Eleventh Edition

Computers: Systems, Terms, and Acronyms

M. Susan Hodges

SemCo Enterprises, Inc.
Winter Park, Florida

Computers: Systems, Terms, and Acronyms

For information, address:

SemCo Enterprises, Inc.
PO Box 147
Winter Park, FL 32790
phone: 407.830.5400
fax: 407.830.0016
email: semco@semcoenterprises.com

Contents

Part One

Computers: Systems

The computer field is incredibly dynamic as new products are released every day, and new terms and acronyms are defined regularly. Working with Information Technology professionals can be overwhelming to someone who is not technical. In fact, even computer professionals are often confused because there are literally tens of thousands of technical skills being used and no one can know them all. While it is impossible to know all of the available technology, everyone, even non-technical people, can develop an understanding of the basic systems that make up Information Technology.

Computer technology consists of the following systems; Computers, Operating systems, Development tools and techniques, Computer languages, Communications systems, Data management, and Applications. These systems have been part of the computer world since the first computers were built, and the number of basic systems neither increases or decreases. Therefore, while it is impossible to understand thousands upon thousands of technical terms, Information Technology can be approached by understanding these seven basic systems and categorizing each technical term as it is encountered.

1. Basic Definitions

The computer industry is an exciting and challenging field that often seems complex and difficult. In reality, the most difficult thing to learn about computer systems is the vocabulary. There are a few basic terms and concepts that must be understood in order to work with information technology.

The first term is "IT." IT stands for Information Technology and refers to any use of computers in processing data. This term is not definitive, terms used synonymously are IS (Information Systems) and MIS (Management Information Systems). All of these names reflect the fact that increasingly more of the processing of data is done by users via computers and/or terminals and that the computer department is concerned more with the collection, currency, security, integrity, and accessibility of the data.

The computer industry encompasses the vendors who make computers and related equipment (hardware) and companies that write programs and systems (software) for the business marketplace. In addition, most mid- and large-size companies hire IT professionals (or computer professionals). IT professionals are people who make their living through computer-related skills and knowledge. This includes programmers, operators, analysts, and other specialists.

The recipients of the work of the IT Department are called users, or end-users. A user is the person who provides the data for a computer system, updates that data, and uses reports from the system in his or her daily work. Many companies don't like the connotations of the term "user" and are replacing it with client, or customer.

Many terms used in the computer industry have multiple definitions and are easily misunderstood out of context. The most obvious of these terms is the simple word "system." Any system is an organized collection of things making up a complex whole. Common examples are the solar system, made up of planets, moons, etc. and the digestive system, made up of various organs. A computer system is a combination of hardware and software that provides the tools for data processing. The hardware is the collection of computers, terminals, printers, disk drives, etc. that comprise the processing environment. The software is the collection of programs and data that provides a function or solves a problem. Software systems fall into two categories: applications (or business) systems and support systems. An applications system is a collection of programs and data that provides a business function for the company. Examples would be the Payroll System, Accounts Receivable, Inventory, etc. These programs solve problems for the company, and relate to administrative (payroll, personnel, etc.) or operational (insurance, banking, etc.) problems. These systems can be either written in-house by the company's own programmers or purchased and modified to fit the company's specific needs.

Support systems fall into three categories; operating systems, DBMS (DataBase Management System), and communications systems. The most important of these are operating systems. An operating system is the collection of programs and data that controls the resources of the computer system. These resources include the computer itself, peripheral devices (disk drives, printers, etc.), programs, and data files. Operating systems vary in complexity, but have one thing in common: these are the programs that handle input/output. This means that a computer is virtually worthless without the operating system, as no programs or data could be entered into the computer system. The term "software system" is sometimes used as a synonym for operating system even though it could actually apply equally to an applications system. When the word system is used alone, most often it refers to an operating system.

The term platform is often used to describe the combination of a computer and the operating system running on it. The platform concept is a critical one, as all software is written for a specific platform. If you change either the machine or the operating system you have changed the platform and a new version of the software must be written or purchased.

In addition to operating systems, most companies have communications and DataBase Management Systems (DBMS). Communications systems contain the software necessary to support online, or real time, programming functions. These systems manage the terminals, programs and data used in online business systems, and must interface with the operating systems and any DBMSs in use. A DBMS is a collection of programs that organizes a collection of data, retrieves and stores the data, and provides security by allowing only approved access to the data. Both communications and DBMS software can be large and complex, and both require special knowledge of the IT staff. Communications software is often included in the DBMS so that online access to the database(s) can be managed. Computer systems are made up of batch and online programs. For a batch program, all the data to be processed is collected and stored, or batched, on tape or disk. The program is loaded into the computer to process the complete batch of data. An example of a batch program is the payroll program. Once loaded, it retrieves data for one employee, produces a paycheck and related data,

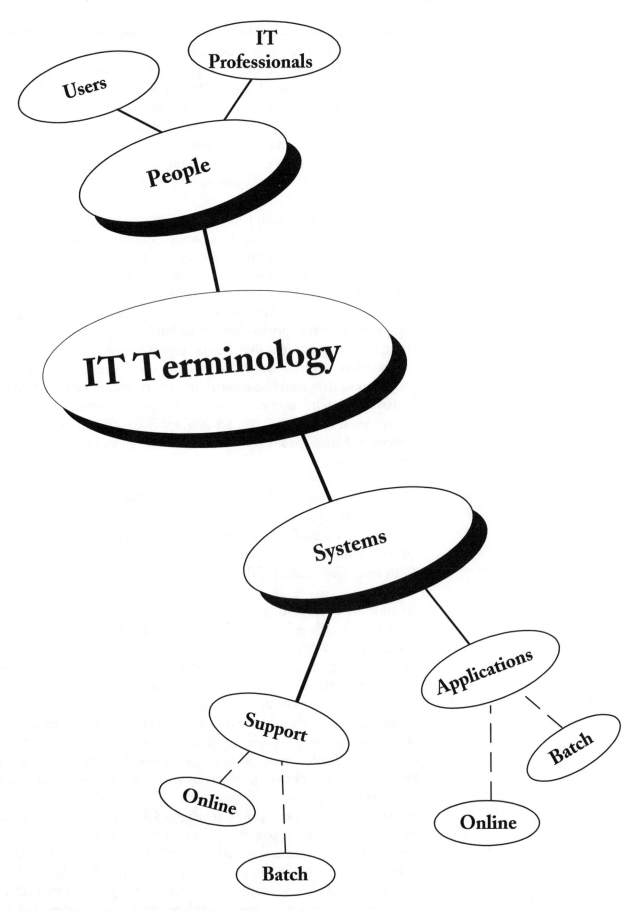

Figure 1.1 Information Technology Overview.

and cycles back to repeat the process for the next employee. This continues until all employees have been processed. Because a batch program runs without human interaction, these programs are often run at night.

Online programs are designed to receive data from a person via a terminal during execution. An example of this would be the program that checks your bank balance. A teller will input your account number, and the program will use this input to search a data file and respond to the terminal with the balance. Terminal is really a catch-all word. It is often a device with a keyboard and screen, but "terminals" are also telephones, the scanners in grocery stores, and even automatic teller machines that get data from plastic identification cards and simple keystrokes. Online systems are also called real-time, interactive, and communications systems. They can operate over great distances because of the hardware and software involved. Online systems use communications lines (telephone, satellite, etc.) and can be defined as any system that receives data from, and/or sends data to, a person working at a terminal. Because of this, online systems run during working hours.

A typical company will have a multitude of computer systems. The hardware system contains the computer(s) and all the peripheral equipment – printers, disk drives, terminals, tape drives, etc. The software systems are many and varied and include both applications and support systems.

2. Computer Systems

Computers are classified into types based on size, processing capabilities, and price. Classification, however, is arbitrary as categories overlap and the most powerful computers in one category exceed the capabilities (and price) of the least powerful computers in the next category. The basic categories are:

Large systems: Mainframes and supercomputers.
Midrange systems: Midsize and RISC machines.
Desktop systems: Microcomputers (PCs) and notebooks.

Large Computer Systems

A large computer system can be identified by three characteristics. First, the computer resides in a data center, or computer room, which provides the necessary dust free, temperature controlled environment. These rooms are generally locked and access is controlled through keys and/or passwords. The second and third characteristics are the people who normally have access to the computer room: the operators and systems programmers. A large computer system has operators, or support personnel, who are responsible for the daily running of the computer and associated equipment. Operators mount tapes, change toner cartridges in printers, perform equipment maintenance functions, and troubleshoot problems that occur. They are concerned mostly with the hardware and its operation, but do get to know something of the software. Systems programmers are the people who work with the support software – the operating systems, database management systems, and communications systems. Systems programmers ensure that all software in the installation works with the hardware and interfaces properly with other software systems. They are responsible for overall performance of the entire computer operation and are the most technical and highest paid of all IT professionals.

Mainframe Computers

A mainframe computer is a large, general-purpose computer. Mainframes are commonly connected to other computers to share the facilities of the data center. The mainframe computer is the dominant computer used in large business today, and the range of capabilities and prices is extensive. A mainframe computer system has a sophisticated operating system and runs both batch and online programs. It allows both multi-users (many users working on terminals), and multi-programming (executing programs concurrently) and often multi-processing (linking more than one computer to execute programs simultaneously as well as concurrently). A mainframe installation has a lot of equipment. Usually there is more than one computer and hundreds of data storage devices and terminals. The computer room, or data center, is temperature-controlled and access is tightly controlled. Typically even applications programmers aren't allowed in the computer room. They work on terminals that are connected to the computer through communications lines and often, in fact, aren't even in the same building.

Mainframe computers are used to execute both batch and online programs in a centralized operating environment. Because many companies are moving towards distributed processing, the role of the mainframe computer is changing in many businesses. Distributed processing is defined as the distribution of data and programs among any number of physical locations, each of which may have a different type of hardware. This means that companies are moving computers and computing functions out to branches and departments that actually need the data and often means applications run on smaller computers. The mainframe can then function more and more as a database manager and do less of the actual processing for online applications. This includes using mainframe computers as server machines in client/server environments.

Supercomputers

Another class of computer, supercomputer, defines a type of computer not commonly used in a business environment. Supercomputers are designed to process complex scientific applications and speed is the most important feature in their design. Today's supercomputers are five times faster than the mainframes used in business and on-going research seeks to increase these speeds.

Midrange Computers

This category actually consists of two types of computers; the midsize computers (which could really be called small mainframes), and the RISC machines, or workstations.

Midsize Computers

A midsize computer is an intermediate size computer that can perform the same kinds of applications as a mainframe, but lacks the speed and storage capacity. They were first produced in the 1970s as machines for medium-size businesses and offer multi-user and multi-programming capabilities. They have sophisticated operating systems and applications programs and also handle both batch and online applications.

Midrange computers have varied functions. One is to serve as the main computer in a medium-size business and, as such, they are similar to mainframes. They can handle a hundred or more terminals, and can have dozens of data storage devices. A second function is to serve as a communications link between mainframes and outlying terminals. Midrange computers can handle much of the necessary communications software, and have proven to be valuable for this function. Yet another function is to act as the server system in a client/server environment. A client/server environment is one in which one computer acts as the server and provides data distribution and security functions to other computers that are independently running various applications.

RISC Machines

Although a RISC (Reduced Instruction Set Computer) is really a large desktop computer, its speed and storage capacity equal that of midsize computers so RISC machines can be categorized as midrange systems. These machines are fast, have high-resolution graphics capabilities, and usually use Unix as the standard operating system. Originally, RISC computers were used in engineering and scientific applications, but are now common in business. In fact, this is the growth area in the computer industry. The name RISC does refer to one of the reasons these machines are so fast. A typical computer has well over 100 instructions. While a complete instruction set handles all operations, it is not necessary. For example, a typical computer has instructions for multiplication. Because multiplication is simply repetitive addition, the multiplication instructions can be left out of a RISC machine. While a multiplication operation will take longer to execute because of the missing instructions, all other operations are faster because of the smaller instruction set. RISC technology is much more complex than this simple example, but the concept is as stated. These machines are desktop computers, but have the capabilities of most midrange systems. The advanced graphics capabilities in these machines are a combination of hardware and software and include both 2-D and 3-D graphics.

RISC machines are heavily used in business environments for software development and are very popular with companies that write software for sale. For example, a company that writes software for a specific industry, such as the retail industry, will use a RISC computer because it is large enough to run multiple systems and software can be developed for multiple platforms. The use of these machines throughout all business programming is also growing because of the proliferation of software development tools. The extra storage capacity and faster speeds of the RISC machine allows them to run the many development tools now available and the RISC environment usually includes a graphic user interface, windowing software, and some CASE products. In addition, object-oriented languages such as Smalltalk and C++ require these capabilities. The proliferation of client/server computing has also increased the popularity of RISC machines; they are commonly used as server machines.

Desktop Systems

A desktop computer requires no special environmental controls, nor does it always require operators and systems programmers. Desktop systems vary from a completely contained single-user microcomputer sitting on someone's desk, to a network of computers linked together sharing printers and data files, to a sophisticated workstation, usually a RISC machine, that is housed in a computer room and supported by a staff of administrators and programmers. Unlike the large systems, the computer room does not have to be air-conditioned and is rarely locked. The title Systems Administrator is used instead of operator because support personnel in desktop systems not only run the hardware, but also provide some software support in terms of running backups for data collections, and maintaining passwords for logons.

Microcomputers, or PCs

A microcomputer, micro, or PC (Personal Computer) is a small computer that includes a terminal, keyboard, and disk and/or diskette storage. Micros are designed primarily as stand-alone machines, but can be connected. They are also called PCs, or Personal Computers. When first introduced in the 1970s, micros were almost ignored by all but small businesses. They were brought into most mid- and large-size companies by users who were frustrated by long backlogs in their IT Departments and decided to handle stand-alone processing problems by themselves. IT Departments soon had to react to provide support for these machines and integrate the work being done on them into overall IT functions.

Microcomputers are desktop machines and have CD and/or diskette drives for data input and output. In addition to these drives, most micros have an installed hard disk and modem. While most function as single-user machines, micros are commonly connected together to provide multi-users and multi-programming. Micros need no temperature control and are found wherever needed throughout a company. While a company will have computer specialists to handle such things as installation and trouble shooting, there usually are no separate staffs of operators or systems people for microcomputers.

Micros can be identified by manufacturer name and identification such as IBM Aptiva, or Apple Macintosh, but computer professionals usually identify the machines by the chip, or processor. A computer chip is the basic building block of the machine. The type of chip determines how large the machine is and how fast instructions execute. There are only a few kinds of processor chips being used, and part of the design of a desktop is choosing the processor.

Processor chips are made up of bits (binary digits). Binary means something that has only two possible states. A light switch is an example of a binary object; a light switch must be either on or off. A computer bit is similar. A bit either has an electronic pulse or it doesn't. To describe the presence or absence of the pulses, numeric values (digits) are assigned. If there is an electronic pulse, the bit has the value 1; if not, it has the value 0 (or vice versa). The chips are differentiated by the number of bits that can be operated one at a time and are 8-, 16-, 32- or 64-bit chips. Some processors have two different bit designations because they move data from storage to the processor in one combination, but process in another. For example, a 16/32 bit machine moves data 16 bits at a time but can combine information and process 32 bits with one operation. The increase in number of bits increases speed and processing capabilities and a 32-bit machine can have both multi-users and multi-programming.

While a sophisticated desktop programmer might use the actual chip, or processor, ID to identify a machine, desktop computers are more often identified by generic names such as Pentium. Pentium refers to the processor, and Pentium computers are the dominant desktop machine today. IBM, Motorola, and Apple have joined together to produce the PowerPC. The name actually refers to a group of processors, and, again, the computer systems are called PowerPCs. DEC (Digital Equipment Corp) is manufacturing the Alpha computer, which is a 64-bit RISC based machine. Alpha, Pentium and PowerPC systems are equivalent in size and capacity to some RISC machines.

Additional PC Types

Notebook computers. A notebook computer is a portable computer that can weigh as little as three pounds. Most notebooks can plug into larger computers which gives the user the capability of working away from the office and simply uploading whatever information is collected or processed while on the road. Most are close to the size of a standard notebook (8 1/2" x 11") and have internal hard disks, modems, CD and diskette drives.

PDAs, or handheld computers. Personal Digital Assistants (PDA). A PDA is a light-weight (1 pound or less) task specific computer. These machines are pen-based, offer voice recognition, fax and modem communication, and often include a pager. A typical use would be to write in "lunch – John." The PDA would send John a fax from a previously entered directory of fax numbers and would enter the lunch date in the user's appointment book. Small as these computers are, they can be programmed for specific needs and can operate standard software such as word processors and games. Palmpilot is an example of a PDA.

Generic Computer Definitions

Server

A server is any computer that runs software that makes programs and/or data available to other computer systems, and does not define the computer. RISC machines are most often used in this capacity, so the two words are often used synonymously, but any machine can act as a server. In fact, one of the most common roles for mainframe computers is to act as a server machine to the rest of the systems in the network.

Network Computer

A network computer is a PC with no disk or CD drives. This ensures that all data and programs must be loaded from a server computer allowing centralized control over software and centralized management of data. The network computer runs graphic programs and allows use of a GUI (the software that allows the user to interface with a mouse rather than keying in commands). Network computers have been introduced for use with the Internet, but can be used internally over Intranets.

SMP (Symmetrical Multiprocessing)

A symmetrical multiprocessing machine is a computer that has more than one processor and can therefore provide parallel processing capabilities. They are commonly used for decision support systems, manufacturing processes, project management, reservation systems, and online transaction processing such as banking and credit uses. These machines are RISC type machines and typically have between two and 64 processors. Unix is the common operating system. They are also called workstations and servers.

MPP (Massively Parallel Processors)

A massively parallel processor, or MPP, is most commonly used with decision support systems, or data warehouses. MPPs are built with literally thousands of RISC processors running independently. They provide the speeds and capabilities of mainframe systems. This provides fast access for complex queries that might have to access dozens of pieces of data to find an answer. A user might, for example, want a list of potential customers for a new product line. The query would have to examine a variety of data such as age, address, sex, hobbies, buying history, income, financial history, and so on. It's hard to picture, but some complex queries can actually take days to run on standard single processing machines. MPP systems are one of the fastest growing areas of computing and are often used in data warehousing. These are complex systems and require experienced people for support.

Analog Computers

Analog computers are used in a scientific, not business, environment. Input to these machines is a measurement of physical values, such as temperature or pressure. Analog computers are special-purpose machines and cannot be programmed. They are often connected to general-purpose computers that convert the measured values to diskrete numbers for processing. One example is the use of analog computers in hospital settings. Analog computers measure temperature, blood pressure, pulse rates, etc. and these values are converted to numbers that are used by a program in a general-purpose computer that monitors the patient.

3. System Development

Programming is, in its simplest definition, writing a sequence of instructions to be executed by a computer to solve a problem. Obviously, there is more to it than this. Programs do not exist alone, but are part of a system. Developing a system is a cyclical activity that has several phases. In order to understand the job of programming, it is necessary to look at the entire project life cycle.

Programming Life Cycle

Developing a computerized system, or even a single program, follows a cycle of activity that is referred to as the project life cycle, the programming life cycle, the system development life cycle. Just about any combination of words can be used with the two important ones: life cycle. There is no definitive breakdown of exact steps, or phases, that make up the project life cycle, but all life cycles contain the same basic phases and activities. The activities in each phase will be discussed separately, but it is important to remember that systems development is a cycle. This means that the work done during each phase overlaps. With large systems, design will start before the detailed analysis is complete, programming starts during the design phase, and so on. Another aspect of the cyclical activity is that completing one phase can result in cycling back to a prior activity. A problem encountered in the programming

Figure 3.1 Progression of activity during the development life cycle.

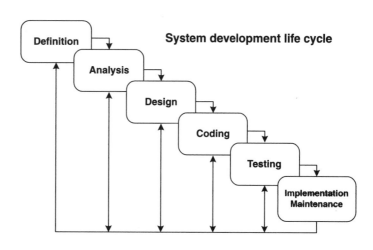

phase often means going back to the original analysis to determine what caused it.

Definition Phase

Defining a system means stating goals and objectives and usually includes a statement of the problems that led to the consideration of creating a new system, or revising an existing one. The most common problems are speed and quantity of data. For example, it is taking too long to produce paychecks, or an increase in business has made it difficult to take inventory. Other problems are caused by changes in government regulations, or changes in business. For example, adding a new product line requires additional data capture and processing.

The definition phase of the programming life cycle is done by the operational departments, or clients, and is completed with a request for programming services.

Analysis Phase

Computer professionals become involved during the analysis phase. Analysis is determining what the problem is. This means carefully examining both the existing system and the problems that exist working with it. It is interesting to note that there is always an existing system. Even if a function has never been automated, there is a manual system that has been used to process the data in question.

The heart of the analysis phase consists of interviews. The analyst spends much time with the client determining not only what the current system does, but what the new system should do. The analyst must work with the client to distinguish between what is needed and what would be nice. At the end of every interview, the analyst should prepare a written summary of what was discussed and decided. Therefore analysts must have strong communications skills, both oral and written. In addition to strong communications skills, an analyst must be logical, organized and be able to keep track of many details and determine how each fits into the total problem. For example, if an analyst were working on a payroll/personnel system, he or she would have to examine all the various pieces of data used throughout the payroll/personnel functions, and understand the relationships between all of them. Understanding the data is necessary in order to understand the processing steps involved in the system. For example, net pay on a paycheck is a piece of data that is calculated from other pieces of data such as gross pay, overtime pay, tax deductions (FICA, federal, state, and city), pension deductions, and unemployment tax. This simple list can be used to illustrate how complex data analysis can be by pointing out that the list would be different for every company and, in fact, would be different for individuals within the same company.

The analysis phase of the project life cycle is critical. If this job is done incorrectly, or incompletely, the results may not show up until the entire project is complete. Analysts have tools to assist them with their work.

These tools are charts and graphs and are used to analyze both the data and the procedures that process the data. One example is data flow diagrams, which chart the movement of data within a company showing both the manual and computerized processing performed on that data. Some CASE (Computer Aided Software Engineering) products have automated analysis functions, particularly data analysis functions.

Design Phase

The design phase consists of determining how the system should be developed. Both the data and the processing must be designed. Designing the data means deciding how the data should be stored and determining such things as whether or not to use a database. It includes defining security measures and setting up backup procedures (maintaining duplicate copies of data). Designing the processing means breaking the problem down into parts that can be handled by a single program. The design of a system might result in a program to input and validate data, another program to handle updating the data, and still other programs to produce reports.

As in analysis, graphic tools exist to help designers with their work. The most popular design tool is a flowchart. A flowchart is similar to an architect's blueprint, and uses symbols and diagrams to describe the processing to be done in the system. Other graphic tools are bubble charts, data flow diagrams, and entity relationship diagrams. Again, there are CASE tools available that automate some of the design

Figure 3.2 Sample of form used by designers during data design.

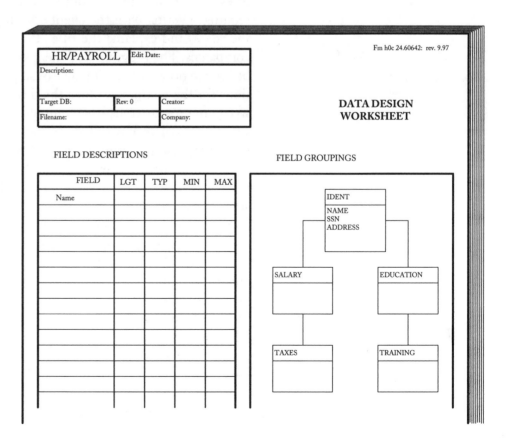

activities. The result of the design step is a diagram of the entire system, and written specifications (program specs) for each program. A program spec contains a description of the input data, the output data (including designs of all reports and/or terminal screens), and the processing steps required to get from the existing input to the desired output. These specs can be formal, typed, multi-page documents, or can be a simple note – "See Ms. Jones in Personnel. She needs a new report." A programmer receives the program specs and the amount of detail will be as extensive as necessary for the experience and skills of the specific programmer. Some companies split the responsibilities for analysis and design and have system designers on staff. Most do not, however, and analysts continue with the design phase. There are usually more people involved in the design phase than there were involved in analysis, and Senior Programmer, Senior Programmer/Analyst, Software Engineer, Project Leader and Team Leader are all common titles for people who perform this function.

Programming Phase

Programming, as previously defined, is writing a sequence of instructions to be executed by a computer to solve a problem. This writing, however, is an involved activity that follows the same cycle as the development of the entire system.

The programmer is given program specs, which are the result of the design phase. This gives him or her the description of the data and a statement of what processing is to be performed. He or she then analyzes the problem to determine exactly what is to be done. This often requires meeting with the client and the designers to clarify points on the specs, and explains the title "Programmer Analyst" used in many companies. The programmer's next step is design. The processing must be broken down into a sequence of steps that can be coded into instructions for the computer. Once the design is complete, the programmer writes the instructions in a computer language. The actual writing of the program is often called coding, and is a fairly simple matter of translating the program design to the particular language to be used. Once the program is written, the programmer tests the program by using actual or simulated data to make sure the program does what it is supposed to do. When testing is complete, the program must be documented. The programming phase, in fact, contains all the activities of the complete project life cycle, only the programmer is working with a single program. Various tools are used by programmers to actually write and test programs. Program generators will actually "write" specific language statements from coded or diagrammatic input. Report generators will generate formatted reports from simple input statements. There are programs to create test data for the testing phase, and programs that will develop an entire test plan. Many of these programs have been incorporated into application development environments.

Testing Phase

Once all the individual programs have been tested, they are combined and executed in sequence as a test of the entire system. Often actual data is used for this final testing and parallel testing (running both the old and new systems for a set period) is done whenever possible.

Testing is often called debugging. Errors in computer programs are called bugs, and fixing the errors is called debugging. Different types of errors can occur. A technical error is one that creates a situation that the computer cannot handle and causes an abend (abnormal end), bomb, or dump. The most common technical error is an instruction in a program that attempts to do some kind of math with data that is not numeric. When the computer encounters this situation, it stops processing and informs the operating system that it cannot continue executing the instructions in this program. The operating system then stops the program and prints a report called a dump, which shows what the program looked like at the time the error occurred. The dump is used by the programmer to determine what the error is in order to fix it. Often programmers will list "dump debugging" as a skill, which indeed it is. Another type of error is a logical error. These bugs are the hardest to correct, and there is no report to help. A logical error is one in which the program does what it should be doing, but it comes up with wrong information. For example, a program to produce paychecks might produce a paycheck for every individual, but close checking of the output shows that the amount withheld for federal tax is not correct in all cases. These errors can be very difficult to find and correct.

The testing phase varies greatly from company to company. First testing is incorporated into the individual program testing, and is handled by the programmers. This is called unit testing. System testing, the testing of all the programs in the system together, can be done by the programmers who wrote the system or, at other times, by a separate group that designs new test data files to eliminate any inadvertent bias in the testing procedure. Systems testing is done by seniors in the IT department, who design new test data and create new test scenarios that deal with the complexities of testing several programs together.

Testing specialists refer to different types of testing such as black box testing, white box testing, and regression testing. Black box testing builds test data and scenarios from the original requirements or program specifications so the testing is not biased by assumptions that were made during development. This testing can be done without knowing the programs, objects or components being tested and is also called behavioral testing, or functional testing. While this is the most popular form of testing, it is not used alone. White box testing is often called structural testing because it depends on the structure of the program. It builds test data and scenarios from the structure of the developed software and test data and scenarios are created using the actual programs. White box testing consists of tests that, i.e., execute every statement at least once, and execute each branch at least once. This form of testing is also called glass-box testing and is used in conjunction with black-box testing. Regression testing is often used with critical systems and, again, is used in addition to other forms of testing. The purpose of regression testing is to protect the integrity of the complete system when making a change to one part of it and to ensure that changes made to a program or system do not affect its performance. All paths of the program and/or system are tested before a change is made and the results are saved. The changes are made and tested. Then the original pre-change tests are rerun to ensure the program or system still functions correctly in all ways.

The final step in testing is called acceptance testing, user testing, or beta testing, and is done by the users of the system. The name "beta test" is used by the computer vendors when they have new software to test. They ask customers to be "beta test sites" where the customer installs the software in a real life environment for a period of time. During this set time period, the system is used in a live environment and is carefully monitored to catch any final problems.

Implementation Phase

Once the system testing is complete, the system can be implemented, or installed. This simply means turning the programs and data over to the operations staff and scheduling regular execution of the programs according to the user's needs. This too, can be quite involved and is often done in phases. For example, if a company were installing a computerized system to handle payroll, the programs that produce the paychecks would probably be implemented before the programs that produce annual W-2 forms were even written. With large systems, testing and implementation overlap with beta testing.

Implementing a system means that the system is no longer considered to be in testing, but is now in production. Production systems are run by the operations staff and follow an established schedule. The programming staff gets involved with the system again only if something goes wrong, or if a change is requested by the user department. Errors don't occur too often, but change is inevitable and frequent. Changes to a production system, whether to fix errors or by request, are called maintenance, and a programmer whose job is described as maintenance is simply implementing changes to a production system. Maintenance programming is not considered to be as challenging as developing new systems but, in fact, it is often more difficult to change an existing program than to create a new one.

Documentation

Documentation is an ongoing process throughout the development cycle. Documentation includes technical documentation for the IT staff which describes the programs and data, and client documentation for the operational departments which describes how to use the reports and run the programs in the system. Technical writers often do the client documentation while the technical documentation is usually done by the programming staff.

Project Management

The management of a programming project is the same as the management of any project; plans and schedules must be made, procedures for implementing changes defined, deadlines established, and checkpoints determined. It is often said that the average development time of a programming system is five years, so management is a critical function. Management starts with the first phase, the definition of the project or problem, and continues throughout the life of the project.

4. Software Engineering

The term "software engineering" was coined as long ago as the 1960s, and is used to refer to the application development process. Software engineering reflects the fact that most application development follows some sort of methodology and uses specific tools for analysis, design, and programming.

Methodologies and Techniques

A methodology is simply an approach to problem solving and is defined by a set of rules and specific development tools. The structured methodologies of the 1970s were the first popularized, and are still commonly used. They have been supplemented, not replaced, by new methodologies such as information engineering. Techniques such as joint application design and rapid application development have also evolved and take part in application development. Knowledge of methodologies and techniques can be part of the skill set required of Information Technology personnel.

Structured Analysis, Design, Programming, and Testing

In the 1970s systems development activity went through what was called the structured revolution. Prior to this time each individual designed and wrote programs according to his or her own logical processes. The programs worked but because so many different approaches were taken, it was difficult for programmers to pick up someone else's program to modify it. Software designers decided that if all IT professionals took the same approach – a structured approach – any program could be easily understood and modified by any programmer.

Structured programming states that all programs are made up of a few definitive logic structures and establishes rules to be followed when analyzing, designing, and writing a program. Many different sets of rules evolved, and most companies do use one of the structured methodologies today. The various methodologies are named after their developers. Nassi-Schneiderman developed the original theory, and Nassi-Schneiderman charts were designed to force adherence to the rules. Yourdon-DeMarco embellished structured theory and developed a very strict programming, testing, analysis, and design style that used pseudo-code heavily. Jackson concentrated on data and developed a methodology based on data flow, and Warnier-Orr replaced flowcharts as a design tool with bubble charts. Other common methodologies are Gane-Sarson, Merise, Ward-Mellor, Hatley, and Chen. Some companies insist on a strict adherence to a given discipline, others have created a unique methodology based on one of the accepted standards, while most simply maintain general guidelines based on structure charts and/or data flow diagrams.

Top-Down Development

A top-down approach to problem solving starts at the broadest and most general level and works down to specific details. For example, the first design of a payroll system would be a diagram of functions to be performed, such as tax routines, benefit deductions, and print routines. Later design would detail routines for federal, state, and FICA taxes; define the specific benefit deductions; and determine how many reports should be printed and what each should look like. In top-down development, a hierarchy is developed and each level is completed before the lower level is started. This is different from a bottom-up approach, which would develop all the tax routines in complete detail before addressing anything about benefits.

This approach to application development is often called functional analysis and design, or functional decomposition because it is based on breaking processing down into the functions (such as taxes, benefits) to be performed. Functional decomposition is a methodology that views an application as a set of functions that can be decomposed, or broken down, into steps that can be easily understood and programmed. The decomposition of a problem takes many steps until the final breakdown produces the program specifications for individual programs. Top-down or functional development was introduced to the Information Technology industry along with structured programming and is used by most developers.

Information Engineering

Information engineering extends the principles of software engineering from a single project to the entire enterprise, or company. Software engineering defines the development methodologies and techniques to be used on a single project; information engineering defines the methodologies and techniques to be used throughout the entire enterprise. It integrates the separate software systems built throughout the company by different people at different times. In order to do this, a repository of planning information, data models, process models, and design information must be built. This repository is called an encyclopedia and allows the identification of common data entities, reusable data, reusable design, and reusable code.

Re-engineering is a term often associated with Information engineering. It is the process of analyzing and redesigning business systems to improve speed, service, and quality before automating the systems. It means designing the manual and non-computerized functions to take advantage of new technologies, not simply automating the existing procedures. A re-engineering effort will result in the creation, elimination, and redefinition of actual jobs as well as the re-development of computer systems. Re-engineering requires a complete commitment from corporate management and takes many years to complete.

Information engineering requires formal methodologies and automated products in order to satisfy the corporate information systems needs. Different methodologies such as JAD and RAD can be used, and businesses have many choices of development tools. Time must be spent making sure the tools picked reflect the needs of the corporation.

JAD (Joint Application Development)

Joint application design is a technique based on group processing. JAD development defines a group of developers made up of both IT professionals and the clients, or users, of the proposed system. A group of between eight and fifteen members meets for a highly concentrated work session to design the system together. This work session often takes several days.

The JAD session is a formal meeting conducted by a trained JAD session leader, or facilitator. This person could be either technical or non-technical and, in fact, does not have to be involved with the system development after the work session. Many companies bring in consultants to fill this position. Another position that can be filled by a disinterested person is the position of scribe. The work session must be thoroughly documented and conclusions recorded.

JAD sessions are often conducted with an EMS (Electronic Meeting System). If an EMS is used, meeting participants each have a workstation so they can participate in a meeting both electronically and verbally. The EMS also makes the job of the scribe much easier. For long distances, the EMS is combined with teleconferencing so remote users can participate without physically being present.

The main benefit of JAD is the heavy involvement of the users during the design phase. Typically the users are interviewed during analysis and some design work is brought to them for review, but they do not actually participate in the process. JAD requires active user participation at the very earliest stages and should result in a better design.

RAD (Rapid Application Development)

Rapid Application Development is a technique which uses iterative prototyping to develop systems, rather than follow formal design and review requirements. It combines development tools with management techniques such as brainstorming to quickly develop prototypes that can be modified to become fully operational. Developers build and present a prototype for each request. A prototype is a model of a program and/or system suitable for evaluation of the design, performance, and potential of the product. It does the work requested by the user. It may, for example, produce a weekly report of the salaries of employees averaged by department and job code. The program that creates this report is, however, probably not a program that would be used on a weekly basis. The prototype would be created by a development tool or a CASE product that would build a model report for the user. The prototype could be easily modified, even expanded, and only when the user was completely satisfied would a production program be written.

RAD development relies heavily on development tools such as application development environments, 4GLs (Fourth Generation Languages), report generators, and program generators. RAD techniques are used with any client/server development as all client/server systems have GUI (Graphic User Interface) front-ends. They are also used to develop applications that run over the Internet. There are specialized development tools available to create Web sites and develop interactive programs that use a Web browser for the front-end.

Application Development Automation

It seems that as soon as people started to develop application programs, software appeared to help the programmer. Some development tools worked with the entire development life cycle, but other tools were developed to work with only one phase. This automation of the development process has been going on for many years and there now are many tools available to developers. Knowledge of one or more of these tools is often a required skill of developers. The tools technically all fall under the header "CASE," which simply stands for Computer Aided Software Engineering, but the term CASE is generally accepted as referring to tools that automate the analysis, design and/or project management phases.

CASE Tools

The tools developed to assist in the automation of the analysis, design, and project management functions are usually referred to as CASE tools and have existed since the 1980s. Several things contributed to the emergence of CASE products. There was (and still is) an overwhelming demand for software. Most IT Departments have a three to ten year backlog of program requests and still spend most of their time maintaining existing systems. Also, while the cost of computers and associated hardware seems high, it doesn't approach the cost of software development. All these are reasons for the industry's look at ways to produce higher-quality programs more quickly and with less effort. The overall goal of the CASE industry is to provide software that, from a specification of what a program is to do, will figure out how it should function and then write and test the program. In other words, CASE products were originally designed to automate the entire systems development cycle. While many of these tools still do this, the automation of the programming and testing phases is commonly done by separate Application Development Environments and tools. Today's CASE products automate analysis, design, managerial functions such as planning and scheduling and provide automatic documentation throughout the systems development.

Technical advances have made the emergence of CASE possible. The advanced graphic capabilities of microcomputers have been critical. Because so much of design work is graphs and charts, until computer tools that could input and modify graphic data were developed, it was not possible to automate design work. The acceptance of standardized design and programming techniques has also had a great effect. CASE systems are based on structured methodologies, so knowledge of these techniques is necessary in implementing CASE. Some products require strict adherence to one set of rules such as Yourdon-DeMarco, or Jackson, but most are adaptable to variations. While CASE products can be purchased to automate a single function, software vendors and CASE users have realized the need for integrated systems. Some CASE products work throughout most of the development cycle while others concentrate on analysis and design functions and integrate with an application development system for the programming and testing functions.

Installing a CASE system can be a major undertaking. For CASE to really work, strict standards must be followed by everyone in the IT Department. This is not easy, especially if a company had fairly lax standards prior to considering CASE. The new tools must be learned, which requires training. And, finally, management must allow for lower productivity during the learning curve. It is the programmers, designers, and analysts who use CASE products; CASE does not significantly affect the systems or operational groups. Programmers often have knowledge in more than one tool as companies have changed their development strategies. The popularity of client/server computing has caused the biggest change in the use of CASE tools. The original tools did not address this environment, so companies either stopped using the tools they had, or replaced them with the newer CASE products that worked with client/server and application development systems.

Application Development Environments

An application development environment, or system, is a software package that enables programmers to write, test, and document computer programs from a terminal. These systems are software packages which provide a programming language which can be used to access existing databases. Some of them also provide report and program generators, DBMSs, CASE tools, and links to other systems. Application development environments provide a GUI (graphic user interface) for the developers and are often called visual languages, or visual development systems. This allows programmers to use point-and-click and drag-and-drop functions rather that writing lines of code. ADEs are also called RAD development tools and GUI builders and are used to develop client/server and Internet applications.

Application development environments are used mainly by applications programmers and often programmers work with more than one system. These systems are very popular in desktop environments and are closely associated with a DBMS (DataBase management system) such as Informix, Ingres and Oracle. Application development systems are also used in the development of client/server systems. Learning to use one of these products is not particularly difficult, and the software is ready to use without involved installation. It is not unusual for a company to use more than one application development environment. These tools are very popular and always used in client/server development. Popular ADEs include Visual Basic and VisualAge.

Application Development Tools

Many programs exist to make the programming function easier and to automate many standard tasks. These programs are called application development tools, and many are contained in both CASE and ADE software. There are several different types of development tools.

Programming Utilities. This category is a catch-all for a variety of different programs that provide specific assistance to programmers. Most programming utilities are written by outside vendors, not by the computer manufacturer. These outside vendors have seen a need for support and often deal with computer systems from many vendors.

Many programming utilities are conversion tools. For instance, these programs can convert applications written to run under one operating system so they can run under another. Some utilities will convert JCL (Job Control Language) from one operating system to the JCL of another. And still others will convert online commands from one online system to another. Other programming utilities provide the needed interfaces so applications programs can access software products (such as databases) from different vendors. These utilities can be important in a company that has more than one database or more than one communications system. A programming utility will, for example, allow the programmer to use the same program to access both a DB2 database and an IDMS/R database.

Still others provide such functions as analyzing programs to make sure they follow company standards, analyzing data descriptions to eliminate redundancy, and defining critical datasets in a system to determine the resources (such as disk space, tape reels) necessary to process the data. Programming utilities provide just about any type of support for the programming function.

Report Generators. A report generator, or report writer, is a program that generates formatted reports from simple input statements, thus eliminating the need to write complete programs to accomplish fundamental print tasks. Report generators exist for all types of computers because they save so much time. Rather than having to write a complete computer program, a programmer can simply provide a description of the data to be used and the way he or she wants the report to look; the generator will do the rest. These programs will generate the instructions necessary to do any desired totaling, averaging or percent calculation. In addition, they will take care of print spacing, header lines and/or trailer lines on reports, and page numbering. All 4GLs, or query languages, contain report generation functions as do all ADEs.

Program Generators. A program generator is a computer program that can generate all, or part, of the code necessary to build a program in a source language. Programmers provide input that describes the data the desired program will work with, how that data should be processed, and what the output should look like. The generator will then produce the source statements that follow the syntax rules and use the vocabulary of the source language. Many of these generators do not produce the entire program, but can only handle part of it. Generators can usually produce source statements for data definitions, but often the statements that manipulate the data must be written by the programmer. The most common program generators produce C, C++, or COBOL statements. Program generators are also called application generators and code generators, and are often part of ADEs.

Screen Editors. Screen editors are computer programs that handle the input and modification of programming statements from a terminal. These programs are similar to word processors and have features such as global changes (i.e. changing the character 'x' to 'x1' every place it appears in a program), block movement of statements, and repetition of statements. In addition to handling programming source statements, screen editors also work with JCL statements. Screen editors also provide access to programming and data libraries so programmers can, through the screen editor, copy programs and/or data files, delete them, rename them, and so on.

Screen Painters/GUI Builders. A screen painter is a program that will generate the code necessary to define a terminal screen for an application system from a sample of the screen. This means that the programmer can design the screen right on the terminal by moving the cursor around. Once he or she is satisfied with the way the screen looks, the program will generate the code to be incorporated in the online program. When a screen painter is also a GUI builder this means the developer can incorporate the GUI interface through the tool. Screen painters/GUI builders are part of application development environments or can be stand-alone tools used for online system development.

5. Object-Oriented Development

Object-oriented programming, or development, differs from the top-down structured approach in many ways. Structured programming deals with the way programs are written and establishes rules to be followed when analyzing, designing, and writing a program. It uses basic definitions of programs and data as separate entities.

Data, in traditional programming, is items of information and descriptions of that information. For example, annual earnings is a piece of data. The information might be $24,000. The description of annual earnings would be: it is a numeric data item (can be manipulated by add, subtract, multiply, and divide); it has a maximum of seven integer places and two decimal places (maximum value of 9,999,999.99); and it must be a positive number.

A program is a series of instructions that manipulate the data, such as: determine annual earnings by adding base salary and overtime. All of the processing, or manipulation, of data is done in the programs. Each program is individually written and determines exactly how a process is to be performed. One program could calculate net salary by adding each weekly overtime amount to the base salary. Another could add all the overtime amounts together and then add the total to the base salary. Both approaches work.

Object-oriented programming theory completely changes this approach. Basically it states that information is made up of objects. An object is a package that contains both a value ($24,000) and methods or descriptions of how to manipulate the value (annual earnings = base salary plus overtime week 1, overtime week 2, and so on). Programming then becomes the simple process of having objects communicate with each other. An object, "W-2 form," would communicate with the object "annual earnings" by sending it a message asking it to return its value ($24,000).

Object-oriented programming develops reusable code. While reusable code can certainly be developed using traditional methodologies, it is not a major goal as it is in object-oriented development. Objects, once defined, can be used over and over again. Writing a new "program" becomes a matter of working with predefined objects. For example, if a company has to incorporate a new city tax in an existing payroll system, the programmer would simply add a new object – city tax – to the class tax and define the methods for calculation. In a completely object-oriented environment, many standard programs such as queries could be done by non-technical people because so much of any given function would already exist in object libraries.

Object-oriented programming is said to be concerned with what needs to be done, not with how it is done. In other words, "W-2 form" does not care how annual earnings are calculated. It only cares that it can receive a value. This means that individual programs can ignore much of the processing they were previously concerned with. The annual earnings figure will be needed many times. It is needed to produce individual W-2 forms; it is needed to produce summary reports for the government; and it is needed to calculate annual bonus figures. With object-oriented programming, any single program need only know that the object "annual earnings" exists and what message should be sent.

As with any theory, object-oriented programming has its own new vocabulary. Object is the first new word, and it is defined as a self-contained entity that contains a piece of information and a description of how to manipulate it. It is correct, however, to think of an object as a single function program. This description of how to manipulate an object is called a method. A method is an action to be performed. In the annual earnings example, the method is to add base salary plus overtime week 1, overtime week 2, and so on. Objects can have more than one method. Another method might be to round annual earnings to the nearest 10,000 figure for purposes of calculating bonuses and/or producing management reports.

A message is the way objects communicate with each other. Determining messages is part of the design and programming activity, and an object can receive more than one message. "W-2 form" could send a message to return annual earnings. "Bonus" could send a message to return annual earnings rounded.

Objects belong to classes. A class is a group of objects that share similar characteristics, and classes have descriptions and methods associated with them. All of the objects discussed above could belong to the class "salary information." This class could have the method, "Carry all math to three decimal places and round to two" associated with it. Any message sent to the objects in the class "salary information" automatically picks up the methods associated with the class. Objects within classes are called instances (simply another new term). And classes follow the principle of "inheritance" – the ability of new classes to use the procedures and data (methods and values) from existing classes. This requires programming only the differences for the new class of objects.

Common terms used with object-oriented are encapsulation, inheritance and polymorphism because an object must have all three of these traits.

Encapsulation means that values and methods should be packaged together as objects for storage, retrieval, and execution. The methods and values are never accessed directly by any of the applications that use them, but are accessible only through the standard message, or interface. Encapsulation is also referred to as hiding data and processing. Encapsulation ensures that no one application can change the data or the processing.

Inheritance is the most obvious and refers to the fact that object inherit characteristics from the class definition. A salary class could be defined as having the characteristics of being numeric and greater than zero. Every salary in the class-gross salary, net salary, year-to-date salary, etc. automatically has those characteristics and the code to validate correctness need be written only once for the entire class.

Polymorphism deals with the standard interface, or message. It simply means that the same message could be sent to different objects which could react in different ways. Taxable salary could be sent to a tax object which would calculate taxes, and reduce the salary. It could also be sent to a benefits program which would simply track the information for retirement benefits. Some of the benefits of object-oriented programming are readily apparent. Duplicate processing is avoided (several programs calculating annual earnings). Consistent results are maintained (all salary figures are rounded to two decimal places). Processing is more efficient because the data and the processing are physically together. A program has to access only one object rather than access two things, a piece of data and a processing statement. Object-oriented systems are designed to work in any computing environment, and programs are written in languages designed especially for this purpose. Java, C++, and Smalltalk, are all object-oriented programming languages. These languages are available in mainframe, midrange, and desktop environments, and programs are transportable.

Object-oriented development has its own standards, rules and methodologies which cover analysis (OOA), design (OOD) and programming (OOP). Analysis and design are very close and use the same tools and methodologies. There are different methodologies which seem to be complementary, therefore more than one methodology is often used. Three popular methods with complimentary strengths include

Jacobsen's Objectory method, Rumbaugh's Object Modeling Technique, and Booch's object-oriented design. Knowledge of the methodologies is often required of senior analysts and designers, but not of programmers. Programmers need to know an object-oriented programming language.

Object-oriented programming languages have been around for a long time. The first, Simula 67 was developed in 1967 and was quickly followed by Smalltalk. Current favorites include C++, CLOS, Eiffel, and Java. While any program can be written in any programming language, it's much easier to use a language specifically designed for object-oriented development. Some of these languages are called hybrids as they are based on another language. C++ is based on C, CLOS is based on Lisp. Others such as Eiffel and Java are pure object-oriented languages.

Component Based Development

Component based architecture is an approach to the development of software systems that builds applications from components, or pre-written building blocks. Manufacturing has been using component-based design and production for a long time. A television consists of many parts: the screen, the front panel, the remote, etc. Some of these parts, i.e. the remote, are complex and are assembled from sub-parts. Sony, for example, buys the working parts of the remote from other companies and encases them in a Sony remote. The size and shape of the remote is controlled by the size, shape and number of the internal parts. Sony makes several different remotes for different models of televisions. For example, one remote might contain the components to operate a stereo system and a VCR while another would contain only components to operate the VCR. Sony might choose to buy the component that operates the VCR from Company X, yet buy the stereo part from Company Y. Panasonic could be buying the same parts for their remotes. Sometimes buying parts is not an option. For example, with new technology parts are not available for purchase, i.e. components might not be available to handle the tuning of the new large flat TV screens. Whether the company buys or builds the components, they must assemble them together so the product works. Often this is difficult when buying parts from different companies, so manufacturers might choose to use a limited number of vendors, or choose to build some of the parts themselves to make the assembly of their final product easier. The assembly of the final product is just as important as the components themselves.

Software can be built the same way. Software developers can create components which can be stored in libraries and reused in any number of applications. If this sounds similar to object-oriented development, it should. While a component is not necessarily an object because it does not have to have inheritance and polymorphism, a component is a pretested, encapsulated object that has a standard interface. The standard interface means that the component will always receive input in the same format. This allows for the component to be pretested. If the input is always the same format, the component can be reused without retesting. Software developers can, with confidence, simply incorporate the component into any application and know it will do exactly the same thing. For example, if a component is built to validate data, it needs two pieces of information: the data and the rules for validation. The component will do the validation and change the color of the field if it is incorrect. The payroll application can use this component and give it someone's salary and the rules "must be a number that is greater than zero and less than 1,000,000." The salary on the screen will change to red if the decimal is omitted from a 100,000.00 salary when inputting. An inventory application can use the same component but give it an item number and rules stating that first character of the number must be alpha and the rest of the characters be numbers. The component would execute properly. Specifying the data and rules must follow a defined standard for this to work.

Encapsulation ensures that the data and processing included in the component are protected from being changed. In other words, the only way the component can be used is through a standard interface. It also means that if the component is changed the changes will be transparent to the application. The importance of this can be shown through a sales tax component. This component is used to do all sales tax calculations and has been included in both the accounts-payables (for items the company purchases) and accounts-receivables (items the company sells) applications. If a state changes the sales tax percent, this change can be made in the component and neither application need be rewritten-both will function correctly with the new tax.

Developing a component can be a major effort. The component that handles sales tax does much more than simply calculate what the tax is on a purchased item. It must also do such things as prepare tax reports, make tax data available to income and expense functions, and predict potential tax liabilities. Components can be complete programs, or parts of programs. In fact, one of the reasons component based development is becoming used more often is that existing programs can be turned into components without having to be rewritten. This means that if companies want to convert systems to client/server, or to a Web based front end, they don't have to rewrite the whole system. This is often referred to as component mining (extracting components from existing code).

Component development has four phases: suitability testing, component adaptation, component assembly, and system evolution.

Suitability testing (also called component qualification). This equates to analysis and is well named. This phase determines whether the component will do the work needed.

Component adaptation. Components can be used as is, extended to handle specifics, or modified. Modifying components will require the most maintenance. Extending them means adding additional functions, but not changing any existing code, while modifying a component does change the original code.

Component assembly. During this phase an infrastructure is built that integrates the components. The most common methodology used is ORBs (Object Request Brokers).

System evolution, or maintenance. All systems will change, and this phase refers to incorporating these changes.

There are commercially available components, and as more vendors offer these products, component development will grow. At present most commercial components are for operating systems, databases, e-mail, and office automation systems. Application (banking, manufacturing, etc.) components are not yet readily available so must be written in-house.

Component based development requires a variety of skills. The components are completely independent and may be written in several different languages and run on different platforms. The developers who create the components use object-oriented languages such as Java and C++ and know tools such as ActiveX, JavaBeans, COM/DCOM and CORBA. The components are then assembled by programmers using RAD tools such as Visual Basic and PowerBuilder or even scripting languages such as VBScript and JavaScript. The most used components are ActiveX and JavaBeans. ActiveX follows COM/DCOM specifications and JavaBeans follows CORBA.

Patterns

Experienced software developers don't think about programs in terms of low level programming language elements, but in higher-order abstractions. Patterns are a way of documenting the abstractions that successful developers use, and the overall goal is to develop a handbook of patterns for software developers, similar to the design handbooks used by other creative building jobs such as architects and automobile designers. Patterns are written as catalog entries with a definitive outline and are have a literary style to provide technical documentation. A group of related patterns forms a "language" for a specific area of development, such as networking or operating systems. Pattern languages are not formal languages, but rather a collection of interrelated patterns, though they do provide a vocabulary for talking about a particular problem

Patterns can be used with any kind of software development, but are usually used with object-oriented development. A pattern is literature, not software and is the written explanation of a solution to a common recurring problem, or an attempt to describe successful solutions to common software problems. Patterns are written by senior developers and are built from experience. Patterns have four elements:

The pattern name. This sounds simple, but in fact can be difficult. Names should be limited to a word or two, but must define the problem sufficiently so that other people will use the pattern. Patterns often have aliases which are really nicknames or synonyms.

The context. A description of the situations in which the problem occurs. The context can be very broad, such as "all inventory systems." The context must define the conditions under which the problem and its solution occur.

The problem. This includes goals and objectives, and is a description of when to apply the pattern. For example, one problem in inventory systems is letting the user update the inventory by changing, adding and deleting items. Often the statement of the problem will include a list of conditions that must be met before using the pattern. Legal issues and constraints can also be included.

The solution. The solution covers not only the design elements that should be used, but also relationships and responsibilities. It provides an abstract description and a general arrangement of elements that should be used. The solution for our inventory update pattern could state that a screen should be built with options for adding, changing and deleting items. The screen must also contain authorization information, and a way of exiting the operation.

There are many types of patterns although design patterns are the most common. These patterns identify the participating classes or modules and describes when the pattern applies. Other pattern types include architectural (describes a set of predefined subsystems), process (describes the development process itself), analysis (used for the analysis function) and organizational (to work with organizational structure).

Frameworks

A framework is not an application, but is an infrastructure, and individual applications are built from one or more frameworks. Frameworks can be built with any kind of development, but are often used with patterns. The relationship between patterns and frameworks is general to specific. A pattern represents the abstract solution to a problem, and the framework is an actual software solution and is written in a programming language. It includes set of software building blocks that programmers can use, extend, or customize for specific computing solutions so the same code doesn't have to be written for each application. These blocks include code for such things as file and edit menus, print reports and pre-defined screens. In the object-oriented world the framework consists of a set of related classes, while with other development the framework consists of reusable modules. A framework must also have a basic infrastructure of code points to plug in the code for a specific application.

6. Programming Languages

Computers are machines, and as intelligent as they might appear, they are devices that only recognize the presence or absence of electronic pulses. Picture a computer as being made up of thousands, even millions, of light switches. Each switch is either on or off – it either has an electronic pulse or it doesn't. These "switches" are called bits (binary digits) and are assigned numeric values. If a switch is on, it has the value 1, if it is off, it has the value 0 (or vice versa).

Each computer is designed with a machine language. Machine language, or machine code, is the combination of bits that the computer recognizes as instructions. For example, the bit combination 1111 1010 could mean add, while 1101 0101 could mean compare. Each computer model has its own machine code. The code for an IBM mainframe is not the same as the code for a Unisys mainframe. In fact, it's not even the same as the code for an IBM microcomputer. Originally, computer programs were written in machine code. The process was not only difficult and time consuming; it was also inaccurate. Simply transposing ones and zeros would change an instruction completely. It's not surprising that computer languages such as COBOL and FORTRAN were quickly developed.

A computer language is an English-like language that has a set vocabulary and syntax rule to cover such things as punctuation and word order. Each language has a language translator, which is a program that takes the English-like instructions and converts them to machine code, thus creating a program the computer can execute. This means that

Figure 6.1 Hierarchy of programming languages.

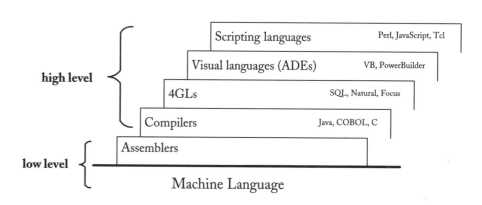

there really are two versions of every program: the English language version used by the developer, and the executable machine language version. There are several different levels of computer languages ranging from low to high level. Low level languages are more difficult to learn and use, but create the most efficient programs. They take more time from the developer, but less time while executing. The higher level languages are easier to learn and use, but the machine code programs generated are less efficient. Low level languages are the closest to the actual machine language and are the most technical. The high level languages are computer independent, can work with any computer system and require less technical knowledge by the programmers. Technical developers are more likely to work with low-level languages and applications developers with high-level languages. Programmers today usually know more than one language, and unless assembler language knowledge is required, it's important to realize that a specific language is the least important skill a programmer has. Languages can be learned quickly by an experienced programmer as most of the activities involved in programming, analysis, design, and testing are independent of the programming language.

Assembler Languages

These languages are also called mnemonic language, or mnemonic code. Assembler languages are closely related to a specific computer, and follow a one-for-one correspondence to its machine language. For example, a computer has as many as a dozen different instructions to add two numbers together. The instructions vary because of the location or size of the numbers. An assembler language must have a corresponding instruction for every one of these adds. Assembler languages have as many as 200 instructions that must be learned by the programmer. Because the instructions are so tied to the machine itself, assembler language is not transferable between machines. If someone knows assembler language for an IBM mainframe, he or she cannot use this language to write a program for a Control Data machine. While most of the assembler programming is done instruction by instruction, programmers can build macros by combining two or more assembler instructions. Working with macros is still low-level programming.

Assembler languages are rarely used in the applications environment, but are used by the systems staff. Because the languages are so close to the actual operation of the machine, more efficient programs can be written although there is a tradeoff in programming time. Actually, extreme efficiency is not always important in an applications environment. When a company runs its payroll, the programs might take hours to run, but they're only run once every two weeks. It makes more sense to save time writing the programs than to save time in execution. The reverse is true with systems programs. A program written by a systems programmer might produce the date on reports. Because this program will be executed hundreds of times every day, speed of execution is critical.

Compiler Languages

Compilers are called third-generation languages and are used for most critical programming. A compiler language is a general-purpose language and is not tied to any machine. Most of these languages were developed outside the computer industry. COBOL, the most popular of all the languages, was developed by the U.S. Navy. PASCAL was developed at a Swiss university. With these languages, the syntax, rules, and vocabulary were defined by themselves and then computer manufacturers wrote translator programs to take the same language input and create machine-specific output. This means that knowledge of a compiler language is transferable. Although minor changes are often needed, a COBOL program written for an IBM machine can indeed be executed on a Control Data machine. All the programmer has to do is use the appropriate translator.

Compiler languages are much easier to learn than assemblers, and programs take less time to write. A compiler, for example, will have a single add instruction. The translator chooses which of the dozen machine instructions to use, and this saves time for the programmer both in learning the language and in writing programs. Compiler languages are heavily used in applications programming and some, such as C and PASCAL, are efficient enough to be used in systems programming. Object-oriented languages fall into this category, and most critical programming is done in compiler languages.

4GLs (Fourth Generation Languages)

4GLs are also called user-friendly, query and non-procedural languages. These languages are almost free-form in structure and have few rules and predefined vocabulary. Because they were designed to be used by non-programmers (hence the title user-friendly) the translator program works from a simple definition of what the output should be and does not require step-by-step instructions to achieve the result.

4GLs concentrate on instructions to query databases. A query to a database can be asking for a response back to a terminal screen (such as

"Show me John Johnson's checking balance") or a complete report ("Produce John Johnson's monthly checking account balance"). These languages are easy to use and programs can be written very quickly using them. They are excellent for queries and reports, but do not have the control of assembler, or even compiler languages. They are used mostly by applications programmers to supplement work done in other languages, to create small systems, and to develop Internet and client/server applications. While learning and using a 4GL is simpler than working with a compiler language, it still requires technical knowledge and these languages did not prove to be "user-friendly."

SQL (Structured Query Language)

SQL is a 4GL but it is worth discussing separately. The language was originally developed to query relational databases, and has become an industry standard. Most databases, from all vendors, are accessible through SQL and it is used in mainframe, midrange, and desktop environments. Its capabilities have been extended to access databases of all types, and even data stored in file structures. More and more, SQL knowledge is becoming part of the necessary skill base for all programmers in business environments. Because SQL has been accepted throughout the entire computer industry, it is dominantly used in any multivendor environment. Client/server computing requires that different systems be able to function together, and all client/server environments use SQL.

Visual Languages (ADEs)

Visual languages are generally part of ADEs (Application Development Environments). There are systems that allow developers to use a GUI interface to create programs rather than writing code statements. ADEs are discussed under Software Engineering

Scripting Languages

These languages are specialized programming languages that process text strings, search files, databases and indexes, and generate reports. They do not have all the features of other languages and would not be used for critical programming, but are excellent for input/output intensive situations. They are easier to use than any other languages and have gained popularity with the growth of the Internet and Web development. Scripting languages have been used with operating systems for many years, particularly with Unix. Unix shell languages are scripting languages, as is the Unix extension Perl.

Interpreters

An interpreter is any programming language that works by translating the English language statements into machine code and immediately executing the code. This creates a slower executing program, as the translation and execution are both done every time the program needs to run, Interpreters are very flexible and allow the developer to immediately see results. Scripting languages are usually interpreted languages, as are some compilers such as Java and Basic.

7. Operating Systems

An operating system is a collection of programs that controls the resources of the computer system. These resources include the computer itself (both the processor and memory), peripheral devices such as disk drives, programs and data files. In order to do this, operating systems use file managers, memory managers and device managers. File managers handle the interactions between the memory and device managers. The memory managers handle both internal memory (the computer's own storage, or RAM) and external memory (storage provided by disks and tape systems). The device managers are concerned with the hardware. Device managers are often called device drivers.

In addition to acting as a resource manager, operating systems have one other thing in common; they contain the programs that handle input/output. This means that a computer is virtually worthless without the operating system, as no programs or data could be entered into the computer system. Beyond these two common functions, there is a variance in complexity in different operating systems for different size systems.

Figure 7.1 Operating system functions.

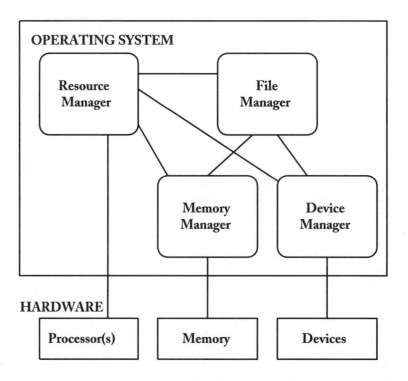

Mainframe Operating Systems

Operating systems for mainframe and midsize systems are written by the computer manufacturer and are purchased or leased along with the computer. The heart of the system is called the control program and includes I/O functions, job scheduling functions, and data management functions. It is the control program that provides multiprogramming (running more than one program in the same computer at one time) capabilities, security functions, cost accounting routines, and reporting and monitoring functions. Some operating systems also provide multiprocessing (supporting more than one processor). Reporting and monitoring are two of the most important functions. While the control program of all operating systems provides these same functions, the level of sophistication varies greatly. The control program keeps track of such things as disk usage and I/O traffic, and produces daily, weekly and monthly reports so that the technical developers and support staff personnel can ensure system resources are available to both applications programmers and the rest of the company. Technical developers are responsible for tuning the operating system to their own company's needs.

In addition to the basic functions provided by the control program, each operating system contains optional support programs. These programs include language translators, data transfer control programs (called Access Methods), and common utilities. Access method programs handle the actual data movement from external devices, such as disk and tape, in and out of the computer. Each operating system has many access method programs so that different types of data files can be handled. Utility programs provide common routines for such tasks as sorts, merges, data file copies, data file prints, and program and data file backup.

Communication with the operating system is handled through Job Control Language (JCL) and commands. JCL is identified by the name of the operating system which often has the letters "OS" somewhere in the name and is used to control the execution of batch jobs. JCL is used mostly by programmers and is a series of statements that define such things as the programs to be executed and the data files the programs will process. JCL, like programming languages, has a predefined vocabulary and syntax (rules for such things as punctuation and word order) that must be learned. After a program is written, a series of JCL statements will be used to tell the operating system the name of the program and name and location of all data files. The operating system will then find the program and data files and control its execution. The operating system works with the JCL for many different programs at one time, and performs its job scheduling tasks by scheduling programs for execution based on information contained in the JCL statements. Then, every time a program is executed, the operating system produces a report showing exactly how the operating system allocated resources to execute the job. Commands are used mostly by operators and are statements that are keyed into a terminal. Commands also have a predefined vocabulary and syntax rules, but are generally simpler and shorter than JCL statements. Commands can be used to control program execution, but are more commonly used to control peripheral devices. Commands are usually used one at a time rather than in a series like JCL. The operating system responds to operator commands by messages direct to the console.

The larger the computer, however, the more complex the operating system. And the more complex the operating system, the harder it is to learn. Therefore, some formal training usually must be provided when an IT professional transfers to a larger operating environment. Applications programmers at most must know the functions of the operating system and the JCL used with it. Because they need not know the internals of the operating system, training time is not extensive. Their knowledge of a company's operation (such as banking or insurance) can be much more valuable. Operators must know the command language, and increasingly are learning JCL. Because their work is solely communicating with the operating system, complete retraining, though not difficult, is necessary. Sometimes an operator's job is more complex in a smaller operating environment, as the operating system will do less. Technical developers usually know the internals of the operating system as it is their job to tune the system to their company's particular needs. While it is true that knowledge of one system makes it easier to learn another, moving a systems programmer from one system to another could involve extensive retraining.

If an IT professional moves from one operating environment to another, some retraining is involved. This training can be as simple as asking a few questions during the first week in the new job, but the new JCL and/or command language must be learned. In general, a move to a smaller operating environment (mainframe to midrange computer system) requires little formal training.

Operating System Enhancements

Many programs have been written that are really extensions to existing operating systems. The general function of all of these programs is to improve the operation of the computer center, and these programs are used by systems programmers and the operations staff. Most installations use enhancement programs to increase system efficiency. Any program that saves computer time or storage space (either disk or computer) falls into this category. Print spoolers are a common efficiency tool. These programs intercept data that is to be printed, store it temporarily on disk, and control the printing of all reports in the system. Other programs monitor disks to fully utilize available space and provide for recovery from hardware problems. Still other programs intercept program failures to maintain system activity. Most installations also use performance monitors and controllers. These programs do not improve the efficiency of the system operation, but they do keep track of the ongoing activity and provide reports on usage of all resources and control activity to eliminate most problems. Performance monitors exist not only for the operating system, but also for database and communications systems. Large companies often have many of these monitors. They could, for example, use one monitor for MVS (the operating system), another to monitor the hardware in the system, another for DB2 (database) and yet another for CICS (communications). Some of these programs also provide capacity planning functions, so an installation can know when its needs outgrow its present system.

Another very common system enhancement is cost accounting software. These programs vary in sophistication, but all provide some kind of chargeback system for using computer center resources. Most computer work is eventually charged back to the user department (such as payroll or accounting) that requested it, so these programs can be very important to the overall financial picture of the entire organization.

The operations staff gets involved with enhancements that provide operator console support. These programs deal with operator commands and messages, and some automate much of the operator's activity. Operations is also involved with disaster control packages. These programs usually run on desktop computers and allow the operations staff and systems programming to input details on all hardware and software locations and priorities. The disaster control program will then provide recommendations on what to do if, for example, a fire destroyed a certain part of the computer room.

Operating system enhancement programs are installed and maintained by the technical developers, and applications developers are usually unaware that they even exist.

Operating System Add-ons

In addition to operating system enhancements, most mainframe installations install software packages that supplement many functions of the operating system. These programs are different from the enhancement programs previously discussed because they do affect the applications programmers. Operating system supplements include data management packages, debugging and testing packages, librarians, and security/auditing software.

Data management packages include programs to monitor disk and tape file storage. These systems provide routines to compress data files, backup and archive data, convert data for accessibility by other systems and/or computers, and organize and optimize disk storage. Systems can also be purchased to translate both data and programs to run on a different computer. The most common type of data management system found in mainframe installations is a tape management system. The programs in a tape management system keep track of what tapes are being used in each applications system and provide reports showing the tape usage, number of scratch (available) tapes, expiration dates of data files on tapes, and information on backup tape storage.

Debugging and testing packages include syntax and logic analyzers and routines to monitor program execution. The most common of these are dump routines. A dump routine will take over when a programming error occurs and produce a report on the probable causes of the problem. Test data generators also fall into this category. All programs must be tested with a sample set of data that closely corresponds to what the actual data will be when the program is put to daily use. A test generator will create this sample data from given parameters.

Librarian packages keep track of programs. They record information about both source programs (the program written in a computer language such as COBOL) and object programs (the corresponding program in its machine code format). They automatically backup all programs and will retrieve the most current version of a program when requested. Reports are provided showing where programs are stored, how often they are changed, and when they were last executed.

Security/auditing packages make sure that only authorized individuals can access data and programs. They all work with IDs (identification codes) and passwords and both IT personnel and users must have valid IDs and passwords in order to use any computer system. Each ID/password combination has a list of programs and data files that the individual is authorized to access. In addition to individual security measures, these programs can provide security measures such as encoding data (change the format of data so it is not readable) and assigning password protection to data terminals (only certain terminals can be used to access some data). Security systems provide reports on any attempt to access unauthorized data.

Unix

Unix is a unique operating system, and is worth discussing separately. It was originally developed by AT&T, and was designed to work in all environments: mainframe, midrange, and desktop. Unix is the standard operating system for RISC machines, is heavily used with midrange systems, and is occasionally used on mainframes. It provides multiprogramming and multiprocessing functions, a graphic user interface (GUI), and support for multiple hardware platforms.

The heart of the Unix, or the control program, is called the kernel and is written in C. It provides the same functions as does the control program in mainframe operating systems. Specifically, the kernel contains the job scheduling, data management and I/O routines. It also supports multiprogramming, provides security functions, cost accounting routines, and reporting and monitoring functions. Different versions of Unix have different kernels. In addition to the kernel, Unix systems also contain optional utility programs including a hierarchical file management system, compilers and program development tools and system management tools.

Unix works completely with commands. Commands are read by a program in the operating system called a shell, so the programmer knows shell language instead of JCL. A Unix shell provides the user interface and reads commands from a terminal or from a prewritten script and performs actions such as executing programs and transferring data between programs. A typical command from a terminal would be to execute the payroll program. A script would be a set of prewritten commands to, for example, execute the validation program, then execute the program to update payroll records, and, finally, execute the payroll program. The shell language is actually a programming language called a scripting language, as it contains statements for logical decisions. For example, commands can be written to execute a program based on what happened during the execution of another program. In the script described above, the commands could be written to execute the payroll program only if the errors found by the validation program were minor.

Figure 7.2 Unix system organization.

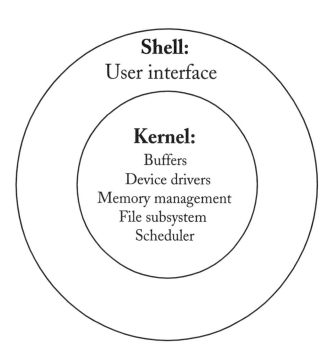

Unix installations can choose which shell program they wish to use. All of the available shells handle both interactive commands from the terminal user and prewritten command scripts. The standard and most popular shell is the Bourne shell. The C shell is also very popular and handles the interactive terminal commands better than does the Bourne shell. Its biggest problem is that it is not available with all versions of Unix. The Korn shell is an extension of the Bourne shell and includes C's superior handling of interactive commands. Most Unix systems run with the Bourne shell as it requires the fewest resources. When reference is made to the "shell language." or "shell programming," or "shell scripts," the absence of any other description means the Bourne shell.

The smaller Unix systems have no operators or systems programmers, but have a position called a System Administrator, which is a mix of the two jobs. The System Administrator is responsible for the integrity of the entire system, for installing new software, adapting software to the system, performing system backups, recovering lost data, and maintaining security. The job is hard to fill, as it is not really a programming job, but does require significant knowledge of Unix. Companies have found that when they assign junior personnel to this job, the person quickly learns, and wants to move into programming.

Unix can be used as the operating system not only for mainframe, midrange and desktop systems, but also for different manufacturers' machines. This means it can be considered a true general-purpose operating system. Many people feel that Unix, or a similar system, will become a universal operating system that is used by everyone, and "Unix" is constantly being upgraded.

Desktop Operating Systems

Desktop computers, like their big brothers, have operating systems that are necessary to their functioning. There are differences, however. Vendors of desktop computers do write proprietary operating systems, but generic systems are dominant. Unlike the large computer operating systems that are written by the hardware vendors and are machine specific, desktop operating systems are written for the computer chip the machine uses, so software vendors can produce operating systems that will work on many different vendors' computers. The most popular desktop computer operating system is the Windows operating system produced by Microsoft for microcomputers built with the Intel computer chip. This is often referred to as the Wintel system.

The interaction with a desktop operating system is different. Because most of the work done on desktop systems is online, desktop operating systems use a GUI (graphic user interface). A GUI is the part of the operating system that lets the user use a mouse and "point and click" to accomplish tasks. By selecting icon or menu choices, the user can ask the operating system to run a program, print a file or display a directory. Internally, desktop operating systems provide the same basic functions as any operating system, but not in as much detail. For example, a mainframe operating system might provide many cost accounting reports while none would be provided by the desktop system. These systems do provide multiprogramming functions, but not multiprocessing. The level of sophistication varies among these systems and there are different systems available to run on a single-user machine and on a larger machine functioning as a server. For example Microsoft has written Windows 95, 98 for the single user system and Windows NT for a server.

Desktop operating systems rapidly change. As the computer chips change, the operating systems must change to take advantage of the newer speeds and functionality. Systems currently take advantage of the 32-bit processing chips; systems for 64-bit machines will follow.

Desktop OS Support Software

Associated with desktop operating systems is operating system support software. These are packages which provide operating system type functions but are not associated with a specific operating system. In fact, much of this software works with multiple operating systems.

Much of the operating system support software works with networks and client/server computing systems. It does such things as balance batch job execution across networks, send all job requests to a central server which matches requests to resources, and provide database performance monitoring and tuning for client/server systems. Much of the software works with networks encompassing both large and desktop computers and extends large system functions to the smaller systems. An example of this would be providing mainframe backup services to desktop systems. In addition to the functions handling networks, operating system support software exists to provide many of the same functions as the operating system enhancements and add-ons for large systems.

8. Data Management

Data, in order to be processed, has to be organized in some manner so that it can be accessed. This is true in a non-computerized environment. Medical records in a doctor's office are stored alphabetically in filing cabinets; post office box numbers are assigned in numeric order; and books in a library are shelved according to the Dewey Decimal System. Electronic records are also stored alphabetically, numerically, or according to a code system. And just as people follow the organizational system to retrieve and store patients' records, computer programs follow the organizational system to retrieve and store records on electronic storage devices.

Files and Databases

Data is basically stored as a file or a database. Accessing data stored in files is simply handled by the operating systems and is relatively simple. Data stored in databases is much more complex and there are different technologies involved.

File Structure

Operating systems contain programs that retrieve and store data in files. A file is a collection of records that can be processed either sequentially (first record, second record, etc.) or randomly (retrieve or store a specific record based on a key value such as social security number or employee number). Each record in the file contains all the information necessary for processing. The file concept, however, has limitations. Payroll/human resources applications provide a good example of the limitations. Each record in a payroll file must contain all the information about each employee to produce paychecks, tax information, and perhaps information on other salary withholding amounts such as medical insurance payments. Chances are this file is too large to contain personnel-type information such as the employee's educational background, non-job-specific skills such as foreign language competency, and prior work history. This information would be stored in the personnel file. Both files would contain common data such as name, social security number, address, number of dependents, and job grade. This type of data is called redundant data because it appears in more than one place.

Sometimes data will be needed from both files. If a report were needed that summarized salaries by educational background, the program that does this would have to input and match records from both files. While this certainly can be done, the program would be long and involved. Redundant data causes problems. Not only is it expensive to enter and store data multiple times, but any updates (changes) to the data must be made more than once. This multiple processing is also expensive, and even more important, introduces errors. To solve these problems, software designers created DBMSs (DataBase Management Systems).

Database Structure

A database is defined as a total collection of data with limited redundancy. In a database structure, the payroll/human resource information discussed above would all be stored in a single database. The data would be grouped according to probable needs, and the database would contain a group of salary information, another group of tax information, one of the common data (name, social security number, etc.) and many more groups. Also, the common data items such as address would have to be stored only once, so redundancy can be controlled. Computer programs then retrieve as many of these groups as needed without having to match data from multiple files. Designers developed three different types of database structures: hierarchical, network, and relational. While the relational design is dominant the older databases have not been dropped. Companies that installed a hierarchical system such as IMS in the 1970s will be using and maintaining these databases for years to come even though new development is being done on

Figure 8.1 File, or flatfile, structure.

FILE DESIGN

PAYROLL FILE	HR FILE
Name	Name
SSN	SSN
Address	Address
Phone	Phone
Empno	Empno
Annual-salary	HS-Diploma
Biweek-gross	AA
Biweek-net	BA
Ins-deduct	BS
Ins-percent	Keyboarding
Biweek-state	OfficeSkills
YTD-state	Access
Biweek-FICA	Progress
YTD-FICA	Paradox
Biweek-disab	dBase
YTD-FICA	Lotus123

relational systems. These older systems are often referred to as legacy systems.

The design of a database is critical. There are two types of design;, the logical design which is done by applications developers, and the physical design which is done by technical developers. The logical design is concerned with the functions, or uses, of the database and consists of determining how many tables are needed, what fields should be grouped together in each table, the relationships between the tables and the probable activity against each table. Database design is done by senior applications developers. A new job and title has evolved with the development of enterprise-wide systems. A Data Modeler, or Information Modeler, is a senior applications person who designs and maintains standard data definitions and routines for the entire company, or enterprise. There are different data modeling processes, the most common of which is entity-relationship modeling. Entity-relationship modeling, or ER, defines entities such as salaries and taxes, and then defines the relationship between the entities. Relationships can be one-to-one (each salary has one tax), one-to-many (each salary can have

Figure 8.2 Database design structure.

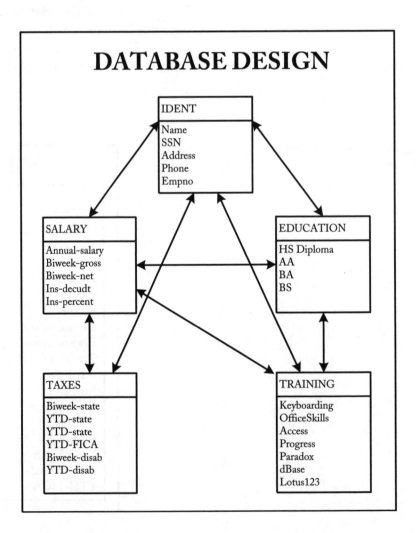

many taxes), or many-to-many (many salaries have many taxes). The entities and relationships are built into ER diagrams which are then used by application developers to create specific systems.

The physical design refers to the hardware and means determining how the actual tables are stored on disks. Physical design requires knowing the hardware, as it takes into account physical facts, i.e. data stored in the middle of a disk will be retrieved faster than data on the outside. Once a database is built, technical developers are responsible for tuning, or performance tuning. Monitor programs report daily on such things as response time and activity against the database so adjustments can be made if necessary. Tuning procedures also must handle the inevitable changes that will occur.

Installing a database system is a major undertaking, and most companies have more than one DBMS. All of these database systems consist of many programs and are often as involved as some operating systems. Not only do they contain routines to manage the data and its relationships, they also contain communications routines for online programming. The key person working with the data environment is the DBA (DataBase Administrator), who is usually a technical developer. This individual is usually technical and understands the internals (the programs that make up the database system) of the system, but could be design oriented and understand the intricacies of the data definitions. Often the DBA is a high-level manager and has technical and design groups reporting to him or her.

Specific database skills are required of applications programmers, analysts and designers, systems programmers and sometimes operators. The systems area commonly has a data group (people who specialize in the database) which reports to the DBA.

Repositories

Databases in many installations now work with data dictionaries and repositories. A data dictionary is actually a specialized database that contains data definitions instead of actual data. For last name, the database would have multiple entries such as Smith, Jones, Brown, etc. The data dictionary would contain a single entry that defined last name as a 23 character, alpha/numeric field. A repository always contains a data dictionary and provides a single point of control for all data management. In addition to the data dictionary, the repository can also contain rules and standards for managing data. Rules would be such things as "carry all math to three decimal places for all salary data," and "put area code in parentheses and have no space between area code and base number" for all U.S. phone numbers. Standards could be complete methodologies, or a simple statement. The data is not necessarily stored in the repository. Data, in fact, could be completely decentralized and be stored in various databases even from different vendors. What would be contained in the repository would be pointers to the data and rules for accessing it. Repositories are used with all enterprise-wide development.

Object-oriented Databases

The growth of object-oriented development has led to changes in the database arena, and there are now object-oriented databases (OODBMS) and object-relational databases, called universal databases. An object-oriented database stores objects which contain both data and methods, and the OODBMS lets developers create new data types containing traditional attributes found in relational data as well as the methods, or processes, that make up objects. New data types are often multi-media and contain video and audio objects.

Object-relational databases, or universal databases, extend existing relational technology, making it easy to convert legacy data and systems to new object-relational databases. Data from relational databases can still be accessed through SQL statements.

Multidimensional Databases

A multidimensional database is built by consolidating multiple databases and letting users share the combined information. This design provides fast response time for complex data queries, and is used with OLAP (OnLine Analytical Processing) systems. The growing popularity of data warehousing has increased the number of multidimensional databases available. Data is stored in cells that exist in multiple layers with a hierarchical structure, i.e. shampoo is part of health & beauty products, which is part of consumer goods. By storing the data in a hierarchy, data can be viewed in three or more dimensions such as time, place and product. Most multidimensional databases are really complete development environments and include tools to create front-end applications which are usually complex queries. the front-end application tools are intended to be used by non-technical people, and all layers can be queried.

Data Warehousing

A data warehouse is a blend of technologies that combine information from multiple databases into a single database designed for analytical processing. It is designed for complex information queries, not operational support, and the query capabilities are just as important as the data storage. A warehouse is designed with end-user access in mind and has a user-friendly interface giving users direct access to corporate data via powerful query tools. A mortgage banker could, for example, ask for a list of potential first-time house buyers from all present customers, or a sales manager could ask which sales region has consistently sold the most over the past six quarters in high ticket items. A data warehouse allows you to keep managerial forecasting and planning queries off the operational databases where they might cause performance problems.

Detail data in the warehouse is collected over time and used for comparisons, trends and forecasting and is migrated from operational systems on a regular basis. Because data is examined differently, the model for building data warehouses demands an outlook and skill set that are different from the skills for building classical operational applications. The detail data that is fed directly from operational databases represents the entire organization, and warehousing is by definition an enterprise-wide activity.

Data is stored in a warehouse at different levels from detail data to highly summarized information called metadata. The detail data is the heart of the warehouse and is stored on a parallel processor. It is organized along subject lines and is maintained for two to five years. It is often refreshed daily. Summary data is maintained for older information and metadata is built for both detail and summary information. Metadata is defined as "data about data" and can be compared to a card catalog in a library. Each card contains some data such as the name, author, publisher, publication date, and number of pages in the book. The card also contains the address, or location, of the actual book. Cards for non-fiction books also contain suggestions for other topics that could be helpful. Metadata is similar and contains basic information such as the description of the data, address or location of the data, and date/time stamps detailing the creation of the metadata. Designers build as many layers of metadata as necessary in order to satisfy the queries. Designing the metadata is the key to successful warehouse development. The warehouse is a historical database and the metadata must be built providing summarization techniques to make it easy for end users to use historical information.

Data modeling for a warehouse is the process of selecting the data from the operational databases and deciding how to build the summary and metadata. A company must decide if it wants a single comprehensive, centralized warehouse, or should it use the data mart concept, which means building several smaller databases aimed at specific users. Data replication is often used to update the warehouse. Replication is automatically copying changes made to one database (an operational database) to another (the warehouse). Because any changes are automatically copied, the warehouse data is synchronized with the operational data. This is complex processing, as operational databases are updated by multiple users and a warehouse receives data from many different DBMS types and can, in fact, be a completely different DBMS from any being used. Whatever the choice, once the warehouse is designed, tools must be chosen that will automatically feed operational data into the warehouse, manage the metadata, and handle the user queries.

Warehousing query tools are often called Decision Support Tools (or Decision Support Systems, DSS) and provide the ability to retrieve, manipulate, and analyze data. Analytical, or complex, queries ask questions such as, "How many cars did we sell in the Midwest during the first quarter of the past two years that had a CD audio system and a base price of less than $25,000?" The presentation of the information is often in the form of charts and graphs varying locations, time periods, price limits, etc. Other queries fall into a category called data mining, which is searching for hidden patterns and associations. Data mining for buying habits of grocery store shoppers turned up the unexpected fact that buyers of beer more often bought diapers than pretzels – a fact which, once known, led a grocery chain to move the pretzels back with the chips and move diapers next to the beer in all their stores.

Data warehousing is an example of OLAP software. OLAP, or On-Line Analytical Processing, is software which creates new business information from existing data. It presents a multidimensional view of data which is independent of how the data is stored. Queries usually involve multiple passes through the data and create summaries and build models for forecasting, spotting trends, and doing statistical analysis. Answers to queries can be by reports and/or 2D or 3D charts and graphs.

Companies have long used computer systems to support their operational systems. Data warehouses take a next step and provide support for managerial functions such as revenue growth, asset management, and market share positioning. Companies are eager to explore the uses of warehousing and it has become one of the fastest growing areas of computer use.

9. Communications

A communications, or online, system is a group of programs that control data transfer between computers, terminals, and other devices such as printers and disk storage. It is a communications system that controls the daily computer activity that is typified by a user working on a terminal. While terminals are thought of as devices with a video display screen and a keyboard, they actually can be any device that can send input and receive output. Telephones are often terminals, as are the sensing devices used in grocery stores to capture item and pricing information. Data communication systems do much more than control these online applications. They also link computers and storage devices to allow transfer of data and even programs between machines.

Communications systems require additional hardware and software. Because the terminals are often long distances from the computers, the system must use telephone lines, satellites, fiber optics, infrared transmission, or even microwaves, for data transmission. Modems (modulator, demodulator) are used to convert parallel signals (signals used by the computer) to serial signals (signals used in communications lines) and back again. Other pieces of equipment are bridges (devices which connect two or more networks), routers (devices which route messages through different communications lines) and multiplexors (devices which consolidate input and output from several terminals).

Networks are the basis for any communications system. The term network refers to both the physical equipment with its connections between the computers and terminals, and the software that supports the communications. All networks are based on protocols, which are standards for the hardware and software involved in the network. Network software is written and maintained by technical developers and they know both the protocols and the networks. Networks are usually transparent to applications developers. Often the people who work with networks are called communications specialists and have expertise in both data processing and communications. While knowledge of networks and protocols is necessary only for technical developers, applications programmers do need to use communications systems. This includes: OLTP (OnLine Transaction Processing), or communications control systems; EDI (Electronic Data Interchange) systems; terminal emulators; and electronic mail systems. Communications also includes the entire field of middleware and servers, client/server systems, and the Internet (including e-commerce).

OLTP (OnLine Transaction Processing)

These packages range from simple data transfer programs to the very complex applications systems such as CICS (Customer Information Control System) that control the interaction of terminal input/output with applications programs. An OLTP, or communications control system has program management functions, data transfer routines, data sharing capabilities, and security routines.

A communications control system becomes an inherent part of a company's operation and runs during working hours. Users will "log on" to the system, which will receive all input from terminals and control the execution of applications programs to handle the requests. For example, if a bank is using CICS as its OLTP system, CICS would receive a teller's request for an account balance, direct the operating system to find the program that handles this type of request, control the execution of the program, receive the appropriate data from the operating system and the user, and transmit the information supplied by the program back to the teller's terminal.

All areas of an IT department can be affected by the communications packages a company uses. Applications programmers and analysts must know how to use communications control systems in order to design and write programs. In other words, programmers must know CICS commands or macros in order to write online programs in a CICS environment. It is these commands that build the screens that are sent to the user at a terminal. The support staff will have operators specially trained in communications packages. A CICS operator will monitor the online system to make sure all authorized terminals can access the system; that response time is quick enough; and that there are no problems with hardware, software, and/or communications lines. The area most affected is the systems area. Companies have technical developers who specialize in the technology of the communications industry, and know the intricacies of a specific communications control system, such as the CICS system. Imagine what happens if a bank teller enters an account number into a terminal and gets no response. It is the systems group that must determine if it is a problem with the terminal, the data lines, other communications hardware, one of the CICS programs, one of the operating system programs, or the application program.

EDI (Electronic Data Interchange)

EDI, Electronic Data Interchange, started as the automation of purchasing, billing and shipping paperwork. In an EDI system, a purchase order is keyed into the system at the purchaser's site. The EDI software translates the information into an acceptable format and transmits the order to the seller, or vendor. The EDI software at the vendor's site accepts the data and translates it to a format that its own software systems can process. The order is now handled electronically; inventories are updated, packing and shipping instructions are produced and invoices are electronically transmitted back to the purchaser. The purchaser may elect to conclude the transaction by using Electronic Funds Transfers (EFTs) to pay the invoice.

The purchaser and vendor in an EDI system are called trading partners, and both must have EDI systems. The systems do not have to be the same, so a system must have the capability of communicating with many others. EDI software must provide a translation of data from the fixed formats necessary for internal, or local, processing to the variable formats necessary for transmission. The software must either have a network or the ability to use existing networks. Third-party networks, called VANs (Value Added Network) are commonly used. These networks are available to companies for fees and add such features as reformatting data messages to the specific requirements of, for example, overseas companies. In addition to the translation and network software, EDI systems must contain security checks. For example, all data transfer is sent with header and trailer information that is checked to ensure no data has been lost. The extent of security provided is one of the major differences among EDI systems. Systems also differ in how extensively business analysis functions are provided. When orders come into a company electronically, it is possible to analyze quickly such things as what product is selling well, and from what geographic region.

There are many reasons to use EDI systems. First, of course, is the time savings afforded by eliminating the mail as the main source of placing orders and sending invoices. The electronic speeds give real-time ordering and billing capabilities. Close in importance is the inventory control offered by these systems. Not only can inventories be updated immediately when an order comes in, but the accuracy of the updating function is greatly improved by receiving the order electronically and eliminating the need to re-input the order information. In fact, because there is only one input point – the original purchase order – errors are both identified and reduced. Error control is one of the main reasons for using EDI. The uses of EDI have expanded beyond handling purchases. EDI is now used for any business function and companies are using EDI software to handle such things as booking and tracking commercial shipments.

The future of EDI is now tied to the Internet, as more and more companies are setting up e-commerce systems to accomplish the same tasks. E-commerce is the overall name for systems that conduct business over the Internet, and putting the functionality of EDI in these systems makes sense. With traditional EDI each company must define and establish links with each business partner. This is not necessary over the Internet as the linkage is already established. Therefore, many EDI systems are being replaced with Internet e-commerce systems. E-commerce is discussed in the Internet chapter.

EFT (Electronic Funds Transfer)

Electronic Funds Transfer may or may not be part of an EDI system, and refers to the electronic transfer of actual money. EFT systems were originally designed to work among banks, and that is still their primary use. Most companies prefer to control money transfer and take advantage of the float time that exists between writing a check and the actual transfer of money to the recipient.

Groupware

Groupware started out as any software that takes common single-user functions such as calendars, word processors, databases, and notepads and incorporates them into a multi-user network. Typical functions would be entering a meeting date and time and letting participants check the system, thus eliminating a lot of telephone scheduling. It has, however, grown much beyond the basic calendaring, scheduling and e-mail functions originally provided. Groupware is now a broad category of software that ranges from sophisticated e-mail systems to entire office automation packages and includes such things as group authoring, calendaring and scheduling, conferencing, information sharing, document and image management, project tracking and work-flow management.

Groupware software is used by workgroups. People in workgroups usually work in the same building or building complex, although recently workgroups are spread across remote locations. The more sophisticated groupware systems have a database associated with them, and communications is usually handled by a LAN. The variety of people and jobs within a workgroup means that the system must handle different operating environments. One workgroup member might be a sophisticated workstation user with needs for multiple databases and windowing applications, while another might be a casual user.

To understand the power of groupware, picture a board in an office that everyone uses to record ideas. Every person writes his or her comments on the board, whenever it is convenient. Comments can include spreadsheets, graphics, drawings – anything your computer system supports. The board might get full and hard to digest, but groupware provides functions that will organize the comments so you can review them by various criteria such as individual and time. Groupware also provides the communications links so that the single board in a single office is now a screen on multiple computers as defined by the workgroup. And, a change on any of these machines will appear on all systems in the group.

Electronic Meeting Systems

Another type of groupware is conferencing, or electronic meetings. EMS (Electronic Meeting System) software provides support for these meetings. Conference rooms can be used, but meeting participants each have a workstation so they can participate in a meeting both electronically and verbally. Over long distances EMS is combined with teleconferencing. Unlike standard workgroup processing, with an EMS participation can be anonymous, so electronic comments on ideas can be more forthcoming. Electronic Meeting Systems are often used in Joint Application Development.

Electronic meetings cannot replace traditional meetings. If the purpose of the meeting is planning and technical work, an electronic meeting is fine and offers advantages of instant minutes of the meeting, equal participation and input, etc. However, when the meeting includes negotiations or sensitive subjects, a face-to-face traditional meeting with subtleties such as body language is more valuable.

Workflow

One of the fastest growing areas of groupware software is workflow. Workflow is defined as the automation and management of business processes. A business process is a sequence of tasks performed in a specific order by specific people in order to meet a business need. An example would be handling business trip expenses. The individual going on the trip might ask for a travel advance. This would require filling out a travel advance request form and taking it to his or her boss. The request, when approved, would be forwarded to the cashier. The cashier would prepare the advance and notify the worker it was available. When the trip ended, an expense account form would be filled out and the travel advance request would be attached. The worker would forward this packet for approval. When approved, the packet would be forwarded to the cashier who would prepare additional money owed, or ask for a refund. Workflow software automates much of this. The worker fills out an electronic form at work or perhaps from a home computer. The workflow system forwards the form through the entire business process and even handles the math involved in deducting the travel advance from the final expenses. Workflow software combines electronic messaging with document management and imaging. In a workflow system the documents are created with routing information to control the distribution of the document. Simultaneous access to documents is allowed and the workflow software controls the eventual document filing and retrieval. Authorized users sign-off at various stages and a sign-off locks all or part of a document from further modification. Processes such as expense processing are well handled by workflow systems.

E-mail (Electronic mail)

Electronic mail systems concentrate on the creation and delivery of messages and even long documents from one terminal to another. In the systems previously discussed, terminal-to-terminal communication was not important. The systems concentrated on terminal-to-computer or even computer-to-computer links. Programs in the systems processed the data being transmitted. In an electronic mail system, however, the data is simply switched from one terminal to another. E-mail is another area that has been greatly affected by the Internet. Many companies use e-mail systems that have been set up over their own or public networks or through groupware systems, but most e-mail is now being conducted over the Internet.

Closely related to electronic mail systems are telephone management systems. These products provide tracking and monitoring of incoming and outgoing calls, call accounting, and automatic call distribution.

Networks

Communications knowledge for technical developers starts with knowing the networks. A network is any link between computers. The network software supports distributed processing, or wide area networks (WANs), networks established for a specific locality called metropolitan area networks (MANs) and local area networks (LANs). Communications is accomplished by having WANs connect LANs.

A distributed processing system, a WAN, is an environment that has computers in several locations that can operate both independently or

Figure 9.1 Local area networks connected by a wide area network.

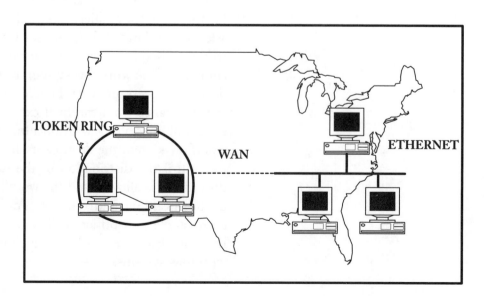

in a link with a main, or host, computer. The host computer is usually a mainframe, while the distributed machines are often midrange computers or, increasingly, desktop computers. The network programs not only control the data transmission between the machines, but also control sharing of disk, printer, and file storage at the main computer site. Often a WAN also provides the long distance linkage necessary to connect LANS from various sites.

A LAN is a network that links computers in the same site although a "campus-area" LAN can be set up. LANs also control the sharing of peripheral devices and data. The way data moves from one system to another in a LAN is controlled by the topology. There are half a dozen different topologies, but only three main structures. A ring topology passes information from computer to computer in a ring format and Token Ring is the most common example. Starlan follows a star topology in which the computers connect through a central node. The most common topology is a bus topology (Ethernet is the best example), where all the computers are connected on a line like a bus route and information is passed along the line. Knowledge of the topology is often a necessary skill for technical developers who have chosen to follow the network specialty.

NOS (Network Operating System)

An NOS is a software package that provides both operating system and LAN (Local Area Network) functionality by doing work such as centralizing file and print services. An NOS can handle diverse systems. It runs on the server in a local area network (LAN) and interfaces with other operating systems. NOSs differ both in the functions offered, and the depth of control for each function. The most common functions are: 1) configuration management, 2) fault management, 3) performance management, 4) security management, and 5) accounting management.

There are different types of network operating systems, peer-to-peer and server based. A peer-to-peer system allows each computer in the network to decide which of its resources (printers, disks, etc.) can be accessed by the other systems. No single computer system is in control of the resources. Peer-to-peer systems are used when security is not a major issue. LANtastic, Windows for Workgroups, and the network operating system functionality included in Windows 95 and Windows 98 are examples of peer-to-peer networks.

A server system designates one computer in the network to function as the server and this single computer has access to the resources that will be shared. The NOS runs on a computer called a server, which is usually a larger computer than the other machines in the network. The NOS, or the server, then controls the sharing of both hardware and software and provides security functions by controlling access on either or both a person/machine basis. NetWare, LAN Manager, LAN Server, and NT Server are all examples of centralized, or server based network operating systems.

Large systems end up with a hierarchy of networking. An NOS can control a single LAN, but often it is necessary to link LANs together. This requires another level of software, network management systems. A network management system controls network operations in a multi-network, multi-operating system environment. This software usually works with networks and operating systems from various

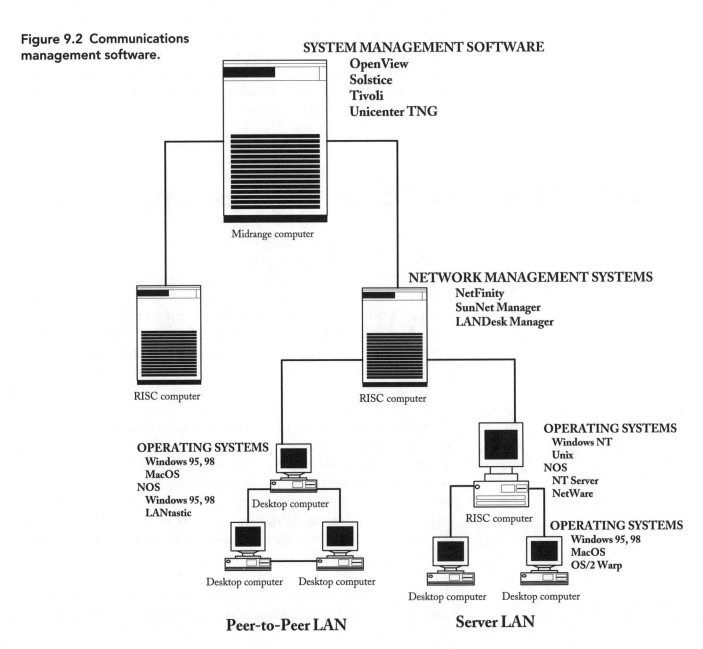

Figure 9.2 Communications management software.

SYSTEM MANAGEMENT SOFTWARE
OpenView
Solstice
Tivoli
Unicenter TNG

Midrange computer

NETWORK MANAGEMENT SYSTEMS
NetFinity
SunNet Manager
LANDesk Manager

RISC computer

RISC computer

OPERATING SYSTEMS
Windows 95, 98
MacOS
NOS
Windows 95, 98
LANtastic

Desktop computer

Desktop computer   Desktop computer

Peer-to-Peer LAN

OPERATING SYSTEMS
Windows NT
Unix
NOS
NT Server
NetWare

RISC computer

OPERATING SYSTEMS
Windows 95, 98
MacOS
OS/2 Warp

Desktop computer   Desktop computer

Server LAN

vendors and includes features such as providing a graphic display of network components, displaying alarms, collecting and graphing statistics, and implementing user-specific management tools.

Still another level of management software, system management software can be used in place of network management systems, or on top of them. This software includes functions from both operating systems and network management systems and provides enterprise-wide management of communications systems, and are used only by companies who need enterprise-wide control.

Companies must have a NOS to control the basic network. Linking computers together doesn't do any good unless that link can be controlled. The other software, network management systems and system management software, is optional.

Protocols

All networks follow protocols. A protocol is simply a set of rules agreed upon for data communications. Each computer vendor developed its own protocols as it developed communications systems. To list just two, IBM's protocols are called SNA (System Network Architecture) and DEC's are called DECnet. Quickly, however, the need for standard protocols became obvious as companies used equipment and software from more than one vendor. CCITT, the Consultative Committee for International Telephone and Telegraph, originated a protocol called X.25. CCITT is part of the United Nations, so this is an international standard. TCP/IP is the dominant standard protocol today. Its popularity has two reasons: it is the standard in Unix systems, and is the standard for the Internet. In fact, the only rule that must be followed to connect to the Internet is that TCP/IP protocols must be followed. SNA, DECnet, X.25, and TCP/IP are most common, but there are literally hundreds of protocols.

OSI (Open Systems Interconnect)

Open Systems Interconnect, or OSI, is a set of standards and definitions by an international organization whose goal is to provide the structure for "open" communication. Open communication means that data may be transferred between hardware and software from different vendors without special programming. While standardized protocols such as X.25 and TCP/IP have provided a measure of "open" communication for several years, OSI goes beyond simply providing protocols. OSI defines the work that is done during the communication process and breaks this work into seven "layers" of communications. It then sets rules, protocols, and definitions for each layer so that each operates independently. Layer 1 defines the work that is closest to the hardware and actual transmission of data, while layer 7 defines rules that work regardless of hardware involved. The lower layers are the most technical and require knowledge not only of computers but also of the communications industry.

Physical Layer. The physical layer is the OSI layer that defines the most basic communications functions, the handshake signals. Handshake signals are the signals sent along communications lines to establish the data transfer. A handshake consists of a request from the originating system, an indication and response from the receiving system, and a confirmation from the originating system prior to any transmission of data. Work at this level of communications requires thorough knowledge of the hardware involved, and knowledge of both the computer and communications industries. Engineers rather than computer specialists do the work at this level.

Data Link Layer. This is the layer that provides standards for such things as bandwidths and error control. The data link layer is responsible for maintaining reliable communication between each node (any

Figure 9.3 OSI communications layers and protocols.

OSI DEFINITION

LAYER	NAME	SAMPLE PROTOCOLS
7	Application	SMTP, FTAM
6	Presentation	AFP
5	Session	APPC, ASP
4	Transport	SPP, SEP
3	Network	X.25, ACMP
2	Data Link	HDLC, Bisync
1	Physical	Set by IEEE

Part One

connection point in a network). HDLC, SDLC and DDCMP are data link protocols.

Network Layer. In communications, the work done at the network layer finds the path in the network, or routes the data. This involves monitoring activity on the networks to make sure overloads do not occur. X.25, and LU6.2 are network protocols.

Transport Layer. The transport layer governs multiple network connections and is responsible for insuring reliable data transfer even though several different networks may be involved. Work done at this layer is similar to the work done at the data link layer except that the transport layer is responsible for the entire communication from source to destination. This layer is also responsible for packet switching, or the breaking apart of long messages into smaller fixed-length packets. Packets are then routed to the destination as independent pieces and may not travel the same route. The packets must be reassembled in the correct sequence at the receiving system.

Session Layer. The session layer provides tools for applications such as , checkpoint/recovery and inquiry/response. This layer establishes and terminates the communications links between the two parties involved in the communication.

Presentation Layer. This layer is concerned with transforming data from the internal representation of one computer to the internal representation of another. The most common transformation would be from EBCDIC to ASCII character code systems.

Application Layer. The final layer, the application layer, provides services to actual application programs. The functions included in this layer are: (1) file transfer and directory operations, such as rename, delete; (2) message handling services, such as ; and (3) job transfer and remote job management. Examples of application layer protocols are: FTP (File Transfer Protocol), SMTP (Simple Mail Transfer Protocol), and NFS (Network File System).

OSI standards are based on SNA, DECnet and TCP/IP. Full agreement by all computer vendors on any standard is not yet a fact so completely "open" communications is still a way off.

10. Middleware and Servers

Middleware is a very broad term. The most basic definition states that middleware is any software that connects heterogeneous computer environments. Put another way, middleware is any software component that sits between end-users at PCs and the database or application system that manages the underlying data. Originally, online systems ran with a single host computer that did all the processing, and dumb terminals allowed the user to communicate with the application running on the host. These terminals were simply I/O devices and did none of the processing. This is no longer how online, or real-time, systems commonly work. Most users now work from a personal computer that does some of the processing. Because the systems have two different computers that need to communicate, middleware is required.

Middleware is confusing because there are so many different types of software that fall into this category. Five categories are generally recognized: database middleware, remote procedure calls (RPC), object request brokers (ORBs), transaction-processing (TP) monitors and message-oriented middleware (MOM).

Database middleware. This software allows client systems to request data from one or more databases through a common access API (Application Programming Interface). This is the most basic form of middleware. Examples are Merant's Data-Direct and Simba Technologies' SimbaExpress.

Remote procedure calls (RPC). This form of middleware allows the client to ask for services, or functions, not just data. For example, the called procedure could check security or convert the data from one form to another. RPCs allow the program logic to be distributed across the network, as any part of the application could run on any system in the network. The calling programming sends a message and data to the remote program, which is executed, and results are passed back to the calling program. Typically the client must wait until the requested operation is completed, and this can result in long program waits. Most RPC software is written following DCE (Distributed Computing Environment) specifications.

Object request brokers (ORB). ORBs are used in object-oriented systems and locate, load and execute objects. It is the link between an object application program and the operating environment. An ORB lets an application request use of another object without knowing where that object is on the system or network. Objects created in different environments (i.e. C++, Smalltalk, Java) cannot work together without some form of intervention. Object request brokers supply the commonality. ORBs allow users to build systems by piecing together objects from different vendors and across different platforms. The two dominant ORB technologies are CORBA (from Object Management Group) and COM/DCOM (from Microsoft). A newly-emerging ORB model is Remote Method Invocation (RMI). RMI is part of the Java virtual machine and allows Java objects to be executed remotely. BEA Systems' ObjectBroker and Iona Technologies' Orbix are examples of ORBs.

Transaction processing monitors (TP). This software starts with the most basic database functions, but adds increased functionality. Some of this functionality includes connecting to many databases at the same time; controlling the completeness of the transfer of all of the data across the multiple databases; controlling the access of hundreds, or even thousands of users; performing load balancing functions; and maintaining a pool of database connections to share with all users. TP monitors are quite complex. Examples include BEA Systems' TUXEDO and IBM's CICS.

Message-oriented middleware (MOM). Most middleware falls into this category, and message-oriented middleware can be the most

Figure 10.1 Message-Oriented Middleware.

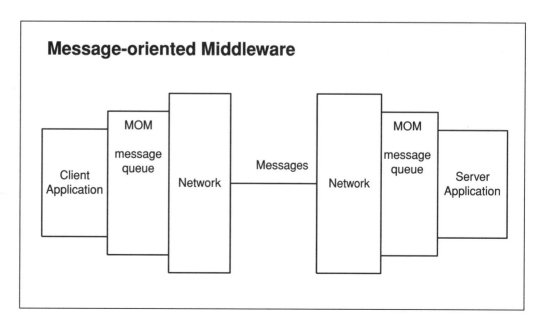

complex of all the types. Its major advantage is that it offers asynchronous processing. This means that the calling program, or client, sends a message to the system but does not have to wait for the response. It also enables applications to exchange messages with other programs without having to know what platform the other application resides on within the network. The messages can contain formatted data, requests for action, or both. IBM's MQSeries is message-oriented middleware.

EAI (Enterprise Application Integration)

Another category of middleware software is EAI (Enterprise Application Integration) software. This software integrates all of the company's applications, including software purchased from vendors and that which is written in-house. Many EAI systems were written to work with ERP systems such as SAP's R/3 and PeopleSoft. EAI includes three types of software – brokers, adapters and agents. Brokers reside on a server and process the requests from the users. The broker is responsible for security and monitors all activity. There are several types of brokers including integration brokers and information brokers. Adapters are written for specific vendor supplied packages including ERP systems, databases, languages, operating systems, and even other middleware systems. This software allows different information resources to seamlessly connect to the broker. The third type, agent, works with the brokers and actually does the integration of the diverse systems. Agents perform conversions, manage processes and handle the business processes.

Servers

Most server software is middleware. The word "server" is another broad term that is used for many things. First of all, it describes both hardware and software. Server software is any program that makes anything – data, other programs, services, devices – available to any other program. A server computer is any computer that runs server software!

Servers are also classified into different types. Some common servers that also fit the middleware definition:

Database server. A database server is a key component in a client/server environment. The term refers to both the hardware and software. It holds the database management system (DBMS) and the databases. Upon requests from clients, it finds selected records and passes them back over the network.

File server. Allows everyone on a network to access files stored in a central place. It includes a computer, disk storage for the data files, and the data management software. It often supports e-mail functions.

Internet server. Provides Internet services including e-mail, news, Web pages, etc. Internet servers include e-mail servers, FTP servers, News servers and Web servers.

Web server. The software that handles requests from Web browsers to download HTML pages. It can also execute scripts that automate functions such as searching databases.

Application Servers

Application servers are growing in popularity because of the Internet. An application server divides the application into different processes that can run in different locations and is used in n-tiered client/server architecture. In these systems the business logic (or the actual processing) resides on a "centralized" middle tier while the database server resides on the server tier and the GUI runs on the client. While this architecture adds additional overhead, it is necessary for large systems. Application servers are also inevitable on the Internet as these Internet applications are automatically partitioned into at least three tiers.

Figure 10.2 Application Servers.

Application Servers

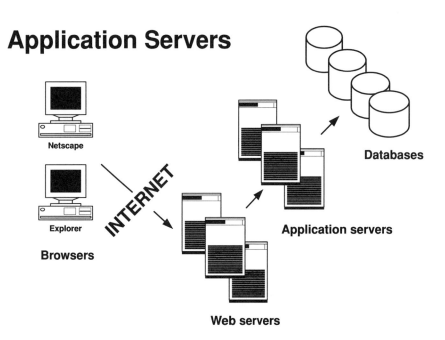

While there are no standards for application servers, there are some features that are usually included. The server must be able to handle a varying access load. Web applications might typically have hundreds of users a day, but a special event such as a sale, or mention of a product on the nightly news, might raise the hits to hundreds of thousands. The application server must be able to scale the application to handle the extra traffic. The application server must also be able to connect to all kinds of data and provide transaction management services such as those provided by TP monitors. In fact, some TP monitors have been enabled as application servers, and some application servers include full TP monitors. Additionally, the server must simplify and manage the multi-tiered architecture and support multiple servers. Finally, many application servers contain complete development environments. Even those that do not include development functions provide interfaces existing development systems.

11. Client/Server

Client/server computing is a type of computer system architecture that encompasses a "server" system that can service many "clients" over a network. The client utilizes the cost/performance of small computers to process the user interface and free up the server, a larger computer, to concentrate on the data-intensive operations. Originally, most of the processing of the data was done on the client system while the server provided data distribution and security functions, but now the processing is more often done on the server system or even a different system altogether. The term client/server is often called cooperative processing or network computing and is a subset of distributed processing.

In a typical client/server environment, desktop computers called clients provide a user interface (which is always a GUI) and perhaps do local data manipulation. Larger computers (large desktops, midrange and even mainframe computers) act as servers to handle database access, security and calculation intensive processing for all the clients. For example, consider a client requesting records for all customers with accounts 30 days overdue. In a typical mainframe system, the host computer would find the requested customers and do whatever processing

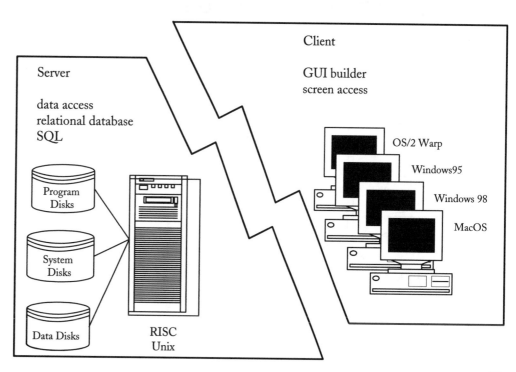

Figure 11.1 Client/server partitioning.

was required. In a local area network (LAN), or a peer-to-peer network, the file server would pass the entire customer file to the requesting computer and the search for the requested customers and all required processing would be done by the computer. In client/server computing, the data access would be done by the server machine, and the user interface would be done by the client. The location of the actual processing, however, varies.

Application partitioning, the splitting of the application logic between client and server nodes in a client/server environment, has no accepted standard. The approaches currently being used include Distributed objects, Multi-tiered (server-based), Two-tiered (client-based), and Three-tiered. The distributed objects method requires knowledge and skills in object-oriented programming and builds objects which can reside in either the client or server. The processing is part of the object, so processing can be in either location. Multi-tiered (server-based) processing splits logic between the server and clients and is defined specifically for each application. The two-tiered (client-based) approach puts all the processing logic on the client systems with the server providing just database access and is often referred to as fat client partitioning. Three-tiered partitioning actually splits the application into three parts with the data manipulation being placed in a third program which often runs on yet a third computer system. In addition to these methods, there are other approaches under development such as Dynamic partitioning and Repository-based code generation.

Figure 11.2 Multi-tiered partitioning.

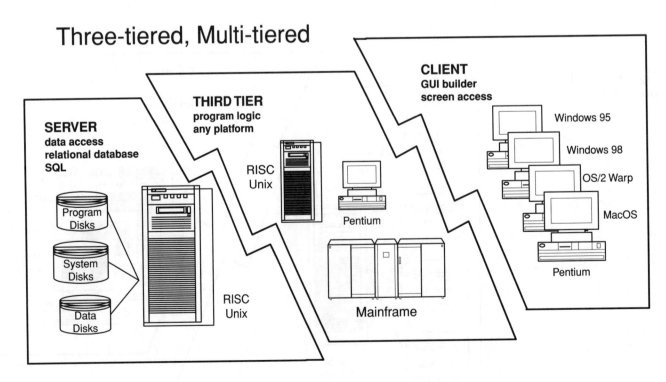

Three-tiered, Multi-tiered

SERVER
data access
relational database
SQL

Program Disks

System Disks

Data Disks

RISC Unix

THIRD TIER
program logic
any platform

RISC Unix

Pentium

Mainframe

CLIENT
GUI builder
screen access

Windows 95

Windows 98

OS/2 Warp

MacOS

Pentium

Part One

Dynamic partitioning locates the processing at run time. Repository-based code generation stores all application requirements in a repository and lets the system generate to code for various platforms.

While original client/server systems followed a two-tier approach, most current systems are either three- or multi-tiered (often called n-tiered). In a multi-tiered architecture, a middle tier, or tiers, handles the needs of many users. This middle tier can take many forms and provides a variety of functions including application execution (actually processing the data) and queuing (keeping and prioritizing a list of data requests. The middle tier can run on any computer system and runs independently of the front-end (the GUI) and the back-end (the data access). If there is a lot of processing, the middle tier could run on a mainframe even though the front-end runs on PCs and the back-end runs perhaps on midsize systems. In object-orient systems, the middle tier is called an ORB (Object Request Broker) and competing standards are CORBA and COM/DCOM. In Internet applications the middle tier is often an application server.

In all the partitioning methods, the client does front-end processing; the server does the back-end work. While client/server computing can be established with any computer and operating system, it requires a relational/SQL DBMS (DataBase Management System) and development tools to build a GUI front-end. This helps make these systems infinitely expandable and produces a computing environment that can grow with a company. It also allows a company to permit multiple platforms for the clients so that individual users can continue in the operating environment they prefer. Within the same system some users could be using Macintosh systems as clients, others Unix systems, and still others Windows.

In all client/server systems controlling the network is done on the server machine, as is handling database management, security functions, e-mail, and batch processing. The client systems then provide the user interface. All client/server systems use a graphic user interface (GUI) and providing the point-and-click interface to users is the main reason for using this design. Because multiple client applications can simultaneously access the same database, client/server systems are developed using SQL-based, relational databases. All databases that use the standard SQL query language follow the same standards, therefore applications can be written across multiple database systems. These systems can then run on most hardware. The relational design allows the system to handle multiple access to the same database.

Finally, there are many development tools available for client/server computing and more are being developed daily. Many of the available tools concentrate on developing client, not server, applications and these tools assist in the development of the screens and menus that make up the user interface, help generate the calls to the server for data and other services. Other tools help develop the application logic and are often called application generators and report generators. Some tools actually automate the partitioning of the application.

Client/server has become the standard architecture for online systems and most new online systems are developed following this architecture. All applications running on the Internet are client/server, as are most online object-oriented systems. In fact, any application that has a GUI front-end is assumed to be a client/server system.

12. The Internet

The Internet is a global collection of networks, or a network of networks. It was started in 1969 in the United States under the name ARPANET and was designed so that information could be exchanged between universities and defense contractors. The original network sites were universities, but the network grew to include government and corporate locations. Growth caused a split and created MILNET, a network to handle military matters. In the mid-80s NSFNET (National Science Foundation Network) was created to increase networking capabilities and by 1990 incorporated the networks that comprised ARPANET and ARPANET was discontinued. In 1991 NREN (National Research and Education Network) was established by the government to connect government and commercial organizations. All of these networks, and many more, are part of the Internet. Corporations, universities, and government functions from all countries have information they wish to share, and information they wish to access.

People use the Internet to share information. E-mail is the most popular use, and simply sending a message to another person will always be a main function. The Internet, however, allows much more than this. Newsgroups and mailing lists provide introductions to, and

Figure 12.1 The Internet.

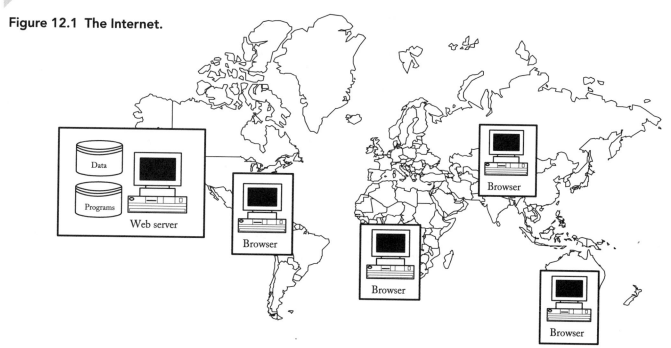

communications with, people who share your interests and concerns on both a personal and professional basis. There are postings for job opportunities and postings of resumes for people looking for work. The government posts legal documents and court decisions. The original purpose of the Internet – the sharing of scientific research – continues to grow, and the fastest growth area on the Internet is the Web where companies conduct business over the Internet.

Any network can be added to the Internet if it follows two technical standards. The network must be a packet-switching network and it must follow TCP/IP (Transmission Control Protocol/Internet Protocol). Packet-switching breaks data down into small "packets" of information, each of which contains header and trailer information identifying destinations and sequencing. These packets can travel through communications media separately and be joined together at the final destination. TCP/IP is the protocol, or rules, that must be followed by all hardware and software on the network. These two requirements aren't complex and it's estimated that a new network attaches itself to the Internet almost every 10 minutes. Every new network brings the possibility of new information, new processes, and new functions to Internet users.

Users can access the Internet through browsers. Most browsers can be freely downloaded or are included in operating systems, and provide access to the Internet functions. E-mail is the most common of Internet programs. E-mail is, of course, sending messages back and forth between individuals and/or computers. The user types a message on his or her computer and sends the message to a mailbox. The recipient checks the mailbox which is on a host computer that contains the e-mail account. The recipient can download the message, forward it to someone else, file it, and/or delete it. E-mail can be sent to anyone who has an Internet address.

Two other common programs are Telnet and FTP. Telnet allows you to interface with a remote computer as if you were directly connected. A user would use Telnet to log onto another system and use the other system's computers to access data. In many cases you need an account and a password to the host computer. NASA SPACElink is an example of a network in the Internet that is accessed through Telnet. Teachers log onto NASA SPACElink to access information that ranges from press releases about the next shuttle launch to computer images from space missions. FTP (File Transfer Protocol) is used to download data from a host system and let your computer process it. For most file transfers you need an account with a logon, but some files can be accessed by anyone through anonymous FTP. Remember, the Internet was created to share information, and universities, government agencies, and even corporations want much of the information they have available to everyone.

World Wide Web

The World Wide Web (also called the Web, WWW, and W3) is part of the Internet. The Web contains information that is accessed through a browser with a GUI. Individuals, companies, schools – anyone who wants to – can establish a Web site which is then accessed through a Web server. A Web site consists of one or more Web pages; a Web page is built using a markup language, usually HTML. HTML (Hypertext Markup Language) formats text, adds graphics, sound, video (any multimedia) and saves the finished document, or page, in a text-only ASCII file that any computer can access and display. The key to HTML is that the author of the page inserts tags that identify areas of the page and builds connections (called links) to other pages. Links allow Web users to jump from page to page, from site to site, by simply clicking on the link. For example, a book store could create a Web page that was divided into two parts. The top of the page could be labeled fiction, and listed under the header could be romance novels, mysteries, science fiction, and so on. The bottom of the page could be headed

Figure 12.2 Web technology.

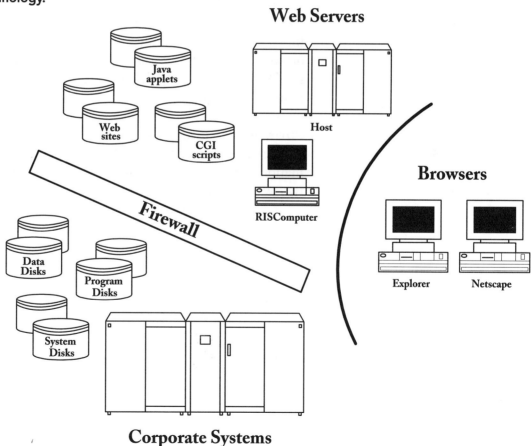

non-fiction, and choices could be history, science, religion, politics, etc. The user would simply click on the topic he or she was interested in, and a page containing available titles for that choice would be automatically displayed. Creating Web sites is called authoring and putting a Web site on a server is called publishing. The information available through the Web is endless. The Encyclopedia Britannica is online, as is AT&T's 800 phone directory.

While HTML can certainly handle multimedia to allow the creation of interesting Web pages complete with video and sound, problems are caused by different systems and retrieval is very slow. Java is a programming language that is often used to write sophisticated Web pages. Java applets are created to handle specific tasks, and during execution the applets are transferred to the user's system where they are translated and then execute. Translating the applet locally addresses the problem of different systems, and executing the multimedia code locally addresses the speed concerns.

Once a Web site is published, it can be accessed using a Web browser through a standard address called a URL (or Uniform Resource Locator). A browser is a program that displays Web files, and the user must have a browser running on his or her machine. The first popular browser was Mosaic which is still in use although Netscape's Navigator and Microsoft's Explorer are dominant today. Web pages are written for specific browsers, and while you can access any Web page from any browser, you won't necessarily be seeing the exact same thing. The links in a Web page can jump the user to another HTML document, or to an interactive program.

Programs specified in links are called gateway scripts, and CGI (Common Gateway Interface) defines how these programs pass information from the HTML page to other programs. CGI is used when a two-way communication is needed to allow the user to not only retrieve information, but also supply information. CGI scripts are written to do things such as track visitors to a Web site, validate user identification to allow access to restricted areas of the site, and add the user's feedback to informational surveys. Because CGI provides a common interface, Web publishers can create interactive documents that will work regardless of the system the user is on. This means Web sites can be developed that link Web front-ends to functions like inventories and customer support thus allowing sales to be conducted over the Internet. CGI scripts are usually written in scripting languages such as the Unix shell languages (Borne, C, Korn), Perl, JavaScript, and VBScript, although compiler languages such as C/C++ and Pascal are also used.

E-commerce (Electronic Commerce)

One of the fastest growing uses of the Internet is commerce – buying and selling over the net. E-commerce sometimes replaces the EDI systems that companies use for business-to-business buying and selling, and is being used to handle purchases from individual consumers. E-commerce systems automate a commercial transaction using Internet functions. A buyer, corporate or individual, gets product information from a Web page. The item is ordered through e-mail, and the e-mail links (CGI scripts) to an purchasing/sales system which checks the inventory, sends a shipping order, and handles payments. E-mail is used to confirm the order with the customer, communicate with the warehouse and send shipping orders or handle backorder situations.

There are several ways to handle electronic payments. Corporate payments are often made by EFTs, (Electronic Funds Transfers) but e-commerce also covers electronic versions of typical consumer payments methods – cash, check, and credit card. Cash and checks over the Internet are simply electronic versions of the paper products. As with paper, a check is made out for a specific amount to a specific recipient, while electronic, or digital, cash is handled by electronic tokens. Digital cash works best with transactions involving small denominations and even includes micropayments, or microcash (amounts under $1). Credit cards are the dominate method used by individuals and many companies to pay for goods, and security is of key importance. Credit cards usage is usually protected by "secure site" functions, which encrypt credit card information so it cannot be accessed by anyone other than the intended receiver.

E-commerce systems typically set up a storefront, which should include a Web-based catalog, a shopping-cart and checkout system, and payment processing and billing. There are several ways of building this storefront. One option is to outsource the entire operation. There are Web commerce providers that will handle the entire storefront for monthly fees, or a percent of sales. Many vendors do not find this attractive because some control is lost, especially over the design of the storefront. A second method is to buy or write each separate function and integrate the pieces. This requires the most work, but gives the company complete control. The third option is to purchase a software suite that integrates all the functions yet leaves the design up to the company. These suites allow you to design your own catalog and storefront, but they still must be integrated with internal systems and databases.

E-commerce systems do more than handle the online purchase, they also provide sales, marketing, order filling and support functions. Companies are building online catalogs and maintaining customer profiles

E-Commerce Systems	
Functions	**Components**
Marketing and Sales	Promotions
	Automated promotions based on buying trends
	Dynamic Web sites that track customer behavior
	Capturing demographic information
	Electronic storefront and/or catalog
	Automatic price quotes
	Shopping cart
	Auction
Procurement and Order Management	Order taking
	Cost calculation including shipping and tax
	Order confirmation
	Accept payment
	Secure transaction processing
	Credit card approval
	Partial orders/backorders
	Online billing
Order filling	Simple shipping
	UPS/FedEX integration
	Freight hauling
	Logistics management
	Freight management
Customer service And Support	Online FAQ
	Searchable knowledge base
	7x24 support (global)
	Return processing/warranty processing
	Order tracking
	Credit and returns
	Customer buying profiles
	Automated proposal/contract approval process

Figure 12.3 E-commerce.

so they can electronically contact customers with new items of interest, or sale items. Internet call centers are being established for support functions to handle both simple returns and complaints and more complex problems associated with products such as software. Internet systems are being built to handle all commercial functions, from advertising to product support.

XML (Extensible Markup Language) has become a very important part of e-commerce systems. XML is an extension to HTML that allows the data definition to be separate from the actual data content. Companies trading together must share common data definitions, i.e. both the buyer and the seller must agree on the number and type of characters that make up a purchase order number.). Because XML is text-based and both people and machine readable, it provides an easy way to do this. XML is also being used to define global and enterprise metadata. Metadata (data about data) is most commonly associated with data warehousing, but it is also very necessary in e-commerce. XML can be used to not only define the data but to also establish security, indicate relationships, and define formats, encryptions, etc. XML was accepted by W3C (World Wide Web Consortium) as a standard in 1998 and is accepted as an e-commerce standard.

Internet Security

Security is important over the Internet and must take into account simply protecting an individual's identity when asking for information through keeping hackers away from corporate systems. Security is extremely important in e-commerce systems where encryption techniques and systems and firewalls are necessary.

Security starts with encryption, which is altering data so that it is meaningful only to the intended receiver, and a "secure site" is one which provides encryption. There are several ways of encrypting, or coding, data. Data is encrypted by keys, and both the sender and the receiver must have the same key. A key is a value associated with a mathematical algorithm, and the longer the key is, the harder it is to break the code. Standard keys vary in length from 56 to 128 bits. Keys are also defined as secret or public. Secret keys have a single key, or formula used to both send and receive the information. Both parties in the transaction have the same key and can encode and/or decode information. Public keys have two keys (formulas), one to send and one to receive. The message is encoded with one key but must be decoded with another and with a public key, even the sender cannot decode a message without the second key. To use public keys, the user generates both a public and a private key through a program that will check for security, commonly a Web browser, or an e-mail program. The private key is kept confidential and the public key is sent to potential correspondents. This public key (also called dual-key or asymmetric) data encryption scheme is used for encryption to provide confidentiality, and authentication.

One key (either one) is used to encrypt a message which is decrypted by using the other key of the pair. Each participant in secure exchanges has his or her own pair of public and private keys. The private key is kept secret, and the public key is distributed to anyone who wants it. When the public key is used to encrypt the message, the message is unreadable to everyone but the holder of the private key, thus the confidentiality is established. When the private key is used to encrypt the message, then anyone (with a copy of the sender's public key) can decrypt the message knowing that only the holder of the private key could have encrypted it. This provides authentication.

Digital certificates use public key encryption. A digital certificate contains the following data: owner name, company, address, owner's public key, owner's certificate serial number, owner's validity dates, certifying company ID, certifying company and digital signature. Digital certificates are sent with messages to ensure that the sender is authentic. Digital certificates are issued by a third party called a certificate authority (CA) which verifies the identity of the site you are connected with. A reliable certificate authority normally charges a fee to the companies it certifies and runs a background check on the company, verifying that it is a legitimate company in the area where it operates.

Security functions also use digital signatures which are used to sign electronic correspondence including digital checks. A digital signature is created by encrypting information which can include but is not limited to, a person's name. Public key encryption can be used to verify signatures.

Firewalls

Firewalls are a necessary security item for companies who build allow Internet access to their own internal systems. A firewall is made up of both hardware and software and provides a single point of control for security. It determines which corporate systems may be accessed from the Internet, and controls who can access them. For a firewall to be effective, all traffic to and from the Internet must pass through the firewall, where it can be inspected. The firewall protects against unauthorized access by which examining each message and blocking those that do not meet specified security criteria. Network managers can then audit and log all significant traffic. Because firewalls provide a central control point, often Web traffic analysis and audit can be provided at the same time. Firewalls are the responsibility of technical developers who use different techniques such as packet filtering, setting up application and/or circuit-level gateways, and building proxy servers. Firewalls can contain all of these options.

13. Application Software

Application software includes all systems that are created to help a company run its business. This software deals with business functions and activity and is often labeled as either "horizontal" or "vertical." Vertical software is created for a single industry or profession, and handles specific functions such as running a law office. This particular software is of interest only to those in the legal profession. Horizontal software, however, supports a business function that is common to many, if not all, businesses. Human Resource software is horizontal – every company, no matter what it does, has HR activity and has software to support this function.

Application software can be separated into several categories although there is some overlap and a lot of interfacing among the categories. Each category has its own functions, and as applications developers become experienced, knowledge of the functions becomes as important a skill as any specific technical skills. For example, experience in manufacturing systems, or even more specifically process control, becomes an important skill. A senior IT person could, in fact, know more about process control than most of the people in the user department.

Accounting. This category includes the basic accounting functions: Accounts Payable (A/P), Accounts Receivable (A/R), General Ledger (G/L), and Fixed Assets (F/A). They are more often referred to by the abbreviation (i.e. G/L) than by the full name. While these are the major programs in an accounting system, software that handles billing, invoicing, and costing also falls into this category.

Financial. Financial systems are many and varied and often overlap with accounting systems. This category includes vertical software used by financial companies such as banks, insurance companies, credit unions, real estate brokers and brokerage houses. It also includes horizontal software that can be used by all companies such as tax services (non payroll), investment management, financial planning and analysis, CPA services, and credit and collections.

Human Resources. These systems are referred to as HRIS (Human Resources Information Systems) and include affirmative action and EEO processing, recruiting and staffing, incentive programs, employment law, hiring and termination processing, job performance evaluations, labor relations, training administration, organization charts, job description processing, time and attendance, competency and skills assessment, workers' compensation, salary administration, discrimination testing, international assignments, and work scheduling. Benefits management, pension administration and payroll are sometimes included as part of the HRIS, but are often stand alone systems or included with a separate payroll system.

Manufacturing. MRP software is probably the most common software in this category and there are alternative definitions for this acronym including Manufacturing Requirements Planning, Material Resource Planning, and Manufacturing Resource Planning. Other software includes MPS (Master Production Scheduling), CRP (Capacity Requirements Planning), MOP (Manufacturing Order Processing), supply chain management, product classification, product configuration, production control, production planning, project budgeting, repetitive manufacturing, job costing, production costing, factory automation (robots), engineering change control, bill of materials, process control, shop floor control and inventory control.

Payroll. Payroll systems can be considered a separate category, but because of the overlap in functionality are often included in human resource or accounting systems. Payroll systems process paychecks and taxes, and, in addition, must either interface with, or include, benefits management, FSA (Fixed Saving Account) programs and pension administration.

Retail. This is often referred to as sales and marketing software. Systems in this category include bar code processing, order entry, order processing, sales and marketing, sales management, sales force automation, distribution management, warehousing, freight and trucking, import/export, mail order, point of sale and wholesale trade.

Supply Chain Management. These systems fall under the manufacturing category, but are treated as a separate function because of their complexity. Supply chain management refers to the process of managing materials from buying supplies through distributing products. These systems are very complex and cover cost analysis and management of inventories, warehouse management, purchasing and procurement, replenishment and resupply, distribution and transportation functions. These systems provide a real-time graphical view of where materials are in the manufacturing process and can cover multiple work sites. Companies use supply chain software to control costs associated with maintaining large inventories, and some, in fact, would like to operate with zero inventory, BTO (Build To Order). Supply chain processing can be very complex because it requires communication with many different companies – all the companies that provide the supplies for the manufacturer to produce their own product. Some supply chain software provides a multi-company interface over the Internet.

Front-office systems. The term "front-office" is used to categorize software that is visible to the user and addresses the business processes that are directly related to customers. This includes the automation of functions such as sales, marketing, help desk, and call center processing. Sales software is often referred to as SFA (Sales Force Automation) systems and allows sales people to check inventories, delivery dates, etc. online to completely process an order. Marketing software includes CIS (Customer Information Systems) which allow sales people and/or office staff to check order status, buying history, etc. The help desk and call center software provides an interface for customers to answer questions and handle problems. Front-office systems integrate with back-office systems which handle the internal, or corporate, processing of the manufacturing operation.

Whether software is horizontal or vertical, applications developers work on these systems. For years, most application software was written internally by each company's technical staff, especially in large and midrange installations. While there is some commonality in all of these systems, every company has its own specific processing that must be handled. For example, while every payroll system calculates taxes from the same tax tables, and withholds the calculated tax amounts from paychecks, companies have very different benefit plans. This means that payroll deductions vary a great deal as one company will deduct for a medical savings account reflecting that benefit, while a company who does not offer it will not. This is obvious by simply reviewing check stubs from different companies; no two are alike. Because of this, companies typically wrote their own application, systems then reflected their particular way of doing business. Software vendors have, however, been able to develop application systems that can handle the different ways of doing business. These systems include the standard functions such as payroll tax deductions. They then provide for variations by making processing such as that handling medical savings accounts an option, or by providing user exits in programs to provide a way for the corporate developers to program any function or situation that is unique to their own company. This flexibility has made it much more viable to purchase applications rather than develop them in-house, and software can now be purchased for every application either as separate programs or complete systems. For example, a company can buy a fixed assets system or a complete accounting package. Software can be purchased for each of the functions described in the individual application categories, and knowledge of specific applications packages in addition to knowledge of the application function becomes important.

Obviously, one of the reasons to buy software is to save time and money. The cost of developing software increases yearly, and companies actively search for ways to control development costs. It is important, however, to realize that whether the software is written in-house or purchased and modified, the developers must thoroughly understand the business function and their own company's implementation of that function. While purchasing software can lessen the amount of programming that needs to be done, it can actually increase the amount of time spent on analysis and design. Two systems must now be analyzed and designed to see if they fit – the internal corporate system and the potential purchase. A company could find that they don't need as many programmers as before, but they need more senior and mid-level people to do the analysis, design, system testing and implementation. Additional knowledge is also needed by the developers in that they not only need to know the business function, they also need to know the specific purchased application package.

ERP (Enterprise Resource Planning)

ERP stands for Enterprise Resource Planning and an ERP system links together systems such as manufacturing, financial, human resources, sales force automation, supply chain management, and data warehousing. The key to ERP systems is the word "Enterprise." These systems combine all business processes in a single application stores all data in a single database and provides a single interface across the entire enterprise. ERP systems are written by software vendors and are most commonly found in large companies. Companies can buy a complete ERP system from a single vendor such as SAP, Oracle and PeopleSoft. These systems do have names, but are often simply referred to by the vendor name.

Installing an ERP system is a major undertaking and requires many specialized people. Because of the additional analysis and design work required in converting a company's internal systems to the new one, consultants specializing in the ERP software are needed through the installation.

SAP's R/3 is the dominant ERP system, and there are many associated skills needed by SAP developers. R/3 is an object-oriented system and uses a relational database, so these skills are required. The technical developers are responsible for building the interfaces and the networks on which the system will run. In R/3 this is called Basis and describes the technical premises behind the system architecture of an R/3 system. This includes a three-tiered client/server design, the networking and the hardware. Consultants are often brought in for Basis skills. Using ERP systems requires both applications and technical developers. Abap/4 is a programming language used with R/3 and both applications and technical programmers use the language, the Abap/4 Repository and Abap/4 Workbench (a set of development tools). The applications developers must know the software and thoroughly understand the business function of the application. R/3 developers must know the separate modules for financial accounting, human resources, manufacturing and logistics, and sales and distribution. With any ERP system, most companies start by installing accounting systems, then human resources and payroll. Financial systems usually follow and front-office systems are the last to be installed. For manufacturing companies, the manufacturing systems can be implemented any place in the order and, in fact, are often implemented first.

ERP does not have to be a single vendor product. Another option is the "best-of-breed" option where the company chooses different applications from different vendors. The single vendor framework works best when highly integrated manufacturing or distribution processes are necessary and makes sense if a company is following a reengineering strategy. The best-of-breed approach obviously is more flexible and can take advantage of software already in place. It does require more attention and support because interfaces must be built to integrate the software from different vendors, but has been the norm for service-oriented businesses like financial services where the interfaces between functions are more standardized.

Figure 13.1 Example of SAP's R/3 ERP system.

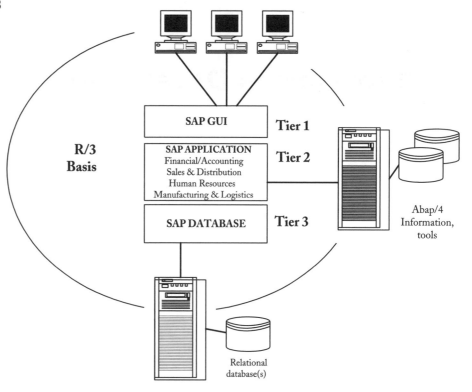

14. Jobs and Titles

There are literally hundreds of different jobs being performed by Information Technology professionals, and it often seems a daunting task to try to understand them all. Every job, and every IT professional, can be defined by a core description and an associated skill set. The core description includes three things: the job type, the computer environment and the seniority required. Job types fall into three categories: application development, technical development and systems support. Applications developers are responsible for creating and maintaining the business software a company needs. Technical developers work with the system software – the operating systems, the database management systems and the network systems. Systems support personnel are responsible for the daily operation of all the computer systems, both hardware and software. The computer environment refers to a mainframe, midrange or desktop system. Seniority is defined as senior, mid, or junior. These levels can be equated to years – a senior has over five years of experience, mid-level developers have from two to five years, and a junior has under two years of experience. The skill set can now be added to this description to have a complete understanding of the job. Skills come from four areas: the platform, development issues, data management and communications skills.

Applications Developers

Most IT jobs fall into this category. These are the people who create and maintain the business, or application, software. This includes, but is not limited to, human resource, payroll, accounting, administration and industry specific software such as MRP (Manufacturing Resource Planning), financial software for banks, insurance companies and brokerages, education packages for schools, and statistical programs for the pharmaceutical industry. Any software that helps a company run its business is produced by these people. Application developers work with the logical functioning of software systems, are concerned with business solutions and know relatively little about hardware.

Junior (up to two years experience) applications developers write and modify programs. They work on several programs at once, but are not

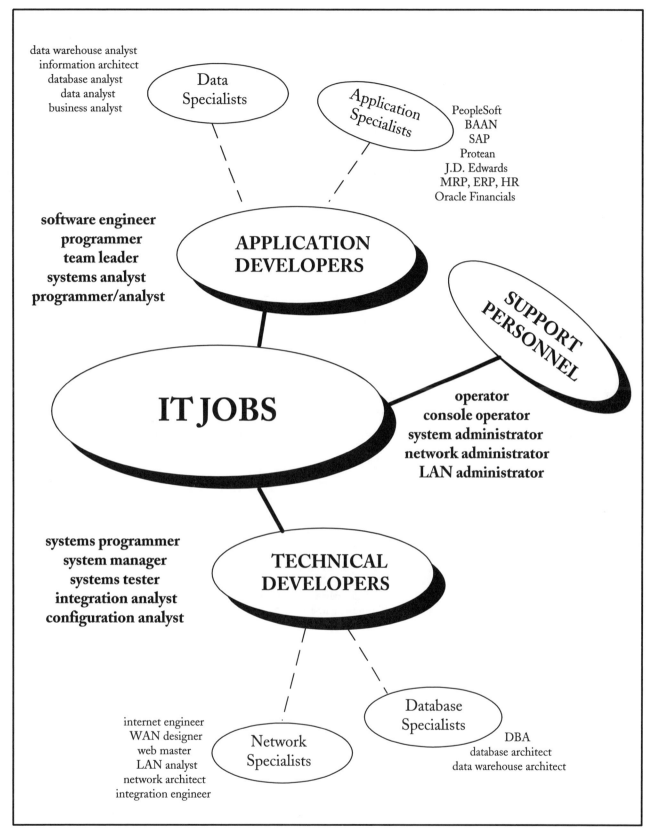

data warehouse analyst
information architect
database analyst
data analyst
business analyst

Data Specialists

Application Specialists

PeopleSoft
BAAN
SAP
Protean
J.D. Edwards
MRP, ERP, HR
Oracle Financials

software engineer
programmer
team leader
systems analyst
programmer/analyst

APPLICATION DEVELOPERS

SUPPORT PERSONNEL

IT JOBS

operator
console operator
system administrator
network administrator
LAN administrator

systems programmer
system manager
systems tester
integration analyst
configuration analyst

TECHNICAL DEVELOPERS

internet engineer
WAN designer
web master
LAN analyst
network architect
integration engineer

Network Specialists

Database Specialists

DBA
database architect
data warehouse architect

Figure 14.1 Overview of IT jobs and titles.

responsible for program interaction. They work under the supervision of mid- or senior-level developers and stay within the confines of the programming department. They have many technical skills, and their jobs are defined exclusively by these skills. Job titles usually include the word "programmer," or perhaps "engineer."

Mid-level (two to five years) applications developers continue to create and maintain software, but do much more. They must work with the interactions and interfaces in the system, so system analysis and design start to be important. The job also requires interpersonal skills as mids have direct contact with the users, or the business men and women within the company. At this point in a development career, programmers often start to specialize. The two specialities in this area are applications and data.

Applications specialists have developed a body of knowledge about an industry, such as banking, a business process, such as benefits processing, or a software product, such as SAP. Often the requirements for mid-level applications developers list the application skill first. In fact, SAP knowledge is often listed as the only skill. There has been such a shortage of developers who know this product that companies don't care about any other skills. Titles for these developers can reflect the specific application, such as EDI Analyst, and SAP Programmer.

Figure 14.2 Application developer jobs.

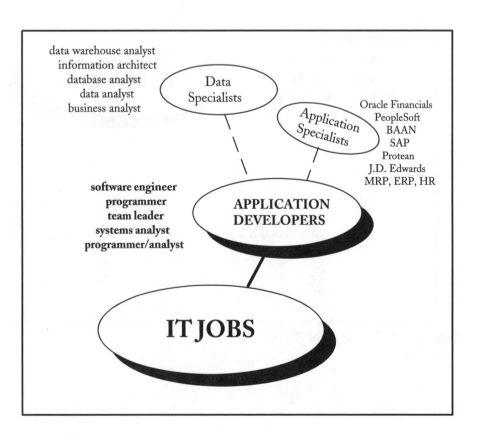

The data specialty requires knowledge of DBMSs (DataBase Management Systems) and data design and analysis tools. These people build the logical data structures that can be specific to a single application or can be enterprise-wide. This logical design requires knowledge of the business environment, as it defines what fields of information will be grouped together, and the data specialists work closely with the user departments. Data specialists can be titled Data Analyst, Database Analyst, or even Information Architect although the noun "architect" is more often used for technical developers. Data Modeler is a title that is used for anyone doing data design on an enterprise-wide basis. Data specialists also work with data warehousing, so Warehouse Architect or Warehouse Analyst are common.

Mid-level applications developers work under the supervision of seniors, but are much more independent than juniors. Job titles still use the word "programmer" and "engineer" is even more popular. New titles include "analyst," "designer," "modeler," and "architect."

Seniors are the most important people in the Information Technology Department. They do everything the mid-level developers do, but are responsible for both the supervision of the mid- and junior-level staff and the correctness of the systems being produced. Those two words – supervise and responsible – are the key to identifying a job or person as senior-level. Titles often simply add the word "senior" to the noun, so we have Senior Programmer, Senior Software Engineer, etc. Seniors also can specialize in applications or data, but there are two jobs that are senior only – Project Leader and Systems Analyst.

A System Analyst is someone skilled and experienced in the analysis phase of the system development cycle. It requires strong interpersonal skills, as analysts spend much of their time with the users determining needs and processing functions. Analysts usually have programming backgrounds, although some companies have moved people from user departments into this position. Analysts have thorough knowledge of the applications systems and often know as much about the company's work, i.e. banking, as anyone in the company. Analysts usually do much of the system design in addition to the analysis work. A minimum of five years' programming experience is usually required. This is not a managerial job, and in many companies is on the same level as project leader.

A Project Leader, or Team Leader, must have managerial skills as well as technical skills. A project leader is responsible for scheduling and planning the systems development. He or she establishes time frames for completion of the project and sets priorities for the work to be done. Occasionally project leaders do some of the analysis, design, and programming work, but usually their work is to provide technical assistance and leadership during these phases. This job is also responsible for evaluating, training, and monitoring the career paths of the programmers. A minimum of five years of programming experience is required, and usually a project leader has more than that.

Technical Developers

Technical developers create, maintain, and develop the system software: operating systems, database management systems, and communications systems. The software has nothing to do with the company's business but rather manages the computer environment. Technical developers know a great deal about the hardware and are responsible for the physical design of data and systems. They are concerned with making sure software runs effectively in the computer environment and often mention tuning, or performance tuning. They are responsible for integrating the different systems within the company. Interpersonal skills are not as important, as there is no contact with the company's businessmen and women and, in fact, many technical developers work alone or in very loose teams. As with applications developers, titles often include the nouns "programmers" and "engineers." In addition, technical developers are likely to have the title "architect."

Figure 14.3 Technical developer jobs.

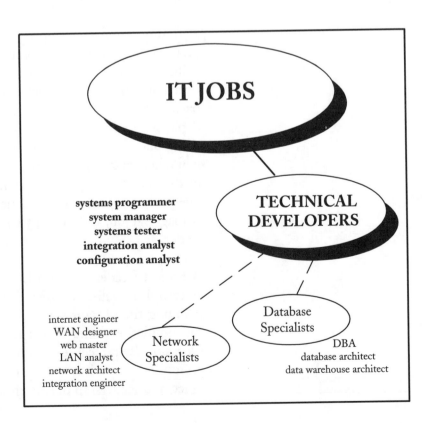

Juniors, mid-level and senior technical developers can be associated with the same time periods as the applications developers (under two years, two to five years, and over five years), but with the technical jobs the specialization is what is important. The technical staff specializes in either operating systems, data, or communications.

Operating systems specialist titles often include the word "systems" as in Systems Programmer or Systems Architect, and the job is to make sure the operating system effectively manages the resources of the company. Each developer is usually responsible for the performance of certain operating system programs. For example, one systems programmer might monitor device handlers, while another would work with the job schedulers. In more senior jobs, technical developers are responsible for such things as capacity planning (making sure the computer environment has hardware and management systems that are large enough and fast enough to effectively run the application systems) and configuration management (keeping track of all the hardware and/or software). Titles can reflect the job duties such as Configuration Manager or can be generic such as Systems Programmer, Integration Analyst, and System Architect.

Data specialists have thorough knowledge of physical database design and implementation. The physical design of a database takes the logical design (the field groupings) and stores the various groups on disk accounting for hardware facts such as: data stored in the center of a disk will be retrieved faster than data stored on the outside. Once the database is built, the technical developers are now responsible for tuning the database to make sure it continues to provide good response time even as situations change. Among many titles, a data specialist can be called Data Architect, or Data Warehouse Designer, or DBA.

The DBA, or DataBase Administrator, is usually a technical developer although some large installations have both technical and applications DBAs. This job administers and controls the organizations' database resources and is responsible for the performance and tuning of the database. Companies have a DBA for each DBMS (i.e. an Oracle DBA) who is responsible for the accuracy, security, and backups of the data.

Network specialists are the highest paid people in Information Technology, and should be, as these jobs require knowledge of two industries – the computer industry and the communications industry. These are the people who create, maintain and develop the networking necessary for all the online services a company provides. An online program is one where a person on a terminal is interacting with a program on a separate computer. This basic definition indicates the level of hardware knowledge required; something is online because of the hardware usage. This specialty includes client/server systems and the Internet access. Because of the growth of both of those areas, communications specialists are much in demand.

Technical developers use the words "network" or "communications" rather than the word "online." They, in fact, have a lot of words they use to describe their work. Topology defines how a LAN (Local Area Network) is set up. A protocol is a set of rules; communications protocols are rules that govern the data transmission. Middleware is software that connects programs following different protocols. A firewall is software that protects a company's internal systems from Internet browsers. Reference to any of these terms identifies the person or the job as a technical developer, specifically a networking specialist. Most of the titles have the word "network" in the title, as in Network Analyst, Network Designer, and Network Manager." Many new titles have appeared that use the word "Internet." There are Internet Programmers, Web Analysts, and, of course, Webmaster.

Another criteria often used for networking jobs is certification. Various vendors provide training in specific skills, then test to confirm that the person has learned what the courses cover. These vendors will certify the participant if he or she passes the test(s). The main certifications are provided by Novell and Microsoft although other companies including Banyan and Cisco certify in the networking area. Some companies are requiring certification for certain jobs, and salaries are higher for at least the Novell and Microsoft certifications.

Systems Support Personnel

The support staff in a company is part of the technical staff, but support personnel have a very different job. Support people do not create, maintain or develop software. They are responsible for running and supporting the software that the developers create. They do not know programming logic, languages or development tools and do no analysis or design. They are responsible for installing hardware and software, running system backups, monitoring the hardware, distributing reports and mounting tapes and toner cartridges, and trouble-shooting problems. Support personnel know how to run both hardware and software. It is the job of the support staff to ensure that the daily production schedule is completed correctly.

Support jobs differ with the computer environment. The support staff in a mainframe installation is called operations, and variations on the title "operator" are the norm. In addition to Operators, Senior Operators, Tape Operators, Console Operators, titles include Tape Handlers, Distribution Managers and Shift Supervisors. The computer runs

24 hours a day, so the operators work in shifts. Juniors perform most of the manual labor – mounting tapes and distributing reports – and seniors do the trouble shooting and supervision. It would be the senior, or console operator who would notice that a certain job was taking longer than usual. He or she would then query the operating system to determine the status of the job. Based on the status, the shift supervisor would call the programmer (at perhaps two a.m.) to come in and fix the problem. Operators sometimes move into technical development as they have necessary hardware knowledge.

Support jobs in midrange environments are usually called administrators – System Administrator and Network Administrator. System Administrator is, in fact, an official title in a Unix environment although it is used in any midrange installation. This job is responsible for such things as installing new software, adapting software to the system, performing system backups, recovering lost data, and maintaining security. System administrators have some of the same duties as mainframe system operators, but have more responsibility. In a mainframe system, the operators run backups. In a midrange system, the system administrator plans and runs the backups. Network administrators will actually install and maintain the networks. They do the actual cabling, respond to problem calls, add and remove devices from the network, and run monitor programs to check hardware and software. These support people often move into technical development. Because they have more responsibility they learn more about the software and many decide the move to development would be a good one.

Another job that is often considered to be part of the support staff is that of help desk personnel. The Help Desk is the department within a company that users go to when they encounter a problem or need assistance. Help Desks were originally set up to provide support for desktop

Figure 14.4 Support jobs.

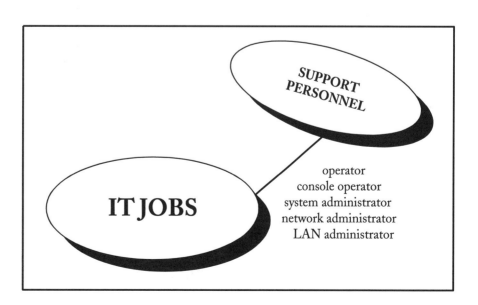

operator
console operator
system administrator
network administrator
LAN administrator

environments and, in general, support included purchasing desktop equipment, training users in desktop software such as Access and Word, establishing networks for desktops, and acting as consultants when problems are encountered. Help desks do have access to technical developers to assist with problems but also most of the personnel are non-programmers who provide training, basic consulting, and purchasing functions. These individuals are skilled in using desktops and the prewritten software packages, but are not programmers. There is no accepted title for Help Desk personnel, and length of experience and specific skills vary greatly. Help desks have been expanded beyond just servicing desktop systems, but this is still their dominant use.

Technical Skill Set

There is a skill set associated with each job. There are literally tens of thousands of products used by IT professionals. In addition, there are sets of rules, methodologies and knowledge sets that appear in job descriptions. The skills can be categorized in four areas: platforms, development, data management, and communications. Some jobs require knowledge of skills in all four areas while others will ask for a single skill.

Skills that are part of the platform reflect the computer systems and the operating systems. Both can be explicitly stated, as in Sun SPARC/Solaris, but they need not be. Often the platform will be referred to simply as Unix. This kind of platform reference means any computer and any flavor of Unix fit the platform requirement for this job. In another situation both machine and operating system might be required.

Development skills include programming languages such as COBOL, C and Java, development tools and techniques. Techniques are the rules and standards defined by developmental methodologies or architectures. Included with techniques are object oriented and component-based development. Protocols and specifications also fit this category. Development tools are programs that automate some function of the development cycle and include CASE, RAD and ADE tools. Development skills are not required of support personnel.

The data management category includes data files, databases, and data warehousing. Data files are usually handled by the operating system, so are considered part of the platform but there are exceptions like IBM's VSAM. Databases should really be referred to as DBMSs and DB/2 and Oracle are examples. Data warehousing is a new use of data and uses DBMS skills and complex query skills. In addition to the product skills, data management required design skills such as data modeling and designing metadata.

The knowledge and skills necessary to develop online applications fall into the communications category. Skills include OLTP monitors such as CICS, LANs, NOS (Network Operating Systems), and communications protocols. In addition, client/server, component based architecture, the Internet and data warehousing all require communications knowledge and skills. This is the most technical area in Information Technology.

Figure 14.5 Information Technology skill sets.

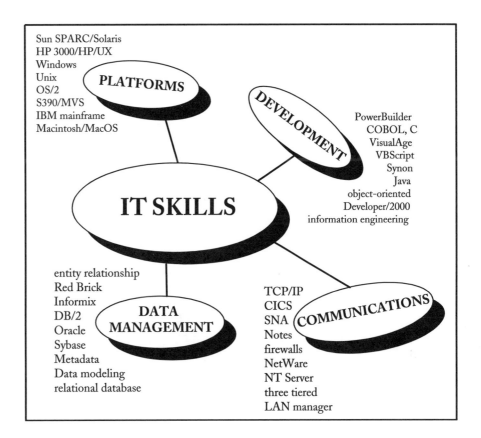

Part Two

Terms and Acronyms

The terms and acronyms are listed in a manner to encompass the varieties of programs, names and synonyms without separately listing every variation. Groups of products that fall in the same category, i.e., specific operating system enhancements, and have an identifying prefix are listed under the prefix. For example, the listing **DRS/** will contain the products DRS/Batch, DRS/Recover and DRS/Update.

Where different versions of a product with slightly different names exist, commas and parenthesis are used to show multiple names. For example **Accell/IDS,SQL** shows two versions: Accell/IDS and Accell/SQL. Parenthesis are used when a version adds something to the name. Commas and parenthesis can be used together. **Asssembler (H,XT)** shows three versions: Assembler, Assembler H and Assembler XT.

The small character 'x' is used in many names where there are too many characters to list. This happens most frequently in the names of computers. Many different numbers or letters can be substituted for all the lower case x's as in the listing. **AS/400 940x, Model xxx.** Wherever an upper case X appears, it is the actual character in the name.

Special characters and case are ignored for purposes of alphabetizing. For example, products and terms starting with "access" are listed in the following order: Numbers appear before letters.

Access	**access method**
Access+	**Access/MVS**
Access/204	**Access SQL**
Access Basic	**Access/Star**
Access-Controller	**Access/VM**
Access/DB,38	**ACCESSMASTER**
Access Executive	**Accessworks**

Also note the importance of spaces and special characters in product names. In this example "Access/VM" will appear before "ACCESS-MASTER" because of the slash. "VisualAge" will follow "Visual SQL."

Product names throughout this glossary are trademarks of the companies listed as vendor.

A+ certification Technical certification. Industry-wide, vendor-neutral certification of service technicians. Two tests verify the technical skills and knowledge to service and support PCs.

A+Edition Operating system software. Extension to Solaris that handles I/O processing in large (over 12 CPUs) networks. Vendor: Sun Microsystems, Inc. and Amdahl Corp.

A/SAP File management utility programs. Runs on DEC, IBM systems. Full name: Access/Star Application Program. Allows users to directly access remote databases. Vendor: Db/Access, Inc.

A/UX Operating system for Apple Macintosh systems. Runs Unix applications. Vendor: Apple Computer, Inc.

A11,14,18 Mainframe computer. Operating systems: Unix, MCP/AS, OS/2, Windows. Vendor: Unisys Corp.

A2B Communications software. Connects computers running Windows to mainframes, any TCP/IP system. Vendor: Simware.

A2x00 Midrange computer. Pentium CPU. Operating system: Windows 95/98. Vendor: Unisys Corp.

AAIS Prolog Artificial intelligence system. Runs on Apple II systems. Vendor: Advanced A.I. Systems, Inc.

Abacus Data Systems Software vendor. Products: ADAMS family of financial, sales, inventory systems for midsize environments.

Abap/4, Abap/4 Workbench Application development tool. Language and tool set used with R/3 applications software. Stands for: Advanced Business Application Programming/4. Vendor: SAP America, Inc.

ABC Activity Based Costing. Determining cost based on specific outputs rather than functions. ABC would determine the cost of i.e. generating a purchase order, rather than budgeting the purchasing department. Used by DOD to isolate critical processes to re-engineer.

abend Stands for abnormal end. Termination of a computer program due to execution error.

Abend-AID Operating system add-on. Debugging/testing software that runs on IBM systems. Versions include: Abend-AID for DB2, Abend-AID for IDMS, Abend-AID/COBOL, Abend-AID/IMS, Abend-AID/SPF, CICS Abend-AID, CICS dBUG-AID, CICS Abend-AID for DB2, DOS Abend-AID. Vendor: Compuware Corp.

ABend-Catcher Operating system add-on. Debugging/testing software that runs on IBM systems. Vendor: Computer Application Services, Inc.

Ability PLUS Desktop system software. Integrated package. Runs on midrange systems. Vendor: Migent, Inc.

Ability VSM Operating system enhancement used by systems programmers. Increases system efficiency in DEC VAX systems. Manages disk storage space. Vendor: Avtech Software, Inc.

AbleCommerce Application development tool used to build Web storefronts. Bundled with Cold Fusion. Vendor: Able Solutions.

ABR Operating system add-on. Data management software that runs on IBM systems. Stands for: Automatic Backup & Recovery. Vendor: Innovation Data Processing.

ABS Business Server Midsize computer. Server. Pentium II processor. Operating system: Windows NT. Vendor: ABS Computer Technologies, Inc.

ABS Notebook Notebook computer. Pentium CPU. Vendor: ABS Computer Technologies.

ABS SCSI Powerhouse Midsize computer. Server. Pentium II processor. Operating system: Windows NT. Vendor: ABS Computer Technologies, Inc.

Abstract/Probe+ Application development tool. Development environment plus documentation generator. Used to development object-oriented applications. Vendor: Advanced Systems Concepts, Inc.

Abstract Window Toolkit Application development tool. Set of Java components including buttons, menus and text areas. Used to build user interfaces for Java applications. Vendor: Sun Microsystems, Inc.

Abstraction The process of determining what is general and what is specific. That which is general can be developed as components, or objects and can be reused in other applications.

ACC Operating system enhancement. Manages disk space for IBM MVS systems. Stands for: Allocation Control Center. Vendor: DTS Software Inc.

AcceleratedSAP See ASAP.

Accelerator Application development tool used to create data marts. Includes data extraction, transformation, and loading functions. Works with SAP and SQL Server. Vendor: Cognos Inc.

Accell/IDS Application development tool. Includes relational database/4GL, program generator, windowing software. Runs on DEC, IBM, Unix systems. Stands for: Integrated Development environment. Vendor: Unify Corp.

Accell/SQL Application development tool. Used for client/server computing. Builds client applications with GUI front-end. Program generator. Interfaces with Unify 2000, Integra, Sybase, Informix, Oracle, Open Look. Runs on Macintosh, Unix systems. Vendor: Unify Corp.

Accell/Workstation Application development tool. Program generator for Unix systems. Supports both Motif and Open Look. Also serves as an enhancement to Accell/SQL. Vendor: Unify Corp.

Accent R Database/4GL for large computer system environments. Runs on DEC systems. Vendor: National Information Systems.

Accent RDM Application development system. End-users create reports and custom applications. Includes 4GL and DBMS. Runs on Unix, VMS, Windows systems. Vendor: National Information Systems, Inc.

acceptance test Part of the testing phase of the development life cycle. A test executed by the users, or clients, of the system.

Access 1. Communications software. Transaction processing monitor. Runs on IBM systems. Interfaces with Com-Plete. Vendor: Software AG Americas Inc. 2. See Microsoft Access.

Access+ Data management system. Runs on IBM systems. Query and reporting system that interfaces with CICS. Vendor: Sonetics Corp.

Access/204 4GL used in mainframe environments. IBM systems. Vendor: Computer Corp. of America.

Access Basic Programming language used with Access databases. Vendor: Microsoft Corp.

Access-Controller Operating system add-on. Security/auditing system that runs on DEC VAX systems. Vendor: Bear Computer Systems, Inc.

Access/DAL See UniPrise Access/DAL.

Access/DB,38 Operating system add-on. Data management software that runs on IBM S/38 systems. Vendor: Lawson Associates, Inc.

Access Executive EIS. Runs on Unix systems. Vendor: Dialogue, Inc.

access method An operating system program that transfers data between storage devices (disks, tapes, etc.) and the computer when requested to do so by a program. Access methods are part of operating systems.

Access/MVS Communications software. Network connecting IBM computers. Vendor: ACC (Advanced Computer Communications).

Access SQL Database server. Utilizes Microsoft Windows graphical user interface for easy access. Allows access to corporate data from single interface. Vendor: Software Products International.

Access/Star Data management system. Runs on DEC, HP, IBM, Tandem, Unix systems. Interfaces with Datatrieve, RAMIS. Vendor: Db/Access, Inc.

Access/VM Operating system add-on. Data management software that runs on IBM systems. Vendor: Goal Systems International, Inc.

ACCESSMASTER Communications software. Security management system. Includes single sign-on for multiple systems, Internet security functions remote access security. Vendor: Bull HN Information Systems, Inc.

Accessworks Communications software. Works with client/server architecture. Allows a DEC VAX system to act as the server for IBM and Macintosh computers while also functioning as a client to IBM mainframes. Vendor: Digital Equipment Corp.

Accolade Application development tools. Includes prototyper, program generator, testing tools for IBM CICS systems. Vendor: Computer Corp. of America.

AccountMate Application software vendor. Produces financial software including account, payroll, sales and purchasing systems.

AccuMAX Operating system enhancement used by systems programmers. Monitors and controls performance in DB2 systems. Vendor: Legent Corp.

Accura Applications Applications software. Suite of client/server business programs including general ledger, accounts payable and receivable, purchase, sales order, payroll, human resources, etc. Runs on Windows systems. Released: 1997. Vendor: Accura Software Inc.

AccuWAN Communications software. Links LANs over WANs. Vendor: AT&T.

ACD Network Communications software. Controls networks in Unix systems. Full Name: ACD Network Knowledge Applications. Vendor: Applied Computing Devices, Inc.

ACE Advanced Computing Environment. An alliance of computer vendors committed to developing standards for desktop computers.

ACE/Server Operating system software. Security package. Runs on most Unix systems. Vendor: Security Dynamics Technologies, Inc.

Ace Timer Support software. Utility that defines performance levels for applications developed via Oracle Forms. Runs on DEC, Unix systems. Vendor: Performance Technologies, Inc.

ACENET General purpose operating system for Unix, Xenix, MS-DOS systems. Vendor: Austec, Inc.

Acer Computer vendor. Manufactures desktop systems.

AcerAltos Midrange computer. Pentium, Pentium Pro CPU. Operating system: Windows NT. Vendor: Acer America Corp.

AcerEntra Desktop computer. AMDK5 CPU. Operating system: Windows 95/98. Vendor: Acer American Corp.

AcerNote Notebook computer. Pentium CPU. Vendor: Acer Computer Corp.

AcerPower Desktop computer. Pentium CPU. Operating system: Windows. Vendor: Acer American Corp.

ACES Application development tool. Program generator for CICS applications. Stands for: Application CICS Environment System. Vendor: Cap Gemini America.

ACF Communications software. Controls networks in IBM systems. Stands for: Advanced Communications Facility. Programs include: ACF/NCP, ACF/TCAM, ACF/VTAM, ACF/VTAME. Vendor: IBM Corp.

ACF2 See CA-ACF2.

ACM/5100 Operating system add-on. Security software which identifies users on DEC VAX systems. Vendor: Security Dynamics, Inc.

ACM Checker Application development tool. Automates programming function. Analyzes source code for such things as adherence to standards and presence of logic flaws (unexecuted code). Languages analyzed: Assembler. Runs on Pentium type desktop computers. Vendor: V-Communications, Inc.

Acma 7200 Notebook computer. Pentium CPU. Vendor: ACMA Computers.

Acma sPower, zPower Desktop computer. Pentium processor. Vendor: ACMA Computers, Inc.

ACMS Operating system enhancement used by systems programmers. Increases system efficiency in DEC systems. Used to define, run and control online applications. Stands for: Application Control and Management System. Vendor: Digital Equipment Corp.

ACMS/AD Application development system. Runs on DEC systems. Provides application development tools for ACMS environment. Vendor: Digital Equipment Corp.

Acom Computer vendor. Manufactures notebook computers.

Acorn Operating system enhancement used by systems programmers. Job scheduler for HP systems. Vendor: Chestnut Data Systems.

ACP 1. Operating system for mainframe Amdahl systems. Vendor: Amdahl Corp. 2. Communications software. Network connecting Apple computers. Stands for: Apple Connectivity Package. Vendor: Pyramid Technology Corp.

Acqua Application management tool. Automatically collects data from existing development and testing tools. Reports status of software projects. Runs on Unix, Windows systems. Released: 1998. Vendor: CenterLine Software, Inc.

Acrobat See Amber.

ACS X12 Communications software. EDI package. Runs on IBM desktop computers. Vendor: Advanced Communication Systems.

ACS2 Communications software. Network connecting IBM desktop computers. Vendor: Network Products Corp.

ACT/1 1. Application development tool. Program that assists in application design in IBM systems. Includes: ACT/1 Design Aid, ACT/1 Production System Option. Vendor: Certified Software Specialists, Ltd. 2. Application development tool. Automates programming and testing functions. Vendor: Mark Jeffry Koch & Associates, Inc.

ACT/Project Operating system enhancement used by systems programmers. Provides system cost accounting for DEC VAX systems. Vendor: UIS.

ActaWorks for SAP Application development tool used to build data marts from SAP R/3 systems. Originally called ActaLink. Released: 1998. Vendor: Acta Technology.

Action 1. Multimedia authoring tool. Mid-level, timeline-based tool. Imports actions created in Director. Applications will run under Windows, Macintosh. Vendor: Macromedia, Inc.
2. Application development tool that works in a client/server architecture. Builds client applications with a GUI-based front end. Interfaces with SQL Server. Vendor: Expertelligence, Inc.

action diagram A design tool that uses nested brackets to show the structure of a program or specification.

Action Plus Groupware. Arranges meetings and client calls. Designed for sales functions. Vendor: Custom Management Systems.

ActionBook Desktop computer. Notebook. AMD processor. Vendor: UMAX Technologies, Inc.

ActionNote Portable computer. Vendor: Epson America, Inc.

ActionWorkflow System Workflow software. Used to re-engineer business processes. Interfaces with Sybase and Microsoft SQL databases. Includes: ActionWorkflow Manager, Analyst, Application Builder. Vendor: Action Technologies, Inc.

Active Data Objects See ADO.

Active Directory Services System software. Tracks everything that is running in Windows 2000 (Windows NT 5.0) networks and connected networks. Vendor: Microsoft Corp.

Active Directory Services Interface See ADSI.

ActiveLink Application development tool used for testing. Works with TrackRecord to provide management functions to development teams. Part of DevCenter. See DevCenter.

Active Server Pages Application development tool. Allows developers to embed programs in HTML pages to create dynamic web pages. Web pages can change for every visitor. Active Server Pages are written in scripting languages like ActiveX, VBScript and JavaScript, which is easier than writing Java programs. Vendor: Microsoft Corp.

Active Storage Manager Operating system software. Provides client backup in client/server systems. System administrators will control the backups, but users can initiate restores. Option with Backup Exec. Vendor: Seagate Technology, Inc.

ActiveWeb Communications, Internet software. Message oriented middleware and development tools. Used to link Internet browsers to corporate databases and applications. Runs on Solaris, Windows NT systems. Vendor: Active Software, Inc.

ActiveWorks EAI software. Middleware. Family of products that includes a broker, agents, adapters, integration and management tools. Vendor: Active Software, Inc.

ActiveCommerce DB Application development tool. Allows users to keep current information from the company's operational databases in Web sites. Interfaces with PowerBuilder, WebObjects, Access Oracle, Sybase. Runs on Unix, Windows NT. Vendor: Open Market, Inc.

ActiveOffice Application development tool. Add-on to Office 97. Extracts text and numbers from any Office application and creates visual representations of the data. Released: 1996. Vendor: Software Publishing Corp.

ActiveProject Application development tool. Allows project developers to use the Web to work with design files, project schedules, spreadsheets, database queries and Web-enabled applications. Includes Web publishing functions. Vendor: Framework Technologies Corp.

ActiveX Software standard. Used in component based architecture. Series of APIs that can activate Internet capabilities and let objects communicate with each other regardless of how they were built. Includes controls, documents and scripts. The part of COM that handles the desktop. ActiveX components are built into windows-based applications and operating systems. Successor to OLE. Competitive with JavaBeans. Vendor: Microsoft Corp.

Actor Professional Application development environment that works in a client/server architecture. Builds client applications with a GUI-based front end. Includes database independent SQL class libraries for object-oriented development. Interfaces with Paradox, dBAse, DB2, Oracle, SQL Server, OS/2 EE Database Manager. Vendor: Symantec Corp.

Actuate Reporting System Application development tool. Web-based report server which provides query and report functions. Interfaces with Oracle, SQLServer, Sybase. Runs on Windows 95/98 systems. Vendor: Actuate Software Corp.

Acu4GL Application development tool. Generates SQL queries for ACUCOBOL programs. Interfaces with Sybase, Informix, SQLServer. Runs on DEC, Unix, Windows systems. Vendor: ACUCOBOL, Inc.

ACUCOBOL-GT Application development tool. Enhances COBOL programs with Windows features such as mouse support, network compatibility. Vendor: ACUCOBOL, Inc.

Acumate ES OLAP software. Multidimensional database. Set of tools that combines 4GL with EIS system. Runs on Unix, VMS, Windows NT systems; supports Windows 95/98 clients. Stands for: Acumate Enterprise Solution. Released: 1995. Vendor: Kenan Systems Corp.

ACX-PM Operating system enhancement used by systems programmers. Increases system efficiency in IBM VM systems. Print monitor. Vendor: ACX Software, Inc.

AD Operating system. Scalable Unix-type system that runs on parallel computers. Vendor: Dascom, Inc.

AD/Advantage CASE product. Automates analysis, design, programming functions. Includes code generator for Mantis. Used to develop client/server applications. Design methodologies supported: Information Engineering, Merise. Databases supported: Supra, DB2, SQL/DS, VSAM. Runs on IBM mainframes. Vendor: Cincom Systems, Inc.

AD/Cycle IBM's CASE environment. A set of standards for the application development cycle from IBM. Both an architecture and a set of development tools. Tools include CASE software from IBM and other vendors such as Bachman Information Systems, Inc. and Index Technology Corporation. Still in the development stage. The first major software written for AD/Cycle is Repository Manager. Vendor: IBM Corp.

Ad-lib Application development tool. Report generator for DEC VAX systems. Vendor: Scientific and Business Systems, Inc.

AD/Method, AD/Method for BPR Methodologies for developing client/server applications. Supports all phases of the development life-cycle. Both work with MAP/Administrator. BPR (Business Process Reengineering) concentrates on business processes. Vendor: Structured Solutions, Inc.

AD/MVS Application development tool. Development environment that allows programmers to develop mainframe systems from a desktop computer. Vendor: Micro Focus, Inc.

AD/VANCE Datamodeler CASE product. Automates analysis and design. Runs on mainframes and interfaces with desktop computer CASE tools. Vendor: On-Line Software International.

Ada, Ada 95 Compiler language used in all computer environments. Language was based on Pascal and developed by U.S. Department of Defense. Language was designed to work with CASE tools and disciplines. Ada 95 is object-oriented version.

ADA RAID Operating system add-on. Debugging/testing software that runs on IBM ADA systems. Vendor: Proprietary Software Systems, Inc.

ADABAS Relational database for large computer environments. Runs on DEC, IBM systems. Uses Natural as compatible 4GL. Vendor: Software AG Americas Inc.

Adabas D Relational SQL database used in desktop environments. Used in client/server systems. Version of mainframe Adabase. Vendor: Software AG Americas Inc.

Adabas SQL Server Database software. Interface to Adabas from SQL. Released: 1994. Vendor: Software AG Americas Inc.

AdaFlow Application development tool. Part of Power Tools. Supports Booch methodology, Ada development. Vendor: ICONIX Software Engineering, Inc.

Adamat/D Application development tool. Automates programming function. Analyzes source code for such things as adherence to standards and presence of logic flaws (unexecuted code). Languages analyzed: ADA. Runs on DEC VAX systems. Vendor: Dynamics Research Corp.

Adams-EDI Communications software. EDI package. Runs on Unix systems. Vendor: Abacus Data Systems, Inc.

Adams family Applications software. Includes financial, sales, inventory products for midsize environments. Vendor: Abacus Data Systems.

adapter Term used with middleware and specifically with EAI software. Software written for specific vendor supplied packages including ERP systems, databases, languages, operating systems, and even other middleware systems. This software allows different information resources to seamlessly connect to the broker.

Adaptfile Systems Image processing system that runs on desktop systems. Vendor: Adaptive Information Systems.

AdaptiveRAID Operating system software. Provides RAID functionality for midsize systems. Runs on Windows NT systems. Released: 1999. Vendor: NStor Technologies, Inc.

Adaptive Replication Engine Application development tool. Uses IBM's MQSeries to handle data updates under Notes. Released: 1995. Vendor: Technology Investments.

Adaptive Server Database architecture from Sybase that unifies SQL Server, SQLAnywhere, and Sybase IQ.

Adaptive Server Anywhere Version of Sybase database for handheld systems. Released: 1998. Vendor: Sybase, Inc.

Adaptive Server Enterprise Original name of Adaptive Server IQ.

Adaptive Server IQ Relational database. Used for enterprise-wide development. Originally called Adaptive Server Enterprise, now called Adaptive Server IQ. Includes DSS software, query functions. Used with data warehousing. Runs on Unix, Windows NT systems. Vendor: Sybase, Inc.

Adaptlications See OEA.

ADAS Operating system add-on. Data management software that runs on IBM systems. Stands for: Automatic Disc Allocation System. Vendor: Universal Software, Inc.

Adasmp Database management software. Add-on to Adabas that allows multiprocessing systems to distribute activity against a single database. One processor does updating while all others do read-only queries. Stands for: Adabas Symmetric Multiprocessor. Vendor: Software AG Americas Inc.

AdaWorld Application development system. Versions available for most desktop systems. Runs on Unix systems. Full name: AdaWorld Cross Development Environment. Vendor: Aonix.

ADB Database. Runs on DEC systems. Is a runtime database. Stands for: Application DataBase. Vendor: Digital Equipment Corp.

ADC/Pro See Aide-de-Camp/Pro.

ADC/Tools Application development tool. Program that assists in developing IDMS applications. Vendor: HSL.

ADC2 See CA-ADC2.

ADCCP Communications protocol standardized by ANSI. Data link level. Stands for: Advanced Data Communication Control Procedure.

ADD system Application development tool. Report generator and program generator. Runs on IBM S/36,34 systems. Vendor: Koala Development.

ADDS Operating system add-on. Data management software that runs on Unisys systems. Stands for: Advanced Data Dictionary System. Works with DMS II databases. Vendor: Unisys Corp.

ADE Application development environment. Object-oriented development tools used to build client/server systems. Stands for: Application Development Environment. Includes C/C++ Workbench, Sablime. Vendor: NCR Corp.

AdeptEditor Application development tool. SGML authoring and editing tool used to publish electronic documents such as CD-ROMs and Web pages. Supports XML. Runs on DEC, OS/2, Unix, Windows systems. Released: 1998. Vendor: ArborText, Inc.

Adesktop computer Desktop computer operating system for Pentium-type desktop computers. Vendor: Wang Laboratories, Inc.

Adesktop computerDOC See CA-Adesktop computerDOC.

ADF Application development tool. Program generator for IMS systems. Stands for: Application Development Facility.

AdHawk Operating system software. Performance monitoring tool for Oracle systems. Allows DBAs to find and correct bottlenecks. Platform independent. Released: 1993. Vendor: Eventus Software, Inc.

ADL Architecture Description Language. The architecture of a system is its components and the connectors between the components. ADLs are used to document the architecture. ADLs include Aesop (client/server architectures), Rapide (component-based designs), MetaH (real-time avionics) and UniCon (component-based designs).

Admins/SQL Application development tool. An SQL interface for Admins/V32 development tools. Vendor: Admins, Inc.

ADMINS/V32 Application development tools. Includes database, screen generator, forms generator, report generator, query language, data management utilities. Vendor: ADMINS, Inc.

ADO Communications software. Middleware. High-level interface that sets up access channels to various databases. Stands for: Active Data Objects. Part of COM. Released: 1997. Vendor: Microsoft Corp.

Adobe PageMaker Desktop system software. Desktop publisher. Runs on IBM, Macintosh compatible systems. Vendor: Adobe Systems, Inc.

Adobe Photoshop Desktop software. Graphics package. Works with images and photos. Runs on Macintosh, Windows systems. Released: 1998. Vendor: Adobe Systems, Inc.

ADP Automatic Data Processing. Term was replaced by EDP (Electronic Data Processing), DP (Data Processing), IS (Information Systems).

ADP-5xxxx, Pxxxxx Desktop computer. Pentium processor. Vendor: Integrated Business Computers.

ADP-GSI Software vendor. Products: TOLAS family of financial, sales, inventory systems for midsize environments.

ADP-P5 Midrange computer. Pentium CPU. Vendor: Integrated Business Computers.

ADPS/P,M Application development tool. Manages and controls development process. ADPS/P uses dialogs to assist developer through process, ADPS/M provides a model for the process. Vendor: IBM Corp.

ADR-ADLIB CASE product. Part of ADR family of CASE products. Vendor: Applied Data Research, Inc.

ADROIT (Plus,II) Authoring language used to create computer-based training packages to run on IBM and compatible desktop computers. Vendor: Computer Associates International, Inc.

ADRS Application Development Tool. Report generator that runs on IBM systems. Vendor: IBM Corp.

ADS 1. Operating system add-on. Debugging/testing software that runs on IBM systems. Stands for: Advanced Debugging System. Vendor: Interactive Solutions, Inc.
2. Application development system. Runs on MAI Basic Four systems. Stands for: Application Development System. Vendor: MAI Basic Four, Inc.
3. Database/4GL for large computer environments. Runs on Convergent systems. Stands for: The Application Development environment. Vendor: Convergent Solutions, Inc.
4. See CA-ADS.
5. See Active Directory Services.

ADS/desktop Computer CASE product. Automates programming and design functions. Vendor: Trinzic Corp.

ADS/Generator See CA-Ads/Generator.

ADS/MVS Operating system enhancement. Checks JCL, generates flowcharts from JCL. Runs on MVS systems. Vendor: A+ Software, Inc.

ADS/Online See ADSO.

ADS/Trace Operating system add-on. Debugging/testing software that runs on IBM systems. Vendor: DBMS, Inc.

ADSI Application development tool. Keeps track of cross-platform applications built using a variety of programming tools such as Visual Basic, Java, or C/C++. Used to build new or adapt existing applications to Windows 2000. Group of COM objects that will reroute database queries to Windows 2000's Active Directory. Converts the old code into LDAP code. Stands for: Active Directory Services Interface. Vendor: Microsoft Corp.

ADSL Communications. Data transfer line. Adjusts transmission to compensate for noise over a circuit. Stands for: Asymetric Digital Subscriber Line.

ADSM Support software. Provides backups on mainframe storage devices for data generated from desktop systems. Includes providing preset backup functions. Stands for: Adstar Distributed Storage Manager. Vendor: IBM Corp.

ADSO Stands for ADS Online environment, which is the online function of IDMS. See CA-ADS(/PC).

ADSP Communications protocol. Session Layer. Part of AppleTalk. Stands for: AppleTalk Data Stream Protocol.

Adstar Distributed Storage Manager See ADSM.

ADVANCE Object-oriented programming language. Includes programming language PAL and an object manager.

Advance II CASE product. Automates programming function. Converts COBOL to COBOL II and provides planning and analysis capabilities. Vendor: Language Technology, Inc.

Advanced Animator See Animator, Advanced Animator.

Advanced Computing Environment See ACE.

Advanced Debugging System See ADS.

Advanced Encryption Standard See AES.

Advanced Function Presentation See AFP.

Advanced Image Management System Image processing system that runs on Hewlett-Packard systems. Vendor: Hewlett-Packard Co.

Advanced Logic Computer vendor. Manufactures desktop computers.

Advanced Peer-to-Peer Networking See APPN.

Advanced Program-to-Program Communications See APPC.

Advanced Revelation (2.0) Application development environment that works in a client/server architecture. Accesses SQL Server, dBase II, III, IV. Includes application and report generators, query language, forms designer, data dictionary. Vendor: Revelation Technologies, Inc.

Advanced Server See Windows NT Server.

Advanced Server for Workgroups Communications software. Network operating system (NOS) connecting LAN Server and Lotus Notes. Vendor: IBM Corp.

AdvanceNET Communications strategy that incorporates IBM's SNA and Open System Interconnect standards for Hewlett-Packard systems. Also supports MAP, StarLAN, Ethernet, X.25 networks.

Advantage Application development system. Includes relational database manager, query tools, report, screen, forms generators. Runs on DEC VAX systems. Vendor: Landmark Software Systems, Inc.

ADVANTAGE-NETWORKS Communications software. Network connecting multi-vendor networks. Interfaces with OSI, TCP-IP, DECnet, SNA. Vendor: Digital Equipment Corp.

Advantage Series Application development system. Runs on IBM systems. Includes: Model 204, User Language, Workshop/204, Imagine, Horizon. Vendor: Computer Corp. of America.

Advantage!xxx Desktop computer. Pentium CPU. Operating system: Windows 95/98. Vendor: AST Research, Inc. (division of Samsung Electronics).

ADvisor Applications development tool. DSS. Provides project management functions for the entire development life cycle. Vendor: PLATINUM Technology, Inc.

Advisor Application development tool used to build component-based CRM (Customer Relationship Management) applications. Java-based, includes rules development environment. Supports CORBA. Runs on Windows systems. Vendor: Neuron Data, Inc.

Advisor Series Desktop software used for budgeting and reporting. Combines relational OLAP with Excel. Uses e-mail reporting over the Internet or intranets. Released: 1996. Vendor: SRC Software.

ADW See COOL: Enterprise.

ADW/Maintenance Workstation CASE product. Reverse-engineering tool. Part of ADW. Vendor: KnowledgeWare, Inc

ADW/Workgroup Coordinator Application development tool. Allows multiple developers to access the ADW encyclopedias and synchronizes their work. Vendor: KnowledgeWare, Inc.

ADW/Workgroup Manager Application development tool. Consolidates separate encyclopedias and handles change management, security. Vendor: KnowledgeWare, Inc.

AeDCSS/Manager Operating system enhancement used by systems programmers. Monitors and controls system performance in IBM VM systems. Vendor: The Adeese Corp.

AeFAST/VMCF Operating system enhancement used by systems programmers. Increases system efficiency in IBM VM systems. Vendor: The Adeese Corp.

AEP Communications protocol. Transport Layer. Part of AppleTalk. Stands for: AppleTalk Echo Protocol.

Aerial xxx Midrange computer. Server. Pentium, Pentium Pro CPU. Vendor: Perifitech, Inc.

AES Security standard. Proposed 128-bit federal encryption standard. Stands for: Advanced Encryption Standard.

Aesop See ADL.

Aether Intelligent Messaging Communications. Middleware for wireless systems. Includes development tools. Handles multiple wireless protocols. Runs on Windows CE devices, handheld devices, pagers. Released: 1999. Vendor: Aether Technologies International LLC.

AF/Operator Operating system enhancement used by Operations staff and systems programmers. Provides operator console support in IBM MVS systems. Vendor: Candle Corp.

AF/Performer Operating system enhancement used by systems programmers. Monitors and controls system performance in IBM MVS, CICS, DB2 systems. Vendor: Candle Corp.

AF/Remote Operating system enhancement used by systems programmers. Provides remote control of MVS systems. Vendor: Candle Corp.

Afficionado Web Surfer Internet browser. Runs on Macintosh, Windows, Unix systems. Vendor: Blackbird Systems (United Kingdom).

Affinity Application development tool. Lets developers manage program interfaces. Vendor: Software Pundits, Inc.

AFP 1. Communications protocol. Presentation and Application Layers. Part of AppleTalk. Stands for: AppleTalk Filing Protocol.
2. Application development tool. Host centered document managing software. Stands for: Advanced Function Presentation. Vendor: IBM Corp.

Agenda Desktop system software. Personal information manager (PIM). Runs on Pentium type systems. No longer actively marketed. Vendor: Lotus Development Corp.

agent A program written to perform specific tasks to access remote sites or run in the background. Commonly used in networking, where agents monitor applications locally and pass information to system management software running at a central site. In Web applications, used for such things as finding the lowest fare airline tickets. In EAI systems, agents work with the brokers and actually do the integration of the diverse systems. They perform conversions and manage the business processes. Also called intelligent agents, network agents.

Ager 2000 Application development tool. Provides data aging functions for files and legacy data. Applies the corporation's own business rules to the aging process. Works with COBOL and PL/1 programs. Vendor: Princeton Softech, Inc. (Division of Computer Horizons Corp.).

aggregation Technique used in data warehousing to improve performance. Summaries of detail data are precomputed and saved. For example, quarterly sales can be calculated and later queries can retrieve the aggregate (the quarterly sales) instead of summarizing the detail data again. Greatly improves performance of the warehouse.

Agile workplace 3 Applications software. Manufacturing system that handles change processing through the entire supply chain. Has Internet interface. Released: 1997. Vendor: Agile Software Corp.

Agora Object-oriented programming language. Developed by Brussels Free University.

AI See artificial intelligence.

AI Flavors Artificial intelligence system. Runs on DEC systems. Object-oriented programming tool for Lisp programmers. Vendor: AI Ware, Inc.

AI/STAGE Artificial intelligence system. Runs on DG systems. Tools for developing expert systems. Vendor: Data General Corp.

Aide-de-Camp/Pro Application development tool. Object based Program that provides configuration management and full life-cycle support for development. Vendor: True Software Inc.

Aim Advantage Application development tool. Used in client/server development. Handles project management, training issues. Vendor: Oracle Corp.

Aim*IT* Application software. Provides asset management functions in a network environment. Collects and maintains an inventory of hardware and software assets. Enterprise and Workgroup editions available. Runs on MacOS, OS/2, Unix, Windows systems. Stands for: Asset and Inventory Management*IT*. Vendor: Computer Associates International, Inc.

Aion Application development environment. Includes GUI and source-level debugger. Provides tools for developing object-oriented systems. Allows integration of existing databases, files, applications. Runs on IBM mainframe and desktop systems. Stands for: AION Development System. Vendor: Computer Associates International, Inc.

Air Mosaic, Air Series See Internet in a Box.

AIX General-purpose operating system for IBM systems. Stands for: Advanced Interactive Executive. Versions include: AIX PS/2, AIX/370, RT desktop computer AIX. Unix-type system. Includes Java, Netscape. Vendor: IBM Corp.

AIX/6000 AIX for IBM's RS/6000 workstation.

Aladin Relational database for desktop environments. Runs on IBM, Apple desktop systems. Vendor: Advanced Data Institute, Inc.

ALANDS Application development tool. Used for developing and debugging Ada applications in DEC VAX environments. Vendor: Tektronix, Inc.

Alcam XL Desktop computer. Notebook. Pentium processor. Vendor: Akia Corp.

ALCIE IV family Application software. Financial, payroll systems for midsize environments. Vendor: Orange Systems.

Aldon/Analyzer Application development tool. Automates programming function. Analyzes source code for such things as adherence to standards and presence of logic flaws (unexecuted code). Languages analyzed: COBOL, RPG-III. Runs on IBM large computer systems. Vendor: Aldon Computer Group, Inc.

Aldus Persuasion for Windows Desktop system software. Graphics package. Interfaces with Lotus 1-2-3, Excel, various word processors. Runs on Macintosh, Windows systems. Vendor: Aldus Corp.

Alert/CICS,MVS,VM,VSE Operating system add-on. Security/auditing system that runs on IBM systems. Vendor: Goal Systems International, Inc.

ALGOL Compiler language used in all computer environments. Versions include ALGOL 60, ALGOL 68. Mathematical language concentrating on algorithms. Stands for: ALGOrithmic Language. More popular in Europe than in the United States.

AlisaMail Communications software. E-mail system. Runs on DEC VAX, Macintosh, Windows systems. Vendor: Alisa Systems, Inc.

AlisaTalk Communications software. Network connecting DEC, Macintosh, IBM desktop computers. Vendor: Alisa Systems, Inc.

ALITE Desktop system software. Spreadsheet. Runs on Pentium type desktop computers. Vendor: Trius, Inc.

All-In-1 Office system software. Includes word processor, appointment calendars, e-mail systems. Used with TeamLinks to create a client/server environment for standard functions. Vendor: Digital Equipment Corp.

Allbase/4GL Application development system. Includes prototyper, report generator, test and maintenance facilities. Runs on Hewlett-Packard systems. Vendor: Hewlett-Packard Co.

Allbase/(SQL) Relational database for large computer environments. Runs on HP systems. Used in client/server computing. Utilizes SQL. Vendor: Hewlett-Packard Co.

Allegris Constructor Enterprise Application development tool. Component-based development tool that allows developers to create, reuse and assemble both visual and non-visual components into client/server applications. Includes MethodScript (language). Runs on Unix, Windows systems. Vendor: INTERSOLV Inc.

Allegris Object Repository Application development tool. Repository for components. Automates reuse across development teams. Runs on Unix, Windows systems. Vendor: INTERSOLV Inc.

Allegris Workshop Application development environment. Object-oriented. Builds components and client/server applications. GUI builder. Runs on Unix, Windows systems. Released: 1996. Vendor: INTERSOLV, Inc.

Allegro 1. Operating system enhancement used by systems programmers. Increases system efficiency in IBM VSE systems. Vendor: Antek Software, Inc.
2. Desktop software. Project management systems with spreadsheet features. Runs on Windows systems. Vendor: Deltek Systems, Inc.

Allegro CL Application development tool. Visual tool used to create applications for Windows 95/98. Dynamic, object-oriented. Based on CLOS. Runs on Unix systems. Stands for: Allegro Common Lisp. Vendor: Franz, Inc.

Allegro family Artificial intelligence system. Provides development tools for AI applications. Runs on Macintosh systems. Includes: Allegro Common Lisp, Allegro Flavors, Allegro Foreign Function Interface, Allegro Stand-Alone Application Generator. Vendor: Coral Software Corp.

Alliance Application development tool. Provides application integration functions. Vendor: Extricity Software Inc.

AllianceSeries Application development tool. Middleware, EAI software. Upgrade to Alliance and includes AllianceInteract (provides packaged templates in five areas: semiconductor manufacturing, electronic systems, consumer packaged goods, third-party logistics, and E-commerce supply-side integration), AllianceFusion (provides connections to internal applications and other middleware products), and AlliancePartner (handles external integration issues). Released: 1999. Vendor: Extricity Software Inc.

ALLINK Operations Coordinator See Operations Coordinator.

Ally Application development tool that works in a client/server architecture. Builds client applications with GUI based front-end. Includes 4GL. Accesses Informix, Oracle Server. Includes application and report generators. Vendor: Ally Software, Inc.

ALM Communications software. Stands for: Application Loadable Module. Building block for AppWare. AppWare contains reusable ALMs for such things as SQL, Imaging, E-mail functions and developers can build their own ALMs. Vendor: Novell, Inc.

Almost TSO Operating system add-on that handles some TSO functions and commands. Runs on IBM MVS systems. Vendor: Applied Software, Inc.

Alpha Computer chip, or microprocessor. 64-bit processor. RISC chip. Next generation of VAX computers. Part of DEC's open systems strategy. Will run Unix type operating systems.

Alpha/Four Relational database for desktop systems. Runs on IBM desktop systems. Vendor: Alpha Software Corp.

ALPHAbook Portable computer. ALPHA CPU. Vendor: Tadpole Technology, Inc.

AlphaBlox Enlighten See Enlighten.

AlphaServer Midrange computer. Alpha RISC CPU. Operating systems: Unix, OpenVMS. Vendor: Compaq Computer Corp.
2. Midrange computer. Alpha CPU. Operating system: Unix. Vendor: *Neko*Tech Inc.

AlphaStation Midrange computer. Alpha CPU. Operating system: Unix. Vendor: *Neko*Tech Inc.

Alpine Midrange computer. Alpha CPU. Vendor: Aspen Systems, Inc.

ALR xxxx Midsize computer. Server. Pentium II processor. Operating systems: Unix, Windows NT. Vendor: Gateway 2000.

AltaVista Internet access, search engine. Used with most browsers. Also available to corporations to use in setting up intranets. Vendor: Digital Equipment Corp.

AltaVista Firewall Communications, Internet software. Firewall. Vendor: Digital Equipment Corp.

Altis Application development tool. Lets users access Windows applications using Web browsers. Users can update applications without having to install the new version on every desktop. Delivers only the part of the application the user needs, so, for example, just the spell checker in a word processor could be downloaded. Vendor: Epicon, Inc.

Altos System xxx Midrange computer. Pentium CPU. Operating systems: Unix. Vendor: Acer America Corp.

AM Application development environment. Generates GUI applications. Used to develop client/server systems. Includes Presentation manager, OS/2 Warp. Builds online forms and connects them to multiple databases. Stands for: Application Manager. Runs on OS/2, Windows systems. Vendor: Intelligent Environments, Inc.

AM:PM Support software. Moves applications and data between IBM mainframes and various workstations including MS-DOS, Windows, OS/2, NetWare, Macintosh and Unix. Vendor: Systems Center, Inc.

AM2000 Application integration tool that provides access to software systems for the AS/400 and IBM System/38. Stands for: Application Manager 2000. Vendor: Software 2000.

AM486 Computer chip, or microprocessor. Vendor: Advanced Micro Devices, Inc.

Amaps/(Q,3000,400) MRP, or CIM software. Runs on HP, IBM systems. Vendor: Dun & Bradstreet Software.

Amber Desktop system software. Document exchange software that allows cross-platform use of documents created under any software. For example, a document created in WordPerfect under MS-DOS/Windows will be readable by a user on a Macintosh system. Can be used instead of HTML to create Web pages. Runs on MacOS, OS/2 Warp, Unix, Windows systems. Formerly called Acrobat. Vendor: Adobe Systems, Inc.

Ambrosia Communications software. Middleware. Used to tie legacy applications to the Internet. Uses Java for client applications. Uses digital signatures. Released: 1997. Vendor: Open Horizons, Inc.

AMD K5,K6 Computer chips, or microprocessors. Vendor: Advanced Micro Devices.

Amdahl Computer vendor. Manufactures large systems that are fully compatible with IBM's mainframes.

America Online Communications service that provides a wide variety of information and services. Includes access to publications, travel and entertainment services, public bulletin boards. Users subscribe for a monthly fee/access charges.

American Business Systems Software vendor. Products: financial, sales, distribution systems.

American Internet The name given the collection of academic networks throughout the United States.

American Software Software vendor. Produces application software for financial, manufacturing, supply chain, accounting functions.

Amiga 3000,4000 Desktop computer. 680X0 CPU. Operating systems: AmigaDOS, Unix. Vendor: Commodore Business Machines, Inc.

AMIGAS II Meteorological system that runs on CDC systems. Stands for: Advanced Meteorological Image and Graphics Analysis System.

AmiPro Desktop system software. Word processor for Pentium-type desktop computers. Vendor: Lotus Development Corp.

AMIS EIS. Runs on IBM AS/400 systems. Vendor: Interactive Software Services, Inc.

Amity CN Notebook computer. Operating system: Windows 95/98. Weighs only 2.4 lbs. Vendor: Mitsubishi Corp.

Amplify Control Application development tool. Provides version control and configuration management for multiple languages, GUIs and platforms. Runs on Unix systems. Vendor: Caseware, Inc.

Amrel Maverick Notebook computer. Pentium CPU. Vendor: Amrel Technology.

Amrel Rocky 2000 Notebook computer. Pentium CPU. Vendor: Amrel Technology.

Amrel Synphony Notebook computer. Pentium CPU. Vendor: Amrel Technology.

AMS 1. Data management utility that handles VSAM files in IBM systems. Manages disk space, key fields, backups. Stands for: Access Method Services. Used synonymously with IDCAMS, which is the JCL and control statements that execute the utility.
2. Computer vendor. Manufactures desktop computers.

AMS-Team CASE product. Automates programming and design functions. Works with IMS, DB2 databases, VSAM files. Vendor: Errico Technologies, Inc.

AMS TravelPro Notebook computer. Pentium CPU. Operating system: Windows 95/98. Vendor: AMS Tech, Inc.

Amsoft Designer Series Application development tools. Used to develop manufacturing, distribution, financial applications on IBM AS/400 systems. Includes: Amsoft Screen Designer, Amsoft Database Designer, Amsoft Inquiry Designer. Vendor: American Software, Inc.

AMTrix Application development tool. Middleware, EAI software. Message oriented middleware and application development tools. Includes prepackaged adapters to integrate front- and back-office applications. Vendor: Frontec AMT.

AMTrix Intelligent Messaging Engine Communications software. EDI. Runs on AS/400, Unix, Windows NT systems. Vendor: Frontec AMT, Inc.

AMX Real-time operating system for Pentium-type computers. Vendor: Kadak Products, Ltd.

Analect/RIM Relational database for large computer environments. Runs on DEC, IBM, Prime systems. Vendor: Dialogue, Inc.

analog computer Computer that takes as input a measurement of physical values such as temperature or pressure.

analysis A phase in the system development cycle. Examining a problem to determine exactly what needs to be done.

analyst Developer, usually application. Determines what should be done to solve the problem. Usually seniors, could be mid-level, not a title used for juniors. See systems analyst.

Analyst 1. Desktop system software. Spreadsheet. Runs on Unix systems. Vendor: Xerox Special Information Systems.
2. CASE product. Part of Re-engineering Product Set. See Bachman Re-engineering Product Set.

Analyst/Designer Toolkit CASE product. Automates analysis and design functions. Vendor: Yourdon, Inc.

Analyst One Desktop system software. Spreadsheet. Runs on Pentium type desktop computers. Vendor: Feldstar Software, Inc.

AnalystStudio Application development tool. Automates analysis functions. Combines requirements management and visual modeling tools. Part of Rational Suite. Runs on Unix, Windows systems. Released: 1999. Vendor: Rational Software Corp.

Analytica Application development tool. DSS. Runs on Macintosh systems. Vendor: Lumina Decision Systems Inc.

Analyzer 1. Application development tool. Programming utility that tracks the times program statements are executed. Part of COBOL/2 Workbench. Vendor: Micro Focus, Inc.
2. Application development tool. Automates programming function. Analyzes source code for such things as adherence to standards and presence of logic flaws (unexecuted code). Languages analyzed: COBOL, CICS. Runs on IBM mainframes. Vendor: Travtech, Inc.
3. Application development tool. Provides statistics for improving data replication under Notes. Vendor: DYS Analytics.

Anatrieve Software that provides access to data stored on film. Vendor: Anacomp, Inc.

Angara Data Server Relational database. In-memory database. Uses main memory as primary data store thus achieves high performance without having to access disks. Runs on a SMP server system with enough memory to store all the data. Vendor: Angara Database Systems.

Animation Works Interactive Multimedia authoring tool. Mid-level, stage-based tool. Applications will run under Windows, Macintosh. Vendor: Gold Disk, Inc.

Animator, Advanced Animator Application development tool. Program utility used in testing. Allows developers to step through program execution monitoring data items in windows. Shows source code on screen while program is running. Part of COBOL/2 Workbench. Vendor: Micro Focus, Inc.

ANSI American National Standards Institute. Private organization that sets standards for various products ranging from computer languages to data communications. Publishes recommendations.

ANSI COBOL Synonym for COBOL. ANSI stands for: American National Standards Institute. See COBOL.

Answer:Architect Case tool. Automates analysis and design. Models business processes. Runs on OS/2 systems. Vendor: Sterling Software (Advanced Systems Division).

Answer:Zim See Zim.

ANT_HTML Application development tool. HTML editor. Works with Microsoft Word. Runs on Macintosh, Windows systems.

AntiSniff Operating system enhancement. Network security software that detects intruders on the network. Released: 1999. Vendor: Lopht Heavy Industries.

ANX Internet commerce network developed by the auto industry. Stands for: Automotive Network Exchange. Automobile companies and business partners connect using certified Internet service providers who are connected at exchange points where security is maintained.

AnyNet Communications software. Middleware. Based on the Multiprotocol Transport Networking architecture. Vendor: IBM Corp.

Anyware Office Software suite. Java programs that include word processing, spreadsheets, database, e-mail. Works with any Java-enabled system. Released: 1997. Vendor: Applix, Inc.

Anywhere Server Communications, Internet software. Lets developers deliver data to Web browsers from corporate databases. Includes a spreadsheet add-on. Vendor: Applix, Inc.

AOCE Groupware. Stands for: Apple Open Collaboration Environment. Supports phone, fax, e-mail, calendar and scheduling functions. Vendor: Apple Computer Inc.

AOEF Operating system enhancement used by systems programmers and Operations staff. Increases system efficiency and provides job scheduling functions in CICS systems. Stands for: Automated Operations Extension Facility. Vendor: Netec International, Inc.

AOS(/VS) Operating system for Data General Eclipse systems. Stands for: Advanced Operating system. Versions include: AOS/DVS, AOS/VS, AOS/VS II, AOS/RT32. Vendor: Data General Corp.

AOS/WS xxx Communications software. Network connecting DG computers. Includes: AOS/WS X.25, AOS/WS XODIAC. Vendor: Data General Corp.

AP:Millennium See Millennium.

AP:Satellite See Satellite.

Apache Communications. Web server. Runs on Unix, Windows NT systems. Vendor: Apache Group and Red Hat Software, Inc.

APAR Authorized Program Analysis Report. A report to IBM for correction of a problem found in an IBM supplied program.

APdesktop computer Communications protocol. SNA based. Allows programs to communicate over local and wide area networks. Vendor: IBM Corp.

APDL Application development tool. Monitors program design and enforces design standards. Stands for: Advanced Program Design Language. Vendor: Advanced Computer Concepts, Inc.

APDS/M, APDS/P Application development tool. Stands for Application Development Project Support/Process Mechanism. ADPS/P is a process for working through the project development tasks detailed in the model. APDS/M is the model that defines application development tasks and their relationships. Runs on IBM mainframe systems. Vendor: IBM Corp.

Aperio Application development tool. Web based OLAP software which allows users to analyze, query, report, and distribute information across the enterprise. Released: 1997. Vendor: Influence Software, Inc.

Aperture Networker Configuration management system that keeps track of everything on a network. Uses graphics to display the network. Runs on Macintosh systems. Vendor: The Graphics Management Group, Inc.

APES Artificial intelligence system. Runs on DEC VAX, IBM desktop computers. Expert system shell. Stands for: Augmented PROLOG for Expert Systems. Vendor: Programming Logic Systems, Inc.

Apex See Rational Apex.

APEX IV Mathematical/statistical programs for CYBER systems. Vendor: Control Data Corp.

APEX Mini-Server, RAID Midsize computer. Server. Alpha, Pentium processors. Operating system: Windows NT. Vendor: ACD Computers, Inc.

API Communications software. Prewritten code that enables a programmer to integrate two pieces of code thus providing the interface between an application program and communications software. Stands for: Application Program Interface.

APILU 6.2 Communications software. Network connecting DG computers. Vendor: Data General Corp.

APL Compiler language. Stands for: A Programming Language. Reputed to be difficult and highly mathematical.

APO Application software. Supply-chain product. Integrates with R/3. Stands for: Advanced Planner and Optimizer. Released: 1998. Vendor: SAP AG.

Apogee Database for Unix environments. Utilizes SQL. Vendor: QNE International.

Apollo Latest version of SuiteSpot. See SuiteSpot.

AppBridge AutoCode, Express, Professional Application development tools. AutoCode runs with Visual Basic or PowerBuilder and builds Windows versions of host screens. Express creates GUI screens for IBM mainframe and AS/400 systems. Vendor: Software Development Tools, Inc.

APPC Communications protocol. Stands for: Advanced Program-to-Program Communications. Session layer. Includes LU 6.2. Allows peer-to-peer communications between programs running on different computers. Vendor: IBM Corp.

AppGen Development System Application development system. Includes documentation generator for both printed manuals and online help for created applications. Runs on Unix systems. Vendor: AppGen Business Software, Inc.

Applaud CASE product. Automates programming and design functions. Vendor: International Consulting.

Applause II Desktop system software. Graphics package. Interfaces with Lotus 1-2-3, Excel, dBase, Framework. Runs on Windows systems. Vendor: Ashton-Tate Corp.

Apple Computer vendor. Manufactures desktop computers.

Apple Open Collaboration Environment See AOCE.

Apple Writer II Desktop system software. Word processor. Runs on Apple II systems. Vendor: Apple Computer, Inc.

AppleShare Communications software. Peer-to-peer network operating system. Connects Macintosh systems. Vendor: Apple Computer Inc.

applet A small program, or object, created by Java that can be sent from the Internet to the user's system. Used for animation, sound, interactive forms, etc. Applets can run on any computer equipped with a Java-enabled Web browser. Applets cannot read or write to local files, so the user's system is protected.

Applet Designer Application development tool. Converts Visual Basic programs to Java applets. Vendor: TV Objects.

Applet Designer Enterprise Application development tool. Generates Java programs and/or applets from existing Visual Basic programs. Runs on Windows systems. Released: 1997. Vendor: TVObjects Corp.

AppleTalk LAN and communications protocols used to connect Apple Macintosh desktop computers. Vendor: Apple Computer, Inc.

AppleWorks Desktop system software. Integrated package. Runs on Apple II systems. Vendor: Claris Corp.

Application Browser Application development tool. Program that analyzes COBOL programs to describe the program logic and control flow. Reverse-engineering software. Runs on DEC, IBM systems. Vendor: Hypersoft Corp.

Application Control Architecture Software that allows separate applications to communicate across a network without requiring each computer to run its own version of that application. Vendor: Digital Equipment Corp.

application development environment 1. Software package that provides functions necessary to develop application systems. Used in RAD development and with client/server architecture. Also used with object-oriented development. Also called visual languages. May include DBMSs, special languages, CASE tools, links to other systems. Applications programmers use the tools provided by these systems to design, write, test programs. Most run on desktop systems even though they can be used to develop large system software. 2. See ADS.

application development tool A computer program that provides support for the development process. Types of tools include: programming utilities, report generators, program generators, screen editors, JCL/SYSOUT maintenance programs. Application development tools can be part of an application development environment.

Application Development Workbench See ADW.

Application Expert Communications software. Provides graphical analysis of distributed systems and measures performance of client/server systems. Runs on Windows systems. Vendor: Optimal Networks Corp.

Application Factory Application development system. Runs on DEC systems. Vendor: CORTEX Corp.

Application Framework See zApp.

application generator See program generator.

Application Integration Architecture Standards that allow programs written for DEC equipment to run in all hardware environments. DEC's version of SAA.

application layer In data communications, the OSI layer that provides services to actual application programs. The functions included in this layer are (1) file transfer and directory operations, such as rename, delete, (2) message-handling services, such as e-mail, (3) job transfer and remote job management. Examples of application layer protocols are: FTP (File Transfer Protocol), SMTP (Simple Mail Transfer Protocol), DNS (Domain Name System), NFS (Network File System).

Application Loadable Module See ALM.

Application partitioning Splitting application logic between client and server nodes in a client/server environment. Building an application and then splitting it into modules that can run on different systems in a network. Various approaches are used: Distributed objects, multitiered, server-based, two-tiered, client-based, three-tiered. Additional approaches are under development.

Application Program Interface See API.

application programmer A programmer that works on the programs that support the company's business.

application server Software that provides both communications and application development tools. Middleware. An application server divides an application into different processes that can run in different locations and is used in n-tiered client/server architecture. In these systems the business logic (or the actual processing) resides on a "centralized" middle tier while the database server resides on the server tier and the GUI runs on the client. While this architecture adds additional overhead, it is necessary for large systems. It is also inevitable on Internet applications where this type of partitioning is automatic. Application servers also include development tools or provide interfaces to existing development environments.

application software Any software used directly by non-technical, or business people in order to perform business functions. This includes functional systems such as insurance and banking systems, and generic tools such as word processors and spreadsheets. Opposed to system software.

Application Software Development Environment CASE product. Automates all phases of the development cycle including project management functions.

Application Structure Database See ASD.

application system Collection of programs that provide a business function for a company.

application tester Application developer. See tester.

ApplicationDEC 400xP, 433MP Midrange computer. Operating systems: Unix. Vendor: Digital Equipment Corp.

Applications Manager See AM.

Applied Digital Computer vendor. Manufactures desktop computers.

Applix Enterprise Suite Application software. Integrated applications that includes Applix Sales, Applix Service, Applix Quality and Applix Helpdesk. Manages the entire sales cycle. Platform independent. Released: 1997. Vendor: Applix, Inc.

ApplixWare EIS. Integrates applications and business data. Includes database access and programming tools. Includes office automation facilities. Runs under Unix. Vendor: Applix, Inc.

AppManager Suite Operating system software. Monitors Windows NT, BackOffice systems across the enterprise. Vendor: NetIQ Corp.

AppMaster Builder Application development tool. Developers define the application once, then generate as needed to different platforms. Accesses data in Oracle, SQL Server, Informix, IMS and DB2 databases. Vendor: MERANT.

AppMaster Renovator Application development tool. Used to move existing COBOL applications to client/server systems. Stores captured legacy objects in a shareable, enterprisewide library and provides an inventory of all application components. Vendor: MERANT.

AppModeler See PowerDesigner.

APPN Communications protocol. IBM protocol used for moving data from IBM SNA mainframe systems through multivendor, multiprotocol networks. Stands for: Advanced Peer-to-Peer Networking.

Approach(96) Relational database used in desktop systems. Designed for use by non-technical people. Includes query and reporting functions. Can access DB2, Oracle, Sybase databases. Interfaces with Notes. Runs on Windows OS/2 systems. Vendor: Lotus Development Corp.

Approval Workflow software. Automates the flow of documents through an approval routing system. Vendor: Marcam Corp.

AppStudio Application development tool. Screen painter. Part of Visual C++. Vendor: Microsoft Corp.

AppScout Communications. Provides access to performance statistics from browsers. Vendor: NetScout Systems, Inc.

Apptivity Application server providing middleware and development tools. Used to build and deploy Web applications. Generates Java code for both server and client tiers. Includes testing tools. Interfaces with multiple databases. Vendor: Progress Software Corp.

AppWare Communications software. Middleware. Application development software. Allows users to write applications across diverse platforms. Interfaces with software following CORBA standards. Vendor: Novell, Inc.

AppWizard Application development tool. Code generator which constructs a program's framework. Part of Visual C++. Vendor: Microsoft Corp.

APRI-0031, 0032, 74 Desktop computer. Pentium Pro CPU. Operating system: Windows NT, NT Workstation. Vendor: DTK Computer, Inc.

Apriori Operating system enhancement. Provides systems management functions for enterprise-wide systems. Help-desk tool. Vendor: PLATINUM Technology, Inc.

APS Application development tool. Application generator for systems to run under OS/2, DOS, OS/400, MVS, VSE. Runs on MVS, OS/2 systems. Vendor: INTER-SOLV, Inc.

APS/desktop computer Workstation Application development tool. Program that allows desktop computers to be used for mainframe program development. Works with CICS, IMS, ISPF, DB2, VSAM. Vendor: INTERSOLV, Inc.

APS Development Center Application development tool. 4GL, GUI, RAD development tools used to develop client/server systems. Uses mainframe terminology which helps mainframe programmers make the transition from character based processing to GUIs. Vendor: INTER-SOLV, Inc.

APS OS/400 Generator Target Application development tool. Program generator. Used to develop COBOL programs on desktop systems to run on the AS/400. Vendor: INTERSOLV, Inc.

APT Workbench Application development tool. Stands for: Application Productivity Tools. Includes APT-Edit, APT-SQL, APT-Build, APT-Execute. Part of SQL Toolset. See SQL Toolset.

APTools Application development tool. Automates programming and testing functions.

Aptiva Desktop computer. Pentium CPU. Operating system: Windows 95/98. Vendor: IBM Corp.

APTuser Application development tool. Query and report functions. Generates reports without writing code. Runs on DEC, Unix systems. Vendor: International Software Group, Inc.

Aquanta Midrange, desktop computers. Pentium, Pentium Pro. Operating systems: Unix, Windows NT, Windows 95/98, NT Workstation. Vendor: Unisys Corp.

Aquanta EN, LN Notebook computer. Pentium CPU. Vendor: Unisys Corp.

Aqueduct Operating system software. Used by software vendors to track usage of their products on the client site. Attached to products such as SAP's R/3 and sends information to SAP about who is using the product, for how long, and for what. Used during beta testing of any product. Vendor: Aqueduct Software, Inc.

AR:Millennium See Millennium.

ARABDOS Desktop computer operating system for IBM and compatible desktop computer systems, Unix systems. Vendor: Gulf Data, Inc.

Arbiter Communications software. Network connecting IBM systems. Micro-to-mainframe link. Vendor: Tangram Systems Corp.

ARC Computer design standard established by ACE for desktop computers and workstations. Based on RISC. Stands for: Advanced RISC Computer.

Arc/Info GIS. Runs on DEC, Windows NT systems. Vendor: Environmental Systems Research Institute, Inc.

Archer Application development tool. Database extraction tool. Lets retail users export customer information to data mining and statistical analysis tools. Released: 1998. Vendor: Retail Target Marketing Systems, Inc.

Archie Communications tool used with the Internet. A search program that will locate files that can be downloaded by anonymous FTP.

Archistrat 4s Midrange computer. Server. Pentium CPU. Vendor Panda Project, Inc.

Architect Application development tools. Suite of products that includes Architect (provides project management functions), The Client/Server Methodology, The Systems Redevelopment Methodology, IE-Expert. Vendor: James Martin & Co.

Architecture Description Language See ADL.

ArchiText Document retrieval system. Works with Web servers. Runs on Unix, VMS systems. Vendor: Interactive Software Engineering Inc.

Archive Operating system add-on. Data management software that runs on DEC VAX systems. Full name: Archive (A File Archival System). Vendor: COSMIC.

Archive-2000 Operating system add-on. Data management software that runs on DEC VAX systems. Vendor: International Structural Engineers, Inc.

Archive·SQL Operating system enhancement used by systems programmers. Increases system efficiency in DEC VAX systems. Manages disk storage space. Vendor: Computertime Network Corp.

Archiver Operating system add-on. Data management software that runs on DEC VAX systems. Vendor: Ziff-Davis Technical Information Co.

ARCNET Communications. LAN (Local Area Network) connecting computers from different vendors. Developed by Datapoint but now available from other vendors. Stands for: Attached Resource Computer Network. Vendor: Datapoint Corp.

ArcServe/Open Operating system software. Provides backup/restore functions for Oracle database systems. Backup Agent for Oracle must be installed. Runs on Unix systems. Vendor: Computer Associates International, Inc.

ARCserve Replication Communications software. Network management system. Provides an automatic backup for a server failure in a multi-server system. Released: 1998. Vendor: Computer Associates International, Inc.

ArcServe*IT* Operating system software. Provides backup/restore functions which can be managed from one central location anywhere on the network. Includes disaster recovery functions. Enterprise, Workgroup and Advanced editions available. Runs on Unix, Windows NT systems. Vendor: Computer Associates International, Inc.

ARCSystem Desktop computer. RISC CPU. Operating systems: RISC/OS, NT. Vendor: MIPS Computer Systems, Inc.

ARCSystem Magnum, Millennium Midrange computer. Operating systems: Unix, MS-DOS. Vendor: MIPS Computer Systems, Inc.

Ardent Software Software vendor. Products include object-oriented databases and development tools. Vmark Software bought Unidata. Unidata bought O2 Technology and named new company Ardent Software, Inc. Company created in 1998.

Arena Internet browser. Runs on Unix systems.

ARES Relational database for large computer environments. Runs on Bull NH systems. Vendor: Bull HN Information Systems, Inc.

Argis Application software. Provides asset management functions for information technology assets. Allows users to manage overall ownership costs including contracts and technical dependencies. Runs on Windows systems. Released: 1998. Vendor: Janus Technologies, Inc.

Aria Recorder Reporter Communications, Internet software. Tracks usage of Web sites/pages. Creates HTML reports. Released: 1997. Vendor: Andromedia.

Aries Computer vendor. Manufactures desktop computers.

Aris Application development software used with SAP's R/3. Automates the process of mapping a company's business processes to R/3 software. Vendor: IDS Professor Scheer.

Aristotle Application development tool. Provides a data analysis front-end to Microsoft's Plato (Multidimensional database and OLAP tool.) Runs on Windows systems. Released: 1998. Vendor: Knosys, Inc.

Armada 1550DMT Notebook computer. Vendor: Compaq Computer Corp.

Armada xxxx Portable computer. Pentium CPU. Vendor: Compaq Computer Corp.

ARMNote Notebook computer. Pentium CPU. Vendor: ARM Computer, Inc.

ARP Communications protocol. Network layer. Used in TCP/IP. Defines addresses in Internet. Used in Vines. Stands for: Address Resolution Protocol.

ARPANET Communications software. A network established by the Advanced Research Projects Agency (ARPA) of the Department of Defense so that information can be exchanged between universities and defense contractors. Obsolete, predecessor to Internet.

Arpeggio Information Publisher Application development tool. Moves data from host to desktop systems. Moves data stored in charts, exports to HTML. Vendor: Concentric Data Systems (division of Wall Data, Inc.)

Arranger Application development software. Uses component-based development. Used to develop client/server applications. Can be used in both large and desktop system development. Can link Visual Basic client applications with Composer. Available in Standard (for users) and Professional (for IS professionals) editions. Vendor: Texas Instruments, Inc.

ARS Operating system enhancement used by systems programmers. Monitors and controls system performance in Unisys systems. Stands for: Activity Reporting System. Vendor: Unisys Corp.

ARSAP Operating system enhancement used by systems programmers. Provides system cost accounting for DEC VAX systems. Full name: ARSAP Resource Management + Chargeback System. Vendor: Gejac, Inc.

ARSAP/Unix Operating system enhancement. Provides system cost accounting for Unix systems. Vendor: Gejac, Inc.

ART Generic term. Stands for: Automated Regression Tester. Software tool that saves the results of test runs and reruns them after a change is made.

ART(-IM) Development tool for artificial intelligence systems. Stands for: Automated Reasoning Tool for Information Management. Expert system building tool. Implemented in LISP. Versions available for mainframe, midrange, desktop systems. SAA compatible. Vendor: Inference Corp.

ART*Enterprise Application development environment. Develops object-oriented, client/server applications. Uses GUI. Develops systems for OS/2, Macintosh, MVS, Unix, Windows. Vendor: Inference Corporation.

ArtBASE Application development environment. Provides Smalltalk environment for VisualWorks and ObjectWorks. Runs on Macintosh, OS/2, Unix, Windows systems. Vendor: ArtInAppleS Ltd.

Artemis (Planner, Project, Prestige) Desktop system software. Project management package. Runs on Pentium type desktop computers. Vendor: Metier Management Systems, Inc.

Artemis 9000/EX Project management software for IBM mainframes. Includes applications development functions. Vendor: Lucas Management Systems.

Artessa Application development tool. Program generator for HP systems. Vendor: Quality Consultants, Inc.

Arthur Enterprise Suite Application software for retail functions. Integrates planning, allocation, and performance. Allows retailers to review performance and plan future seasons. Runs on Windows systems. Released: 1997. Vendor: Comshare, Inc.

artificial Intelligence Programs that allow a computer to perform functions that are normally associated with human intelligence, such as reasoning, learning, self-improvement.

artificial worlds See virtual reality.

Artist Dream Machine Desktop computer. Pentium CPU. Operating system: Windows 95/98. Vendor: Comtrade Electronics, U.S.A. Inc.

ARTour Communications software. Middleware. Connects wireless systems with enterprise networks. Includes: ARTour Mobile, ARTour Gateway, ARTour Emulator Express Client, ARTour Emulator Express Server. Vendor: IBM Corp.

ARTOS Real-time operating system for desktop computers. Vendor: Locamation.

Arts & Letters Graphics Editor Desktop system software. Graphics package. Interfaces with Lotus 1-2-3, Excel. Runs on Windows systems. Vendor: Computer Support Corp.

As-Easy-AS Desktop system software. Spreadsheet. Runs on Pentium type desktop computers. Vendor: Trius, Inc.

AS/400 Midrange computer. Operating systems: OS/400, OS/2, AIX. Stands for: Applications System/400. Some models can be configured as the server in a client/server system. Vendor: IBM Corp.

AS/400 Portable Portable computer. Vendor: IBM Corp.

AS/Center Operating system enhancement used by systems programmers. Increases system efficiency in IBM midrange systems. Manages disk storage space. Vendor: Systems Center, Inc.

As/Set UCI, Integrator CASE product. Automates analysis, design, programming functions. Includes code generator for RPG. Design methodologies supported: Information Engineering. Databases supported: OS/400 relational database. Runs on IBM AS/400 systems. Vendor: System Software Associates, Inc.

ASAP Application development tool. Used for fast installation of SAP's R/3 ERP system. Valid for companies who have limited modifications to make. Released: 1998. Stands For: AcceleratedSAP. Vendor: SAP AG.

Ascend Groupware. Time management software. Released: 1996. Vendor: Franklin Quest Co.

Ascentia Notebook computer. Vendor: AST Research, Inc. (division of Samsung Electronics).

ASCII Character code system; the binary codes assigned to each character. Used in desktop systems and some midrange computers. Stands for: American Standard Code for Information Interchange.

ASCIIBridge Application development tool. Part of Power Tools. Merges, Imports, and Exports ASCII information from PowerTools repository. Vendor: ICONIX Software Engineering, Inc.

ASD Repository software. Stands for: Application Structure Database. Vendor: Microsoft Corp.

ASE Artificial intelligence system. Runs on IBM AS/400 systems. Stands for: Application Software Expert. Vendor: Software Artistry, Inc.

AshWin Operating system software. Provides mainframe type job scheduling functions for Windows systems. Vendor: Creative Interaction Technologies.

ASI-ST Database for large computer environments. Runs on IBM systems. Interfaces with IMS and TOTAL databases. Vendor: Applications Software, Inc.

ASK/1000 Data management system. Runs on HP systems. Database reporting and inquiry system. Vendor: Corporate Computer Systems, Inc.

Ask Jeeves Application software. Search engine that allows users to state queries in plain English instead of key words. The vendor has a Web site used for general questions and other vendors use Ask Jeeves software to handle such things as customer service. Vendor: Ask Jeeves, Inc.

ASK/Windows 4GL with object-oriented functions. Includes support for class libraries. Interfaces with Ingres, Oracle, Sybase. Vendor: ASK Group, Inc.

askSam Application development tool. Query and reporting tool that works with flat files and relational databases. Runs on Windows systems. Vendor: askSam Systems.

ASM2 See CA-ASM2.

ASP 1. See Active Server Pages. 2. Communications protocol. Session Layer. Part of AppleTalk. Stands for: AppleTalk Session Protocol.

Aspects Groupware. Includes EMS (electronic meeting system). Runs on Macintosh systems. Vendor: Group Logic, Inc.

Aspen See CA-Visual Objects.

Aspen/2 Data management system. Runs on Bull HN systems. Includes data dictionary and screen formatting tools. Vendor: Tekton Systems, Inc.

Aspire Desktop computer. Pentium CPU. Operating system: Windows. Vendor: Acer American Corp.

Assembler (H, XT) Assembler language used in IBM systems.

assembler (program) A computer program that translates English language codes into machine-readable codes. Also called assemble, assemble program.

Asset Insight Application software. Provides asset management functions. Vendor: Tangram Enterprise Solutions Inc.

asset management Identifying, tracking and managing hardware and software resources.

AssetCenter Application software. Provides asset management functions. Vendor: Peregrine Systems.

AssetExplorer, AssetPRO Application software. Asset management tools for IT departments. Collects information from network management systems, system management software, help desk software. Determines cost of purchase and cost of ownership. Tracks physical assets, purchasing data, warranties, contracts and leases. Runs on MacOS, OS/2, Windows systems. Vendor: Asset Software International Corp.

AssetView Operating system software. Stores information about computers, software licenses, printers, other technical assets. Vendor: Hewlett-Packard Co.

Assist/GT Application development tool. Program that assists in the development of CICS applications. Vendor: GT Software, Inc.

Assistance center Communications. Network management software. Includes multiprocessing support, remote support, network monitors. Part of OS/2 Warp. Vendor: IBM Corp.

Association of Web Professionals See AWP.

Assure Policy Networking Communications software. Manages policy based networking. Vendor: Cisco Systems, Inc.

AST-5250 File Transfer Communications software. Network connecting IBM systems. Micro-to-midrange link. Vendor: AST Research, Inc. (division of Samsung Electronics).

AST-FTSII Communications software. Network connecting IBM systems. Micro-to-mainframe link. Vendor: AST Research, Inc. (division of Samsung Electronics).

AST Research Computer vendor. Manufactures desktop and pen-based computers. Purchased by Samsung Electronics in 1997.

AsteaObjects Library Application development tool kit. Lets developers create the same look and feel across diverse applications. Vendor: Astea International.

Astex Operating system enhancement used by systems programmers. Increases system performance in IBM MVS systems. Manages DASD storage. Stands for: The Automated Storage Expert. Vendor: LEGENT Corp.

Astex Migration Manager Operating system enhancement used by systems programmers. Manages inactive data. Vendor: Legent Corp.

Astound Multimedia authoring tool. Entry-level, slide-based tool. Runs on Windows systems. Vendor: Gold Disk, Inc.

ASTR-IX Desktop computer operating system for NEC desktop computer systems. Vendor: NEC Information Systems, Inc.

Astra QuickTest Application development tool. Used to test e-commerce applications. Allows developer to record and monitor tests through any browser. Interfaces with TestDirector. Runs on Windows systems. Vendor: Mercury Interactive Corp.

Astra SiteManager Application development tool used to test Internet and intranet applications. Analyzes Web sites, tests links, and tracks usage. Runs on Windows systems. Released: 1998. Vendor: Mercury Interactive Corp.

Astra SiteTest Application development tool used to test Internet and intranet applications. Generates real Web traffic to conduct load testing. Interfaces with Astra SiteManager. Runs on Windows systems. Vendor: Mercury Interactive Corp.

AstraNet Data communications software. Connection to the Internet offered through Prodigy. Includes paid services. Vendor: Prodigy.

Astute Operating system enhancement used by systems programmers. Increases system efficiency in IBM systems. Manages disk storage space. Vendor: Astco Ltd.

ASVD Standard communications network for which a variety of products is available. Uses existing analog, modem connected systems. Used with multimedia communication. Alternative to ISDN. Stands for: Analog Simultaneous Voice/Data Technology.

AT/RTX See RTX.

AT&T 3270 Emulator Communications software. Network connecting IBM systems. Midrange-to-mainframe link. Vendor: AT&T Information Systems.

AT&T Global Computer vendor. Manufactures midrange and desktop computers. Owns NCR Corp.

AT&TEDI Communications software. EDI package. Runs on mainframe, midrange, desktop systems. Vendor: AT&T.

At Work for Handhelds See WinPad.

Atari Computer vendor. Manufactures handheld computers.

ATF 1. Application development tool. Allows testing of client/server software. Stands for: Automated Test Facility. Vendor: Softbridge, Inc.
2. Application development tool. Provides regression testing for KBMS systems. Vendor: Trinzic Corp.

Athene Capacity planning tool. Software that forecasts performance of applications systems. Runs on MVS, Unix systems. Vendor: Metron Systems, Inc.

Atlantra AS/xxxx Midrange computer. Pentium CPU. Operating system: Unix. Vendor: Mobius Computer Corp.

Atlas Application development tool. Used for business process modeling in client/server systems. Vendor: Avalon Software, Inc.

Atlas AMD Desktop computer. AMD processor. Vendor: ET Technology.

Atlas Geocoder Application development tool. Geocoding software. Vendor: Strategic Mapping.

AtlasX Desktop computer. Pentium CPU. Operating system: Windows 95/98. Vendor: ProGen Technology Inc.

ATM 1. In communications, backbone technology for high-speed, switched data networks. Used in both local and wide area networks. Mixes video, voice, data transmission. Used fixed length packet to increase speeds. Stands for: Asynchronous Transfer Mode. Replacing many T1/T3 systems. Alternative technologies are SMDS, frame relay. Widely used by IBM.
2. Automatic Teller Machine. Specialized communications terminal that allows banking transactions from a walk-in or drive-in terminal.

ATMizer Communications software. Creates VLANs. Vendor: Agile Networks.

ATMS Communications software. Controls networks in IBM, DEC systems. Full name: ATMS voice/data network management system. Vendor: The Info Group.

ATP Communications protocol. Transport Layer. Part of AppleTalk. Stands for: AppleTalk Transaction Protocol.

ATPK Utility program for DEC VAX systems.

Atriom Application development tool. Used to create, browse, manage object-oriented applications. Vendor: Semaphore.

Attachmate Software vendor. Produces communications software.

attribute querying Analyzes information without having to define rules or set up filters. Lets users determine trends in real time. Without attribute querying, users would have to define filters such as: men age 20-30, men age 31-40, men age 41-50, etc. Often used with data warehousing.

ATX Powerstation Desktop computer. Pentium processor. Vendor: ACE Computers.

Audit 1. Operating system add-on. Security/auditing system that runs on DEC VAX systems. Vendor: Clyde Digital Systems.
2. Application development tool. Locates problem modules responsible for 80% of the errors. Produces information and statistics to allow the developers to further test or rewrite the problem modules. Part of Logiscope. Runs on Unix, Windows systems. Vendor: CS Verilog.

Auditor Operating system add-on. Security/auditing system that runs on DEC VMS systems. Vendor: Braintree Technology, Inc.

Auditre Operating system enhancement used by systems programmers. Monitors and controls system performance in Adabase systems. Vendor: Treehouse Software, Inc.

AuditTrack Operating system software. Audit tool that tracks use of LAN resources. Vendor: On Technology Corp.

AuditWare Operating system software. Security and reporting tool for NetWare and AS/400 systems. Runs on Windows systems. Vendor: Preferred Systems, Inc.

Aurora Operating system. Version of OS/2 Warp Server. Includes JVM. Release date: 1999. Vendor: IBM Corp.

AurumFrontOffice Application software. Front-office system that uses the Internet to interface to Baan, Oracle, and SAP ERP systems. Includes customer-oriented functions such as sales, call-center operations and configuration. Vendor: AurumSoftware, Inc. (division of Baan).

authoring tool Multimedia term. High-level, programming facility using English or graphical commands specifically designed to build multimedia applications.

Authorware Professional, Interactive Studio Multimedia authoring tool. High-level, icon-based. Applications will run under Windows, Macintosh. Vendor: Macromedia, Inc.

Auto-Answer Operating system enhancement used by systems programmers and Operations staff. Job scheduler for IBM systems. Vendor: Star Tech Software.

Auto/Command Operating system enhancement used by systems programmers. Automates the control of distributed computers. Vendor: Boole & Babbage, Inc.

Auto Intelligence Artificial intelligence system. Builds knowledge base that can then be imported into Intelligence/Compiler, desktop computer Plus, LISP, PROLOG systems. Runs on desktop systems. Vendor: IntelligenceWare, Inc.

AUTO-MATE Plus CASE product. Automates design function. Part of IDMS/Architect. Vendor: Computer Associates International, Inc.

Autobox Application development tool. Statistical forecasting tool. Allows user to make predictions based on past patterns. Runs on Windows systems. Vendor: Automatic Forecasting Systems, Inc.

AutoClient Operating system software. Provides management of desktop systems from server. Runs on Solaris systems. Vendor: Sun Microsystems, Inc.

Autocoder Early programming language. Obsolete.

AutoController Application development tool used to test distributed applications. Tests the stress on the network. Runs on OS/2, Windows systems. Released: 1996. Vendor: AutoTester, Inc.

AUTODIN The worldwide communications network of the U.S. Defense Communications Systems. Stands for: AUTOmatic DIgital Network.

AutoInstall Operating system software. Automated software distribution and installation for PCs. Released: 1999. Vendor: 20/20 Software, Inc.

Automate/MVS,XC,VM Operating system enhancement used by operations staff and systems programmers. Provides operator console support in IBM systems. Vendor: LEGENT Corp.

Automater QA Application development tool. Provides automated software testing. Runs on Windows systems. Vendor: Direct Technology, Ltd.

Automatic Backup and Restore Operating system enhancement used by systems programmers. Increases system efficiency in IBM systems. Manages disk storage space. Vendor: Innovation Data Processing.

Automatic Operator II Operating system enhancement used by operations staff and systems programmers. Provides operator console support in IBM MVS, VM and VSE systems. Vendor: Blueline Software, Inc.

Automotive Network Exchange See ANX.

AutoNomy Application development tool used for information filtering. Finds articles of interest on the Internet. Runs on Windows 95/98 systems. Vendor: Autonomy Inc.

AutoOPERATOR Operating system enhancement used by operations staff and systems programmers. Provides operator console support in IBM systems. Vendor: Boole & Baggage, Inc.

AutoPLAN CASE tool. Automates project management. Automates Gantt bar charts PERT flow diagrams. Runs on Unix systems. Vendor: Digital Tools, Inc.

AUTOPRO Database for large computer environments. Runs on DEC systems. Vendor: Smith, Abbott and Co., Inc.

AutoSecure ACX Operating system enhancement. Security management system for distributed Unix environments. Used to protect Unix-based Web servers and MVS systems connected to Unix platforms. Vendor: PLATINUM Technology, Inc.

AutoSecure SSO Operating system enhancement. Security management system for heterogeneous environments including mainframes, distributed systems and PCs. Vendor: PLATINUM Technology, Inc.

AutoSys Operating system enhancement. Provides systems management functions for enterprise-wide systems. Job scheduling tool. Interfaces with SAP, Oracle, PeopleSoft application systems. Runs on MVS, Unix, Windows NT systems. Vendor: Computer Associates International, Inc.

Autotester (Web) Operating system add-on. Debugging/testing software that runs on IBM System/34, 36, 38 systems. Vendor: Software Recording Corp.

AutoTester Application development tool. Debugging and testing product for GUI applications. Web version available. Runs on OS/2, Windows systems. Released: 1996. Vendor: AutoTester.

AutoTune Operating system enhancement used by systems programmers. Monitors and controls system performance in IBM S/36, S/38 systems. Vendor: Help/38 Systems.

AV 2650(R) Desktop computer. Pentium II processor. Released: 1998. Vendor: Data General Corp.

AV Object Office Communications software. Allows Windows applications to run with Unix servers on a Novell network. Vendor: Data General Corp.

AV20000 Midrange computer. Parallel processor. Up to 64 Pentium II processors. NUMA architecture. Vendor: Data General Corp.

Availability Command Center Operating system software. Manages a broad range of server platforms including SunOS, HP/UX, AIX, Windows NT, OS/400, NetWare. Vendor: Candle Corp.

Avalanche Midrange computer. Alpha CPU. Vendor: Aspen Systems, Inc.

Avanta Desktop computer. Pentium processor. Vendor: Unicent Technologies.

AvantGo Communications software. Links handheld computers to corporate databases. Works with Windows CE, PalmPilot. Released: 1998. Vendor: AvantGo, Inc.

Aventail IPN Communications. Networking software. Allows corporations to manage and authenticate access to Internet and internal network resources. Prevents direct connection between internal and external systems and supports any application that runs over TCP/IP. Runs on Unix, Windows systems. Released: 1997. Vendor: Aventail Corp.

Aventail VPN Communications, Internet software. Provides security measures for VPNs. Enables users to share information with customers, suppliers, remote employees, etc. Protects Java, ActiveX and custom corporate applications. Uses 128 bit encryption, user-based authentication, key/certificate management. Runs on Unix, Windows systems. Released: 1998. Vendor: Aventail Corp.

Aviator Wireless Network NOS. Windows-based, peer-to-peer system. Used with wireless LANs. Released: 1998. Vendor: WebGear, Inc.

AViiON AV xxxx Midrange computer. RISC machine. 88xx0 CPU. Operating systems: DG/UX, Unix. Vendor: Data General Corp.

AVOS Operating system for midrange Computer Designed Systems. Vendor: Computer Designed Systems, Inc.

AVS/Express Application development tool. OLAP software. Visualization tool that can color-code groups or clusters of data and display trends and patterns as graphs. Used for analysis type queries and with data warehousing. Vendor: Advanced Visual Systems, Inc.

Aware/MVS,VSE Application development tool. Program that runs on IBM systems. Vendor: Phoenix Software Co.

AWK Programming language used in Unix systems. String processing language used for writing simple programs. Stands for: Aho, Weinberger and Kernighan (the authors of the language).

AWP Association of Web Professionals. Non-profit consortium of both individuals and corporations working on developing standards for Web software development. Provide vendor independent certification for Web developers. Currently have defined the Certified Web Technician. Two additional certifications, Certified Web Designer and Certified Web Manager will follow.

Axcess Internet software. Provides administration and security functions. Provides centralized and secure access to Web-based information from Intranets. Runs on Unix, Windows NT systems. Released: 1997. Vendor: AXENT Technologies, Inc.

aXcess/400 Application development tool. Query and report functions. Runs on AS/400 systems. Vendor: Glenbrook Software, Ltd.

Axiant Application development tools. Used for developing client/server applications and includes the visual tools Powerplay and Impromptu, methodologies including RAD, automated maintenance for analysis reports, multiuser object-oriented repository. Runs on HP, IBM, Unix, VMS, Windows systems. Vendor: Cognos, Inc.

Axil Ultima family, UPX 1000 Midrange computer. UltraSPARC CPU. Operating systems: Solaris. Vendor: Axil Computer, Inc.

AxilNet! Midrange computer. SPARC CPU. Operating systems: Unix. Vendor: Axil Computer, Inc.

AxilServer Midrange computer. SPARC CPU. Operating systems: Solaris. Vendor: Axil Computer, Inc.

Axilxxx Midrange computer. SPARC CPU. Operating systems: Unix, Solaris. Vendor: Axil Computer, Inc.

Axiom(/mx) MRP (or CIM) software. Runs on DEC systems. Vendor: Avalon Software, Inc.

AXP See Alpha AXP.

Axum Application development tool used for data mining. Provides graphs of different subsets or summaries of data. This allows user different perspectives of the data. Runs on Windows 95/98 systems. Vendor: Mathsoft, Inc.

AXViews Application development tool. Used to display real-time data in Windows or Web-based applications. Works with Visual Basic, Visual C++ and other integrated development environments. Runs on Windows systems. Vendor: DataViews Corp.

AZ7 Application development tool. Report generator for Bull HN systems. Accesses IDS, Total databases. Vendor: Azrex, Inc.

Aztec-Cxx-c,d,p,r Compiler language used in desktop computer environments. Combination of C and assembler language.

B2B Integration Server Communications. Enables data exchange among applications, and Web sites. Integrates Java, JavaScript, C, C++, Visual Basic and Active X applications. Released: 1998. Vendor WebMethods, Inc.

Baan Software vendor. Produces ERP software also called Baan. Includes GUI and project management applications. Includes Orgware, tools which automatically configure enterprise software to the company's way of doing business. Includes Informix DBMS. 1998 version is Baan V, called BaanSeries.

BaanFrontOffice Application software. Front-office systems that uses the Internet to interface to Baan ERP systems. Includes customer-oriented functions such as sales, call-center operations and configuration. Vendor: AurumSoftware, Inc. (division of Baan).

Baby 4XX Application development tool. Migrates AS/400 applications to run on microcomputers. Vendor: California Software Products, Inc.

Bac/Plus Operating system add-on. Tape management system that runs on DEC VAX systems. Vendor: ProSoft Computer Systems.

Bachman/Ellipse Communications software. Transaction processing monitor. Builds and manages client/server on-line transaction processing systems. Runs on IBM mainframe, Unix systems. Vendor: Cayenne Software (formerly Bachman Information Systems, Inc. and Cadre Technologies, Inc.)

Bachman/Re-engineering Product Set CASE product. Automates analysis, design and programming functions. Includes code generator for SQL. Design methodologies supported: Bachman DM entity-relationship models. Databases supported: DB2. Used to move applications from mainframe systems to Unix. Supports client/server computing. Runs on IBM systems. Includes: Bachman/Analyst, Data Analyst, Database Administrator, DBA Catalog Extract, DBA Enabler, DBA Repository Services, Designer, Shared Work Manager. Vendor: Cayenne Software (formerly Bachman Information Systems, Inc. and Cadre Technologies, Inc.)

Bachman/Windtunnel Operating system support software. Provides capacity planning for client/server systems. Provides analytic modeling. Runs on OS/2, Windows systems. Vendor: Cayenne Software (formerly Bachman Information Systems, Inc. and Cadre Technologies, Inc.)

back-end The part of the program that contains the data access. In client/server, the server portion.

back-office software Application software that supports the business. Provides record keeping and data management functions. Includes automation of functions such as inventory control, supply-chain processing, human resource, accounting, manufacturing control and payroll. ERP systems are back-office systems.

backbone Communications term. The part of the network that is connected by routers and switches. The backbone connects the subnetworks. Backbone technologies are ATM, FDDI, switched Token Ring, Fast Ethernet, Gigabyte Ethernet.

backdoor A hidden program that confers remote access and control over a PC to unauthorized persons. Backdoors are used by hackers to get into a system as an authorized user. Other destructive programs are called viruses, worms, Trojan Horses and logic bombs.

BackOffice See Microsoft BackOffice.

BackPack Operating system add-on. Data management software that runs on HP systems. Vendor: Tymlabs Corp.

backup The copying of programs and data onto disk or tape for safekeeping.

Backup CMS Operating system add-on. Data management software that runs on IBM systems. Vendor: Boole & Babbage, Inc.

Backup Exec Operating system software. Provides automatic backups and data restoration for desktop systems. Runs on OS/2, Windows systems. Vendor: Seagate Technology, Inc.

Backup Express Operating system enhancement. Backs up data on Unix, NetWare, Windows NT systems. Released: 1996. Vendor: Syncsort, Inc.

Backup Unet Operating system enhancement used by systems programmers. Increases system efficiency in Unix systems. Manages disk storage space. Vendor: Systems Center, Inc.

BackWeb Communications, Internet software. Push technology. Users select from content channels. Focuses on Intranets, Extranets. Vendor: BackWeb Technologies.

BACP Communications protocol. Used in ISDN. Coordinated addition and dropping of channels on an as-needed basis. Stands for: Bandwidth Allocation Control Protocol.

backdoor A hidden program that confers remote access and control over a PC to unauthorized persons. Backdoors are used by hackers to get into a system as an authorized user. Other destructive programs are called viruses, worms, Trojan Horses and logic bombs.

Baikonur Internet/Intranet Suite, Super-Server Application server providing middleware and development tools. Develops, deploys and administers Web-based enterprise-wide systems. Interfaces with Delphi, C++Builder and JBuilder. Runs on Unix, Windows NT systems. Available in Russian only until 1999. Vendor: Epsylon Technologies.

BAL Basic Assembler Language. Assembler language for IBM mainframes.

balanced scorecard A business measurement model in which companies set goals for business directions and then assign values to performance that can be measured against these goals.

Balans Communications. Network management software that balances application loads across different systems. Works with Sun, OSF, Hewlett-Packard network environments. Vendor: VXM Technologies, Inc.

bandwidth The capacity of electronic lines including communications networks and computer I/O functions. It is expressed in bits per second, bytes per second or in Hertz (cycles per second).

banner Communications. Internet function. An advertisement on a Web site. Usually linked to the advertiser's site.

Banner xxxx Application software. Includes many applications including human resources, finance, customer information, and sales. Runs on Unix systems. Vendor: System and Computer Technology Corp.

Banyan Communications supplier. See VINES.

Baronet Desktop system software. Project management package. Runs on Pentium type desktop computers. Vendor: Computerline, Inc.

BART Operating system enhancement. Backup and recovery package for Unix systems. Vendor: Raxco.

Base/OE Data management system. Runs on DEC systems. Handles updating of relational files. Includes query facility and report generator. Vendor: Information Structures, Inc.

BaseWorX Object Services Package Application development environment. Used to develop object-oriented client/server applications. Includes GUI builder and C++ class libraries. Runs on Unix systems. Vendor: TCSI Corp.

BASIC Compiler language. Stands for: Beginner's All-purpose Symbolic Instruction Code. Mainly used in desktop environments. Versions include: BASIC II, Extended BASIC (EBASIC), Business BASIC, BASIC-Plus, Turbo BASIC, GW-BASIC.

BASIC-2 Desktop operating system for Wang systems. Vendor: Wang Laboratories, Inc.

BASIC/260 Operating system for Hewlett-Packard systems. Vendor: Hewlett-Packard Co.

Basic Mapping Support See BMS.

Basis Applications software. Part of SAP's R/3 ERP (Enterprise Resouce Planning) system. Describes the technical premises that make up the system architecture. Used by technical developers to build the networks and infrastructure.

BASISplus Text information management system that manages data stored as documents. Runs on most large computer systems. Includes BASISplus/DMXRV, an image processing module. Vendor: Information Dimensions, Inc.

BASYS Operating system add-on. Library management system that runs on Unisys systems. Vendor: ESI.

Batch Accelerator See SmartBatch.

batch processing Processing accumulated data at one time with little or no human interaction.

BatchPipes See SmartBatch.

Bay Networks Hardware vendor. Manufactures equipment used in communications networks.

BBASIC See UBASIC/BBASIC.

BBEdit Application development tool. Web authoring tool. Vendor: Bare Bones Software.

BBN/Cornerstone Application development tool. Data analysis package that runs in client/server systems. GUI based. Vendor: BBN Software Products.

BBN/StatsWise Communications. Network management software that collects and analyzes network data. Interfaces with SunNet Manager, OpenView, Spectrum. Vendor: Bolt, Beranek and Newman, Inc.

BC Series Desktop computer. Pentium CPU. Operating systems: OS/2, Windows 95/98. Vendor: Cubix Corp.

BDAM Access method used in IBM mainframe systems.

BDBF Operating system enhancement used by systems programmers. Provides system cost accounting for IBM MVS systems. Stands for: Billing Database Facility. Vendor: Duquesne Systems, Inc.

BEA Builder Application development tool. Used to develop OLTP applications. Supports multiple systems and protocols. Full name: BEA Builder-Active Expert. Runs on Unix, Windows NT systems. Released: 1997. Vendor: BEA Systems, Inc.

BEA Connect Communications. Family of connectivity products used to scale online applications. Provides a single standard interface across diverse systems. Runs on DEC, MVS, OS/400, Unix, Windows NT systems. Released: 1998. Vendor: BEA Systems, Inc.

BEA Jolt See Jolt.

BEA M3 See M3.

BEA MessageQ Communications. Message oriented middleware. Provides call interfaces to allow messages to be delivered between distributed applications across different platforms. Runs on DEC, OS/2, Unix, Windows systems. Released: 1997. Vendor: BEA Systems, Inc.

BEA TopEnd See TopEnd.

BEA TUXEDO See TUXEDO.

BEACON Application development tool. Generates Ada, C, FORTRAN code for Unix operating systems. Creates documentation and test schemes. Vendor: Applied Dynamics International.

Beamit Communications software. Network connecting IBM systems. Mainframe-to-micro link that moves NOMAD database information to desktop software systems. Vendor: MUST Software International.

bean dipping Technology that allows developers to add licensing and security type functions to JavaBeans. Developed by IBM.

Bean Extender Application development tool. Used to enhance JavaBeans components without the Java source code. Based on JDK. Vendor: IBM Corp.

BeanBuilder Application development tool. Visual JavaBeans assembly tool used by advanced site builders and rapid component developers. Released: 1998. Vendor: NetObjects, Inc.

Beaverton Computer vendor. Manufactures desktop computers.

Bedrock Application development technology. Used to develop a single version of an application that will run on both Macintosh and Windows systems. Technology is used in OpenDoc. Vendor: Apple Computer, Inc. and Symantec Corp.

Being There EMS (Electronic Meeting System). Vendor: Intelligence at Large.

benchmark A test of performance. Usually refers to hardware performance, but can be used for software. Also used to indicate the acceptable standards against which performance is measured. Equipment is rated according to how close to accepted benchmarks it performs.

Benchmark Factory 97 Application development tool used in testing. Provides industry standard benchmarks for database, Web applications. Runs multiple benchmarks and detects bottlenecks. Runs on Windows systems. Released: 1997. Vendor: Client/Server Solutions, Inc.

BeOS Operating system for Macintosh systems. Released: 1996. Vendor: Be, Inc.

Berard Object & Class Specifier See BOCS.

Berkeley Software Distribution V4.3 See BSD V4.3

Best/1 Operating system support software. Provides capacity planning for client/server systems. Provides analytic modeling, real-time monitoring. Runs on Unix systems. Vendor: BGS Systems, Inc.

Best/1-MVS,SNA,VM,MVS Operating system enhancement used by systems programmers. Monitors and controls system performance in IBM systems. Vendor: BGS Systems, Inc.

Best/1-VAX Operating system enhancement used by systems programmers. Monitors and controls system performance in DEC VAX systems. Vendor: BGS Systems, Inc.

Best/1-Visualizer for Lans Systems management tool. Gathers and analyzes data for heterogeneous networks. Released: 1996. Vendor: BGS Systems, Inc.

BEST/AOS Operating system for Qantel systems. Vendor: QANTEL Business Systems, Inc.

Best Buy Desktop computer. Pentium CPU. Operating systems: DOS/Windows, Windows 95/98. Vendor: Micron Computer Electronics, Inc.

Bestnet Boundary,MSNF Communications software. Network connecting IBM computers. Vendor: BGS Systems, Inc.

Beta Object-oriented programming language.

BETA 91 Operating system enhancement used by systems programmers. Increases system efficiency in IBM systems. Automates batch restarts. Vendor: BETA Systems Software, Inc.

Beta 92 Application development tool. JCL and/or SYSOUT maintenance program for IBM systems. Stands for: Online Sysout Control Archival and Retrieval System. Vendor: BETA Systems Software, Inc.

BETA 93 Operating system enhancement used by systems programmers. Increases system efficiency in IBM systems. Handles print disbursement. Vendor: BETA Systems Software, Inc.

beta test A test of a computer system where users run the system under normal conditions. A beta test follows regular testing by the Information Systems staff.

BeyondMail Communications software. E-mail system. Includes Beyond Notes that connects to Lotus Notes databases. Internet version available. Vendor: Banyan Systems, Inc.

BFAS Operating system add-on. Data management software that runs on Bull HN systems. Stands for: Basic File Access System. Vendor: Bull HN Information Systems, Inc.

BH-3000 Communications software. Network connecting Bull HN and DEC computers. Vendor: Biles & Associates.

BI/Broker, BI/Query, BI/Web Application development tool. Java based. Includes services such as queries, data models, reports, and graphics and a repository of information objects. BI/Broker works with BI/Query which allows users to build queries and create reports from enterprise systems, and BI/Web which works with queries through Web browsers. Stands for: Business Intelligence. Released: 1998. Vendor: Hummingbird Communications, Ltd.

BIBLOS Application development system. Runs on IBM systems. Includes COBOL code generator and links with ADABAS, IDMS and IMS databases. Enables development on desktop systems with testing on mainframe. Vendor: BIBLOS, Inc.

BigScreen3 Desktop computer. Notebook. Pentium processor. Vendor: The Brick Computer Company, Inc.

BIM- Operating system enhancements used by systems programmers. Increases system efficiency in IBM systems. Group of programs that handle different aspects of system performance improvement. Vendor: B. I. Moyle Associates, Inc.

BIND Berkeley Internet Name Domain. A service used in public networks to allow client systems to obtain the names and addresses of network hosts. Used with Unix.

BindView Communications software. Network management software. Includes BindView Network Control System, BindView for Netware Directory Services (NDS). Works with Netware systems. Vendor: BindView Development Corp.

BIOS Basic Input-Output System. The code in an IBM desktop computer that controls data movement from diskette, disk, or keyboard to the computer.

BIS See Business Intelligence System.

BISE Application development tool. Used to create an integrated set of sales analysis and OLAP reporting functions. Stands for Business Intelligence for the Small Enterprise. Vendor: Cognos Inc.

Bisync Communications protocol. Data Link layer protocol being replaced by HDLC.

bit BInary Digit. One of the values 0 or 1. The basic building block of computers.

bitmap indexing Application development technique used in data warehousing. An efficient method for data indexing where almost all of the operations on database records can be performed on the indices without resorting to looking at the actual data underneath. By performing operations primarily on indices, the number of database accesses is reduced significantly.

Bitwise Computer vendor. Manufactures notebook and portable computers.

BitWise 433,466 Portable computer. Vendor: Bitwise Designs, Inc.

BizCase Application software. Commercial version of TurboBPR. Released: 1996. Vendor: SRA International, Inc.

BL300, BL700 Relational database machine. A dedicated computer that contains the DBMS necessary to process the database. These databases are compatible with most mainframe operating systems and databases. Vendor: Sharebase Corp.

black box testing Type of testing that builds test data and scenarios from the original requirements or program specifications so the testing is not biased by assumptions that were made during development. Also called behavioral testing, or functional testing. This testing can be done without knowing the programs, objects or components being tested. Contrast with white box testing.

Black Hole Communications, Internet software. Firewall. Vendor: Milkyway Networks, Inc.

BlackSmith Developer Application development system. Menu driven environment to develop screens, menus, reports, and update processes. Runs on Unix, VMS, Windows systems. Vendor: BlackSmith Corp.

blast BLocked ASynchronous Transmission. Used in name of many communications software packages.

BLIS/COBOL Operating system for Data General Eclipse systems. Vendor: Information Processing, Inc.

BLISS 36 Language used to develop systems software. Stands for: Basic Language for Implementing System Software. Runs on DEC systems. Vendor: Digital Equipment Corp.

BLOb Binary Large Object. Also called multimedia data. Includes textual, graphics, sound, video data.

BLOCKADE... Communications software. Includes security systems for LANs WANs and the Internet. Vendor: Blockade Systems Corp.

BLOCKMASTER See CA-BLOCK-MASTER.

BLT Desktop system software. Operating system add-on. Runs on Pentium type systems. Stands for: Bloomington Library Tool. Manages library of Turbo Pascal programs. Also available, Son of BLT. Vendor: dogStar Software.

Blue Object-oriented programming language. Developed as a teaching tool.

Blue Tooth Communications. Technology for wireless links. Allows notebook, handheld and mobile telephones to use radio links to connect to the Internet. Developed by Ericsson, IBM, Intel and Toshiba.

Bluebird Operating system software. Version of OS/2 Warp Server that builds client/server environment that will run any Java application. Vendor: IBM Corp.

BlueVision Communications. Network management software that manages SNA networks from an SNMP based platform. Vendor: Cabletron Systems, Inc.

BMP Batch Message Processing Program. Type of program in an IMS online environment.

BMS Basic Mapping Support. An interface between CICS and applications programs that allows programmers to define how data should appear on terminal screens.

BNA Communications software. Network connecting Unisys computers. Full name: BNA (Network Architecture). Vendor: Unisys Corp.

Bobcat Desktop computer. Alpha CPU. Operating system: Windows NT. Vendor: *Neko*Tech Inc.

BOCS Application development tool. Used in object-oriented development to document and model systems. Contains libraries to manage requirements, instance and class specifications, graphical models. Stands for: Berard Object & Class Specifier. Vendor: Berard Software Engineering, Inc.

Boeing CALC Desktop system software. Spreadsheet. Runs on Pentium type systems. Vendor: Garrison Software Corp.

Boeing Graph Desktop system software. Graphics package. Runs on Pentium type systems. Vendor: Garrison Software Corp.

Boeing Rim Relational database for large computer environments. Runs on CDC CYBER, DEC, IBM systems. Vendor: Boeing Computer Services.

Boks Communications. Controls user logons to access multiple systems. Uses real-time tokens rather than passwords. Vendor: Securix, Inc.

BOLT xx Notebook computer. Pentium CPU. Vendor: Altura Computer Systems.

Bolt, Beranek, Newman Computer vendor. Manufactures large computers.

Bon Object-oriented development methodology. Used for analysis and design. Stands for: Business Object Notation. Associated with Nerson & Walden.

BONeS Designer Application development tool. Used to design networks. Runs on Unix systems. Vendor: Alta Corp.

Booch Object-oriented development methodology. Provides analysis and design techniques.

Bookmanager family Software that allows users to read electronic versions of books and manuals. Products include: Bookmanager Read/DOS 1.2, Bookmanager Read/DOS 2 1.2, Bookmanager Build/VM 1.2, Bookmanager Read/VM 1.2. Vendor: IBM Corp.

BorderManager Communications. Integrated family of directory based network services that manage, secure and accelerate data access. Includes firewall, VPN and remote access functions. Runs on windows systems. Released: 1997. Vendor: Novell, Inc.

BorderManager FastCache Communications. Internet software. Caching tool that accelerates Web access. Runs on Macintosh, Unix, Windows NT. Vendor: Novell, Inc.

BorderWare Communications, Internet software. Firewall. Runs on Macintosh, Unix, Windows systems. Full name: BorderWare Firewall Server. Vendor: Secure Computing Corp.

Border Services Communications, Internet software. Includes firewall software, Wolf Mountain clustering technology. Works with IntranetWare, Unix, Windows NT. Vendor: Novell, Inc.

Borland C++ Application development environment. Object-oriented visual language, debugging tool. Develops code for Windows, Windows NT, OS/2 platforms. Includes OWL. Vendor: Inprise Corp.

Borland C++ Development Suite Application development environment. Combines Together/C++ object modeling tool set and Borland C++. Programs can be written in C++ code or by creating object models. Vendor: Inprise Corp.

Borland MIDAS See MIDAS.

Borland Office Software suite. Desktop system software that includes WordPerfect (word processor), Quattro Pro for Windows (spreadsheet), Paradox for Windows (database). Vendor: Inprise Corp. and WordPerfect Corp.

BOS Desktop computer operating system for Stride desktop computer systems. Vendor: Stride Micro.

BOS/Apex Operating system enhancement used by systems programmers. Increases system efficiency in Unix systems. Allows simultaneous running of BOS and Unix. Vendor: BOS National, Inc.

BOS/CBOS/5 Desktop computer operating system for IBM and compatible desktop computer systems. Versions include: BOS/5, MBOS/5. Vendor: BOS National, Inc.

BOS-Complement Operating system enhancement. Creates on-line documentation and pop-up help screens under VTAM. Vendor: B.O.S. Software Technologies.

BOS/LAN Communications software. Network operating system connecting IBM desktop computer and DEC PDP. Vendor: BOS National, Inc.

BOSaNOVA Desktop computer. Server. Pentium processor. Operating system: Windows 95/98. Vendor: BOS (Better On-line Solutions).

BOSS/(IX,VS,VX) Operating system for MAI systems. Vendor: MAI Systems.

Boston Code name for an integrated application development environment. Allows developers to use multiple languages and tools from the single environment. Vendor: Microsoft Corp.

bot See spider.

bottom up Application development technique used in data warehousing. A data warehouse strategy based on building incremental data marts to test products, methodologies and designs first, then using these data marts to justify the construction of the enterprise data warehouse.

Boundless Network Computer Network computer. Vendor: Boundless Technologies, Inc.

BoundsChecker Application development tool used for testing. Detects run-time errors in Visual C++ code. Part of DevCenter. See DevCenter.

Bourne Shell The standard shell program in Unix environments. See shell.

bozo filter Internet program. Type of snippet. Blocks access to a Web site by the URLs specified by the site owner.

BPAM Access method used in IBM mainframe systems.

BPCS Application software. Manufacturing, financial systems. Includes accounting, payroll, manufacturing, sales, budgeting and supply chain software. Stands for: Business Process Control System. Client/Server software. Can run over the Internet. Runs on AS/400, Unix, Windows systems. Released: 1995. Vendor: System Software Associates, Inc.

BPCS EDIPath Communications software. EDI package. Runs on AS/400 systems. Vendor: System Software Associates, Inc.

BPR Business Process Redesign. Another term for re-engineering.

Bpwin Application development tool. Business process modeling tool. Allows managers to model and re-engineer business processes such as order processing, inventory tracking and personnel changes. Interfaces with ERwin to input process models into application development. Available in Standard or Pro versions. Runs on Windows systems. Released: 1998. Vendor: Computer Associates International, Inc.

BQM E-mail specification. Sets standards for message-oriented middleware for e-mail. Stands for: Business Quality Messaging. Introduced by IBM, Intel and Microsoft.

BrainMaker Application development tool used for data mining. Uses neural networks to discover and predict relationships in data. Runs on Macintosh, Windows systems. Vendor: California Scientific Software.

Brainstorm Object eXecutive Object-oriented, real-time operating system. Vendor: Brainstorm Engineering Co.

Branch Validator See CASEdge.

Bravo Application programming interface. Creates two-dimensional images for Web pages. Allows developers to create images, line art and text that can run on any platform. Integrates with Java. Vendor: Adobe Systems, Inc.

Bravo LC,MS 1. Desktop computer. Pentium CPU. Operating system: Windows 95/98. Vendor: AST Research, Inc. (division of Samsung Electronics).

Breur Software vendor. Products: financial systems for DEC environments.

BrewMaster Application development tool. Used to develop Java applications. Includes version control, syntax checking, word processing, debugging tools. Runs on Solaris, Windows systems. Released: 1997. Vendor: Objectsoft, Inc.

bridge Communications interface used to connect two networks that operate under the same data-link layer protocols. Term used for both hardware and software.

BridgePoint Application development tool. Object-oriented code generator. Generates C, C++, SQL, Ada. Based on Shlaer-Mellor methodology. Vendor: Project Technology, Inc.

BridgeWare Application development tool. Automatically generates code to link desktop PowerBuilder and Visual Basic systems to mainframe applications. Vendor: Micro Focus, Inc.

Brio Enterprise Communications. Internet server. Includes both push and pull technologies. Includes OnDemand application server for ad-hoc queries and Broadcast Server for push reports. Runs on Unix, Windows. Released: 1998. Vendor: Brio Technology, Inc.

Brio.Insight Application development tool. Provides analysis functions and allows user to create 3D charts, comprehensive reports and aggregates. Runs on Macintosh, Unix, Windows systems. Released: 1998. Vendor: Brio Technology, Inc.

Brio.Web.warehouse Application development tool used in data warehousing. Allows Web browser tools to access predefined warehouse reports. Vendor: Brio Technologies, Inc.

BrioQuery Application development tool. Query and reporting tool used by end-users. Interfaces with Oracle, Sybase, Red brick, Microsoft and multidimensional databases. Includes Explorer Designer. Used in data warehousing. Runs on Macintosh, Unix, Windows systems. Released: 1998. Vendor: Brio Technology, Inc.

BriteLite Portable computer. Vendor: RDI Computer.

Broadbase Server Application development tool used to create data marts. Includes complex query functions. Runs on Windows NT systems. Released: 1998. Vendor: Broadbase Information Systems, Inc.

Broadcast Agent Application development tool. Used to push reports via e-mail, pager, and fax. Released: 1998. Vendor: Business Objects, Inc.

broadcast Communications. Sending a single message to every user on a network or communications system.

BroadQuest EIP, corporate portal. Provides single browser interface corporate information including ERP packages and front-end CRM systems. Released: 1999. Vendor: BroadQuest Inc.

Brock Activity Manager Relational database for a desktop systems environment. Runs on AT&T, Compaq, DEC, HP, IBM RT systems. Utilizes SQL. Vendor: Brock Control Systems, Inc.

broker Term used with middleware and specifically with EAI software. The broker resides on a server and processes the requests from the users. It provides security and monitors activity. There are several types of brokers including integration brokers, information brokers.

brouter Communications device that can operate as either a bridge or a router.

browser Communications, Internet software. Program used to view sites on the Web.

BRS/Search Application development tool. Provides text retrieval functions for client/server systems. Runs with Unix, VMS, Windows NT server systems. Vendor: Dataware Technologies, Inc.

BSA Business Software Alliance. Association of companies. Purpose is to prevent software piracy.

BSAM Access method used in IBM mainframe systems.

BSC-3270 Communications software. Terminal emulator. Vendor: Computer Logics, Ltd.

BSCLIB Mathematical/statistical programs. Vendor: Boeing Computer Services.

BSD (V4.3) Operating system. Evolution of the original Unix system. Stands for: Berkeley Software Distribution. Ultrix (DEC's version of Unix) is very close to BSD V4.3.

BSD/OS Operating system. Version of Unix easily adapted to small systems. Vendor: Berkeley Software Design, Inc.

BTAM Access method used in IBM mainframe systems. Stands for: Basic Telecommunications Access Method.

BTAS/1 Operating system add-on. Data management software that runs on IBM Series/1 systems. Vendor: Business Management Systems, Inc.

BTG AXP275 Midrange computer. Alpha RISC CPU. Operating system: Windows NT. Vendor: BTG Technology Systems.

BTnet Communications network. Public packet switching network set up by the British government. Uses X.75 protocols.

BTOS II Operating system for mainframe Unisys systems. Vendor: Unisys Corp.

Btrieve(/N) Data retrieval subroutines used in IBM and compatible desktop computers. Vendor: Novell, Inc.

BTS Communications software. Network connecting IBM mainframes and terminals. Stands for: Binary Synchronous Communications Transport System. Vendor: Digital Communications Associates, Inc.

bubble chart A tool used during the analysis and design phases of the project life cycle. Uses bubble-like symbols and shows the movement of data within a processing system.

Budget DASD Operating system enhancement used by systems programmers. Increases system efficiency in IBM systems. Manages disk storage space. Vendor: Empact Software.

buffer A temporary holding place for data as it is being transferred from an external source (disk, terminal) to or from a computer. This allows a lot of data to be transferred and held for processing. Caching is a type of buffering.

bug A program error. Can also be used to refer to a hardware error.

BugTracker Operating system add-on. Debugging/testing software that runs on DEC VAX, Unix, IBM desktop computers. Vendor: NAS Software Associates.

Bug Trapper Application development tool. Automatically tracks bugs and generates a log of how and where bugs occurred, providing a more direct path to the cause of problem behavior. Runs with Microsoft's Visual Studio. Runs on Windows NT systems. Vendor: MuTek Solutions, Inc.

Build-IT Application development tool. Used to build large Web sites. Coordinates work done by many developers. Logs usage and connections between large numbers of files. Part of Web Sphere Studio 3. Vendor: IBM Corp.

Build Momentum Application development environment. Object-oriented tool for creating client/server applications with a variety of GUIs. Vendor: Sybase, Inc.

BuilderXcessory System software. Deploys Unix applications to Windows and the Internet. Works with C, C++ programs. Vendor: Integrated Computer Solutions, Inc.

BuildProfessional Application development environment. Graphical, object-oriented. Supports Oracle, Informix, ODBC. Runs on Unix, Windows systems. Released: 1996. Vendor: TODAY Systems, Inc.

Bull HN Computer vendor. Manufactures desktop computers.

BUNDL Operating system enhancement used by systems programmers. Increases system efficiency in IBM systems. Automates report distribution. Vendor: LEGENT Corp.

BURCOM Communications software. Network connecting DEC and Unisys computers. Vendor: Applied Information Systems, Inc.

Burroughs Old computer vendor. Merged with Sperry to form Unisys.

bus topology A network configuration where each computer is connected to a central cable that passes through the network.

BusiGEN Program and report generator for DG systems. Vendor: Data General Corp.

Business 1,2 WaveStation Desktop computer. Pentium processor. Vendor: Directwave Inc.

business analyst Application developer. Similar to systems analyst, but business analysts usually do not have a technical background. See systems analyst.

Business BASIC Compiler language. Version of BASIC.

Business Design Facility Application development tool. Builds graphical business models. Interfaces with IEF. Vendor: Texas Instruments, Inc.

Business Engineering Workbench Application development tools used with SAP's R/3 application systems. Automates the configuration of R/3 systems. Released: 1996. Vendor: SAP AG.

Business Intelligence for the Small Enterprise See BISE.

Business Intelligence System Application development tool. Contains twenty predefined business analysis functions. Works with Express, Express analyzer, Express Objects. Released: 1998. Vendor: Oracle Corp.

Business Information Warehouse Data warehouse software. Gathers information from R/3 systems and suppliers to make business forecasts. Released: 1998. Vendor: SAP AG.

Business Partner See FX-600,800.

business portal See EIP.

business process A business process is a sequence of tasks performed in a specific order by specific people in order to meet a business need. Term is used in workflow technology.

Business Process Redesign See BPR.

Business Quality Messaging See BQM.

Business Station, Server, Workstation Desktop computer. Pentium processor. Vendor: ABS Computer Technologies, Inc.

Business Traveler Video conferencing system. Vendor: RSI Systems, Inc.

BusinessMachine Desktop computer. Pentium CPU. Operating system: Windows 95/98. Vendor: Comtrade Electronics, U.S.A. Inc.

BusinessMAX Desktop computer. Pentium CPU. Operating system: Windows 95/98. Vendor: CyberMax Computer, Inc.

BusinessMiner Application development tool used in data warehousing. Data mining tool for end-users. Automatically detects patterns in data. Runs on Windows systems. Vendor: Business Objects.

BusinessObjects Application development tool. OLAP software. Data query and reporting software. Allows end-users to query relational and multidimensional databases without knowing SQL or the database structure. Runs on , Macintosh, Unix, Windows systems. Vendor: Business Objects, Inc.

BusinessQuery Query language used with Excel. Vendor: Business Objects, Inc.

BusinessWare EAI (Enterprise Application Integration) software that integrates disparate applications by allowing the end-user to design the solution. Works with Clarify, PeopleSoft, SAP, CICS, MQSeries, MSMQ, Informix, Oracle. Includes Connector SDK (Software Development Kit) to build connections between legacy and custom applications. Released: 1998. Vendor: Vitria Technology, Inc.

Butterfly Nickname for IBM's ThinkPad notebook. Weighs 4.5 lbs, has expandable keyboard so is only 10″ x 8″ x 2″. Vendor: IBM Corp.

BuyerXpert See CommerceXpert.

BW-MultiConnect Communications. Network management software. Connects Windows NT, NetWare systems. Vendor: Beame & Whiteside Software.

BX, LX Supreme Desktop computer. Pentium processor. Vendor: ABS Computer Technologies, Inc.

BX Pro Application development environment. Visual tool. Interfaces with Java and C++. Includes user interface objects that can be selected, positioned and combined to create an application's front-end. Runs on MacOS, Unix, Windows systems. Released: 1999. Vendor: Integrated Computer Solutions, Inc.

byte String of eight bits operated on as a unit. Byte is the basic unit used to describe size of storage units, including computer storage, programs, data files. Stated as kilobytes, megabytes, etc.

C Compiler language. Language was developed by AT&T in conjunction with Unix operating systems. It is not, however, limited to Unix environments.

C++ Object-oriented version of C language.

C++Builder Application development environment. Visual, object-oriented. Includes templates to build web applications and generates HTML code. Has forms-oriented RAD functions for designing front end applications and assembler and API tools for system developers. Includes project management functions, Turbo Debugger (debugging tools), and database development tools. Standard, Professional and Client/Server versions available. Released: 1997. Vendor: Borland Tools, division of Inprise Corp.

C++ Coder Application development tool. Generates C++ code for applications to run under Windows, OS/2. Vendor: Superbica.

C++ POWERBench See POWERBench.

C++ Professional Application development environment. Used to develop object-oriented applications. Compatible with Visual C++ and Frameworks. Vendor: Symantec Corp.

C++/Softbench Object-oriented extension of CASE environment provided by CASEdge. Vendor: Hewlett-Packard Co.

C++/Views Application development tool. Object-oriented, cross-platform development toolkit. Allows users to develop code that will work in diverse environments including Windows, Presentation Manager, Motif, Macintosh. Runs on OS/w, Unix. Windows systems. Vendor: Liant Software Corp.

C.A.P. System Desktop computer. AMD processors. Vendor: Austin Computer Systems, Inc.

C-A-T Relational database for desktop computer environments. Runs on Apple Macintosh systems. Stands for: Contacts-Activity-Time. Vendor: Chang Labs, Inc.

C/C++ Workbench Application development tool. Object-oriented programming system that eases transition to C++. Runs on Unix systems. Included with ADE. Vendor: NCR Corp., Lucid, Inc.

C-Cover Application development tool. Provides testing, debugging functions. Analyzes source code to determine what code has not been tested. Runs on OS/2, Unix, Windows systems. Vendor: Bullseye Testing Technology.

C Executive Real-time operating system for CISC and RISC systems. Vendor: JMI Software Systems, Inc.

C.I.P. Database for desktop environments. Runs on IBM desktop computer systems. Includes report generator. Full Name: The Concentric Information Processor. Vendor: Concentric Data Systems, Inc.

C/S Composer, SES/Workbench Operating system support software. Provides capacity planning for client/server systems. Simulation tools. Runs on Unix systems. Vendor: SES, Inc.

C/S Elements Application development tool. Builds cross platform GUIs including Presentation Manager, Open Look, Motif, Windows. Runs on most desktop systems. Vendor: Neuron Data, Inc.

C Set++ Application development environment. Includes C/C++ compiler, KASE:Set, debugging and testing tools and development utilities. Develops object-oriented applications for OS/2, AIX. Vendor: IBM Corp.

C shell A Unix shell program that efficiently handles interactive commands. See shell.

C/Spot/Run Application development tool. Automates analysis functions and helps developers find and debug errors. Runs on HP, Sun systems. Vendor: Procase Corp.

C-Telex Communications software. E-mail system. Runs on Concurrent systems. Vendor: Concurrent Computer Corp.

c-tree Plus Data management software. Controls sequential and random file access for programs written in C. Based on B+tree routines. Runs on DEC, Macintosh, Unix, Windows systems. Released: 1998. Vendor: FairCom US Corp.

C-XIX Operating System for Counterpoint systems. Vendor: Acer Counterpoint, Inc.

C/XL Compiler language used in midrange and desktop environments. Runs on HP systems. Vendor: Hewlett-Packard Co.

C090, C100, C120 Notebook computer. Pentium CPU. Vendor: Hitachi PC Corp.

C3 Series Mainframe computer. Operating system: Unix. Vendor: Convex Computer Corp.

C3ADA Compiler language used in midrange and desktop environments. Allows use of ADA on Concurrent systems. Vendor: Concurrent Computer Corp.

C4 Application development environment. Object-oriented. Runs on Unix systems. Full name: C4 Object-oriented Programming Language. Vendor: Axon Development Corp.

C4/XA Model C46xx Mainframe computer. Operating systems: ConvexOS, Unix. Vendor: Convex Computer Corp.

C7 Compiler language. Shorthand for Microsoft's C/C++ compiler.

C9xx, EL9xx, J9xx Supercomputer. Operating systems: Unix, COS, Unicos. Vendor: Cray Research, Inc.

CA-1 Operating system add-on. Tape management system that runs on IBM systems. Part of CA-UNIPACK/SRM. Vendor: Computer Associates International, Inc.

CA-11 Operating system enhancement used by systems programmers. Increases system efficiency in IBM systems. Handles job reruns without JCL changes. Part of CA-UNIPACK/APC. Vendor: Computer Associates International, Inc.

CA-7(/RPT) Operating system enhancement used by systems programmers. Manages the entire data center for IBM systems. Part of CA-UNIPACK/APC. Vendor: Computer Associates International, Inc.

CA-9 Reliability Plus Operating system enhancement used by systems programmers. Monitors and controls disk storage in IBM systems. Vendor: Computer Associates International, Inc.

CA-Accpac/2000 Accounting software that runs in client/server environments. Runs on Windows systems. Vendor: Computer Associates International, Inc.

CA-Accuchek Application development tool. Used during testing. Compares data on different files and prepares a variety of reports so developers can compare results from different test scenarios. Runs on IBM mainframe systems. Vendor: Computer Associates International, Inc.

CA-ACF2 Operating system add-on. Security/auditing system that runs on IBM systems. Stands for: Access Control Facility. Vendor: Computer Associates International, Inc.

CA-ADC2 Operating system enhancement used by systems programmers. Job scheduler for IBM systems. Vendor: Computer Associates International, Inc.

CA-ADS/Generator Application development tool. Automates programming functions. Vendor: Computer Associates International, Inc.

CA-ADS(/PC) Application development environment. Includes 4GL, prototyping facilities. Interfaces with IDMS. Runs on IBM mainframe systems; desktop computer version available. Formerly called ADSO (ADS online). Vendor: Computer Associates International, Inc.

CA-Agentworks System management software. Provides load balancing, performance monitoring, some network management. Monitors databases and networks in Unix systems. Based on SNMP, CMIP protocols. Part of CA-Unicenter. Vendor: Computer Associates International, Inc.

CA-APCDOC Operating system enhancement used by systems programmers. Increases system efficiency in IBM systems. Part of CA-UNIPACK/APC. Vendor: Computer Associates International, Inc.

CA-ASM2 Operating system add-on. Data management software that runs on IBM systems. Part of CA-UNIPACK/SRM. Stands for: Automated Storage Management. Vendor: Computer Associates International, Inc.

CA-BLOCKMASTER Operating system enhancement used by systems programmers. Increases system efficiency in IBM systems. Reduces I/O activity. Part of CA-UNIPACK/SRM. Vendor: Computer Associates International, Inc.

CA-CAS MRP software. Includes sales analysis and performance measurements. Runs on IBM MVS systems. Vendor: Computer Associates International, Inc.

CA-Clipper C compiler language for dBase/XBase programs. Includes data dictionary. Runs on Windows on IBM and compatible desktop computers. Part of xBase environment. Works with CA-Visual Objects. Vendor: Computer Associates International, Inc.

CA-Commonview Application development tool. A framework, or class library, for developing GUI applications in C++. Runs on OS/2, Windows systems. Vendor: Computer Associates International, Inc.

CA-Converter Application development tool. Used to convert COBOL and Assembler programs from VSE to MVS operating systems. Will create applicable JCL and create documentation. Vendor: Computer Associates International, Inc.

CA-Cricket Presents Desktop system software. Graphics package. Interfaces with Lotus 1-2-3, Excel. Runs on Windows systems. Vendor: Computer Associates International, Inc.

CA-Culprit Application development tool. Information retrieval and management tool that produces user-designed reports from IDMS databases. Runs on IBM mainframe systems. Vendor: Computer Associates International, Inc.

CA-Datacom/DB Relational database/4GL for Unix environments. Utilizes SQL. Includes report generator. Runs on IBM systems. Includes CA-Datadictionary and CA-Dataquery. Vendor: Computer Associates International, Inc.

CA-Datadictionary See CA-Datacom/DB.

CA-DataMacs(/II) Operating system add-on. Data management software that runs on IBM systems. Interfaces to IMS and IDMS databases. Vendor: Computer Associates International, Inc.

CA-Dataquery(/PC,VAX) Application development tool. Provides ad hoc query, data migration and reporting functions for CA-Datacom databases. Vendor: Computer Associates International, Inc.

CA-DB:cBASE Application development system. Includes data dictionary, online testing and debugging tools, prototyping tools. Interfaces with CA-Datacom/DB, DB2, CA-ACF2, and CA-Top Secret. Runs on IBM mainframe systems. Vendor: Computer Associates International, Inc.

CA-DB:Generator/PC, DB:Architect CASE product. Automates analysis, design, and programming functions. Includes code generator for C, C++, Pascal, Fortran. Design methodologies supported: Chen, entity-relationship models. Databases supported: IDMS. Runs on Pentium type desktop computers. Vendor: Computer Associates International, Inc.

CA-DB/Unix Database for Unix environments. Utilizes SQL. Vendor: Computer Associates International, Inc.

CA-dBFast Application development tool. Development environment for dBase applications. Allows users to migrate dBase applications to Windows systems. Includes compiler, screen editor. Interacts with Clipper. Vendor: Computer Associates International, Inc.

CA-Deliver Operating system enhancement. Handles printing and report distribution. Can deliver specific pages or combine pages and/or complete reports for printing and distribution in real-time. Runs on IBM mainframe systems. Vendor: Computer Associates International, Inc.

CA-Dispatch Operating system enhancement used by systems programmers. Increases system efficiency in IBM systems. Controls print distribution. Part of CA-UNIPACK/APC. Vendor: Computer Associates International, Inc.

CA-DUO Application development tool used to convert programs from VSE to MVS. Provides an interface that allows VSE applications to runs under MVS without alteration. Vendor: Computer Associates International, Inc.

CA-DYNAM Operating system add-on. Data management software that runs on IBM systems. Programs include: CA-DYNAM, CA-DYNAM/B VM, CA-DYNAM/D, CA-DYNAM/FI, CA-DYNAM/T(VM), CA-DYNAM/VM. Vendor: Computer Associates International, Inc.

CA-DYNAM/TLMS Operating system add-on. Tape management system that runs on IBM systems. Part of CA-UNIPACK/SRM. Vendor: Computer Associates International, Inc.

CA-Dynamic Scheduling Engine Object-oriented, client/server manufacturing software. Provides real-time scheduling functions. Vendor: Computer Associates International, Inc.

CA-EasyPROCLIB Application development tool. JCL and/or SYSOUT maintenance program for IBM systems. Vendor: Computer Associates International, Inc.

CA-Easytrieve 4GL. Vendor: Computer Associates International, Inc.

CA-e-mail+ Communications software. E-mail system. Runs on IBM systems. Vendor: Computer Associates International, Inc.

CA-Endevor for xxx Application management tool. Software configuration manager. Manages both in-house and vendor-supplied software. Provides configuration management, inventory management, change control, and release management functions. Versions available for MVS, Unix. Vendor: Computer Associates International, Inc.

CA-Estimacs CASE product. Automates project management function. Supplies cost accounting information for project development and maintenance. Vendor: Computer Associates International, Inc.

CA-EXAMINE Operating system enhancement used by systems programmers. Monitors and controls system performance in IBM systems. Reports on status and modifications to system components. Part of CA-UNIPACK/SCA. Vendor: Computer Associates International, Inc.

CA-EZTEST/CICS Operating system add-on. Debugging/testing software that runs on IBM systems. Vendor: Computer Associates International, Inc.

CA-FastDASD Operating system enhancement used by systems programmers. Increases system efficiency in IBM systems. Improves DASD performance. Part of CA-UNIPACK/PMA. Vendor: Computer Associates International, Inc.

CA-Gener/OL Application development tool. Runs on IBM systems. Vendor: Computer Associates International, Inc.

CA-HRISMA Applications software. Human resources package that interfaces with DB2. Runs on OS/2, Windows systems. Client/server technology. Vendor: Computer Associates International, Inc.

CA-ICMS Application development tool. Delivers data in relational format to user workstations. Provides support for end-user application development. Runs on IBM mainframe systems. Vendor: Computer Associates International, Inc.

CA-IDEAL Application development system. Includes report generator, data dictionary, screen painter. Runs on IBM systems. Integrates with CA-Datacom/DB and DB2. Vendor: Computer Associates International, Inc.

CA-IDMS Relational database/4GL for mainframe IBM, Unix systems. Utilizes SQL. Includes report generator. Vendor: Computer Associates International, Inc.

CA-IDMS/Architect CASE product. Automates analysis, programming, and design functions. Works with IDMS and RMS databases. Associated product names: Knowledge BUILD, AUTO-MATE Plus, Application Development System. Vendor: Computer Associates International, Inc.

CA-IDMS-DC Communications software. Transaction processing monitor. Runs on IBM systems. Communications monitor for online IDMS applications. Vendor: Computer Associates International, Inc.

CA-IDMS/DDS Maintains IDMS databases at separate locations. Used by system programmers. Stands for: Distributed Database System. Vendor: Computer Associates International, Inc.

CA-IDMS/SQL Relational database/4GL for large computer environments. Runs on DEC VAX systems. Utilizes SQL. Includes report generator. Vendor: Computer Associates International, Inc.

CA-INGRES/Gateways Communications software. Gateways connecting Sun, DEC, HP and IBM workstations. Vendor: Computer Associates International, Inc.

CA-Ingress II See Ingress II.

CA-INGRES/Windows Application development environment. Includes windowing software, 4GL, and GUI. Compatible with Windows, Open Look, and OSF/Motif. Vendor: Computer Associates International, Inc.

CA-InterTest, CA-InterTest/Batch Application development tool. Debugging/testing programs.. Runs on IBM mainframe systems. Vendor: Computer Associates International, Inc.

CA-ISS/Three Operating system enhancement used by systems programmers. Monitors and controls system performance in IBM systems. Capacity management tool. Part of CA-UNIPACK/PMA. Vendor: Computer Associates International, Inc.

CA-JARS/ Operating system enhancement used by systems programmers. Provides system cost accounting for IBM systems. Includes: CA-JARS/CICS, CA-JARS/DSA, CA-JARS/IMS, CA-JARS/MVS, CA-JARS/SMF, CA-JARS/VSE, CA-JARS/VM. Vendor: Computer Associates International, Inc.

CA-JCLCHECK Application development tool. JCL and/or SYSOUT maintenance program for IBM systems. Part of CA-UNIPACK/APC. Vendor: Computer Associates International, Inc.

CA-Jobtrac Operating system enhancement. Manages system workload. Provides automatic job submission and tracks the progress of each job. Runs with Unicenter TNG. Vendor: Computer Associates International, Inc.

CA-LDM Application development tool used in data warehousing. Used to extract, convert and move large volumes of data from host-based systems to client/server systems. Stands for Legacy Data Mover. Vendor: Computer Associates International, Inc.

CA-LIBRARIAN Operating system add-on. Library management system that runs on IBM systems. Vendor: Computer Associates International, Inc.

CA-LOOK Operating system enhancement used by systems programmers. Monitors and controls system performance in IBM systems. Part of CA-UNIPACK/PMA. Vendor: Computer Associates International, Inc.

CA-Manager Operating system enhancement used by systems programmers. Job scheduler for IBM systems. Supports TSO, CICS, RJE, ROSCOE, BTAM, VTAM. Vendor: Computer Associates International, Inc.

CA-Masterpiece, Masterpiece/Net Application software. Financial management package for multinational, multi-organizational companies. Euro compliant. Masterpiece/Net has Web interface. Vendor: Prestige Software, a division of Computer Associates International, Inc.

CA-Mazdamon Communications. Network management software. Controls networks in IBM systems. Vendor: Computer Associates International, Inc.

CA-MetaCOBOL Compiler language used in mainframe environments. Version of COBOL that supports structured programming and enhances conversion from COBOL to COBOL II. Includes program generation functions by creating program shells and validating generated code. Vendor: Computer Associates International, Inc.

CA-MINDOVER MVS Operating system enhancement used by systems programmers. Monitors and controls system performance in IBM systems. Capacity planning and performance tuning system. Vendor: Computer Associates International, Inc.

CA-NETMAN Operating system enhancement used by systems programmers. Manages the entire data center for IBM systems. Part of CA-UNIPACK/DCA. Vendor: Computer Associates International, Inc.

CA-NetSpy Communications. Network performance monitor. Measures response time and network traffic. Performs real-time problem analysis, recommends problem solutions. Integrates with Unicenter TNG. Runs on MVS systems. Vendor: Computer Associates International, Inc.

CA-OpenRoad See OpenRoad.

CA-OPERA Operating system enhancement used by Operations staff and systems programmers. Provides operator console support in IBM systems. Part of CA-UNIPACK/APC. Vendor: Computer Associates International, Inc.

CA-Optimizer Application development tool. Automates programming function. Analyzes source code for such things as adherence to standards and presence of logic flaws (unexecuted code). Languages analyzed: COBOL. Runs on IBM mainframes. Vendor: Computer Associates International, Inc.

CA-Panvalet Operating system add-on. Library management system that runs on IBM systems. Vendor: Computer Associates International, Inc.

CA-PHIPS Image processing system that runs on desktop computer systems. Vendor: Computer Associates International, Inc.

CA-Planmacs CASE product. Automates project management function. Supplies cost accounting information for project development and maintenance. Vendor: Computer Associates International, Inc.

CA-Prevail/XP-Paradigm Operating system enhancement used by systems programmers. Problem management software. Tracks actions and time taken to solve system problems. Runs on Unix systems. Part of CA-Unicenter. Vendor: Computer Associates International, Inc.

CA-PRMS MRP software. Includes manufacturing, distribution and financial software. Runs on IBM AS/400. Vendor: Computer Associates International, Inc.

CA-Quick-Fetch Operating system enhancement. Used to optimize DASD and I/O control units. Runs on MVS systems. Vendor: Computer Associates International, Inc.

CA-QuickD Operating system software. Software delivery system. Packages, distributes and installs software on both clients and servers on an enterprise basis. Integrates with CA-Unicenter. Vendor: Computer Associates International, Inc.

CA-RAMIS Database/4GL for large computer environments. Runs on IBM systems. Vendor: Computer Associates International, Inc.

CA-RAPS Operating system enhancement used by systems programmers. Increases system efficiency in IBM CICS systems. Controls printers. Part of CA-UNIPACK/APC. Vendor: Computer Associates International, Inc.

CA-Realia CICS Mainframe-compatible version of CICS that runs on desktop computers. Vendor: Computer Associates International, Inc.

CA-Realia COBOL Compiler language used in desktop computer environments. Compatible with mainframe COBOL. Vendor: Computer Associates International, Inc.

CA-Realia DL/1 Mainframe-compatible version of DL/1 that runs on desktop computers. Vendor: Computer Associates International, Inc.

CA-Realia II Workbench Application development tool. Allows developers to develop mainframe COBOL/CICS applications from desktop systems. Vendor: Computer Associates International, Inc.

CA-Realia IMS Mainframe-compatible version of IMS that runs on desktop computers. Vendor: Computer Associates International, Inc.

CA-Realizer Application development environment. Works in client/server architecture. Builds client applications with a GUI-based front end. Interfaces with SQLBase. Allows developers to write applications for either Windows or OS/2 and recompile for the other platform. Vendor: Computer Associates International, Inc.

CA-Roscoe Application development tool used to develop IBM mainframe systems. Runs on IBM online programs. Supports CICS, IMS/DC, and DB2 development. Vendor: Computer Associates International, Inc.

CA-RSVP Operating system enhancement. Storage management software that manages system catalogs and allows administrators to define installation storage criteria. Runs on IBM mainframe systems. Released: 1992. Vendor: Computer Associates International, Inc.

CA-SCHEDULER Operating system enhancement used by systems programmers. Job scheduler for IBM systems. Part of CA-UNIPACK/APC. Vendor: Computer Associates International, Inc.

CA-SORT Operating system enhancement used by systems programmers. Increases system efficiency in IBM systems. Efficient sort program. Part of CA-UNIPACK/SRM. Vendor: Computer Associates International, Inc.

CA-SRAM Operating system enhancement used by systems programmers. Increases system efficiency in IBM systems. Allows multiple concurrent sorts. Stands for: Sort Reentrant Access Method. Part of CA-UNIPACK/SRM. Vendor: Computer Associates International, Inc.

CA-SuperCalc Desktop system software. Spreadsheet. Runs on Pentium type desktop computers, Apple systems. Version also available for IBM mainframes and DEC VAX systems. Vendor: Computer Associates International, Inc.

CA-Superimage (for VAX) Desktop system software. Graphics package. Runs on DEC VAX, Pentium type systems. Vendor: Computer Associates International, Inc.

CA-SuperProject Desktop system software. Project management package. Runs on Pentium type desktop computers, DEC VAX systems. Versions include: SuperProject Expert, SuperProject Plus, SuperProject for Windows. Vendor: Computer Associates International, Inc.

CA-SYSVIEW/E Operating system software. Real-time system monitoring and analysis package. Provides a graphical view of current and historical resource utilization for MVS, CICS, JES2 and CA-DATACOM. Runs on MVS, OS/390 systems. Released: 1998. Vendor: Computer Associates International, Inc.

CA-Telon CASE product. Automates analysis, design, and programming functions. Includes code generator for COBOL, PL/1. Databases supported: DB2, IDMS, IMS, VSAM. Design methodologies supported: Chen, Constantine, DeMarco, Yourdon. Runs on IBM, Unix systems. Includes: Telon Teamwork, Telon Plus. Vendor: Computer Associates International, Inc.

CA-Textor Desktop system software. Word processor. Runs on Windows systems. Vendor: Computer Associates International, Inc.

CA-Top Secret for xx Operating system add-on. Security/auditing system that runs on IBM systems. Versions include: Top Secret for VM, Top Secret for MVS, Top Secret for VSE. Vendor: Computer Associates International, Inc.

CA-Transit Application development tool. Converts COBOL programs written for Unisys, CDC, Bull, AT&T, HP, Prime and DEC to run under IBM VSE or MVS. Vendor: Computer Associates International, Inc.

CA-TSO/MON Operating system enhancement used by systems programmers. Monitors and controls system performance in IBM systems. Controls TSO performance. Vendor: Computer Associates International, Inc.

CA-UCANDU Operating system add-on. Data management software that runs on IBM systems. Part of CA-UNIPACK/SRM. Vendor: Computer Associates International, Inc.

CA-Unicenter See Unicenter TNG.

CA-Unicenter/Star Operating system software. Runs on OS/2 systems and manages client/server environment. Manages diverse systems from a single workstation. Versions for Windows, MVS will be available. Vendor: Computer Associates International, Inc.

CA-UNIPACK/APC Operating system enhancement used by systems programmers. Increases system efficiency in IBM systems. Stands for: Automated Production Control. Includes: CA-7, CA-11, CA-SCHEDULER, CA-OPERA, CA-RAPS, CA-DISPATCH, CA-JCLCHECK, CA-APCDOC. Vendor: Computer Associates International, Inc.

CA-UNIPACK/DCA Operating system enhancement used by systems programmers. Manages the entire data center for IBM systems. Stands for: Data Center Administration. Includes: CA-NETMAN software. Vendor: Computer Associates International, Inc.

CA-UNIPACK/PMA Operating system enhancement used by systems programmers. Increases system efficiency in IBM systems. Stands for: Performance Measurement and Accounting. Includes: CA-FastDASD, CA-ISS/Three, CA-MAZDAMON, CA-JARS(/CICS), CA-LOOK, CA-MINDOVER. Vendor: Computer Associates International, Inc.

CA-UNIPACK/PPS Operating system add-on. Testing and debugging tools. Vendor: Computer Associates International, Inc.

CA-UNIPACK/SCA Operating system add-on. Security/auditing system that runs on IBM systems. Stands for: Security Control & Auditing. Includes: CA-ACF2, CA-Top Secret, CA-VMAN, CA-EXAMINE. Vendor: Computer Associates International, Inc.

CA-UNIPACK/SRM Operating system enhancement used by systems programmers. Increases system efficiency in IBM systems. Stands for: Storage and Resource Management. Includes: CA-DYNAM/TLMS, CA-BLOCKMASTER, CA-SORT, CA-SRAM, CA-ASM2. Vendor: Computer Associates International, Inc.

CA-Universe Relational database for large computer environments. Runs on IBM systems. Includes data dictionary, query and reporting facilities. Allows application development through screen painting. Vendor: Computer Associates International, Inc.

CA-UpToDate Groupware. Vendor: Computer Associates International, Inc.

CA-Verify Application development tool. Debugging/testing programs. Vendor: Computer Associates International, Inc.

CA-View Operating system enhancement. Allows administrators to view output from the operating system in real-time. Runs on MVS, OS/390 systems. Vendor: Computer Associates International, Inc.

CA-Visual Express Application development tool. Windows based, client/server query and reporting tool. Accesses over 20 standard databases. Incorporates Easytrieve and Ramis 4GLs. Vendor: Computer Associates International, Inc.

CA-Visual Objects Application development environment for xBase systems. Object-oriented system that includes visual development techniques, class libraries, and SQL access. Works with Clipper. Originally called Aspen. Vendor: Computer Associates International, Inc.

CA-Visual Realia Application development tool. Provides a GUI for COBOL programs. Used to move legacy applications into client/server systems. Supports CA-Ingres, Oracle, Informix, Sybase databases. Vendor: Computer Associates International, Inc.

CA-VMAN Communications. Network management software. Controls networks in IBM systems. Part of CA-UNIPACK/SCA. Vendor: Computer Associates International, Inc.

CA-Vollie Application development environment. Runs on IBM online systems. Includes JCL and COBOL syntax checkers, full-screen editing capabilities, job-submission and status-checking routines. Vendor: Computer Associates International, Inc.

CA-XCOM Operating system enhancement. Used to transfer files across different platforms. Vendor: Computer Associates International, Inc.

CA90s An open-system architecture from Computer Associates. Integrates applications across different computers and operating systems. Vendor: Computer Associates International, Inc.

cable The medium used to connect systems in a LAN. Twisted-pair, coaxial, and fiber-optic are common cables. Cables can be baseband (one message at a time) or broadband (multiple messages).

Cabletron Hardware vendor. Manufactures equipment used in communications networks.

Cabletron Specialist Technical certification. Can take courses and pass tests in competency areas for Cabletron products. Can certify in multiple areas.

Cabletron Systems Engineer Technical certification. Achieved by passing all Cabletron specialist courses.

CABS Accounting system that runs on IBM RS/6000 systems. Stands for: Client Accounting and Bookkeeping System. Vendor: Briareus Corp.

Cache Object database. Uses a transactional multidimensional data model. Has backwards compatibility mode for relational databases. Interfaces with Web or GUI based applications. Runs on OpenVMS, Unix, Windows systems. Released: 1997. Vendor: InterSystems.

CacheFlow Application development tool. Provides web caching functions. Vendor: CacheFlow, Inc.

caching A performance technique that improves network performance by storing frequently requested programs and/or data locally. Often used with client/server.

Cactus Application server providing middleware and development tools. Includes 4GL, supports Java, ActiveX. Develops applications for client/server systems or Web front ends. Interfaces with RDBMSs, legacy databases. Released: 1996. Vendor: Information Builders, Inc.

CAD/CAM Computer-aided Design/Computer-aided Manufacturing. Programs that allow computers to design and develop products such as circuit boards and other computers.

CADD A CAD system with additional features for the drafting function. Stands for: Computer-Aided Design and Drafting.

CADIS Operating system add-on. Data management software that runs on Unisys systems. Stands for: Cache Disk Interface Software. Vendor: Unisys Corp.

CAFC Operating system enhancement used by systems programmers. Increases system efficiency in IBM CICS systems. Stands for: CICS Application File Control. Vendor: NETEC International, Inc.

Cafe, Visual Cafe Application development tool. Used to generate Java code, JavaBeans components. Runs on Macintosh, Windows systems. Creates applications that use a Web browser as a user interface. Incorporates JavaBeans. Formerly called Espresso. Released: 1996. Vendor: Symantec Corp.

Caffeine Application development tools. RAD tool user for Internet applications. Provides CORBA interfaces. Released: 1996. Vendor: Netscape Communications Corp, Visigenics, Inc.

CAI Computer-Assisted Instruction. Computer programs that operate in a dialogue with users and "teach" by providing information on whether responses to questions are correct or not. Uses graphics extensively and can also incorporate such aids as interactive video disks. Also called CBT.

Cairo Operating system for midrange, desktop computers. Development name for version of Windows NT, although it actually is an object-oriented layer on top of NT that includes a file system, object management and user interface. Vendor: Microsoft Corp.

Caldera Network Desktop Operating system providing access to the Internet. Based on Linux (Unix based public domain operating system). Includes GUI. Vendor: Caldera, Inc.

Caliper Communications software. Used to design complex networks. Can be used for both WAN and LAN. Released: 1997. Vendor: Network Tools, Inc.

Calisto Midrange computer. Pentium Pro CPU. Operating system: Windows NT. Vendor: NeTpower, Inc.

CallPilot Network management tool. Messaging tool that manages e-mail, voice mail and fax messages. Interfaces with MS Outlook, Lotus Notes, Netscape Messenger, Eudora Pro. Vendor: Nortel Networks.

CallSPONSOR Midsize computer. Server. UltraSPARC Pentium processor. Operating system: Unix. Vendor: Periphonics Corp.

Calout Plus Communications software. Transaction processing monitor. Runs on DEC systems. Vendor: Clyde Digital Systems.

CALS Computer-Aided Acquisition and Logistic Support. An image-processing and document-management program designed by the U.S. Department of Defense. The program was intended to implement existing standards for the electronic transfer of data.

Calypso Application development tool. Used for object-oriented analysis, design and modeling. Interfaces with GainMomentum. Vendor: Sybase, Inc.

Camelot Code name for VisualAge C++. See VisualAge C++.

CAMS 1. Operating system enhancement used by systems programmers. Monitors and controls system performance in IBM CICS systems. Vendor: Universal Software, Inc.
2. Application development tool. Program that converts COBOL programs and JCL from a DOS/VSE environment to MVS. Stands for: Conversion Analysis and Management System. Vendor: Comp Act Data Systems, Inc.

CANDE Application development system. Runs on Unisys systems. Vendor: Unisys Corp.

Candle Command Center See CCC.

Candle Command Center for S/390 Unix System Services Operating system enhancement. Manages Unix applications running on MVS systems. Released: 1998. Vendor: Candle Corp.

CANE System software. Graphic, object-oriented tool for designing, installing and maintaining networks. Runs on Windows NT systems. Stands for: Computer Aided Network Engineering. Vendor: ImageNet, Ltd.

Canon Computer vendor. Manufactures desktop computers.

Canonizer Application development tool. Automates programming function. Assists programmers in converting database models from one environment to another. Vendor: Six Sigma Case, Inc.

CAP Communications software. Provides interface between Unix and SNA networks. Stands for: Communications Access Processor. Vendor: Unisys Corp.

capacity planner Technical developer. Determines what hardware and software is needed to keep the computer system functioning at maximum capacity. Makes sure the system has enough storage and speed to handle the workload. Senior title most often used in mainframe computer systems.

Capacity Requirements Planning See CRP.

Caplink Application development tool. Links Netron/Cap with Bachman's Bachman/Analyst. Vendor: Netron, Inc.

Capone Communications software. Messaging software that allows the operating system to handle workgroups. Works with the Chicago version of Windows (Windows 95). Renamed Touchdown. Vendor: Microsoft Corp.

Caprera Communications software. Middleware. Runs on Unix, Windows NT systems. Vendor: Tactica Corp.

Captain Application development tool. Automates programming and testing functions. Works with IMS, IDMS, DB2, VAX Rdb/VMS databases. Vendor: Parker Shannon Associates, Inc.

Capture/MVS,SNA,VM,VMS Operating system enhancement used by systems programmers. Monitors and controls system performance in IBM and DEC VAX systems. Vendor: BGS Systems, Inc.

CARE Communication. Links mainframe systems to the Web. Includes Java-based terminal emulation development tools to create E-commerce applications. Vendor: Computer Network Technology Corp.

Carleton Passport Application development tool. Data warehouse software. Concentrates on the development of metadata. Runs on IBM systems. Runs on OS/2, Windows. Vendor: Carleton Corp.

Cars 5 Operating system add-on. Security/auditing system that runs on most major computer systems. Vendor: Triton Systems.

CART Application development tool used for data mining. Decision-tree software used to find trends and relationships within data warehouses. Runs on Unix, Windows NT systems. Stands for: Classification and Regression Trees. Released: 1998. Vendor: Salford Systems, Inc.

cartridge Object-oriented architecture used to integrate relational and object databases. Interface modules are independent and connect to the database server across a network. Alternative to DataBlades. Developed by Oracle Corp.

CARTS-TM Operating system enhancement used by systems programmers. Increases system efficiency in IBM systems. Manages disk storage space. Vendor: On-Line Documentation, Inc.

Cascade Operating system software that will provide Windows NT services along with the Solaris operating system. Will allow applications written for Windows to run under Solaris. Cascade is a code name. Release date: 1999. Vendor: Sun Microsystems, Inc.

CASE Computer Aided Software Engineering. A program that automates any phase of the development cycle.

CASE 2000 CASE product. Automates analysis and design functions. Associated product name: NRUNOFF. Vendor: Nastec Corp.

CASE*Designer CASE product. Automates design function. Part of Oracle Tool Kit. Vendor: Oracle Corp.

CASE*Dictionary CASE product. Automates testing and project management functions. Part of Oracle Tool Kit. Vendor: Oracle Corp.

CASE:PM,W Application development tool. Automates programming and testing functions. Program generator for applications running under windows. Vendor: Caseworks, Inc.

CASE Project Pack Software that aids organizations in evaluating the benefits of CASE. A part of the decision-making process. Package includes consulting, training, and support services as well as rental of Excelerator. Vendor: Index Technology Corp.

CASEdge CASE product. Name for family of CASE tools. Products include: Soft-Bench, Encapsulator, Branch Validator. Vendor: Hewlett-Packard Co.

CASEGenerator for SQL*Forms/SQL*Menu CASE Tool. Automates programming phase. See SQL*Forms and SQL*Menu. Vendor: Oracle Corp.

CASEGenerator for SQL*ReportWriter\ SQL*Plus CASE Tool. Automates programming phase. See SQL*ReportWriter and SQL*Plus. Vendor: Oracle Corp.

CasePac CASE product. Automates analysis, programming, and design functions. Works with DB2 databases. No longer being sold. Vendor: On-Line Software International.

CASETools CASE product. Interfaces with CASE*Designer, CASE*Dictionary, CASE*Generator. SQL*Forms. Vendor: Oracle Corp.

CASEVision/Workshop Application development tool. Automates programming, testing functions. Used to build applications in both business and technical environments. Runs on Unix systems. Interfaces with ToolTalk. Vendor: Silicon Graphics, Inc.

Cassiopeia Handheld computer. Operating system: WinCE. Discontinued 1997, Vendor: Casio.

CAST/FRAMEWORK-SYNERGY CASE tool. Automates full life cycle. Companies customize to their own methodologies. Runs on OS/2, Windows systems. Vendor: CMD Corp.

CAST Workbench CASE tool. Full life cycle development modules for SQLServer and Sybase applications. Generates documentation, includes testing functions. Runs on Windows systems. Vendor: CAST Software, Inc.

Castanet CAST Communications. Internet software. Distributes and manages application software over the Internet. Updates only the portion of the application that has changed since the last download. Released: 1997. Vendor: Marimba, Inc.

Catalog Server Internet software. Search engine. Vendor: Netscape Communications Corp.

Catalysis Object-oriented development methodology. Associated with D'Souza and Wills.

Catamount Midrange computer. Code name for latest release of VAX systems. Vendor: Digital Equipment Corp.

Catapult Internet software. Firewall. Vendor: Microsoft Corp.

Cathode ray tube See CRT.

CATI Computer-Aided Testing and Implementation. Generic name for coordinated and integrated tools for testing and debugging. Provide such diverse functions as dump debugging and network simulators that can simulate, for example, a CICS environment.

CATS 1. Operating system enhancement used by systems programmers. Increases system efficiency in IBM CICS systems. Vendor: Fischer International Systems Corp.
2. Consortium for Audiographics Teleconferencing Standards, Inc. Works on standards to incorporate graphics into videoconferencing.

Caucus Groupware. Vendor: Camber-Roth.

Cayenne Software Software development company formed by Bachman Information Systems and Cadre Technologies in 1996. Purchased by Sterling Software in 1998.

CB Cocomo CASE product. Automates project management function. Supplies cost accounting information for project development and maintenance. Vendor: Market Engineering Corp.

CBConnect See Component Broker Connect.

CBE Technical certification. Certified Banyan Engineer. CBS certification is a prerequisite. Requires passing two tests. NT option requires passing additional tests.

CBLDOC Word processor for IBM mainframes. Vendor: Compute (Bridgend), Ltd.

CBLVCAT Operating system add-on. Data management software that runs on IBM systems. Vendor: Compute (Bridgend), Ltd.

CBR Express Applications development tool. Front-end processor for ART-IM. Allows users to add to an ART-IM information collection without any knowledge of object-oriented languages. Stands for: Case Based Reasoning Express. Vendor: Inference Corp.

CBS(/NT) Technical certification. Certified Banyan Specialist. Requires passing three tests. NT option requires more tests. CBS certification is a prerequisite for CBE certification.

CBT Computer-Based Training. See CAI.

CBToolkit Application development tool. Used for component based development. Extends VisualAge using JavaBeans as the component model. Supports object models from third party vendors such as Rational Rose, Select OMT. Based on CORBA. Runs under AIX, MVS, Windows NT systems. Released: 1998. Vendor: IBM Corp.

CBUG/3000 Operating system add-on. Debugging/testing software that runs on HP systems. Debugs COBOL programs. Vendor: Systems of the Future, Inc.

CC:Mail(/MHS) Communications software. E-mail system. Runs on local area networks. Vendor: Lotus Development Corp.

CCC Communications. Network management tools to monitor and control distributed Oracle and Sybase databases. Stands for: Candle Command Center. Versions available for: MQSeries, IMS, OS/390, Sysplex, Distributed Systems (Unix), CICS, DB2plex. Runs on IBM mainframe, OS/2, Unix, Windows systems. Released: 1995. Vendor: Candle Corp.

CCC/Bridge Application development tool. Links Desktop and mainframe development. Assists in developing mainframe applications on a desktop. Vendor: Softool Corp.

CCC(/DM) Operating system add-on. Library management system that runs on DEC VAX systems. Stands for: Change and Configuration Control(/Development and Maintenance). Vendor: Softool Corp.

CCC/Harvest Application development tool. Provides configuration management and version control for cross platform client/server and Internet systems. Vendor: PLATINUM Technology, Inc.

CCC/Life Cycle Manager Application development tool. Manages configuration and change to synchronize development activities under MVS. Vendor: PLATINUM Technology, Inc.

CCCA Application development tool. Program utility that converts old macro-level CICS programs to the new command-level interface. Vendor: IBM Corp.

CCDA Technical certification in Cisco products. Cisco Certified Design Associate.

CCDP Technical certification in Cisco products. Cisco Certified Design Professional.

CCE Groupware. Includes: SoftSolutions, InForms, GroupWise, MHS. Stands for: Collaborative Computing Environment. Vendor: Novell, Inc.

CCI Net Communications software. Network operating system (NOS) connecting Concurrent system. Vendor: Concurrent Controls, Inc.

CCIE Technical certification. Certifies knowledge in internetworking devices and concepts and Cisco routers. Requires a written test and a two day hands-on lab exam. Stands for: Cisco Certified Internetwork Expert.

CCITT Consultative Committee for International Telephone & Telegraph. Developed X.25 protocol.

CCMS Operating system enhancement used by systems programmers. Monitors and controls system performance in IBM systems. Maintains history of program changes. Full name: Change Master. Vendor: System Connections, Inc.

CCNA Technical certification in Cisco products. Cisco Certified Network Associate.

CCNP Technical certification in Cisco products. Cisco Certified Network Professional.

CCO System Operating system enhancement used by systems programmers. Controls communications networks in IBM systems. Vendor: Telco Research Corp.

CCP Technical certification. Certified Computing Professional. Must pass a core exam and two specialty exams. One specialty exam can be replaced by two language exams. Stands for general certification.

CCS Common Communications Support. The component of IBM's SAA that sets standard controls on how systems store, retrieve and move data through networks.

CD Net Plus Operating system software. Allows network access to CD-ROM storage. Vendor: Meridian Data, Inc.

CD Notebook TFT Notebook computer. Pentium CPU. Operating system: Windows 95/98. Vendor: MAXIMUS COMPUTERS, Inc.

CD-PowerBrickx Desktop computer. Notebook. Pentium processor. Vendor: The Brick Computer Company, Inc.

CDA Software that supports the use of compound documents. Stands for: Compound Document Architecture. Includes: CDA Toolkit, CDA Viewers, CDA Converter Library. Vendor: Digital Equipment Corp.

CDB/Fast Copy Operating System enhancement used by systems programmers. Increases system efficiency in IBM systems. Copies DB2 database tables. Vendor: CDB Software, Inc.

CDB/Rexx Operating System add-on. Runs on IBM systems and interface between DB2 and Rexx. Vendor: CDB Software, Inc.

Cdb Toolkit Relational database for midrange and desktop environments. Runs on Unix systems. Vendor: Jaybe Software.

CDC Computer vendor. Manufactures large and midrange computers.

CDC 4680 InfoServer Midrange computer. RISC machine. MIPS R6000 CPU. Operating system: EP/ix. Vendor: Control Data Corp.

CDCDLC Communications protocol. Data link level. Proprietary, developed by CDC. Stands for: Control Data Corporation Data Link Control.

CDD Central storage of the descriptions of data in DEC systems. Stands for: Common Data Dictionary. Vendor: Digital Equipment Corp.

CDD/Administrator CASE software. A windows-based tool that allows users access to CDD/Repository. Vendor: Digital Equipment Corp.

CDD Plus CASE software. Automates programming function. Vendor: Digital Equipment Corp.

CDD/Repository, CDD/Plus Repository CASE software. A distributed repository based on object-oriented interfaces. Allows the user to run both technical and commercial applications organized in an object-oriented manner. Allows the user to change part of a program. Vendor: Digital Equipment Corp.

CDDF Operating system enhancement used by systems programmers. Increases system efficiency in IBM CICS systems. Stands for: CICS Dump Display Facility. Vendor: NETEC International, Inc.

CDDV Application development tool. Program that verifies and fixes entries in CDD in DEC systems. Stands for: Common Data Dictionary Verify. Vendor: Digital Equipment Corp.

CDE 1. Common Desktop Environment. Specifications for standard Unix to give a common desktop to Unix systems. Includes: Visual User Environment (Hewlett-Packard), Motif (OSF), Common User Access shell (IBM), Desktop Manager (Novell), ToolTalk (SunSoft).
2. See Oracle CDE.

CDF Internet technology. Standard for automated client calls and push software. Stands for: Channel Definition Format.

CDK Application development tool. Allows users to create OCX controls. Part of Microsoft's C++. Stands for: Control Development Kit. Vendor: Microsoft Corp.

CDL 1. Application development tool. Stands for: Change Definition Language. Language developed to define changes to IBM DB2 systems. CDL generates DDL statements (required by DB2). Part of Change Manager. Vendor: BMC Software, Inc.
2. Data management system. Builds an intranet system by indexing all distributed data and creating a central repository. Web browsers can be used to locate, annotate, and monitor usage of documents. Stands for: Corporate Digital Library. Runs on Sequent systems. Vendor: Sequent computer Systems, Inc.

CDPD Communications protocol. Uses cellular networks transmitting data during the pauses occurring in voice transmission. Stands for: Cellular Digital Packet Data.

CDS-1 Mainframe computer. CMOS technology. Supercooled. Release date: 1998. Vendor: Commercial Data Servers, Inc.

CDS-2000 Mainframe computer. Vendor: Commercial Data Servers, Inc.

CDS/Acorn Operating system enhancement used by systems programmers and Operations staff. Job scheduler for Hewlett-Packard systems. Vendor: Chestnut Data Systems, Inc.

CDUX Application development tool. Program that shows linkages between COBOL programs. Runs on Tandem systems. Stands for: COBOL/SCOBOL Data Utility X-Reference. Vendor: Menlo Business Systems, Inc.

CE Pro Desktop computer. Handheld, although they have full notebook-size screens. Operating system: Windows CE. Vendor: Microsoft Corp.

Cedar Application development tool. Used to link mainframe and PC applications. Provides a gateway between Visual Basic applications with COBOL mainframe programs. Vendor: Microsoft Corp.

Celebris (5xxx) Midrange computer. Pentium CPU. Operating system: Windows 95/98. Vendor: Digital Equipment Corp.

Celeron Computer chip, or microprocessor. 266-300 MHz Pentium machines based on the same architecture as the Pentium II, but lacking some high-performance features Celeron is the low-end (and low cost) member of the Pentium family of microprocessors. Vendor: Intel Corp.

Cello Graphical interface for the Internet. Internet browser. Runs with Windows. Public domain software written by the Legal Information Institute (Cornell University).

cellular communications Wireless communications. Geographical areas are divided into cells which each have a transmitter and controller. Communications via radio waves. Used in mobile phone systems, and in data communications.

CENTIX Operating system for midrange Unisys systems. Vendor: Unisys Corp.

Central processing unit See CPU.

Centralan Midrange computer. Server. SuperSPARC CPU. Vendor: AST Research, Inc. (division of Samsung Electronics).

Centura for Java Application development tool. Used to generate Java code. Runs on Windows systems. Creates applications that use a Web browser as a user interface. Vendor: Centura Software Corp.

Centura net.db Application development tool. Reporting tool that provides access to data through Web browsers. Runs on Windows systems. Released: 1998. Vendor: Centura Software Corp.

Centura SApplication Builder Application development system. 4GL facilities and three tiered architecture for SAP systems. Released: 1996. Vendor: Centura Software Corp.

Centura Software Software vendor. Formerly called Gupta Corp.

Century Tap, LAN Analyzer Communications software. Protocol analyzer used with Fast Ethernet. Software and analyzer cards. Released: 1996. Vendor: Shomiti Systems, Inc.

Centura Team Developer Application development environment. Used to develop component-based applications accessing multiple databases. Applications can run on Windows or the Web. Runs on Windows systems. Released: 1998. Vendor: Centura Software Corp.

Centura Web Developer Application development environment. Version of Centura Team Developer used for Web development only (no Windows programs created). Runs on Windows systems. Released: 1997. Vendor: Centura Software Corp.

CEO Desktop system software. Integrated package. Vendor: Data General Corp.

CEO/MHS Communications software. Gateway for e-mail systems. Runs Data General systems. Vendor: Relational Data Systems, Inc.

CEO Object Office A software package that provides basic office functions such as e-mail, filing, and print services. Runs on Data General systems. Vendor: Data General Corp.

CERBERUS Artificial intelligence system. Runs on DEC systems. Provides basis for building expert systems. Vendor: COSMIC.

certificate authority Third party organizations that will authenticate both parties in an online transaction. Used with the Internet. Leading authorities: Entrust, VeriSign, GTE Cybertrust.

Certified Banyan Engineer See CBE.

Certified Banyan Specialist See CBS(/NT).

Certified Internet Professional See CIP.

Certified Novell Engineer See CNE.

Certified Novell Administrator See CNA.

Certified Web Technician Certification proposed by AWP (Association of Web Professionals) that is vendor independent. Supported by IBM, Microsoft, Netscape, among others.

CEX Communications. Network management software. Controls networks in DEC systems. Stands for: Communications Executive. Vendor: Digital Equipment Corp.

Cezanne Application development tool. Object-oriented, rapid application development environment. Vendor: Netlogic, Inc.

CF-M31 Notebook computer. Released: 1998. Vendor: Panasonic Personal Computer Co.

CFML Cold Fusion Markup Language. See Cold Fusion.

CFMS Relational database for mainframe large computer environments. Runs on NCR systems. Stands for: Central File Management System. Vendor: Century Analysis, Inc.

CFO Vision Application development tool. OLAP software used to analyze business figures. Vendor: SAS Institute, Inc.

CGI Common Gateway Interface. A set of application programming interfaces for linking a Web server to other systems. Used to link Web front-ends to functions like inventories, customer support, etc. to process Internet sales. Allows visitor on Web page to access databases.

CGI script A program written in any number of languages that provides an interface between users on the Internet and internal corporate systems. This enables people on the Web to access resources not directly on the Web. A CGI script could send a form to a user on a browser asking, i.e., what type of books the user would be interested in. The user could fill in "science fiction." The script would then convert the information to an SQL query against a corporate database. When the list of science fiction books was returned to the script, it would convert the information to HTML and return the Web page to the browser. Scripts can be written in any language, but most commonly are written in Perl, VBScript and JavaScript.

CGOS 4/OS 3 Operating system enhancement used by systems programmers. Increases system efficiency in Bull HN systems. Vendor: Bull HN Information Systems.

Challenge Server Tools Application development tool. Visualization tool that can color-code groups or clusters of data and display trends and patterns as graphs. Used for analysis type queries and with data warehousing. Vendor: Silicon Graphics, Inc.

Chameleon Communications software. Software suite that includes terminal emulation, e-mail, and an Internet interface. Vendor: NetManage Inc.

Chameleon Hostlink Communications software. Allows access to mainframe applications through the Internet. Released: 1997. Vendor: NetManage, Inc.

Chameleon Open Communications software. Protocol analyzer which provides simultaneous testing of LAN and WAN interfaces and applications. Vendor: Tekelec.

Chameleon UNIXLink Communications software. Provides PC to Unix link through Web browsers. Vendor: NetManage, Inc.

Champs/Case Application development tool. Program generator that works with Rdb and runs on VMS systems. Generates C programs. Integrates with DECforms. Vendor: Champs Software, Inc.

Change Action Operating system enhancement used by systems programmers. Monitors and controls system performance in IBM MVS systems. Monitors all system changes. Vendor: Action Software, International.

Change Agents Groupware. Document management system. Sends users e-mail notification when documents change. Vendor: OpenText Corp.

Change Man Application development tool. Programming utility that allows changes to mainframe source libraries from workstations. Vendor: Optima Software, Inc.

Change Manager 1. Operating system add-on. Library management system that runs on IBM systems. Vendor: Software Controls, Inc.
2. Application development tool. Manages changes to DB2 systems and synchronizes changes across multiple application systems. Vendor: BMC Software, Inc.

Channel Definition Format See CDF.

channels 1. See communications lines.

Channels Groupware. Interactive discussion database that allows multiple users to log in to a discussion, post comments, add documents through e-mail, etc. Vendor: Team Software, Inc.

CharacterEyes OCR software. Vendor: Ligature Software.

Chargeback Operating system enhancement used by systems programmers. Provides system cost accounting for HP systems. Vendor: Operations Control Systems.

Chariot Communications software. Network performance testing tool. Allows user to evaluate performance and capacity of almost any network device such as routers, switches, adapters and network software. Runs under: MVS, OS/2, Unix, Windows. Released: 1997. Vendor: Ganymede Software, Inc.

Charisma Desktop system software. Graphics package. Used for presentation graphics. Vendor: Micrografx, Inc.

Charles River Computer vendor. Manufactures midrange computers.

Chart.J Application development tool. Java based. Used to build and share dynamically generated charts. Vendor: Rogue Wave Software, Inc.

ChartWorks Application development tool. Java based. Used to build and share dynamically generated charts. Vendor: ChartWorks, Inc.

Chatterbook Desktop computer. Notebook. Pentium processor. Vendor: Talkto Computers, Inc.

ChatterBox Midrange computer. Server. Pentium CPU. Vendor: Chatcom, Inc.

Check-Out Operating system add-on. Debugging/testing software that runs on Unisys systems. Vendor: Programming Aids, Inc.

Check Plus for DB2 Operating system enhancement. Ensures DB2 integrity and increases data availability. Vendor: BMC Software, Inc.

Checkout/VM Operating system enhancement used by systems programmers. Increases system efficiency in IBM VM systems. Checks for system failures. Vendor: Legent Corp.

Checkpoint 1. Application development tool. Assists in managing software development by analyzing development cost, quality, etc. and suggesting alternate approaches. Provides performance benchmarks. Runs on MS-DOS, Unix, Windows systems. Vendor: Software Productivity Research, Inc.
2. CASE product. Automates project management function. Vendor: Software Productivity Research, Inc.

checkpoint restart The processing of establishing checkpoints that can be used to restart a computer job other than at the beginning in case of problems.

CheetaRack Desktop computer. SPARC Pentium processor. Vendor: Marner International, Inc.

ChemBook 2600, 22700, 6000, 9750 Notebook computer. Pentium II CPU. Operating system: Windows 95/98. Vendor: CHEM USA Corp.

Chembookxxx Notebook computer. Pentium CPU. Vendor: Hiquality Computer Systems.

Chen Structured programming design methodology named for its developer. Based on information modeling and considered to be a precursor to object-oriented analysis and design. Uses entity-relationship diagrams. Accepted as a standard design methodology by some companies, and used by some CASE products.

Chess MRP, or CIM software. Runs on many systems. Vendor: Xerox Computer Services.

Chicago Development name for Windows 95.

Chili!Reports Application development tool. Query and report functions. Runs on Windows NT systems. Vendor: Chili!Soft Inc.

ChiliASP Application development tool. Used to allow non-Microsoft Web servers to run Active server pages. Can be used with Netscape, Unix servers. Vendor: Chilisoft Inc.

Chimera Real-time operating system. Developed for research purposes. Vendor: Carnegie Mellon University.

chip Basic building block of the computer. An electronic circuit that carries electrical pulses from one point to another.

Chisel Operating system software. Measures performance in client/server systems by simulating user activity to measure response time. Includes scheduling functions. Released: 1998. Vendor: Network Tools, Inc.

Choreographer Application development tool that works in a client/server architecture. Accesses OS.2 EE Database Manager, SQL Server, Oracle Server. Builds client applications with a GUI based front-end. Includes program generator. Vendor: Guidance Technologies, Inc.

Chorus/ClassiX Real-time operating system for SPARC workstations. Interfaces with Unix systems. Vendor: Chorus Systems, Inc.

Chorus/COOL ORB Application development tool. Used to develop real-time object-oriented applications. Corba compliant. Runs on Unix, Windows systems. Vendor: Chorus Systems, Inc.

Chorus/Fusion Real-time operating system for RISC systems. Includes Unix features. Vendor: Chorus Systems, Inc.

Chorus/OS Family of real-time operating system products. Vendor: Chorus Systems, Inc.

Chosenlan Communications software. Network operating system. Runs on midrange computers. Vendor: Moses Computers, Inc.

Chrome Development technology and tool. Three-dimensional, interactive media tool used to provide an easier PC interface. Will be used in Windows 98 and Windows NT. Available for developers to build 3-D images for Windows applications. Vendor: Microsoft Corp.

CHRP Hardware standard. Stands for Common Hardware Reference Platform. Defines PowerPC systems. CHRP PowerPC computers will be able to run a variety of operating systems including Windows NT, AIX, MacOS, OS/2, Solaris thus providing cross-platform compatibility and access. Standards are being defined by IBM Corp, Apple Computer, Inc., and Motorola, Inc. CHRP computers are being manufactured by IBM, Apple, Radius, and PowerComputing Corp.

Churn/Customer Profiling System See CPS.

CI-Link Communications. Middleware. Connects SQL databases to Macintosh and PC applications. Interfaces with Oracle, Ingres, Informix, Progress. Runs on Unix systems. Released: 1992. Vendor: Cornut Information.

CICS Communications software. Transaction processing monitor. Runs on IBM systems. Data can be accessed through CICS Command Level statements, or CICS Macro Level statements. Command Level is most often used by Applications Programmers, while Macro Level is a more technical method of access. Latest version of CICS utilizes only Command Level statements. Versions of CICS available for CMS, VS, DOS/VSE, AIX, OS/2, and MVS. Can be used on HP systems. Can be used in client/server systems with an OS/2 server and Dec, OS/2, Macintosh and Sun clients. Interfaces with Java, Internet. Stands for: Customer Information Control System. Vendor: IBM Corp.

CICS/6000,9000 Communications software. Versions of CICS that run under Unix. CICS/6000 works with IBM's RS/6000; CICS/9000 works with Hewlett-Packard's HP 9000 systems. Vendor: IBM Corp. and Hewlett-Packard Co.

CICS Abend-AID See Abend-AID.

CICS Automation Operation/MVS Operating system enhancement used by systems programmers. Monitors and controls IBM CICS systems. Vendor: IBM Corp.

CICS Central Operating system enhancement used by systems programmers. Monitors and controls system performance in IBM CICS systems. Vendor: Computer Corp. of America.

CICS Command Level See CICS.

CICS Connection Operating system enhancement used by systems programmers. Allows users to update same VSAM files in a CICS environment without compromising data integrity. Vendor: Beacon Software International.

CICS dBUG-AID See Abend-AID.

CICS-EDGE Operating system add-on. Debugging/testing software that runs on IBM systems. Vendor: IBS Corp.

CICS/ESA Communications software. Transaction processing monitor. Runs on IBM systems. Latest release of CICS. Takes advantage of MVS/ESA features. Programmers must use Command Level statements. Includes SAA Common Programming Interface for communications. Vendor: IBM Corp.

CICS Integrity Series Operating system enhancement used by systems programmers. Increases system efficiency in IBM MVS and VSE CICS systems. Includes: Recovery for CICS, Recovery Plus for VSAM, Journal Manager Plus, Data Vault. Vendor: BMC Software, Inc.

CICS Liberator Communications software. Network connecting IBM CICS and DEC VAX systems. Vendor: Advanced Systems Concepts, Inc.

CICS Macro Level See CICS.

CICS Manager Operating system enhancement used by systems programmers. Monitors and controls system performance in IBM CICS systems. Vendor: Boole & Babbage, Inc.

CICS OS/2 Communications software. Version of CICS that runs on OS/2, Unix systems. Vendor: IBM Corp.

CICS Playback Operating system add-on. Debugging/testing software that runs on IBM systems. Vendor: Compuware Corp.

CICS RADAR Operating system add-on. Debugging/testing software that runs on IBM CICS systems. Interprets CICS dumps. Vendor: Compuware Corp.

CICS/Recorder Application development tool. Program that allows programmers to decipher CICS/COBOL code. Vendor: Language Technology, Inc.

CICS/Replay Operating system add-on. Debugging/testing software that runs on IBM systems. Vendor: Interactive Solutions, Inc.

CICS-TDS Operating system add-on. Debugging/testing software that runs on IBM systems. Vendor: Startech Software Systems, Inc.

CICSORT Operating system enhancement used by systems programmers. Increases system efficiency in IBM CICS systems. Provides online sort capability. Vendor: Syllogy.

CICSPARS/MVS,VSE Operating system enhancement used by systems programmers. Monitors and controls system performance in IBM systems. Vendor: IBM Corp.

CICSplex Operating system software. Allows CICS applications to work in a parallel sysplex environment without reprogramming. Vendor: Neon Systems, Inc.

CICSPlex SM/ESA Operating system enhancement. Administer multiple CICS systems from a single point of control. Runs on IBM CICS systems. Stands for: CICSPlex System Manager/ESA. Vendor: IBM Corp.

CID Software distribution manager. Allows Information System managers to centrally manage the distribution and installation of software to workstations throughout the company. Stands for: Configuration, Installation and Distribution. Vendor: IBM Corp.

CIIM CIM (Computer Integrated Manufacturing) system. Interfaces with Oracle, SQL*Forms. Runs on Unix, DEC VAX, OS/2, Windows systems. Client/server technology. Stands for: Computer Interactive Integrated Manufacturing. Vendor: Avalon Software.

CIM 1. Computer Integrated Manufacturing. Systems that integrate plant-floor computers with front office administrative machines.
2. Corporate Information Management. Information technology initiatives by the Department of Defense intended to reform and standardize all computer systems used by the military.
3. Common Information Model. Standard that defines a model for network devices, systems and applications to display information about themselves and pass this information to management tools. Information includes static information from desktop systems such as serial numbers to dynamic data such as traffic on router ports.

Cimpro MRP, or CIM software. Runs on Unisys, IBM, DEC, Unix systems. Vendor: Datalogix International, Inc.

CIMS Operating system enhancement used by systems programmers. Provides system cost accounting for IBM systems. Full name: CIMS Job Accounting and System Performance. Vendor: PLATINUM Technology, Inc.

CIO-Vision EIS. Runs on Pentium type midrange computers. Vendor: Computer Associates International, Inc.

CIP Technical certification. Certified Internet Professional. Five tracks to follow: Internet Business Strategist, Web Designer, Web Developer, Intranet Manager, Internet Architect. Some tracks apply to applications personnel. Newest program offered by Novell.

Cirrus Relational database for midrange computer environments. Runs on Pentium type systems. Vendor: Microsoft Corp.

CIS Customer Information System. See front-office software.

CISC Complex Instruction Set Computer. Architecture used in large computer systems. Contrast to RISC, Reduced Instruction Set Computer, architecture used in many desktop systems.

Cisco Hardware vendor. Manufactures equipment used in communications networks.

Cisco Certified Design Associate See CCDA.

Cisco Certified Design professional See CCDP.

Cisco Certified Internetwork Expert See CCIE.

Cisco Certified Network Professional See CCNP.

Cisco Certified Network Associate See CCNA.

CiscoBlue Communcations hardware and software. Road map of products used to build switched networks and manage SNA performance. Vendor: Cisco Systems, Inc.

CiscoWorks Communications. Network management software. Allows users to monitor and model networks. Vendor: Cisco Systems, Inc.

CISLE family CASE product. Automates analysis, design, programming functions. Runs on most midrange and desktop systems. Vendor: Software Systems Design, Inc.

CIT Computer Integrated Telephony. Computer software control of telephone applications. Includes such things as automatically routing customer calls to the appropriate salesperson and automatically connecting a caller with data files.

Citrix Multiuser Communications software. Network operating system. Runs on multiple midrange computers. Vendor: Citrix Systems, Inc.

CL Command language. Used with OS/400 Operating System. Comparable to JCL.

CL/Conference Communications software that provides data conferencing to allow people in different locations to share the same data screen. Runs on IBM MVS systems. Combines telephone and data conferencing. Vendor: Candle Corp.

CL/GATEWAY Communications software. Network connecting IBM computers. Includes CL/GATEWAY for IMS, CL/GATEWAY for MVS. Vendor: Candle Corp.

CL/SuperSession Communications software. Network connecting IBM computers. Vendor: Candle Corp.

Clarify Software vendor. Produces front office application software. Products included start with "Clear" and provide automation of call center, contract, helpdesk, sales and support functions.

Clarion Application development environment. Includes screen builder, 4GL, report writer. Provides Internet support. Runs on Windows systems. Vendor: TopSpeed Corp.

Claris Home Page Application development tool. Web authoring tool. Available on multiple platforms. Development code name: Loma Prieta. Vendor: FileMaker, Inc.

Clarus Application software. Includes financial software (G/L, F/A, A/R, A/P), procurement software for buying supplies over the Web, a budgeting application, and HRMS software which includes a program which lets employees sign up for benefits through the software. Released: 1998. Vendor: SQL Financials International, Inc.

Classic/AL Application development tool. Program that works with CICS applications in IBM environments. Vendor: Goal Systems International, Inc.

ClassiX See Chorus/ClassiX.

class In object-oriented development, the definition, or template, of an object. Object, or class libraries have multiple objects which have the same methods (or processing) but different values.

ClassWizard Application development tool. Connects the screen painter to the compiler in Visual C++. Vendor: Microsoft Corp.

cleanroom software engineering An approach to software development emphasizing defect prevention. Uses rigorous engineering-based practices based on mathematical principles. Team-based, puts greatest emphasis on design.

Clear Access See VISION:Clearaccess.

Clear Manager See VISION:Clearmanage.

Clear... Application software. Front office packages that include: ClearCallCenter, ClearContracts, ClearHelpdesk, ClearLogistics, ClearSales, ClearSupport, ClearTelebusiness. Runs on Unix, Windows systems. Vendor: Clarify, Inc.

ClearBasic Application development tool. Provides scripting language, allows users to customize Clarify application software. Released: 1996. Vendor: Clarify, Inc.

ClearCase Case product. Automates project management functions. Includes configuration management. Vendor: Rational Software Corp.

ClearCase Attache, MultiSite Application development tool. Client/server configuration manager. Manages workspaces, parallel development, software reuse. MultiSite is used with projects being developed from multiple sites. Runs on most RISC systems. Vendor: Rational Software Corp.

ClearConnect Communications. Provides two-tier connectivity from Windows to DB2 databases. Vendor: Sybase, Inc.

ClearGuide Application development tool. Extension to ClearCase. Automates development processes, including parallel development and client/server development. Runs on Unix, Windows NT systems. Vendor: Rational Software Corp.

ClearPath (2200) Mainframe computer. Operating systems: proprietary, Unix, Windows NT. Vendor: Unisys Corp.

ClearPath HMP ix, LX500, mx Midrange computer. Proprietary CPU. Operating systems: OS 2200, MCP/AS, Unix, Windows NT. Vendor: Unisys Corp.

ClearPath SMP Midrange computer. RISC CPU. Operating systems: Unix, Windows NT. Vendor: Unisys Corp.

ClearSales Application software. Sales force automation software. Included with FrontOffice98. Vendor: Clarify, Inc.

Clementine Application development tool used for data mining. Used to determine which of hundreds of relationships among variables are significant. Runs on Unix, Windows NT systems. Vendor: ISL Decision Systems.

client See user.

Client Access/400 Operating system software. Connects Windows 94 users to AS/400 systems. Vendor: IBM Corp.

Client/Builder Application development tool. Allows users to simultaneously merge applications from diverse large computer platforms into a single graphical interface. Reengineers AS/400 and mainframe systems for client/server computing. Vendor: ClientSoft, Inc.

client/server A type of computer system architecture that encompasses a "server," or host, system that can service many "clients" over a network. The clients are personal computers and the client software uses a GUI (Graphic User Interface). The server accesses data from any relational database. Term is often used interchangeably with cooperative processing and network computing. Client/server architecture is common with database management systems and is a subset of distributed processing.

Client/Server Ready Application development tool. Migrates midrange applications to client/server computing. Vendor: Cognos Inc.

ClientPack Application development tool. Provides run-time testing tools for Windows-based client software. Part of Cyrano Suite. Vendor: Cyrano, Inc.

ClientPro Desktop computer. Pentium II CPU. Operating system: Windows 95/98, NT Workstation. Vendor: Micron Electronics, Inc.

Clientele, ClienteleNet Application software. Front office customer information system. Organizes, tracks and shares prospect, sales, and post-sales information. ClienteleNet provides Internet access through common browsers. Vendor: Platinum Software Corp.

ClientWorks Desktop software. Personal information manager (PIM). Vendor: Information Management Services.

CLIII/95-xx Midrange computer. CMOS processor. Operating system: MAXIV. Vendor: Modular Computer Systems, Inc. (MODCOMP).

CLIP Communications software. Transaction processing monitor. Runs on most midrange systems. Stands for: Communications Link Interface Program. Vendor: Packaged Solutions, Inc.

Clipper See CA-Clipper.

clipper chip Government backed security standard which would provide data encryption through a device called a "clipper chip." Uses Skipjack encryption, an 80-bit key. Government security agencies would control the devices. Controversial because it includes a second key that would allow authorized government agents to decrypt information.

Clips Artificial intelligence system. Expert system building tool. Stands for: C Language Integrated Production System. Vendor: Cosmic (Computer Software Management & Information Center).

Cliqcalc Desktop system software. Spreadsheet. Runs on Unix systems. Vendor: Quadratron Systems, Inc.

CLIST Command list. A set of TSO commands.

CLNP Communications protocol. Transport layer. Developed by OSI. Stands for: Connectionless Network Protocol.

cloning Also called disk imaging. Technology that uses snapshot imaging to copy the operating systems, programs and data from one hard disk to another in a single repeatable procedure.

CLOS Object-oriented version of LISP. Stands for: Common LISP Object System. See LISP.

CLOSQL Object-oriented database. Research database designed for prototyping. Based on CommonLISP.

Close-up Communications software. Allows remote access control of midrange computers. Supports TCP/IP. Vendor: Norton-Lambert, Inc.

Clout 4GL used in midrange computer environments. Works with R:BASE databases. Vendor: Microrim.

Cloverleaf EAI Suite EAI software. Middleware. Includes components, adapters and utilities that integrate in-house and external applications. Runs on Unix, Windows systems. Vendor: Healthdyne Information Enterprises Inc. (HIE).

cluster A group of computers linked together to provide multiprocessing capabilities. High-availability clustering links a second computer which acts as a backup in case of system failure. Performance clustering is also called parallel processing and links many computer systems together. During application execution, queries are divided into work units which are handled separately by each system. Performance clustering is used when many items of data are necessary to answer the query. Because each system can retrieve one piece of data, the query is satisfied much faster than if a single system had to do all the data retrieval.

Cluster Server Nxxxx Midrange computer. Server. Pentium, Pentium Pro CPU. Vendor: NetFrame Systems, Inc.

ClusterView Version of OpenView that handles clustered systems including Window NT clusters. See OpenView.

ClusterX Operating system software. Cluster management system that handles MSCS (Microsoft Cluster Server) clusters. Consolidates multiple clusters into a single interface that lets network administrators manage the entire enterprise. Performs automatic load balancing across multiple clusters. Released: 1999. Vendor: NuView Inc.

CM/1 Groupware. EMS system. Vendor: Corporate Memory Systems, Inc.

CM-2 Connection Massively parallel processor. Supports up to 64,000 processors. Operating system: VM/370. Vendor: Thinking Machines Corp.

CM-2a Model 4,8 Massively parallel processor. Supports up to 8192 processors. Operating system: VAX/VMS. Vendor: Thinking Machines Corp.

CM-5 Connection Massively parallel processor. Supports up to 16,384 processors. Operating system: Unix. Vendor: Thinking Machines Corp.

CM:Millennium See Millennium.

CM-Tsoplus Operating system add-on. Provides JCL and/or SYSOUT maintenance for IBM systems. Vendor: Symark International, Inc.

CMAP Operating system enhancement used by systems programmers. Monitors and controls system performance in IBM VM systems. Vendor: VM/CMS Unlimited, Inc.

CMAS Applications software. Stands for: Construction Management Accounting Software. IBM uses this name, as does Real Estate Information Tracking Systems. Programs include construction accounting, construction payroll, job costing, etc.

CMC API for messaging in client/server systems. Provides the interface between different servers. Stands for: Common Messaging Call. Other common messaging APIs are VIM, MAPI.

CMDS Operating system software. Security package that alerts users to suspicious activity to provide real-time detection of intrusion. Detects virus attacks. Stands for: Computer Misuse Detection System. Runs on Unix systems. Vendor: Science Applications International Corp.

CMF Monitor 1. Operating system enhancement used by systems programmers. Monitors and controls system performance in IBM systems. Vendor: Boole & Babbage, Inc.
2. Operating system enhancement. Suite of products that provides historic analysis, performance modeling for Cray Unicos systems. Vendor: TeamQuest Corp.

CMIP Communications protocol. Application layer. Stands for: Common Management Information Protocol. Provides a common format for devices such as modems, routers, and bridges. Developed by ISO. Adhered to by IBM and DEC.

CMM Capability Maturity Model. Method for evaluating the effectiveness of software development. Defines the principles and practices that should be followed and defines five levels of effectiveness; initial (ad hoc and chaotic), repeatable, defined, managed, optimizing (piloting new development ideas and technologies.) Defined by Software Engineering Institute (SEI).

CMOS Computer chip design. Uses transistors in a complementary fashion that requires less power to operate. Can use in air-cooled machines. Stands for: Complementary Metal Oxide Semiconductor.

CMOS-390 Superserver computer. Operating systems: AIX, MVS/Open Edition. Vendor: IBM Corp.

CMOS 7S Computer manufacturing technology that uses copper instead of aluminum to build transistors. Developed by IBM Corp. in 1997.

CMS 1. Communications software. Provides environment for application development. Includes programming and testing utilities. Works with VM operating system and usually seen as VM/CMS. Vendor: IBM Corp.
2. Operating system for midrange Unisys systems. Vendor: Unisys Corp.
3. See Customer Messaging System.

CMS CodeReview Application development tool used for testing. Provides automatic source code analysis of Visual Basic programs. Part of DevCenter. See DevCenter.

CMS/EDAC Operating system enhancement used by systems programmers. Increases system efficiency in IBM VM systems. Stands for: CMS Extended Data Array Capability. Vendor: IBM Corp.

CMX Real-time operating system for desktop computers. Vendor: CMS Co.

CNA Technical certification. Certified Novell Administrator, or Certified NetWare Administrator. First and least technical certification. Pass one exam in NetWare, IntraNetWare. Certification for support personnel.

CNE Certified Novell Engineer, or Certified NetWare Engineer Technical developers who work with Novell's NetWare can take a course and pass a test to receive certification from Novell. Banyan and Microsoft also offer training and certification in their LANs (VINES and LAN Manager).

Coad/Yourdon Object-oriented development methodology. Provides analysis and design techniques.

coaxial cable Communications lines connection used in many LANs. Provides high speed transmission. Also commonly used by cable TV.

COBOL Compiler language. Stands for COmmon Business-Oriented Language. Dominant language for business applications in large computer environments.

COBOL/2 Compiler language used in midrange computer environments. Runs on IBM midrange systems under OS/2. Compatible with mainframe COBOL and COBOL II. Includes COBOL/2 and COBOL/2 Workbench. Vendor: Micro Focus, Inc.

COBOL/2 Workbench Application development system that runs on IBM systems under OS/2 and Xenix. Has interfaces for IMS and CICS. Interfaces to Excelerator. Used for cross-platform development. Supports Windows, OS/2, Unix. Includes Animator, Analyzer, CSI, Session Recorder. Also called Micro Focus COBOL. Vendor: Micro Focus, Inc.

COBOL Analyst Application development tool. Analyses and maintains COBOL applications in a midrange computer environment. Vendor: Seec, Inc.

COBOL for OS/2 Object-oriented visual language. Version of COBOL that allows object-oriented COBOL development. Vendor: IBM Corp.

COBOL II Compiler language used in mainframe environments. Enhancement to COBOL. Includes structured programming facilities and allows programmers to work with efficiency considerations.

COBOL/METRICS Application development tool. Program that provides quality assurance for COBOL programs. Vendor: Computer Data Systems, Inc.

COBOL POWERbench See POWERbench.

COBOL SET Application development tool. Tools including object-oriented support, remote and local data access, debugging tools, and online documentation for AIX systems. Released: 1995. Vendor: IBM Corp.

COBOL/SF Application development tool. Program utility that converts COBOL code to COBOL II code and restructures the program. Vendor: IBM Corp.

COBOL Softbench Application development tool. Migrates COBOL applications to Unix. Vendor: Hewlett-Packard Co.

COBOL Source Analyst Application development tool. Interactively analyzes COBOL code. Runs on SPF/PC systems. Vendor: Command Technology Corporation.

COBOL Source Intelligence See CSI.

COBOL VisualSet Object oriented COBOL. Used on mainframe, midsize, and desktop systems. Interfaces with DB2, IMS, VSAM, Oracle, Sybase, Informix. Vendor: IBM Corp.

Cobra(200,275),EV5 Midrange computer. Alpha RISC CPU. Operating system: Windows NT. Vendor: Carrera Computers, Inc.

CocoBase Application development tool. Database access modules used with Internet-based Java applications accessing relational databases. Interfaces with Oracle, Sybase, Informix. Released: 1997. Vendor: Thought, Inc.

CoCoPro Application development tool. Part of Power Tools. Used to estimate costs of software projects. Vendor: ICONIX Software Engineering, Inc.

Codan Application development tool. Automates programming function. Analyzes source code for such things as adherence to standards and presence of logic flaws (unexecuted code). Language analyzed: C. Runs on Pentium type midrange computers. Vendor: Implements, Inc.

Code Check Application development tool. Automates programming function. Analyzes source code for such things as adherence to standards and presence of logic flaws (unexecuted code). Languages analyzed: C, C++. Runs on Pentium type midrange computers, RS/6000, Macintosh, DEC VAX. Vendor: Abraxas Software, Inc.

code generator See program generator.

Codebase Application development tool. Library of routines that allows C programs to access xBase files. Used for cross-platform development. Supports Windows, Windows NT, OS/2, Unix. Vendor: Sequiter.

CodeCenter Application development system. Used to develop C programs. Includes debugging functions, graphical browsers. Interfaces with Informix, Oracle. Runs on Unix systems. Vendor: Centerline Software, Inc.

CodeGuard Application development tool. Debugging tool. Locates and diagnoses memory management bugs. Works with Borland C++. Vendor: Inprise Corp.

CodeWarrior Application development tools. Used to develop C/C++ programs for various environments including Windows, Java, PowerPC, PalmPilot. Released: 1997. Vendor: Metrowerks, Inc.

Codewright Professional Application development tool. Language editor for most desktop languages. Vendor: Premia Corp.

Coesco Operating system enhancement used by Operations staff and systems programmers. Disaster control package. Vendor: Parnassus, Inc.

CoFac Application development tool. Automates programming and testing functions. COBOL code generator. Vendor: Coding Factory.

Cogen Application development tool. Program, or code generator. Generates COBOL statements. Includes data dictionary and system documentation. Runs on Pentium type PCs. Vendor: Bytel Corp.

Cognos BI Applications Application development tool used in data warehousing. Used to build data marts using pre-defined extraction definitions. Includes predefined reports. Stands for: Cognos Business Intelligence. Vendor: Cognos Inc.

COGNOSuite Application development tools. Includes PowerPlay, Scenario, Impromptu. Vendor: Cognos Inc.

COGO Programming language designed for coordinating geometry problems in civil engineering.

Coherent General purpose operating system. Unix type system. Vendor: Mark Williams Co.

Cohesion The umbrella name for DEC's CASE software and strategy. Includes CDD/Repository, CDD/Administrator.

Cohesion Team/See Application development tools. Manage data and development across diverse Unix environments. Vendor: Digital Equipment Corp.

CohesionworX Application development system for Unix systems. Provides GUI interface, editing and debugging tools. Vendor: Digital Equipment Corp.

CoHost Application software. Provides Windows front-end to mainframe systems including Dun & Bradstreet's Smart-Stream software. Written in PowerBuilder and can be customized to individual user requirements. Vendor: Dun & Bradstreet Software.

Coinserv Document retrieval system. Runs on Unix systems. Vendor: INSCI Corp.

COLD Computer output to Laser Disk. Writes and indexes mainframe reports to laser disks. Searches, selects and outputs data faster than microfiche.

Collabra Server Communications, Internet software. Groupware application. Supports discussion groups based on NNTP standards. Vendor: Netscape Communications Corp.

Collabra Share Groupware. Includes database, runs on multiple e-mail systems including CC:Mail, Mail, All-in-1, Profs. Used for electronic discussions, meetings. Vendor: Collabra Software, Inc.

Colonel CASE product. Automates design function. Works with IMS, IDMS, DB2, VAX Rdb/VMS databases. Vendor: Parker Shannon Associates, Inc.

Columbine Desktop computer. Pentium CPU. Operating systems: Windows 95/98, NT. Vendor: Aspen Systems, Inc.

Com-plete Communications software. Transaction processing monitor. Runs on IBM systems. Vendor: Software AG Americas Inc.

Com-pose Communications software. Transaction processing monitor. Runs on IBM systems. Vendor: Software AG Americas Inc.

COM 1. Middleware technology for Windows-based software. Used to access data in different databases running on Windows clients and servers. Consists of ActiveX, which handles the desktop, and DCOM, which handles the servers and databases. Stands for: Component Object Model. New features allow access to data on non-Windows servers. Components include: OLE DB, ADO, ODBC. Contrast with CORBA, which is platform independent. Vendor: Microsoft Corp. 2.Computer Output Microfilm (or Microfiche). Software and hardware that send reports to microfilm (or fiche) instead of to paper. Used for large volumes of printing.

Comboard Communications software. Network connecting IBM and DEC computers. Vendor: Software Results Corp.

Comet/11 Communications software. E-mail system. Runs on DEC systems. Vendor: Maxcom, Inc.

Comfort/36-38 Communications software. Network connecting IBM System/36,38. Vendor: Comware International.

ComGate/36-38 Communications software. Terminal emulator. Used in IBM System/36,38 environments. Vendor: Comware International.

Command Configuration management system that keeps track of everything on a network. Uses graphics to display the network. Runs on Unix systems. Vendor: Isicad, Inc.

Command Center (Plus) EIS. Runs on IBM mainframe, Macintosh, Unix systems. Vendor: Pilot Executive Software.

Command File Scheduler Operating system enhancement used by systems programmers and Operations staff. Job scheduler for DEC VAX systems. Vendor: Morss Software Development, Inc.

Command/MQ Communications software. Manages message queues from middleware systems that handle communications from Unix and Windows to mainframe systems. Interfaces with MQSeries. Released: 1996. Vendor: Boole & Babbage, Inc.

Command/Post Operating system software. Monitors systems and correlates alerts and status messages from diverse systems. Released: 1998. Vendor: Boole & Babbage Inc.

Commander Decision (Web) Application development tool. OLAP software that provides multidimensional analysis of business information. Interfaces with Oracle, express, Essbase, TM[1]. Web version uses Java-enabled browsers. Released: 1997. Vendor: Comshare, Inc.

Commander EIS EIS. Runs on Pentium type midrange computers, Macintosh. Utilizes SQL. Vendor: Comshare, Inc.

Commander OLAP OLAP software. Multidimensional database. Provides data scrubbing, query, and reporting functions. Runs on OS/2, Unix, Windows NT systems. Vendor: Comshare, Inc.

Commander Prism Desktop software. Multidimensional spreadsheet. Runs on Windows systems and is a follow-up to One-Up. Vendor: ComShare, Inc.

Commax Computer vendor. Manufactures midrange computers.

CommBuilder Application development tool. Works with Easel to build graphical applications. Runs on Pentium type desktops. Vendor: Easel Corp.

Commence Desktop system software. Personal information manager (PIM). Runs on Pentium type systems under Windows. Vendor: Jensen-Jones, Inc.

Commerce Builder Communications software, Web server. Provides access to the Web from Windows NT systems, Windows 95/98. Lets users view documents. Vendor: The Internet Factory.

Commerce:Doculink/BSC,MNP,OS,SNA Communications software. EDI package. Runs on AS/400 systems. Vendor: Sterling Software, Inc. (Network Services Division).

Commerce Server Communications software, Web server. Provides access to the Web from Windows NT, Unix, OSF/1 systems. Lets users view documents. Vendor: Netscape Communications Corp.

Commerce Station Application software for e-commerce. Provides complete storefront for both business-to-business and business-to-consumer sales. Interfaces with Sybase, Oracle, SQLServer. Runs on Unix, Windows NT systems. Runs on Unix, Windows NT. Vendor: ATG Inc. (Art Technology Group).

CommerceXpert Applications software. E-commerce software. Family of component-based products used to buy, sell, merchandise, and deliver product over the Internet. Includes: ECXpert and DeveloperXpert (backbone and software development kit), SellerXpert (handles business-to-business commerce), BuyerXpert (automates procurement process), PublishingXpert (includes advertising, billing functions), MerchantXpert (handles business-to-individual commerce). Released: 1998. Vendor: Netscape Communications Corp.

Commercial Data Servers, Inc. Computer vendor. Builds mainframes. Founded by Gene Amdahl in 1997.

Commodore Computer vendor. Manufactures microcomputers (PCs).

Common Ground Desktop system software. Document exchange software that allows cross-platform use of documents created under any software. For example, a document created in WordPerfect under Windows 95 will be readable by a user on a Macintosh system. Vendor: No Hands Software Inc.

Common Ground Web Publishing Application development tool. Creates digitized versions of any print document. Uses Java applets that run on any platform. Vendor: Hummingbird Communications, Ltd.

Common Hardware Reference Platform See CHRP.

Common Knowledge Groupware. Allows project management, document collaboration, bulletin board and administrative functions for Windows and Macintosh users. Vendor: On Technology.

Common Server Database management. Refers to versions of DB2 that run on desktop systems. Runs on OS/2, Unix, Windows NT systems. Vendor: IBM Corp.

Common User Access See CUA.

CommonLISP Version of LISP that was developed to set a standard for the LISP language after multiple versions were created.

CommonPoint Application development environment. Object-oriented development environment. Certification and testing program available to ensure that software developed under CommonPoint will run on multiple operating systems including OS/2, Macintosh, and HP/UX. Vendor: Taligent, Inc.

CommonPoint for OS/2, AIX Application development environment. Object-oriented, includes pre-generated code for commonly used functions. Versions available for OS/2 and AIX systems. Vendor: IBM Corp.

communications Sending data over communications lines to allow for processing data over distances.

Communications Builder Communications software, Web server. Provides access to the Web from Windows NT, Windows 95/98 systems. Lets users view documents. Vendor: The Internet Factory.

communications control system A system that contains data transfer routines, data sharing capabilities, and security routines. Used to control online business applications.

Communications Integrator Communications software. Middleware. Connects Unisys, Apollo, IBM and DEC networks. Vendor: Covia Technologies.

communications layers Communications activity is described in layers, and protocols state the work done in each layer. SNA is a seven layer protocol, DECnet is a five layer protocol. OSI defines seven layers. These layers, in order of most basic to most complex, are: physical, data link, network, transport, session, presentation, and application. There are protocols defined for each layer. Layers can have sublayers (which would also have protocols).

communications lines Long-distance lines such as telephone, satellite, radio, etc. Also called channels.

Communications Server Communications software, Web server. Provides access to the Web from Windows NT, Unix, OSF/1 systems. Lets users view documents. Vendor: Netscape Communications Corp.

communications specialist Technical developer. Usually works with WANs and supports the networking for host-based long distance systems. Senior or mid-level title.

Communicator Communications, Internet software. Browser. Includes messaging and groupware functions. Vendor: Netscape Communications Corp.

CommWorks for Windows Communications software. Package of five communications programs. Includes: LapLink, LapLink Remote Access, TS Fax, TS On-Line, LapLink Alert. Vendor: Traveling Software, Inc.

Comnet III Communications. Network management software. Allows users to build models of networks and proposed changes to determine effect of changes. Runs on AIX, IRIX, OS/2, OSF, SunOS, Solaris, Windows systems. Vendor: CACI Products Co.

Compaktor Operating system add-on. Data management software that runs on IBM systems. Vendor: Innovation Computer Corp Data Processing.

Compaq Computer vendor. Manufactures microcomputers (PCs); purchased DEC (Digital Equipment Corp.) in 1998.

COMPAREX Operating system add-on. Debugging/testing software that runs on IBM systems. Vendor: Sterling Software (Systems Software Marketing Division).

Compas Operating system enhancement used by Operations staff and systems programmers. Disaster control package. Vendor: Comdisco Disaster Recovery Services.

Compel Multimedia authoring tool. Entry-level, slide-based tool. Applications will run under Windows, Macintosh. Vendor: Asymetrix Corp.

Compete EIS. Vendor: Manageware, Inc.

:Compfile Operating system add-on. Data management software that runs on HP systems. Vendor: Systems Express.

Compile PRF Application development tool. Companion to PRF. Converts reports, queries, and SQL statements into COBOL programs. Called PRF. Vendor: PLATINUM Technology, Inc.

compiler, compiled language A programming language that works by translating the English language statements into machine code and creating an executable machine code program which can be run at any time, and as many times as needed. Contrast with interpreter, interpreted language. Compiler programs execute faster than interpreted programs. Some languages, i.e. Basic, allow the English language statements to be either compiled or interpreted.

compiler (program) A program that translates high-level language into machine executable language.

complex data types Unstructured, nonrelational data types. Includes voice, video, fingerprints, text, etc.

component A software building block. In object-oriented development, software that is similar to an object in that it is encapsulated (contains data and processing), but does not necessarily support inheritance and polymorphism. Applications are built by putting together components. JavaBeans are components.

component based architecture
Development of software systems that builds applications from components, or pre-written building blocks. Based on object-oriented development. JavaBeans and ActiveX are used to build components.

Component Broker Connector Communications technology. Middleware, includes an object request broker (ORB). Manages transactions and applications throughout distributed systems. This allows, i.e., mainframe applications to be accessed from Internet browsers without modification. Uses CORBA standards. Released: 1997. Vendor: IBM Corp.

component mining Application development technology. Term used with component based architecture. Refers to extracting components from existing code. This allows existing programs to be used, at least in part, and entire systems do not have to be rewritten.

Comport Application development tool. Assists in developing input/output functions. Vendor: GT Software, Inc.

Compose II Application development tool. Report generator for DG systems. Vendor: Evolution 1, Inc.

Composer See COOL:Gen

compound document A document that contains a number of integrated components which may include text, graphics, and scanned images.

Compound Document Architecture See CDA.

Compress Operating system add-on. Data management software that runs on Bull HN systems. Vendor: Scientific and Business Systems, Inc.

Compress/ Operating system add-on. Data management software that runs on IBM systems. Programs include: Compress/DB2, Compress/IDMS, Compress/IMS, Compress/PDS, Compress/VSAM-VSE. Vendor: DataBase Technology Corp.

COMPstation Midrange computer. SPARC RISC CPU. Operating system: Solaris. Vendor: Tatung Science & Technology Inc.

Compstation U10-360 Desktop computer. SPARC IIi 64-bit processor. Operating system: Unix. Vendor: Tatung Science & Technology, Inc.

Compu-Tek family Midrange computers. 486, Pentium CPU. Operating systems: DOS/Windows, Windows 95/98. Includes: Compu-Tek 486xxx/xx, Compu-Tek EISA File Server, Compu-Tek MPC Multimedia, Compu-Tek PCI Power Server, Compu-Tek PCI Power Station. Computek-VBL. Computek-Windows Power. Vendor: Compu-Tek International.

CompuBook 650 Notebook computer. Pentium CPU. Vendor: Impulse Computer Corp.

CompuLert Communications software. Network management. Allows network administrators to monitor and control Windows NT networks, multiplatforms distributed systems, and hardware through the enterprise. Full name: CompuLert Enterprise Management Solution. Vendor: Tone Software Corp.

CompuSeries Desktop publishing software. Version of CompuSet. Vendor: Document Sciences Corp.

CompuServe Communications service that provides a wide variety of information and services. This includes online editions of computer publications, travel and entertainment data, various bulletin boards, and e-mail functions. Users subscribe to CompuServe for a monthly fee.

CompuSet Publishing software that runs on mainframes. Vendor: Document Sciences Corp.

CompuSource Computer vendor. Manufactures desktop systems. Full name: Compusource International, Inc.

Computer Machine that processes data.

Computer-Aided Methodology CASE product. Automates project management function.

computer-aided telephony Integration of midrange computers and telephones. Will allow user to dial into an e-mailbox. System would read back messages. Through the phone, the user could collect data from databases, cause reports to be generated, and fax the reports as requested from the messages.

Computer-Assisted Instruction See CAI.

Computer Associates Software vendor. Products include financial, database, and systems software.

computer center See data center.

Computer Integrated Telephony See CIT.

computer language Also called machine language. The binary language used to program computers. The output of language translators.

Computer Masters Computer vendor. Manufactures midrange computers.

Computer Misuse Detection System See CMDS.

Computrend Computer vendor. Manufactures midrange computers.

Comquest Computer vendor. Manufactures midrange computers.

COMS Communications software. Transaction processing monitor. Runs on Unisys systems. Stands for: Communication Management System. Vendor: Unisys Corp.

COMSAT A private organization that launches and operates communications satellites. Stands for: COMmunications SATellite Corp.

Comshare BudgetPLUS Application software. Financial application used for budgeting. Combines spreadsheet functions and data is entered through Excel or Lotus 1-2-3 screens. Runs on OS/2, Windows systems. Released: 1998. Vendor: Comshare, Inc.

Comshare FDC Application software. Financial application that handles budgets, financial plans, forecasts and operating financial data. Users can enter data from Excel or Lotus 1-2-3 screens. Stands for: Financial Data Control. Runs on Windows systems. Released: 1997. Vendor: Comshare, Inc.

concentrator Communications hardware. Joins several communications lines together.

ConceptBase Object-oriented database. Multi-user system used for conceptual modeling in design situations. Freely distributed and used for research and teaching.

Concert Series family Application software. Financial, HR, payroll systems for AS/400 environments. Vendor: NewGeneration Software.

Concerto Operating system software. Allows a single Desktop server to run up to four operating systems concurrently. Operating systems include Unix, NetWare, OS/2, Windows NT. Vendor: NetFrame Systems, Inc.

Concerto 4/25,4/33 Handheld, pen-based computer. Vendor: Compaq Computer Corp.

Concurrent Computer vendor. Manufactures midrange computers.

Concurrent DOS Operating system for IBM and compatible midrange computer systems. Versions include: Concurrent DOS 386, Concurrent PC DOS XM. Vendor: Digital Research, Inc.

concurrent engineering Manufacturing approach that overlaps analysis, design and product functions by having several functions (design, development, marketing, manufacturing, service and sales) work together as a team.

Condor Application development tool. Screen editor. Runs on IBM mainframe systems. Vendor: Phoenix Software International.

Condor 3 Relational database for midrange computer environments. Runs on IBM, HP midrange systems. Vendor: Condor Computer Corp.

Conductor Groupware. Workflow software. Used to define the process flow by end-users without working with the application logic. Vendor: Forte Software, Inc.

configuration analyst, manager Technical developer. Manages hardware and/or software and analyses uses, interactions, and cost benefit. Senior title most often used in mainframe computer systems.

configuration management Identifying and controlling devices in a computer system. Controlling changes to computer systems. In client/server, the function that keeps the client and server parts of applications in sync with each other and with the other resources in the computer environment. In operating systems, the function that balances and controls hardware and software resources.

CONFIGURE Application development tool. Program that automatically formats COBOL programs. Vendor: Computer Data Systems, Inc.

CONMAN-E Operating system enhancement used by systems programmers. Monitors and controls system performance in IBM systems. Vendor: Jason Data Services, Inc.

Connect... Series of system management software that supports message-oriented middleware. Includes Lotus Connect, SAP Connect, IMS Connect, and Internet Connect. Includes Command Center management software. Vendor: Candle Corp.

Connect:Direct Communications software. Transaction processing monitor. Handles whole file transfers between NetWare and IBM and Unix servers. Vendor: Sterling Software, Inc.

CONS Communications protocol. Transport layer. Developed by OSI. Stands for: Connection-Oriented Network Service.

console Terminal used by the Operations staff to communicate with the operating system. Sometimes used as a synonym for terminal.

CONSOLE-Master Operating system enhancement used by Operations staff and systems programmers. Provides operator console support in IBM systems. Vendor: Xenos Computer Systems, Inc.

console operator Support personnel. Person who monitors the daily computer activity and is responsible for making changes when required to the production schedule and/or the equipment assignments. Senior level. Title used in mainframe installations.

Constantine Structured programming design methodology named for its developer. Based on functional decomposition. Accepted as a standard design methodology by some companies, and used by some CASE products.

Constellar Hub Communications. Message oriented middleware. Data transformation and application development tools. EAI software that provides data transformation, data movement and interface management for ERP systems. Uses Oracle database. Vendor: Constellar Corp.

Constellar WareHouseBuilder Application development tool used in data warehousing. Builds scalable data warehouses and data marts. Runs on Unix, Windows systems. Vendor: Constellar Corp.

Constellation xxx Midsize computer. Server. Pentium CPU. Operating systems: Windows 95/98. Vendor: Technology Advancement Group (TAG).

Construction Site Application development tool. Links Web front end systems to client/server systems. Creates applications that use a Web browser as a user interface. Vendor: W3.Com

Context Spider Application development tool. Web authoring tool. Automatically inserts HTML tags into text to create Web pages. Vendor: InContext Systems.

ConText Option Application development tool. Manages and retrieves data on Oracle7 databases. Interfaces with the Web. Vendor: Oracle Corp.

Continuum xxxx Mainframe, midrange computers. Operating systems: Unix, Stratus FTX. Vendor: Stratus Computer, Inc.

Continuus/CM Application management tool. Software configuration manager. Includes standard process life cycle models, and process management, version control, release management and change management functions. Integrates with SoftBench, CodeCenter, ObjectCenter, FrameMaker and Software Through Pictures. Runs on Unix, Windows NT systems. Released: 1997. Vendor: Continuus Software Corp.

Continuus/PT Application development tool. Problem Tracker. Manages software change cycle

Interfaces with Continuus/CM. Runs on Unix systems. Vendor: Continuus Software Corp.

Contrl Communications software. Transaction processing monitor. Runs on DEC systems. Vendor: Clyde Digital Systems.

Control 1. Operating system add-on. Library management system that runs on Tandem systems. Vendor: Network Concepts, Inc.
2. Application development tool. DSS. Provides access to data stored in multidimensional databases. Vendor: KCI Computing.

ControlIT Communications software. Allows remote control of PCs from anywhere on a network. Allows users to view multiple systems, conduct chat sessions and execute remote applications from any Windows system. Runs on Windows systems. Released: 1998. Vendor: Computer Associates International, Inc.

Control Data Corp See CDC.

Control-O Operating system enhancement used by systems programmers. Used to develop automation of the data center. Vendor: 4th Dimension Software, Inc.

Control SA Operating system software. Security package that handles end-user security issues. Released: 1995. Vendor: New Dimension Software, Inc.

Control-x Operating system enhancements used by systems programmers in IBM systems. Job scheduling and production systems. Includes Control-D,D/PC,O,M,R. Vendor: 4th Dimension Software, Inc.

Controlmanager CASE product. Part of Manager family. Vendor: Manager Software Products, Inc.

ControlShell Application development environment. Used to develop real-time applications from reusable code objects. Runs on Unix systems. Vendor: Real-Time Innovations, Inc.

Convectis Application development tool used for information filtering. Finds articles of interest according to user defined specifications. Runs on Solaris, Windows NT systems. Vendor: Aptex.

conversation Communications term. Continuous dialog between two or more systems. Can consist of overlapping executions in a distributed environment. Useful for updating distributed databases. Some middleware uses conversations.

Convert Series Application development tool. Migrates programs, data, data definitions from one database to another. Vendor: Forecross Corp.

Convex Computer vendor. Manufactures large computers. Division of Hewlett-Packard Co.

Convex Exemplar SPP1000, SPP1200 Mainframe computer. Parallel processor. Supports up to 128 processors. Operating systems: Unix, HP-UX. Vendor: Convex Computer Corp.

ConvexOS Operating system for Convex mainframe systems. Vendor: Convex Computer Corp.

cookie (counter) Internet program. Type of snippet. A file kept by each browser that visits your Web page. The cookie is a small amount of information that is sent from a server to your computer and stored on your hard disk. When you visit the site later the server can recall the information (the cookie) and you can pick up where you left off. Cookies record information about each visitor to a Web site such as such as the visitor's identification (URL), page preferences, and what they've requested (or bought) from you before. Cookies raise privacy concerns.

COOL:2E CASE product. Automates analysis, design, and programming functions. Includes code generator for COBOL, RPG languages. Used to build client/server applications. Design methodologies supported: Information Engineering. Uses IBM AS/400 as server and connects client desktop computers via Token Ring. Database supported: OS/400 relational database. Runs on AS/400, Unix systems. Formerly known as Synon/2E. Vendor: Sterling Software, Inc.

COOL:Biz Application development tool. Business and data modeling toolset. Uses diagrams that can be understood by both the users and the developers. Works in a multi-user, team-based environment allowing multiple users to access the same model at the same time. Formerly known as COOL:DAT and KEY:Model. Runs on Windows systems. Released: 1998. Vendor: Sterling Software, Inc.

COOL:BusinessTeam Application development tool. Allows users to build both data and process models. Runs on Unix, Windows systems. Formerly known as GroundWorks. Vendor: Sterling Software, Inc.

COOL:Cubes Application development tool. Used to develop component based applications by assembling software from predefined and pretested components. Vendor: Sterling Software, Inc.

COOL:Dat See COOL:Biz.

COOL:DBA Application development tool used to create, maintain, and migrate databases. Visual environment used to design relational databases. Supports Oracle, Sybase, DB2. Runs on OS/2, Windows systems. Formerly known as Terrain. Vendor: Sterling Software, Inc.

COOL:Enterprise Application development environment. Used to develop both host-based and client/server applications. OS/2 based. Formerly called: KEY:Enterprise. Released: 1996. Vendor: Sterling Software, Inc.

COOL:Gen Application development environment. Provides full life-cycle support. Used for client/server, component based development. Uses visual development tools to develop client/server systems. Provides application partitioning and shows how balanced the code will be and what potential network problems could result. Provides data modeling functions. Used for enterprise development. Generates COBOL, C code. Generates error-free code for multiple applications. Formerly called: Composer. Runs on DEC, IBM mainframes, OS/2, Unix, Windows systems. Released: 1997. Vendor: Sterling Software, Inc.

COOL ICE Communications. Middleware used over the Internet. Integrates multiple and different database systems for access over the Internet. Released: 1998. Stands for: COOL Internet Commerce Enabler. Vendor: Unisys Corp.

COOL:Jex Application development tool. Automates all phases of the development cycle. Generates C++, ADA, Smalltalk, embedded SQL, PowerBuilder, Java, VisualBasic NewEra code. Interfaces with Versant, Objectstore, Objectivity, Ontos, Raima. Includes ObjectTeam Application Factory (suite of development tools) and ObjectTeam Enterprise (application development environment). Formerly known as ObjectTeam. Released: 1998. Vendor: Sterling Software, Inc.

COOL:Plex Application development tool. Used for enterprise development. Object-oriented approach to development. Uses pattern-based technology to build scalable applications for AS/400, Unix and Windows NT environments. Works with Synon/2E. Formerly called Obsydian. Vendor: Sterling Software, Inc.

COOL:Quik Application development tool. Reusable pattern procedures that can be copied to create procedures for data items. Runs on Unix, Windows systems. Released: 1998. Vendor: Sterling Software, Inc.

COOL:Spex Application development tool. Allows developers to analyze business rules and convert them directly into interfaces for business components. This means development teams can assemble applications from reusable components. These component-based systems are compatible with any object framework including COM, CORBA and Java. Released: 1998. Vendor: Sterling Software, Inc.

COOL:Stuff Application development suite. Includes: COOL:Biz, COOL:Dat, COOL:Gen, COOL:Jex, COOL:Cubes. Vendor: Sterling Software, Inc.

COOL:TeamWork CASE product. Automates analysis, design, and programming functions. Supports modeling and development of real-time and embedded systems. Works with rigorous requirements specifications which must be adhered to during the development. Includes code generator for Ada, C. Design methodologies supported: Constantine, DeMarco, Ward-Mellor, Yourdon. Runs on DEC VAX, HP/Apollo, IBM RT, Sun, Unix systems. Vendor: Sterling Software, Inc.

COOL:Xtras Application development tools. Set of utilities and add-on products to the COOL development set. Includes Advise (tool that provides on-line methodology for project development) and Webview (tool that publishes current versions of development models over the Internet or corporate intranets). Vendor: Sterling Software, Inc.

Cooperation Groupware. Runs on OS/2, Unix systems. Designed to build mission critical software systems, replacing mainframe applications. Vendor: NCR Corp.

Cooperative Development Environment See CDE.

cooperative processing A type of client/server architecture where most of the processing is done on the host system.

Cooperative Server See Oracle7.

cooperative-server database A database which allows applications to access data located on multiple computers as if the data were on a single computer.

Coordinator Groupware. Vendor: Action Technologies.

Copernicus Communications software. Message oriented middleware, data transformation tool. Uses a translator to store modules of code in an internal database to link disparate applications. Can run queries against multiple databases on different platforms. Runs on Unix, Windows systems. Vendor: Vie Systems.

Copics/CIMapps MRP, or CIM software. Runs on IBM VSE, MVS systems. Vendor: IBM Corp.

Copland Operating system for PowerPC Macintosh computers. Includes multiprocessing support, microkernel design. Combined with NextOS to make Rhapsody. Vendor: Apple Computer, Inc.

COPS Operating system add-on. Security/auditing system that runs on IBM CICS systems. Stands for: CICS-On-Line-Protection System. Vendor: TelTech.

COPY PLUS See DB2 COPY PLUS.

Coral 66 Compiler language. Follows British standards. Language is based on Algol and incorporates aspects of FORTRAN. Used to develop real-time applications. Vendor: Concurrent Computer Corp.

CORBA Communications standards. Stands for: Common Object Request Broker Architecture. Standards set by OMG. Provides a common platform for application developers in object-oriented environments to write middleware adhering to CORBA standards. Allows diverse applications and databases to communicate with each other regardless of platform. Based on IDL. Contrast with COM which is Windows-based middleware. Supported by IBM, Oracle, Netscape, Sun.

Core 2000 Foundation Software Application development tools. Suite of software that provides reusable modules for such things as database interaction, error correction, and user interfaces. Object-oriented tools. Uses VisualWorks. Runs on OS/2, Unix systems. Vendor: American Management Systems, Inc.

CorelCentral Desktop software. Provides desktop management, PIM functions. Vendor: Corel Corp.

CorelDraw Desktop system software. Graphics package. Interfaces with Lotus 1-2-3, Excel. Runs on OS.2, Unix, Windows systems. Vendor: Corel Systems Corp.

Coronado Application development tool. Provides a front-end to Microsoft's SQL Server database. Allows users to look at more than two data variables simultaneously, and can assign data variables such as size, color and animation. Released: 1998. Vendor: Portola Dimensional Systems.

Corporate MyYahoo EIP, corporate portal. Uses Yahoo's existing information services, including news, weather, stocks, etc. and also accesses internal corporate information such as employee directories, benefits, supply-chain status, and accounts payable. Released: 1999. Joint venture: Tibco and Yahoo.

Corporate Online Repository Groupware. Document management system. Links DOCS Open to Lotus Notes. Vendor: PC Docs.

corporate portal See EIP.

Corporate Tie Communications software. Network connecting IBM systems. Vendor: Pansophic Systems, Inc.

Corridor Communications software. Allows a Web browser to emulate a dumb terminal to access mainframe systems. Vendor: Teubner & Associates, Inc.

Corsair Graphical interface for the Internet. Internet browser. Runs on Macintosh, NetWare, Windows systems. Vendor: Novell, Inc.

Cortex Application development tool. Program that is used by systems programmers during VSE to MVS conversions. Vendor: Computer Task Group, Inc.

Corvet Application development tool. Includes program generator and produces program documentation. Runs on DG systems. Vendor: Analysts International Corp.

CorVision CASE product. Automates design and programming functions. Runs on DEC VAX, IBM 286 midrange systems. Interfaces with IEW, Excelerator. Vendor: CORTEX Corp.

COS 1. Operating system for SCS, Cray, and Harris systems.
2. An organization of most of the computer vendors whose charter includes monitoring OSI and ISDN standards development. Stands for: Corporation for Open Systems.

COSACS Operating system enhancement used by Operations staff and systems programmers. Provides operator console support in Xenix systems. Stands for: Computer Operations Surveillance and Control Systems. Vendor: NYNEX Service Co.

COSE Common Operating System Environment. Initiative by Unix vendors (including IBM, Sun, DEC, HP, Univel) to adhere to common APIs and produce Unix software that maintains similar user interfaces. Produce CDE (Common Desktop Environment) standards and programs.

Costar CASE product. Automates project management function. Supplies cost accounting information for project development and maintenance. Vendor: Softstar International, Inc.

Cougar Desktop computer. Alpha CPU. Operating system: Windows NT. Vendor: *Neko*Tech Inc.

Council EMS (Electronic Meeting System). Vendor: CoVision.

CP Control Program. Major part of VM/CMS. Controls all hardware specific operations and directs all I/O operations.

CP-6 Operating system for mainframe Bull HN systems. Suited to engineering and scientific environments. Vendor: Bull HN Information Systems.

CP/M Operating system for midrange computers. The first general operating system that ran on a variety of midrange computers. Obsolete.

CP:Millennium See Millennium.

CP/OS Operating system for NCR systems. Vendor: NCR Corp.

CPF Operating system for IBM System/38. Includes a relational DBMS. Stands for: Control Program Facility. Vendor: IBM Corp.

CPG Application development tool. Program generator for IBM CICS systems. Stands for: Communications Program Generator. Vendor: THORN EMI Computer Software.

CPI Common Programming Interface. The component of IBM's SAA that specifies how programs are to be written, and how programs are to be incorporated into the SAA structure.

CPK Operating System enhancement used by systems programmers. Reorganizes data and free space on disk. Vendor: Innovation Data Processing.

CPR Packaged software designed to assist sales and marketing. Assists with telemarketing, field sales, customer support and quality management. Interfaces with Oracle, Sybase, Unify, Informix. Runs on Unix systems. Stands for: Customer Resource Planning. Vendor: Aurum Software, Inc.

CPS Application software. Sales functions. Identifies customers who are likely to defect so they can be approached with special offers. Stands for: Churn/Customer Profiling System. Vendor: SLP InfoWare, Inc.

CPSR Computer Professionals for Social Responsibility. Group that focuses on issues such as public access to government information and women in computing.

CPU Central Processing Unit. The part of the computer that does the actual processing of data. Used as a synonym for computer.

CPX 1. Latest version of Prograph. See Prograph.
2. Operating system enhancement used by systems programmers. Monitors and controls system performance in IBM systems. Stands for: Capacity Planning Extended. Vendor: IBM Corp.

CQCF Operating system enhancement used by systems programmers. Increases system efficiency in IBM CICS systems. Stands for: CICS Queue Command Facility. Vendor: NETEC International Inc.

CQCS 4GL tool kit for IBM, DEC, Data General, Sun, and Unix systems. Contains mail merge, transaction processing and graphic presentation functions as well as a query language and report writer. Stands for: Cyberquery/Cyberscreen. Vendor: Cyberscience Corp.

cQLLC Communications software. Network connecting IBM systems. Vendor: Systems Strategies, Inc.

CQS-Convert2/DB2 Operating system add-on. Data management software that runs on IBM systems. Converts IDMS/R, Adabase, and Supra applications to DB2. Vendor: Carleton Corp.

Cradle CASE product. Automates analysis, programming, and design functions. Runs on HP, Sun systems. Vendor: Yourdon, Inc.

Cray Former computer vendor. Manufactures supercomputers. Now owned by Silicon Graphics.

Cray J90, T3D, T3E, T90 See appropriate listing.

CREDO Operating system add-on. Data management software that runs on Bull HN systems. Stands for: Comprehensive Report Examination/Display Option. Creates reports in an online environment. Vendor: Bull HN Information Systems.

CRI Common Repository Interface. Incorporates the application programming interfaces that define how third party software can communicate with IBM and DEC repositories. Strategic plan of Computer Associates International, Inc.

Cricket Presents See CA-Cricket Presents.

CRL See KnowledgeCraft.

CRM Customer Relationship Management. Applications software providing front-office functions. Automates sales, marketing and customer service functions. Sales force automation systems fall into this category.

Cross Platform Services Communications. Network management software. Runs the core services of NetWare on various platforms. Vendor: Novell, Inc.

Cross Platform Toolset Application development tool. Allows developers to create interfaces between Motif and Windows applications. Vendor: Visual Edge Software.

Cross System Product See CSP.

Cross Target Builder Multidimensional database. Interfaces with Data Integrator. Runs on MacOS, MVS, OS/2, Unix, Windows systems. Vendor: Dimensional Insight.

CrossAccess Communications software. Middleware. Used to create client/server applications. Builds access to both SQL and nonrelational data stores. Allows queries against multiple databases on different platforms. Vendor: Cross Access Corp.

CrossCode Application development tool. Makes programs accessing Ideal databases run on IBM systems. Vendor: On-Line Software International.

CrossGraphs Application development tool used for data mining. Provides graphs of different subsets or summaries of data. This allows user different perspectives of the data. Runs on Windows 95/98 systems. Vendor: Belmont Research.

CrossRoads Software vendor. Changed name to CrossWorlds Software in 1998.

Crosstalk XVI Communications software. Network connecting IBM midrange computers. Vendor: Crosstalk Communications/DCA.

CrossTarget Multidimensional database and application development tool. Object-oriented data access and analysis software. Users can view data through charts and graphs and can access additional data by selecting any part of the current display. Creates a model that users can access through Diver (GUI). Used in data warehousing. Runs with Windows, Macintosh, and Motif. Vendor: Dimensional Insight, Inc.

CrossTies Data management software. Provides user interface to access data from spreadsheets, word processors, and/or graphics packages. Runs on MS/DOS, Windows systems. Vendor: CrossTies Software Corp.

CrossView Application development tool. Report generator for DEC VAX systems. Vendor: Ross Systems, Inc.

CrossWorlds Software vendor. Creates EAI integration software (also called processsware). Main product: CrossWorlds Customer Interaction.

CRP Application software. Part of manufacturing systems. Stands for: Capacity Requirements Planning.

CRT Cathode Ray Tube. The terminal screen.

CruisePad Handheld computer. Wireless. Vendor: Zenith Data Systems.

Crunch Desktop system software. Spreadsheet. Runs on Apple Macintosh systems. Vendor: VisiCorp.

Cryptopak Operating system add-on. Security/auditing system that runs on IBM systems. Vendor: Computation Planning, Inc.

CryptoWall Internet software. Firewall. Vendor: Radguard Ltd.

Crystal Info Application development tool. Tool for extracting, analyzing and distributing information from most relational databases. Version available for Essbase multidimensional database. Queries can run from Web browsers or Windows systems. Vendor: Seagate Technology, Inc.

Crystal Performance Evaluator Performance evaluation software. Evaluates use of computer system resources by checking for such things as code execution redundancy. Vendor: BGS Systems, Inc.

Crystal Reports Application development tool. Query tool and report writer. Accesses over thirty data sources including databases and files. Queries can run from either Web browsers or Windows systems. Runs on Windows systems. Vendor: Seagate Technology, Inc.

CS-150 Midsize computer. Server. Pentium Pro processor. Operating system: Windows NT. Vendor: Tandem Computers, Inc.

CS/7000,8000, CS/1000 Application development tool. Expert system used to develop client/server applications. Controls the development cycle. Modules include a methodology selector, a task manager, project planner, resource manager, documentation generator. CS/7000,8000 are used to develop workgroup and departmental systems; CS/1000 is used for large-scale development. Vendor: Client/Server Connection, Ltd.

CS-COMP Operating system add-on. Data management software that runs on IBM systems. Vendor: C-S Computer Systems, Inc.

CS Master Operating system add-on. Security/auditing system that runs on DG systems. Vendor: Data General Corp.

CS PROXI Application development tool. Program generator for DG systems. Also generates reports. Vendor: Data General Corp.

CSA 1. Application development tool. Analyzes COBOL programs and stores data in a dictionary. Used with CSM. Runs on IBM mainframe systems. Stands for: Current Systems Analyzer. Vendor: Texas Instruments, Inc.
2. Application development tool. Program that reformats COBOL code according to company-defined standards. Stands for: COBOL Structuring Aid. Runs on IBM systems. Vendor: MARBLE Computer, Inc.
3. See COBOL Source Analyst.

CSAR 1. Operating system enhancement used by systems programmers. Monitors and controls system performance in IBM systems. Stands for: Computer Scheduling and Reporting System. Vendor: Stockholder Systems, Inc.
2. Operating system enhancement used by systems programmers. Job scheduler for IBM systems. Vendor: Software Engineering of America, Inc.

CSG Application development tool. Program generator. Full name: Synon/Client Server Generator. Generates applications using AS/400 systems as the server and IBM PS/2 systems as clients with a Token Ring local area network. Vendor: Synon Corp.

CSI Application development tool. Program utility used for maintenance and re-engineering. Part of COBOL/2 Workbench. Stands for: COBOL Source Intelligence. Vendor: Micro Focus, Inc.

CSM Application development tool. Analyzes impact of changes to COBOL programs and generates updated COBOL code. Uses CSA for analysis. Stands for: Current Systems Modification. Runs on IBM mainframe systems. Vendor: Texas Instruments, Inc.

CSMA/(CD) Communications protocol. Data link layer. Used in Ethernet. Stands for: Carrier Sense Multiple Access/(Collision Detection).

CSobject See IMS CSobject.

CSOS Operating system for IBM midrange computer systems. Vendor: IBM Corp.

CSP Application development tools. Includes program generator, testing tools. Runs on IBM systems. Works with CICS, TSO, and CMS. Interfaces with DL/1, DB2, SQL. Stands for: Cross System Product. Includes CSP/AD (Application Development) and CSP/AE (Application execution). Midrange computer versions called: EZ-Prep (application development), EZ-Run (application Execution). Vendor: IBM Corp.

CSP/ADE Program development tool. Toolset for IBM's AD/Cycle which provides a set of reusable applications models for CSP development. Stands for: Cross System Product/Application Development Enabler. Vendor: K-C Computer Services.

CSPP Computer Systems Policy Project. An association of computer vendors including IBM, Apple, Hewlett-Packard and Cray doing research and development in conjunction with the Department of Energy.

CSRL Artificial intelligence system. Runs on Apollo, DEC VAX systems. Stands for: Conceptual Structures Representation Languages. Vendor: Battelle.

CST Software vendor. Produces application development software. Full name: Client/Server Technology Ltd.

CT-Mail Communications software. E-mail system. Runs on Convergent systems. Vendor: Unisys Corp.

CT-NET Communications software. Network connecting Convergent computers. Vendor: Unisys Corp.

CT*OS/IT*OS Word processor for DEC systems. Vendor: Intermation Corp.

CTIX SNA Communications software. Network connecting Convergent computers. Vendor: Unisys Corp.

CTL Midsize computer. Server. Pentium Pro processor. Operating system: Windows NT. Vendor: Computer Technology Corp.

CTL A-400, Internet Server Midrange computer. Server. Pentium CPU. Vendor: Computer Technology Link.

CTOP(III) Operating system enhancement used by systems programmers. Increases system efficiency in IBM CICS systems. Vendor: H & W Computer Systems, Inc.

CTOS Operating system for Unisys systems. Vendor: Unisys Corp.

CTOS/BTOS Operating system for mainframe Burroughs systems. Vendor: Convergent Technologies, Inc.

CTOS/VM Operating system for IBM and compatible midrange computer systems. Vendor: Convergent Technologies, Inc.

CTS-300 Operating system for DEC systems. Stands for: Commercial Transaction Operating System. Vendor: Digital Equipment Corp.

CUA Common User Access. The component of IBM's SAA that sets standards for how end-users interact with applications systems.

Cubicalc Application development tool used for data mining. Provides fuzzy analysis of multiple criteria, giving user results close to stated criterion. Runs on OS/2, Windows systems. Vendor: HyperLogic Corp.

Cubix Computer vendor. Manufactures microcomputers (PCs).

CubixConnect Desktop computer. Pentium CPU. Operating system: Windows 95/98. Vendor: Cubix Corp.

CubixNet Operating system. Unix-type systems. Vendor: Cubix Corp.

CULPRIT Application development tool. Report generator for IDMS systems. Vendor: Cullinet Software, Inc.

Current Desktop system software. Personal Information Manager (PIM). Vendor: Jensen-Jones, Inc.

Current System Analysis, Modification Application development tool. Used to redesign mainframe systems and add network access. Vendor: Texas Instruments, Inc., Price Waterhouse.

CUSP Operating system enhancement used by systems programmers. Increases system efficiency in Concurrent systems. Print spooler. Stands for: Commercial User SPooler. Vendor: Concurrent Computer Corp.

customer See user.

customer callback Techniques which allow the consumer to ask for a return call from a salesperson by clicking on a Web icon.

Customer Interaction Application development tools. EAI software that integrates disparate applications by allowing the end-user to design the solution. Works with front office packages from Aurum, Clarify, Siebel, Trilogy and Vantive and back office packages from Baan, PeopleSoft and SAP. Released: 1997. Vendor: CrossWorlds Software, Inc.

Customer Messaging System Applications software. Customer service and messaging software. Enables companies to manage, monitor and restore high-volume e-mail. Contains Java-based application server. Runs under Windows 95/98, NT. Also called CMS. Vendor: Kana Communications Inc.

Customer Relationship Management See CRM.

CustomerFirst Notes Remote Help desk software that keeps a database of problems and solutions available to remote users. Vendor: Repository Technologies.

CustomGantt Desktop system software. Project management package. Runs on Pentium type systems. Vendor: SofTrak Systems, Inc.

Customizer CASE product. Part of Excelerator system. Vendor: Index Technology Corp.

CX/UX,SX,RT Operating system for Harris computer. Vendor: Harris Corp.

Cxdb Operating system add-on. Debugging/testing software that runs on DEC VAX, IBM midrange, Unix systems. Full name: Cxdb: C Source Level Cross Debugger. Vendor: Whitesmiths, Ltd.

CXML Application development tool. Extension to XML that provides definitions of metadata used in e-commerce. Developed as a collaborative effort to standardize commerce over the Internet by defining the content of business catalogs and business transactions such as purchase orders, change orders, shipping notices, etc. Stands for: Commerce XML. Implemented in 1999.

CXOS Operating system for Motorola systems. Vendor: Motorola Computer Systems, Inc.

Cyber 920C Server Midrange computer. RISC machine. MIPS R3000 CPU. Operating systems: IRIX OS. Some models can be configured as the server in a client/server system. Networks: LAN Manager. Client systems can run Unix. Vendor: Control Data Corp.

Cyber xxxx, CyberPlus Mainframe computers. Operating system: NOS/VE, VX/VE. Vendor: Control Data Corp.

CyberAgent SDK Application development tool. Used to build applications for the Internet. Contains scripting language. Creates applications that use a Web browser as a user interface. Stands for: CyberAgent Software Development Kit. Vendor: FTP Software, Inc.

Cybercop Communications. Internet software. Firewall. Vendor: Network Associates, Inc.

Cyber Graf Desktop computer. Pentium processor. Vendor: Royal Electronics Inc.

CyberGuard (Firewall) Internet software. Firewall. Vendor: CyberGuard Corp.

Cyberprise Host Communications/Internet software. Accesses mainframe systems through Web browsers. Supports Microsoft's Visual Studio. Runs on Unix, Windows systems. Released 1998. Vendor: Wall Data, Inc.

Cyberjack Internet browser. Runs on Windows 95/98. Vendor: Delrina Corporation.

Cyberleaf Development tool used to connect multimedia content (i.e. graphics, text images). The coding that allows files to appear as formatted pages on the World Wide Web. Includes GUI and format conversions from standard word processors. Easier to use than SMGL or HTML. Vendor: Interleaf, Inc.

Cybermax Computer vendor. Manufactures midrange computers.

Cyberprise Portal EIP, corporate portal. Provides control of information access and encryption functions. Runs on Unix, Windows NT systems. Vendor: Wall Data, Inc.

Cyberprise Server Application server providing middleware and development tools. Supports ActiveX, Java and HTML. Runs on Windows NT systems. Vendor: Wall Data, Inc.

Cyberquery/Cyberscreen See CQCS.

cyberspace See virtual reality.

Cyborg Systems Software vendor. Produces Solution Series/ST which includes HR, Payroll and Time and Attendance modules.

Cyc Artificial intelligence system that contains millions of basic facts about everyday existence. Used as the basis for applications systems.

Cyclone Desktop computer. Pentium CPU. Operating systems: Windows 95/98, Windows NT Workstation. Vendor: Tangent Computer, Inc.

Cyclone II Midsize computer. Server. PA-RISC Pentium processor. Operating system: IRIX. Vendor: Colorbus, Inc.

Cyclone Office Midsize computer. Server. Pentium II processor. Operating system: IRIX. Vendor: Colorbus, Inc.

Cyrano Suite Application development tools. Automates testing functions. Includes Cyrano Robot which executes testing and Cyrano Manager which sets criteria for tests. For specific applications includes: ClientPack, DBPack, ServerPack, VTPack (see separate listings). Vendor: Cyrano, Inc.

Cyrix Hardware vendor. Manufactures computer chips, or microprocessors. Full name: Cyrix Corp.

D&B Software Software vendor. Products include: SmartStream series, McCormack & Dodge systems. Purchased by Gaec Computer Corp, Ltd.

D-Fast/VSE See TD-Fast/VSE.

D-MON Disk Monitor Operating system enhancement used by systems programmers. Increases system efficiency in DEC VAX systems. Manages disk storage space. Vendor: Bear Computer Systems, Inc.

D-NIX General purpose operating system. Unix type system. Conforms to POSIX standards. Follows TCP/IP, NFS, X.25, SNA/SDLC protocols. Vendor: Diab Data, Inc.

D'OLE Application development tool. Allows Windows applications to connect with NextStep applications. Vendor: Apple Computer Inc.

D The Data Language Relational database/4GL for desktop computer environments. Runs on IBM desktop computer systems. Vendor: Caltex Software, Inc.

D3M Operating system add-on. Data management software that runs on Apollo systems. Stands for: Domain Distributed Data Management System. Vendor: Apollo Computer, Inc.

DA 1 Database for large computer environments. Runs on IBM systems. Vendor: Consolidated Business Systems, Inc.

Dacom CASE product. Automates design functions. Vendor: D. Appleton Company, Inc.

DADS Operating system enhancement used by Systems Programming. Increases system efficiency in IBM CICS systems. Stands for: Dynamic Allocation/Deallocation Subsystem. Vendor: On-Line Software International.

DAIS Application development tools. CORBA based. Used to create and run distributed applications. Users independently create components which can be linked and distributed throughout a network. Runs on Unix, Windows systems. Vendor: ICL, Inc.

Dais Com²CORBA Application development tool. Provides a bridge between COM and CORBA technologies. This allows applications to access both Unix and Windows systems. Vendor: ICL Inc.

DAISys CASE product. Automates analysis, design, and programming functions. Includes code generator for C, COBOL. Interfaces with DB2, VSAM, CICS. Runs on Pentium type desktop systems. Full name: Developer's Assistant for Information Systems. Vendor: S/Cubed, Inc.

DAL Computer language based on SQL that allows Macintosh users to access database information on large computer systems. Stands for: Data Access Language. Vendor: Apple Computer, Inc.

DAP Communications protocol. Proprietary applications layer protocol used with DECnet.

DAP Technologies Computer vendor. Manufactures handheld computers.

Daprex Software vendor. Products: financial, payroll HR, sales systems for AS/400 environments.

Darwin 1. Application development tool. Data mining software. Originally created by Thinking Machines Corp. Acquired by Oracle Corp in 1999.
2. Midrange computer. UltraSPARC processor. Operating system: Solaris. Vendor: Sun Microsystems, Inc.

DASCOMP/B,M Communications software. Transaction processing monitor. Runs on IBM systems. Increases speed of data transmission. Vendor: Information Management Software, Inc.

DASD Direct Access Storage Device. A data storage device that allows random access to data. Includes disks and drums.

DASD Advisor Operating system enhancement used by Systems Programming. Increases system efficiency in IBM systems. Detects bottlenecks. Vendor: Boole & Babbage, Inc.

DASD Inventory Operating system enhancement used by systems programmers. Increases system efficiency in IBM systems. Manages disk storage space. Vendor: Design Strategy Corp.

DASD manager for DB2 Operating system enhancement used by systems programmers. Increases system efficiency in IBM systems. Manages disk storage space. Vendor: BMC Software, Inc.

DASD/Plus Operating system enhancement used by systems programmers. Increases system efficiency in IBM systems. Manages disk storage space. Vendor: A+ Software, Inc.

DASDMON Operating system enhancement used by Systems Programming. Increases system efficiency in IBM systems. DASD tuning software. Vendor: Legent Corp.

DASDR Operating system add-on. Data management software that runs on IBM systems. Stands for: Direct Access Storage Dump Restore. Used by Systems Programming. Vendor: IBM Corp.

DASDtrak Operating system enhancement used by systems programmers. Increases system efficiency in IBM systems. Manages disk storage space. Vendor: Advanced Software Products Group, Inc.

Dasher II Midrange computer. Pentium CPU. Operating systems: Unix, Windows. Vendor: Data General Corp.

data Information.

DATA Application development tool. DSS. Runs on Windows systems. Vendor: TreeAge Software Inc.

data, database architect Developer, could be application or technical. Designs the database(s) and the interfaces between the database software and applications. Senior or mid-level title. Often interchanged with data or database analyst.

Data Accelerator 1. Operating system enhancement used by systems programmers. Improves I/O efficiency and reduces elapsed time between batch jobs. Runs on IBM MVS systems. Vendor: BMC Software, Inc.
2. See SmartBatch.

Data Access Language See DAL.

Data-Ace Relational database for desktop computer environments. Runs on IBM desktop systems. Vendor: Advanced American Technology, Inc.

Data Analyst CASE product. Part of Re-engineering Product Set. See Bachman/Re-engineering Product Set.

Data Base Management Database for midrange and desktop computer environments. Runs on DEC, IBM VM/CMS systems. Includes query facility and report generator. Vendor: SEED Software Corp.

Data Base-Plus Relational database for midrange and desktop computer environments. Runs on AT&T, DEC, DG, IBM, NCR, Xenix systems. Includes query facility, report generator, and utilities. Vendor: Tominy, Inc.

data center The hardware, software, and personnel used to support the company's data processing activity. Also called computer center.

data communications See communications.

Data Content Card Catalog Application development tool used with data warehousing. Used to do content searches of warehouses. Vendor: Pine Cone Systems, Inc.

data cube Application development design tool. A proprietary data structure used to store data for an OLAP end user data access and analysis tool.

Data Detective Application development tool. Provides data access across multiple data formats and media. Vendor: Data Discovery, Inc.

Data Desk Application development tool used for data mining. Provides graphs of different subsets or summaries of data. This allows user different perspectives of the data. Runs on Macintosh systems. Vendor: Data Description, Inc.

data dictionary A computerized glossary containing descriptions of data fields and their relationships. Some databases contain data dictionary facilities; some data dictionaries stand alone.

Data Dictionary/Solution Data repository, that runs on IBM's DB2 systems. Utilizes SQL and interfaces with QMF, IMS, and IEW. Vendor: Brownstone Solutions, Inc.

Data Dumper See Toolset-DB2.

Data Encryption Standard See DES.

Data Expediter Application development tool. Program that analyzes data descriptions to reduce redundancy and ensure standards are followed. Runs on IBM COBOL systems. Vendor: Data Administration, Inc.

Data Express Computer vendor. Manufactures desktop systems.

data flow diagram A tool used during the analysis and design phases of the project life cycle. A data flow diagram charts the movement of data within a system and shows both the manual and computerized processing performed on that data.

Data General Computer vendor. Manufactures large and desktop computers.

data geocoding Finding geographic and demographic patterns in data that can be used for processes such as targeting direct marketing campaigns, choosing proper retail outlets and appraise locations.

Data Interchange/2 Communications software. EDI package. Runs on IBM PS/2 systems. Vendor: IBM Corp.

Data Interpretation System See DIS.

Data Junction Application development tool. Converts data from different Windows environments. Vendor: Data Junction.

data library A dataset that contains sets of related data.

data link layer In data communications, the OSI layer that provides standards for such things as bandwidths and error control. HDLC, SDLC and DDCMP are data link protocols.

data locking Database technology. DBMS software provides data locking in order to allow multiple users to update the same database. Means protecting data being updated from access by any other person/program until the change is completed. Most common form of data locking is record locking in which only the record being changed is protected. Other possibilities are page locking and table locking.

Data Mail Communications software. EDI package. Runs on Pentium type PCs. Vendor: Advanced Communications Systems.

data management software Software that manages data by providing such functions as: compression of data files, backing up and archiving data, converting data for accessibility by other systems and/or computers, organizing and optimizing disk storage.

Data Manipulation Language See DML.

data marts A set of data designed and constructed for optimal decision support end user access. Data marts created from a data warehouse are called dependent or architected data marts while those created directly from legacy systems are called independent data marts.

Data Mart Solution Application development tool. Used to build data marts. Copies operational data from mainframe DB2 and Unix systems to Windows NT systems. Provides query capabilities. Vendor: Sagent Technology, Inc.

Data Mart Suites Application development tool. Rule-driven software used to load data warehouses. Vendor: Oracle Corp.

data mining Using large, unstructured datasets to analyze data. Automatically detects trends and associations. Often used to refer to using data from legacy systems for current management decisions. Querying data collections with no expectations of the results. Used with data warehousing and for extracting data from legacy system reports so it can be accessed by desktop query tools.

data modeler Application developer. Builds a model of the data relationships to be used for enterprise-wide applications development. Often involved in data warehousing. Senior level title.

data modeling A design technique used heavily in systems that need to integrate data from different databases. A data model is built that describes all the data and the relationships between data. The model also identifies redundant data and describes the processes that generate the data. The model is converted into physical database designs and interfaces. All maintenance is done through the model.

Data Packer/DB2, IMS, VSAM Operating system add-on. Data management software that runs on IBM systems. Vendor: BMC Software, Inc.

data propagation See replication.

data refresh The process of continuously updating the data warehouse's contents from its data sources. In addition, depending on the archive criteria the current data warehouse data will be archived for historical purposes.

data replication See replication.

data scrubbing Validating data for accuracy. Includes eliminating duplicates and inconsistencies. Often used with data warehousing, migrating legacy systems to newer technologies.

Data-Station CASE product. Automates analysis and design functions. Works with Oracle database. Vendor: Charles River Development.

data structure diagram Application design tool. Chart of the data types, linkages, access rules for all the data used in a specific system.

Data+Trac Relational database for desktop computer environments. Runs on IBM desktop systems. Vendor: Datanetics.

Data Transfer Services Application development tool. Downloads mainframe SAS data to desktop computers. Vendor: SAS Institute, Inc.

Data Transport Computer Massively parallel processor. Operating systems: Unix. Vendor: Wavetracer, Inc.

Data Usage Tracker Application development tool used in data warehousing. Monitors and analyzes growth and usage of warehouses. Vendor: Pine Cone Systems.

Data Vault See CICS Integrity Series.

data visualization Application development function used in data design. The process of displaying data in a graphical form (i.e. pie charts, scatter charts, bar graphs, etc.) to ease analysis.

data warehouse Enterprise-wide data access that interfaces current and legacy data. Data can be moved to a common data collection and integrated into a consistent format or can be replicated from one system to another to provide common access. Common usage is to take data from a production database and load it into an end-user database. Warehousing then includes the query and reporting tools to access the data.

data warehouse architect Database architect that works with data warehouses. See database architect.

Data Warehouse *Plus* Data warehouse framework. Strategy and product selection intended to offer smooth integration between the warehouse and the operational systems. Vendor: IBM Corp.

Data Warehousing Initiative Data warehouse framework. Strategy and product selection intended to offer smooth integration between the warehouse and the operational systems. Vendor: SAS Institute, Inc.

DATA Workbench See SQL Toolset. Includes VQL.

Data-XPERT Application development tool. ISPF-like facility. Runs on IBM mainframe systems. Vendor: Compuware Corp.

DataArchitect See PowerDesigner.

DataAtlas Application development tool. Data design and modeling tool for relational databases. Part of VisualAGE. Vendor: IBM Corp.

databank A public database to which companies or individuals can subscribe containing general-interest information. An example of a databank would be a collection of Stock Exchange information such as stock prices, trading volume, and high and low daily prices.

database A total collection of data with limited redundancy. Databases have amultiple groupings of fields in each record.

Database Relational database for large computer environments. Runs on DEC systems. Interfaces to INGRES. Vendor: HOK Computer Service Corp.

database/4GL A database that has a 4GL associated with it.

database administrator See DBA.

Database Administrator CASE product. Part of Re-engineering Product Set. See Bachman/Re-engineering Product Set.

Database Analyzer (for Oracle) Application development tool. Application development tool. Manages configuration and change to synchronize development activities under MVS. Vendor: PLATINUM Technology, Inc.

Database Designer Application development tool. Used to build and migrate databases. Provides graphical modeling and reverse engineering functions. Supports Oracle, DB2 SQLServer and any ODBC compliant database. Runs on Windows systems. Released: 1996. Vendor: Oracle Corp.

database management system. See DBMS.

database middleware Communications, middleware. Software that allows client systems to request data from one or more databases through a common access API (Application Programming Interface). This is the most basic form of middleware.

Database Mining Application development tool. Data mining tool. Runs on Unix, Windows NT. Vendor: Ascent Technology, Inc.

DataBase Mining Marksman Application development tool used for data mining. Used to determine which of hundreds of realtionships among variables are significant. Runs on Windows NT systems. Vendor: HNC Software Inc.

database server The software that runs in the host computer of a distributed processing system, or in the server computer of a client/server system that holds and manages the database. Provides data retrieval, storage, protection, and security functions.

DataBasic Application development tool. Creates stored procedures for IBM DB2 databases. Procedures can be invoked through Visual Basic. Runs on AIX, OS/2 systems. Vendor: IBM Corp.

Databasic II IMS database utility. Provides automatic creation of test databases. Vendor: Consumer Systems Corp.

DataBench Application development tool. Graphical point-and-click interface to query and manipulate data. Vendor: Minx Software, Inc.

DataBlade Object-oriented architecture used to integrate relational and object databases. Builds plug-in modules and stores them on the database server. Supports video, audio, text objects. Alternative to cartridge technology. Developed by Informix Software, Inc.

DataCentral Mainframe-class database server computer. Interfaces with IBM, DEC, Unix systems. Multiprocessing machine (includes up to 32 processors). Vendor: Unisys Corp.

Datacom/DB See CA-Datacom/DB.

Datadex/3000 Application development tool. Program that provides query facility and prototyping function for Image databases. Vendor: Dynamic Information Systems Corp.

Datadictionary See CA-Datacom/DB.

DataDirect Connect Communications. Middleware. OLE version provides connections to leading relational databases such as Oracle and Informix, and other sources including Lotus Notes. ODBC version contains drivers that leverage existing DBMS middleware to deliver ODBC-enabled client, web and server applications to more than 25 databases. Vendor: MERANT.

DataDirect Explorer, SmartData See Virtual Data Warehouse.

DataDirect SequeLink Communications. Database middleware. Connects multiple clients and platforms to multiple database servers. Developer's toolkit available. Versions available for ODBC, Java, OLE. Runs on Unix, Windows systems. Released: 1998. Vendor: MERANT.

DataDirector for Java Application development tool. Drag and drop tool used to build Java applets without writing code. Vendor: Informix Software, Inc.

DataDirector for Visual Basic Application development tool. Develops systems by prototyping. Generates code for all data access and data management functions. Vendor: Informix Software, Inc.

DataEase Application development environment. Includes menu system, report generator. Interfaces with SQL Server, Oracle, OS/2 EE database manager, DB2. Associated products: DataEase Developer, DataEase Personal. Runs on Pentium type desktop systems. Vendor: DataEase International, Inc.

DataEase Connect Communications software. Network connecting IBM systems. Micro-to-mainframe link. Vendor: DataEase International, Inc.

DataEase Express Visual end-user database. Integrates with DataEase. Runs on Windows systems. Vendor: DataEase International, Inc.

DataEase SQL Application development tool that works in a client/server architecture. Accesses SQL Server. Includes application and report generators. Vendor: DataEase International, Inc.

DataEdit Application development tool. Used for RAD development of forms-based client/server applications. Vendor: Brio Technology, Inc.

Datafast Database for desktop computer environments. Runs on Macintosh systems. Vendor: Statsoft, Inc.

Dataflex Application development system. Includes relational database/4GL for large computer environments. Runs on DEC, IBM, Unix systems. Vendor: Data Access Corp.

DataFountain Application development tool. Used to analyze and report on data over intranets. Includes Java applets for display functions. Runs on Unix, Windows NT. Vendor: Dimensional Insight, Inc.

DataGate Application development software. EAI product. Released: 1997. Vendor: Software Technologies Corp.

DataGuide Data management software. Information catalog tool that provides a table of contents to the information stored in a data warehouse. Manages the metadata. Runs on AIX, OS/2, OS/400, AIX, Windows NT systems. Vendor: IBM Corp.

DataHub Data management software. Administers multiple databases from a single point. Works with DB2 (all versions). Runs on OS/2, Unix systems. Unix version will also work with Oracle, Sybase. Vendor: IBM Corp.

Datainterchange/MVS,AS/400 Communications software. EDI package. Runs on IBM MVS, AS/400 systems. Vendor: Advantis.

DataJoiner Communications software. Middleware. Can access data from multiple databases in response to a single SQL query. Vendor: IBM Corp.

Datalan XA Communications software. Network operating system. Runs on Datapoint, IBM desktop systems. Vendor: Datapoint Corp.

DataLib Operating system add-on. Library management system that runs on DEC VAX, DG systems. Vendor: Centel Federal Services Corp.

DataMacs See CA-DataMacs.

Dataman Application development tool. DSS. Provides access to data stored in multidimensional databases. Vendor: SLP Infoware.

DataManager CASE product. See Manager family.

DATAMAT Relational database for midrange and desktop computer environments. Runs on DEC, IBM, Unix, Xenix systems. Includes DSS. Vendor: Transtime Technologies Corp.

DataMart Manager Application development tool used in data warehousing. Works with Intelligent Warehouse. Vendor: PLATINUM Technology, Inc.

Datamatch Database for midrange and desktop computer environments. Runs on Sperry, CDC, IBM desktop systems. Vendor: Inter Systems, Inc.

DataMerchant Communications. E-commerce application. Lets users build "storefronts," maintains customer registration, collects payments. Interfaces with multiple databases. Controls access to internal databases from the Web. Runs on Windows systems. Released: 1998. Vendor: Cognos Inc.

DataMind Application development tool used for data mining. Used to determine which of hundreds of realtionships among variables are significant. Runs on Unix, Windows NT systems. Vendor: DataMind Corp.

DataModeler Application development tool. Part of PowerTools. Supports Chen, Martin, data modeling methodologies. Vendor: ICONIX Software Engineering, Inc.

DataMover See NDM.

Datapac Communications network. Public packet switching network set up by the Canadian government. Uses X.75 protocols.

Datapacker/II Operating system enhancement used by Systems Programming. Increases system efficiency in IBM CICS systems. Vendor: H & M Systems Software, Inc.

Datapair 1000 Operating system enhancement used by Systems Programming. Increases system efficiency in HP systems. Protects against data loss from hardware malfunctions. Vendor: Hewlett-Packard Co.

DataPass Communications software. Network connecting IBM computers. Micro-to-mainframe link. Vendor: DTSS, Inc.

DataPerfect Database for desktop computer environments. Runs on IBM desktop computer systems. Vendor: WordPerfect Corp.

DataPrism Application development tool. Used to create ad hoc queries. Runs on IBM AS/400 systems. Used in data warehousing. Vendor: Brio Technology, Inc.

DataPropagator (Relational, Nonrelational) Data management software. Replicates (copies) data from its original database into a database on another platform. Works with IBM databases. Vendor: IBM Corp.

Datapulse See Group Four Datapulse.

Dataquery See CA-Dataquery(/PC,VAX).

DataRamp Communications software. Middleware. Provides access to relational databases over the Internet. Runs on Windows NT systems. Vendor: Working Set, Inc.

DataRight Data warehousing software. Prepares data for warehouses by standardizing data items such as names. Runs on VMS, Unix, Windows systems. Vendor: PostalSoft, Inc.

DATARUN Application development tool. Provides a project management framework which works with existing development tools. Runs on Windows systems. Released: 1996. Vendor: Silverrun Technologies Inc.

Datasage Application development tool used for data mining. Used to determine which of hundreds of realtionships among variables are significant. Runs on IBM mainframe systems. Vendor: Cirrus Recognition System.

Datascan Relational database for large computer environments. Runs on DG systems. Includes query facility and report generator. Vendor: Vista Computers, Inc.

DataServer Analyzer Database and query system designed for EIS activity. Uses Express query language to provide multiple views of information. Vendor: IRI Software.

dataset See file.

DataShare Operating system enhancement. Provides mainframe based backups of data on all systems. Vendor: Encore Computer Corporation.

DataShark Application development tool. Creates test data for applications using data stored in Oracle databases. Extracts live data from production systems based on developer supplied criteria. Released: 1998. Vendor: HardBall Software, Inc.

DataShopper Application development tool used in data warehousing. Used to access and analyze metadata in PLATINUM repositories. Runs on OS/2, Windows. Vendor: PLATINUM Technology, Inc.

DataStage Application development tool. Used to design and build data warehouses. Automates the extraction, transformation, integration and maintenance of data from multiple sources. Runs on Windows NT systems. Released: 1996. Vendor: Ardent software, Inc.

DataStar Software vendor. Products: Uniquest family of financial, payroll systems for midsize environments.

DATASTORE:lan,pro Relational database for desktop computer environments. Runs on IBM, HP desktop systems. Vendor: LanQuest Group.

Datatec-DS Application development tool. Programming utility that converts COBOL data definitions into standard data elements and record definitions. Re-engineering tool. Runs on IBM MVS systems. Vendor: XA Systems Corp.

Datatec-EA Application development tool. Analyzes production systems and stores information about such things as JCL and DDL statements in a relational repository. Re-engineering tool. Runs on IBM MVS systems. Vendor: XA Systems Corp.

Datatek Application development tool. Analyzes existing programs to determine what applications need restructuring to fit into a new system. Vendor: XA Systems Corp.

DataTracker Application development tool. Object-oriented, used to create data marts and sales and marketing applications. Includes data import, data mart creation and loading, and data access functions. Released: 1996. Vendor: Silvon Software, Inc.

Datatran Communications software. EDI package. Runs on Unix systems. Vendor: St. Paul Software.

DataTree Communications software. Network connecting IBM mainframe, Cray, DEC, and IBM desktop computers. Vendor: General Atomics.

Datatrieve Application development tool. Program that manipulates data from DEC files and databases. Includes query language and report generator. Vendor: Digital Equipment Corp.

DataVantage Operating system add-on. Debugging/testing software that runs on IBM IMS systems. Vendor: On-Line Software International.

DataView Application development tool. Navigation tool included with Visual C++ Enterprise Edition. Runs on Windows NT systems. Vendor: Microsoft Corp.

DataViews Application development tool. Graphic tool used to create 2D static or animated drawings of such things as instrument panels and control systems. Creates applications that monitor and control real-time processes. Includes DV-Draw (allows users to design graphics of the processes the applications will monitor) and DV-Tools (allows users to manage animated screens). Runs on Unix, Windows NT systems. Released: 1996. Vendor: DataViews Corp.

DataWeb See Empress DataWeb.

DataWindows Application development tool. Part of PowerBuilder. Programming object that lets developers manipulate data from relational databases without coding SQL statements. Vendor: PowerSoft Corp.

Datellite 300L, 400L Pen-based computer. Vendor: Microslate.

Dauphin 1050 Portable computer. Vendor: Dauphin Technology, Inc.

Dauphin DTR Pen-based computer. Vendor: Dauphin Technology, Inc.

Dauphin Pentop Pen-based computer. Vendor: Dauphin Technology, Inc.

Dauphin Technology Computer vendor. Manufactures portable and pen-based computers.

DaVinci E-mail Communications software. E-mail system. Vendor: DaVinci Systems.

DaynaNet Communications. LAN (Local Area Network) connecting up to 100 Apple Macintosh and/or IBM desktop computers. Vendor: Dayna Communications.

Daytona 1. Desktop computer. Pentium processor. Vendor: Racer Computer Corp. 2. See Windows NT.

Dazel Output Server Operating system software. Distributes print jobs to fax machines, e-mail, printers, or the Web. Includes MetaWeb which handles Internet e-mail. Runs on Unix, Windows NT systems. Vendor: Dazel Corp.

DAZEL SDK Application development tool. Used to provide network management functions such as configuration and queue management from client/server applications. Runs on Unix, Windows systems. Vendor: Dazel Corp.

DB/Analyzer See DBA Tool Kit.

DB/Assist Application development tool. Allows users to create and access SQL logic. Vendor: Easel Corp.

DB/Audit See DBA Tool Kit.

DB/AUDITOR Operating system enhancement used by Systems Programming. Monitors and controls system performance in IBM DB2 systems. Vendor: Systems Center, Inc.

DB:cBASE See CA-DB:cBASE

DB/Center Operating system enhancement used by systems programmers. Monitors and controls SQL/DS database performance in IBM VM systems. Vendor: VM Systems Group.

DB/CLIST Application development tool. Program that allows users to access DB2 through TSO CLISTS. Vendor: Updata Software Co.

DB/Compress See DBA Tool Kit.

DB/DASD for DB2 Operating system enhancement used by systems programmers. Increases system efficiency in IBM systems. Manages disk storage space. Vendor: Candle Database Tools Division.

DB-Delivery Database management tools. Provides change management services for IBM DB2 systems. Vendor: Goal Systems International, Inc.

DB Designer CASE product. Reverse-engineering. Used to migrate IMS, VSAM, RMS and IDMS data to Oracle. Runs on Unix systems. Vendor: Cadre Technologies, Inc. (now Cayenne Software).

DB/EDITOR Application development tool. Program that provides table editing in SQL/DS environments. Vendor: Systems Center, Inc.

DB/Enable Application development tool. Used to develop 3-tier, object-oriented applications. Allows traditional 2 tier client logic to be migrated into middle tier objects that can be shared among multiple clients. Provides interface between relational databases and object-oriented systems. Includes CORBA based adapters to access and retrieve many different data structures. Databases supported include Oracle, Sybase, Informix, SQL Server. Supports simultaneous connections to multiple dissimilar databases. Runs on Unix, Windows systems. Released: 1997. Vendor: Black & White Software, Inc.

DB Excel Application development tool. Program that lets users define and manage corporate data. Includes: Plan Manager, DDL Manager. Vendor: Reltech Products, Inc.

DB/Explain Operating system enhancement used by Systems Programming. Improves system efficiency and monitors and controls IBM DB2 systems. Vendor: Candle Corp's Database Tools Division.

DB Express Application development tool. Visualization tool that can color-code groups or clusters of data and display trends and patterns as graphs. Used for analysis type queries and with data warehousing. Vendor: Computer Concepts Corp.

DB Magic Relational database for midrange and desktop computer environments. Runs on HP, IBM systems. Utilizes SQL. Can be used in client/server computing. Vendor: Advanced MicroSolutions.

DB/Migrator for DB2 Operating system enhancement. Moves IBM DB2 applications from test to production environments. Vendor: Candle Corp.

DB/OPTIMIZER Operating system enhancement used by Systems Programming. Increases system efficiency in IBM DB2 systems. Vendor: Systems Center, Inc.

DB/ProEdit Application development tool. Program that works with DB2 databases. Vendor: Updata Software Co.

DB Publisher Application development tool. Query and report functions. Java based ad hoc query tool. Accesses multi-databases. Runs on multi-platforms. Vendor: Xense Technology Inc.

db-Query 4GL used in desktop computer environments. Vendor: Raima Corp.

DB/Quickcompare Application development tool. Utility that compares DB2 databases to identify differences so users can synchronize them. Vendor: Candle Corp.

DB/Reorg See DBA Tool Kit.

DB/REORGANIZER Operating system enhancement used by Systems Programming. Increases system efficiency in IBM SQL/DS systems. Vendor: Systems Center, Inc.

DB/Reporter Application development tool. Report generator for IBM systems. Accesses SQL/DS, DB2 databases. Vendor: Systems Center, Inc.

db-Revise Application development tool. Program that restructures desktop computer databases. Part of db-VISTA III. Vendor: Raima Corp.

DB/SECURE Operating system enhancement used by Systems Programming. Monitors and controls system performance in IBM DB2 systems. Vendor: Systems Center, Inc.

DB/SMU for DB2 Operating system enhancement. Provides online repair of IBM DB2 databases. Vendor: Candle Corp.

DB/Text Intranet Spider Application development tool used with intranets. Allows users to search on words or contents. Spider will search both HTML and non-HTML documents and return information such as personnel titles and URLs associated. Runs on Windows systems. Released: 1997. Vendor: Inmagic, Inc.

DB/Text WebPublisher Application development tool used to build Web sites. Allows users to add material to Web sites without using HTML. Vendor: Inmagic, Inc.

DB/TextWorks Application development tool. Database management system designed to manage text residing on PCs, networks. Builds an index for retrieval. Runs on Windows systems. Vendor: Inmagic, Inc.

db-UIM/X Application development environment. Visual programming tool. Users base systems on C or C++. Supports Informix, Oracle, Sybase. Interfaces with UIM/X. Vendor: Bluestone Consulting, Inc.

DB/Unix See CA-DB/Unix.

DB-Vision Applications development tool. Package that monitors and tunes Oracle databases. Full name: DB-Vision for Oracle. Vendor: Aston Brooke.

db-VISTA III Database for midrange and desktop computer environments. Runs on DEC, IBM, Macintosh, Unix, Xenix systems. Includes: db-Query, db-Revise. Utilizes SQL. Vendor: Raima Corp.

DB/Workbench for DB2 Operating system enhancement. Provides catalog and query functions for IBM DB2 systems. Vendor: Candle Corp.

DB2, DB2/400 Relational database for large computer environments. Runs on IBM systems. Stands for: Database 2. Utilizes SQL. Can be used in client/server computing. Vendor: IBM Corp.

DB2/2, DB2/6000 Relational database. DB2/2 runs on Pentium type desktops under OS/2, DB2/6000 runs on the RS/6000. Desktop version of DB2. Handles complex data types including video, voice, image and fingerprint data. Vendor: IBM Corp.

DB2 ACTIVITY MONITOR Operating system enhancement used by Systems Programming. Monitors and controls system performance in DB2 systems. Vendor: BMC Software, Inc.

DB2 ALTER Operating system enhancement used by Systems Programming. Increases system efficiency in IBM DB2 systems. Vendor: BMC Software, Inc.

DB2 AM Operating system enhancement used by Systems Programming. Monitors and controls system performance in IBM DB2 systems. Stands for: DB2 Activity Monitor. Vendor: BMC Software, Inc.

DB2/BATCH Operating system enhancement used by Systems Programming. Increases system efficiency in IBM DB2 systems. Runs DB2 programs in batch through JCL. Vendor: Relational Architects, Inc.

DB2 CATALOG MANAGER Operating system enhancement used by Systems Programming. Increases system efficiency in IBM DB2 systems. Vendor: BMC Software, Inc.

DB2-CICS Access Server Operating System enhancement used by systems programmers. Used to integrate DB2/CICS applications in client/server applications. Vendor: Micro Decisionware, Inc.

DB2 COPY PLUS Operating system enhancement used by Systems Programming. Increases system efficiency in DB2 systems. Reduces copy time and provides copies in one pass. Vendor: BMC Software, Inc.

DB2-DASD Operating system enhancement used by Systems Programming. Monitors DASD usage in DB2 systems. Vendor: CDB Software, Inc. Marketed by Candle Corp.

DB2 DASD MANAGER Operating system enhancement used by Systems Programming. Increases system efficiency in IBM DB2 systems. Vendor: BMC Software, Inc.

DB2-DBX Application development tool. Program that provides ISPF interface to DB2 databases. Vendor: IBS Corp.

DB2 Everywhere for Windows CE and PalmOS Version of DB2 to run on handheld systems. In development. Vendor: IBM Corp.

DB2 LOADPLUS Application development tool. Programming utility that improves speed of loading data into DB2 databases. Runs on IBM DB2 systems. Vendor: BMC Software.

DB2 MASTERMIND Operating system add-on. Data management software that runs on IBM DB2 systems. Includes: DB2 ALTER, DB2 CATALOG MANAGER, DB2 DASD MANAGER. Vendor: BMC Software, Inc.

DB2 NOMAD See NOMAD(2).

DB2 Parallel Edition Communications software. Transaction processing monitor. Runs on RS/6000, MPP systems. Uses parallel processing. Vendor: IBM Corp.

DB2 REORG PLUS Operating system add-on. Data management software that runs on IBM DB2 systems. Vendor: BMC Software, Inc.

DB2 Universal Database See Universal Server.

DB2 Universal Database Satellite Edition Version of DB2 designed for large-scale laptops. In development. Vendor: IBM Corp.

DB2 UNLOAD PLUS Application development tool. Programming utility that increases speed of unloading data from DB2 databases. Offers selection capabilities. Runs on IBM DB2 systems. Vendor: BMC Software.

DB2-XPERT Operating system add-on. Data management software that runs on IBM DB2 systems. Vendor: XA Systems Corp.

DB2PM Operating system enhancement used by Systems Programming. Monitors and controls system performance in IBM DB2 systems. Stands for: DB2 Performance Monitor. Vendor: IBM Corp.

DB2PRT Operating system enhancement used by Systems Programming. Monitors and controls system performance in IBM DB2 systems. Stands for: DB2 Performance Reporting Tool. Vendor: IBM Corp.

DB2SAM Operating system add-on. Add-on to popular security packages such as RACF and Top Secret. Works with DB2 databases. Vendor: Optima Software, Inc.

DBA Developer, usually technical. Administrates and controls the organization's database resources. Responsible for performance and tuning of the database. Companies have a DBA for each DBMS (i.e. an Oracle DBA). Responsible for security, backups, and accuracy of the data. Usually part of the technical staff, but some mainframe installations have DBAs in both the application and technical areas. Senior title. Stands for: DataBase Administrator.

DBA 1000 Relational database machine. A dedicated computer that contains the DBMS necessary to process the database. These databases are compatible with most mainframe operating systems and databases. Vendor: Accel Technologies, Inc.

DBA Advantage System software. Provides integrated database management and monitoring. Provides automated SQL reports, point-and-script generation and table data editing for DBAs, project managers and application developers. Vendor: PLATINUM Technology, Inc.

DBA Catalog Extract CASE product. Part of Re-engineering Product Set. See Bachman/Re-engineering Product Set. (now Cayenne Software).

DBA Enabler CASE product. Part of Re-engineering Product Set. See Bachman/Re-engineering Product Set. See Bachman/Re-engineering Product Set.

DBA Master Application development tool. Programming utility that monitors and reports on multiple databases from mixed platforms in a desktop environment. Runs on Windows systems. Vendor: Management Information Technology, Inc.

DBA Repository Services CASE product. Part of Re-engineering Product Set. See Bachman/Re-engineering Product Set.

DBA Tool Kit System that manages and controls IDMS environments. Includes: DB/Analyzer, DB/Audit, DB/Reorg, DML/Online, Formula/1, Journal Analyzer, Log Analyzer, Online Log Display, and Schema Mapper products. Vendor: DBMS, Inc.

DBA-XPERT Application development tools. Provide data management functions for Oracle databases. Includes Security-Xpert, Query-Xpert, Reorg-Xpert, Change-Xpert. Vendor: Compuware Corp.

DBAPort, DBA Prep Application development tools. Application generators. Automates conversion of data definitions and indexes between DB/2, DB2/2, Oracle, SQL Server, Ingres, Informix, XDB. Runs on Windows systems. Vendor: N Systems.

DBApp Developer, DBApp Publisher Application development tools. Used to connect Web servers with database applications. Released: 1998. Vendor: Wall Data, Inc.

dBase 5 Database for desktop computer environments. Runs on IBM desktop systems. Part of xBase. Vendor: Inprise Corp.

dBASE DIRECT Communications software. Allows dBase applications to communicate with mainframe applications. Vendor: Inprise Corp.

dBase Mac Database for desktop computer environments. Runs on Macintosh systems. Vendor: Inprise Corp.

dBase Mac RunTime Database for desktop computer environments. Runs on Macintosh systems. Execute-only version of dBase Mac. Includes dBase Mac. Vendor: Inprise Corp.

dbBRZ/xxx Operating system software. Backup and recovery system. Versions available for Oracle, R/3, Informix. Runs on Unix systems. Released: 1996. Vendor: SCH Technologies.

DBC Operating system enhancement used by Systems Programming. Monitors and controls system performance in IBM IMS systems. Stands for: Database Control. Vendor: Schumann Consulting Group, Inc.

DBC/1012 See Teradata.

DBDA Operating system add-on. Data management software that runs on IBM systems. Stands for: Data Base Design Aid. Vendor: IBM Corp.

DBExpert Relational database. End-user database. Supports dBase IV, DB2/2, Oracle, SQL. Runs on AS/400, OS/2, MVS systems. Vendor: Sundial Systems.

:DBEXPRESS Operating system add-on. Data management software that runs on HP systems. Works with Image databases. Vendor: Systems Express.

DBFRAME Artificial intelligence system. Runs on IBM, DEC, Unix systems. Extension of EXSYS expert shell system. Vendor: California Intelligence.

DBM 1. Operating system add-on. Data management software that runs on Bull HN systems. Stands for: DataBase Manager. Works with I-D-S/II databases. Includes DDL (Data Description Language) and DML (Data Manipulation Language). Vendor: Bull HN Information Systems, Inc.
2. Database for large computer environments. Runs on Harris systems. Includes report generator and utilities. Vendor: Harris Corp.

DBMAN V Database for Unix environments. Vendor: Versasoft Corp.

DBMGR Image database utility for HP Image systems. Vendor: Dynamic Information Systems Corp.

DBMS Database Management System. The software system that controls database creation and modification. DB2, Informix and Oracle are examples of DBMSs.

DBN 54xxA Notebook computer. Pentium CPU. Vendor: DTK Computer, Inc.

DBNet Application development tool used with data warehousing. Lets users access data marts over the Internet. Vendor: Silvon Software, Inc.

DBOMP Database management system used in manufacturing. Stands for: DOS Bill Of Materials Processor. Old system. Runs on IBM DOS/VSE systems. Vendor: IBM Corp.

DBPack Application development tool. Provides run-time testing tools for SQL performance. Part of Cyrano Suite. Vendor: Cyrano, Inc.

DBPAS Software that generates dBbase II and III databases and small Pascal programs. Part of Insight 2+.

DbPower Application development tool. Query and report functions. Creates 3D business graphics. Runs on Unix systems. Vendor: Db-Tech Inc.

dbProbe Application development tool. Summarizes database information into interactive graphs which users can analyze to see underlying trends. Creates GUI front-end to database information from anywhere on the intranet or Internet. Released: 1998. Vendor: InterNetivity, Inc.

dbProphet Application development tool used for data mining. Uses neural networks to discover and predict relationships in data. Runs on Windows NT systems. Vendor: Trajecta Inc.

DBQ Query language used with DEC databases. Includes: Callable DBQ (used by programs) and Interactive DBQ. Stands for: Database Query. Vendor: Digital Equipment Corp.

DBRAD Operating system add-on. Data management software that runs on IBM systems. Full name: DBRAD for MVS and VM. Used by DBA. Vendor: IBM Corp.

DBRC Operating system add-on. Data management software that runs on IBM systems. Stands for: IMS/VS Data Base Recovery Control Feature. Vendor: IBM Corp.

:DBReport Application development tool. Report generator for HP systems. Works with Image/3000 databases. Vendor: Systems Express.

dBRun III Plus Database for desktop computer environments. Runs on IBM desktop systems. Execute-only version of dBase III+. Vendor: Ashton-Tate Corp.

DBS/IRX Database for large computer environments. Runs on NCR systems. Vendor: NCR Corp.

DBtools.h++ Application development tool used in object-oriented development. Encapsulates relational data to be used as C++ objects. Object-oriented library interface to Ingres, Informix, Oracle, Sybase, Microsoft SQL Server databases. Vendor: Rogue Wave Software, Inc.

DBTools.h++ XA Application development tool. Allows developers to create applications that simultaneously access multiple, heterogeneous databases, and ensure data integrity. Provides support for Encina TPM, giving you cross-database recoverability and distributed transactions. Runs with DBTools.h++. Vendor: Rogue Wave Software, Inc.

:DBTrans Application development tool. Program that transfers data between KSAM files, Image databases. Runs on HP systems. Vendor: Systems Express.

DBVision Application development tool. Monitors and manage performance of Oracle and Sybase databases in distributed Unix environments. Vendor: PLATINUM Technology, Inc.

dBXL Relational database for desktop computer environments. Runs on IBM desktop systems. dBase lookalike. Vendor: Wordtech Systems, Inc.

DC-OSx Operating system for desktop computers. Unix-type system. Vendor: Siemens Pyramid Information Systems, Inc.

DCAS/3000 Operating system enhancement used by Systems Programming. Provides system cost accounting for HP systems. Vendor: Robert-Andrew Associates, Inc.

DCAS Operating system enhancement used by Systems Programming. Provides system cost accounting for HP systems. Vendor: Unison Software.

DCATS Operating system add-on. Debugging/testing software that runs on IBM systems. Stands for: Data Communication Applications Test System. Vendor: Systems Design & Development Corp.

DCC Operating System enhancement. Increased system efficiency in IBM DB2 systems. Manages the entire DB2 environment. Stands for: Database Control Center. Includes: DCC-Assist, DCC-Compact, DCC-SQL. Vendor: Tone Software Corp.

DCC-Compact Operating system enhancement. Compresses data in DB2 databases. Runs on MVS systems. Vendor: Tone Software Corp.

DCD II, III Application development tool. Program that generates narrative documentation of programs and systems. Runs on IBM systems. Stands for: Data Correlation and Documentation System. Vendor: MARBLE Computer, Inc.

DCE Distributed Computing Environment. Middleware. A set of technologies used in distributed processing. Includes naming services, remote procedure calls, distributed file services and security functions. Part of OSF. Allows diverse systems to share resources and applications. Supports desktop computers, Unix, IBM mainframes, Dec VAX systems. Vendor: Open Software Foundation.

DCE Developers Kit Application development tool. Allows users to make existing applications portable across various platforms. Vendor: Transarc Corp.

DCF Application development tool. Programming utility that provides the interfaces necessary to automate data collection. Stands for: Data Control Facility. Vendor: NCR Corp.

DCL Data Control Language. Subset of SQL.

DCLE Application development tool. Data conversion tool. Runs on Unix, Windows systems. Pronounced "diesel." Vendor: Reliant Data Systems, Inc.

DCM Operating system enhancement used by Systems Programming. Monitors and controls system performance in IBM systems. Vendor: Allen Systems Group, Inc.

DCM/I Operating system enhancement used by Systems Programming. Monitors and controls system performance in IBM systems. Stands for Data Center Manager. Vendor: Systemware, Inc.

DCM/Pak Operating system enhancement used by Systems Programming and Operations staff. Job scheduler for Hewlett-Packard systems. Vendor: Unison Software.

DCOM Object-oriented middleware technology. Connects Windows clients and various servers including Unix and mainframes. Similar to CORBA, but clients must be Windows. The part of COM that handles the server and databases. Stands for: Distributed Component Object Model. Vendor: Microsoft Corp.

DCP/Exec Operating system for IBM and compatible desktop computer systems. Vendor: Emulex Corp.

DDCMP Communications protocol. Data Link protocol used in DEC's DNA networks. Stands for: Digital Communications Message Protocol. Vendor: Digital Equipment Corp.

DDCS/2 Communications software. Data manager that runs on IBM desktops with DB2/2 and allows access to DB/2, SQL/DS and OS/400 databases. Stands for: Distributed Database Connection Services/2. Vendor: IBM Corp.

DDE Dynamic Data Exchange. Automatically moving data from one application to another in a windows environment. For example, DDE will update a word processing document when a number in a spreadsheet is changed.

DDFF Operating system add-on. Data management software that runs on IBM systems. Stands for: Distributed Disk File Facility. Vendor: IBM Corp.

DDL Data Definition Language. The language used to define data in a database. Used by both IBM (in DB2 systems) and DEC. Subset of SQL.

DDL Manager Application development tool. Program that supports data definitions in DB2 environments. Part of DB Excel. Vendor: Reltech Products, Inc.

DDM Operating system enhancement used by systems programmers. Increases system efficiency in DEC VAX systems. Manages disk storage space. Stands for: Distributed Disk Manager. Vendor: Demax Software, Inc.

DDP Communications protocol. Network layer. Part of AppleTalk. Stands for: Datagram Delivery Protocol.

DDT Operating system add-on. Debugging/testing software that runs on DEC systems. Used with DIBOL; stands for: DIBOL Debugging Technique. Vendor: Digital Equipment Corp.

DE-Light Communications software. Connects Internet users to DCE networks. Based on Java. Runs on a Web server and downloads applets written in Java to Web browsers. Vendor: Transarc Corp.

Debug Operating system add-on. Debugging/testing software that runs on Unix systems. Vendor: Language Processors, Inc.

Debug/32 Operating system add-on. Debugging/testing software that runs on Concurrent systems. Vendor: Concurrent Computer Corp.

debugging/testing software A computer program that automates either a debugging or testing activity such as: creating test data; producing dump summaries; monitoring program execution.

DEC Computer vendor. Manufactures both large and desktop computers under the generic name VAX. Most systems use the VMS operating system. DEC stands for: Digital Computer Equipment Corporation and was purchased by Compaq in 1998.

DEC 10000 AXP Mainframe computer. Operating systems: OpenVMS, OSF/1. Vendor: Digital Equipment Corp.

DEC/EDI Communications software. EDI package. Runs on DEC VAX systems. Part of NAS. Vendor: Digital Equipment Corp.

DEC Fuse Application development tool. Automates programming function. Provides integrated group of tools to be used in debugging, coding, and testing. Vendor: Digital Equipment Corp.

DEC Object/DB Object-oriented database. Links to Rdb. Vendor: Digital Equipment Corp.

DEC SNA 3270 Communications software. A library of routines that allows users to build IBM applications on DEC VAX systems. Vendor: Digital Equipment Corp.

DeCaf Application development tool. Used to protect Java programs. Encrypts Java code so users can't copy it. Vendor: KRDL.

DECdecision A DECwindows application that provides decision support services. Includes query functions, transforms data into graphic format for inclusion in compound documents, provides spreadsheet analysis and records, and automated repetitive tasks. Vendor: Digital Equipment Corp.

Decdesign CASE software. Automates analysis and design functions. Vendor: Digital Equipment Corp.

DECdtm Operating system enhancement used by Systems Programming. Provides support for online applications in DEC VMS systems. Stands for: DEC Distributed Transaction Monitor. Vendor: Digital Equipment Corp.

DECforms A set of software development tools and run-time environment for creating and implementing a forms-based (fill-in-the-blanks) user interface. Replaces VAX FMS and VAX TDMS.

DecideRight Application development tool. DSS. Runs on Windows systems. Vendor: Avantos Performance Systems Inc.

DECintact Communications software. Transaction processing monitor. Provides environment for application development. Includes programming and testing utilities. Includes terminal management, file management, network management, and security functions. Supports DECforms, Rdb, and RMS. Vendor: Digital Equipment Corp.

Decision Analyzer Application development tool. Query and report functions. Interfaces with ADABAS, DB2, IMS, VSAM. Runs on MVS, Windows systems. Vendor: Decision Technolog.

Decision DB Relational database. Runs on MP-2 MPP systems. Vendor: MasPar Computer Corp.

Decision Frontier Solution Suite Application development tools used in data warehousing. Includes: Informix Dynamic Server, MetaCube, DataStage and Seagate Crystal Info. Vendor: Informix Software, Inc.

Decision Series Application development tool used for data mining. Used to determine which of hundreds of realtionships among variables are significant. Runs on Unix systems. Vendor: NeoVista Solutions Inc.

Decision Support Suite See Pilot Decision Support Suite.

Decision Support System See DSS.

decision table Application design tool. A diagram that sets conditions and actions to be taken upon the combination of conditional happenings.

DecisionBase Application development tool used in data warehousing. Used to build and deploy data warehouses and data marts. Uses Platinum Repository to ensure that data defined in data marts is compatible with a full enterprise warehouse. Runs on windows NT systems. Released: 1998. Vendor: Computer Associates International, Inc.

DecisionEdge Application software. Lets users manage marketing campaigns and analyze future sales activity. Runs on AS/400, RS/6000 systems. Vendor: IBM Corp., BSG Corp.

Decisionhouse Applications software. Used to run behavioral analysis on customer databases for the banking, insurance and retail industries. Analyzes customer credit, customer responses, and retention traits. Released: 1999. Vendor: Quadstone, Ltd.

DecisionMaker Application development tool. DSS. Runs on Macintosh, Windows systems. Vendor: Palo Alto Software Inc.

DecisionMaster Application software. Decision support system designed specifically for retail industry. Includes retail data model. Includes application workbenches for Merchandising, Category Management, Store Operations, Marketing and Executive Management. Released: 1997. Vendor: Intrepid Systems.

DecisionPoint for Financials Data warehouse software. Integrates data stored in Oracle Financials with other internal or external data. Vendor: Sequent Computer Systems, Inc.

DecisionPro Application development tool. DSS. Provides risk analysis and forecasting functions. Runs on Windows systems. Vendor: Vanguard Software Corp.

DecisionView Application development tool. DSS. Provides access to data stored in multidimensional databases. Vendor: Knosys.

Decisions/Decisions Decision support software. Runs on MS-DOS systems. Vendor: Dalton Dialogic, Inc.

DecisionStream Application development tool. Used to build data warehouses and to conduct analysis on existing relational databases. Restructures operational data into formats suitable for general reporting and OLAP analysis. Supports both Star and Snowflake schemas. Runs on DEC, Unix and Windows NT servers with Windows clients. Acquired from Relational Matters in 1999. Vendor: Cognos Inc.

DecisionSuite (Server) Decision support software. An OLAP engine which provides multidimensional views of most major relational databases. Runs on Unix systems. Vendor: Information Advantage, Inc.

DECmail Communications software. E-mail system. Runs on DEC systems. Vendor: Digital Equipment Corp.

Decmcc Director Part of DEC's EMA (Enterprise Management Architecture). Program that manages equipment from other vendors. Vendor: Digital Equipment Corp.

DECMessageQ Communications software. Middleware. Provides high-level programming interface that handles the communications links between different applications. Originally written by DEC. Vendor: BEA Systems, Inc.

DECmpp 12000-xx Massively parallel processor. Supports up to 16,384 processors. Operating systems: Ultrix. Vendor: Digital Equipment Corp.

DECnet Communications software. Network connecting DEC computers. Includes: DECnet Router, DECnet-20, DECnet/SNA, DECnet/OSI. Supports OSI standards, TCP/IP and proprietary protocols. Vendor: Digital Equipment Corp.

DECprint Print utility that prints DEC's CDA documents. Vendor: Digital Equipment Corp.

DECquery, DECdecision Application development tool. Query and report functions. Generates SQL statements. Runs on VMS systems. Vendor: Touch Technologies, Inc.

DECrpc Set of tools for developing distributed applications that operate over a multi-vendor network. Is part of NCS. Stands for: DEC Remote Procedure Call. Vendor: Digital Equipment Corp.

Decscheduler for VMS Operating system enhancement used by Systems Programming and Operations staff. Job scheduler for DEC VMS systems. Vendor: Digital Equipment Corp.

DECspin Multimedia software. Runs on DECstation systems. Vendor: Digital Equipment Corp.

DECsystem 5xxx Midrange computer. RISC machine. Operating systems: Ultrix. Vendor: Digital Equipment Corp.

DECwindows Software that allows the development of windowing applications. Provides common user interface for different hardware and software platforms. Based on X-Windows. Available with Ultrix, VMS, and MS-DOS. Vendor: Digital Equipment Corp.

DECwrite Word processor for DEC's CDA documents. Vendor: Digital Equipment Corp.

DEF Operating system add-on. Security/auditing system that runs on IBM systems. Stands for: Data Encryption Facility. Vendor: Applied Software, Inc.

Defensor family Operating system software. Security package that includes client, server, gateway components as well as a mainframe toolkit. Works with diverse networks and geographic locations. Vendor: CyberSafe Corp.

Defrag Operating system add-on. Data management software that runs on DEC VAX systems. Vendor: H & E Concepts.

Deft CASE product. Automates analysis, design, programming and project management functions. Works with Ingres, Oracle, Rdb, Sybase, Informix, ADABAS, DB2, Non-stop SQL databases. Uses Gane-Sarson, Chen/Bachman, and Yourdon methodologies. Runs on DEC VAX, Macintosh systems. Vendor: Sybase, Inc.

Dell Computer vendor. Manufactures desktop computers.

Delphi Application development environment. Used for RAD development of departmental client/server systems. Uses visual design and debugging tools. Includes code optimizing. Interfaces with dBase, Oracle, Sybase, SQL Server, Informix, InterBase. Smaller version available that interfaces with Paradox, dBase. Runs on Windows systems. Supports both COM and CORBA. Vendor: Borland Tools, division of Inprise Corp.

Delta IMS Operating system enhancement used by systems programmers. Used to make changes to IMS systems without disrupting operations. Vendor: BMC Software, Inc.

Delta Solutions Application development tool. DSS. Provides access to data stored in multidimensional databases. Vendor: MIS AG.

Deltamon Operating system enhancement used by Systems Programming. Monitors and controls system performance in IBM systems. Vendor: Candle Corp.

Deluxe Multimedia Desktop computer. Pentium CPU. Operating system: Windows 95/98. Vendor: Compu-tek International.

DeMarco Structured programming design methodology named for its developer. DeMarco worked with Yourdon and developed a methodology often referred to as Yourdon-DeMarco methodology. Based on functional decomposition. Accepted as a standard design methodology by some companies, and used by some CASE products.

DEN Communications. Specification used to simplify administration of complex networks. Stands for: Directory Enabled Networks. Proposed by Microsoft and Cisco Systems in 1998.

departmental development The development of small applications that support from one to a few users, usually within the same department. Also called workgroup development. Projects concentrate on queries and reporting functions and are usually GUI/SQL based. Contrast with enterprise development.

Depot/J Application development tool. Allows developers to combine business rules and fields from mainframe and Unix databases and files into objects that can be linked to Java applets. Users could view and update the data as if it came from a single database. Based on Odapter. Runs on HP systems. Vendor: Hewlett-Packard Co.

derived data Data that results from calculations or processing applied by the data warehouse to incoming source data.

DES Encryption methodology. 56-bit key, secret key. Current standard is Triple DES. Stands for: Data Encryption Standard.

Deschutes Computer chip. Fast chip for notebook and portable computers. Released: 1998. Vendor: Intel Corp.

DeScribe Desktop system software. Word processor. Runs on Windows systems. Vendor: DeScribe, Inc.

Descrypt/EDI(+) Communications software. EDI package. Runs on IBM, DEC systems. Vendor: Prime Factors, Inc.

Descrypt(+)/BS,FS,CS,MS Operating system add-on. Security/auditing system that runs on IBM, DEC VAX systems. Vendor: Prime Factors, Inc.

design A phase in the system development cycle. Determining how to solve a problem. Deciding how both data and programs should be structured.

Design/1 CASE product. See Foundation.

Design Aid CASE product. Automates analysis and design functions. Vendor: Nastec Corp.

Design Data Systems Software vendor. Products: SQL* family of financial, sales systems for midsize environments.

Design Machine CASE product. Automates design functions. Vendor: Ken Orr and Associates.

Design/Monitor Application development tool. Works with Lotus Notes. Tracks changes made to the design of Notes applications. Vendor: DSSI.

Design Recovery Re-engineering tool. Creates design diagrams from COBOL source code. Used to analyze and redesign code in existing mainframe systems. Runs on Pentium type desktop computers. See APS, Excelerator. Design Recovery. Vendor: INTERSOLV, Inc.

Design Recovery for Windows Application development tool. Analyzes COBOL programs and produces design information. Interfaces with Excelerator. Vendor: INTERSOLV, Inc.

DesignAid II CASE product. Automates analysis, design functions. Runs on Pentium type desktop computers. Vendor: CGI Systems, Inc.

Designer CASE product. Part of Re-engineering Product Set. See Bachman/Re-engineering Product Set.

Designer/2000 CASE product. Used to design and model applications. Provides application partitioning, business process reengineering support. Runs on RISC systems. Interfaces with Developer/2000. Vendor: Oracle Corp.

Designer Workbench Application development tool. Windows based GUI. Vendor: Unisys Corp.

DesignMachine Application development tool. Automates programming, testing, and project management functions. Vendor: Optima, Inc.

DesignManager CASE product. See Manager family.

DesignVision CASE product. Automates design function. Vendor: Optima, Inc.

Deskpro Desktop computer. Pentium CPU. Operating systems: Windows 95/98. Vendor: Compaq Computer Corp.

Deskstation Computer vendor. Manufactures desktop computers.

DeskMan/2 Operating system enhancement. Controls OS/2 desktops and allows users to customize, enhance, protect OS/2 Workplace Shell objects. Runs on OS/2 systems. Development Technologies.

desktop computer A computer which requires no special environment or controls. Commonly used to refer to single user computers. Desktop computers include all of the following: handheld computers, laptops, micros, microcomputers, PCs, PDAs, portables, RISC machines, workstations.

Desktop Conferencing Groupware. Conferencing software. Runs on NetWare systems. Vendor: Fujitsu Industry Networks, Inc.

Desktop DBA System software. Multiserver data administration and migration tool for Oracle, Sybase, SQL Server and Informix databases. Vendor: PLATINUM Technology, Inc.

Desktop DNS Operating system software. Allows desktop managers to configure software and settings when moving or upgrading desktops thus retaining the look and feel of the old machines. Released: 1999. Vendor: Miramar systems, Inc.

Desktop Management Interface See DMI.

Desktop Management Task Force See DMTF.

desktop publishing The use of personal computers to prepare high-quality documents comparable to typeset documents.

DeskWorks Spreadsheet package that runs on AS/400 systems. Vendor: Generic Software, Inc.

Desqik Operating system add-on. Security/auditing system that runs on IBM systems. Vendor: Computation Planning, Inc.

Desqview/X Application development tool. Graphic user interface for MS-DOS systems. Converts DOS and Windows programs into X based programs. Vendor: Quarterdeck Office Systems.

Destination Desktop computer. Pentium II CPU. Operating system: Windows 95/98. Vendor: Gateway 2000, Inc.

Destiny 1. Database/4GL for large computer environments. Runs on DEC systems. Vendor: Software Engineering of America.
2. Operating System for desktop computers. Unix type system. Utilizes either Motif or Open Look GUI. Vendor: Unix System Laboratories, Inc.
3. Document management system. Automates distribution of reports. Users can get a print copy, browse an Internet page, or check e-mail or fax. Released: 1998. Vendor: Tivoli Systems, Inc.

DevCenter Application development tools used for testing and project management. Automates the detection and tracking of program errors and provides managers with information about the project's status automatically. Integrates with QACenter and Microsoft's Visual Studio. Includes: CodeReview, SmartCheck, BoundsChecker, Jcheck, TrueTime, TrueCoverage, FailSafe, TrackRecord, ActiveLink. Released: 1999. Vendor: NuMega Technologies, division of Compuware.

Developer CASE product. Automates design function. Works with MDBS, MDBS III, DB2 databases. Vendor: DataEase International, Inc.

Developer/2000 Applications development environment. Upgrade to Oracle CDE. Supports enterprise-wide development. Includes Web interface. Used to develop client/server systems. Addresses application partitioning by splitting application into client/server parts through drag and drop features. Interfaces with Oracle databases, Designer/2000. Generates Java code. Released: 1996. Vendor: Oracle Corp.

Developer Connection Application development tool. Consolidates existing development programs. Vendor: Sun Microsystems, Inc.

Developer's Toolkit Application development tool. Builds complete reporting applications. Contains its own data manipulation language. Part of Focus Six for Windows. Vendor: Information Builders, Inc.

Developer Studio See Pro*IV*.

Developers Suite See zApp.

DevelopMate Application development tool. Used to build enterprise models and prototypes of business functions. Works during design phase of application development. Works with AD/Cycle. Runs on OS/2, MVS systems. Released: 1990. Vendor: IBM Corp

DevelopmentStudio Application development tool. Used to build Web-based, e-commerce and enterprise applications. Interfaces with Microsoft's Visual Studio. Part of Rational Suite. Runs on Unix, Windows systems. Released: 1999. Vendor: Rational Software Corp.

DevPartner See NuMega DevPartner xxx.

DEXAN Operating system enhancement used by Systems Programming. Monitors and controls system performance in IBM systems. Reports on system bottlenecks. Stands for: Degradation Exception Analyzer. Full name: DEXAN for MVS. Vendor: Candle Corp.

DFDP CASE product. Automates design function. Full name: Diagraphics for Data Processing. Vendor: ADPAC Corp.

DFDS Communications software. Transaction processing monitor. Runs on IBM systems. Stands for: Data Facility Device Support. Vendor: IBM Corp.

DFDSS Operating system enhancement used by Systems Programming. Increases system efficiency in IBM systems. Protects and recovers data stored on disks. Stands for: Data Facility/Data Set Services. Vendor: IBM Corp.

DFHSM Operating system enhancement used by Systems Programming. Monitors and controls system performance in IBM systems. Works with data files stored on disk. Stands for: Data Facility Hierarchical Storage Manager. Vendor: IBM Corp.

DFI 66xx Notebook computer. Pentium CPU. Vendor: DFI, Inc.

DFP Operating system add-on. Data management software that runs on IBM systems. Stands for: Data Facility Product. Vendor: IBM Corp.

DFR Operating system add-on. Data management software that runs on Unisys systems. Stands for: Data File Recovery System. Vendor: Unisys Corp.

DFSMS/VM Operating system enhancement used by systems programmers. Increases system efficiency in IBM systems. Manages disk storage space. Vendor: IBM Corp.

DFSORT Operating system add-on. Data management software that runs on IBM systems. Vendor: IBM Corp.

DG See Data General.

DG/'1 Communications software. Network connecting DG systems. Vendor: Data General Corp.

DG/L Compiler language. Application development language for DG systems. Vendor: Data General Corp.

DG/RDOS Operating system for Data General midrange and desktop computer systems. Vendor: Data General Corp.

DG/SQL Relational database for large computer environments. Utilizes SQL. Includes data dictionary, query tool. Runs on DG systems. Vendor: Data General Corp.

DG/UX Operating system for midrange Data General systems. Unix-type system. Vendor: Data General Corp.

DG/Vision Desktop computer. Pentium CPU. Operating systems: Windows 95/98. Vendor: Data General Corp.

DGEN/36 CASE product. Automates design function. Vendor: Iris Software.

Dharma Integrator Communications software. Middleware. Allows queries to multiple databases on different platforms. Runs on Windows systems. Vendor: Dharma Systems, Inc.

DHCP Communications protocol. Used in client/server systems to assign IP addresses to network devices. Can reassign IP addresses dynamically. Stands for: Dynamic Host Configuration Protocol.

DHTML Dynamic HTML. Extended version of HTML and JavaScript. Includes improved animation. Allows the content of a page to change each time the user clicks, drags or points to elements on the page.

DI+ Operating System enhancement used by systems programmers. Protects data integrity in IBM IMS systems. Stands for: Database Integrity Plus. Vendor: BMC Software, Inc.

DI-Atlantis Application development tool. Provides business analysis and reports on large volumes of data. Accesses most databases. Runs on AS/400, MVS, Unix systems. Released: 1997. Vendor: Dimensional Insight Inc.

Diagnostics Operating system add-on. Debugging/testing software that runs on Pentium type desktop computers. Vendor: Hewlett-Packard Co.

dialog In an online system, a series of related inquiries and responses between the individual and the program.

Dialog System Application development tool. Provides the user interfaces to be incorporated into COBOL programs to run on desktop computer systems. Supports multiple GUIs and character user interfaces. Full name: Micro Focus Dialog System. Vendor: Micro Focus, Inc.

Dialog System Professional Application development tool. GUI used with RAD (rapid application development). Allows developers to create GUIs for DOS/Windows and OS/2. Vendor: Micro Focus, Inc.

Diamond CM Application management tool. Provides software configuration management from a single point of control. Includes automatic documentation, version control, project and problem management, and software distribution functions. Runs on Unix, Windows systems. Released: 1998. Vendor: Diamond Optimum Systems, Inc.

DIBOL Compiler language used in large computer environments. Stands for: Digital Interactive Business Oriented Language. Runs on DEC systems. Vendor: Digital Equipment Corp.

Dictionary/3000 Operating system enhancement used by Systems Programming. Monitors and controls system performance in HP systems. Vendor: Hewlett-Packard Co.

DictionaryManager CASE product. See Manager family.

DIDOS Operating system for midrange Siemens Nixdorf systems. Vendor: Siemens Nixdorf Computer Corp.

Diffusion Information Delivery Server Communications. Internet software. Uses push technology to transmit information from Web sites and business applications to users via the Internet, e-mail, fax, pager, printer, etc. Concentrates on financial businesses. Runs on Solaris, Unix, Windows NT. Vendor: Diffusion, Inc.

Digging for Diamonds Application development tool used in data warehousing. Data Mining tool. Vendor: SAS Institute, Inc.

Digibase Data management system. Runs on DEC systems. Includes report generator and utilities. Vendor: Timeline, Inc.

digital certificates Security feature. Uses encryption technology. Can be used instead of passwords. Easier to use because the system manages and authenticates digital signatures.

Digital Communications Message Protocol See DDCMP.

digital computer General-purpose computer used in a business environment. Mainframes, midrange computers, and desktop computers are all digital computers.

Digital Nervous System Communications strategy of integrating applications to ensure that employees have access to the current, accurate data. Proposal from Microsoft Corp.

Digital Personal Workstation au series Desktop computer. Alpha CPU. Operating systems: Unix, Windows NT. Vendor: Digital Equipment Corp.

Digital Personal Workstation i,a series Desktop computer. Pentium II CPU. Operating systems: Windows NT Workstation. Vendor: Digital Equipment Corp.

digital signature Encrypted code that can be attached to e-mail or documents to verify the origin of the data. Also used to prove that documents haven't been altered in transit. Used in digital certificates.

Digital Subscriber Line See DSL.

digital wallet Internet software. Allows users to pay for purchases over the Web. Holds credit-card numbers, shipping addresses, phone numbers, etc. plus a password. The information is held by a financial institution and can be used in later purchases because it will be automatically entered into order fields across e-commerce sites. Uses digital certificates for security. Also called electronic wallet.

Digitalk Smalltalk See Smalltalk/V.

Dimension Desktop computer. Pentium CPU. Operating systems: Windows 95/98. Vendor: Dell Computer Corp.

Dimension 5 Application development tool. Program that builds menu-based systems. Vendor: VAXCalc Co.

dimension tables Application design tool used in data warehousing. A table used in a star schema to store descriptive, hierarchical and metric information about an aspect of the business that is used for an analytical perspective (i.e. time, product, and customer).

Dimons 3G Communications software. Object-oriented distributed software that allows users to get information on any object from anywhere in a network. Includes application programming interface. Called OverLord when first released for Sun workstations. Runs on Sun, NCR, Siemens/Nixdorf systems. Vendor: NetLabs, Inc.

DIOSS Application development tool. Used to build client/server interfaces. Stands for: Distributed Interface Object Server System. Runs on Unix systems. Released: 1995. Vendor: DIOSS Corp.

DIPOS Operating system for midrange Siemens Nixdorf systems. Vendor: Siemens Nixdorf Computer Corp.

Direct access storage device See DASD.

Direct:Mailbox Communications software. Uses a Unix server to hold EDI transmissions. Vendor: Sterling Software, Inc.

DirectAdmin System software. Directory management system which reads, updates and deletes objects in many directories including NDS, Windows NT, Exchange and Active Directory. Vendor: Entevo Corp.

DirectConnect Communications. Connects MVS, AS/400 and LANs. Vendor: Sybase, Inc.

Direction PC Desktop computer. Pentium II CPU. Operating system: Windows NT. Vendor: NEC Computer Systems.

DirectMigrate 2000 Operating system software used to migrate to Windows 2000. Creates a new hierarchy for objects and users. Released: 1999. Vendor: Entevo Corp.

Director Multimedia authoring tool. Mid-level, timeline-based tool. Applications will run under Windows, Macintosh. Vendor: Macromedia, Inc.

Directory Listing of locations of network items such as objects, files, devices and user IDs. Includes metadirectories, which keep track of all references to each entry, i.e. noting that "Jane Doe" is listed in an e-mail directory, an application directory, and a network operating system. Also includes dynamic directories, which are enhanced metadirectories which search all directories if the item asked for is not in the first directory listed.

DirectoryAnalyzer Operating system software used to migrate to Windows 2000. Troubleshoots the infrastructure of the migration. Released: 1999. Vendor: netPro Computing, Inc.

directory services System software. Software that tracks information stored in electronic directories. Part of e-mail services and often a function of network operating systems (NOS).

DIS Decision support system. Stands for: Data Interpretation System. Interfaces with Sybase, Oracle. Runs on MS-DOS, OS/2 systems. Vendor: Metaphor, Inc.

DISCOVER CASE tool. Used to develop C/C++, Unix systems. Vendor: Software Emancipation Technology, Inc.

Discover 3D Desktop computer. Pentium CPU. Operating system: Windows 9598. Vendor: ProGen Technology Inc.

Discover MPEG II Desktop computer. Pentium CPU. Operating system: Windows 95/98. Vendor: ProGen Technology Inc.

Discoverer Application development tool. Query tool used for multidimensional queries. Interfaces with Oracle Reports. Can be used with non-Oracle databases. Vendor: Oracle Corp.

//Discovery Application development tool. Suite of data mining tools. Vendor: HYPERparallel.

Discovery for Developers Application development tool. Visualization tool that can color-code groups or clusters of data and display trends and patterns as graphs. Used for analysis type queries and with data warehousing. Vendor: Visible Decisions, Inc.

Discovery Suite Groupware. Workflow software. Combines document management, imaging and workflow software. Vendor: FileNet Corp.

disk A storage device that allows random access.

Disk Compressor Operating system enhancement used by systems programmers. Increases system efficiency in IBM midrange systems. Manages disk storage space. Vendor: Pac, Inc.

disk imaging See cloning.

Disk-Pac Operating system add-on. Data management software that runs on DG systems. Vendor: Eagle Software, Inc.

Diskeeper Operating system enhancement used by systems programmers. Consolidates fragmented files on disk in VMS systems. Vendor: Executive Software, Inc.

diskette A flexible disk mainly used in desktop computer environments.

DISKIT/RSTS,2000 Operating system enhancement used by Systems Programming. Increases system efficiency in DEC VAX. Upgrades basic DEC utilities. Vendor: UIS.

DiskLock Operating system software. Provides security functions for desktop systems. Full name: Norton DiskLock Administrator. Vendor: Symantec Corporation.

Diskman Operating system enhancement used by systems programmers. Increases system efficiency in Wang systems. Manages disk storage space. Vendor: Software Extraordinaire, Inc.

DISOSS/VSE,370 Communications software. E-mail system. Runs on IBM systems. Stands for: Distributed Office Support/370. Stands for: Distributed Office Support. Vendor: IBM Corp.

Dispatch See CA-Dispatch.

Dispatch/3000 Operating system enhancement used by Systems Programming. Job scheduler for HP systems. Vendor: Computer Consultants & Service Center, Inc.

Dispatcher Operating system enhancement used by Systems Programming. Job scheduler for HP systems. Vendor: Operations Control Systems.

DISPF/VSE Application development tool. Program that works with applications in IBM VSE systems. Vendor: IBM Corp.

DisplayWrite/36,370 Word processor for IBM large computer computers. Vendor: IBM Corp.

DisplayWrite 4 Desktop system software. Word processor. Runs on IBM systems. Versions available for large computer systems. Vendor: IBM Corp.

DistribuLink Operating system software. Manages application software and data files. Runs on IBM MVS systems using OS/2 workstations. Interfaces with Network Navigator to move programs across LANs. Includes: Warehouse Manager, Profiler, Scheduler, InSure. Vendor: Legent Corp.

DistribuLink-Unix DistribuLink for Unix systems. See DistribuLink.

Distributed Computing Environment See DCE.

Distributed Disk Manager See DDM.

Distributed Management Environment See DME.

Distributed Object Maitre'd Knowledge management software. Provides access to document management systems over the Internet. Vendor: Altris Software, Inc.

Distributed Object Management Facility Application development tool. Tool kit for developing object-oriented systems. Vendor: Hewlett-Packard Co.

Distributed Object Transactions/Messaging Middleware. See DOT/XM.

distributed objects Part of client/server computing. One type of application partitioning. Requires knowledge of object-oriented programming and builds objects which can reside in either the client or server.

Distributed Objects Environment See DOE.

Distributed Office Support See DISSOS/370.

distributed processing The distribution of data and programs among any number of physical locations, each of which may have a different type of hardware. A distributed processing environment can have each location independently running applications and maintaining all necessary data, or it can be set up to have each location independently running applications against a central repository of data.

Distributed Smalltalk Application development environment. Extends Visual-Works. Used to develop peer-to-peer distributed object systems. Supports Unix, Windows, Windows NT, OS/2. Vendor: ParcPlace, Digitalk.

Distributed SmartMode Application development tool. Query analyzer. Interfaces with EDA/SQL. Vendor: Information Builders, Inc.

DISTRIX Operating system for Convergent systems.

DITTO Operating system add-on. Data management software that runs on IBM VSE systems. Vendor: IBM Corp.

Diver Application development tools used for business access and analysis. OLAP tools that provide access to large data volumes. Vendor: Dimensional Insight, Inc.

Diversi-DOS Operating system for Apple II desktop computer systems. Vendor: Dynacomp Inc.

DIY Communications. Software that analyzes network performance by condensing thousands of network statistics into trends and problem situations. Software to collect data from the network is downloaded from the Internet, then is uploaded to the DIY site for analysis whenever user desires. Stands for: Do-It-Yourself. Vendor: NetOps Corp.

DL/I Database language for IBM IMS environments. Full name: Data Language/I. DL/I actually is the language used to access IMS databases, but it's often used as a synonym for IMS. Vendor: IBM Corp.

DLL Application development tool. Library of executable routines or data that can be called by multiple applications. These routines are called while an application is running and do not have to be compiled with any application. If the library routine is changed, the change is then in effect for all applications that call it. DLLs are used in Windows operating systems, with some databases, and in client/server technology. Stands for: Dynamic Link Library.

DLM Operating system enhancement. Used with Unix systems to build Unix clusters. Stands for: Distributed Lock Manager.

DLSw Communications standard used to connect SNA systems with TCP/IP. Vendor: IBM Corp.

DLSw+ Data communications software. Used to connect SNA systems to TCP/IP systems. Is IBM's DLSw, but also builds a hierarchy of connections to handle networks with thousands of users. Vendor: Cisco Systems, Inc.

DLT Digital Linear Tape. Technology used for tape backups.

DM Relational database for large computer environments. Runs on DEC, CDC systems. Has text capabilities. Vendor: Information Dimensions, Inc.

DM/Manager Operating system software used to migrate to Windows 2000. Moves users from Windows NT domains to Active Directory domains. Released: 1999. Vendor: Fast Lane Technologies, Inc.

Dmart Data warehouse. Used to build data marts. Combines Esperant query tool and Adabas relational database. Vendor: Software AG Americas Inc.

DMCL Special-purpose language used to define the relationship between physical files and database areas. Used with DEC databases. Vendor: Digital Equipment Corp.

DME Distributed Management Environment. Standards established by OSF to provide the ability to manage multivendor networked systems via a broad range of network management applications and a common GUI, Motif.

DMF Communications software. Network connecting Wang systems. Stands for: Distributed Management Facility. Vendor: Wang Laboratories, Inc.

DMG/Net Communications software. Network connecting DEC systems. Vendor: Digital Management Group, Ltd.

DMI 1. Desktop Management Interface. Standard for networking desktop computers. Defines how various desktop systems make status and configuration information available to network management systems. Most vendors including IBM, Novell, DEC, Hewlett-Packard and others have committed to implementing DMI. 2. See NetShare.

DML Data Manipulation Language. Language used to access data in databases. Used for IDMS, DB2 and DEC databases. Subset of SQL.

DML/Online See DBA Tool Kit.

DMS 1. Operating system add-on. Debugging/testing software that runs on IBM systems. Full name Development Management System Debug II. Vendor: IBM Corp. 2. Program development tools for DEC systems. Stands for: Data Management Software. Vendor: Digital Equipment Corp.

DMS/CMS Communications software. Transaction processing monitor. Runs on IBM VM systems. Vendor: IBM Corp.

DMS/DX Application development environment. Runs on IBM VSE systems. Vendor: IBM Corp.

DMS II Data management system. Runs on Unisys systems. Provides data definition, security facilities, and interfaces to user languages. Vendor: Unisys Corp.

DMS-Micro,Plus Database for midrange and desktop computer environments. Runs on DEC, IBM systems. Includes report generator, word processor, inquiry processing, and text editing. Vendor: Campus America/POISE.

DMS/OS Data management system. Runs on IBM systems. Storage management system that can be operated by non-technical users. Stands for: DASD Management System. Vendor: Sterling Software (Systems Software Marketing Division).

DMTF Desktop Management Task Force. A standards body working with network management standards, including WBEM.

DMU Data management utility that maintains the CDD in DEC environments. Vendor: Digital Equipment Corp.

DNA 1. See Windows DNA 2. Digital Network Architecture. Set of protocols that controls all communication through DEC networks.

DNA-4 Relational database for large computer environments. Runs on DG systems. Vendor: Exact Systems and Programming Corp.

DNDS Communications. LAN (Local Area Network) connecting desktop computers. Vendor: Connections Telecommunications.

DNOS Operating system for Texas Instruments systems. Vendor: Texas Instruments, Inc.

DNOS DBMS Database for desktop computer environments. Runs on TI desktop computer systems. Vendor: Texas Instruments, Inc.

DNS Communications protocol. Applications layer protocol that deals with naming standards and control for the parts of a network. Used with TCP/IP, the Internet. Maps user friendly names to IP addresses. The user friendly name has two parts separated by an @. To the left of the sign is the user name and to the right is the domain name identifying the computer where the user has a mail account. Domain names must be registered, and typically companies, groups, schools, etc. have their own domain names. The rightmost part of the domain name contains broad identifiers: com (commercial), net (Internet organizers and providers), mil (military), edu (education), org (organizations), gov (government), and, outside the United States, two character country identifiers: uk (United Kingdom), fr (France), etc. Stands for: Domain Name Services.

DNS 300 Communications software. Network operating system for mainframe Bull HN systems. Vendor: Bull HN Information Systems, Inc.

Doc-Aid Operating system enhancement. Provides JCL maintenance for IBM systems. Vendor: Allen Systems Group, Inc.

Doc-To-Help Application development tool. Takes documentation created in Microsoft's Word and converts it to on-line help format. Vendor: Wextech Systems, Inc.

DocEXPRESS Application development tool. Document manager. Generates documents from repositories of CASE tools. Runs on Unix, Windows systems. Vendor: ATA, Inc.

docking Plugging a laptop into a workstation. Also see "hot docking."

DocRoute Application development tool. Integrates document management systems with workflow. Vendor: Action Technologies, Inc.

DOCS Operating system enhancement used by Operations staff and Systems Programming. Provides operator console support in IBM VSE systems. Stands for: Display Operator Console Support. Vendor: Smartech Systems, Inc.

DOCS Open Document management system. Versions available for investment, construction, medical, legal applications. Vendor: PC Docs, Inc.

Docu/Master Document retrieval system. Runs on MVS, Unix, VSE, Windows systems. Vendor: Document Systems, Inc.

DocuAnalyzer Application development tool used for data mining. Filters and summarizes data according to predefined statements. Runs on AS/400 systems. Released: 1996. Vendor: Mobius Management Systems, Inc.

document management system A form of groupware that allows multiple authors to work on the same document simultaneously.

Document Type Definition See DTD.

documentation Part of the development process. Textual description of a computer program, or system. Also refers to the manuals prepared for people who use the programs and/or systems. Documentation is often produced by technical writers.

Documents See Oracle Groupware.

DocumentDirect Document management system. Provides a single point of access to all enterprise records, documents and data, regardless of how they were created or where they are stored. Runs on OS/2, Windows systems. Released: 1995. Vendor: Mobius Management Systems, Inc.

Documentum Workflow software. Document management system. Object-oriented, client/server system. Runs on Windows, Macintosh, Unix systems. Vendor: Documentum, Inc.

DOE Application development environment with an object layer on top of a Unix operating system, a distribution method for objects, and an application development environment. Handles both development and run-time environments. Runs on Solaris systems. Includes NextStep/OpenStep development environment. Stands for: Distributed Objects Environment. Vendor: SunSoft, Inc.

domain In artificial intelligence, the area of knowledge covered by an expert system.

Domain/OS Operating system for midrange Apollo systems. Vendor: Hewlett-Packard Co.

Domain Migrator Operating system software used to migrate to Windows 2000. Moves and consolidates domains from Windows NT to Windows 2000. Included in Windows 2000. Released: 1999. Vendor: Mission Critical Software, Inc.

Domain Name Service See DNS.

Dome Application development system. Builds client/server applications where messages can be passed without knowing the physical destination of the message. Runs on DEC, Unix, Windows systems. Vendor: Suitesoftware.

DOMF Distributed Object Management Facility. Technology from Hewlett-Packard used to create interoperable applications.

Domino Combined Lotus Notes/Web server. Groupware, e-mail system. Runs on mainframe, midrange and desktop systems. Allows both Notes clients and Web browsers. Can access Notes applications and documents. Translates Notes documents to HTML and stores HTML documents. Used to build intranets. Supports IMAP, LDAP protocols. Vendor: Lotus Development Corp.

Domino.Action Application development tool. Communications, Internet software. Used to create Web sites. Used with Lotus Notes. Vendor: Lotus Development Corp.

Domino.Broadcast Application development tool. Communications, Internet software. Used to send text, programs, video to other Domino users. Used with Lotus Notes. Vendor: Lotus Development Corp.

Domino.Connect Application development tools. Server based tools that link Domino to legacy systems including databases, online systems, and business applications. Domino provides a Web link. Interfaces with DB2, Oracle, ODBC. Vendor: Lotus Development Corp.

Domino.Doc Document management software. Allows documents to be spread over several locations. Documents to be viewed from Web browsers, Notes clients. Vendor: Lotus Development Corp.

Domino Extended Search Application development tool. Allows users to search across Notes databases, relational databases and the Internet at the same time. Runs on Windows NT systems. Vendor: Lotus Development Corp.

Domino Go WebServer Communications. Web server. Runs on OS/2, S/390, Unix, Windows systems. Vendor: Lotus Development Corp.

Domino.Merchant Application development tool. Communications, Internet software. Used to build online commerce applications. Used with Lotus Notes. Vendor: Lotus Development Corp.

DOOS Object-oriented development methodology. Associated with Wifts-Brock et al.

DOS 1. Old desktop operating system. Versions of DOS are available; MS-DOS from Microsoft, PC -DOS from IBM, DOS from Novell. All provide the same basic functions, but are competitive products. DOS systems have been replaced by Windows operating systems.
2. Sometimes used to refer to VSE/SP, an IBM mainframe operating system that used to be named DOS.

DOS Abend-AID See Abend-AID.

DOS/VSE See VSE.

DOSTOPS 3.0 Communications software. Network operating system. Runs on Pentium type desktop computers. Vendor: Sitka Corp.

DOT/XM Communications software. Message oriented middleware. Used to create applications that span PC and mainframe systems. One client can access multiple back-end systems. Stands for: Distributed Object Transactions/Messaging Middleware. Vendor: Level 8 Systems, Inc.

Double Helix II Relational database for desktop computer environments. Runs on Macintosh systems. Includes DSS. Vendor: Odesta Corp.

Double Impact Desktop computer. Pentium CPU. Operating systems: Windows NT. Vendor: Mega Computer Corp.

DoubleBax Operating system software. Security backup system for NetWare. Vendor: Oneac Corp.

DoubleSpace Operating system utility. Compresses data for storage. Used in MS-DOS. Vendor: Microsoft Corp.

download Transferring data files from large computer systems to smaller ones. Possibilities include downloading from mainframe to midsize and/or PC, from a server to a PC, and more and more common, from the Internet to PC.

downtime The time a computer system is inoperable due to hardware problems or scheduled hardware maintenance.

DP Data Processing. Original term for computer departments and systems. Anything to do with the computer industry. Also called ADP, EDP, IS, IR, IT, MIS.

DP 6200 Desktop computer. Pentium Pro CPU. SMP multiprocessor. Vendor: Cubix Corp.

DP-AID-2001 Operating system enhancement used by Operations staff and Systems Programming. Disaster control package. Vendor: Target Marketing Group.

DP Manager Operating System enhancement used by systems programmers. Monitors and controls system performance in IBM MVS and VSE systems. Tracks hardware and software inventories, change requests and problem reports. Vendor: MacKinney Systems.

DPA Operating system enhancement used by Systems Programming. Provides system cost accounting for IBM systems. Full name: Data Processing Accounting for IMS/VS. Provides accounting information for IMS environments. Vendor: IBM Corp.

DPEX Operating system for midrange Siemens Nixdorf systems. Stands for: Distributed Processing Executive. Vendor: Siemens Nixdorf Computer Corp.

DPL Application development tool. DSS. Runs on Windows systems. Vendor: Applied Decision Analysis.

DPPX/370,370 SP Operating system for midrange IBM 9373, 9379, 8100 systems. Stands for: Distributed Processing Programming Executive/System Product. Vendor: IBM Corp.

DPPX Assembler Assembler language used in IBM 8100 systems.

DPS/7000 Midrange computer. RISC machine. IBM PowerPC CPU. Operating systems: AIX, BOX/X. Vendor: Bull HN Information Systems, Inc.

DPS 9000/xxx Midrange computer. Multiprocessor. Vendor: Bull HN Information Systems, Inc.

DPWa, DPWau Desktop computer. Alpha processor. Vendor: Compaq Computer Corp.

DPX/20 Midrange computer. PowerPC RISC CPU. Operating systems: Unix. Vendor: Bull HN Information Systems, Inc.

DR DOS Operating system for IBM and compatible desktop computer systems. Vendor: Digital Research, Inc.

drag and drop Phrase used with object linking and embedding. Lets users move embedded objects from one application to another.

DragonDictate Voice recognition. Runs on Windows systems, uses 16-bit sound cards, 8M of RAM. Vendor: Dragon Systems, Inc.

Draw Perfect Desktop system software. Graphics package. Interfaces with Lotus 1-2-3, Excel, Quattro, Plan Perfect. Runs on MS-DOS systems. Vendor: WordPerfect Corp.

DRDA Standards to allow data access from different databases. Stands for: Distributed Relational Data Architecture.

Dreamedia Station Desktop computer. Pentium processor. Vendor: ACE Computers.

DriverStudio Application development tool used for testing. Tool set used to develop, debug, tune and test device drivers for the Microsoft's Windows 2000 operating system. Released: 1999. Vendor: NuMega Technologies, division of Compuware.

DRP-EZ Operating system enhancement used by Operations staff and Systems Programming. Disaster control package. Stands for: Disaster Recovery Planning. Vendor: Strohl Systems.

DRS 1. Operating system enhancement used by Operations staff and Systems Programming. Disaster control package. IBM systems. Stands for: Disaster Recovery System. Vendor: MAI Basic Four, Mainframe Software Products Corp.
2. Operating system enhancement used by Systems Programming. Increases system efficiency in IBM online systems. Print spooler. Stands for: Dynamic Report System. Vendor: Levi, Ray & Shoup, Inc.

DRS/ Operating system add-on. Data management software that runs on IBM systems. Programs include: DRS/Batch, DRS/Recover, DRS/Update. Vendor: Integrity Solutions, Inc.

drum A storage device that allows random access.

Drumbeat Application development tool. Allows users to connect Web pages to ODBC accessible databases without coding. Uses drag and drop, menus and dialog boxes instead of writing code. Generates HTML, JavaScript code. Released: 1998. Vendor: Elemental Software.

DS/1000-IV Communications software. Transaction processing monitor. Runs on HP systems. Vendor: Hewlett-Packard Co.

DS/3000 Communications software. Network connecting HP computers. Vendor: Hewlett-Packard Co.

DS Standard Communications software. Management and migration programs used to upgrade versions of NetWare. Vendor: Preferred Systems, Inc.

DSA Database architecture designed to use parallel processing in Informix parallel databases. Used for data warehousing. Stands for: Dynamic Scalable Architecture. Vendor: Informix Software, Inc.

DSA/ Communications software. Network connecting Bull HN computers. Includes: DSA 300, DSA-NCF, DSA/SNA, DSA/6. Vendor: Bull HN Information Systems, Inc.

DSDM Consortium of users and vendors that sets guidelines for the management of RAD projects. Stands for: Dynamic Systems Development Method. UK based.

DSEE (III) Application development tool. Program that coordinates programming tasks in Unix systems. Stands for: Domain Software Engineering Environment. Vendor: Apollo Computer, Inc.

DSF Operating system enhancement used by Systems Programming. Monitors and controls system performance in IBM MVS systems. Stands for: Device Support Facilities. Handles disk operations. Vendor: IBM Corp.

DSL Digital Subscriber Line. Technology that uses copper telephone lines to transmit digital data without the need for repeater devices. Can use existing twisted pair wires so T1 lines don't have to be installed. Often used in e-commerce. Includes ADSL (Asymmetric) SDSL (Single line), HDSL (High data rate), VDSL (Very high speed), UDSL (Universal, for home use).

DSM Communications. Network manager that performs network printer/spooler management, security auditing, disk/file system management, and logical resource pooling Runs on DEC, Unix, Windows systems. Released: 1998. Stands for: Distributed Systems Manager. Vendor: Enlighten Software Solutions, Inc.

DSM-11 Operating system for DEC, MDB systems.

DSN/DS Communications software. Network connecting HP computers. Vendor: Hewlett-Packard Co.

DSN/MRJE Communications software. Transaction processing monitor. Runs on HP systems. Vendor: Hewlett-Packard Co.

DSNL/370 Communications software. Network connecting IBM computers. Used in banking environment. Used with MERVA/370. Vendor: IBM Corp.

DSOM Distributed SOM. Allows objects to be exchanged across platforms. See SOM.

DSR Database for midrange and desktop computer environments. Runs on IBM, Wang systems. Full name: Data Storage and Retrieval System. Vendor: Valuation Systems Co.

DSS Decision Support System. Software that provides tools for making operational management decisions. Used by analysts and managers, a DSS performs strategic planning, budgeting and forecasting, consolidated reporting, what-if analysis, goal-setting, and optimization on large volumes of data. OLAP software.

DSS Administrator Application development tool. Used in data warehousing. Manages and monitors warehouse applications. Uses DSS Server. Released: 1997. Vendor: MicroStrategy, Inc.

DSS Agent Application development tool. OLAP softwareused in data warehousing. Provides query and workflow automation functions. Vendor: MicroStrategy Inc.

DSS Architect Application development tool. Defines multidimensional models for relational databases. Used in data warehousing. Vendor: MicroStrategy, Inc.

DSS Executive EIS. Also used in data warehousing. Vendor: MicroStrategy, Inc.

DSS for OS/2 Operating system enhancement. Gives users access to enterprise data from OS/2 Warp. Uses Kerberos encryption. Stands for: Directory and Security Server. Released: 1996. Vendor: IBM Corp.

DSS Objects Application development tool. Allows developers to build customized interfaces with Visual Basic, PowerBuilder, C++, Excel. Vendor: Microstrategy, Inc.

DSS Server Application development tool. OLAP software. Data warehouse scheduler and administrator. Interfaces with relational databases. Vendor: MicroStrategy Inc.

DSS Web Application development tool. OLAP software that provides a Web interface to data in relational databases. Vendor: Microstrategy, Inc.

DSSI Operating system enhancement used by Systems Programming. Increases system efficiency in DEC VAX systems. Handles external storage devices. Stands for: Digital Storage Systems Interconnect. Vendor: Digital Equipment Corp.

DSX Operating system for midrange IBM systems. Stands for: Distributed Systems Executive. Vendor: IBM Corp.

DTA/ Operating system enhancements used by systems programmers. Runs on IBM VSE systems. Includes: DTA/Copy, DTA/COS, DTA/ICCF, DTA/LMON, DTA/Print, DTA/QCopy, DTA/RECOV, DTA/VTOC. Vendor: Davis, Thomas & Associates, Inc.

DTA/COBOL Operating system add-on. Debugging/testing software that runs on IBM systems. Vendor: Davis, Thomas & Associates, Inc.

DTD Document Type Definition A file that defines the format codes, called tags, embedded within it. Part of SGML, so also part of HTML and XML. DTDs provide an automated validation and exchange of information.

DTK Computer vendor. Manufactures desktop computers.

DTK Quin Desktop computer. Pentium CPU. Operating systems: Windows 95/98. Vendor: DTK Computer, Inc.

DTK Quin-38 Desktop computer. Pentium CPU. Vendor: DTK Computer, Inc.

DTN 5xxxx Notebook computer. Pentium CPU. Vendor: DTK Computer, Inc.

dtSearch Document retrieval system. Works with Web servers. Runs on Windows systems. Vendor: DT Software Inc.

Dual NT Enterprise Server Midsize computer. Server. Pentium Pro processor. Operating system: Windows NT. Vendor: Ace Computers.

dumb terminal A terminal without processing capabilities. Most mainframe terminals are dumb terminals.

dump 1. A report produced when an error occurs during program execution. The report contains a copy (or dump) of what the program looks like in the computer when the problem occurred. The report is used to debug the problem.
2. A printout of the contents of a tape or disk.

Dump-Edge Operating system add-on. Debugging/testing software that runs on IBM systems. Vendor: IBS Corp.

Dun & Bradstreet See D&B Software.

Duracom Computer vendor. Manufactures desktop computers.

Durango(II) Midrange computer. Alpha CPU. Vendor: Aspen Systems, Inc.

DV-Centro Application development tool used in RAD development. Used to develop complex graphical systems such as tracking telecommunications equipment or configuring factory systems. Runs on Windows systems. Released: 1996. Vendor: DataViews Corp.

DV-Draw, DV-Tools See DataViews.

DV-Proto Application development tool. Programming utility that enables application developers to evaluate and present an application without any coding. Vendor: V.I. Corp.

DVD Digital Versatile Disk. Successor to CD (Compact Disk). Records audio, video, data. Used in IT for distributing software, transporting files, creating backup and archival files. DVD-ROM holds read only data. There are two types of rewritable DVD technology: DVD-RAM and DVD+RW.

DX10 Operating system for Texas Instruments systems. Vendor: Texas Instruments, Inc.

DX10 DBMS Database for desktop computer environments. Runs on TI desktop systems. Vendor: Texas Instruments, Inc.

DX500 Open Directory Operating system software. Provides directory services. Follows LDAP protocols. Based on X.500 standards, but uses an SQL database to store names and associated information instead of files. Vendor: Datacraft Ltd.

DXT 4GL used in mainframe environments. IBM systems. Stands for: Data Extract. Works with DB2 databases. Vendor: IBM Corp.

DYL-260 Application development tool. Report generator for IBM systems. Vendor: Sterling Software (Dylakor Division).

DYL-270 Operating system add-on. Data management software that runs on IBM systems. Vendor: Sterling Software (Dylakor Division).

DYL-280 See VISION:results.

DYL-Audit See VISION:Audit.

DYL-Inquiry 4GL used in mainframe environments. Uses Natural language capability. Accesses IMS databases. Vendor: Sterling Software (Dylakor Division).

DYL-IQ Express(/OEM) 4GL used in mainframe environments. Accesses VSAM files and DB2 and IMS databases. Vendor: Sterling Software (Dylakor Division).

Dyl-link Communications software. Network connecting IBM systems. Vendor: Sterling Software (Dylakor Division).

DYL-Online TSO Communications software. Provides environment for application development. Includes programming and testing utilities. Used to develop DB2 applications. Vendor: Sterling Software (Dylakor Division).

DYL-Security See VISION:Security.

Dylan Object-oriented programming language.

DynaBase Application development tool used for Web publishing. Supports XML. Vendor: Inso Corp.

DYNAM See CA-DYNAM.

DYNAM/TLMS See CA-DYNAM/ TLMS.

Dynameasure Operating system software. Measures capacity and performance of Windows NT server networks. Released: 1996. Vendor: Bluecurve, Inc.

Dynamic 4GL Application development tool. Used to converts 4GL applications to graphics. Recompiles existing code to build Windows and X-Windows GUI front-ends to legacy applications. Released: 1998. Vendor: Informix Software, Inc.

dynamic data exchange See DDE.

Dynamic Data Store See Enterprise Data Mart.

dynamic directory See directory.

Dynamic HTML See DHTML.

Dynamic partitioning Part of client/server computing. One type of application partitioning. Partitions applications at run time.

Dynamic Scalable Architecture See DSA.

Dynamic Server See Informix Dynamic Server.

Dynamic SQL Tracker Application development tool. Speeds up loading and querying databases. Used in data warehousing. Runs on MVS systems. Vendor: Softbase Systems Inc.

DynamicCube Application development tool. DSS. Provides access to data stored in relational databases. Runs on Windows systems. Vendor: Data Dynamics, Ltd.

Dynamo 1. Application server providing both middleware and development functions. Used with Java software including servlets, JavaBeans, EJBs and CGI scripts. Integrates with legacy systems and relational databases. Released: 1997. Runs on Unix, Windows NT. Vendor: ATG Inc. (Art Technology Group).
2. Simulation language used to model business, social, economic, biological, and engineering systems. Runs on most systems. Vendor: Pugh-Roberts Associates, Inc.

Dynamo e/BSD Application development tool. Developer's Kit which includes a developer's license and source code for BSD/OS, including the kernel, networking stack, link layer, and file systems, as well as a complete development environment. Vendor: Berkeley Software Design, Inc.

Dynasty Application development environment. Used for enterprise development. Includes object-oriented functions. Generates C, SQL. Used for cross-platform development. Supports Windows, OS/2, Macintosh, Unix. Interfaces with Oracle, Sybase. Generates codes for specific platforms from one specification. Used to develop client/server applications. Runs on Macintosh, OS/2, Unix, Windows systems. Vendor: Dynasty Technologies, Inc.

DynaText Document retrieval system. Works with Web servers. Runs on Macintosh, Unix, Windows systems. Vendor: Electronic Book Technologies, Inc.

Dynax Resources Software vendor. Products: financial systems for AS/400 environments.

Dynix(/ptx) Operating system. Unix-type system. Vendor: Sequent Computer Systems, Inc.

(E)JES Operating system enhancement used by systems programmers. Manages spooling in an IBM MVS JES3 environment. Vendor: Phoenix Software International.

E.Advantage Suite Application software. E-commerce package that uses an Internet interface to financial management systems. Includes online order entry and access to accounting data. Released: 1998. Vendor: Accpac International, Inc.

e-COBOL Programming language used to build e-commerce applications. Allows developers to create e-commerce without knowledge of CGI / NSAPI / ISAPI APIs. Vendor: MERANT.

e-commerce Electronic commerce. Conducting financial transactions over the Web. EDI using Web browsers. Many vendors use DSL (Digital Subscriber Line) for e-commerce applications.

e-commerce developer Application developer. Develops systems to do business over the Internet. Requires Internet skills such as HTML, Java, CGI Script, etc. and business knowledge and skills. EDI knowledge often part of the job. Senior or mid-level title.

e-mail A communications system which concentrates on the creation and delivery of messages and even long documents from one terminal to another.

e-mail See CA-e-mail.

E-mail Communications software. E-mail system. Runs on Wang, Windows systems. Vendor: Integrated Custom Software.

E-Mon Operating system enhancement used by systems programmers. Controls communications networks in DEC VAX systems. Monitors Ethernet performance. Vendor: Bear Computer Systems, Inc.

E-Portal Suite EIP, corporate portal. Provides access to business information in unstructured documents such as word processing documents, Web pages, and other text based systems, as well as structured information in files and databases. Vendor: Viador, Inc.

e-Vantage Communications software. Allows users to use a Web front end to access mainframe systems. Funnels access to the host through web servers. Released: 1998. Vendor: Attachmate Corp.

E-Z Order Communications software. EDI package. Runs on Pentium type desktop computers. Vendor: CAN/AM TECH (USA), Inc.

E1000 200 PC, or micro computer. Pentium CPU. Operating system: Windows 98. Released: 1998. Vendor: Gateway 2000, Inc.

E100D, E133T Notebook computer. Pentium CPU. Vendor: Hitachi PC Corp.

E1CICS Application development tool. Program that converts Environ/1 applications to CICS. Vendor: Business Information Systems, Inc.

EA/400 Application development tool. Used to develop client/server systems with AS/400 servers and Windows clients. Stands for: Easy Access to the AS/400. Vendor: Remora Development Corp.

EA Server See Enterprise Application Server.

EAI Enterprise Application Integration. Part of ERP (Enterprise Resource Planning). Provides integration of products from different vendors and different applications. Includes middleware, messaging, and database software. Provides the infrastructure for ERP.

Eagle 1. Application development methodology. Used to develop client/server, object-oriented systems. Goal is to develop applications with 60% to 80% reusable components. Vendor: Andersen Consulting.
2. A database application language. Includes a report generator and data dictionary. Vendor: Migent, Inc.
3. Communications, Internet software. Firewall. Runs on Windows NT systems. Vendor: Raptor Systems, Inc.

EAS Application development tool. Internet software. Combines a graphical development tool with an e-mail server to create e-mail applications. Stands for: E-mail Application server. Released: 1999. Vendor: Delano Technology Corp.

Easel/(2,Win,DOS) Application development environment. Develops client/server applications. Runs on OS/2, Windows systems. Screen scraper, used to renovate existing systems with GUIs. Associated product: Easel Workbench. Vendor: Easel Corp.

Easel Workbench Application development tool that works in a client/server architecture. Builds client applications with a GUI-based front end. Interfaces with IBM, Oracle, Sybase, SQL Server. Vendor: Easel Corp.

Easy-Archive Operating system add-on. Data management software that runs on Bull HN systems. Vendor: Scientific and Business Systems, Inc.

Easy Base Relational database for desktop computer environments. Runs on IBM desktop computer systems. Vendor: Dac Software, Inc.

Easy-Base II Database for Unix environments. Vendor: Automated Systems.

Easy-Connect Communications software. EDI package. Runs on IBM AS/400, Windows systems. Vendor: EDI Support, Inc.

Easy*SQL Simplified version of SQL used with Oracle databases. Runs on Pentium type desktops, DEC VAX, AT&T, Unix systems. Vendor: Oracle Corp.

Easy Tools Operating system software. Works with OS/2 Warp, Windows. Includes security measures, NetFinity (network management software). Vendor: IBM Corp.

Easy Word II Desktop system software. Word processor. Runs on Pentium type systems. Vendor: Dac Software, Inc.

EasyCASE CASE product. Automates design and programming functions. Includes EasyCASE Professional and EasyCASE System Designer. Runs on Pentium type desktops. Vendor: Evergreen CASE Tools, Inc.

EasyDBG Application development tool. Debugging software for applications written in Pascal and Fortran. Runs on Pentium type desktops. Vendor: Silicon Valley Software.

EasyER/EasyOBJECT Application development tool. Data modeling tool used to build entity-relationship diagrams and class diagrams based on object-oriented methodologies. Supports desktop databases including Access, dBase, FoxPro, Paradox, SQLServer and Oracle. Interfaces with Visual Basic. Runs on Windows systems. Released: 1998. Vendor: Visible Systems Corp.

EasyHelp/Web Application development tool. HTML editor. Used to create HTML documents and Windows Help files. Vendor: Eon Solutions.

Easynet NOS/2 Plus Communications software. Network operating system. Runs on Pentium type desktop computers. Vendor: Lanmark Distributing Corp.

EasyPROCLIB See CA-EasyPROCLIB.

EasySAA Object-oriented CASE tool used to develop SAA compatible interfaces and applications. Vendor: Multi Soft, Inc.

EASYSPEC Application development tool. Program that assists in database design. Works with Ingres, Oracle, DB2, SQL, NOMAD2, IDMS databases. Vendor: Easyspec, Inc.

Easytalk 4GL. Designed for end-users to access Oracle databases. Vendor: Servio Logic Development Corp.

Easytrieve See CA-Easytrieve.

EasyWriter II Desktop system software. Word processor. Runs on Pentium type systems. Vendor: Computer Associates International, Inc.

EbaServiceMonitor Internet software. Collects data on activity at Web sites through a Java applet that travels with the Web page and reports back to the server. Eliminates the need for cookies. Released: 1999. Vendor: Candle Corp.

EBASIC Compiler language. Full name: Extended BASIC. Version of BASIC.

EBCDIC Character code system; the binary codes assigned to each character. Used in mainframe and some midrange systems. Stands for: Extended Binary Coded Decimal Interchange Code.

Ebusiness System Communications software. Internet product used with e-commerce. Provides customer service over the Web. Vendor: Silknet Software Inc.

ECCO Desktop software. Personal Information Manager (PIM). Vendor: Arabesque Software, Inc.

ECF Servers-Requesters Communications software. Network connecting IBM computers. Provides micro-to-mainframe link. Vendor: IBM Corp.

Echos Pxxx, Pro Notebook computer. Pentium CPU. Vendor: Olivetti Personal Computers USA, Inc.

Eclipse Code name for GroupWise XTD. See GroupWise (XTD).

Eclipse MV/xxxx Mainframe and midrange computers. Operating systems: AOS/xxx, DG/UX, DG/RDOS. Some models can be configured as the server in a client/server system. Networks: Portable Netware. Client systems can run DOS, OS/2, Unix, Macintosh. Vendor: Data General Corp.

ECMA PCTE European CASE standard designed to integrate application development tools so data integration and sharing can be established across different tools. Stands for: European Computer Manufacturer's Association Portable Common Tools Environment. Standard is gaining acceptance in the U.S.

ECML Electronic Commerce Modeling Language. Used in e-commerce. Provides a standard form for Web shoppers to fill out customer and credit information. The form is stored on the individual's PC as an icon and can be dragged to ECML compliant sites thus eliminating the need to key in any information.

ECO Application development tool. Program that manages change in program design and development. Stands for: Engineering Change Order. Vendor: Team One Systems, Inc.

EcoCHARGEBACK Operating system enhancement. Provides cost accounting for Unix systems. Vendor: Compuware Corp.

EcoCLIENT Operating system software. Monitors client systems in distributed processing. Vendor: Compuware Corporation.

ECON Communications software. EDI package. Runs on Unix systems. Stands for: Electronic Commerce Operational Network. Vendor: Lawrence Livermore National Laboratory.

EcoNet Communications. Network management software. Monitors traffic from each server and workstation on the network and shows where applications are being processed. Reports on usage by software name so administrators can see how software is being used. Runs on Windows systems. Vendor: Compuware Corp.

Ecop-Logical Operating system enhancement used by systems programmers. Disaster recovery system that provides backups and writes JCL to restore the environment when a problem occurs. Runs on IBM mainframe systems. Vendor: Advanced Software Products Group, Inc.

ECOS Operating system for midrange Harris systems.

EcoSCHEDULER Operating system software. Provides management functions for client/server environments. Lets users schedule batch jobs based on calendar events and also records status of jobs. Vendor: Compuware Corp.

EcoSCOPE Communications software. Network management tool. Tracks such things as what applications are running on the network, how many users are accessing the applications, etc. Vendor: Compuware Corp.

EcoSNAP Operating system enhancement. Application monitor and client/server fault management tool that strengthens the recovery of Oracle applications. Provides real-time detection of application failures, immediately notifies key personnel that a failure has occurred, and captures critical failure data to speed recovery from your application failure. Runs under Unix, Windows systems. Vendor: Compuware Corp.

EcoSYSTEMS Communications. Network management software. Includes Eco-TOOLS. Vendor: Compuware Corp.

EcoTOOLS Operating system software. Management system that monitors heterogeneous databases, operating systems, and network systems. Includes event management, performance monitoring, capacity planning, configuration management, and security functions. Can be used with Baan, PeopleSoft and SAP ERP systems. Released: 1997. Vendor: Compuware Corp.

ECXpert and **DeveloperXpert** See CommerceXpert.

EDA/Copy Manager Application development tool. Provides replication functions through scheduled copies of any EDA/SQL source to OS/2 databases. Part of EDA/SQL. Vendor: Information Builders, Inc.

EDA for Enterprise E-Commerce Communications. Extension to EDA middleware products. Transfers data among data warehouses. Vendor: Information Builders, Inc.

EDA Application development tool. Database middleware. Family of products that works in a client/server architecture to allow access to data stored in nonrelational files or databases. Runs on most systems and on all computer types. Stands for: Enterprise Data Access. Vendor: Information Builders, Inc.

EDE-PC Communications software. EDI package. Runs on Pentium type desktop computers. Vendor: American Business Computer.

Edgar Database of corporate financial statements. Maintained by the Securities and Exchange Commission (SEC), available over the Internet.

Edge xx Notebook computer. Pentium CPU. Vendor: Austin Computer Systems, Inc.

EDI 1. Electronic Data Interchange. Computer-to-computer exchange of commercial transactions between trading partners. 2. Operating system enhancement used by systems programmers. Increases system efficiency in IBM MVS systems. Protects stored data files. Stands for: Enhanced Data Set Integrity. Vendor: Duquesne Systems, Inc.

EDI/36,38,400 EDI/e Communications software. EDI package. Runs on S/36, S/38, RS/6000, AS/400 systems. Vendor: Premenos Corp.

EDI 400 Communications software. EDI package. Runs on AS/400 systems. Vendor: JBA International, Inc.

EDI analyst Application developer. Expert in EDI systems. Senior, or mid-level title.

EDI/ANK, IDK,SCK,WDK Communications software. EDI package. Runs on AS/400 systems. Stands for: Advanced Networking Kernel, Integrated Development Kernel, Standard Compliant Kernel, Workstation Kernel. Vendor: System Software Associates, Inc.

EDI-Answer Communications software. EDI. Runs on Unix, Windows NT systems. Vendor: Datacom Global Corp.

EDI Application Integrator Communications software. Message oriented middleware, data transformation tool. Performs EDI mapping, translation and management. Runs on Unix systems. Vendor: GE Information Services, Inc.

EDI*Asset Communications software. EDI package. Runs on Pentium type desktop computers. Vendor: Electronic Data Systems Corp.

EDI/Batch, EDI/Developer, EDI/Entry Communications software. Provide support functions for EDI systems. Runs on AIX, Windows systems. Vendor: DNS Worldwide.

EDI BIZIBIX ECMS System Communications software. EDI package. Runs on Pentium type PCs. Vendor: Synergistic Systems, Inc.

EDI Business Partner Translator Communications software. EDI package. Runs on Pentium type desktop computers. Vendor: American Custom Software.

EDI/comm Communications software. EDI. Runs on AS/400 systems. Vendor: Fischer EDI, Inc.

EDI/Edge Communications software. EDI package. Runs on Pentium type desktop computers. Vendor: DNS Associates, Inc.

EDI Excel Communications software. EDI package. Runs on Pentium type desktop computers. Vendor: American Business Computer.

EDI*Expert Communications software. EDI package. Runs on Pentium type desktop computers. Vendor: Electronic Data Systems Corp.

EDI Link 1. Communications software. EDI package. Runs on Unix, Windows systems. Vendor: Dunn Systems, Inc. 2. Communications software. EDI package. Runs on DEC, IBM, Unix systems. Vendor: Cincom Systems, Inc.

EDI-Link-M,S,SB Communications software. EDI package. Runs on Pentium type PCs. Vendor: Intercoastal Data Corp.

EDI Manager Communications software. EDI package. Runs on OS/2, Unix, Windows systems. Vendor: EDI Able, Inc.

EDI*Net Communications software. EDI package. Runs on multiple networks. Vendor: BT North America.

EDI/Open Communications software. EDI. Runs on Unix, Windows NT systems. Vendor: Premenos Technology Corp.

EDI*PC Communications software. EDI package. Runs on Pentium type desktop computers. Vendor: GE Information Services, Inc.

EDI/Synapse Communications software. EDI package. Runs on Pentium type desktop computers. Vendor: Data Dispatch Corp.

EDI*T System Communications software. EDI package. Runs on Bull HN, IBM, Tandem, Unisys systems. Vendor: GE Information Services, Inc.

EDI-Ware Communications software. EDI package. Runs on Pentium type desktop computers. Vendor: Grace Computer Resources, Inc.

EDI Windows Communications software. EDI package. Runs on DEC, Unix systems. Vendor: Trinary Systems, Inc.

EDI/WINS-3000 Communications software. EDI package. Runs on HP systems. Vendor: Logistic Systems.

EDIeasy Communications software. EDI package. Runs on Pentium type PCs. Vendor: EDI Solutions, Inc.

Edify Software vendor. Produces workflow software. Full name: Edify Corp.

Edify Electronic Banking System Application software. Links proprietary banking systems to the Internet to allow customers to check account balances, pay bills, and transfer funds over the Net. Vendor: Edify Corp.

EDIpl Translator Communications software. EDI package. Runs on MS-DOS, Unix systems. Vendor: LEK Product Marketing, Inc.

EDISIM Communications software. EDI package. Runs on IBM AS/400 systems. Vendor: Foresight Corp.

edit To check data for validity, i.e. a zipcode can be edited to make sure all the characters are numbers (one of the characters 0–9). Term is used interchangeably with validate.

Edit+ Application development tool. Screen editor for NCR systems. Vendor: Century Analysis, Inc.

Edit 1000 Application development tool. Screen editor for HP systems. Vendor: Hewlett-Packard Co.

EDITool Application development tool. Screen editor for DEC VAX systems. Vendor: Software Partners/32, Inc.

Editor Desktop system software. Word processor. Runs on Pentium type systems. Vendor: Condor Computer Corp.

EDItran; EDIfast Communications software. EDI package. Runs on Unix systems. Vendor: EDI Solutions, Inc.

EDITXS See EXSYS.

EDLIN An elementary text editor that is part of MS-DOS operating system.

EDM 1. Electronic document management.
2. Operating system software. Object-based. Tracks data about applications and upgrades. Electronically distributes the software for client/server applications. Runs on MVS, Sun, Unix systems. Interfaces with Web servers and Internet browser clients. Stands for: Enterprise Desktop Manager. Vendor: Novadigm, Inc.

EDMS EAI software. Integrates data from heterogeneous, distributed environment. Extracts and transforms legacy data into applications systems and/or data warehouses and datamarts. Stands for: Enterprise Data Management System. Requires Oracle7 or higher. Vendor: Oracle Corp.

EDP Electronic data processing. Term was replaced by DP.

EDP AUDITOR Operating system add-on. Security/auditing system that runs on IBM systems. Vendor: Cullinet Software, Inc.

Edwards/Odell/Martin Object-oriented development methodology. Provides analysis and design techniques.

Edword Word processing software that runs on IBM mainframes. Vendor: Trax Softworks, Inc.

EDX Operating system for midrange IBM Series 1 systems. Vendor: IBM Corp.

EFT Electronic Funds Transfer. The exchange of money over communications lines.

Eiffel Application development environment for developing object-oriented systems. Includes: EiffelBench (compiler/interpreter), EiffelBuild (application builder). EiffelVision (used for GUI), EiffelBase (development libraries), EiffelStore (database libraries), EiffelCase (analysis/design tools), ArchiText (document builder). Vendor: Interactive Software Engineering Inc.

EIM Electronic Image Management. Any system pertaining to the storage, transfer and processing of images.

EIP Enterprise Information Portal. A Web site that provides access to a company's information base. Classified as knowledgeware. Provides access to business information in unstructured documents such as word processing documents, Web pages, and other text based systems, as well as structured information in files and databases. Includes customizable home pages where users can, i.e., list reports they want to access. Handles security for both Internet and extranet applications. Includes a metadata repository, search engine, and publish and subscribe engine. Also called business portal, corporate portal.

EIS 1. Executive Information System. Term used for software designed to assist senior management. EIS packages typically contain access to external databases (such as stock exchange information), interfaces to other software (such as spreadsheet packages), access to internal databases, multiple methods of locating information, integration of data from many sources, and good security measures. EIS packages are more presentation systems than analytical systems such as DSS. OLAP software.
2. EIS package that runs on any system running Windows. Interfaces with spreadsheets, databases, e-mail, communications packages, and word processors. Vendor: Compuserve/Collier Jackson.

EIS/Corporate Performance Analysis EIS. Runs on Unix systems. Vendor: Cogent Information Systems, Inc.

EIS/G Application development tool. Automates programming function. Stands for: Executive Information System. Vendor: PILOT Executive Software.

EIS II EIS. Vendor: Information Resources, Inc.

EIS Toolkit EIS. Runs on Pentium type desktop computers. Vendor: Ferox Microsystems, Inc.

EIS-Track EIS. Runs on Pentium type desktop computers. Vendor: Intelligent Office Co.

EISA Extended Industry Standard Architecture. Design for microcomputers.

EISToolKit (for Windows) EIS. Runs on Macintosh, Windows systems. Vendor: Microstrategy, Inc.

EISTrack EIS. Runs on AS/400, Unix systems. Vendor: Productivity Software Resources, Inc.

EJB See Enterprise JavaBeans.

EKASYS Knowledge management software. Accessible through the Internet. Vendor: Psytep Corp.

EKS/Empower Application development tool. DSS used in data warehousing. Provides access to data stored in multidimensional databases. Vendor: Metapraxis, Inc.

Elan Traveler Desktop computer. Notebook. Pentium processor. Vendor: Census Computer Inc.

Elan Workstation Desktop computer. Pentium processor. Vendor: Austin Computer Systems, Inc.

Electric InterConnect Communications service which connects LAN based e-mail to the Internet. Vendor: The Electric Mail Co.

electronic brainstorming Another name for electronic meeting system. See EMS.

electronic commerce See e-commerce.

Electronic Commerce Modeling Language See ECML.

Electronic Data Interchange See EDI.

electronic directory An electronic phone book. Includes e-mail addresses, social security numbers, digital signatures and security information. Directories are built following protocols such as X.500 and LDAP.

Electronic Forms Designer Desktop software. Electronic forms software. Vendor: Microsoft Corp.

electronic forms software Desktop system software. Builds, revises, and edits forms. Most scan in existing forms. Forms can be filled in electronically, and the information automatically written to a database. Interface with most databases.

Electronic Funds Transfer See EFT.

Electronic Meeting System See EMS.

electronic paper See gyricon.

Electronic Publishing Edition Communications software, Web server. Provides access to the Web from OS/2 systems. Lets users view documents. Vendor: IBM Corp.

electronic signature Computerized capture of a signature.

Electronic Software Distribution See ESD.

electronic wallet Part of e-commerce. Storing information about customers including credit cards so a payment form can be filled out with a single key stroke. Electronic wallets can include cybercash. Also called digital wallet.

Electronic Workforce Application development tool used to develop customer service systems. Supports COM and connects to back office systems. Includes runtime environment. Interfaces with phone, fax, e-mail and databases. Enables users to call customer service call centers via the World Wide Web. Vendor: Edify Corp.

Electronic World Strategy from Hewlett-Packard defining Internet plans. Includes range of products in the Electronic Commerce area.

Elements Application development environment. Object-oriented, includes GUI development, data access. Generates C++, Java code. Can be used for enterprise development. Vendor: Neuron Data, Inc.

ELF Programming language. Stands for: Extension Language Facility. Used with ApplixWare development tools. Vendor: Applix, Inc.

ELIZA Artificial intelligence system. Runs on IBM systems. Psychology applications. Vendor: Artificial Intelligence Research Group.

Ellipse Application development tool that works in a client/server architecture. Builds client applications with a GUI-based front end. Interfaces with SQL Server. Vendor: Bachman Information Systems. (now Cayenne Software).

ELS NetWare Operating system for IBM and compatible desktop computer systems. Stands for Entry Level Solution NetWare. Vendor: Novell, Inc.

ELSI Specification for distributing software over the Internet. Stands for: Electronic Licensing and Security Initiative. Sets rules and procedures for controlling the transfer of software over the Net. Supported by Microsoft, IBM AT&T and other software vendors.

EMA Application software. Provides marketing automation by responding to and tracking customer information requests, analyzing when a lead changes to a prospect, and analyzing which marketing programs are most effective. Stands for: Enterprise Marketing Automation. Released: 1998. Vendor: Rubric, Inc.

EMACS Application development tool. Screen editor for Prime systems. Vendor: Prime Computer, Inc.

Emacs-W3 Internet browser. Runs on OS/2, Unix, VMS, Windows NT systems.

Emanager Suite Communications. E-mail application for the Internet that provides traffic management, load balancing, blocks spamming and filters out unwanted e-mail. Released: 1998. Vendor: Trend Micro, Inc.

embedded systems Software built in machines, i.e. the programs in cars that tell the driver when lights are left on are part of an embedded system. Embedded systems usually contain both operating and application system programs.

EMBOS Operating system for mainframe ELXSI Systems. Vendor: ELXSI.

EMC Data Manager Operating system enhancement. Provides backup and restore functions for Windows NT data stored on Symmetrix subsystems. Vendor: EMC Corp.

Emc2/TAO Communications software. E-mail system. Runs on IBM systems. Vendor: Fischer International Systems Corp.

EMD See EMC Data Manager.

Emedia Application development software. Combines documents from imaging systems, legacy systems, and Web servers and makes them available to a single user. Released: 1998. Vendor: Optika Imaging Systems.

Emerald Bay Relational database for a desktop computer environment. Designed to handle multiple users over a network. Interfaces with Eagle and Summit. Vendor: Migent, Inc.

Emeraude PCTE Application development tool. Integrates heterogeneous development and project management tools. Stands for Emeraude Portable Common Tool Environment. Runs on Unix systems. Vendor: Transtar Software, Inc.

EMerchant Internet software. E-commerce applications. Provides templates for processing high volume sales. Handles repeat business by storing past orders. Used for business-to-business commerce. Released: 1999. Vendor: Magic Software Enterprises Inc.

Emissary Internet browser. Runs on Windows systems. Part of Power Desktop.

EML See XML.

Emperor Operating system enhancement used by systems programmers and Operations staff. Job scheduler for Hewlett-Packard systems. Vendor: Carolian Systems International, Inc.

EmphaSys/36 Application development tool. Supports migration from System/36 to Unix systems. Vendor: Emphasys Software, Inc.

Empress 4GL Application development environment. Includes 4GL, screen painter. Creates applications to access relational databases without coding. Runs on Unix systems. Vendor: Empress Software, Inc.

Empress(/32) Relational database/4GL for midrange and desktop computer environments. Runs on Cray, DEC, HP, Sun, Unix systems. Utilizes SQL. Runs on DEC, Unix, Windows systems. Released: 1998. Vendor: Empress Software, Inc.

Empress DataWeb Application development tool. Used to develop Web pages and collect data on customers who visit a Web site. Runs on Unix systems. Released: 1996. Vendor: Empress Software, Inc.

EMS 1. Electronic Meeting System. Meeting participants each have a workstation so they can participate in a meeting both electronically and verbally. Electronic participation is anonymous, so electronic comments on ideas can be more forthcoming. Often part of groupware software. Used in Joint Application Development.
2. Operating system for mainframe ELXSI systems. Vendor: ELXSI.
3. See Enterprise Messaging Server.

EMS3 Family of CAD, engineering, and manufacturing products. Stands for: Engineering Modeling System. Vendor: Intergraph Corp.

emulation The imitation of any part of a system by another. For example, terminal emulators are programs written to allow desktop computers to act as terminals on a large computer system.

EMX Communications software. E-mail system. Stands for: Enterprise Mail Exchange. Uses X.400 protocols. Vendor: Lotus Development Corp.

Enable Desktop system software. Integrated package. Runs on Pentium type systems. Vendor: The Software Group.

encapsulation Term used with object-oriented development. The principle that values and methods should be packaged together as objects for storage, retrieval, and execution.

Encapsulator See CASEdge.

Encina, Encina++ Communications software. Transaction processing monitor for DEC, Hewlett-Packard, IBM systems. Also called OLTP system. Supports TCP/IP, SNA LU 6.2. Licensed by both Hewlett-Packard and IBM to use in software connecting IBM mainframes with Unix systems. Encina++ is a distributed object development environment that allows reuse of software objects. Interfaces with Java, the Internet . Runs on AIX, DEC, HP, OS/2, MVS, Sun systems. Vendor: Transarc Corp (subsidiary of IBM).

Encompass Application development environment. Used to develop client/server applications. Follows DCE specifications. Supports C, C++, COBOL for server applications, Visual Basic, PowerBuilder for client development. Vendor: Open Environment Corp.

Encore Computer vendor. Manufactures large and desktop computers.

Encore EIS Toolkit See EIS Toolkit.

Encore RSX Midrange computer. Operating systems: Unix, Umax, MicroMPX, MicroARTE. Vendor: Encore Computer Corp.

encryption Altering data so that it is meaningful only to the intended receiver. There are several ways of encrypting, or coding, data. Data is encrypted by keys, and both the sender and the receiver must have the same key. A key is a value associated with a mathematical algorithm, and the longer the key is, the harder it is to break the code. Standard keys vary in length from 56 to 128 bits. Keys are also defined as secret (or symmetric), or public (asymmetric). Secret keys have a single key, or formula used to both send and receive the information. Public keys have two keys (formulas), one to send and one to receive, With a public key, even the sender cannot de-crypt a message.

encyclopedia A repository of planning information, data models, process models, and design information. Used in information engineering.

end-user See user.

Endevor-C1,DB,DB2 Operating system add-on. Library management system that runs on IBM systems. Versions include: Endevor-DB, that works with IDMS databases, and Endevor-DB2. Vendor: Business Software Technology, Inc.

Endevor for xxx Application management tool. Software configuration manager. Manages both in-house and vendor-supplied software. Provides configuration management, inventory management, change control, and release management functions. Versions available for MVS, Unix, Windows NT. Vendor: Computer Associates International, Inc.

Energize Programming System Application development tool. Used to develop object-oriented, C++ applications. Runs on IBM, HP Apollo, NCR, Sun systems. Vendor: Lucid, Inc.

Energizer PME for R/3 System enhancement for SAP R/3 systems. Includes OptiTrak. Stands for: Energizer Performance Management Environment. Vendor: OptiSystems Solutions Ltd.

eNetwork Communications software. Provide the infrastructure to build extranet and intranet applications. Provides secure access to the Internet. Includes software for network computing, mobile communication and host communications. Allows PC to mainframe access. Vendor: IBM Corp.

Enfin/2,3 Application development environment. Used to develop object-oriented client/server systems. Builds client applications with a GUI-based front end. Interfaces with IBM, SQL Server, Oracle, Centura. Vendor: Enfin Software Corp.

ENFIN for AIX Application development environment based on SmallTalk object-oriented language. Builds applications for AIX that can be ported to OS/2 and Windows. Full name: ENFIN SQL Edition for AIX Motif. Vendor: Easel Corp.

ENFORCER I, II Application development tool. Program that checks programs against company-defined standards and reports on violations. Runs on IBM systems. Vendor: Clarity Concept Systems Co.

Engage.Fusion, Engage.Discover Application development tools. Used in data warehousing. Used to track and analyze online traffic. Developed by Engage Technologies. Vendor: Red Brick Systems, Inc.

Engineer See programmer.

Engineering Document Management System Image-processing system that runs on large computer systems. Vendor: Formtek, Inc.

Engineering Modeling System See EMS3.

Enhanced NCSA Mosaic Improved commercial version of Mosaic. See Mosaic. Vendor: Spyglass, Inc.

Enhanced Spectrum Communications. Network management software. Manages applications and devices in addition to the network resources. Vendor: Cabletron Systems, Inc.

Enhydra Application development environment and server which includes a servlet-based run-time environment, XML compiler, application development wizard, and utility called Enhydra Jolt which lets developers use embedded Java code for dynamic HTML. Runs under Mac OS, Unix, and Windows 95/98/NT. Available as a free download from the vendor. Vendor: Lutris Technologies.

Enix Operating system. Unix-type system. Vendor: Everex Systems, Inc.

Enlighten Application development tool. Used to build Java applications. Vendor: AlphaBlox Corp.

Enovia Application software. PDM software for large systems. Vendor: Dassault Systems SA.

EnQuiry Application development tool. Graphically builds Visual Basic forms and queries. Vendor: Progress Software Corp.

ENS Communications software. Services include a directory (StreetTalk), messaging, security, and administrative routines. Runs with Novell NetWare under HP/UX, Solaris, AIX operating systems. Stands for: Enterprise Network Services. Vendor: Banyan Systems, Inc.

Ensemble Graphics user interface. Includes windowing functions. Runs on Pentium type desktop computers. Vendor: Geoworks.

Ensemble Test Application development tool. Toolset for debugging C programs. Runs on Unix systems. Vendor: Cayenne Software, Inc.

Ensemble Viewer Application development tool. Provides graphical view of program flow, data structures, etc. for C programs. Vendor: Cayenne Software Inc.

ENTAO Extended Enterprise Suite Communications. Internet software. Used to set up intranets. Includes e-mail, Text Chat, Visual Chat, Forums and Contact Management. Released: 1998. Vendor: Skunk Technologies, Inc.

Entark Application development environment. Develops and deploys enterprise-wide multi-user, heterogeneous client/server systems. Runs on Unix, Windows NT systems. Vendor: Infosys Technologies, Ltd.

Entera Communications, remote procedure call (RPC) middleware. Used to build three-tiered client/server applications. Runs on Macintosh, MVS, OS/2, Windows, Unix systems. Released: 1995. Vendor: Inprise Corp.

enterprise The entire company. Software developed for the enterprise considers all aspects of the company as opposed to departmental software development which only considers a specific department or function.

Enterprise/Access, Enterprise Connect Application development tools. Connects Internet browsers to mainframe systems. Vendor: Apertus Technologies.

Enterprise Application Integration See EAI.

Enterprise Application Server Application server providing middleware and development tools. Used to develop Web and client/server applications. Includes Jaguar CTS and PowerDynamo. Supports HTML, Java, C, ActiveX, JavaBeans, CORBA, COM. Also called EA Server. Vendor: Sybase, Inc.

Enterprise Application Studio Application development tools. Used to develop distributed Web and client/server Java applications. Includes: PowerBuilder, PowerJ and Enterprise Application Server. Vendor: Sybase, Inc.

Enterprise:Builder Application development tool. Automates programming and testing functions. Interfaces with Rdb, RMS, and Oracle. Vendor: Cullinet Software, Inc.

Enterprise Builder Application development environment. Creates a single layer that combines information from disparate corporate databases to allow single access including functions such as drag and drop from one database to another. Can be used on the Internet. Part of TopTier. Vendor: TopTier, Inc.

Enterprise Connect Communications software. Middleware. Provides access to heterogeneous platforms and databases. Vendor Sybase, Inc.

Enterprise Console System management tool. Allows users to manage system from one console. Included in POEMS. Vendor: PLATINUM Technology, Inc.

Enterprise ControlStation Operating system enhancement. Automates management of the entire data center. Runs on RS/6000, Sun systems. Vendor: 4th Dimension Software, Inc.

Enterprise Data Access/SQL See EDA/SQL.

Enterprise Data Mart Data warehousing software. Used to combine data marts into an enterprise data warehouse. Includes Dynamic Data Store (database created of data common to all the data marts), Global Data Mart Repository (Manages metadata to ensure that queries are routed to the correct data mart). Released: 1997. Vendor: Informatica Corp.

Enterprise:DB Relational database for large computer environments. Runs on DEC, Unix systems. Utilizes SQL. Vendor: Cullinet Software, Inc.

Enterprise DBA Application development tool used in data warehousing. Manages RDBMSs (Oracle, Sybase, SQLServer) across multiple Unix and Windows platforms. Vendor: PLATINUM Technology, Inc.

Enterprise Desktop Manager See EDM.

Enterprise Developer Application development environment. Used for enterprise wide client/server development. Includes SCALE (a repository which includes data models, entity-relationship diagrams, and business rules), SCALEScript (a 4GL), Team Enterprise Developer (allows development team to share objects and business models within SCALE repository), report writers, debuggers. Vendor: Symantec Corporation.

Enterprise Direct Communications software. Middleware. Allows queries to multiple databases on different platforms. Runs on IS/2, Unix, Windows NT systems. Vendor: Neon Systems, Inc.

Enterprise Edition Version of Visual Basic intended for the larger corporate user. Includes remote procedure call technology. Vendor: Microsoft Corp.

enterprise, enterprise-wide development The development of large scale projects that apply to the entire company, or enterprise. Projects include mission critical, transaction oriented systems with emphasis on data updating, security, and response time. These systems often operate in multiplatform environments and follow formal developmental methodologies. Enterprise systems that give users access to all information needed across the entire "enterprise." Contrast with departmental development.

Enterprise Expert Application development tool. Rules-driven development environment for automating business processes and procedures. Used for financial applications. Runs on Windows systems. Vendor: Neuron Data, Inc.

Enterprise FrameWork See FrameWork.

Enterprise:Generator Application development tool. Automates programming and testing functions. Vendor: Cullinet Software, Inc.

Enterprise Information Factory Data warehouse framework. Strategy and product selection intended to offer smooth integration between the warehouse and the operational systems. Vendor: AT&T GIS.

Enterprise Information Portal See EIP.

Enterprise Inquisit Communications service. News filtering service that links with corporate intranets. Monitors hundreds of news sources and delivers articles according to user defined criteria. Vendor: Inquisit, Inc.

Enterprise JavaBeans Application development specification. Defines an object-oriented API that extends the component model enabling developers to build platform independent Java applications. Used to build blocks of reusable code that runs on the server, not the browser. Specifications published in 1998 by Sun Microsystems, Inc.

Enterprise Knowledge Server Application development tool used for business access and analysis. OLAP database used in Windows applications. Vendor: Metapraxis, Ltd.

Enterprise Management Architecture See EMA.

Enterprise Messaging Server Communications software. Expands the capabilities of LAN e-mail systems allowing thousands of users. Is an extension of Lotus Notes. Runs with Windows. Uses Access relational database. Also called EMS. Vendor: Microsoft Corp.

enterprise messaging software Communications software. Expands LAN e-mail systems to handle thousands of users.

Enterprise Miner Application development tool. Used in data warehousing for data mining. Includes a database to allow users to analyze detail data from existing data warehouses or data marts. Runs on DEC, Unix, Windows systems. Released: 1998. Vendor: SAS Institute Inc.

Enterprise Momentum Application development tool. Data and process modelers that would be managed by a repository that would unite all Sybase development tools. Vendor: Sybase, Inc.

Enterprise Network Services See ENS.

Enterprise Object Library Engine Application development tool. Allows developers to store PowerBuilder objects in relational databases. Provides distributed objects to let users divide application processing among different systems. Part of PowerBuilder. Vendor: PowerSoft (division of Sybase, Inc.).

Enterprise Objects Framework Application development tool. Allows object-oriented applications to encapsulate and use data stored in relational databases. Accesses Oracle and Sybase databases. Vendor: Apple Computer, Inc.

Enterprise Production Management Application software. Provides production scheduling and automated systems operations. Runs on mainframe and desktop systems. Vendor: New Dimension Software.

Enterprise Reporter Application development tool. Report generator for end-users. Allows non-technical people to design a report and then select data for the report without understanding the underlying database. Runs on Unix, Windows systems. Released: 1997. Vendor: SAS Institute Inc.

Enterprise ResponseAgent Application software. Manufacturing software handling supply chain functions. Runs on Unix systems. Released: 1996. Vendor: Red Pepper Software Co. (division of PeopleSoft, Inc.).

Enterprise Server Communications software. Web server. Runs on Unix, Windows NT systems. Vendor: Netscape Communications Corp.

Enterprise/Solver Communications. Network management software. Provides planning and troubleshooting functions. Works with networks using SNA and TCP/IP protocols. Vendor: NETSYS Technologies, Inc.

Enterprise Storage Manager Operating system software. Provides mainframe data backup and recovery functions for desktop/LAN systems. Vendor: Legent Corp.

Enterprise-Wide Work Management System Project management package. Manages multiple projects simultaneously and gives a complete picture of all activity underway at any time. Also called Multitrak. Runs on IBM MVS systems. Vendor: Multitrak Software Development Corp.

Enterprise Workbench Application development tool. Provides point-and-click access to SQL relational databases. Integrates CASE tools. Includes DB/Assist. Vendor: Easel Corp.

Enterprise X-D Midrange computer. Pentium II CPU. Released: 1998. Vendor: Tangent Computer, Inc.

EnterpriseCONNECT Communications software. Middleware. Provides cross-platform, multi database connectivity. Vendor: Sybase, Inc.

EnterprisePro Communications. Software that analyzes network performance by condensing thousands of network statistics into trends and problem situations. Vendor: International Network Services Inc. (INS).

EnterpriseWeb/MVS,VM Communications/Internet software. Web server. Allows users to access variety of information on MVS-class mainframes. Lets users view text, binary files, images, sound clips, video and Java applets stored on MVS or VM systems. Vendor: Beyond Software, Inc.

EnterpriseWeb/Vision Communications/Internet software. Enables standard Web browsers to be used as client front-ends for sending and receiving e-mail and attachments from OfficeVision host. Supports enterprise-wide calendars and scheduling. Runs on IBM systems. Vendor: Beyond Software, Inc.

Entire Access Application development tool. Used with Natural to develop client/server applications that are hardware, operating system, and DBMS independent. Applications can work with multiple databases. Runs on Unix, Windows systems. Vendor: Software AG Americas Inc.

Entire Broker Communications software. Message-oriented middleware. Provides the client/server communication agents for the parts of an application that reside on different systems. Vendor: Software AG Americas Inc.

Entire Net-Work Communications software. Middleware. Part of Unix Productivity Pack. Vendor: Software AG Americas Inc.

Entire Security SAF Gateway Operating system software. Extends mainframe security systems to Windows NT, HP-UX and Web applications. SAF stands for: System Authorization Facility. Vendor: Software AG Americas Inc.

Entire Transaction Propagator Software tool that will send updates to Adabas databases to remote diverse sites. Updates are sent at user specified intervals and to user selected sites. Vendor: Software AG Americas Inc.

EntireX Application development tool. Converts systems and integrates legacy systems into client/server and Internet/intranet applications. Includes Developers Kit and most Entire programs. Runs on DEC, Unix, Windows systems. Released: 1998. Vendor: Software AG Americas Inc.

entity relationship diagram A tool used during the analysis and design phases of the project life cycle. A chart that lists attributes of data and then defines the relationships between the attributes.

entity-relationship models Design tool used in some structured programming methodologies.

Entrada 1000 Desktop computer. Pentium processor. Vendor: Duracom Computer Systems.

Entrust Certificate authority. (NASDAQ: ENTU)

Entrypoint 90 Application development tool for designing custom data entry screens. Includes editing, data validation and export options. Interfaces with most data communications networks. Runs on MS-DOS, Windows systems. Vendor: Datalex, Inc.

Entrypoint 90/Plus Application development tool. Includes editing, data validation, export options, audit trail facility. Interfaces with most data communications networks. Runs on MS-DOS, Windows systems. Vendor: Datalex, Inc.

EnView See Sentinel.

Environ/1 Operating system enhancement used by systems programmers. Monitors and controls system performance in IBM systems. Works with online systems. Vendor: Cincom Systems, Inc.

Envision CASE product. Automates programming, testing, and project management functions. Vendor: Future Tech Systems, Inc.

Envision3D Communications. Web based software which allows visual collaboration across a wide geographic area. Vendor: Adaptive Media, Inc.

enVISN Communications software. Adds high speed switching functions to existing networks. Vendor: Digital Equipment Corp.

EnVista Midrange computer. RISC system; Pentium Pro processor. Operating system: Windows NT. Vendor: Amdahl Corp.

EnVista Gateway for DW/DSS Communications software. Provides bulk data downloading from operational databases to data warehouses. Runs on MVS, Unix, Windows NT systems. Vendor: Amdahl Corp.

EnVista Gateway for OLTP Communications software. Connects databases to user applications. Runs on MVS, Unix, Windows NT systems. Vendor: Amdahl Corp.

Envive Inspector for R/3 System management software for SAP R/3. Performance monitor.Released:1997. Vendor: Envive Corp.

Envoy Desktop system software. Document exchange software that allows cross-platform use of documents created under any software. For example, a document created in WordPerfect under Windows 95 will be readable by a user on a Macintosh system. Vendor: WordPerfect.

Envy/Developer Application development tool. Used for team development. Provides change management of source and object modules. Runs on most workstations. Written in Smalltalk. Vendor: Object Technology International, Inc.

EO Personal communication system. Vendor: AT&T.

EOA Enterprise Object Architecture. Strategy designed to migrate to client/server computing. Vendor: Digitalk.

EP/IX Operating system for midrange Control Data computers. Vendor: Control Data Corp.

EPAT Operating system add-on. Tape management system that runs on IBM systems. Vendor: SDI.

EPC EDB Application development tool. Interactive debugger that allows the developer to work with either the machine or source code. Works with programs written in any of the EPC languages (C, C++, Fortran, Modula-2, Pascal). Runs under Unix. Vendor: Edinburgh Portable Compilers.

EPC xxxxx Compiler languages, including: C, C++, Fortran, Modula.2, Pascal. Vendor: Edinburgh Portable Compilers.

EPDM Operating system enhancement used by systems programmers. Collects information from various IBM mainframe components including the MVS, CICS, VM, and RACF. Stands for: Enterprise Performance Data Manager. Vendor: IBM Corp.

Epic EIS. Runs on Pentium type desktop computers, Unix systems. Vendor: Epic Software, Inc.

EPIC/CMS-Console Operating system enhancement used by systems programmers. Monitors and controls system performance in IBM VM systems. Vendor: Tower Systems International.

Epic/CMS,MVS,VSE Operating system add-on. Data management software that runs on IBM systems. Products include: Epic/VSE-Disk, Epic/VSE-Tape. Vendor: Tower Systems International.

Epic for VSE,CMS Operating system enhancement used by systems programmers. Increases system efficiency in IBM systems. Manages disk storage space. Vendor: Goal Systems International, Inc.

Epicentric Portal Server EIP, corporate portal. In addition to standard internal access has established partnerships with external data sources to provide information on such things as weather and stock prices. Released: 1999. Vendor: Epicentric, Inc.

Epilog Operating system enhancement used by systems programmers. Monitors and controls system performance in IBM MVS systems. Includes: Epilog for MVS, Epilog 1000 for VM. Vendor: Candle Corp.

Epoc/16 Operating system for handheld computers. Vendor: Psion, Inc.

Epoch Enterprise Backup Operating system software. Client/server backup/restore program that works with microcomputer based systems, relational databases, and Unix file systems. Vendor: Epoch Systems, Inc.

EpochBackup Application development tool. Automatically backs up and restores files on heterogeneous computer systems. Used in client/server environments. Runs on Sun systems. Vendor: Epoch Systems, Inc.

EpochMigration Application development tool. Migrates client files in diverse environments to the EpochServ system. See EpochServ. Vendor: Epoch Systems, Inc.

EpochServ Application development tool. Provides complete data management functions for a client/server environment encompassing diverse systems. Vendor: Epoch Systems, Inc.

EPOS CASE product. Automates analysis, design, and programming functions. Includes code generator for Ada, C, Fortran, Pascal. Runs on mainframe, midrange, and workstation systems. Vendor: SPS Software Products and Services, Inc.

EPS See Energize Programming System.

EPS Apex Notebook computer. Pentium CPU. Vendor: EPS Technologies, Inc.

EPT Data management product. Automatically replicates updates to Adabas databases to systems across heterogeneous networks. Works with mainframes, OS/2, Unix, OpenVMS. Stands for: Entire Transaction Propagator. Vendor: Software AG Americas Inc.

EQL 4GL. Accesses FOCUS databases. Stands for: English Query Language. Vendor: Elliott Bay Computing, Inc.

EQM Query language used with MAI Origin ADS databases. Stands for: Executive Query Manager.

Equium Desktop computer. Pentium II, Pro CPU. Vendor: Toshiba America Information Systems, Inc.

ER Designer CASE product. Automates design function. Works with DB2, Oracle, Ingres, and other databases. Vendor: Chen & Associates.

ERBD Relational database. Interfaces with standard development tools in a desktop environment. Vendor: Automated Technology Associates, Inc.

Ergo PowerBrick See CD-PowerBrick.

Ergo PowerBrickx Desktop computer. Notebook. Pentium processor. Vendor: The Brick Computer Company, Inc.

ERI/CICS Operating system enhancement used by systems programmers. Integrated CICS management tools including a single sign-on option and an application profile editor. Runs on MVS RACF systems. Vendor: Enterprise Research, Inc.

Eroom Communications software. Teamware. Allows users to share information and files over the Web. Includes a common calendar and the ability to edit and trade documents. Vendor: Instinctive Technologies.

EROOS Object-oriented development methodology. Stands for: Entity-Relationship Object-Oriented Specifications.

ERP Enterprise Resource Planning. Software that links together systems such as manufacturing, financial, human resources, sales force automation, supply chain management and data warehousing. These systems combine all business processes in a single application to be used throughout the entire enterprise.

ERwin Application development tool. Used for data modeling, building data warehouses. Works with PowerBuilder, Visual Basic, SQL Windows. Integrates with Passport. Vendor: Computer Associates International, Inc.

ES/9000 Model xxx Mainframe and midrange computer systems. Operating systems: MVS, VSE, VM. This is IBM's newest line of computers. Stands for: Enterprise System/9000. Utilizes fiber-optic channels. Vendor: IBM Corp.

ES RE/Vision See RE/Vision.

ESA Stands for Enterprise System Architecture. See MVS.

ESA/390 Mainframe architecture from IBM. High-end design of large business computers (large 3090 and ES/9000 machines) offering improved system performance.

Escala PowerCluster Midrange computer. RISC system. Operating system: AIX. Vendor: Bull HN Information Systems, Inc.

Escape Operating system for DEC systems, ADA compatible. Vendor: Proprietary Software Systems, Inc.

ESCAPE Data management system. Runs on IBM systems. Allows programs written in non-IDMS environments to access IDMS databases. Vendor: Computer Associates International, Inc.

Escon architecture The architecture used by IBM in the ES/9000 large computer systems. Uses fiber optics for communications channels. Escon channels could become popular as other vendors develop compatible hardware and software. Stands for: Enterprise Systems connections.

ESD Electronic Software Distribution. Purchasing and distributing software over the Internet .

ESDS Entry Sequence Data Set. A type of VSAM file.

EService 98 Communications software. Internet customer service system. Interfaces with help desk software packages. Vendor: Silknet Software, Inc.

ESF 1. Operating system enhancement used by systems programmers. Increases system efficiency in IBM Series 1 systems. Vendor: IBM Corp.
2. External Source Format. Part of IBM's CSP that will work with AD/Cycle and allow data collected from CASE analysis and design tools to be passed to CSP for code generation.

ESIE Artificial intelligence system. Runs on Pentium type desktop computers. Expert system shell. Stands for: Expert System Inference Engine. Vendor: Lightwave.

ESM See Enterprise Storage Manager.

ESnet The Department of Energy's network.

ESP Operating system enhancement used by systems programmers and Operations staff. Job scheduler for IBM systems. Stands for: Execution Scheduling Processor. Vendor: Cybermation, Inc.

ESP Advisor Artificial intelligence system. Runs on DEC, IBM desktop computer. Expert system shell. Provides base of over 3000 rules. Vendor: Expert Systems International.

ESP Frame-Engine Artificial intelligence system. Runs on Pentium type desktop computers. Expert system shell. Vendor: Expert Systems International.

ESP II Communications software. EDI package. Runs on 286 and larger model desktop computers. Vendor: Foretell Corp.

ESP-Tempd Operating system enhancement used by systems programmers. Increases system efficiency in IBM systems. Manages disk storage space. Vendor: Enhanced Software Products, Inc.

Esperant Application development tool. Data query and reporting software. Allows end-users to do database queries without knowing SQL or the database structure. Vendor: Software AG Americas Inc.

Esplanade Application development tool. Web server that links Web applications with relational databases. Runs on Windows NT systems. Used in data warehousing. Released: 1996. Vendor: SPL Software, Inc.

Espresso See Cafe.

ESQL-C,COBOL,Ada See Informix-ESQL(Ada,C,Cobol).

ESS 1. Spreadsheet for mainframes. Stands for: Electronic Spread Sheet. Vendor: Trax Softworks, Inc.
2. Operating system enhancement used by systems programmers. Monitors and controls system performance in Unisys systems. Stands for: Extended System Software. Vendor: Unisys Corp.

Essbase OLAP software. Multidimensional database. Links OLAP processing with data in relational databases. Used in data warehousing. Runs on OS/2, Unix, Windows NT systems; supports OS/2, Windows, Macintosh, Unix clients. Vendor: Hyperion Solutions Corp.

Essbase/400 Application development tool. DSS. Provides access to data stored in multidimensional databases. Vendor: Hyperion Solutions Corp.

Essbase Analysis Application development tool used for business access and analysis. OLAP database. Provides multiuser access to data. Used with data warehousing. Runs on Unix, OS/2, Windows NT systems. Vendor: Hyperion Solutions Corp.

Essentia Pxxx Desktop computer. Pentium processor. Vendor: Census Computer Inc

Essential Tools Application development tool. Collection of utility programs to handle administrative functions such as adding and deleting users to a Notes environment. Transfers users' e-mail files across systems. Released: 1998. Vendor: InfoImage, Inc.

Estimacs See CA-Estimacs.

Estimator CASE product. Automates project management function. Vendor: Spectrum International, Inc.

Estrella 300-xxx Midsize computer. Server. PowerPC processor. Operating system: AIX. Vendor: Bull HN Information Systems Inc.

eSuite DevPack Application development tool. Used on network computers to develop Internet applications. Runs on any JVM. Includes applets for: SQL\JDBC data access, CGI data access, project scheduling, word processing, spreadsheets, and presentation graphics. Released: 1998. Vendor: Lotus Development Corp.

eSuite Workplace Communications. Provides user interface for network computers. Runs on network computers, any JVM. Functions include browsing, file management, e-mail, PIM features. Includes Java applets for: calendar, mail, address book, word processor, spreadsheet, presentation graphics. Released: 1998. Vendor: Lotus Development Corp.

ESW See Existing Systems Workbench.

ETA System V Operating system for supercomputers. Unix-type operating system. Vendor: Compuware Corp.

ETC Communications protocol for cellular systems. Stands for: Enhanced Throughput Cellular. Written by AT&T/Paradyne.

ETEwatch Operating system software. Measures performance in client/server systems by monitoring actual response time experienced by the user. Vendor: Candle Corp.

Ethernet Communications. LAN (Local Area Network). Follows TCP/IP protocols. Uses bus topology. Commonly used with Unix systems. Developed by Xerox, DEC, and Intel.

ETI Extract Tool Suite Application development tool. Automates and expedites the movement of large volumes of data in complex environments. Used to load data warehouses. Runs on RISC systems, IBM mainframes. Vendor: Evolutionary Technologies International.

ETL Term used for a category of software used in data warehousing. Stands for: Extraction, Transformation and Loading and refers to the software that is used to get data into a warehouse. which includes data loading, replication, scrubbing and extraction programs. Other data warehouse software categories are DBMS (DataBase Management Systems) and DSS (Decision Support Systems).

EToolkit/400 Communications software. EDI. Runs on AS/400 systems. Vendor: Impro International, Inc.

ETOS Communications software. Provides environment for application developmen. Includes programming and testing utilities. Runs on DEC systems. Program development timesharing option. Vendor: Federated Consultants.

Eudora (Pro) Communications. Internet e-mail software. Includes support for multiple mail accounts, voice mail support and handling HTML messages. Supports IMAP, LDAP protocols. Vendor: Qualcomm, Inc.

euro, euro compliant Reference to the new European currency. Euro compliant systems have incorporated the new monetary system.

EuroCom 3500 Portable computer. Vendor: Europak International.

Europak Computer vendor. Manufactures desktop computers.

Event Control Server Operating system software. Scheduling and batch processing software for client/server environments. Supports most mid-size, RISC, and desktop systems. Vendor: Vinzant, Inc.

event-driven programming Programming that reflects action, i.e. clicking on an icon will trigger a series of steps. The program does not solve a problem, but rather creates an environment in which the user solves the problem. Different from traditional procedural programming in which the program solves the problem. Often used in object-oriented development.

event response Design methodology associated with structured programming. Views an application as a black box that responds to events occurring outside it. Each event results in a program specification for a single program. Methodologies designed on event response are Ward-Mellor.

EventIX Application development tool. Allows network administrators to build automation and management applications for Unix systems. Interfaces with NetView, SunNet Manager, SNMP managers. Vendor: Bridgeway Corp.

Everest 1. Code name for SCO OpenServer 5. See SCO OpenServer 5. 2. Web server jointly built by Netscape Communications Corp and Silicon Graphics, Inc. In development.

Everex Computer vendor. Manufactures desktop computers.

Evolution Desktop computer. Pentium CPU. Operating system: Windows 95/98. Includes MP II, PRO, 2XL, Q-SMP. Also includes multiprocessors: Revolution DUAL6, QUAD6, 6x6. Vendor: Advanced Logic Research, Inc.

EWorld Communications service that provides a wide variety of information and services. Includes user forums, news groups, Apple customer support, access to the Internet . Vendor: Apple Computer, Inc.

EX, GX Model xx See HDS EX,GX Models xx.

EX3278 Operating system add-on. Debugging/testing software that runs on IBM systems. Vendor: Systems Strategies, Inc.

EXceed Communications software. Used in client/server computing. Links Windows clients to Unix servers. Vendor: Hummingbird Communications.

Excel Desktop system software. Spreadsheet. Runs on Macintosh, IBM systems. Includes VBA. Vendor: Microsoft Corp.

Excelan Communications software. Network connecting IBM, Apple, DEC, Sun systems. Vendor: Excelan.

Excelerator CASE tool. Automates analysis, design. Supports development for diverse platforms, object-oriented development. Vendor: INTERSOLV, Inc.

Excelerator for Windows Application development tool. Provides functions for prototyping, reporting, chart and diagram maintenance. Runs on Windows systems. Vendor: INTERSOLV, Inc.

Excelerator II CASE product. Automates analysis, design, programming functions. Integrates with relational/SQL databases. Runs on OS/2 systems. Vendor: INTERSOLV, Inc.

Excellink (Host/V) Communications software. Network connecting IBM systems. Micro-to-mainframe link. Vendor: OBS Software.

eXcelon Application development tool used to develop e-commerce applications. Used to build XML-based Internet applications that work with all data sources and application server. Includes a data server and a toolset used to build and store XML data. Runs on Windows systems. Released: 1999. Vendor: Object Design, Inc.

Exchange (Server) Communications software. Groupware. Client/server messaging product. Includes e-mail, file handling, MAPI support, EMS, workflow. Supports MS-Mail, NNTP, IMAP, LDAP. Runs on Windows NT systems. Various releases called Osmium, Platinum. Platinum is the code name for version four. Vendor: Microsoft Corp.

Excite for Web Servers Communications, Internet software. Search engine. Vendor: Excite, Inc.

EXCP Execute Channel Program. A program that contains the actual commands that control data transfer in or out of the computer. Written by systems programmers.

ExDiff Operating system add-on. Debugging/testing software that runs on Unix systems. Vendor: Software Research Associates.

EXEC procedure A set of CMS commands.

Execumate II EIS. Runs on multiple desktop computer and workstation platforms. Vendor: Southware Innovations, Inc.

Executive Decisions EIS. Runs on IBM PS/2 systems. Vendor: IBM Corp.

Executive Edge EIS. Vendor: Execucom Systems Corp.

Executive Information System 1. See EIS 2. EIS. Vendor: Meta Media, Inc.

Executive Management System EIS. Runs on Macintosh systems. Vendor: Softouch Software, Inc.

Executive NOMAD Application development tool. Provides visual access to information for managers at all levels. Displays information in charts, drawings, images and text. Runs with NOMAD and interfaces with Lightship. Runs on IBM mainframe systems. Released: 1996. Vendor: Aonix.

Executive Partner See FT-70,80.

ExecuTrieve/36 Communications software. E-mail system. Runs on IBM systems. Vendor: Computer Corp. of America.

Exemplar Linking and embedding code that allows applications across Apple, NetWare, and Unix systems to be mixed. Vendor: Apple Computer Inc.

Exemplar X,S-class See Hp Exemplar.

Existing Systems Workbench Application development tool. Includes re-engineering support, program editing, interactive testing functions. Available for COBOL, PL/1 systems. Works in all IBM environments. Vendor: Viasoft, Inc.

EXLU6.2/CICS,S36 Operating system add-on. Debugging/testing software that runs on IBM systems. Vendor: Systems Strategies, Inc.

EXODUS Object-oriented database developed by the University of Wisconsin used for research.

Expandable/MRP MRP, or CIM software. Runs on DEC, IBM, systems. Vendor: Expandable Software, Inc.

ExpEDIte Communications software. EDI package. Runs on IBM systems. Vendor: IBM Corp.

Exper family Artificial intelligence system. Runs on Macintosh systems. Includes: ExperLisp, Exper-Common Lisp, ExperFACTS, ExperMulti OPS5, ExperProlog II. Vendor: ExperTelligence.

Expert 1. Desktop system software. Project management package. Runs on IBM, Macintosh systems. Vendor: Decision Science Software, Inc.
2. Data management system. Runs on HP systems. Allows end-users access to database. Vendor: Cognos, Inc.

Expert Choice Application development tool. DSS. Runs on Windows systems. Vendor: Expert Choice, Inc.

expert system In artificial intelligence, a system that can solve problems by drawing inferences from a collection of information that is based on human experience. Synonymous with knowledge-based system.

ExpertR Artificial intelligence system. Runs on Wang systems. Builds expert systems. Vendor: Coyne Kalajian, Inc.

eXplora Desktop computer. AMD CPU. Operating system: Windows 95/98. Vendor: Everex Systems Inc.

Explora Network computer. Vendor: Network Computing Devices (NCD).

Explore OnNet Communications software. Internet kit which facilitates the use of basic Internet services. Vendor: FTP Software.

Explore/ Operating system enhancement used by systems programmers. Monitors and controls system performance in IBM systems. Includes: Explore/CICS, Explore/VM, Explore/VSE. Vendor: Goal Systems International, Inc.

Explora Network computer. Vendor: Network Computing Devices, Inc.

Explorer See Internet Explorer.

Express 1. Operating system enhancement used by systems programmers and Operations staff. Job scheduler for Hewlett-Packard systems. Vendor: Operations Control Systems.
2. Communications software. Gateway for e-mail systems. Runs with Netware networks. Vendor: Immedia Telematics Corp.
3. Query language. Works with DataServer Analyzer. Vendor: IRI Software.
4. See Oracle Express.
5. See Forte Express.

Express5800 Midrange computer. Pentium Pro CPU. Parallel processor with up to four processors. Operating system: Windows NT. Vendor: NEC, Inc.

Express Analyzer Application development tool. Query tool used with Express databases. Released: 1995. Vendor: Oracle Corp.

Express Objects Application development tool. Graphical development environment used for OLAP. Interfaces with Express Analyzer. Access data stored in Express multidimensional databases. Vendor: Oracle Corp.

Express Publisher Desktop system software. Desktop publisher. Runs on Pentium type desktop computers. Vendor: Power Up Software Corp.

Express Server 1. Desktop computer. RISC machine. NEC VR4400 CPU. Operating system: Windows NT. Vendor: NEC Computer Systems.
2. EIS, multidimensional database. Includes query functions. OLAP software. Runs on IBM mainframes, DEC, Prime, Sun, HP, Unix, Windows NT systems. Vendor: Information Resources, Inc., Oracle Corp.

ExQc Quality control system for manufacturing. Vendor: Automated Technology Associates, Inc.

Exsys Application development system. Includes tools to assist with analysis, design, testing, and documentation functions. Runs on DEC VAX systems. Vendor: Jordan-Webb Info Systems, Ltd.

EXSYS 1. Artificial intelligence system. Expert system building tool. Runs on AT&T, DEC, IBM, VAX, Unix systems. Includes: EXSYS, EDITXS, SHRINK. Vendor: EXSYS, Inc. (NY).
2. See XSYS/EXSYS.

Extend Application development tool. Provides process modeling functions. Runs on Macintosh, Windows systems. Vendor: Imagine That, Inc.

Extended Enterprise Strategy that uses the Internet /Web for all corporate communications including employees, customers, business partners and suppliers. Uses Pervasive Solutions framework. Vendor: Hewlett-Packard Co.

Extensa Notebook computer. Pentium CPU. Vendor: Acer America Corp.

Extensible Markup Language See XML.

Extension Builder Application development tool. Allows users to write custom extensions to interface enterprise application with the Internet. Runs on Solaris, Windows systems. Vendor: Netscape Communications Corp.

Extol EDI Integrator Communications software. EDI package. Runs on AS/400 systems. Vendor: Extol, Inc.

EXTRA! Communications software. Gateway connecting LANS to IBM's SNA networks. Vendor: Attachmate Corp.

EXTRACT Tool Suite Application development tool. Used to automate the migration of data between dissimilar systems. Used in data warehousing. Builds a bidirectional interface between the warehouse and the operational databases. Vendor: Evolutionary Technologies International.

Extraction, Transformation and Loading See ETL.

extranet A network that allows non-employees to have Web access to internal corporate applications.

Extreme xxxx Desktop computer. Pentium processor. Vendor: PC Importers, Inc.

Extricity AllianceSeries See AllianceSeries.

Eyewitness Operating system enhancement used by systems programmers. Increases system efficiency in IBM CICS systems. Processes CICS dumps. Vendor: Landmark Systems Corp.

EZ EDI Communications software. EDI package. Runs on AS/400, Windows systems. Vendor: Intercoastal Data Corp.

EZ/IQ Query language used with VSAM files. Stands for: EZ/Interactive Query. Vendor: Oracle Corp.

EZ-Prep Part of CSP. Program Generator for IBM desktop computers. Vendor: IBM Corp.

EZ-RPC Operating System add-on. Used in client/server environments. Allows applications on one computer to use procedures residing on another. Generates C code. Stands for: EZ-Remote Procedure Call. Vendor: Noblenet, Inc.

EZ-Run Part of CSP. Application Execution environment for IBM desktop computers. Vendor: IBM Corp.

EzBook Notebook computer. Pentium CPU. Vendor: CTX International, Inc.

ezBridge Transact Communications software. Middleware. Vendor: Systems Strategies, Inc.

EZshell Operating system enhancement. Front end user interface to Unix systems. Upwards compatible with Bourne, Korn, and C shells so existing applications will work. Vendor: Touch Technologies, Inc.

EZTEST/CICS See CA-EZTEST/CICS.

EzX Application development tool used with Motif. Allows users to link to databases for prototyping without actually writing programs. Can interface with LISP, Basic. Includes: EzXdraw (draws windows, menus, etc. to build the GUI), EzXtalk (links scripts to windows), EzXpresent (demonstrates GUI before writing code), and EzXcode (generates C, Ada code. Vendor: Sunrise Software International, Inc.

F-Secure VPN Communications, Internet software. Security system that provides encryption to set up Virtual Private Networks. Provides secure tunnels between corporate sites. Runs on Windows systems. Vendor: Data Fellows.

F1 Manager Operating system software. Disk manager. Locates and reports file allocation. Allocates disk space by department. Vendor: Kisco Information Systems.

F3 Forms Software Includes: F3 Pro Designer which automates the design of forms, F3 Design & Mapping which creates stand alone forms to work with F3 Fill, a program which fills out, edits, prints, and faxes forms. Vendor: BLOC Development Corp.

F90 Application development tool. Language converter. Converts old Fortran programs to newer versions of Fortran. Vendor: Parasoft Corp.

FAAP Operating system enhancement used by systems programmers. Monitors and controls system performance in Harris systems. Vendor: Rimsco Software, Inc.

Fabs Plus Data management program. Maintains indexes for up to twenty files at a time. Runs on MS-DOS, Unix systems. Vendor: Computer Control Systems, Inc.

Face to Face Electronic meeting system that allows users to share entire documents. Most systems only allow users to share the page of the document that is on the screen. Vendor: Crosswise Corp.

Facelift Application development tool. Programming utility that allows the user interface on a program to be changed without having to rewrite code. Vendor: On-Line Software International.

Facets Application development environment. Object-oriented system. Interfaces with most major databases. Includes 4GL. Integrates with Smalltalk. Runs on RISC systems. Vendor: Reusable Solutions, Inc.

FacetTerm Operating system support. Runs multiple sessions on single terminal. Runs on Unix systems. Vendor: Structured Software Solutions, Inc.

Facilitator Groupware. EMS system. Provides anonymity. Records and analyzes votes. Vendor: McCall, Szerdy & Associates.

fact table A table used in a star schema to store the detail transaction level data.

Factfinder 1. Data management system. Runs on Macintosh systems. Vendor: ICON Review.
2. Data management system. Runs on IBM S/36 systems. Provides query and report generation facilities. Vendor: Pansophic Systems, Inc.

Factory See zApp.

Facts Data management system. Runs on DEC, Prime, Unix systems. Vendor: Database Systems Corp.

FailSafe Application development tool used for testing. Automatically recovers from Visual Basic code errors. Part of DevCenter. See DevCenter.

Falcon Code name for MSMQ. See MSMQ.

Falcon Gateway Communications software. Connects MQSeries and MSMQ. Vendor: Level 8 Systems, Inc.

FalconMQ Communications. Message-oriented middleware. Connects Windows NT applications with Unix, MVS, AS/400, VMS and Unisys systems. Implements MSMQ. Vendor: Level 8 Systems, Inc.

Family Pak Desktop computer. Pentium processor. Vendor: Royal Electronics Inc.

FamilyMAX Desktop computer. Pentium CPU. Operating system: Windows 95/98. Vendor: CyberMax Computer, Inc.

FAQS/MVS,VSE,VM Operating system enhancement used by systems programmers. Monitors and controls system performance in IBM systems. Vendor: Goal Systems International, Inc.

FAQS/PCS for VSE Operating system enhancement used by systems programmers and Operations staff. Job scheduler for IBM VSE systems. Vendor: Goal Systems International, Inc.

FAS 1100 Operating system add-on. Data management software that runs on Unisys systems. Vendor: Unisys Corp.

Fast Operating system add-on. Data management software that runs on Prime systems. Vendor: Database Systems Corp.

FAST Desktop system software. Spreadsheet. Runs on Pentium type systems. Stands for: Financial Advisory Support Techniques. Has Lotus 1-2-3 and Turbo Pascal formats available. Vendor: Financial Proformas, Inc.

FAST/DFAST Operating system enhancement used by systems programmers. Increases system efficiency in Bull HN systems. Vendor: Bull HN Information Systems, Inc.

Fast Dump Restore Operating system enhancement used by systems programmers. Increases system efficiency in IBM systems. Manages disk storage space. Vendor: Innovation Data Processing.

Fast Ethernet In communications, backbone technology which transfers data at 100mbps. Also called 100BASE-T. See Ethernet.

Fast Forward Application development tool. Reverse engineering. Extracts processes and data definitions to be used in client/server computing. Vendor: Seer Technologies, Inc.

Fast Load Operating System enhancement used by systems programmers. Increases system efficiency in IBM DB2 systems by loading tables faster than a standard load. Vendor: PLATINUM Technology, Inc.

Fast Path Function in IMS that allows faster response in an online system.

Fast Unload Operating system enhancement. Quickly unloads data files. Versions available for DB2, SQL Server, Sybase, Oracle. Vendor: PLATINUM Technology, Inc.

Fastback (Plus) Operating system enhancement. Improves system performance in IBM and compatible desktop computers. Handles file backups. Runs on Pentium type desktop computers. Vendor: Fifth Generation Systems.

FASTCPK Operating system enhancement. Provides DASD reorganization functions. Vendor: Innovation Data Processing.

FastDASD See CA-FastDASD.

FastForward Application software. Client/server system which includes financial, resource planning functions. For mid-size companies. Vendor: Oracle Corp.

FastFrEDI See FrEDI, FastFrEDI.

FASTGENR Operating system enhancement used by systems programmers. Increases system efficiency in IBM systems. Improves speed of IBM data utility IEBGENER. Vendor: Software Engineering of America.

FasTran Communications software. EDI package. Runs on AS/400 systems. Vendor: Intercoastal Data Corp.

FastStart Application development tool. Used to build datamarts. Part of Universal Warehouse. Released: 1997. Vendor: Informix Software, Inc.

FastTask Application development tool. Part of Power Tools. Used in object-oriented development for real-time systems. Vendor: ICONIX Software Engineering, Inc.

FASTVSAM Operating system add-on. High-speed copy program. Includes file and catalog management functions. Runs on IBM mainframe systems. Vendor: Software Engineering of America.

fat client In client/server computing, a system where most of the processing is done on the client system. "Fat client" systems require more powerful client machines.

FATAR Operating system add-on. Data management software that runs on IBM systems. Stands for: Fast Analysis and Recovery. Vendor: Innovation Data Processing.

FATS Operating system add-on. Data management software that runs on IBM systems. Stands for: Fast Analysis of Tape Surface. Vendor: Innovation Data Processing.

fault management Documenting and reporting network errors. One of the functions of network management software.

fault tolerant A system that continues to operate if a failure occurs. Fault-tolerant computers typically have redundant processors that automatically take over in the event of a failure.

Fault-XPERT Application development tool. Debugging tool. Client/server version of Abend-Aid. Released: 1995. Vendor: Compuware Corp.

Faver/VSE,MVS for DB2 Operating system add-on. Data management software that runs on IBM systems. Stands for: FAst Virtual Export and Restore. Vendor: Goal Systems International, Inc.

FCP See Foundation for Cooperative Processing.

FCS Communications standard for high speed communications channels, particularly fiber optics. Stands for: Fiber Channel Standard. Adhered to by many hardware vendors.

FCS AM,DS Notebook computer. Pentium CPU. Vendor: First Computer Systems, Inc.

FCS Pentium Desktop computer. Pentium CPU. Operating system: Windows 95/98. Options include: Business Pak, Express Pak, Green Pentium, MultiMedia Plus, Power Pak, President Pak. Vendor: First Computer Systems, Inc.

FDD Imaging software. Includes network printing, text retrieval, multimedia annotations, OCR, bidirectional faxing. Stands for: Feith Document Database. Vendor: Feith Systems and Software, Inc.

FDDI In communications, backbone technology. Stands for: Fiber Distributed Data Interface. Standard for high-speed fiber-optics communications. Used in local area networks.

FDL Special-purpose language used to write specifications for data files. Stands for: File Definition Language. Vendor: Digital Equipment Corp.

FDR Operating system add-on. Data management software that runs on IBM systems. Stands for: Fast Dump Restore. Vendor: Innovation Data Processing.

FDR/ABR Operating system enhancement used by systems programmers. Increases system efficiency in IBM systems. Complete DASD management system including disaster recovery. Vendor: Innovation Data Processing.

FDR/Upstream Operating system add-on. Mainframe system that provides automatic PC backup. Runs on MVS, interfaces with OS/2, Unix, Windows NT systems. Vendor: Innovative Data Processing.

FDRCLONE Operating system enhancement. Provides testing functions including cloning datasets from backups to reproduce the testing environment. Runs on MVS systems. Released: 1999. Vendor: Innovation Data Processing.

FDREPORT Operating system software. Provides real-time reporting functions. Released: 1997. Vendor: Innovation Data Processing.

FDRQUERY Operating system enhancement used by Operations staff and systems programmers. Monitors DASD usage. Vendor: Innovation Data Processing.

FDRREORG Operating System add-on. Data management software that runs on IBM systems. Automates the reorganization of VSAM, IAM and PDS datasets. Vendor: Innovation Data Processing.

FDRSOS Operating system enhancement. Provides backup functions for large databases at high speeds. Stands for: FDR Safeguarding Open Storage. Vendor: Innovation Data Processing.

Fdump Operating system add-on. Data management software that runs on Unisys systems. Vendor: Information Systems Corp.

FEDI Communications software. EDI package. Runs on IBM mainframe systems. Stands for: Financial Electronic Data Interchange. Vendor: Servantis Systems, Inc.

FETCH(/XA) Operating system enhancement used by systems programmers. Monitors and controls system performance in IBM CICS systems. Monitors CICS load time. Vendor: Axios Products, Inc.

Fiber Channel Standard See FCS.

fiber optics Cable used in communications that transmits signals using light beams. Provides high speed digital data transmission.

fibre channel Communications. High speed interconnect technology. Hardware interface that allows for the connection of peripheral devices. Increases both speeds and distances over SCSI.

field A piece of information. For example, information in a human resource file or database would contain fields for last-name, phone-number, street-address, city, state, etc. Each piece of information is a separate field.

FIFO A method of processing data in that the oldest item is processed first. Stands for: First In, First Out.

fifth-generation computer The next generation of computers that are designed around artificial intelligence methods and applications. Still in the design stage.

Figaro(+) Implementation of graphics standard (PHIGS) designed for advanced graphics. Runs on IBM mainframe, DEC VAX, Unix systems. Vendor: Template Graphics Software, Inc.

Filcon Operating system add-on. Data management software that runs on Bull HN systems. Vendor: Scientific and Business Systems, Inc.

file A collection of records processed as a unit. For example, a record for each employee containing personal and salary information is grouped together to make up the payroll file. Each record has only one group of fields, so each access to the file retrieves or stores all the information. These records are stored, retrieved, and processed together. Also called flat file, dataset.

File Data management system. Runs on Macintosh systems. Vendor: Microsoft Corp.

File-AID Operating system add-on. Data management software that runs on IBM systems. Programs include: File-AID for IMS, File-AID for ROSCOE, File-AID/Batch, File-AID/SPF, File-AID/XE, FILE-AID/PC. Vendor: Compuware Corp.

File Explorer Operating system software. File management system. Part of Windows 95/98. Vendor: Microsoft Corp.

file server A device in a network (usually a computer and one or more disks) on which files and applications are stored to be shared throughout the network.

File Transfer Communications software. Network connecting IBM systems. Micro-to-mainframe link. Vendor: IBM Corp.

File Transfer Protocol See FTP.

File Transfer System Communications software. Network connecting IBM systems. Micro-to-mainframe link. Vendor: Mackensen Corp.

Filebase Database for desktop computer environments. Runs on IBM desktop systems. Vendor: EWDP Software, Inc.

Filecomp Operating system add-on. Data management software that runs on IBM systems. Vendor: Dataware Conversion Services.

FileMaker Pro (for Windows) Relational database that runs on Macintosh, Windows systems. Developer edition includes tools and programming interfaces. Multi-user DBMS. Vendor: FileMaker, Inc.

filePRO 16 (Plus) Relational database for midrange and desktop computer environments. Runs on IBM, Unix, Xenix systems. Vendor: The Small Computer Co., Inc.

Filer Database for desktop computer environments. Runs on Apple II systems. Vendor: Spinnaker/Hayden.

Filesave/(Archive,RP) Operating system add-on. Data management software that runs on IBM systems. Vendor: Computer Corp. of America.

Filetab-D Application development tool. Program that includes debugging, maintenance, and documentation support. Runs on DEC systems. Vendor: EEC Systems, Inc.

Final Exam Internet Application development tool. Testing and debugging software that lets Web developers repeatedly test Web pages and analyze Web server performance under high user load levels. Includes WebLoad, Internet Test, and C/S Test. Released: 1996. Vendor: PLATINUM Technology, Inc.

Financial Electronic Data Interchange See FEDI.

Financial Stream, HR Stream Applications software. Financial tracking and human resource systems. Runs on Unix, Windows. Client/server technology. Part of SmartStream. Vendor: Dun & Bradstreet Software.

Find Application development tool. Provides full text search for Exchange-based information. Vendor: Vendor: Fulcrum Technologies, Inc.

Finder Desktop system software. Operating system utility that manages the desktop on Apple Macintosh systems. MultiFinder is another version. Vendor: Apple Computer, Inc.

FindOut, FindOut Builder Application development tools used for business access and analysis. Data access and information modeling tools for client/server environments. Vendor: Open Data Corp.

Finest Hour Desktop software. Scheduling package. Vendor: Primavera Systems, Inc.

finger Type of communications software. A program in the Internet that lets you see if a specific user is online, or which users are online at a specific site.

fire-and-forget Transaction queuing technique that allows users to run multiple requests against a database without waiting for the results of each query.

Fire-Wall-1 Communications, Internet software. Firewall. Provides three levels of access for programs written in Java. Vendor: CheckPoint Software Technology, Ltd.

Firebox II Communications software. Firewall. Released: 1998. Vendor: Watchguard Technologies, Inc.

Firefly Passport System software. Allows users to define what information about themselves can be accessed over the Web by defining a passport. Written by Firefly, but purchased by Microsoft in 1998. Vendor: Microsoft Corp.

firewall Security system. The software and/or hardware used to block certain kinds of traffic between the Internet and corporate information systems thus protecting corporate systems and information.

Firewall-1 Communications. Firewall that includes policy-based security management. Vendor: Cisco systems, Inc.

firewall appliance Hardware and software bundled to provide Internet security.

FireWire Communications connection used with multimedia equipment. Used in electronics industry; starting to be used in peripheral/computer connections. Known as IEEE standard 1394. Used in PowerMac G3 systems.

firmware Programs permanently stored in the computer. Firmware programs are always available and cannot be changed by the user. Firmware programs are specific to each computer and are written by technical programmers who know the hardware.

First Aid Operating system add-on. Set of utilities to improve the operation of the system. Includes WebScan X. Vendor: CyberMedia, Inc.

First Computer Computer vendor. Manufactures desktop computers.

first-generation computer Computers built with vacuum tubes. Commercially available in the 1950s.

firstCASE Case product. Automates the entire project life cycle including project management. Runs on Pentium type desktop computers. Vendor: AGS Management Systems.

FirstClass Communications software. Messaging and groupware. Includes e-mail, discussion databases, remote access, forms processing, EMS. Provides Internet access. Vendor: SoftArc, Inc.

FirstSense Enterprise Operating system software. Measures performance in client/server systems by monitoring actual response time experienced by the user. Vendor: FirstSense Software, Inc.

FirstSTEP Application development tool. Provides process modeling functions. Runs on Windows systems. Vendor: Interfacing Technologies Corp.

FIS Application development tool. Report generator for DEC VAX systems. Vendor: Park Software, Inc.

Fiscal Application development tool. DSS. Provides access to data stored in relational databases. Runs on Macintosh, OS/2, Unix, Windows systems. Vendor: Lingo Computer Design, Inc.

Flagship Application development tool. Enables Clipper programmers to develop under Unix. Vendor: WorkGroup Solutions.

FLASHER Operating system enhancement used by systems programmers. Job scheduler for IBM systems. Vendor: Tone Software Corp.

flashpoint See VISION:Flashpoint.

flat file See file.

flat panels Hardware. Flat, LCD display monitors. Provide excellent pictures, lightweight, present more information than traditional CRT screens. Drawback—expensive.

Flavors Object-oriented extension to LISP. See LISP.

Flecs Application development tool. Program, or code generator. CICS generator and information retrieval system. Runs on IBM mainframes. Vendor: DEK Software International.

Flee/VSE Operating system add-on. Library management system that runs on IBM VSE systems. Vendor: Goal Systems International, Inc.

Flex Operating system add-on. Data management software that runs on Unisys systems. Vendor: Software Clearing House, Inc.

Flex xxx Desktop computer. Pentium CPU. Operating system: Windows 95/98. Vendor: USA Flex Inc.

Flexelint Application development tool. Analyzes C language programs looking for redundancies, inconsistencies and bugs. Runs on MVS, Unix, VMS systems. Vendor: Gimpel Software.

FLEXexpress Internet software. E-commerce system that supports software distribution and licensing. Runs on DEC, Unix, Windows systems. Released: 1998. Vendor: Globetrotter Software, Inc.

FlexiPayables Applications software. Accounting package. Runs on OS/2, Unix, Windows systems. Interfaces with NetWare, SQLBase. Client/server technology. Vendor: FlexiWare Corp.

FlexLink-VAX,Gould,IBM,Sun Communications software. Network connecting various systems. Vendor: FlexLINK International Corp.

FlexOS Operating system for IBM and compatible desktop computer systems, Unix systems. Vendor: Digital Research, Inc.

FlexQL Application development tool. Report writer. Accesses data stored in DataFlex, dBase, Paradox, Lotus, SYLK, DIF, and ASCII . Vendor: Data Access Corp.

FloodGate-1 Communications software. Controls bandwidth allocations to ensure that vital business needs have priority. Vendor: CheckPoint Software Technologies Ltd.

floppy disk See diskette.

Flow CASE product. Automates design function. Vendor: Digital Management Group, Ltd.

flow manufacturing Manufacturing design that integrated production lines that schedule work only when orders are placed. Replaces traditional stand-alone manufacturing departments. Produces large cost savings, especially in inventory management.

FloWare and Map/Builder Workflow software. Object-oriented package. Vendor: Recognition International.

flowchart A graphical representation of the design of a program and/or system. Flowcharting skills indicate knowledge of program and/or system design.

FlowLogic Application development tool. Models and manages inventory control, document management, project management. Provides the interface for these systems. Vendor: Workflow Systems.

FlowMark Workflow software. Runs on AIX, OS/2, Windows systems. Vendor: IBM Corp.

FlowModel Workflow software. Depicts, analyzes, and communicates complex business processes. Runs on Windows systems. Vendor: Arcland, Inc.

FLUFA 770 Notebook computer. Pentium CPU. Vendor: Chaplet Systems USA, Inc.

FMS Data management system. Runs on DEC systems. Stands for: File Management System. Vendor: NSI, Inc.

FMS II family Application software. Financial systems for midsize environments. Vendor: Mitchell Humphrey & Co.

FMU Application development tool. Program that manages form descriptions. Stands for: Forms Management Utility. Runs on DEC systems. Vendor: Digital Equipment Corp.

FOCAL Compiler language. Used for scientific and mathematical applications.

Foccalc Desktop system software. Spreadsheet. Runs on Unix systems. Vendor: Information Builders, Inc.

Focnet Communications software. Network connecting FOCUS systems. Vendor: Information Builders, Inc.

FocTalk Communications software. Network connecting IBM computers. Micro-to-mainframe link. Vendor: Information Builders, Inc.

FOCUS 4GL/database. Application development environment. Includes 4GL, database management system, report writer, business graphics, statistics, reporting functions. Versions available for most platforms. Vendor: Information Builders, Inc.

Focus/EIS for Windows EIS software that combines Lightship with Focus to produce an EIS that can access mainframe databases such as DB2 and Focus and works on the Windows desktop. Vendor: Jointly produced by Pilot Executive Software and Information Builders, Inc.

FOCUS for Open Environments Application development environment. Supports end-user computing. Provides relational database interfaces, data migration tools. Includes parallel query technology. Runs on Unix systems. Vendor: Information Builders, Inc.

FOCUS Fusion Multidimensional database. Used for decision support systems using complex queries. Allows users to access relational and non-relational data and build software using vendor-supplied or user-written applications. Runs on Unix systems. Released: 1996. Vendor: Information Builders Inc.

Focus/Reporter EIS tool. Query and reporting tool for users. Vendor: Information Builders, Inc.

Focus Six for Windows Application development tool. Family of query tools. Includes: Reporter Edition, Managed Reporter Edition, Managed Reporter Administrator's Kit, Developer's Toolkit, Report Server. Vendor: Information Builders, Inc.

FolderView Imaging system. Runs on Windows systems. Vendor: FileNet Corp.

Folio Views Document retrieval system. Works with Web servers. Runs on Macintosh, Windows systems. Vendor: Open Market Inc.

font A set of characters of a particular design and use. Used in printing with word processors and desktop publishers.

footprint The amount of geographic space covered by an object. A computer footprint is the desk or floor surface it occupies. The word is most often used to refer to the amount of RAM a program occupies while executing. Especially important in small systems such as handhelds, and for imbedded systems. Effective programs for these systems are said to have "small footprints."

FOR_C Application development tool. Converts Fortran code to ANSI C. Runs on DEC, Unix systems. Released: 1997. Vendor: Cobalt Blue, Inc.

For:Pro Operating system for midrange Fortune systems. Vendor: Fortune Systems Corp.

FOR_STRUCT Application development tool. Builds structure into old FORTRAN programs. Runs on DEC, Unix systems. Released: 1996. Vendor: Cobalt Blue, Inc.

ForComment Groupware. Allows group members to annotate and revise documents. The author can review and incorporate revisions with a single keystroke. Vendor: Computer Associates International, Inc.

Forecast Pro Application development tool. Statistical forecasting tool. Allows user to make predictions based on past patterns. Runs on OS/2, Windows systems. Vendor: Business Forecast Systems, Inc.

Foresight CASE product. Automates design and project management functions. Vendor: Computer & Engineering Consultants Ltd.

ForeSite Application development tool. Used to convert existing Visual Basic, PowerBuilder, C++, COBOL, Centura and other applications to the Web. Runs on Windows systems. Released: 1997. Vendor: Centura Software Corp.

Forest & Trees EIS. Runs on Pentium type desktop computers. Vendor: PLATINUM Technology, Inc.

FORGO-77 Compiler language used in midrange environments. Fortran compiler for Harris systems. Vendor: Harris Corp.

FORMAC An extension of PL/1 designed for non-numeric manipulation of mathematical expressions. Stands for: Formula Manipulation.

Formatrix Applications development tool. 4GL program generator used in DEC VAX systems. Vendor: Iskra Software International.

Formida Fire (for the Web) Application development environment. Used to develop software in Formida source code, which can be executed directly. Interfaces with multiple databases and can work with textual, graphical, spatial, numerical or 3D data. Formida Fire for the Web includes Web deployment capabilities. Runs on DEC, Unix, Windows systems. Vendor: Formida Software Corp.

FormFlow Workflow software. Includes electronic forms. Runs on Windows systems. Client/server software. Vendor: Delrina Technology, Inc.

Forms-Plus/400 Application development tool. Creates forms, labels, and checks on Desktops and allows them to be uploaded to AS/400 systems. Vendor: Eclipse Corp.

FormTalk Electronic forms software. Uses WorkPlace Shell. Vendor: IBM Corp.

FORMULA/1 See DBA Tool Kit.

Formula One Net Application development tool. DSS. Provides spreadsheet capabilities for Web pages. Vendor: Visual Components, Inc.

Forte Application development environment. Used for enterprise development of enterprise systems. Includes 4GL, GUI, application partitioning, interactive debugger. Object-oriented. Creates three-tiered applications. Applications are built independent of platform. A logical application definition is stored in a repository and the system automatically maps the application to the desired platform(s). Released: 1994. Vendor: Forte Software, Inc.

Forte Application Environment Application development tool. Integrates CORBA and DCOM to allow users to connect systems that run the two technologies.

Forte Express Application development tool. Automatically generates multitier applications. Handles networking requirements and supports applications management. Runs with Forte's development environment. Released: 1996 Vendor: Forte Software, Inc.

Forte Fusion Application development tool. Integrates applications from major ERP and application software vendors. XML-based. Works with SAP's R/3, Siebel's and Vantive's front-office applications, PeopleSoft's ERP system and others. Released: 1999. Vendor: Forte Software, Inc.

Forte WebEnterprise Application server providing middleware and development tools. Used to create Web applications. Includes a graphical wizard based Internet application generator. Vendor: Forte Software.

FORTH Compiler language used in desktop computer environments. Designed for applications involving mathematical calculations. Used for astronomy, robotics, and graphics.

Fortiva Desktop computer. Pentium CPUs. Operating system: Windows 95/98. Vendor: Leading Edge Hardware Products, Inc.

FORTRAN Compiler language. Stands for Formula Translator. Designed for applications using mathematical calculations and used primarily in scientific, engineering and mathematical applications. There are many versions of FORTRAN available.

FORTRAN POWERBench See POWERBench.

FOS Operating system for Hewlett-Packard systems. Vendor: Hewlett-Packard Co.

Foundation 1. CASE product. Automates analysis, design, and programming functions. Includes code generator for C, COBOL. Design methodologies supported: Chen, DeMarco, Information Engineering, Merise, Yourdon. Databases supported: DB2, Rdb, Sybase, Oracle. Runs on IBM, DEC, Bull HN systems. Includes: Method/1, Plan/1, Design/1, Install/1. Vendor: Andersen Consulting. 2. Database for desktop computer environments. Runs on Apple II systems. Includes spreadsheet and word processing functions. Vendor: Foundation Corp.

Foundation for Cooperative Processing Application development tool that works in a client/server architecture. Builds client applications with a GUI-based front end. Includes object-oriented repository. Interfaces with IBM, SQL Server. Vendor: Andersen Consulting.

Foundation Vista CASE product. Automates analysis and design functions. Uses Gane-Sarson, Jackson, and Yourdon methodologies. Runs on Macintosh systems. Vendor: Menlo Business Systems, Inc.

Fountain Communications, Internet software. Includes VRML browser and world-building tool. Runs on Windows systems. Vendor: Caligari.

FourGen Application software. Manufacturing software handling inventory control, order entry, distribution, supply chain processing, warehouse management. Runs on Unix systems. Released: 1996. Vendor: FourGen Software, Inc.

FourGen Visual Application development tool. Provide decision support functions. Runs on Windows systems. Vendor: FourGen Software, Inc.

fourth-generation computer Computers built with chips. Commercially used in the 1980s and 1990s.

fourth-generation language See 4GL.

FoxBASE (+,Plus), FoxPRO Relational database for desktop computer environments. Runs on IBM, Macintosh, Unix, Xenix systems. FoxPRO includes development tools. Part of xBase. Vendor: Microsoft Corp.

Foxfire Application development tool. Query and report functions. Runs on Macintosh, Windows systems. Vendor: Micromega Systems, Inc.

Foxgraph Desktop system software. Graphics package. Interfaces with Lotus 1-2-3, Excel, dBase. Runs on MS-DOS systems. Vendor: Fox Software, Inc.

frame relay Communications technology for wide area networks. Used in both private and public networks. Alternative technologies are SMDS, ATM.

FrameMaker Desktop software. Integrated software that combines word processing with desktop publishing. Runs on Macintosh, Unix, Windows systems. Vendor: Frame Technology Corp.

framework A set of software building blocks that programmers can use, extend, or customize for specific computing solutions. The basic infrastructure of applications. Provides developers with such things as file and edit menus, print reports and pre-defined screens so the same code doesn't have to be written for each application. A framework used in object-oriented development is a set of related classes.

Framework Workflow software. Object-oriented process modeling tools. Runs on Unix, Windows NT systems. Vendor: Ptech, Inc.

Framework IV Desktop software. Integrated package. Runs on Pentium type desktop computers. Vendor: Inprise Corp.

FrEDI, FastFrEDI Communications software. EDI package. Runs on Pentium type PCs. Vendor: EDI Able, Inc.

FreeFlow Application development tool. Part of Power Tools. Supports DeMarco structured analysis. Vendor: ICONIX Software Engineering, Inc.

FreeHand Desktop system software. Graphics package. Runs on Macintosh computers. Vendor: Aldus Corp.

Freelance Graphics for OS/2, DOS Desktop system software. Graphics package. Interfaces with Lotus 1-2-3, Excel, dBase. Vendor: Lotus Development Corp.

Freelink Communications software. Network connecting IBM systems. Micro-to-mainframe link. Vendor: On-Line Software International.

Freeloader Communications, Internet software. Off-line browser. Vendor: Freeloader, Inc.

FreeSpeech 98 Speech recognition software. Released: 1998. Vendor: Philips Electronics N.V.

freeware Software that is free. Also called public domain software. Available over the Internet through anonymous FTP.

Frequent Flier Image processing software. Interfaces with most SQL databases. Vendor: Optika Imaging Systems.

Fresco Group of hardware and software products used to deploy Java applications over networks. Includes Fresco Designer (RAD development tool), Fresco Information Server (Internet server), Fresco Adapters (adapters). Released: 1997. Vendor: Infoscape, Inc.

front-end The part of a program that contains the user interface. In client/server, the client portion of the application.

front-office software Application software that is visible to the user. Addresses the business processes that are directly related to customers. Includes automation of functions such as sales, marketing, help desk, call center, and configuration management. SFA (Sales Force Automation) and CIS (Customer Information systems) are front-office systems. Integrates with ERP systems which are then the back-office systems.

Frontline Manager System management tool designed for small businesses. Keeps track of inventories and manages PCs. Vendor: Manage.Com, Inc.

Frontline Viewbuilder Application development tool. Program that allows development of applications that can run on both mainframes and desktop computers. Runs on DEC, IBM systems. Vendor: Dun & Bradstreet Software.

FrontMind Communications. Internet software. Artificial intelligence software that monitors Web site visitors actions and adjusts the site to the projection of what the visitor wants in real-time. Runs on Linux, Solaris and Windows systems. Vendor: Manna, Inc.

FrontOffice See AurumFrontOffice.

FrontOffice98 Application software. Integrated front-office suite that includes call-enter, field-service and sales force automation systems. Released: 1998. Vendor: Clarify, Inc.

FrontPage (98) Application development tool. Web authoring tool. Automatically inserts HTML tags into text to create Web pages. FrontPage 98 generates dynamic HTML and works with Explorer. Runs on Windows systems. Vendor: Microsoft Corp.

FRT Operating system enhancement. Speeds IMS database recovery. Stands for: Fast Reorg Facility. Vendor: BMC Software, Inc.

FSA Applications software. Part of HR, Payroll, and/or Accounting systems. Stands for: Fixed Savings Account.

FSE Operating system add-on. Data management software that runs on CDC systems. Stands for: Full Screen Editor. Vendor: Control Data Corp.

FSE+ Application development tool. Screen editor for IBM MVS mainframe systems. Stands for: Full Screen Editor +. Vendor: Applied Software, Inc.

FT-5xx Active, Passive Color Notebook computer. Pentium CPU. Vendor: Wynn Data. Ltd.

FT/Express TSO,CMS Communications software. Network connecting IBM systems. Micro-to-mainframe link. Vendor: Digital Communications Associates, Inc.

FTAM 1. Communications protocol. Applications-level protocol that controls transfer of files, particularly between different vendors. OSI accepted protocol. Stands for: File Transfer Access and Management.
2. Communications software. Network connecting DG computers. Stands for: File Transfer, Access, and Management. Vendor: Data General Corp.

FTP 1. Communications protocol. Applications-layer protocol that handles file transfer. Not as general-purpose as FTAM. Stands for: File Transfer Protocol. Used with TCP/IP.
2. Operating system add-on. Data management software that runs on IBM systems. Stands for: File Transfer Program. Vendor: IBM Corp.

FTS Operating system add-on. Data management software that runs on Prime systems. Vendor: Prime Computer, Inc.

FTS/ Communications software. Network connecting IBM computers. Stands for: File Transfer System. Micro-to-mainframe link. Includes: FTS/CICS, FTS/IDMS, FTS/IMS, FTS/VTAM, FTS/Entry Level, FTS Send/Receive, FTS/PassTHRU. Vendor: Mackensen Corp.

ftSPARC Midrange computer. SuperSPARC RISC CPU. Operating system: Solaris. Vendor: Sun Microsystems, Inc.

FTX Operating system. Fault-tolerant version of Unix. Vendor: Stratus Computer Inc.

Fujitsu Computer vendor. Manufactures pen-based computers.

Fulcrum SearchServer Document management software. Indexes documents as soon as they are added to the database. and reindexes them if they are modified. SearchRuns on Unix, Windows systems. Released: 1996. Vendor: Fulcrum Technologies Inc.

Full Moon Operating system software. Provides clustering capabilities allowing Sun users to link multiple SPARC servers. Vendor: Sun Microsystems, Inc.

functional decomposition, analysis, design Design methodology associated with structured programming. Views an application as a set of functions that can be decomposed, or broken down, into steps that can be easily understood and programmed. The decomposition of a problem takes many steps until the final breakdown produces the program specifications for individual programs. Also called top-down development. Methodologies designed based on functional decomposition are Constantine, DeMarco, Gane and Yourdon.

FUP Operating system add-on. Data management software that runs on Tandem systems. Vendor: Tandem Computers, Inc. (a Compaq company).

Fusion 1. Object-oriented development methodology. Associated with Coleman et al.
2. Artificial intelligence system. Expert system building tool. Runs on desktop systems. Interfaces with Lotus 1-2-3. Vendor: 1st-Class Expert Systems, Inc.
3. See Focus Fusion.
4. See NetObjects Fusion.
5. See Netron Fusion.
6. See Chorus/Fusion.

Fusion FTMS Communications software. Transfers files from LANs to mainframes. Runs on MVS, Windows NT systems. Stands for: Fusion File Transfer Management System. Vendor: Proginet Corp.

Fusion OneBook Desktop computer. Pentium processor. Vendor: Akia Corp.

FutureMate Notebook computer. Pentium CPU. Vendor: Futuretech Systems, Inc.

FuziCalc Application development tool used for data mining. Provides fuzzy analysis of multiple criteria, giving user results close to stated criterion. Provides best and worst case results. Runs on Windows systems. Vendor: FuziWare, Inc.

fuzzy logic Logic based on imprecise parameters. Allows such queries as "select employees who are tall and young." Traditional queries would have set values and state "select employees where height is greater than six feet and age is less than 30. Fuzzy logic is used in artificial intelligence.

Fuzzy System Standard Environment Data structure to represent fuzzy logic systems. Attempt to create public-domain standards by Motorola Microprocessor, Memory Technologies Group, and Aptronix.

FW5xxx, FW7000 Portable computer. Pentium CPU. Vendor: Fieldworks, Inc.

FX32 System software. Emulates Intel CISC chip from an Alpha RISC/CPU. Vendor: Digital Equipment Corp.

FYI Communications, Internet software. Security system that provides public key encryption to set up Virtual Private Networks. Includes firewall technology, handles digital certificates. Released: 1997. Vendor: Internet Dynamics, Inc.

FYISuite Application development tool used for business access and analysis. DSS for forecasting and sales analysis. Includes FYI Planner. Vendor: Think Systems Corp.

FYPlan Desktop system software. Spreadsheet. Enhancements to spreadsheets specifically for budgeting. Runs on Windows systems. Vendor: Pillar Corporation.

G, GB Gig. Stands for gigabyte, approximately one billion bytes.

G-Logis Programming language that combines LISP and Prolog. Used in artificial intelligence. Vendor: Graphael, Inc.

G-Series Desktop computer. Pentium II CPU. Operating system: Windows 95/98. Vendor: Gateway 2000, Inc.

G/SNA Gateway Communications software. Network connecting HP systems. Micro-to-midrange link. Vendor: Gateway Communications Corp.

G4, G5 Mainframe computer. Includes air cooled processors. Part of S/390 family. Vendor: IBM Corp.

GA Workbench Application development tool. Used to add a graphical front-end to existing OS/2 and/or Windows applications. Stands for: Graphical Application Workbench. Vendor: INTERSOLV, Inc.

Gaec Computer Corp Ltd. Software vendor. Acquired D&B Software in 1996. Two divisions: Gaec Enterprise Server which produces Expert and Millennium accounting, HR, and payroll systems for mainframes; and Gaec SmartStream which does the same for client/server environments.

Galilelo Pentium Desktop computer. Pentium processor. Vendor: ET Technology.

Gain, GainMomentum Application development environment. Used to develop object-oriented systems. Handles multimedia objects, European and Japanese character sets. Allows users to combine video, voice, data. Runs on Unix systems, interfaces with Sybase, Oracle. Uses SQL. Vendor: Sybase, Inc.

Galaxy 1100 Portable computer. Pentium CPU. Vendor: Science Applications International Corp.

Galaxy(/C++) Application development environment. Used for enterprise development. Includes workbench software. Allows for development of object-oriented multi-platform, client/server applications. Builds GUI front-ends. Developers can create systems under Windows NT, Sun/Open Look, Motif/Unix, or Macintosh. Interfaces with Oracle, Sybase. C, C++ versions available. Vendor: Visix, Inc.

Galaxy Model xxx Midrange computer. Server. Pentium, Pentium Pro CPU. Vendor: Technology Advancement Group, Inc.

GAM Access method used in IBM systems.

GameBreaker 3D Desktop computer. Pentium CPU. Operating system: Windows 95/98. Vendor. ProGen Technology Inc.

Gamma Application development tool. Program generator for IBM systems. Vendor: KnowledgeWare, Inc.

Gane-Sarson Structured programming design methodology named for its developers. Based on functional decomposition. Accepted as a standard design methodology by some companies, and used by some CASE products.

gateway In communications, the connection of dissimilar communications systems. A mechanism used to connect two networks that operate under dissimilar protocols. Can refer to hardware or software.

Gateway 2000 Computer vendor. Manufactures desktop computers.

Gateway*Express Communications software. EDI package. Runs on IBM, Unix systems. Vendor: SMS Corp.

Gateway family Desktop computer. Pentium CPU. Operating system: Windows 95/98. Vendor: Gateway 2000, Inc.

Gateway PC Communications software. Network connecting IBM systems. Micro-to-mainframe link. Vendor: Software Corp of America.

Gateway Professional Desktop computer. Pentium CPUs. Operating system: Windows 95/98. Vendor: Gateway 2000, Inc.

Gateway Solo See Solo S90.

Gator Mail(-M,Q) Communications software. Gateway for e-mail systems. Runs Macintosh systems. Vendor: Cayman Systems, Inc.

Gauntlet Internet Firewall Communications, Internet software. Firewall. Vendor: Trusted Information Systems, Inc.

GBB Application development environment for developing artificial intelligence programs on HP/Apollo workstations. Vendor: Blackboard Technology Group, Inc.

GC-LISP Stands for: Golden Common LISP. See LISP.

GCOS General-purpose operating system for Bull HN large computer systems. Versions include: GCOS 6, GCOS 64, GCOS 7, GCOS 8. Vendor: Bull HN Information Systems, Inc.

GDBS Database for large computer environments. Desktop computer version available. Runs on Bull HN, DEC, HP, IBM, Prime systems. Full name: General Data Base System. Vendor: DTRO, Inc.

GDC/SCON Operating system enhancement used by Operations staff and Systems Programming. Provides operator console support in IBM MVS systems. Consolidates console messages from multiple systems. Vendor: Duquesne Systems, Inc.

GDDM Graphics package that runs on IBM mainframes. Stands for: Graphical Data Display Manager.

GD*Draw* Application development tool. Provides engineering drawing facilities including drawing templates. Part of Graphical Designer. Vendor: Advanced Software Technologies, Inc.

GD*MethodBuilder* Application development tool. Used to create corporate methodologies, or customize standard methodologies. Includes scripting language—GDL. Part of Graphical Designer. Vendor: Advanced Software Technologies, Inc.

GD*Pro* Application development tool. Used for object-oriented design and re-engineering. Supports Rumbaugh, Booch, Shlaer/Mellor methodologies. Includes Java, C++, Smalltalk code generators. Part of Graphical Designer. Vendor: Advanced Software Technologies, Inc.

GDX 1. Application development system. Includes tools to develop online systems. Allows for development of mainframe systems on a desktop computer. Runs on IBM systems. Vendor: General Data Systems, Ltd.
2. Application development tool. Automates programming and testing functions. IBM, Tandem systems. Stands for: General Data Expansion. Includes: desktop computer version, GDX Application Generator, GDX Menu-Driven Interface, GDX Programming Language. Vendor: Electronic Data Systems Corp.

Gecomo Plus CASE product. Automates project management function. Supplies cost accounting information for project development and maintenance. Vendor: GEC-Marconi Software Systems.

GEM System management software. Integrates tools for managing IBM mainframe system with Tivoli Management Environment (TME 10). Stands for: Global Enterprise Manager. Vendor: Tivoli Systems, Inc.

GemBase Application development system. Includes applications generator, report generator, data dictionary, menu processor, and 4GL. Runs on DEC, Unix systems. Vendor: Ross Systems, Inc.

GEMbase Database for large computer environments. Runs on NCR systems. Vendor: Software Clearing House, Inc.

GEMCOS Communications software. Transaction processing monitor. Runs on Unisys systems. Stands for: Generalized Message Control System. Vendor: Unisys Corp.

Gemini Code name for a Unix operating system developed by Hewlett-Packard, Santa Cruz Operation, and Novell.

GEMMS Application software. Process manufacturing system. Includes production management, sales order processing, MRP, financial applications, production scheduling and other modules. Stands for: Global Enterprise Manufacturing Management System. Runs on DEC, Unix systems. Released: 1995. Vendor: Oracle Corp.

GemORB Communications. Middleware, ORB. CORBA compliant. Includes CORBA development tools and system administration tools. Runs on Unix, Windows systems. Released: 1997. Vendor: GemStone Systems, Inc.

GemStone Object-oriented database. Can be accessed through Smalltalk, C++, Pascal, Ada, COBOL, Fortran, LISP. Uses Opal. Runs on DEC, IBM, Sun, Unix systems. Vendor: GemStone Systems, Inc.

GemStone GeODE Application development environment. Used for object-oriented development. Used to visually develop code-free applications. Runs on IBM RS/6000, DEC, Hewlett-Packard, Sun systems. Vendor: GemStone Systems, Inc.

GemStone/J Application server providing middleware and development functions. Used to build and deploy enterprise-wide Java business-to-business e-commerce applications. Provides an all-Java CORBA ORB. Supports? a wide choice of development tools, components, Web servers, client browsers and databases. Vendor: GemStone Systems, Inc.

GemStone Object/Web Server Application development environment. Used to build, deploy and manage enterprise-wide, scalable, client/server applications. Runs on Unix, Windows systems. Includes GemORB. Vendor: GemStone Systems, Inc.

GemStone/S Application server providing middleware and development functions. Used to develop and partition Smalltalk three-tiered applications. Provides concurrent access to shared objects, data distribution, data replication, on-line services for continuing operations, security. Runs on Macintosh, Unix, Windows systems. Vendor: GemStone Systems, Inc.

Gen/C++ Application development tool. Part of Object Development Workbench. Generates C++ code for Windows. Vendor: System Software Associates, Inc.

Gen/X Application development environment. Generates applications for X databases. Includes GUI builder. Runs on Unix systems. Vendor: Agetek, Inc.

Gen81 Application development tool. Automates programming and testing functions. Vendor: Evansville Data Processing Corp.

Gener/OL Application development tool. Runs on IBM systems. Vendor: Pansophic Systems, Inc.

GenerAda Application development tool. Program generator for Ada applications. Vendor: Oracle Corp.

General Automation Computer vendor. Manufactures desktop computers.

General Manager II Database for desktop computer environments. Runs on Apple II systems. Vendor: Sierra On-Line, Inc.

Generation 5, 6 Mainframe computers. Versions of IBM's S/390 parallel enterprise server.

Generation Five Database for large computer environments. Runs on Bull HN systems. Vendor: Generation Five and Services, Inc.

Generic Lint See Lint.

Genesis LT,MP Midrange computer. PowerPC RISC CPU. Operating system: System 7. Vendor: Daystar Digital, Inc.

Genesis V Application development tool. Automates programming and testing functions. Vendor: Help/38, Inc.

Genesys Enterprise Series Applications software. Human resource, payroll systems. Client/server. Runs on Sun systems. Vendor: Genesys Software Systems, Inc.

Geneva Application development tool. EAI software used to integrate legacy, e-commerce, Web and Windows application systems. Works with most message-oriented middleware systems. Incorporates COM/DCOM, MTS, IIS and MSMQ. Works with VB or Java Script and XML. Runs on Windows NT systems. Released: 1999. Vendor: Level 8 Systems, Inc.

Geneva V/T Application development tool used for data mining. Used to determine which of hundreds of relationships among variables are significant. Runs on MVS systems. Vendor: Price Waterhouse LLP.

GeneXus CASE tool. Automates analysis, design, programming function. Runs on OS/2, Unix, Windows systems. Vendor: GeneXus, Inc.

Genie Desktop system software. Word processor. Runs on Pentium type systems. Vendor: Foresight Technologies Corp.

Genifer Application development tool. Program generator for IBM desktop computers. Vendor: Bytel Corp.

Genisys I Database for large computer environments. Runs on DG systems. Database for non-technical users. Vendor: DMS Systems, Inc.

Genpulse Application development tool. Program generator for Unisys systems. Part of Group Four application development environment. Vendor: ESI.

Gentia, GentiaDB Application development tool used in data warehousing. Client/server system that provides complex analysis and reporting tools. Can be used over the Web. Includes GentiaDB, a multidimensional database that stores up to 16 terabytes. Vendor: Gentia Software.

Gentia WebSuite Application development tool. Used to deploy DSS and OLAP applications through Web browsers. Users create pages from drag-and-drop and HTML is generated. Runs on Macintosh, OS/2, Unix, Windows systems. Released: 1996. Vendor: Gentia Software.

Gentium EIS (Executive Information System). Uses object technology to allow users to reformat data without technical help. Includes database. Vendor: Planning Sciences.

Gentran Communications software. Message oriented middleware, data transformation tool. EDI package. Runs on IBM mainframe systems. Family of products including Gentran:Control, Dataguard, Examiner, Plus, Realtime, Structure, Viewpoint. Vendor: Sterling Software (Interchange Software Division).

GeoBook NB-xx Desktop computer. Notebook. AMD processor. Vendor: Brother International Corp. Inc.

geocoding Process of assigning map coordinates to data such as customer or store addresses. Used to develop GIS software.

Geode Application development tool. Automates programming and testing functions. Vendor: Verilog USA.

GeODE See GemStone GeODE.

geographic information system See GIS.

Geos Operating system for handheld computer systems. Runs on Zoomer PDAs. Vendor: Geoworks, Inc.

Geoworks Bindery Application development tool. Used to develop applications for handheld computers. Vendor: Geoworks, Inc.

Gescan Relational database machine. A dedicated computer that contains the DBMS necessary to process the database. These databases are compatible with most mainframe operating systems and databases. Vendor: Gescan International, Inc.

GESCAN Data management system. Runs on DEC systems. Provides full-text retrieval (entire text will be searched for any character string and requires no indexes). Includes report generator and thesaurus. Vendor: Netscan Technology Corp.

GetAccess Communications. Internet software that provides Web security. Allows companies to let all applications work from a single password. Runs behind the firewall. Vendor: EnCommerce, Inc.

GFX Internet Firewall System Communications, Internet software. Firewall. Vendor: Global Technology Associates, Inc.

Giant/8 Database for large computer environments. Runs on DEC systems. Vendor: Solutions Unlimited.

Gibraltar Communications software, Web server. Provides access to the Web from Windows NT systems. Lets users view documents. Vendor: Microsoft Corp.

gigabyte Approximately one billion bytes.

Gigabyte Ethernet Communications. Backbone technology. Fiber optics cables that can transmit data up to 1G per second. Also called gigabit enternet. Contrast with Fast Ethernet, which transfers at 100M per second.

gigaPoPs Communications. The type of connectivity between the universities and other organizations in Internet2. Also the connection between Internet2 and NGI networks. Stands for: Gigabit per second Points of Presence.

Giles Operating system enhancement. Maintains a directory of MVS components without manual input. Automatically extracts information from JCL, COBOL, CICS tables and libraries. Vendor: Global Software, Inc.

GIS Geographical Information System. Programs that combine such things as demographics, corporate revenues, and taxes with maps. The maps allow geographically pertinent information to be incorporated into standard company information processing. Software designed to present and analyze spatial information. Used for such things as demographics, sales potential and performance, routing and scheduling functions.

GKS A graphics system used to create 2-D and 3-D images. Has been adopted as a standard. Stands for: Graphical Kernel System.

GL:Millennium See Millennium.

GL:Satellite See Satellite.

Glacier (II) Desktop computer. Pentium CPU. Operating systems: Windows 95/98, NT. Vendor: Aspen Systems, Inc.

GLE Application development system. Runs on Unisys systems. Stands for Generative Language Environment. Includes: Irgent, Quikfacts, Simon, Lexicon. Vendor: PROGENI Systems, Inc.

Global Data Manager for NetBackup Operating system software. Provides centralized management of backup and recovers across distributed systems. Runs with NetBackup. Released: 1999. Vendor: Veritas Software Corp.

Global Data Mart Repository See Enterprise Data Mart.

global directory Type of directory services that tracks information across multiple databases and directories. Directory information includes logon name, network address, security clearance and job title, and can be located in many different places in a large company.

Global Enterprise Manager See GEM.

Global Enterprise Manufacturing Management System See GEMMS.

Global Enterprise Manufacturing System MRP, or CIM software. Runs on DEC VAX, IBM RS/6000, HP systems. Vendor: Datalogix International, Inc.

Global Information Manager EIS. Runs on Pentium type desktop computers. Vendor: Global Software, Inc.

Global MHS Communications software. Message handling services. Shares directories in NetWare and e-mail systems. Vendor: Novell, Inc.

Global Trust Organization See GTO.

GlobalView Application development tool. Document generator. Runs on Unix systems. Can be used in electronic meetings. Vendor: XSoft (division of Xerox Corp).

Go/ISDN Communications software. Network connecting desktop computers to IBM midrange computer systems. Vendor: Trisystems.

GoBook Notebook computer. Up to eleven hours of battery life. Vendor: Micron Electronics, Inc.

Gold-spreadsheet Desktop system software. Spreadsheet. Runs on Pentium type desktop computers. Vendor: Dynacomp, Inc.

Goldatabase Relational database for desktop computer environments. Runs on IBM desktop computer systems. Vendor: Goldata Computer Services, Inc.

Golden MailBridge Communications software. E-mail. Allows exchange of e-mail over diverse systems. Released: 1997. Vendor: NBS Systems, Inc.

Goldengate Desktop system software. Integrated package. Runs on Pentium type systems. Vendor: Computer Associates International, Inc.

GoldMine Application software. Sales force automation. Runs on Windows systems. Released: 1996. Vendor: GoldMine Software Corp.

Goldrun Application development tool. Automates design, programming and testing functions. Vendor: Peat Marwick Advanced Technology.

GoldWorks Artificial intelligence system development tool. Written in LISP. Runs on Pentium type desktop computers. Vendor: Gold Hill Computers, Inc.

Gopher Document retrieval system used on the Internet. Through Gopher, users can view information spread out on many hosts. Developed by University of Minnesota. No longer in common use.

GOSIP Government's application of the OSI communications model. Stands for: Government OSI Profile. Refines standards and limits options in each layer of the OSI communications model.

Governor Facility Application development tool. Monitors and controls the use of PRF and QMF. Vendor: PLATINUM Technology, Inc.

GNOME Project planning to build a complete user-friendly desktop with free software. Part of the GNU project. Uses GTK+ as a GUI toolkit for all GNOME compliant software. Is the GUI included in many versions of Linux. Stands for: GNU Network Object Model Environment.

GNU Project sponsored by the Free Software Foundation that is developing a software environment that includes an operating system kernel, utilities, compilers and debuggers. The GNU kernel is used in most versions of Linux. The software provides a GNU General Public License (GNU GPL) that gives everyone the right to use and modify the source code as long as they make the modifications available to everyone else with the same licensing stipulation. This license is also called "copyleft."

GNUPro Toolkit Application development tools. Includes C, C++ compilers and debugging tools. Used to develop software for desktops and embedded systems. Runs on DEC, Unix, Windows. Released: 1997. Vendor: Cygnus Solutions.

GPAR(/DOS) Operating system enhancement used by Systems Programming. Monitors and controls system performance in IBM systems. Stands for: General Performance Analysis Reporting. Vendor: IBM Corp.

Gpf Application development tool. Interactive tool designed for developing graphical user interfaces. Stands for: GUI programming facility. Runs on OS/2 systems. Vendor: Gpf Systems, Inc.

GPSS Programming language used to build models for simulation. Stands for: General Purpose System Simulator.

GQL Application development tool. Used for ad hoc queries. GUI interface, accesses most relational databases. Stands for: Graphical Query Language. Platform independent. Released: 1996. Vendor: Andyne Computing.

Graffiti Handwriting recognition software. Runs on GEOS, Magic CAP, NewtOS systems. Vendor: Palm Computing, Inc.

Grafika 4xxx Desktop computer. Pentium CPU. Operating systems: OS/2, Windows. Vendor: DTK Computer, Inc.

Grafsman Application development tool. Programming utility that allows users to display data from multiple databases in a graphical format. Part of Uniface Application Development toolset. Vendor: Uniface Corp.

Graham Object-oriented development methodology. Provides analysis and design techniques.

Grail Internet browser. Runs on Unix systems.

Grandview Desktop system software. Personal Information Manager (PIM). Runs on Pentium type systems. Vendor: Symantec Corp.

granularity Application development term used in data warehouse design. Granularity expresses the level of detail in a data warehouse. The higher the granularity, the more detailed the data is (the higher of level of abstraction).

GrapeVINE for Notes/Domino, Web Application development tool. Requires users to build an interest profile. Then searches multiple Notes databases and sends a message to the user when anything that fits the profile changes. Vendor: GrapeVINE Technologies, Ltd.

Graph Pro MMX Desktop computer. Pentium CPU. Operating system: Windows 95/98. Vendor: Royal Electronics, Inc.

graphic user interface GUI. Software that allows access to computer systems via graphical methods such as icons and pull-down menus. Includes window functions. GUIs are found in operating systems, application development systems and database systems. Both system and application software can be developed with a graphic user interface.

Graphical Designer family Application development tools. Used for re-engineering. Includes: GD*Pro*, GD*MethodBuilder*, GD*Draw*. Support team design. Runs on Unix, Windows NT systems. Vendor: Advanced Software Technologies, Inc.

Graphical Query Language See GQL.

Graphical Systems Manager Application development tool. GUI for enhancing DEC job control language and Unix shell commands to manage systems in a client/server environment. Vendor: Integrated Solutions, Inc.

graphics The creation and processing of picture images in a computer.

Graphics Networker Configuration management system that keeps track of everything on a network. Uses graphics to display the network. Runs on MS-DOS systems. Vendor: The Graphics Management Group, Inc.

graphics package A desktop computer program that produces pictures, graphs, charts, and diagrams, which may or may not be accompanied by text.

Graphicway Desktop system software. Graphics package. Interfaces with Lotus 1-2-3, dBase. Vendor: Tilcon Software, Ltd.

Great Gantt! Desktop system software. Project management package. Runs on Macintosh systems. Vendor: Varcon Systems, Inc.

Great OS Communications software. Network operating system which links MS-DOS systems. Vendor: Gateway Communications, Inc.

GreatCircle Application development tool. Eliminates memory bugs in C and C++ programs. Vendor: Geodesic Systems, Inc.

Green Power Workstation Desktop computer. Pentium CPU. Operating system: Windows 95/98. Vendor: EPS Technologies, Inc.

Green River Development name for NetWare release. See NetWare.

green-screen Text-based, dumb terminals. Used in large computer environments.

GRiDPAD Handheld, pen-based computer. Discontinued 1997. Vendor: AST Research, Inc. (division of Samsung Electronics).

GRIP Application development tool. Integrates host-based systems in Novell networks. Used with client/server development. Stands for: Guaranteed, Reliable, Interoperable, Processing. Vendor: Itautec/America.

GRMS Software vendor. Products: manufacturing, financial systems for midsize environments.

GroundWorks See COOL:BusinessTeam.

Group Four Application development system. Runs on Unisys systems. Full name: Group Four Application Generation Information Retrieval System. Includes: Datapulse, Genpulse, Impulse, MicroStation, Micropulse. Vendor: ESI.

Group Four Datapulse Database/4GL for large computer environments. Runs on Unisys systems. Part of Group Four application development environment. Vendor: ESI.

Group Four Micropulse Communications software. Network connecting Unisys systems. Micro-to-mainframe link. Part of Group Four application development environment. Vendor: ESI.

GroupMaster Application development tool used for information filtering. Informs users of changed Internet and intranet information. Runs on Windows systems. Vendor: Revnet Systems.

Groupscape Application developers toolkit. Allows developers to create Notes applications that can be used with Navigator Web browser. Lets Netscape users run Notes applications. Vendor: Brainstorm Technologies.

GroupSystems Groupware. Includes EMS functions. Provides anonymity for participants. Called TeamFocus and marketed under that name by IBM through a licensing agreement. Vendor: Ventana Corp.

groupware Software that takes common single-user functions such as calendars, word processors, databases, and notepads and incorporates them into a multi-user network. Typical functions would be entering a meeting date and time and letting participants check the system, thus eliminating a lot of telephone scheduling. Groupware also can include EMS software. Groupware systems can have an internal DBMS and/or provide interfaces to existing DBMSs.

GroupWise WebAccess Communications software. Adds Web access to GroupWise to let users access e-mail, calendars, etc. from Web browsers. Released: 1997. Vendor: Novell, Inc.

GroupWise (XTD) Groupware. Also considered a major e-mail product. Includes electronic messaging, calendaring, scheduling, document management, fax processing, Internet messages, workflow, remote services. Replaced WordPerfect Office. XTD is client/server version. Supports IMAP, LDAP protocols. Code name of GroupWise XTD is Eclipse. Vendor: Novell, Inc.

Gryphon Code name for release of Sybase's SQL Server 11. Vendor: Sybase, Inc.

GS File Data management system. Runs on Apple II systems. Vendor: SoftWood Co.

GSAM Access method used in IBM systems.

Gsharp Application development tool. Graphics tool that produces precise 2D and 3D graphics. Web edition available. Runs on DEC, Unix systems. Released: 1995. Vendor: Advanced Visual Systems, Inc.

GSSP Generally Accepted System Security Principles. Developed by the Information Systems Security Association. Defines a set of principles and security measures that can be used as guidelines for developing client/server and Internet applications. Follows the lead of the medical, legal, and accounting professions.

GTE CyberTrust Certificate authority.

GTFPARS Operating system enhancement used by Systems Programming. Monitors and controls system performance in IBM systems. Stands for: Generalized Performance Analysis Reporting. Vendor: IBM Corp.

GTK+ Application development tool. Open source toolkit used primarily to develop x-windows systems. Basis for the GUI in most versions of Linux. Stands for: Gimp ToolKit. GNOME is built on top of the GTK (also called GTK +) library.

GTO Global Trust Organization. Formed by eight international banks to provide a set of common rules for digital signatures and certificates and to provide backing for financial transactions enacted with GTO identifications.

Guardian 1. Communications, Internet software. Firewall. Vendor: Netguard. 2. Operating system software. Security package that provides password control and restricts number of concurrent users according to specified login groups. Runs on Unix systems. Released: 1991. Vendor: DataLynx, Inc.

GUARDIAN 90, 90XF Operating system for Tandem midrange and desktop computer systems. Vendor: Tandem Computers, Inc. (a Compaq company).

Guardian, Guardian Plus Portable computer. Pentium CPU. Vendor: Modgraph, Inc.

GuardIT Communications. Internet software. Firewall. Controls access to all enterprise servers at all points of entry from both internal intranets and the Internet. Available in enterprise and workgroup editions. Vendor: Computer Associates International, Inc.

Guest Application development tool. Runs on CICS systems. Vendor: Adrem, Inc.

Guest-IDT Application development environment. Includes testing and prototyping functions. Used to develop on-line applications. Runs on IBM mainframe systems. Vendor: Allen Systems Group, Inc.

GUI See graphic user interface.

Guide 1. Application development tool. Programming utility that allows object-oriented development for Open Look applications. Vendor: Sun Microsystems, Inc.
2. Users group for information technology professionals. Focuses on management issues.

Gulliver Handheld computer. Operating system: GEOS. Vendor: Hyundai Electronics America.

Gupta Software vendor now called Centura Software.

GW-BASIC See BASIC.

gyricon A silicon rubber sheet that can hold digital images for months without power. Can display the images using AA batteries. Can updates the images. Also called Electronic Paper. Information can be transferred to the sheets through wireless technology. Vendor: Xerox Corp.

HA1000 Midrange computer. SPARC CPU. Vendor: Integrix, Inc.

HabaCalc (1-2-3) Desktop system software. Spreadsheet. Runs on Apple II systems. Vendor: Haba Systems, Inc.

Habafile Database for desktop computer environments. Runs on Apple II systems. Vendor: Haba Systems, Inc.

HabaGraph Desktop system software. Graphics package. Runs on IBM systems. Vendor: Haba Systems, Inc.

HabaWord Desktop system software. Word processor. Runs on Apple II systems. Vendor: Haba Systems, Inc.

Habitat Database for large computer environments. PC version available. Runs on DEC systems. Full name: Habitat Integrated Database and User Interface Management System. Vendor: Esca Corp.

HACMP for AIX Operating system enhancement. Handles clusters of up to eight processors. Stands for: High Availability Cluster Multi-Processing Runs on IBM AIX systems. Released: 1996. Vendor: IBM Corp.

HAHTsite Application server providing middleware and development tools. Used to develop Web applications and works with any Web server, browser, or database. Supports distributed servers so applications can run simultaneously on multiple servers. Runs on Unix systems. Full name: HAHTsite Integrated Internet Development System. Vendor: Haht Software, Inc.

handheld computer See PDA.

handshake In data communications, the signals sent along communications lines to establish the data transfer. A handshake consists of a request from the originating system, an indication and response from the receiving system, and a confirmation from the originating system prior to any transmission of data.

HAPSE CASE product. Automates all phases of the development cycle including project management functions. Works with applications developed in Ada on Harris systems. Stands for: Harris Ada Programming Support Environment. Vendor: Harris Corp.

Harbor Operating system enhancement. File manager providing automatic backup/restore functions for file servers and workstations in a mainframe environment. Runs on MVS systems. Vendor: New Era Systems, Inc.

HarborLight Application development tool. Query and report functions. Interfaces with relational databases. Runs on Unix systems. Vendor: Harbor Software Inc.

HarborView Application development environment. Allows both technical and non-technical people to create applications using a visual programming language. Runs on Unix systems. Interfaces with Oracle. Vendor: Harbor Software, Inc.

hard copy A permanent, usually printed, copy of electronically stored data. Can also be microfilm or microfiche.

hardware Physical equipment used in a computer system. Includes the computer itself, storage devices such as tapes and disks, and input/output devices such as printers and terminals.

Harmony Real-time operating system. Developed for control of robotics experiments. Used for industrial manufacturing and assembly. Vendor: National Research Council Laboratories.

Harmony Spreadsheet Desktop system software. Spreadsheet. Runs on Pentium type desktop computers. Vendor: Open Systems, Inc.

Harris INFO Relational database for large computer environments. Runs on Harris systems. Includes query facility, report generator and utilities. Vendor: Harris Corp.

HarrisData Software vendor. Products: financial, payroll systems for AS/400 environments.

Harry EIS. Runs on IBM mainframe systems. Vendor: Adviseurs, Inc.

Harvard Graphics Desktop system software. Graphics package. Interfaces with Lotus 1-2-3, Excel, dBase systems. Runs on MS-DOS systems. Vendor: Software Publishing Corp.

Harvard Project Manager Desktop system software. Project management package. Runs on Pentium type systems. Vendor: Software Publishing Corp.

Harvest software Communications software. Receives faxed forms and extracts data which it enters into host application systems. Vendor: Harvest Software, Inc.

HASP Houston Automatic Spooling Priority System. Performs job spooling and controls communication between local and remote processors.

Hatley Structured programming design methodology named for its developer. Based on functional decomposition. Concentrates on real-time applications. Accepted as a standard design methodology by some companies, and used by some CASE products.

Hawkeye Application development tool. Program that standardizes COBOL programs. Runs on most large computer systems. Vendor: Blackhawk Data Corp.

HC 100,110,120 Handheld computer. Vendor: Psion, Inc.

HD Computers Computer vendor. Manufactures desktop systems to order.

HDAM Access method used in IBM systems.

HDLC Communications protocol used in X.25 networks. Data Link level protocol. Stands for: High-level Data Link Control.

HDM Operating system enhancement that monitors and controls system performance in IBM CICS systems. Allows end-users to communicate with an online help desk. Vendor: Main Frame Software Products.

HDS EX Mainframe computer. Operating system: MVS, VM. Vendor: Hitachi Data Systems Corp.

HDS GX Mainframe computer. Operating system: VM. Vendor: Hitachi Data Systems Corp.

HDS Skyline Series Mainframe computer. Operating system: MVS. Vendor: Hitachi Data Systems Corp.

HDSL See DSL.

Headliner (Professional) Communications, Internet software. Push technology for news. Users select from content channels. Professional, Enterprise and Workgroup editions. Released: 1996. Vendor: Lanacom.

Helios Real-time operating system for desktop computers. Vendor: Perihelion Distributed Software.

Helix VMX Relational database for midrange and desktop computer environments. Runs on Apple II, DEC systems. Vendor: Odesta Corp.

Helm Application development tool. DSS. Provides access to data stored in multidimensional databases. Vendor: Codework.

Helmsman Application software. Includes software for financial analysis, budgeting and reporting. Vendor: Helmsman Group, Inc.

Help Desk Operating system enhancement used by Systems Programming. Monitors and controls system performance in IBM systems. Monitors status of data center problems. Vendor: Remedy Corp.

help desk personnel Support personnel. Provides user telephone support for personal computer systems. Some help desks also install software and software upgrades and provide training. Can be any experience level.

HelpBreeze Application development tool. Authoring tool used to create help files. Add-on to Word. Vendor: Microsoft Corp.

HelpMate Application development tool. Creates business knowledge repositories. Graphical software that provides process components, operating procedure definitions, job responsibilities, quality requirements, etc. This allows users to do such queries as: "show me how to enter a sales order" and "define my learning processes." Vendor: The Cobre Group.

Hercules 200 Midrange computer. RISC system. Operating system: Windows NT. Vendor: Carrera Computers, Inc.

Hermes 1. See System Management Server.
2. Joint European network dedicated to crossing borders with railway cable systems. This saves users from having to deal with telephone providers in each country.

Hertz Computer vendor. Manufactures desktop computers.

Hertz NT WebServer Midrange computer. Pentium CPU. Operating system: Windows NT. Vendor: Hertz Computer Corp.

Hertz PCI Dual P-xxx Desktop computer. Pentium CPU. Operating system: Windows NT. Vendor: Hertz Computer Corp.

Hertz Z-Pentium Desktop computer. Pentium CPU. Operating system: Windows NT. Vendor: Hertz Computer Corp.

Heuristic Optimized Processing System See HOPS for Unix, for Macintosh.

Hewlett-Packard Computer vendor. Manufactures desktop computers.

HFAS Operating system add-on. Data management software that runs on Bull HN systems. Full name: HFAS File Maintenance Utility Set. Vendor: Bull HN Information Systems, Inc.

Hi -Life EDI Communications software. EDI package. Runs on Windows systems. Vendor: SAI Software Consultants, Inc.

HIBOL Application development tool. Program generator for IBM CICS systems. Vendor: Matterhorn, Inc.

HIDAM Access method used in IBM systems.

Hierarchical Storage Manager Operating system enhancement. Migrates inactive files on microcomputer hard disks to tape or optical disk. Runs on NetWare, Unix systems. Vendor: Network Imaging Corp.

HIERS Operating system add-on. Data management software that runs on Harris systems. Stands for: Harris Information Entry and Retrieval System. Vendor: Harris Corp.

Higgins Groupware. Includes e-mail functions. Vendor: Enable Software, Inc.

Higgins Gateways Communications software. Gateway for e-mail systems. Runs with Netware, Vines, 3+Open, LAN Manager networks. Vendor: Enable Software Inc.'s Higgins Group.

High C/C++ Toolset Application development tool. Allows users to create objects directly from C++ source code. Cross-platform compiler that creates code for Windows, Windows NT, OS/2, Unix platforms. Runs on OS/2 systems. Vendor: MetaWare, Inc.

High-Level Language Application Program Interface See HLLAPI.

High Productivity System See Seer HPS.

HighPoint Application development tool. Application generator. Runs on both large and desktop systems. Vendor: IBM Corp.

HighVIEW/SQL for Windows Application development system. Used to develop imaging applications. Vendor: Highland Technologies, Inc.

Hindsight Application development tool. Program that analyzes, tests, and debugs existing C language programs. Reverse-engineering tool. Runs on Apollo, DEC, IBM, Sun systems. Vendor: Advanced Software Automation, Inc.

HiNote Notebook computer. Pentium CPU. Vendor: Digital Equipment Corp.

Hiperfocus 4GL. High performance version of Focus. Vendor: Information Builders, Inc.

HIPO A graphical representation of the design of a computer program and/or system. Stands for: Hierarchy: Input, Process, Output.

HiQ xxxxx Midrange computer. Server. Pentium, Pentium Pro CPU. Vendor: Hi-Quality Computer Systems.

HISAM Access method used in IBM systems.

Historian Plus Operating system add-on. Library management system that runs on Prime systems. Vendor: Opcode, Inc.

Hit List Enterprise Communications, Internet software. Tracks usage of Web sites/pages. Supports SQL. Vendor: Marketwave.

Hitachi Computer vendor. Manufactures large computers.

HLLAPI IBM API used to write applications that can manipulate 3270 terminal screen output. Allows users to capture screen output from one or more mainframe applications and process the data into a format for PC screens. Stands for: High-Level Language Application Program Interface.

HMX(Pro) Network computer. Vendor: Network Computing Devices (NCD).

HNS Computer vendor. Manufactures mainframe computers.

HNSX Computer vendor. Manufactures large computers.

HOD Communications software. Allows a Web browser to emulate a dumb terminal to access mainframe systems. Stands for: Host-on-Demand. Vendor: IBM Corp.

Hollywood Desktop system software. Graphics package. Interfaces with Lotus 1-2-3, Excel. Runs on Windows systems. Vendor: IBM Corp.

Holos 1. Multidimensional database used with data warehousing. Runs on DEC, HP, IBM RS/6000, Sequent systems. Vendor: Holistic Systems, Inc.
2. See Seagate Holos.

Home Office Pro Desktop computer. Pentium processor. Vendor: ACE Computers.

HOPS Data warehouse. Loads and indexes data from operational databases for ad hoc querying and decision support. Supports parallel data access. Includes query tools, report generator, scripting language. Stands for: Heuristic Optimized Processing System. Vendor: HOPS International Inc.

HOPS for UNIX, for Macintosh Application development tool used in data warehousing. Handles all data access functions including query tools, report generators, scripting languages and random data generator. Handles files in the multiple terabyte range. Stands for: Heuristic Optimized Processing System. Released: 1995. Vendor: Hops International, Inc.

Horizon Communications software. Network connecting IBM systems. Part of the Advantage Series. Vendor: Computer Corp. of America.

Horizon/3000 Communications software. Network connecting HP and IBM desktop computers. Vendor: Datasoft International.

Horizon MPC Desktop computer. Pentium CPU. Operating system: Windows 95/98. Vendor: Mega Computer Systems.

Host-on-Demand See HOD.

Host Operating System PC operating system for Apple II PC systems. Vendor: Panasonic Industrial Co.

HostOffice Communications software. When combined with SNA, integrates applications running on mainframe systems with Web based Windows NT systems. Released: 1997. Vendor: Proginet Corp.

HostView Server Communications. Links mainframe systems to the Web and provides security features such as user authentication and encryption. Vendor: Attachmate Corp.

hot docking Plugging a running laptop into a workstation.

Hot Dog (Pro) Application development tool. HTML editor. Automatically inserts HTML tags into text to create Web pages. Runs on Macintosh, Unix, Windows systems. Vendor: Sausage Software.

HotJava Internet browser. Can be downloaded from Sun's home page on the Internet. Includes animated icons. Handles multimedia information. Written in Java, runs Java applets. Runs on Solaris, Windows systems. Vendor: Sun Microsystems, Inc.

HotJAVA Views Operating system software. Adds a GUI to MS-DOS. Part of Project Rescue. Vendor: Sun Microsystems, Inc.

Hotmail Free e-mail service. Accessible through wireless modems.

HoTMetaL (Pro) Application development tool. HTML editor. Automatically inserts HTML tags into text to create Web pages. Runs on Macintosh, Unix, Windows systems. Vendor: SoftQuad International, Inc.

Hotshot Presents Desktop system software. Graphics package. Interfaces with Lotus 1-2-3, Excel, dBase. Runs on MS-DOS systems. Vendor: Symsoft Corp.

HotSpot Application development tool. Compiles Java code to create faster execution of Java applications. Full name: HotSpot Java Virtual Machine. Released: 1998. Vendor: Sun Microsystems, Inc.

HP See Hewlett-Packard.

HP 200LX palmtop PC Handheld computer. Operating system: MS-DOS. Vendor: Hewlett-Packard Co.

HP 300LX, 320LX100LX, palmtop Handheld computer. Operating system: WinCE. Vendor: Hewlett-Packard Co.

HP 3000,9000 Midrange computer. RISC machine. PA-RISC CPU. Operating systems: HP-UX, MPE/IX, Unix. Vendor: Hewlett-Packard Co.

HP 620-LX Handheld computer. Runs common office applications. Vendor: Hewlett-Packard Co.

HP 660LX Palmtop PC Handheld computer. Operating system: Windows CE. Vendor: Hewlett-Packard Co.

HP 9000 EPS21, EPS30 Mainframe computer. Operating system: HP-UX. Vendor: Hewlett-Packard Co.

HP Brio Desktop computer. Pentium II processor. Vendor: Hewlett-Packard Co.

HP Domain Enterprise Midrange computer. PA-RISC CPU. Operating system: HP-UX. Vendor: Hewlett-Packard Co.

HP Enterprise Link Application development tool. Integrates SAP manufacturing software with other manufacturing products. Released: 1996. Vendor: Hewlett-Packard Co.

HP Exemplar S-Class Midrange computer. PA-RISC CPU. Operating system: SPP-UX. Vendor: Hewlett-Packard Co.

HP Exemplar X-Class Mainframe computer. Operating system: SPP-UX. Vendor: Hewlett-Packard Co.

HP Fusion Object-oriented development methodology. Provides analysis and design techniques.

HP Lan Communications software. Network operating system which links different systems. Interfaces with LAN Manager. Vendor: Hewlett-Packard Co.

HP LaserRX Operating system enhancement used by Systems Programming. Monitors and controls system performance in HP systems. Monitors CPU performance. Vendor: Hewlett-Packard Co.

HP Net Vectra Network computer. Vendor: Hewlett-Packard Co.

HP NetServer Midrange computer. Pentium CPU. Operating systems: Windows 95/98. Vendor: Hewlett-Packard Co.

HP OmniBook Notebook computer. Pentium CPU. Operating systems: Windows 95/98. Vendor: Hewlett-Packard Co.

HP OmniGo Handheld computer. Operating system: MS-DOS. Vendor: Hewlett-Packard Co.

HP ORB Plus Application development tool. Creates and runs CORBA compliant C++ applications. Includes ORB, IDL. Runs on Unix, Windows NT systems. Vendor: Hewlett-Packard Co.

HP Pavilion Desktop computer. Pentium CPU. Operating system: Windows 95/98. Vendor: Hewlett-Packard Co.

HP-UX Operating system for midrange Hewlett-Packard systems. Unix-type system. Vendor: Hewlett-Packard Co.

HP Vectra 500 Desktop computer. Pentium CPU. Operating systems: OS/2, Windows. Vendor: Hewlett-Packard Co.

HP Windows Client Desktop computer. Pentium CPU. Operating system: Windows 95/98. Vendor: Hewlett-Packard Co.

HP X.400 Communications software. Gateway for e-mail systems. Runs on HP systems; follows OSI standards. Vendor: Hewlett-Packard Co.

HPCC High Performance Computing and Communication. Government program designed to build hardware and software to handle major processing problems such as climate modeling and drug design. Is developing parallel processors. Government functions such as the Census Bureau and the Patent Office are looking at the program to help tabulate census data and make patent information available to the public through HPCC's network.

HPQS Massively parallel processor. Used to off-load DB2 databases from mainframe systems and allow parallel query processing. Stands for: Highly Parallel Query System. Later models will support additional databases. Vendor: IBM Corp.

HP-RT Real-time operating system for Hewlett-Packard desktop systems. Vendor: Hewlett-Packard Co.

HPS See Seer HPS.

HPSQL/V Database for mainframe and/or midsize environments. Runs on HP systems. Utilizes SQL. Vendor: Hewlett-Packard Co.

HPW10E2 Handheld computer. Vendor: Hitachi Home Electronics (America), Inc.

HR:Millennium See Millennium.

HR Stream See Financial Stream, HR Stream.

HSAM Access method used in IBM systems.

HSM 1. Hierarchical Storage Management. Software that automatically migrates unused files from disk to tape and retrieves these files when necessary. Common part of large computer systems operating systems and starting to be used on desktop systems.
2. Operating system add-on. Data management software that runs on IBM systems. Stands for: Hierarchical Storage Manager. Vendor: IBM Corp.

HTML Development tool used to connect multimedia content (i.e. graphics, text, images). The coding that allows files to appear as formatted pages on the World Wide Web. Stands for: HyperText Markup Language. Some word processors offer the option of creating HTML documents which could then be accessed through the Internet. Based on SGML. Dynamic HTML lets users create simple animation without extensive programming and enables developers to build interactive Web pages.

HTML Author Application development tool. HTML editor.

HTML converters Application development tool. Used to develop Web pages. Converts documents created from standard word processors to HTML code. Available for major work processors and desktop publishers.

HTML editors Application development tools. Used to develop Web pages. Similar to word processors, but handle HTML tags. Some editors can be added to standard word processors. Also called Web authoring tools.

HTML Help The format used for Help screens by Microsoft. This format must be used with Explorer browsers.

HTML Writer Application development tool. HTML editor.

htmSQL Application development tool. Uses Web browsers to access data warehouses. Vendor: SAS Institute.

HTTP HyperText Transfer Protocol. Protocol which defines data representation across diverse systems. Used on Web to transfer HTML documents.

hub Communications hardware. Connects several communications lines together in a star topology. To a great extent replaced by switches.

Hugo/ISPF Application development tool. Works with Datamanager in IBM ISPF environments. Vendor: Global Software, Inc.

Huron Applications development system for mainframes. Contains a database called Metastor that contains all software components C rules (processing statements), data attributes and actual data. No source code need be maintained, and changes are made to rules or data attributes. Vendor: Amdahl Corp.

Hurricane 1. Desktop computer. Pentium CPU. Operating systems: Windows 95/98, Windows NT Workstation. Vendor: Tangent Computer, Inc.
2. Communications. Web server. Supports Enterprise JavaBeans. Accesses data in legacy systems and used on intranets and e-commerce applications. Interfaces with VisualAge for Java, Fusion. Runs on OS/2, OS/400, S/390, Unix, Windows NT. Released: 1998. Vendor: IBM Corp.

Hurricane NetStar Midrange computer. Pentium Pro CPU. Operating system: Windows NT. Vendor: Tangent Computer, Inc.

Husky FC-486, FS/2 Handheld computer. Operating system: MS-DOS. Vendor: Husky Computers, Inc.

HVS 6 Plus Operating system for Bull HN systems. Vendor: Bull HN Information Systems, Inc.

HyBase Object-relational database. Runs on Macintosh, Windows systems. Released: 1996. Vendor: Answer Software Corp.

Hydra Original name of WTS. See WTS.

Hyper-Buf Operating system add-on. Data management software that runs on IBM systems. Vendor: Goal Systems International, Inc.

Hyper Note Pad Desktop system software. Word processor. Runs on windows on IBM and compatible desktop computers. Vendor: Maxthink, Inc.

Hyperbook Handheld, Pen-based computer. Vendor: HyperData Technology Corp.

Hyperbook 700 Pro Portable computer. Pentium CPU. Vendor: HyperData Technology Corp.

hyperbolic tree Design technology that provides a multi-dimensional browsing and display that offers comprehensive views of complex data. The main area of interest is focused in the center of the screen with logically related information appearing on the edges. The user can drag a remote element to the center of the screen, which will magnify it, bring out more detail and provide links to still more information. Also called "wide widgets." Developed by Xerox Parc in 1998.

Hypercard Software for Apple computers. Presents screens as index cards and users can make notes, type, or even draw on these cards. Cards are then grouped in stacks for easy retrieval. Vendor: Apple Computer, Inc.

hyperchart Application design tool. A collection of diagrams defining interrelated processes. A change to any single diagram will change the hyperchart; a change to the hyperchart could cause changes to any single chart. Also called hyperdiagram. Used in information engineering.

Hyperdata Computer vendor. Manufactures pen-based computers.

HyperHelp Application development tool. Provides programming interface for full text searches. Works with Motif and Open Look. Vendor: Bristol Technology, Inc.

Hyperion Enterprise Application development tool. Analyzes financial information for global environments. Supports Essbase, Oracle, Sybase. Runs on Unix, Windows systems. Vendor: Hyperion Solutions Corp.

Hyperion OLAP Applications software. Allows users to analyze sales results and other financial information from a Windows based multidimensional database. Released: 1997. Vendor: Hyperion Solutions Corp.

Hyperion Pillar Application software used for budgeting, planning and forecasting. Interfaces with Essbase. Vendor: Hyperion Solutions Corp.

Hyperion Solutions Software vendor. Company formed by Arbor Software Corp's acquisition of Hyperion Software Corp in 1988. Software includes multidimensional database and application analysis tools. Products include financial and accounting systems.

Hyperkernel Real-time operating system for desktop computers. Vendor: Imagination Systems, Inc.

HyperProject Application development tool. Customizes Navigation System project development for specific projects. See Navigation System Series. Vendor: Ernst & Young LLP.

Hypersource Maintenance Application development tool. Analyzes COBOL programs and assists in debugging COBOL programs. Mainframe and Desktop versions available. Vendor: Software Control.

HyperStar Communications. Message-oriented middleware used with object-oriented systems. Provides access to most databases from most development environments including PowerBuilder, Visual Basic. Runs on DEC, Unix, Windows systems. Released: 1995. Vendor: Ardent Software, Inc.

hypertext A text file with words or pictures identified by color or a change in font as hot links. Clicking on a hot link causes a jump to another place in the file—or even to another file. Used in the Web (Internet access).

Hypertext Markup Language See HTML.

Hypertracs Communications software. EDI system. Based on LU6.2 protocols. Vendor: Sterling Software.

Hypertree Communications, Internet software. File management system. Allows users to organize e-mail and files in a single window. Vendor: Netscape Communications Corp.

Hyundai Computer vendor. Manufactures handheld and notebook computers.

I*B*I*S System Communications software. EDI package. Runs on Windows systems. Vendor: Integrated Business Image Solutions, Inc.

I-CASE Integrated-CASE. CASE tools that work with multiple phases of the development cycle; typically at least analysis, design, and programming.

I-Comm Internet browser. Runs on Windows systems.

I/CSCS PC Desktop system software. Project management package. Runs on Pentium type desktop computers. Vendor: Metier Management Systems, Inc.

I-D-S/II Database for large computer environments. Runs on Bull HN systems. Full name: Integrated Data Store/II. Entry-level database. Vendor: Bull HN Information Systems, Inc.

I Hate Algebra Desktop system software. Spreadsheet. Runs on Windows systems. Vendor: T.Maker Research Co.

I-MON Operating system enhancement used by Systems Programming. Monitors and controls system performance in DEC VAX systems. Monitors machine time consumption. Vendor: Bear Computer Systems, Inc.

I/O Input/Output. Transferring data in or out of the computer.

I/O Express Operating system enhancement used by systems programmers. Speeds up response time and eliminates I/O bottlenecks on DEC VAX systems. Vendor: Executive Software, Inc.

I-O Netstation Network computer. Vendor: I-O Corp.

I.Q. EIS. Vendor: New Generation Software, Inc.

I Seek You See ICQ.

I Spy Operating system enhancement. Performance monitor for CA-Ideal programs. Runs on MVS, VSE systems. Released: 1995. Vendor: Princeton Softech, Inc. (Division of Computer Horizons Corp.).

I-Spy Application development tool. Data warehouse monitoring and optimization tool. Allows administrators to assess and optimize usage by reviewing queries. Provides Web-based interface. Released: 1999. Vendor: Informix Software, Inc.

I/Watch System control software. Monitors the operating system and Oracle databases. Alerts the DBA to potential problems and provides a step-by-step resolution. Can be programmed to automatically take corrective action. Vendor: Quest Software, Inc.

IA-64 Computer chip, or microprocessor. 64-bit architecture developed by Intel and Hewlett-Packard which merges RISC and CISC technology. Operating systems: Unix, Windows NT. Stands for: Intel Architecture 64-bit chip. Code name: Merced. Release date: 1999.

IA-SPOX Operating system software used with multimedia systems. Real-time environment that allows multimedia functions to run on the processor without requiring separate processors. Stands for: Intel Architecture SPOX. Vendor: Intel Corp.

iAcquire See Interchange2000.

IAM Operating system add-on. Data management software that runs on IBM systems. Alternative to VSAM file structure. Vendor: Innovation Computer Corp Data Processing.

IAS Operating system for DEC systems. Stands for: Interactive Application System. Vendor: Digital Equipment Corp.

IAST1 Application development tool. Automates programming function. Generates code for DB2, Oracle databases. Runs on CDC systems. Stands for: Information Analysis Support Tool. Vendor: Control Data Corp.

IBM Computer vendor. Manufactures large and desktop computers. Names associated with IBM: MVS, VSE, and VM are large operating systems. Dominant vendor in business computer market. Full company name: International Business Machines Corp.

IBM 1401 Obsolete. Early business computer.

IBM 2488 Model 300, 800 Handheld, pen-based computer. Vendor: IBM Corp.

IBM C++ Application development system. Includes open class library. Supported by VisualAge. Vendor: IBM Corp.

IBMFAL Communications software. Network connecting IBM and DEC computers. Stands for: IBM File Access Listener. Vendor: Interlink Computer Sciences, Inc.

IBMxxx/DECnet 3711S Gateway Communications software. Network connecting IBM and DEC computers. Vendor: Interlink Computer Sciences, Inc.

IBS Flowmaster Workflow software. Includes: FlowManager, FlowController. Runs on IBM RS/6000 systems. Vendor: Image Business Systems Corp.

IBS/Solution 2000 Tools Application development tool. Used to convert data processing for year 2000. Vendor: IBS Conversions, Inc.

ICA Communications. Protocol for network computers. Used to develop thin-client client/server systems. Data link layer. Includes components for server software, client software and the network. Used in MetaFrame, WinFrame. Stands for: Independent Computing Architecture. Vendor: Citrix Systems, Inc.

iCalendar Standards set for calendars in groupware and workflow systems. This will allow users on different groupware systems to share the same calendar.

Icare Communications software. Internet product used with e-commerce. Provides customer service over the Web. Vendor: NetDialog Inc.

ICC/ Communications software. Network connecting Unisys computers. Micro-to-mainframe link. Includes: ICC/Link, ICC/Xtract II+, ICC/DataXpress, ICC/Intercom 102, ICC/File-Xpress. Vendor: Intercomputer Communications Corp.

ICCF Communications software. Provides environment for application development, Includes programming and testing utilities. Runs on IBM systems. Also contains program library functions. Stands for: Interactive Computing Control Facility. Synonym: VSE/ICCF. Vendor: IBM Corp.

ICCP Institute for Certification of Computer Professionals. Provides CCP general certifications for computer professionals. Old certifications were CDP (Certificate in Data Processing) and CSP (Certified Systems Professional).

ICD Operating system add-on. Debugging/testing software that runs on Unisys systems. Stands for: Interactive COBOL Debugger. Vendor: University Computing Services Corp.

ICE Information and Content Exchange. Proposed standard to simplify the transmission of data between businesses over the Internet.

ICE.Block Internet software, firewall. A simple GUI that requires no Unix knowledge is used for installation, administration and monitoring. Runs on Unix systems. Released: 1997. Vendor: J. River, Inc.

ICE.TCP (PRO) Communications. Terminal emulation software that links Windows PCs to UNIX hosts . Features automated network installation. Runs on Unix, Windows systems. Released: 1997. Vendor: J. River, Inc.

ICL Computer vendor. Manufactures desktop computers.

ICMP Communications protocol. Network Layer. Part of TCP/IP. Stands for: Internet Control Message Protocol.

ICMS See CA-ICMS.

icon A graphical representation of an object, such as a data file. Instead of typing the object name, the user points to the picture with a device such as a mouse, lightpen, touch screen, or moves through a menu with arrow keys.

icon-based Multimedia term. Used to describe multimedia tools that create applications by connecting groups of graphical icons that represent the actions the computer will take.

IconAuthor Multimedia authoring tool. High-level, icon-based, includes scripting. Allows non-technical multimedia users to produce applications combining graphic, text, animation, audio, and video. Supports OLE, Windows, Macintosh, Unix, OS/2. Vendor: Asymetrix Learning Systems, Inc.

Iconic Query Application development tool. Allows access to databases through graphical icons rather than through query languages. Interfaces with Paradox, Oracle, Sybase, DB2. Runs on Pentium type desktop computers. Vendor: IntelligenceWare, Inc.

ICP Communications protocol. Network Layer. Part of Vines. Stands for: Internet Control Protocol.

ICPL Application development environment. Includes 4GL, screen builder, graphics, interactive debugger, word processor. Runs on Pentium type desktop computers. Stands for: Interactive Control and Programming Language. Vendor: XYZT Computer Dimensions, Inc.

ICQ Communications, Internet software. Establishes contact with other ICQ users allowing one person to page and/or send notes. Stands for: I Seek You. Vendor: Mirabilis Ltd.

IDAPI API. Provides a standard access to relational/SQL databases. Simplifies connection between front-end (user interface) and back-end (database access) programs. defined by Borland, IBM and Novell. Stands for: Integrated Database API.

IDB Object Database Object database for Macintosh, RISC systems. Runs on NextOS, Unix, Windows systems. Vendor: Persistent Data Systems, Inc.

IDCAMS See AMS.

IDD IDMS database utility. Stands for: Integrated Data Dictionary. Vendor: Cullinet Software, Inc.

Idea 1. Artificial intelligence system. Runs on IBM desktop computers. Vendor: AI Squared, Inc.
2. Application development tool. Runs on DEC systems. Vendor: Sapiens International Corp.

Idea PreView Application development tool used for business access and analysis. Provides end-user access for queries, reports, charts. Vendor: Idea Corp.

IDEAL See CA-IDEAL.

IDF Data management utility that controls online access to data files. Stands for: Interactive Development Facility. Runs on Bull HN systems. Vendor: Bull HN Information Systems, Inc.

IDL 1. Interface Definition Language. Language used with CORBA to define components. IDL is used to describe what the component does, and how information is to be passed to it. IDL is independent of the programming language and can work with C++, Smalltalk, Object COBOL, Java programs.
2. Application development environment for scientific computing. Includes image processing, two and three dimensional plotting. Stands for: Interactive Data Language. Runs on DEC, MacOS, Unix, Windows. Released: 1997. Vendor: Research Systems, Inc.

IDM Groupware. Workflow system which accesses documents and images through a Web browser. Runs on Windows systems. Stands for: Integrated Document Management. Released: 1998. Vendor: Filenet Corp.

IDM 500x Relational database machine. A dedicated computer that contains the DBMS necessary to process the database. These databases are compatible with most mainframe operating systems and databases. Vendor: Sharebase Corp.

IDMS/Architect See CA-IDMS/Architect.

IDMS-DC See CA-IDMS-DC.

IDMS(/R) See CA-IDMS(/R).

IDOL-IV Application development system which includes relational database, 4GL, report writer, centralized dictionary. Can be used for RAD development. Runs on DEC, Unix systems. Includes: Dictionary-IV, Report-IV, Script-IV. Vendor: Thoroughbred Software International, Inc.

IDRIS PC operating system for Hewlett-Packard desktop computer systems. Unix-type system. Vendor: Whitesmiths, Ltd.

IDS I,II Database/4GL for large computer environments. Runs on Bull HN systems. Vendor: Bull HN Information Systems, Inc.

IDSL Communications channels. Service that combines ISDN with Internet servers. Vendor: Uunet Technologies, Inc.

IE:Expert Application development tool. Includes a data modeler, a report generator, and an SQL generator. Runs on MS-DOS, Windows systems. Vendor: Information Engineering Systems Corp.

IEEE Institute of Electrical and Electronic Engineers. Membership organization that includes a Computer Society and is involved with setting standards for the computer and communications industries.

IEF See Information Engineering Facility.

IEF for Client/Server Implementation, IEF for Client/Server Encyclopedia Versions of IEF used in client/server development. Vendor: Texas Instruments, Inc.

IEW See Information Engineering Workbench.

IEW/GAMMA Application development tool. Program generator for IBM systems. Part of Information Engineering Workbench. Vendor: KnowledgeWare, Inc.

IFS 1. Operating system add-on. Data management software that runs on IBM systems. Stands for: Interactive File Sharing. Vendor: IBM Corp.
2. Data management system. Runs on DEC systems. Allows creation of new database files. Vendor: VAXCalc Co.
3. See Oracle8i.

IFS Applications Applications suite. ERP software. Includes distribution, engineering, finance, human resource, manufacturing, and maintenance modules which can be purchased separately. Interfaces with Oracle databases, Rational Rose development tools. Runs under OpenVMS, Unix, and Windows NT systems. Vendor: IFS Industrial and Financial Systems.

iHTML Scripting language used for Internet applications. Used to create interactive Web sites. Vendor: Inline Internet Systems, Inc.

IIOP Communications. Protocol connecting distributed objects following CORBA standards to TCP/IP networks including the Internet. Stands for: Internet Inter-ORB Protocol.

IIS Communications. Web server. Manages Web sites with multiple Web servers. Used to deploy intra- and inter-net sites. Stands for: Internet Information Server. Bundled with Windows NT. Integrates with Microsoft BackOffice. Vendor: Microsoft Corp.

IIS/DESTINY Database/4GL for large computer environments. Runs on DEC systems. Includes data dictionary, report generator. Vendor: Intel Corpligent Information Systems, Inc.

IKP Communications. Security protocol for e-commerce. Provides secure credit card payments over the Internet. Includes 1KP, 2KP and 3KP models. These models define five parties: a Certificate Authority, a financial institution holding public and secret keys, the issuer of the credit card (usually a bank), the merchant and the customer. Part of the basis for SET, the current standard protocol. Designed by IBM-Zurich.

Illustra Object-relational DBMS. Handles video, audio, images, and two- and three-dimensional modeling. Vendor: Illustra Information Technologies, Inc.

Illustrator Desktop system software. Graphics package. Vendor: Adobe Systems, Inc.

ILOG JViews Application development tool. Includes a set of application specific objects which can be customized to create graphics that are more detailed than standard GUIs. Used to create Java applications. Platform independent. Released: 1997. Vendor: ILOG, Inc.

ILOG Rules Application development tool used to develop artificial intelligence systems. Works with C++. Runs on DEC, Unix, Windows systems. Released: 1998. Vendor: ILOG, Inc.

ILOG Schedule, Solver, Server, Broker Application development tool. Creates complex scheduling applications such as manufacturing systems. Runs on Unix systems. Vendor: ILOG, Inc.

Ilog Solver Application development tool. C++ development tools used for scheduling, configuration, and management applications. Vendor: ILOG, Inc.

Ilog Views Application development tool. C++ development tools used in object-oriented system development. Vendor: ILOG, Inc.

iLUFA 770 Notebook computer. Pentium CPU. Vendor: Chaplet Systems USA, Inc.

IM/Control Stand-alone data dictionary. Runs on CDC systems. Vendor: Control Data Corp.

IM/DM Data management system. Runs on CDC systems. Set of programs that defines, creates, and manages databases. Includes query language report generator, screen editor, and utilities. Full name: Information Management/Data Management. Vendor: Control Data Corp.

IM/FAST Application development tool. Runs on CDC systems. Vendor: Control Data Corp.

IM/Personal Relational database for desktop computer environments. Runs on Pentium type desktop computer systems. Vendor: Control Data Corp.

IM/QUICK 4GL used in mainframe environments. Runs on CDC systems and is designed for end-users. Vendor: Control Data Corp.

IM/VE Data management system. Runs on CDC systems. Vendor: Control Data Corp.

iMac Desktop computer. G3 processor. Constructed as a single unit including processor, monitor and hard disk. Released: 1998. Vendor: Apple Computer, Inc.

Image Operating system add-on. Data management software that runs on DEC VAX systems. Vendor: Edison Software Systems.

Image/ Database for large computer environments. Runs on HP systems. Versions include: Image/SQL, Image/1000, Image/1000-II, Image/3000, Image/260. Vendor: Hewlett-Packard Co.

image editing software Part of image processing. Software that allows retouching of images including editing color brightness.

Image Filer Image processing system that runs on desktop systems. Vendor: Optika Imaging Systems, Inc.

Image Flow Image processing system that runs on large computer systems. Vendor: Recognition Equipment, Inc.

Image-In Color Image editing software. Runs on MS-DOS, Windows systems. Vendor: Image-In, Inc.

Image Master Image processing system that runs on large computer systems. Vendor: Cimage Corp.

image processing Storing, retrieving, and processing graphic data such as pictures, charts, and graphs instead of, or in addition to, textual data. Includes techniques that can identify levels of shades and colors that cannot be differentiated by the human eye.

Image Search Plus Image processing system that runs on desktop systems. Vendor: Bell & Howell Document Management Products Co.

Image Way Image processing system that runs on midrange Prime systems and integrates with Pick operating system. Vendor: Prime Computer, Inc.

ImageBasic 1. Image processing software. Allows customization of images through Visual Basic. Interfaces with dBase, SQL Windows, Lotus Notes. Vendor: Diamond Head Software, Inc.
2. Application development tool. Used with imaging systems. Used to develop imaging and workflow systems. Works with Visual Basic. Vendor: Diamond Head Software, Inc.

ImageExtender Image processing system that runs on various platforms. Client/server architecture. Vendor: GeneSys Data Technologies, Inc.

ImageFast Document retrieval system. Runs on Windows systems. Vendor: Compusearch Software Systems, Inc.

Imagelink Image processing software. Works with CDs, optical disks, microfilm. Vendor: Eastman Kodak Co.

ImageMover Workflow software. Object-oriented package. Vendor: IMC, Inc.

ImagePlus(/2) Image processing system that provides distributed processing capabilities for documents, including both text and graphic images, and runs on IBM large computer systems. Imageplus/2 is a desktop computer version that runs in a client/server environment. Vendor: IBM Corp.

Imagesystem Image processing system that runs on desktop systems. Vendor: Image Business Systems Corp.

Imageworks Document image management system that runs on Bull HN Bull systems. Vendor: Bull HN Information Systems, Inc.

Imagic Image management system. Vendor: Westbrook Technologies, Inc.

Imagine 1. Application development tool. Report generator for IBM systems. Generates reports in both batch and online environments. Part of The Advantage Series. Vendor: Computer Corp. of America. 2. Operating system add-on. Debugging/testing software that runs on HP systems. Vendor: Technalysis Corp.

Imagine GIS Geographic Information System. Includes imaging software. Allows users to view images at any scale. Runs on DEC workstations. Vendor: Erdas, Inc.

imaging See image processing.

Imaging for Windows Professional
Imaging software. Includes software to convert scanned or faxed documents to word processor or HTML documents. Vendor: Wang Laboratories, Inc.

Imaging Software for Windows 95/98
Imaging software. Allows users to scan, view, annotate and store images. Can be downloaded free from the Internet. Users need a full imaging system with document management functions to build production systems. Vendor: Wang Laboratories, Inc.

IMAN Application software. PDM package. Maintains product data in a tree structure which can contain different types of data, such as 3-D Cad images, stress analysis and digital photos. The system uses electronic sign-offs to ensure parts pass smoothly through the processes of design, testing, and manufacturing. Vendor: Unigraphics Solutions.

iManage (Internet) Document management software. Allows users to create and access documents over the Internet. Released: 1996. Vendor: iManage Inc.

iManage Network Document management software. Has three tier design, with a middle tier that eliminates data drivers on client machines. Released: 1998. Vendor: iManage Inc.

IMAP Communications protocol used for Internet e-mail. Provides standards for offline processing (storing the e-mail on a server and having each user download e-mail to his or her own machine and at the same time deleting it from the server), but concentrates on online operations which enable the use to search the mail on the server and process each piece of mail separately. Stands for: Internet Mail Access Protocol.

Imation Media Performance Manager
Operating system software. Monitors and manages media and storage performance. Released: 1997. Vendor: Imation Corp.

IMF Operating system enhancement used by Systems Programming. Monitors and controls system performance in IBM IMS systems. Stands for: IMS Management Facilities. Vendor: Boole & Babbage, Inc.

IMF(/1000) Level I,II Image database utilities. Vendor: Corporate Computer Systems, Inc.

IMFT Operating system add-on. Data management software that runs on Bull HN systems. Stands for: Inter-Multics File Transfer Facility. Vendor: Bull HN Information Systems, Inc.

IML Initial Machine (or Microcode) Load. The procedure that prepares a device for use. Function of the Operations staff.

Immune System for Cyberspace Virus detection technology for the Internet. Combination of hardware and support services. Will detect viruses across enterprise networks. Released: 1999. Vendor: IBM Corp.

iMonitor EIS. Runs on Unix systems. Vendor: BayStone Software Inc.

IMOS Operating system for midrange NCR systems. Stands for: Interactive Multiprogramming Operating system. Versions include: IMOS III, IMOS V. Vendor: NCR Corp.

Impact Operating system enhancement used by Systems Programming. Monitors and controls system performance in IBM DB2 systems. Stands for Information Management Problem And Change Tracking. Vendor: Infolink Software, Inc.

Impact PCI Desktop computer. Pentium CPU. Operating system: Windows NT. Vendor: Mega Computer Systems.

ImpactChecker Application development tool. Allows the developer to determine the impact of the use of resources such as files, data types, parameters, etc. Part of Logiscope. Runs on Unix, Windows systems. Vendor: CS Verilog.

Impromptu Application development tool. OLAP software. Data query and reporting software. Interfaces with Sybase, Oracle, Rdb, Interbase. Runs on Windows systems. Integrates with PowerPlay. Vendor: Cognos Inc.

Impromptu Web Query Application development tool. Query tool that allows end-users on browsers to build queries. Vendor: Cognos, Inc.

Improv Desktop system software. Multidimensional spreadsheet. Runs on Windows systems. Vendor: Lotus Development Corp.

IMPRS Relational database/4GL for midrange and desktop computer environments. Runs on DEC, IBM, Tandem desktop systems. Includes query facility, report generator and DSS. Stands for: Information Management Process Reporting System. Vendor: Ruf Corp.

Impulse Application development tool. Report generator for Unisys systems. Part of Group Four application development environment. Vendor: ESI.

IMRS Ontrack EIS. Runs on Pentium type desktop computers. Vendor: IMRS Co.

IMS Database for large computer environments. Runs on IBM systems. Stands for: Information Management System. Interactive system is called IMS/DC (IMS Communications). Associated language is DL/I. Vendor: IBM Corp.

IMS/ADF (II) Application development tool. Provides skeleton programs and databases to build specific applications. Runs on IBM mainframe systems. Stands for: IMS Application Development Facility. Vendor: IBM Corp.

IMS/AutoOPERATOR Operating system enhancement used by Operations staff and Systems Programming. Provides operator console support in IBM IMS systems. Vendor: Boole & Babbage, Inc.

IMS Client Server/2 Application development tool. Allows access to IMS mainframe databases through OS/2 users. Vendor: IBM Corp.

IMS Client Server Toolkit Application development tool. Set of utilities and application generators which allow developers to add graphical front ends to IMS systems. Licensed by IBM under the name IMS CS Toolkit. Vendor: Multi Soft, Inc.

IMS CSobject Object-oriented software that allows online IMS systems to be distributed to OS/2 systems. Included with IMS Version 5. Stands for: IMS Client/Server Object. Vendor: IBM Corp.

IMS DB/DC Way of referring to both batch (DB) and online (DC) processing of IMS databases.

IMS/DC Communications software. Transaction processing monitor. Runs on IBM systems. Communications monitor for online IMS applications. Vendor: IBM Corp.

IMS/DC Version 5 Latest release of IMS/DC that enables pieces of IMS processing to be distributed to OS/2 systems. End users can access and update data without using the mainframe. Includes IMS CSobject. Vendor: IBM Corp.

IMS-XPERT Application development tool. ISPF-like facility for IBM IMS systems. Vendor: XA Systems Corp.

IMSAM/3000 Application development tool. Program that works with Image databases. Stands for: Image Sequential Access Method. Vendor: Dynamic Information Systems Corp.

IMSAO Operating system enhancement used by systems programmers. Increases system efficiency in IBM IMS systems by controlling multiple IMS systems on one or more MVS systems. Runs on Netview systems. Stands for: IMS Automated Option. Vendor: IBM Corp.

IMSASAP Operating system enhancement used by Systems Programming. Monitors and controls system performance in IBM IMS systems. Stands for: IMS Monitor Summary and System Analysis. Vendor: IBM Corp.

IMSConnect Communications software. Links IBM's MQSeries middleware with IMS systems. Vendor: Candle Corp.

IMSL Libraries Mathematical and statistical routines that run on Macintosh systems. Vendor: IMSL, Inc.

IMSPARS Operating system enhancement used by Systems Programming. Monitors and controls system performance in IBM IMS systems. Stands for: IMS Performance Analysis and Reporting System. Vendor: IBM Corp.

IMX 700 Communications software. Transaction processing monitor. Runs on DEC systems. Vendor: Incotel, Inc.

IN/ix General-purpose operating system for Wang, Unix systems. Vendor: Interactive Systems Corp.

in-memory database Database which stores all the data in the computer's memory using disks only for storing log and backup information. Can increase database performance by as much as 50 times.

InCharge for the Internet Systems management software. Installs and manages Internet services including the Web, file transfer protocol, e-mail on multiple servers. Maintains profiles of devices and applications on the network and identifies failures and the effect of the failure. Users interface through a Web browser. Interfaces with Tivoli. Released: 1996. Vendor: System Management Arts, Inc.

INCISA Communications, Internet software. Push technology. Users select from content channels. Released: 1997. Vendor: Wayfarer Communications.

InCompare Operating system add-on. Data management software that runs on IBM systems. Vendor: Incepts, Inc.

InConcert Workflow software. Interfaces with Oracle, Sybase. Runs on Sun, IBM RS/6000 systems. Vendor: Xsoft (division of Xerox Corp.).

Independent Computing Architecture See ICA.

Index Server Communications, Internet software. Search engine. Vendor: Microsoft Corp.

Indigo Midrange computer. RISC machine. Operating system: Unix. Vendor: Silicon Graphics, Inc.

INDISY(/PC) Communications software. Network connecting IBM computers. Vendor: INDISY Software, Inc.

Indy GUI used with Windows 95/98. Will also be used with Cairo (object-oriented Windows NT) so both operating systems will have the same interface. Vendor: Microsoft Corp.

inference The process of learning from experience by generalizing new facts and rules from existing ones. A computer program must be able to do this in order to be said to have artificial intelligence.

Inference Find Knowledge management software. Accessible through the Internet. Vendor: Inference Corp.

Inferno Application development tool used to build distributed applications for multiple devices including microcomputers, network computers, televisions, and telephones. Includes Limbo (programming language), supports Java. Works on any network including the Internet. Can run as a stand alone operating system on microcomputers, or can run under Unix, Windows 95/98, Windows NT. Released: 1996. Vendor: Lucent Technologies, Inc.

Infinia Desktop computer. Pentium CPU. Operating system: Windows 95/98. Vendor: Toshiba America Information Systems, Inc.

Infinity 90 Series Mainframe computers. Parallel processor. Supports up to 16 processors. Operating systems: Unix, Umax. Vendor: Encore Computer Corp.

Infinity(/Query) Database for large computer environments. Runs on MODCOMP systems. Vendor: MODCOMP.

Infinity R/T Series Mainframe computers. Operating systems: Umax, Unix. Vendor: Encore Computer Corp.

Infinium Applications software. Object-oriented human resource, financial, and materials systems that run on AS/400 systems. Vendor: Software 2000, Inc.

Influence Knowledge Warehouse for SAP Application development tool. Prepackaged data warehouse for SAP's R/3. Web based analysis tool used to access data collected by R/3. Released: 1998. Vendor: Influence Software, Inc.

INFO Relational database/4GL for midrange and desktop computer environments. Runs on most midrange computer systems. Vendor: Henco Software, Inc.

Info 7 Application development tool. Query and reporting system. Vendor: Seagate Software Inc.

INFO-DB+ Database/4GL for midrange and desktop computer environments. Runs on DEC systems. Includes IQL. Vendor: Henco Software, Inc.

Info-XL Desktop system software. Personal Information Manager (PIM). Runs on Pentium type systems. Vendor: Valor Software.

InfoAdvisor Application development tool. Client/server DSS (Decision Support System). Runs on DEC Alpha servers with PC clients. Vendor: PLATINUM Technology, Inc.

InfoAssistant Application development tool. Query and report functions. Supplies predefined queries. Runs on Windows systems. Vendor: Asymetrix Corp.

Infobase Data management system. Runs on Datapoint systems. Vendor: Recognition Equipment, Inc.

InfoBeacon Application development tool. DSS. Provides access to data stored in relational databases. Runs on Unix, Windows systems. Vendor: Computer Associates International, Inc.

Infocenter Knowledge management software. Collects information from Internet sources and corporate databases and makes it available through push technology. Accessible through the Internet. Vendor: BackWeb Technologies.

InfoConnect Communications software. Allows a Web browser to emulate a dumb terminal to access mainframe systems. Vendor: Attachmate Corp.

InfoCube Application development tool. Provides pre-written HTML-based templates that will allow any user to generate a wide variety of reports that will meet most business analysis needs. Requires WebFocus. Runs on Unix, Windows NT systems. Vendor: Information Builders, Inc.

Infocus Application development tool. A query and report tool for mainframe human resource management software. Vendor: Integral Systems, Inc.

InfoExpress Application development tool. Data replicator. Vendor: PLATINUM Technology, Inc.

Infoflex Application development system. Uses SQL, WYSIWYG syntax for developing screens. Runs on Macintosh, Unix, VMS systems. Vendor: Infoflex, Inc.

Infogate Communications software. Network connecting IBM computers. Micro-to-mainframe link. Vendor: Cullinet Software, Inc.

InfoGOLD Midrange computer. Server. Pentium CPU. Vendor: American Multisystems, Inc.

InfoGuard Operating system add-on. Security/auditing system that runs on Unisys systems. Vendor: Unisys Corp.

InfoHub Communications software. Middleware. Allows relational database users to query nonrelational data. Runs on IBM mainframe systems. Vendor: Sybase, Inc.

InfoImage Imaging system. Used as a pilot configuration for companies evaluating imaging technology. Vendor: Unisys Corp.

Infolio 160 Pen-based computer. Vendor: PI Systems.

InfoMagnet Application development tool used for information filtering. Searches text of real-time news feeds. Runs on Windows NT systems. Vendor: CompassWare Development Inc.

InfoMaker Application development tool. Used to design, build and deploy workgroup applications. Includes extensive reporting and analysis functions. Interfaces with sQL Anywhere, Oracle, SQLServer, PowerBuilder. Runs on Windows 95/98 systems. Vendor: Powersoft Corp (division of Sybase).

InfoManagement Operating system enhancement used by Systems Programming. Monitors and controls system performance in IBM systems. Vendor: IBM Corp.

InfoManager EIS. Runs on AS/400 systems. Vendor: Ferguson Information Systems Inc.

InfoModeler Application development tool. Visual relational database design tool. Includes browser, report generator. Relational database design tool that maps English statements into the database model. Works with Visual Basic. Generates code for FoxPro, Access, Paradox, dBase. Runs on Windows systems. Vendor: Asymetrix Corp.

INFOPAK Operating system enhancement used by Systems Programming. Increases system efficiency in IBM systems. Compresses data on DASD. Vendor: InfoTel Corp.

INFOPAC-xxxxxxx Operating system enhancements used by Systems Programming and Operations staff. Job scheduler for IBM mainframe systems. Includes: INFOPAC-JCL Management, INFOPAC Report Distribution System, INFOPAC VTAM Remote Output, INFOPAC Automatic Balancing system, INFOPAC-OPMAN, INFOPAC-Scheduler, INFOPAC-TapeSaver. Vendor: Mobius Management Systems, Inc.

InfoPilot Application development tool used for information filtering. Informs users of changed intranet information. Runs on Solaris, Windows NT systems. Vendor: FirstFloor, Inc.

InfoPortal EIP, corporate portal. Supports ODBC, XML and the WebDAV protocol recently approved by the World Wide Web Consortium. In addition to standard internal access has established partnerships with external data sources to provide information on such things as weather and stock prices. Released: 1999. Vendor: Glyphica.

InfoPump Application development tool used in data warehousing. Includes database and middleware. Provides bi-directional data movement. Used for replication, integration of data warehouses and operational data. . Moves data from Unix, OS/2, Windows NT and mainframe platforms. Runs on OS/2, Unix, Windows NT systems. Vendor: Computer Associates International, Inc.

InfoRefiner Application development tool. Takes data from legacy databases and integrates it into relational databases. Runs on MVS systems. Vendor: Computer Associates International, Inc.

InfoReports Application development tool. Report generator for Web applications. Supports multi-database reporting with parameter passing, scheduling, thin client/web browser report viewing and HTML output. Vendor: Computer Associates International, Inc.

Inform/3000 See Rapid/3000.

Information Access Application development tool. Provides visual programming environment. Used for prototyping, queries and reports. Vendor: Hewlett-Packard Co.

Information and Content Exchange See ICE.

Information Center The department within a company that provides support to the desktop computer users within the department.

Information Engineering Structured programming design methodology. Accepted as a standard by some companies and used by some CASE products.

Information Engineering Facility CASE product. Automates analysis, design, and programming functions. Includes code generator for C, COBOL. Design methodologies supported: Information Engineering. Databases supported: DB2, OS/2 DBM, Rdb, Oracle. Runs on Unix systems, IBM and compatible desktop computers. Vendor: Texas Instruments, Inc.

Information Engineering Workbench CASE product. Automates analysis, programming, and design functions. Includes: IEW/GAMMA. Interfaces with ADW. Runs on IBM systems. Uses Chen, DeMarco, Gane-Sarson, Martin, and Yourdon methodologies. Vendor: KnowledgeWare, Inc.

Information Management Communications. Network management software that provides a connection between its internal database and NetView. Provides problem, change, configuration and inventory functions. Runs on MVS systems. Released: 1994. Vendor: IBM Corp.

Information Master Database for desktop computer environments. Runs on Apple II systems. Vendor: High Technology Software Products, Inc.

Information Model Library CASE product. Part of Re-engineering Product Set. Vendor: Bachman Information Systems, Inc. (now Cayenne Software).

Information/pc Relational database for desktop computer environments. Runs on IBM desktop systems. Vendor: Prime Computer, Inc.

Information(Plus) Relational Database/4GL for large computer environments. Runs on Prime systems. Includes Information EXL, Information Escape. Vendor: Prime Computer, Inc.

Information Systems Network See ISN.

Information Warehouse Set of standards from IBM that addresses issues of access to corporatewide data wherever it is in the company. Includes standards for communicating with other vendor systems. Part of IBM's SAA. Vendor: IBM Corp.

Informer Application development tool. DSS. Provides access to data stored in multidimensional and relational databases. Vendor: Reportech.

Informix Entire development environment including Dynamic Server database and development tools. Vendor: Informix Software, Inc.

Informix 4GL,4GL++ 4GL. 4GL++ has object-oriented extensions. Vendor: Informix Software, Inc.

Informix CLI Application development tool. Used to develop applications to access heterogeneous databases following ODBC standards. Runs on Unix, Windows systems. Stands for: Informix Call Level Interface. Vendor: Informix Software, Inc.

Informix Client SDK Application development tool. Includes APIs that allow developers to write applications in Java, C, C++ or ESQL. Includes Informix Connect. Released: 1998. Vendor: Informix Software, Inc.

Informix Connect Communications. Contains the runtime libraries used by applications accessing Informix databases and running on client systems. Works with Informix Client SDK. Vendor: Informix Software, Inc.

Informix-Data Director Application development tool. Generates data access code thus reducing amount of code that must be written. Versions available for Web and for Visual Basic. Vendor: Informix Software, Inc.

Informix DataStage Application development tool used in data warehousing. Extracts data from operational databases and transforms and loads into the warehouse. Runs on Unix, Windows systems. Vendor: Informix Software, Inc.

Informix-DSA See DSA.

Informix Dynamic Server Relational database. Runs on all size systems. Supports parallel processing and Web interfaces. Handles all data types. Used for data marts and data warehouses. Works with OS/2, Unix, VMS, Windows systems. Vendor: Informix Software, Inc.

Informix-Enterprise Gateway Communications software. Allows access to non-Informix databases. Runs on Unix, Windows systems. Vendor: Informix Software Inc.

Informix-ESQL(/Ada,C,Cobol,Fortran) Application development tool. Creates imbedded SQL statements for programs written in the appropriate language. Accesses Informix databases. Runs on Unix systems. Vendor: Informix Software, Inc.

Informix-Gateway with DRDA Communications. Connects IBM databases to Unix systems. Conforms to DRDA standards. Runs on Unix systems. Released: 1994. Vendor: Informix Software, Inc.

Informix i.Reach Application development tool. Provides a single repository for Web site content. Vendor: Informix Software, Inc.

Informix i.Sell Internet software. Electronic storefront which integrates database and e-commerce software. Vendor: Informix Software, Inc.

Informix/Illustra See Universal Server.

Informix Internet Foundation 2000 Object-oriented database. Incorporates Web-based, geospatial, video and image data. Includes Datablades technology. Supports Java, COM, ActiveX, C, C++. Released 1999. Vendor: Informix Software, Inc.

Informix-JWorks Application development tool. Used to build Web client/server applications using Informix databases. Automatically generates Java code from drag-and-drop environment. Runs on Unix, Windows systems. Vendor: Informix Software, Inc.

Informix-Net Communications. Connects Informix databases on servers with Informix development tools running on client systems. Runs on DEC, Unix systems. Released: 1994. Vendor: Informix Software, Inc.

Informix-NewEra Application development environment. Provides graphical development tools. Includes object-oriented language, interactive debugger, repository, Java code generator. Runs on Unix, Windows systems. Vendor: Informix Software, Inc.

Informix-OnLine (7.0) Database server that runs on Unix systems. Version 7.0 uses PDQ to provide parallel processing. Vendor: Informix Software, Inc.

Informix Red Brick Warehouse Data warehouse. Used to build warehouses or data marts. Loads and indexes data from operational databases for ad hoc querying and decision support. Supports parallel data access. Vendor: Informix Software, Inc.

Informix-SE Relational database. Version of Informix for laptops. Runs on OS/2, Windows systems. Vendor: Informix Software, Inc.

Informix-SQL 4GL used in large computer environments. Accesses Informix databases. Vendor: Informix Software, Inc.

Informix Universal Server See Universal Server.

Informix Vista Application development tool used in data warehousing. Provides aggregate computations to improve warehouse performance. Determines what aggregates to create by analyzing existing queries. Automatically evaluates the effectiveness of existing aggregates. Vendor: Informix Software, Inc.

InForms Groupware. Electronic forms software. Runs with GroupWise. Interfaces with dBase, Paradox, FoxPro, DataPerfect and most SQL databases. Vendor: Novell, Inc.

INFOS II Operating system add-on. Data management software that runs on DG systems. Vendor: Data General Corp.

InfoScout Application development tool used for information filtering. Finds articles of interest according to user defined specifications. Runs on Windows NT systems. Vendor: Mayflower Software.

InfoSession Application development tool. Integrates legacy application data with client/server applications by linking mainframe systems to OS/2, Unix, Windows clients and Internet browsers using SQL. Vendor: Computer Associates International, Inc.

InfoTransport Application development tool. Moves data from heterogeneous systems without requiring additional gateways, intermediate servers or custom conversion code. Vendor: PLATINUM Technology, Inc.

InfoWizard Application development tool used for information filtering. Finds articles of interest according to user defined specifications. Uses Web and non-Web sources. Runs on Windows systems. Vendor: Amulet Inc.

InfraSet Application development tool used in component-based development. Allows complex data retrieval and analysis to be performed against high-level reusable components. Follows COM and Microsoft's DNA. Vendor: Methods Bay, Inc.

Infrastructure Manager Application software. Provides asset management functions. Vendor: Fortress Technologies, Inc.

INFRONT/ Development tools that contain micro-to-mainframe links to allow programs developed on a desktop computer to run on, and/or access data on, host computers. Includes: INFRONT/DS, INFRONT/RT, INFRONT/HPO, INFRONT/BCF, INFRONT/DB2, INFRONT/SDF. Vendor: Multi Soft, Inc.

INGRES See CA-Ingres.

Ingres II Relational database. Links Open-Road development environment and C, C++, Visual Basic. Used for enterprise-wide development. Vendor: Computer Associates International, Inc.

inheritance In object-oriented programming, the ability of new classes of objects to use the procedures and data (methods and values) from existing classes. This requires programming only the differences for the new class.

Innovation Access Method Operating system enhancement used by systems programmers. Increases system efficiency in IBM systems. Manages disk storage space. Vendor: Innovation Data Processing, Inc.

Innovation Team Project Groupware. Project management software that allows scheduling and resource management across multiple groups. Vendor: Workflow Technologies, Inc.

InPerson Groupware. Desktop conferencing system. Runs on Unix systems. Vendor: Silicon Graphics, Inc.

InPower HR Applications software. Human resource system. Runs on OS/2 systems. Interfaces with NetWare, SQL-Base. Client/server technology. Vendor: Integral, Inc.

Inprise New name for Borland, a software vendor. Name created when Borland (creates development tools) purchased Visigenic (creates middleware products) in 1998. Full name: Inprise Corporation.

Inprise Application Server Application server providing middleware and development tools. Used to build enterprise and Web applications. Integrates with Delphi, Jbuilder, C++Builder. Includes CORBA based transaction server and supports Enterprise JavaBeans. Released: 1998. Vendor: Inprise Corp.

input/output See I/O.

Inquire Query language used on IBM mainframes.

InQuizative Application development tool. Report generator. Part of PowerHouse. Designed for use by non-technical people. Runs on Unix systems. Vendor: Cognos Inc.

inSight Application development tool. Query tool for SAP R/3, MetaCube, and most databases. Supports team development. Vendor: ARCPLAN.

Insight 1. Operating system enhancement used by Systems Programming. Monitors and controls system performance in IBM systems. Vendor: Universal Software, Inc. 2. Application development system. Runs on HP systems. Vendor: Unison Software. 3. Communications, Internet software. Tracks usage of Web sites/pages. Measures delivery of data, download time, and interrupted transfers. Vendor: Vendor: Accrue Software. 4. Document retrieval system. Works with Web servers. Runs on Windows systems. Vendor: Enigma Information Retrieval Systems, Inc.

InSight 1. Operating system add-on. Debugging/testing software that runs on DEC VAX, IBM systems. Vendor: Intellisys, Inc. 2. Application development tool. Provides a multiuser analysis and design modeling environment that works with Visual Basic and relational databases. Vendor: LBMS. 3. Application software. Provides client/server financial and human resource functions. Vendor: Lawson Software.

Insight 2+ Artificial intelligence system. Expert system building tool. Includes PRL (Production Rule Language) and DBPAS. Latest version of Insight 2+ is called LEVEL5 and marketed by Information Builders, Inc. (parent company of Level Five Research, Inc.). Vendor: Level Five Research, Inc.

Insight/DB2 Operating system enhancement used by Systems Programming. Increases system efficiency in IBM DB2 systems. Monitors system and improves response time. Vendor: Database Utility Group, Inc.

INSPEC Operating system software. Manages Unix systems. Audits, monitors and reports on both the operating and application system performance. Runs on Unix systems. Released: 1996. Vendor: Elegant Communications, Inc.

Inspector 1. CASE product. Automates analysis, design, and testing functions. Vendor: Language Technology, Inc. 2. Operating system software. Monitors SAP R/3 systems. Finds bottlenecks and excessive access time and suggests corrections. Released: 1997. Vendor: Envive Corp.

Inspiration Notebook computer. Tillamook CPU. Vendor: Dell Computer Corp.

Inspire Systems management software. Suite of client/server storage management applications. Runs on Unix systems. Vendor: Alphatronix, Inc.

Inspiron Notebook computer. Operating system: Windows 95/98. Vendor: Dell Computer Corp.

InSQRIBE Application Development tool. Report generator. Interfaces with Developer/2000, Visual Basic, PowerBuilder, etc. Works with most databases. Vendor: SQRIBE Technologies (acquired by Brio Technology in August, 1999).

Install/1 CASE product. Part of Foundation system. Vendor: Arthur Andersen & Co.

InstallShield Operating system enhancement. Deploys enterprise and large-scale applications. Versions: Professional, Express, Objects. Vendor: InstallShield.

Instance Monitor Application development tool. Real-time monitoring and diagnostic tool for Oracle databases. Displays real-time database activity and identifies bottlenecks using flows, graphs and visual icons. Released: 1999. Vendor: Quest Software, Inc.

Instant Update Groupware. Includes EMS (Electronic meeting system). Vendor: On Technology Corp.

InstaPlan Desktop system software. Project management package. Runs on Pentium type systems. Vendor: InstaPlan Corp.

InSure See DistribuLink.

Insure++ Application development tool. Run-time debugging tool that allows developers to identify bugs such as memory errors and library problems. Includes InVision (displays patterns of memory and data use) and InUse (displays memory-use information). Supports Microsoft's Visual Studio. Runs on DEC, Unix, Windows systems. Released: 1998.Vendor: ParaSoft Corp.

INTACT Database utilities used to verify the integrity of Hewlett-Packard databases. Vendor: Carolian Systems International, Inc.

Integra Relational database for midrange and desktop computer environments. Runs on Unix systems. Utilizes SQL. Can be used in client/server computing. Vendor: Coromandel Industries.

Integrated Document Management See IDM.

Integrated Document Processing System Image processing system that runs on Tandem systems. Vendor: Tandem Computers, Inc. (a Compaq company).

integrated package A desktop computer program that provides word processing, spreadsheet, database, graphics, and communications functions in one.

integration analyst, architect, engineer Technical developer. Determines what is needed to integrate various software packages such as databases, communications programs, and application software. Usually senior level title, could be mid-level.

Integration Server Application development tool. Moves analysis data from relational databases to Essbase. Used to build analysis models. Released: 1998. Vendor: Arbor Software Corp.

Integration Systems Application development tool. EAI software. Includes prepackaged adapters to integrate front- and back-office applications. Vendor: Active Software.

integration software Software used to integrate diverse systems such as MVS and Unix. Allows companies to use the same applications to access different environments.

integration test A test to make sure that the interfaces between programs in a system work.

Integrator See PLATINUM Integrator.

Integrion Business-to-business extranet set up for the banking industry. Provides home-banking services for member banks.

Integrity ADE Application development environment. Object-oriented. Used to develop component based applications. Uses a three-tier architecture (the user interface, the data access, and the business logic) and each object belongs to one and only one of the tiers. Can import ActiveX or Java object. Works with both COM and CORBA. Interfaces with Oracle, Informix, Sybase, SQL Server, runs under OS/2, Unix, Windows systems. Released: 1996. Vendor: Cleyal Ltd.

Integrity Data Re-engineering Tool Application development tool. Integrates data from multiple operational databases, outside sources, and historical data sources into a data warehouse. Runs on IBM MVS systems. Vendor: Vality Technology, Inc.

Integrity family Midrange computer. RISC machines. MIPS Rxxxx CPUs. Operating systems: NonStop-UX. Vendor: Tandem Computers, Inc. (a Compaq company).

Integrity Programming Environment Application development tool. Analyzes, matches, reformats data extracted from legacy systems. Used in re-engineering, data warehousing. Used to build relational databases from flat files. Vendor: Vality Technology, Inc.

Integrix Computer vendor. Manufactures desktop computers.

Intel Hardware vendor. Manufactures Pentium, II, Pro, MMX computer processor chips.

Intel Supercomputer Computer vendor. Manufactures supercomputers. Division of Intel Corp.

IntelaGen Application development tool. Program generator for IBM systems. Vendor: On-Line Software International.

Intellect 400 Application development tool. Natural language system that accesses DB2, IMS and IDMS databases. Also works with KBMS. Vendor: AI Corp., Inc.

Intellect/Focus Application development tool. Program that translates English into FOCUS statements. Vendor: AI Corp., Inc.

Intellect/SQL/DS and DB2 Application development tool. Program that accesses SQL/DS and DB2 databases by AI-based natural language. Vendor: AI Corp., Inc.

Intellecte Document management software. Client/server tool that provides workflow, viewing and distribution. Used for document intensive processes such as research. Used in the scientific community. Runs on Windows systems. Vendor: Interleaf, Inc.

Intelligence/Compiler Artificial intelligence system. Expert system building tool. Intended for use by programming personnel. Runs on desktop computers. Vendor: IntelligenceWare, Inc.

intelligent agent See agent.

Intelligent Miner Application development tool used for data mining. Used to determine which of hundreds of relationships among variables are significant. Runs on AIX systems. Vendor: IBM Corp.

Intelligent Notebook Computer vendor. Manufactures notebook computers.

Intelligent OOA Application development tool. Provides Object-Oriented Analysis functions. Supports Shlaer/Mellor methodology. Vendor: Kennedy Carter.

Intelligent Planner Desktop system software. Project management package. Runs on Pentium type desktops. Vendor: PlanView, Inc.

Intelligent Query Application development tool. Query tool for end-users. Includes report generator. See IQ 5.0 family. Vendor: IQ Software Corp.

intelligent terminal A terminal with processing capability. Intelligent terminals can provide such functions as simple data validation when data is input. Also can run systems developed to run under windows.

Intelligent Warehouse Warehouse software that supports the modular warehouse concept that builds department size warehouses that can be incorporated into a total enterprise warehouse at a later date. Builds information catalog describing metadata. Vendor: PLATINUM Technology, Inc.

IntelligentPad Application development environment. Breaks software into components that can be reused. Components are called Pads and can be assembled graphically into specific applications. Runs on Windows systems. Released: 1996. Vendor: Fujitsu Software Corp.

IntelliScope Document retrieval system. Runs on Unix, Windows systems. Vendor: Inso Corp.

IntelliServ Application development tool used for information filtering. Finds articles of interest according to user defined specifications. Runs on Windows NT systems. Vendor: Verity, Inc.

Intellisys Applications development tool. Develops custom database applications for DEC VAX systems. Vendor: Genex Systems, Inc.

IntelliView MPS-110,210 Portable computer. Pentium CPU. Vendor: Intellimedia Corp.

Intelliwatch Tracer Application development tool. Used for testing Lotus Notes applications. Released: 1998. Vendor: Candle Corp.

InTempo Groupware. Workflow system that works with Exchange e-mail systems. Users can be Windows, Web browsers. Released: 1998. Vendor: Jetform Corp.

Interact Integrator Data management software that provides the framework for integrating CASE products in a network environment. Vendor: Interact Corp.

interactive Synonym for online. See online.

Interactive Query EIS. Runs on IBM AS/400 systems. Vendor: New Generation Software, Inc.

InterBase Relational database/4GL for midrange and desktop computer environments. Runs on most computer systems. Utilizes SQL. Can be used in client/server computing and embedded systems. Vendor: Inprise Corp.

Intercase Knowledgeware Gateway. Application development tool. Programming utility that provides an interface between Knowledgeware's CASE tools and Intercase Reverse Engineering Workbench. Loads COBOL programs into a repository and automatically transfers information about the programs to the CASE software. Vendor: Interport Software Corp.

Interceptor Firewall appliance. Provides a single path for data between corporate networks and the Internet. Vendor: Technologic, Inc.

Interchange2000 EIP software, corporate portal. XML based. Includes iAcquire which allows diverse systems, databases and applications to make data available to Interchange2000, and iPresent, which provides the Web-based front-end to Interchange2000. Vendor: Sequoia Software Corp.

Interconn Communications software. EDI package. Runs on MS-DOS systems. Vendor: St. Paul Software.

InterConnect for Notes Groupware. Public network that small and medium size companies can use. Vendor: IBM Corp.

Interel Relational database/4GL for large computer environments. Runs on Bull HN systems. Utilizes SQL. Can be used in client/server computing. Includes report generator, CASE tools. Vendor: Bull HN Information Systems, Inc.

Interface Builder Application development tool. Object-oriented program generator for Next systems. Accesses data in multidimensional databases. Part of NeXTStep. Uses prepackaged objects. Vendor: Next, Inc.

Interface Definition Language See IDL.

Interfacing Studio Application development tool used to build application interfaces and maintain interfaces so data can be mapped from target to source. Works with CORBA. Runs on Unix, Windows NT systems. Vendor: SmartDB Corp.

Intergraph Computer vendor. Manufactures desktop systems.

Intergy See OEA.

Interix Operating system software. Provides a UNIX system environment for Microsoft Windows NT systems. Includes utilities such as KornShell, Bourne shell, C shell, and awk. Runs on Windows NT systems. Released: 1998. Vendor: Software Research, Inc.

Interleaf Publisher Desktop system software. Desktop publishing package. Runs on Pentium type systems. Vendor: IBM Corp.

InterLISP(-D) See LISP.

InterMart Toolkit Application development tool. Used to build links between relational databases and Web browsers. Vendor: NetScheme Solutions.

Internet Communications network that originally linked computer systems at universities and government facilities. Now includes commercial users. Follows TCP/IP protocols.

Internet Adapter Application development tool. Connects Versant databases to Web. Increases speed for Web servers. Vendor: Versant.

internet applications developer, web applications developer See web programmer.

Internet Assistant Application development tool. Text editor. Allows users to create and edit HTML documents from within Microsoft Word. Runs on Macintosh systems. Vendor: Microsoft Corp.

Internet Chameleon Communications software. Internet kit which facilitates the use of basic Internet services. Vendor: NetManage.

Internet Client Station Network computer. Vendor: Idea Associates, Inc.

Internet Developer Toolkit Application development tool. Used to develop client/server applications that can be accessed from a Web browser. Vendor: Powersoft Corp.

internet engineer, web engineer Developer, usually technical. Builds the interfaces between the Internet user interface and internal corporate systems. Knowledge of TCP/IP, firewalls. Maintains connectivity between Internet and internal networks. Can be any experience level.

Internet Explorer(VR) Internet browser. Runs on Unix, Windows systems. VR is an add-on which allows creation of VRML documents. Part of Microsoft Plus. Vendor: Microsoft Corporation.

Internet File System See Oracle8i.

Internet in a Box Communications software. Internet kit which facilitates the use of basic Internet services. The browser is Air Mosaic. Based on NCSA Mosaic. Known as the Air Series. Vendor: Spry, Inc.

Internet Information Server See IIS.

Internet Inter-ORB Protocol See IIOP.

Internet Mail Access Protocol See IMAP.

Internet Mail Server Communications. Internet e-mail software. Includes support for multiple mail accounts, voice mail support and handling HTML messages. Supports IMAP, LDAP protocols. Also called Solstice Internet Mail Server. Vendor: Sun Microsystems, Inc.

Internet Protocol See IP.

internet software developer, web software developer Developer. Used for a variety of job skills and levels, so has no real meaning other than requiring Internet skills such as Java, CGI scripts, HTML, etc.

Internet Terminal Server Communications, Internet software. Allows access to the Internet through mainframe terminals. Vendor: Idea Corp.

Internet Web Server Communications. Internet software. Web server. Vendor: Novell, Inc.

Internet Workshop Application development environment. Creates client/server applications based on Java. Includes Java WorkShop, JOE. Vendor: SunSoft.

Internet2 Internet2 is a collaborative effort by more than 150 U.S. universities, working with partners in industry and government, to develop advanced Internet technologies and applications to support the research and education missions of higher education. Internet2 is a project of the University Corporation for Advanced Internet Development (UCAID).

InternetConnect Communications software. Links IBM's MQSeries middleware with the Internet. Vendor: Candle Corp.

InterNotes News Application development tool. Converts Notes databases and documents to HTML format for use on the Web. Runs on OS/2, Windows NT systems. Released: 1995. Vendor: Lotus Development Corp.

InterOFFICE Communications software. E-mail system. Runs on DEC, Wang systems. Vendor: The Boston Software Works.

InterOffice (Suite) Groupware. Client/server package that includes Oracle DBMS, Fax, voice mail, calendaring, and document sharing. Supports Java, TCP/IP, Internet protocols. Runs on Solaris, Unix, Windows NT systems. Vendor: Oracle Corp.

interoperable The principle that different systems must be able to communicate and exchange data easily. Part of open systems. Or, running software and exchanging data in a heterogeneous network made up of several different LANs with different platforms.

interpreter, interpreted language A programming language that works by translating the English language statements into machine code and running them as they are translated. The translation is done each time the program runs. Scripting languages are usually interpreted languages, as are Java and Basic. Contrast with compiler, compiled language.

InterQ See Smartstream, Financial Stream Analysis, InterQ.

InterServe Midrange computer. RISC machine. Operating systems: Windows. Vendor: Intergraph Corp.

INTERSOLV Company developed by the merger of Index Technology and Sage Software. Concentrates on CASE technology.

InterViso/IVQuery, IV Build Application development tools used in data warehousing. Used to build virtual data warehouses. InterViso/IVBuild manages metadata. Runs on Unix systems. Released: 1997. Vendor: Data Integration, Inc.

InTEXT Document management system. Includes APIs to handle desktop-to-mainframe communication and end-user GUIs to issue search and display queries. Runs on IBM mainframe, OS/2, Unix systems. Released: 1995. Vendor: InTEXT Systems.

InText WebPack See WebPack.

Intouch Application development tool. Includes online testing and debugging tools. Runs on DEC VAX systems. Vendor: Touch Technologies, Inc.

Intouch*EDI Communications software. EDI package. Runs on Pentium type desktop computers. Vendor: Harbinger*EDI Services.

InterViso Communications software. Middleware. Allows queries to multiple databases on different platforms. Runs on Unix systems. Vendor: Data Integration, Inc.

Intra.doc Application development tool. Document management system used over intranets. Runs on Solaris, Windows NT systems. Vendor: Intranet Solutions, Inc.

IntraBuilder Application development tool. Visual tool used to build intranets. Vendor: Borland International, Inc.

IntraExpress Application development tool used for information filtering. Finds articles of interest according to user defined specifications. Runs on Windows systems. Vendor: Diffusion, Inc.

intranet An internal corporate network that uses Internet standards and terminology.

Intranet Genie Communications, Internet software. Search engine. Used on Internet and intranets. Vendor: Frontier Technologies.

IntranetWare Communications software. Network operating system. NetWare plus software providing Web server functions. Vendor: Novell, Inc.

IntrNet See SAS/IntrNet.

IntRprise See Passport IntRprise.

Intuity Communications, Internet software. Provides unified messaging services. Allows users to access messages from a laptop. Includes: Intuity Message Manager, Intuity AUDIX, Intuity Interchange Server. Vendor: Lucent Technologies, Inc.

Invircible Anti-virus software. Looks for viruses in production systems. Vendor: NetZ Computing Ltd.

Invisible LAN Communications software. Network operating system (NOS) connecting Windows systems. Vendor: Invisible Software, Inc.

Involv Communications software. Teamware. Allows users to share information and files over the Web. Includes a common calendar and the ability to edit and trade documents. Vendor: Changepoint Corp. and US West Inc.

Involv Intranet Groupware. Includes project collaboration, discussion management and software management functions. Interfaces with Domino. Vendor: Changepoint International Corp.

INword Desktop system software. Word processor. Runs on DEC, IBM, Unix systems. Vendor: Interactive Systems Corp.

InXight Software Software vendor. Marketing company of Xeror Parc. See Xerox Parc.

IOC Software Application development tool. Program that facilitates the development of IOC applications. Runs on DEC VAX systems. Vendor: APTEC Computer Systems, Inc.

ION Communications network that combines voice, data and video in an integrated, network. Stands for: Integrated On-demand Network. Vendor: Sprint Corp.

IOS 1. Communications software. Network operating system. Connects diverse networks. Handles multimedia data. Stands for: Internetworking Operating System. Vendor: Cisco Systems, Inc.
2. An office automation system that provides such things as e-mail and appointment calendars. Was designed to work on all IBM systems. Vendor: IBM Corp.

IP Communications protocol. Transport layer. Part of TCP\IP. Handles packet forwarding. Stands for: Internet Protocol.

IP address A 32-bit number that specifies a network address and a host ID. Used on TCP/IP networks, and can identify any device—computers, phones, fax, etc. An IP address can access any other IP address.

IP-Watcher Communications. Network security systems. Allows administrator to detect, monitor and control intruders in real-time. Runs on Unix systems. Released: 1996. Vendor: En Garde Systems, Inc.

IPADI API (Application Programming Interface). Standard for integrating databases. Allows programmers to write to a single interface that will access SQL-relational databases and flat files. Used in Desktop environments. Stands for: Integrated Database Application Programming Interface. Written by Borland.

IPC Communications protocol. Transport Layer. Part of Vines. Stands for: Interprocess Communications Protocol.

IPCF Operating system add-on. Testing software that runs on Bull HN systems. Stands for: Interactive Program Checkout Facility. Vendor: Bull HN Information Systems, Inc.

IPCS Operating system enhancement used by Systems Programming. Increases system efficiency in IBM VM/CMS systems. Stands for: Interactive Problem Control System Extension. Handles on-line problem management, including debugging functions. Vendor: IBM Corp.

IPF 1. Application development tool. Runs on CDC online systems. Vendor: Control Data Corp.
2. Operating system enhancement used by Systems Programming. Increases system efficiency in IBM VSE, VM systems. Stands for: Interactive Productivity Facility. Vendor: IBM Corp.

IPL Initial Program Load. The initialization procedure that starts the operating system. Function of the Operations staff.

Ipos Operating system for midrange International Parallel Machines computers. Vendor: International Parallel Machines, Inc.

iPresent See Interchange2000.

IPS/OPT Operating system enhancement used by systems programmers. Monitors and controls system performance in IBM MVS systems by analyzing workloads. Runs on desktop computers. Vendor: Chicago-Soft, Ltd.

iPSC/860 Mainframe computer. Supports up to 128 processors. Vendor: Unix. Vendor: Intel Supercomputer Systems (Division of Intel Corp).

Ipsec Communications. Protocol covering encryption, authentication, and key functions for Internet security. Used with VPNs. Stands for: Internet Security Protocol.

Ipsys Engineer CASE tool. Supports the development of large client/server and Internet applications requiring no code to be written. Object based. Supports multiple developers. Vendor: Lincoln Software.

Ipsys Hood Application development tool. Used for developing mission-critical applications which must be 100% reliable. Vendor: Lincoln Software.

IpTeam Application software. Web-based engineering tool that allows collaboration by design teams. Includes decision making function that automates and documents design decisions. Vendor: NexPrise, Inc.

IPTrack Communications software. Manages TCP/IP addresses throughout a Novell network. Vendor: On Technology Corp.

IPX Communications protocol. Used by Novell in NetWare. Works with Token-ring, ARCnet and Ethernet topologies.

IPX/SPX Communications protocol. Transport layer. Stands for: Internet Packet Exchange/Sequence Packet Exchange. Created by Novell and used in Netware networks. Also used in Nextstep.

IQ Application development tool. Indexing and data retrieval tool which optimizes data queries. Used with data warehousing. Vendor: Sybase, Inc.

IQ 5.0 family Application development tools. Provides query and reporting functions. Used in data warehousing. Includes graphic and charting tools for business users. Includes IQ/SmartServer, IQ/Access, IQ/Objects. Vendor: IQ Software Corp.

IQ/Live Web Application development tool. Allows the user to query a database, create a report, and post it to an Internet server. The report can be accessed through a browser. Runs on Unix, Windows NT systems. Vendor: Information Advantage, Inc.

IQ/Objects, IQ/SmartServer Application development tool. Query and report functions. Object-oriented. Runs on Windows systems. Vendor: Information Advantage, Inc.

IQ/Vision Application development tool. DSS. Provides access to data stored in relational databases. Runs on Unix, Windows systems. Vendor: Information Advantage, Inc.

IQL 4GL. Works with INFO-DB+ databases. Vendor: Henco Software, Inc.

IQS 4GL used in mainframe environments. Stands for: Integrated Query System. Used in Bull HN systems. Vendor: Bull HN Information Systems, Inc.

IR Information Resources. New term being used by some in place of DP, although IT is more common. See IT.

IRAD Internet Rapid Application Development. Tools used to access data stored in legacy systems over the Internet.

IrDA Communications. Specification for short range, wireless channel for LANS.

IRE-DDS Database for large computer environments. Runs on IBM systems. Full name: Database Development System. Includes DSS and data dictionary. Vendor: International Research & Evaluation.

IRE Marketing Warehouse Application development tool. Speeds up loading and querying databases. Used in data warehousing. Runs on Unix systems. Vendor: Mercantile Software Systems, Inc.

Irgent Application development tool. Report generator for Unisys systems. Part of GLE development environment. Vendor: PROGENI Systems, Inc.

Iridium Communications system that will allow users to send data and voice messages from pocket-size portable telephones to anyplace in the world. Low orbit satellites will relay messages in areas where no communications infrastructure exists.

IRIS Operating system for midrange Point 4 and Data General systems. Stands for: Interactive Real-time.

Iris Explorer Applications development tool. Builds object-oriented applications. Vendor: Silicon Graphics, Inc.

IRIS Indigo Midrange computer. RISC CPU. Operating Systems: IRIX. Vendor: Silicon Graphics, Inc.

IRIX Real-time operating system for various computer systems. Unix-based system that includes some OSF/1 functions. Supports up to 128 processors. Includes system management software. Supports CORBA, COM WTS, Unix/NT integration software. Conforms to Ace specifications. Vendor: Silicon Graphics, Inc.

IRIX OS Operating system for midrange Control Data computers. Unix-type system. Vendor: Silicon Graphics, Inc.

IRMA CASE product. Automates design function. Works with dBase III, IMS, IDMS/R, and DATACOM databases.

IRMALAN Communications software. Gateway connecting LANs to IBM's SNA networks. Vendor: Digital Communications Associates, Inc.

Irmalink Communications software. Network connecting IBM systems. Micro-to-mainframe link. Vendor: Digital Communications Associates, Inc.

iRMX Real-time operating system for Intel systems. Vendor: Intel Corp.

IRU (II,III) Operating system add-on. Data management software that runs on Unisys systems. Stands for: Integrated Recovery Utility. Vendor: Unisys Corp.

IRX Operating system for midrange NCR systems. Stands for: Interactive Resource Executive. Vendor: NCR Corp.

IS Information Systems. New term being used by some in place of DP, although IT is more common. See IT.

IS/3 General-purpose operating system for various systems. Unix-type system. Vendor: Interactive Systems Corp.

IS-CS,desktop computer,PC/PDN Communications software. Network connecting Unisys, IBM PCs. Vendor: Unisys Corp.

ISA Integrated Software Architecture. A set of standards and interfaces that allows users to write software that can run on IBM, DEC, Wang systems. Similar to IBM's SAA. Vendor: Software AG Americas Inc.

ISA/Task Master Operating system enhancement used by Systems Programming and Operations staff. Job scheduler for DEC VAX systems. Vendor: ISA Solutions.

ISAM Access method that allows indexed access of data. Replaced in IBM systems by VSAM.

ISDN A standard communications network for which a variety of products is available. Used instead of analog, modem connected systems. Integrates voice, video, and data signals into digital lines. Provides faster response times, multiple channels. Stands for: Integrated Services Digital Network.

ISE Eiffel See Eiffel.

Isee/Accell CASE product. Automates design and programming functions. Interfaces with Oracle, Sybase, Informix. Runs on Unix systems. Vendor: Westmount Technology.

ISEE, TSEE, RTEE CASE product. Automates analysis, design, and programming functions. Includes code generator for C. Design methodologies supported: Chen, Constantine, DeMarco, Jackson, Ward-Mellor, Yourdon. Databases supported: Ingres, Informix, Unify, Sybase, Oracle. Runs on most workstations under Unix. Vendor: Westmount Technology.

ISG Midrange computer. SPARC CPU. Vendor: Integrix, Inc.

ISG Navigator Communications software. Middleware. Allows queries to multiple databases on different platforms. Runs on Windows 95/98 systems. Vendor: International Software Group, Inc.

Isis Communications software. Used in client/server systems. Groups servers in cluster-like settings to spread databases among the machines. Originally public-domain software from Cornell University. Used by Stratus Computer, Inc. in Unix systems.

ISIS Application software. Manufacturing software handling supply chain, inventory control and order entry functions. Runs on Windows systems. Stands for: Integrated Supply Chain Information System. Released: 1998. Vendor: Prescient Systems, Inc.

Isis for Database Application development environment. Builds fault-tolerant database applications. Includes replication functions, automatic online recovery. Runs on Unix systems. Vendor: Isis Distributed Systems, Inc.

Island InTEXT Document retrieval system. Works with Web servers. Runs on Unix, Windows systems. Vendor: Island Software Corp.

ISM System management software. Used with mainframe, Unix systems. Has object-oriented interface. Stands for: Integrated System Management. Vendor: Bull Worldwide Information Systems.

ISN A fiber optic network system developed by AT&T used in LANs and MANs. Stands for: Information Systems Network.

IsoEnet Networking topology standard for multimedia systems. Stands for: Isochronous Ethernet. Handles voice, video, data traffic. Vendor: National Semiconductor.

Isoplex Communications software. Connects divergent mail systems and provides access to the Internet. Runs on Unix systems. Full name Scalable Isoplex Message Server. Vendor: Isocor.

ISP Operating system enhancement used by Systems Programming. Increases system efficiency in IBM systems. Handles print spooling. Stands for: Inter-system Spool Processor. Vendor: Tone Software Corp.

ISPF Communications software. Provides environment for interactive application development. Runs on IBM systems. Stands for: Interactive System Productivity Facility. Vendor: IBM Corp.

ISPF/PDF Application development tool. Screen editor for IBM systems. Stands for: ISPF Program Development Facility. Contains text editor and program library management routines. Vendor: IBM Corp.

ISS/Three See CA-ISS/Three.

ISSC Integrated Systems Solutions Corp. Subsidiary of IBM that provides outsourcing functions.

IT Information Technology. Anything to do with the computer industry. The department within a company that operates and maintains computer systems including programming, operations, and systems programming groups. Also called IS, MIS.

IT Charge Manager Operating system enhancement. Provides charge back system for IT resources. Runs on DEC, Unix, Windows systems. Released: 1997. Vendor: SAS Institute Inc.

IT DecisionGuru Communications. Network management software. Allows users to simulate adding new applications, users and technologies to networks, thus predicting network needs and avoiding failures. Developed jointly with Hewlett-Packard who markets it under the name OpenView Service Simulator. Vendor: MIL-3 Inc.

IT Director System management tool designed for small businesses. Keeps track of inventories and manages PCs. Vendor: Tivoli systems, Inc.

IT Ledger Application software. Provides asset management functions. Vendor: NetBalance Inc.

IT Service Vision System management software. Accesses performance data from computers, networks, phone systems, the Internet, applications and data warehouses. Locates bottlenecks, peak traffic periods, main users of any resource. Released: 1996. Vendor: SAS Institute Inc.

Itasca See Orion.

ithink Application development tool. Provides process modeling functions. Runs on Macintosh, Windows systems. Vendor: High Performance Systems Inc.

ITMS Database for large computer environments. Runs on DEC systems. Vendor: Information Access Systems.

ITOS Operating system for NEC systems. Vendor: NEC Information Systems, Inc.

Itronix Computer vendor. Manufactures desktop computers.

ITS-OS Real-time operating system for desktop computers. Vendor: In Time Systems Corp.

ITX Operating system for midrange NCR systems. Stands for: Interactive Transaction Executive. Vendor: NCR Corp.

Ivan-Submit Operating system enhancement used by Systems Programming and Operations staff. Job scheduler for NCR systems. Vendor: Carlisle Systems Group.

IView System Manager Operating System enhancement. Monitors and controls software and hardware in networks. Works with products from different vendors. Vendor: Independence Technologies.

IX/370 General-purpose operating system for IBM 370 and 9370 systems. Unix type system. Stands for: Interactive Executive for System 370. Vendor: IBM Corp.

IX Informix Database for large computer environments. Runs on IBM systems. Includes query language and report generator. Vendor: IBM Corp.

IX UltraCalc Desktop system software. Spreadsheet. Runs on IBM systems. Vendor: IBM Corp.

IXL Artificial intelligence system. Analyzes databases and builds rules from this analysis for the knowledge base. Used with Intelligence/Compiler. Runs on desktop computers. Vendor: IntelligenceWare, Inc.

iXpress Application development tool. Allows users to build Web front ends to existing enterprise applications. Runs on Unix, Windows NT systems. Vendor: Software AG Americas Inc.

J.B.Muncer & Associates Software vendor. Products: JBM family of financial, payroll, sales systems for AS/400 environments.

J.D. Edwards Software vendor. Products include application software named OneWorld. Runs on AS/400, mainframe systems, Unix, Windows NT systems. Includes financial, manufacturing and ERP systems.

J/Direct Application development tool. Connects Java built applications to Windows systems. Uses Windows APIs. Vendor: Microsoft Corp.

J++ Builder Application development tool. Used to build Java applications. Released: 1997. Vendor: Inprise Corp.

J90 Series Supercomputer. Parallel processor. Supports up to 32 processors. Operating systems: UNICOS. Vendor: Cray Research, Inc.

Jacada Application development tool. Used to create Java front end systems for online transaction processing systems. Developers import terminal screens and Jacada will generate the Java code. Runs on Windows systems. Released: 1997. Vendor: Client/Server Technology, Ltd.

Jackson Structured programming design methodology named for its developer. Accepted as a standard design methodology by some companies, and used by some CASE products.

Jacobsen Object-oriented development methodology. Also called Objectory which stands for the object factory for software development.

JAD Joint Application Design. Design by a group of people in meeting settings rather than single design efforts that are later merged.

Jade Application development environment. Allows users to create multimedia applications that will run without change on browsers when users connect to a Web site. Interfaces with Jasmine DBMS. Vendor: Computer Associates International, Inc.

Jaguar, Jaguar CTS Communications software. Middleware. Handles distributed applications for SQL Server applications. Interfaces with other databases, Internet. Allows users to run three tiered applications on the Web. CTS version used for component applications and stands for: Jaguar Component Transaction Server. Released: 1997. Vendor: Powersoft (division of Sybase, Inc.)

Jakarta See Visual J++.

JAM Application development environment that works in a client/server architecture. Used for enterprise development. Builds client applications with a GUI based front-end. Includes visual object repository, application and report generators, debugging tools, flat file management functions. Interfaces with most relational/SQL databases. Supports three-tiered application partitioning by having an application server between the client and database server systems. Stands for: JYACC Applications Manager. Vendor: Prolifics, a JYACC Company.

JAM/CASE Application development tool. Interface between JAM's development tools and Teamwork CASE tool. Vendor: JYACC, Inc. and Cadre Technologies, Inc. (now Cayenne Software).

JAM(/Dbi) Communications software. Network connecting IBM and compatible desktop computers and DEC VAX computers. Vendor: Prolifics, a JYACC company.

JAM/ReportWriter Application development tool. Report writer. Add-on to JAM. Allows users to create complex reports using familiar components of JAM's graphical development environment. Released: 1995. Runs on Unix, Windows systems. Vendor: Prolifics, a JYACC company.

JAM/WEB Application development environment. Used to develop Web applications. Allows users to create and deploy server based programs that automatically generates HTML for display on any Web browser. Vendor: Prolifics, a JYACC company.

Jamba Communications, Internet software. Java authoring tool that allows users to add animation, sound and interaction to Web sites. Vendor: Aimtech Corp.

Jampack Operating system add-on. Data management software that runs on Unisys systems. Vendor: Software Clearing House, Inc.

JARS/ See CA-JARS/.

JAS Operating system enhancement used by systems programmers. Provides system cost accounting for Tandem systems. Vendor: DND, Inc.

Jasmine Object-oriented DBMS and visual application development environment. Used to develop applications with graphics, animation, audio and video data. Includes JAVA support. Interfaces with Oracle, Informix, Sybase, DB2, IMS, VSAM. Released: 1997. Vendor: Computer Associates International, Inc.

Jasmine SDK Application development environment. Provides visual, drag-and-drop authoring tools. Used with Jasmine object-oriented DBMS. Stands for: Jasmine Software Development Kit. Released: 1997. Vendor: Computer Associates International, Inc.

Java Object-oriented programming language. Developed as a subset of the C language by Sun Microsystems, and has quickly become the industry's primary cross platform development language. Java is used extensively for WWW, applet, and thin client application development. Java's strength is its ability to run on any computer, provided the computer has a JVM (Java Virtual Machine) available. Its greatest weakness is that it is an interpreted language, meaning the JVM must translate the universal Java code to the native computer's operating code as the application executes. This leads to slow performance. This weakness has been mitigated by Java compilers, which generate machine specific code allowing quick performance, albeit at the cost of the universal cross platform nature of Java byte code. Created by Sun Microsystems in 1996.

Java 2 Platform Application development environment. Used to develop Java applications. Available in three editions: Enterprise (for large server applications); Standard (for PC clients and workstations); Micro (small devices such as appliances, pagers, PDAs). Formerly called JDK. Released: 1998. Vendor: Sun Microsystems, Inc.

Java Activator System software. Allows users to decide whether to use either Sun or Microsoft's JVM to view Web pages. Vendor: Sun Microsystems, Inc.

Java Blend Communications software. Middleware. Allows Java applications to access data from disparate databases without using database code. Vendor: Sun Microsystems, Inc.

Java compiler Application development tool. Creates a machine code version of Java programs and applets. The machine code program can then execute over and over again without further translation. This speeds up execution time.

Java-Designer Application development tool. Converts C++ systems to Java for use on the Internet. Version of X-Designer. Runs on Windows systems. Vendor: Imperial Software Technology Ltd.

Java Development Kit See JDK.

Java Dynamic Management Kit Communications. Network management tool. JavaBeans tool that creates distributed agents that can detect and possibly fix network problems. Vendor: Sun Microsystems, Inc.

Java Foundation Classes See JFC.

Java Objects Everywhere See JOE.

Java Media Framework Application development tool. Allows developers to add multimedia elements to Java applets. Released: 1999. Vendor: Sun Microsystems, Inc.

Java Naming and Directory Interface Application development tool. API that provides a unified interface from Java to multiple directory services. Released: 1998. Vendor: Sun Microsystems, Inc.

Java Server Pages Application development tool. Used to create dynamic web pages which can change for every visitor. Built on servlets and uses embedded tags executed on the server side, along with server-side Java code. Competitive with Active Server Pages and Coldfusion. Released: 1998. Vendor: Sun Microsystems, Inc.

Java Studio Application development tool used for component based development. Allows developers to assemble JavaBeans components to create Java applets and applications without programming. Includes over 50 JavaBeans for such things as animation, sound, database processing, GUI controls. Supports JDK. Can use the included JavaBeans, create new JavaBeans, and use JavaBeans from other sources. Released: 1998. Vendor: Sun Microsystems, Inc.

Java Virtual Machine See JVM.

Java Workshop Application development tool. Used to generate Java code. Provides drag-and-drop interface. Creates applications that use a Web browser as a user interface. Incorporates JavaBeans. Supports JDK. Runs on Solaris, Windows systems. Released: 1998. Vendor: Sun Microsystems, Inc.

JavaBeans Application development tool. Creates JAVA components. Applets and full applications are then created by combining JavaBeans components. Used over the Internet. Work with various platforms, including most flavors of Unix. Competitive with ActiveX. Vendor: Sun Microsystems, Inc.

JavaBeans Development Kit Application development tool. Used to build JavaBeans components. Released: 1997. Vendor: Sun Microsystems, Inc.

JavaEnterprise Application development tool. Adds Java front-ends to legacy applications thus adding a GUI to older systems. Converts AS/400 or mainframe applications into Java applets. Runs on Windows NT systems. Released: 1998. Vendor: ClientSoft Inc.

JavaOS Operating system for network computers. Executes Java applications on Sun SPARC and Intel microcomputers. Released: 1997. Vendor: Sun Microsystems, Inc.

JavaOS for Business Operating system used for Java-based network computing. Vendor: IBM Corp and Sun Microsystems, Inc.

JavaPlan Application development tool. Used for analysis, design and generation of Java applications. Provides drag-and-drop modeling. Supports UML. Generates Java, C++, Objective-C code. Runs on Solaris, Windows systems. Released: 1996. Vendor: Lighthouse Design, Ltd.

JavaScope Application development tool. Ensures that test data is used for every line of code. Released: 1997. Vendor: Sun Microsystems, Inc.

JavaScript Scripting language used with Netscape Navigator. Allows Web page developers to incorporate GUI actions into Web pages. Based on HTML. Used to develop CGI scripts, interactive Web pages. Vendor: Netscape Communications Corp.

JavaSpec Application development tool. Tests program interfaces. Released: 1997. Vendor: Sun Microsystems, Inc.

JavaStar Application development tool. Tests Java applications. Released: 1997. Vendor: Sun Microsystems, Inc.

JavaStation Network computer. Vendor: Sun Microsystems, Inc.

JavaSuite See Pro*IV*.

Javelin Application development tool. Visual program used to develop Java Internet applets and corporate client\server applications. Runs on Windows systems. Released: 1997. Vendor: Step Ahead Software.

Javelin (Plus) Desktop system software. Spreadsheet. Runs on Pentium type systems. Vendor: Javelin Software Corp.

JAZ Operating system enhancement used by systems programmers. Monitors and controls system performance in IBM systems. Provides daily reports of data center activity. Vendor: Universal Software, Inc.

Jazz Desktop system software. Spreadsheet. Runs on Macintosh systems. Vendor: Lotus Development Corp.

JBA Software vendor. Produces ERP software under System 21 name. Full name: JBA International Ltd.

JBA System 21 EDI Communications software. EDI. Runs on AS/400 systems. Vendor: JBA International, Ltd.

JBM family Application software. Financial, payroll, sales systems for AS/400 environments. Vendor: J.B. Muncer & Associates, Inc.

JBridge Application development tool. Allows Unix, MacOS and 16-bit Windows systems to run 32-bit Windows programs using Java as a bridge. Runs on Windows NT. Vendor: Corel Corp.

Jbuilder 3 Application development tools. RAD tools used for Internet applications. Incorporates JavaBeans. Interfaces with Java 2. Standard, Enterprise and Professional editions available. Released: 1997. Vendor: Borland Tools, division of Inprise Corp.

Jbusiness Application server providing middleware and development tools. Used to build and deploy and manage distributed Java applications. Uses Enterprise JavaBeans. Vendor: Novera Software, Inc.

JC/Cheetah Desktop computer. Pentium processor. Vendor: JC Information Systems Corp.

JC/Kiwi Midrange Computer. Server. Pentium Pro CPU. Vendor: JC Information Systems Corp.

JC/Lion Desktop computer. Pentium processor. Vendor: JC Information Systems Corp.

JC/Mango Midrange Computer. Server. Pentium Pro CPU. Vendor: JC Information Systems Corp.

JCentral Application development tool. Assists Java developers. Web search engine used to find code, Java applets, JavaBeans, articles about Java, etc. Vendor: IBM Corp.

JCheck Application development tool used for testing. Analyzes Java code. Part of DevCenter. See DevCenter.

JCL Job Control Language. The language used to communicate with the operating system. Developers write a set of JCL to execute batch programs. The JCL statements tell the operating system what resources are needed (memory size, datasets, etc.) for the program to execute. Called DCL in DEC systems, CL in AS/400 systems. JCL is used with mainframe and midsize operating systems.

JCL/Convert Operating system enhancement. Provides JCL maintenance for IBM systems. Allows for global changes to JCL streams. Replaces JCL/Cross-Reference and PDS/Manager. Vendor: MB Solutions, Inc.

JCL/Cross Reference Operating system add-on. Automatically documents production JCL. Runs on IBM MVS systems. Vendor: MB Solutions, Inc.

JCL maintenance program A computer program that keeps track of JCL and allows for easy updating.

JClass Chart Application development tool. A JavaBean, or component used to build and share dynamically generated charts. Vendor: KL Group.

JCLCHECK See CA-JCLCHECK.

JCLCLEAN Operating system enhancement used by systems programmers. Provides JCL and/or SYSOUT maintenance for IBM systems. Vendor: Software Engineering of America.

JCLMan/VSE Application development tool. JCL and/or SYSOUT maintenance program for IBM VSE systems. Vendor: Goal Systems International, Inc.

JCLWTR Application development tool. JCL and/or SYSOUT maintenance program for IBM systems. Vendor: Applied Information Development, Inc.

jConnect Communications software. Provides Java developers with data access. Runs on Unix, Windows systems. Vendor: Powersoft Corp (division of Sybase, Inc).

JDBC Communications, Internet software. Java API. Middleware. Specification to allow users to access data from multiple vendor databases. The database vendors must prepare database drivers to provide the connectivity. Stands for: Java DataBase Connectivity. Vendor: Sun Microsystems, Inc.

JDesignerPro Application development environment. Visual environment that generates Java code. Vendor: Bulletproof Corp.

Jdeveloper Application development tools. Suite of products to build and manage component-based database applications for the Internet. Used to develop Java applications to run on the server. Developers write Java code and generate HTML for client systems. Supports JDBC, JavaBeans, Enterprise JavaBeans, JFC, RMI, CORBA, and IIOP, Oracle Application Server 4.0, and Oracle Database Server. Released: 1998. Vendor: Oracle Corp.

JDK Application development environment. Used to develop Java applications for personal and network computers. Includes run time engine and class libraries. Latest release called Java 2. Stands for: Java Development Kit. Released: 1996. Vendor: Sun Microsystems, Inc.

JeeVan Object-oriented database. Used to develop Web-enabled Java systems. Platform independent. Allows users to retrieve objects by specifying values. Vendor: W3apps, Inc.

JEF Operating system enhancement used by Operations staff and systems programmers. Disaster control package. IBM systems. Maintains journal of data changes. Vendor: Davis, Thomas & Associates, Inc.

Jefferson Project Code name for document management software to be included with Novell's GroupWise. Vendor: Novell, Inc.

JEP Operating system enhancement used by systems programmers. Increases system efficiency in IBM VSE systems. Stands for: Job Entry Program. Handles remote operating units. Vendor: IBM Corp.

JES-Master Operating system enhancement used by systems programmers. Increases system efficiency in IBM systems. Handles print distribution. Vendor: Xenos Computer Systems, Inc.

JES2,3 Operating system enhancement used by systems programmers. Job scheduler for IBM systems. Stands for: Job Entry Subsystem. Also handles output disbursement. Vendor: IBM Corp.

JetBook xxxx Desktop computer. Notebook. Pentium processor. Vendor: Jetta Computers Ltd.

JetForms Desktop system software. Electronic forms software. Runs on Macintosh, MS-DOS, Unix, Windows systems. Vendor: JetForm.

JFactory Application development environment. Generates Java code to create applications that will run unchanged on Windows, Unix platforms. Vendor: Rogue Wave Software, Inc.

JFC Application development tool. Library of GUI components used to develop Internet and desktop applications. Stands for: Java Foundation Classes. Vendor: Sun Microsystems, Inc.

JHS II Application development tool. JCL and/or SYSOUT maintenance program for IBM systems. Stands for: Job History System. Vendor: Systemware, Inc.

Jikes Application development tool. Standalone Java compiler, which turns Java programs into code that runs on standard Java virtual machines. Available under open source standards. Released: 1999. Vendor: IBM Corp.

Jini Communications software. Technology that will create spontaneous networks from diverse equipment. The software includes a series of Java class libraries that work with Java Virtual Machines. The Jini code is licensed to hardware vendors who "Jini enable" their products. The hardware can then be plugged in and run without configuration or integration with the operating system. Released: 1998. Vendor: Sun Microsystems, Inc.

JMail-MHS Communications software. Gateway for e-mail systems. Runs with Netware networks. Vendor: Joiner Associates, Inc.

JMP Application development tool. Statistical forecasting tool. Provides summary information and determines significance of degree of relationship between two factors. Runs on Macintosh, Windows systems. Vendor: SAS Institute, Inc.

JMS/SWITCH Operating system add-on. Data management software that runs on IBM CICS systems. Vendor: Integrity Solutions, Inc.

JNDI See Java Naming and Directory Interface.

Jnet NJE Communications software. Network connecting DEC and IBM computers. Includes Jnet NJE BSC, Jnet NJE SNA. Vendor: Joiner Associates, Inc.

Job Alert Operating system enhancement. Monitors and controls system performance in IBM MVS systems. Vendor: Allen Systems Group, Inc.

Job control language See JCL.

Job Flow Operating system enhancement used by systems programmers. Job scheduler for Unisys systems. Vendor: ESI.

job library A dataset that contains JCL datasets.

Job/Master Operating system enhancement used by systems programmers and Operations staff. Job scheduler for IBM mainframe systems. Vendor: Mantissa Corp.

Job/Scan Operating system enhancement. Provides JCL and/or SYSOUT maintenance in IBM systems. Vendor: Diversified Software Systems, Inc.

Jobform Application development tool. JCL and/or SYSOUT maintenance program for HP systems. Vendor: Operations Control Systems.

Jobnet Operating system enhancement used by systems programmers and Operations staff. Job scheduler for IBM mainframe systems. Vendor: Allen Systems Group, Inc.

Jobscope (, Jobscopejr) MRP, or CIM software. Runs on HP, IBM systems. Vendor: Jobscope Corp.

Jobsys Operating system enhancement used by systems programmers. Job scheduler for DEC VAX systems. Vendor: Software Partners/32, Inc.

Jobtrac Operating system enhancement used by systems programmers and Operations staff. Job scheduler for IBM systems. Vendor: Goal Systems International, Inc.

JobTrack 2 MRP, or CIM software. Runs on IBM RS/6000, Sun workstations. Vendor: JobTrack Systems, Inc.

JOE Application development tool. Links Java applications with corporate networks. Uses CORBA middleware standards. Stands for: Java Objects Everywhere. Vendor: Sun Microsystems, Inc.

Jolt Application development tool used to make applications accessible from the Intranet or intranets without additional programming. Interfaces with TUXEDO. Platform independent. Released: 1997. Vendor: BEA Systems, Inc.

Journal Analyzer See DBA Tool Kit.

Journal Manager Plus See CICS Integrity Series.

JOVIAL A multipurpose programming language. Stands for: Jules' Own Version of International Algorithmic Language. Used to develop real-time applications.

JPadPro Application development environment. Used to develop Java applications. Interfaces with JDK. Vendor: Modelworks Software.

JPU/E Operating system enhancement used by Operations staff and systems programmers. Disaster control package. IBM systems. Recovers lost data. Vendor: Softsystems, Inc.

Jsafe Application development tool. Encryption package for Java. Released: 1998. Vendor: RSA Data Security, Inc.

Jserver Operating system software. A Java Virtual Machine (JVM) that runs within the Oracle8i database. In effect, this means the database can also be an application server. Vendor: Sun Microsystems, Inc.

JSP See Java Server Pages.

JSS Operating system enhancement. Provides data center management for DEC VAX systems. Full name: JSS-Job Scheduling and Control System. Vendor: ICAM Technology Corp.

JumpStart 1. Application development tool. Set of 80 JavaBeans used in e-commerce. Released: 1999. Vendor: The theory Center.
2. Application development tool. Converts Notes data, hyperlinks, and objects to Exchange. Released: 1996. Vendor: The Mesa Group.

Junior Operator See Entry-level Operator.

Junior Programmer See Entry-level Programmer.

Junior Systems Programmer See Entry-level Systems Programmer.

Juno Free e-mail service. Accessible through wireless modems.

Jupiter CAD software. Object-oriented technology. Runs on Windows systems. Vendor: Intergraph Corp.

Justwrite Desktop system software. Word processor. Runs on Windows systems on IBM and compatible desktop computers. Vendor: Symantec Corp.

JVM Operating system software. Java interpreter code that is usually embedded in operating systems and Web browsers to run Java programs. Can also be embedded in DBMSs. Stands for: Java Virtual Machine.

K, KB Kilo. Usually stands for kilobyte, which is approximately 1000 bytes.

K2 Toolkit Application development tool. Scalable knowledge management tool that divides large queries across a network of servers. Accesses large volumes of unstructured data through the Internet or intranets. Vendor: Verity, Inc.

K2000, K2100, K2500 Handheld, pen-based computer. Vendor: Kalidor.

K4 Desktop computer. Pentium CPU. Operating systems: Windows 95/98, NT. Vendor: Aspen Systems, Inc.

K5 Computer chip. Clone of Intel's Pentium chip. Also called AMD K5. Vendor: Advanced Micro Devices, Inc.

Kaffe Application development tool. Clone of Java used to create embedded systems. Released: 1999. Vendor: Transvirtual Technologies, Inc.

Kalidor Computer vendor. Manufactures desktop computers.

Kameleon Application development tool. Provides process modeling functions. Runs on Windows systems. Vendor: P-E International PLC.

Kappa Application development environment. Visual programming system used to develop object-oriented Unix applications. Runs on Sun systems. Vendor: IntelliCorp, Inc.

Karat System management software. Object-oriented. Provides both network and systems management functions. Based on NetView (network management). Will support MVS, OS/2, AIX, OS/400. Vendor: IBM Corp.

KASE:Set Application development tool. Visual design and code generator. Vendor: Kaseworks, Inc.

KASE:VIP Application development tool. Code generator for Windows, OS/2 client/server applications. Includes SQL designer. Generates C, C++, COBOL code. Vendor: Kaseworks, Inc.

Kaspia Automated Monitoring System Communications. Network management software. Monitors device statistics and reports to provide an early warning system. Runs on Windows NT systems. Released: 1997. Vendor: Kaspia Systems, Inc.

Katmai Computer chip, or microprocessor. Upgrade to Pentium II. Provides speeds up to 500 MHz. Released: 1999. Vendor: Intel Corp.

KAWA Application development environment used to build Java applications, applets, and JavaBeans. Vendor: Tek-Tools, Inc.

Kayak Desktop computer. Operating systems: Unix, Windows NT. Released: 1997. Vendor: Hewlett-Packard Co.

KBMS(/PC,VAX,For Windows) Artificial intelligence system development tool. Stands for: Knowledge Base Management System. Vendor: Trinzic Corp.

KE Texpress Object-oriented database. Handles structured and textual data. Uses query-by-example so non-technical users can access the database. Has Internet interface. Runs on Unix systems. Vendor: KE Software, Inc.

KDES Operating system add-on. Security/auditing system that runs on IBM systems. Provides file compression and encryption. Vendor: Kolinar Corp.

KEDIT Application development tool. Screen editor for IBM desktop computers that emulates XEDIT. Vendor: Mansfield Software Group, Inc.

KEE Artificial intelligence system. Expert system building tool. Stands for: Knowledge Engineering Environment. Implemented in InterLISP. Vendor: IntelliCorp, Inc.

Kerberos Encryption system used by financial institutions, universities, and government agencies.

Kermit Communications protocol developed by Columbia University. Used in several public-domain communications systems to transfer files between desktop systems and host computers.

kernel The control programs in any operating system. Term originated with Unix and most often refers to Unix systems although it can be used with any operating environment.

Kestrel Code name for application development tool in development. Will add debugging functions, performance tuning and optimizations. Vendor: Sun Microsystems, Inc.

KEY:Advise See COOL:Xtras.

KEY:ADW See Application Development Workbench.

KEY:Analyze Application development tool. Automates analysis and design functions. Builds data and process models. Runs on OS/2 systems. Vendor: Sterling Software (Applications Development Division).

KEY:Assemble Application development environment. Visual environment. Includes BASIC-like scripting language. Runs on OS/2, Windows systems. Vendor: Sterling Software, Inc.

Key Chart 2000 Desktop system software. Graphics package. Interfaces with Lotus 1-2-3, Excel, dBase. Runs on MS-DOS systems. Vendor: Softkey Software Products, Inc.

KEY:Construct-400 Application development tool. Uses layouts and design from Design Workstation to generate COBOL, RPG code for AS/400 systems. Vendor: Sterling Software (Applications Development Division).

KEY:Construct-GUI Application development tool. Automates programming functions. Includes code generator. Interfaces with DB/2, Oracle, Sybase. Runs on OS/2 systems. Vendor: Sterling Software (Applications Development Division).

KEY:Construct-MVS Application development tool. Automates programming functions. Uses specifications developed with design Workstation to generate COBOL, COBOL II code for IMS/MVS systems. Interfaces with DB/2, VSAM, Oracle, Sybase. Runs on OS/2 systems. Vendor: Sterling Software (Applications Development Division).

KEY:Design Application development tool. Automates design functions. Runs on OS/2 systems. Vendor: Sterling Software (Applications Development Division).

KEY:Plan Application development tool. Analyzes and models business goals with technology. Runs on OS/2 systems. Vendor: Sterling Software (Applications Development Division).

KEY:Team Application development tool. Used with Knowledgeware's ADW CASE tools. Allows access to LAN based encyclopedia. Runs on OS/2 systems. Vendor: Sterling Software (Applications Development Division).

Keyfile Workflow software. Object-oriented package. Concentrates on document management. Vendor: Keyfile Corp.

Keyflow Groupware. Workflow system that works with many e-mail systems. Users can be Windows, Web browsers. Released: 1998. Vendor: Keyfile Corp.

KeyKOS/370 Operating system for mainframe IBM systems. Vendor: Key Logic.

Keynote 8660 Notebook computer. Pentium CPU. Vendor: Keydata International, Inc.

Keynote xxxx Desktop computer. Notebook. Pentium processor. Vendor: Keydata International, Inc.

KEYS Operating system enhancement used by systems programmers. Monitors and controls system performance in IBM MVS, VSE systems. Hardware/software inventory and help desk management system. Vendor: Software Engineering of America.

Keys/(MVS,VSE) Operating system enhancement used by systems programmers. Manages the computer room help-desk by providing analysis and information about the entire data center. Runs on CICS systems. Vendor: Software Engineering of America.

KEYview Operating system utility. Used with e-mail and the Internet, allows Windows users to view, convert, and print most file types without needing the original software. Runs on Lotus Notes, Windows systems. Vendor: FTP Software.

Keyworks Keyboard utility for IBM and compatible systems. Allows user definition of keyboard. Vendor: Alpha Software Corp.

Khalix Application development tool. DSS. Provides access to data stored in relational databases. Runs on Unix, Windows systems. Vendor: Longview Solutions, Inc.

Khoros Pro Application development environment. Used to develop scientific software. Visual environment, GUI builder. Runs on Unix systems. Released: 1996. Vendor: Khoral Research, Inc.

Kicks/SQL Application development tool. Programming utility that allows users to submit SQL queries to DB2 databases through CICS. Vendor: Cone Software Laboratory, Inc.

KIKS400 Application development tool. Programming utility that migrates CICS, VSAM, IMS, and DB2 programs to AS/400 systems. Vendor: Access to Information, Inc.

KIMS System (4000,5000) Image processing systems. Vendor: Eastman Kodak Co.

Kinetix Application development software. DSS system. Provides a point-and-click interface to SQL databases and flat files. Produces charts, reports and spreadsheets. Interfaces with Oracle, Informix, Ingres, Sybase, Progress, Allbase. Runs on Unix systems. Released: 1993. Vendor: Hilco Technologies, Inc.

KIVA NET Communications software. Network connecting IBM Series 1 computers. Vendor: Anasazi, Inc.

KNET Communications software. Network connecting IBM computers. Vendor: Fibronics International, Inc.

Knowledge Base Management System See KBMS(/PC).

knowledge based system See expert system.

Knowledge BUILD CASE product. Part of IDMS/Architect. Vendor: Cullinet Software, Inc.

Knowledge Center EIP, corporate portal. Supports data creation and mining, and also offers self-publishing from desktop applications. Allows corporate users to search on key words and find documents across the Web. Integrates with application servers, including MTS, SilverStream and NetDynamics. Knowledge Center can be clustered across multiple NT and Unix machines to handle the varying traffic loads. Released: 1999. Vendor: KnowledgeTrack, Inc.

Knowledge Engineering Environment See KEE.

Knowledge Gallery Application development tool. Provides a central repository for corporate reports that can be distributed over intranets using Web browsers. Interfaces with Aperio. Released: 1998. Vendor: Influence Software, Inc.

Knowledge Kiosk Knowledge management software. Accessible through the Internet. Used for help desks. Vendor: Molloy Group, Inc.

knowledge management Building software systems to manage organizational knowledge including processes, procedures, patents, reference works, formulas, fixes, forecasts. Includes intranets, groupware, data warehousing, bulletin boards, video conferencing.

Knowledge Management System Knowledge management software. Accessible through the Internet. Includes search and retrieval, groupware and database technologies. Vendor: Intraspect Software.

KnowledgeCraft Artificial intelligence system. Runs on DEC systems. Integrates object-oriented programming and develops expert systems for large environments. Includes CRL (Carnegie Representation Language). Implemented in LISP. Vendor: Carnegie Group, Inc.

KnowledgeMaker Artificial intelligence system. Builds rule database for many expert system building tools including KnowledgePro. Runs on desktop computers. Vendor: Knowledge Garden, Inc.

Knowledgeman/2 Relational database for desktop computer environments. Runs on DEC systems. Utilizes SQL. Can be used in client/server computing. Vendor: mdbs, Inc.

KnowledgePro Artificial intelligence system. Expert system tool. Uses KnowledgeMaker to build the rule database. Runs on desktop systems. Vendor: Knowledge Garden, Inc.

KnowledgePro Windows See KPWIN.

KnowledgeShare Knowledge management system. Vendor: Cambridge Technology Partners, Inc.

KnowledgeX Application development tool. Graphically displays data relationships from Internet, database sources. Uses e-mail to acquire, publish, and distribute the data. Runs on Windows systems. Released: 1997. Vendor: KnowledgeX, Inc.

KOMAND III/ Operating system enhancement used by systems programmers. Provides system cost accounting for IBM systems. Includes: KOMAND III/DAS, KOMAND III/DAMS, KOMAND III/DBI (for DB2 systems), KOMAND III/DIS, KOMAND III/ICI (for IMS systems), KOMAND III/IDCI (for IDMS systems), KOMAND III/RBS, KOMAND III/UCI, KOMAND III/VMCI (for VM systems), KOMAND III/CCI (for CICS systems.) Vendor: Pace Applied Technology, Inc.

Kona WorkPlace Software suite. Includes word processing, presentation graphics, spreadsheets. Java based. Runs on text-only terminals, network computers. Release date: 1997. Vendor: Lotus Development Corp.

KOPE Application development system. Runs on IBM, Prime systems. Stands for: KOBOL On-line Programming Environment. Vendor: KOS & Associates, Inc.

Korn Shell An extension to the Bourne shell program that is used in Unix environments. Enhances processing of interactive commands. See shell.

Kprobe Operating system add-on. Debugging/testing software that runs on IBM systems. Vendor: VM Systems Group, Inc.

KPWIN Artificial intelligence system development tool. Runs on Windows systems and includes object-oriented programming techniques. Intended to be used by non-professionals. Stands for: KnowledgePro Windows. Vendor: Knowledge Garden, Inc.

Krypton Midrange computer. Server. Pentium CPU. Vendor: Xediom Corp.

KSDS Keyed Sequence Data Set. A type of VSAM file.

Kurzweil Voice Voice recognition system. Runs on Windows systems. Vendor: Kurzweil Applied Intelligence.

KyberPass Communications, Internet software. Firewall. Vendor: Devon Software.

L2F Communications. Protocol used to set up open tunneling over the Internet. Used with VPNs. Stands for: Layer 2 Forwarding. Supplanted by L2TP.

L2TP Communications. Protocol used to set up open tunneling over the Internet. Used with VPNs. Developed in 1998 and intended to supplant PPTP and L2F. Stands for: layer 2 Tunneling Protocol.

La Mans Desktop computer. Pentium processor. Vendor: Racer Computer Corp.

LAN Local Area Network. A computer network located on a single premise. Usually refers to a network of desktop computers.

LAN/3000,9000 Communications. LAN (Local Area Network) connecting HP computers. Vendor: Hewlett-Packard Co.

LAN administrator See Network Administrator.

LAN II Communications software. EDI package. Runs on IBM systems. Micro-to-mainframe link. Vendor: Network Systems Corp.

LAN Manager Communications software. Network operating system connecting IBM and compatible desktop computers. Also called OS/2 LAN Manager. Used in client/server environments. Vendor: Microsoft Corp.

LAN NetView Communications software. Client/server management system which manages Windows, DOS, OS/2 clients from an OS/2 server. Interfaces with LAN Server, NetWare, DB2/2. Vendor: IBM Corp.

LAN Operating System See Network Operating System.

LAN Server Communications software. Peer-to-peer network operating system connecting IBM and compatible desktop computers. Versions available for OS/2, AIX, OS/400, VM, MVS. Used in client/server environments. Vendor: IBM Corp.

LAN WorkPlace Communications software. Allows access to diverse systems. Accesses NetWare, TCP/IP systems, Internet. Works with Windows, Unix, OS/2, Macintosh. Vendor: Novell, Inc.

LANAlert Communications software. Network management tool that allows users to monitor Windows NT Server and Novell NetWare Servers from the same console. Vendor: Network Computing, Inc.

Lanalyzer Communications software. Network analysis tool. Works with Netware and analyzes Netware, Banyan, SNA, and OSI networks. Versions also available for Ethernet and Token Ring. Vendor: Novell, Inc.

*LANDesk Manager Communications. Network management software. Runs on Windows systems. Manages NetWare and Windows NT systems from a single console. Vendor: Intel Corp.

LANDesk Virus Protect Communications software. Scans networked clients and servers for viruses. Works with Windows clients and NetWare, Windows NT servers. Released: 1997. Vendor: Intel Corp.

Landmarq Ipx Desktop computer. Pentium processor. Vendor: DFI, Inc.

LANexpress Communications software. Connects remote users to centralized LANs. Vendor: Microcom, Inc.

LANfocus Communications. Network management software providing LAN to LAN connections. Monitors, configures, and downloads software to LANs from a single OS/2 server. Vendor: IBM Corp.

LANLink Communications. LAN (Local Area Network) connecting IBM desktop computers. Vendor: The Software Link, Inc.

Lansa Application development environment. Client/server environment. Uses same 4GL to create both client and server software. Integrates with SQL Anywhere, Watcom C/C++, Crystal Reports. Available for Windows, the Web, AS/400. Vendor: Lansa USA, Inc.

LANsmart Communications software. Network operating system (NOS) connecting IBM and compatible PC systems. Also has e-mail program. Runs on Novell networks. Vendor: D-Link Systems, Inc.

LanSoft Communications software. Network operating system which links MS-DOS systems. Provides multi-lingual versions. Vendor: Accton Technology Corp.

LANSpy Communications software. Monitors performance on Token-Ring LANs. Vendor: Legent Corp.

LanStation Pro Network computer. Vendor: Accton Technology Corp.

LANstep Communications software. Network operating system which links MS-DOS systems. Vendor: Hayes Microcomputer Products, Inc.

LANtastic (NOS) Communications software. Peer-to-peer network operating system. Runs on Macintosh, IBM and compatible desktop computers. Versions include: LANtastic for Macintosh, NetWare, Windows. Vendor: Artisoft, Inc.

LANtegrity Operating system software. Lets users upgrade NetWare network operating system without disturbing daily operations. Released: 1995. Vendor: Network Integrity, Inc.

LANVIEW/Windows Communications. Network management software used with Ethernet and Token Ring networks. Vendor: Cabletron Systems.

LAPB, LAPD, LAPM Communications protocols. All are extensions to data link protocol HDLC.

LAPDOS Operating System for IBM and compatible desktop computer systems. Vendor: Traveling Software, Inc.

LapLink (for Windows) Communications software. Allows remote access control of desktop computers. Included with CommWorks for Windows. Vendor: Traveling Software, Inc.

laptop computer Desktop computer. Generally weights between three and six pounds. Also called notebook computers.

Laser Optic Filing System Image processing system that runs on desktop systems. Vendor: TAB Products Co.

LaserData Image processing system. Runs on desktop systems. Vendor: LaserData, Inc.

Laserview Image processing system that runs on desktop systems. Vendor: Laserdata, Inc.

LaserWare Operating system add-on. Data management software that runs on DEC VAX systems. Vendor: Perceptics Corp.

Lassisnet Communications. Network management software linking multiple networks. Vendor: Synoptics Communications, Inc.

Latte Application development tool. Converts C++ systems to Java for use on the Internet. Runs on Windows systems. Vendor: Inprise Corp.

LattisWare Communications. Network management software providing network management services for diverse LANs. Vendor: Connects LANs with SNA systems. Vendor: SynOptics Communications, Inc.

Lattitude Desktop computer. Notebook. Pentium processor. Vendor: Dell Computer Corp.

Launchpad Operating system software. Part of OS/2 Warp. Allows users to easily call up frequently used files or applications. Vendor: IBM PC Corp.

Lawson Insight Application software. Products include financial packages, human resources, procurement, and supply chain process functions for midsize environments. Interface with Informix databases. Runs on Unix systems. Released: 1996. Vendor: Lawson software.

Layer 2 Forwarding See L2F.

Layer 2 Tunneling Protocol See L2TP.

Lazer DBMS Database for large computer environments. Runs on DEC systems. Vendor: Canalta.

LBMS Systems Engineer See Systems Engineer.

LCN Software Communications software. Network connecting CDC, IBM, DEC, and other compatible computers. Vendor: Control Data Corp.

LCS See Lotus Communications Server.

LCS/CMF Operating system add-on. Library management system that runs on IBM systems. Stands for: Library Control System/Change Management Facility. Vendor: Pansophic Systems, Inc.

LDA CASE product. Automates design function. Stands for: Linc Development Assistant. Vendor: Unisys Corp.

LDAP Communications protocol. Stands for: Lightweight Directory Access Protocol. Standardization for communications directory products. Allows users to use a single desktop to access multiple directories from e-mail, database and network operating systems.

LDJ Messenger Communications software. EDI package. Runs on Pentium type desktop computers. Vendor: LDJ, Inc.

LDM See CA-LDM.

Leading Edge Computer vendor. Manufactures microcomputers (Pcs).

Legacy Desktop system software. Word processor. Runs on Windows on IBM and compatible desktop computers. Vendor: NBI, Inc.

Legacy Data Mover See CA-LDM.

legacy middleware Communications software that allows client/server applications to access legacy systems.

legacy system An old system still in use. Uses flat files, or non-relational databases.

Legacy Workbench (for Windows) Application development tool. Maintains, redesigns, and migrates mainframe applications. Runs on MVS, OS/2 systems. Vendor: Sterling Software (Application Engineering Division).

Legato Networker Portable computer. Vendor: Legato Systems, Inc.

Legend Supreme Desktop computer. Pentium processor. Vendor: Packard Bell NEC Inc.

LENS Object-oriented programming language. Stands for: Late-bound Encapsulated Name Spaces.

LEVEL5 Artificial intelligence system development tool. Runs on IBM, VAX systems. Builds rule-based expert systems. LEVEL5 is the newest version of Insight 2+. Vendor: Information Builders, Inc.

Level 5 Object Object-oriented application development tool that allows programs to be developed on desktop computers and then used on mainframes. Runs on DEC, IBM systems. Vendor: Information Builders, Inc.

Level5 Quest Application development tool used for data mining. Provides fuzzy analysis of multiple criteria, giving user results close to stated criterion. Runs on Unix, Windows systems. Vendor: Information Builders Inc.

Lex-Graph Desktop system software. Graphics package. Interfaces with Lotus 1-2-3. Runs on MS-DOS, VMS systems. Vendor: Trajectory Software, Inc.

LEXCALC Desktop system software. Spreadsheet. Runs on Pentium type desktop computers. Vendor: Trajectory Software, Inc.

Lexicon Data management system. Runs on Unisys systems. Data dictionary system. Vendor: PROGENI Systems, Inc.

LGHPC Handheld computer. Operating system: WinCE. Includes e-mail, fax, PIM. Vendor: LG Electronics USA.

LGS-III Operating system enhancement used by systems programmers. Terminal simulator and performance monitor in Bull HN systems. Stands for: Load Generator System-II. Vendor: Information Systems Consultants, Inc.

Liana Application development tool. Object-oriented programming language. Develops GUI applications for Windows systems. Includes run-time library. Runs on Windows systems. Vendor: Base Technology.

Lib Operating system add-on. Library management system that runs on NCR systems. Vendor: Ivan Software, Inc.

LibC/Inside Application development tool. Analyzes object code. Tracks library calls, finds date routines, etc. Vendor: Electris Software, Ltd.

Liberty Desktop software. Electronic forms software. Vendor: NCR Corp.

Liberty Web Publisher Application development tool used to build Internet and e-commerce applications. Runs on Windows systems. Released: 1997. Vendor: Liberty Integration Software.

Libr8 Application development tool. Program that assists in moving VMS applications to Unix. Vendor: Accelr8 Technology Corp.

Librarian Operating system add-on. Library management system that runs on HP systems. Vendor: Operations Control Systems.

LIBRARIAN See CA-LIBRARIAN.

library 1. A file, or set, of related files. Libraries include: program libraries, test libraries, job libraries, production libraries, and data libraries.
2. In object-oriented development a collection of classes or orjects.

Libretto Handheld computer. Runs Windows 95/98. Vendor: Toshiba America Information Systems, Inc.

Librex 386SX Notebook computer. Vendor: Librex Computer Systems, Inc.

Life-Cycle Productivity System CASE product. Automates analysis, design, and programming functions. Includes code generator for COBOL. Design methodologies supported: DeMarco, Yourdon. Databases supported: Adabas, DB2, IMS. Runs on IBM, DEC systems. Vendor: American Management Systems, Inc.

LifeBook xxx Notebook computer. Pentium CPU. Vendor: Fujitsu PC Corp.

LifeKeeper Communications software. Middleware. Used in client/server systems for hardware and software fault detection. Integrates SAP's R/3 applications with Informix database software. Vendor: NCR Corp.

LIFELINE L200, L275 Midrange computer. Server. ALPHA CPU. Vendor: Invincible Technologies Corp.

LIFELINE NFS Server Midrange computer. Server. PA-RISC CPU. Vendor: Invincible Technologies Corp.

LIFO A method of processing data in which the newest item is processed first. Stands for: Last In, First Out.

Lightning Computers Computer vendor. Manufactures desktop computers.

LightShip EIS. Client/server system. Used in data warehousing. Runs on Pentium type desktop computers. Utilizes SQL, multidimensional databases. Vendor: Pilot Software, Inc.

Lightship Professional Application development tool. Windows based GUI. Vendor: Pilot Software, Inc.

Lightship SMIS Application software. Sales and marketing system. Stands for: Lightship Sales and Marketing Intelligence System. Provides 80/20 analysis and ranking functions. Vendor: Pilot Software, Inc.

Lightweight Internet Person Schema See LIPS.

Limbo Programming language based on C and Pascal. Used to develop Internet applications. Part of Inferno. Vendor: Lucent Technologies, Inc.

Linc Development Assistant See LDA.

Linc Environment CASE product. Automates analysis, design, and programming functions. Includes code generator for COBOL. Design methodologies supported: Linc, data-flow diagrams, entity-relationship models. Databases supported: Oracle, proprietary. Runs on Unisys systems. Vendor: Unisys Corp.

LinguistX Application development tool. Natural language processing software. Analyzes massive text repositories and provides abstracts and summaries of the data. Includes LinguistX Platform, LinguistX Summarizer. Available in many languages. Released: 1997. Vendor: InXight Software (subsidiary of Xerox Corp).

Linkage Editor An IBM operating system support program.

LinkWorks Groupware. Includes workflow software. Used for e-mail, work-flow, document management. Object-oriented, client/server software that supports multiple platforms. Vendor: Digital Equipment Corp.

LinoServer Midrange computer. Server. Pentium CPU. Vendor: Linotype Hell Co.

Lint Operating system add-on. Debugging/testing software that runs on DEC VAX, IBM, Unix systems. Full name: Generic Lint. Used with C language programming. Versions include: Lint, PC-Lint. Vendor: Gimpel Software.

LINUS Database for large computer environments. Runs on Bull HN systems. Functions as subsystem of MULTICS. Stands for: Logical INquiry and Update System. Vendor: Bull HN Information Systems.

Linux Operating system. Unix based, public domain software. Developed on the Internet by volunteers, provides multi-user, multi-tasking access to the Internet.

LIONL Operating system enhancement used by systems programmers and Operations staff. Job scheduler for Unisys systems. Full name: LIONL/Job Flow. Vendor: ESI.

LIPS Communications, Internet software. Specification for retrieving information such as names and e-mail addresses over the Internet. Based on LDAP. Stands for: Lightweight Internet Person Schema.

LISP Compiler language. Language designed for list processing and used extensively in artificial intelligence. Versions include: CommonLISP, InterLISP(-D), ZetaLISP, GC-LISP, MacLISP, VAX LISP.

LISP machine Special single-user computer built to optimize development and running of software written in LISP. Used for AI applications; also called AI machines.

LispWorks Application development environment. Object-oriented. Includes CLOS, Prolog. Accesses SQL databases, interfaces with C, C++. Runs on Unix, Windows systems. Vendor: Harlequin, Inc.

ListManager Application development tool used in data warehousing. Extracts data from warehousing for use in managing customer lists. Vendor: Group 1 Software, Inc.

Live Commerce Application software for e-commerce. Includes tools for taking and tracking orders, handling shipping and processing payments and taxes. Runs on Windows NT systems. Released: 1998. Vendor: Open Market.

<Live Markup>PRO Application development tool. HTML editor. Has WYSIWYG interface. Runs on Windows systems.

Live Model for R/3 Application development software used with SAP's R/3. Simulation tool that will allow users to see how R/3 configurations will execute once the system is in production. Vendor: IntelliCorp, Inc.

Live200 Video conferencing system. Released: 1997. Vendor: PictureTel Corp.

LiveAnalyst Application development tool. Provides data transfer between legacy systems and SAP R/3 systems. Interfaces with LiveModel. Generates Abap programs. Runs on Windows systems. Released: 1997.Vendor: IntelliCorp, Inc.

LiveCache Application development tool. Transfers complex and calculation-intensive data and processing into a memory cache to reduce database access time. Part of SAP R3. Released: 1998. Vendor: SAP AG.

LiveContact Communications software. Internet product used with e-commerce. Provides customer service over the Web. Vendor: Balisoft Technologies, Inc.

LiveExchange Application software. Web-based auction software that allows businesses to sell off excess assets and inventories. Vendor: Moai Technologies, Inc.

LiveLink (Intranet) Communications software. Suite of programs that includes a Web server, Web browser, management utilities, HTML, and document management function. Used to build an intranet. Vendor: Open Text Corp.

LiveLink Search Communications, Internet software. Search engine used to corporate intranets. Released: 1996. Vendor: OpenText Corp.

LiveMeeting Suite Groupware. Electronic meeting system. Vendor: OnLive Technologies, Inc.

LiveModel Application development environment. Visual system. Allows users to build GUI system with window painting tool. Runs on Unix, Windows systems. Vendor: IntelliCorp. Inc.

LiveWire (Pro) Application development tool. Visual tool suite which allows developers to create dynamic web pages. Applications are developed in JavaScript. Runs under Unix, Windows systems. Vendor: Netscape Communications Corp.

LMU Communications. Network management software for Unix systems. Stands for: LAN Manager for Unix. Vendor: UniPress Software, Inc.

load balancing Communications. A function of some middleware and some network management systems. Allows distributed applications to spread their work-loads through two or more duplicated applications parts.

Load/INGRES INGRES database utility. Vendor: Db/Access, Inc.

Load/SYBASE SYBASE database utility. Vendor: Db/Access, Inc.

LOADPLUS See DB2 LOADPLUS.

LoadRunner Application development tool. Testing tool for client/server development. Vendor: Mercury Interactive Corp.

LoadTest See SQA Suite.

LOCO Operating system enhancement used by Operations staff and systems programmers. Provides operator console support in IBM systems. Vendor: NETEC International, Inc.

Log Analyzer Operating system enhancement used by systems programmers. Provides auditing information and information on activity for DB2 systems. Vendor: PLATINUM Technology, Inc.

logic bomb destructive program routine that destroys data, but does not affect other programs. Also, a resident computer program that lies dormant for a period, and then triggers an unauthorized act when a certain event, such as a date, occurs. Other destructive programs are called viruses, worms, backdoors and Trojan Horses.

Logical Decisions Application development tool. DSS. Runs on Windows systems. Vendor: Logical Decisions.

logical processing Term used for any processing that is hardware independent. Applications programmers solve problems logically; their solutions should run on any computer and data could be stored on any storage device. Contrast with logical processing.

Logility Software vendor. Creates manufacturing software covering demand planning, distribution, manufacturing, supply chain processing, transportation, and warehousing. Full name: Logility, Inc. (subsidiary of American Software, Inc.).

Logiscope Application development tool. Automates programming function. Analyzes source code for such things as adherence to standards and presence of logic flaws (unexecuted code). Languages analyzed: Ada, assembler, C, COBOL, Fortran, Pascal, Pl/1. Includes RuleChecker, Audit, TestChecker, ImpactChecker. Runs on DEC, HP/Apollo, IBM, Sun systems. Vendor: CS Verilog.

LOGIX Relational database for midrange and desktop computer environments. Runs on Unix systems. Vendor: Logical Software, Inc.

LOGO Compiler language used in midrange and desktop computer environments. Developed for desktop computers as a tool to teach procedural logic.

Lone-Tar Operating system software. Performs backups on Unix systems. Used for data archiving and recovery. Vendor: Lone Star Software Corp.

LOOK See CA-LOOK.

Look! Application development tool. Debugging tool for C++ applications. Vendor: Power Software.

Looking Glass Graphics user interface for Unix systems. Specifically designed for workstation use and adheres to Posix, OSF, and X/Open standards. Vendor: Visix Software, Inc.

Loops Object-oriented programming language.

Loox, Loox++, LooxWin Application development tools. Object-oriented graphics development tool. Works with LooxLib (a C programming library). Runs on Unix, Windows systems. Vendor: Loox Software, Inc.

Lotus 1-2-3 Desktop system software. Spreadsheet. Runs on Pentium type systems. Vendor: Lotus Development Corp.

Lotus Communications Server Communications software. Expands the capabilities of LAN e-mail systems allowing thousands of users. Is an extension of Lotus Notes. Has built-in database. Also called LCS. Vendor: Lotus Development Corp.

Lotus Components Application development tool. Allows users to embed, spreadsheets and project management templates in Notes documents. Uses ActiveX. Released: 1997. Vendor: Lotus Development Corp.

Lotus Forms Desktop software. Electronic forms software. Vendor: Lotus Development Corp.

Lotus Notes See Notes.

Lotus Notes: Document Imaging Imaging software. Stores images in a Notes database. Represents documents with icons. Vendor: Lotus Development Corp.

Lotus Notes VIP Application development system. Used to create Notes applications. Stands for: Lotus Notes Visual Programming. Runs on Windows systems. Vendor: Revelation Technologies, Inc.

Lotus Soft-Switch Communications software. Messaging system. Vendor: Lotus Development Corp.

LotusScript Computer language included with Notes. Object-oriented derivative of Visual Basic. Generates macros which can be invoked by Lotus applications running under Windows, MS-DOS, Unix, Macintosh. Vendor: Lotus Development Corp.

LPS CASE Product. Automates all phases of the development cycle including project management functions. Full name: Lifecycle Productivity System.

LSE Computer vendor. Manufactures desktop computers.

LSF Operating system software. Assigns job processing in a series of Unix systems in a client/server environment. Keeps a database of system resources and, i.e., moves Unix jobs to idle computers on the system. Stands for: Load Sharing Facility. Vendor: Platform Computing Corp.

LTE 5400 CTFT Portable computer. Pentium CPU. Vendor: Compaq Computer Corp.

LTE Elite, 5000 Notebook computer. Vendor: Compaq Computer Corp.

LU 6.2 Communications protocol devised by IBM. Used to connect PCs to mainframes; Lans to WANS.

Lucid 3D Desktop system software. Spreadsheet. Runs on Pentium type desktop computers. Vendor: Daceasy, Inc.

Luminate for SAP R/3 Operating system software. Measures performance in client/server systems by simulating user activity to measure response time. Released: 1998. Vendor: Luminate Software Corp.

Lynx 1. Software distribution tool. Used to install and maintain multiple versions of a single application running on a network. Vendor: Lotus Development Corp. 2. Text only Internet browser.

Lynx(OS) Real-time operating system for various RISC machines. Unix-type system. Provides real-time multitasking. Ascribes to Posix standards. Vendor: Lynx Real-Time Systems, Inc.

MB Meg. Stands for Mega, which usually stands for megabyte: approximately one million bytes.

M.1 Artificial intelligence system. Expert system building tool. Implemented in PROLOG, but rewritten in C. Runs on desktop computers. Vendor: Tecknowledge.

M-Bridge Communications software. Gateway for e-mail systems. Runs with Netware, MS-NET networks. Vendor: Computer Mail Services, Inc.

M&D Usually refers to McCormack & Dodge Corp. McCormack & Dodge have written financial and administrative software used by many corporations. See Millennium. Company was purchased by Dun & Bradstreet Software, who was, in turn, purchased by GEAC Computer Corp. The software is referred to by both M&D and D&B designations.

M-Link Communications software. Network connecting DEC, CDC, and Unix computers. Vendor: Century Analysis, Inc.

M-Note Pentium Notebook computer. Pentium CPU. Vendor: Mikon, Inc.

M/P/E Application development tool. Provides module foundation to develop new applications and integrate existing applications into a single system. Runs on DEC VAX systems. Vendor: Unicad, Inc.

M:PDL Procedural language used with Millennium:SDT. Vendor: Dun & Bradstreet Software Services.

M-Power Desktop computer. RISC processor. Vendor: APS Technologies.

M/SQL Relational database for desktop computer environments. Runs on DEC, IBM, Unix systems. Utilizes SQL. Can be used in client/server computing. Vendor: Intersystems Corp.

M/Text Mainframe word processor. Works with IDMS databases. Vendor: Cincom Systems, Inc.

M1 Computer chip, or microprosessor. Faster than the 486 processor. Vendor: Cyrix Corp.

M100x, M120T, M133T Notebook computer. Pentium CPU. Vendor: Hitachi PC Corp.

M³ Communications. Middleware. Combines Tuxedo with an object request broker to allow transaction management software to work with object technology. Supports both COM and CORBA. Part of WebLogic Enterprise. Released: 1998. Vendor: BEA Systems, Inc.

Mac-EDI Communications software. EDI package. Runs on Macintosh systems. Vendor: Digit Software, Inc.

Mac OS X Operating system for Macintosh computers. Upgrade to Mac OS 8, replaces Rhapsody. Release date: 1999. Vendor: Apple Computer, Inc.

Mac-schedule Desktop system software. Project management package. Runs on Macintosh desktop systems. Vendor: Mainstay.

Mac2Win Application development tool. Converts Macintosh applications to Windows platforms. Vendor: Altura Software, Inc.

Mac3270 Communications software. Terminal emulator for Apple Macintosh desktop computers. Vendor: The Vermont Software Co., Inc.

MacAnalyst(/Expert) CASE product. Automates analysis and design functions. Uses data flow diagrams, control flow diagrams, process specifications, screen prototyping. Includes data dictionary and supports data modeling with entity-relationship diagrams. Vendor: Excel Software.

MacApp Application development environment. Used for object-oriented program development and runs on Macintosh systems. Includes some object libraries. Interfaces with Object Pascal.

MacAPPC Communications software. Network connecting Apple Macintosh desktop computers to mainframes, midrange systems, and other desktop computers. Vendor: Apple Computer, Inc.

MacBrain Artificial intelligence system. Runs on Macintosh systems. Imitates the brain. Performs diagnostic, robotic, and analytic processing. Vendor: Neuronics.

MacCalc Desktop system software. Spreadsheet. Runs on Macintosh systems. Vendor: Bravo Technologies, Inc.

MacDesigner(/Expert) CASE product. Automates design functions. Creates structure charts. Vendor: Excel Software.

MacDOS Operating system for Apple Macintosh systems. Vendor: Traveling Software, Inc.

MacDraw II Desktop system software. Graphics package. Runs on Macintosh systems. Vendor: Claris Corp.

MacDSS Decision support software that runs on Macintosh systems. Vendor: Apple Computer, Inc.

MacExcel See Excel.

MACH General-purpose operating system based on Unix, and considered a possible replacement of Unix. Basis for NeXT's proprietary operating system, and used as the cornerstone for OSF/1. Originally developed by Berkeley and now produced by Carnegie Mellon. Used on NeXT, RS/6000 systems.

Mach 1 Database for Unix environments. Utilizes SQL. Vendor: Tominy, Inc.

Mach Micro-Kernel See microkernel.

machine language The language of the computer. Binary language. Also called computer language.

Macintosh Desktop/midrange computers. PowerPC CPU. Operating systems: MacOS, A/UX System 7, BeOS, Rhapsody. Vendor: Apple Computer, Inc.

Macintosh PowerBook Notebook computer. Vendor: Apple Computer, Inc.

Macintosh Programmers Workshop See MPW.

MacIRMA Communications software. Terminal emulator for Apple Macintosh desktop computers. Vendor: Digital Communications Associates, Inc.

MacLISP See LISP.

MacNFS Data communications software. Connects Macintosh and Unix systems. Lets Unix, DOS and Windows applications share data. Runs on Macintosh systems. Vendor: Thurisby Software Systems, Inc.

MacOS Operating system for Apple Macintosh systems. Vendor: Apple Computer, Inc.

MacPac Application software. Manufacturing, distribution, and financial software. Runs with diverse hardware. Vendor: Andersen Consulting.

MacPaint Desktop system software. Graphics package. Runs on Macintosh systems. Vendor: Claris Corp.

MacProject II Desktop system software. Project management package. Runs on Macintosh systems. Vendor: Claris Corp.

macPROLOG Artificial intelligence system. Runs on Macintosh systems. Vendor: Programming Logic Systems, Inc.

macro A single predefined statement that will cause the generation of more than one machine statement. Used mostly with Assembler languages.

Macro Assembler Assembler language used in IBM desktop computer systems.

MacroTrap System software. Anti-virus software used on networks. Vendor: Trend Micro Devices, Inc.

MacScheme Artificial intelligence system. Runs on Macintosh systems. Programming environment for Lisp. Vendor: Semantic Microsystems, Inc.

MACSYMA A programming language designed for non-numeric manipulation of mathematical expressions.

MacTCP Communications software. Network connecting Apple Macintosh desktop computers to other vendors' computers. Vendor: Apple Computer, Inc.

MacTerminal Communications software. Terminal emulator for Apple Macintosh desktop computers. Vendor: Apple Computer, Inc.

MacWeb Internet browser. Runs on Macintosh systems.

MacWrite Desktop system software. Word processor. Runs on Macintosh systems. Vendor: Claris Corp.

MacX Operating system extension to MacOS that runs X Windows applications. Supports X Windows, Motif, and Open Look. Vendor: Apple Computer, Inc.

Macyacc Application development tool. Program (or code) generator. Generates ANSI C code. Runs on Macintosh systems. Vendor: Abraxas Software, Inc.

Madman MIB Communications. Stands for: Mail and Directory Management (Madman) Management Information Base (MIB). Defined by EMA (Electronic Messaging Association).

MAE Emulation software that allows applications written for Macintosh systems run on Sun and Hewlett-Packard systems. Stands for: Macintosh Application Environment. Vendor: Apple Computer, Inc.

Maestro Operating system enhancement used by systems programmers. Job scheduler. Manages jobs under Unix, Windows NT. Interfaces with SAP financial applications. Vendor: Unison Software.

Maestro II CASE product. Automates analysis, design, and programming functions. Design methodologies supported: Merise. Includes code generator for COBOL. Used for re-engineering and moving systems from mainframe to Unix. Used for client/server computing. Runs on Bull HN, DEC VAX, HP, IBM, Unisys systems. Vendor: Softlab, Inc.

Maestro II RTW Application development tool. Used to maintain systems. Provides change management, source control, mainframe interface. Links to COBOL Workbench. Runs on OS/2, Unix, Windows systems. Stands for: Redevelopment Workstation. Vendor: Softlab, Inc.

MAG/base(2) Database for desktop computer environments. Runs on IBM desktop systems. Includes report generator. Vendor: Rocky Mountain Software Systems.

MAGEC for MVS, VSE, VM/CMS Application development tool. Includes program generator, security functions, documentation tools, prototyping tools, and window facilities. Runs on IBM mainframe systems. Stands for: Mask and Application Generator and Environment Controller. Vendor: Magec Software.

Magic(7) Application development environment. Used in RAD for client/server and host environments. Used in enterprise development. Includes functions for concurrent multiuser development, SQL code generation, relational database processing. Allows visual development of programs to run under MS-DOS, Unix, CTOS, VMS. Vendor: Magic Software Enterprises, Inc.

Magic * Calc Desktop system software. Spreadsheet. Runs on DEC systems. Vendor: Data Control Systems, Inc.

Magic Cap Operating system for handheld computers. Includes Telescript. Vendor: General Magic, Inc.

Magic/L Compiler language used in midrange and desktop computer environments.

Magic Link Handheld computer, or PDA. Uses Magic Cap operating system. Vendor: Sony Electronics, Inc.

Magic PC,LAN Relational database for desktop computer environments. Runs on Pentium type desktop computer systems. Vendor: Aker Corp.

Magic Words Desktop system software. Word processor. Runs on Apple II systems. Vendor: ARTSCI, Inc.

Magicalc Desktop system software. Spreadsheet. Runs on Apple II, IBM systems. Vendor: ARTSCI, Inc.

Magix Application development system. Includes relational DBMS, programming language, communications and network software, screen and report formatters. Runs on IBM desktop systems. Vendor: Advanced Software Technologies, Inc.

Magna 8/TP, Magna Case, Magna View Application development system. Magna 8/TP includes relational database, code generators, and data management utilities. Magna Case allows development of mainframe applications from a desktop computer and integrates with Excelerator. Magna View provides window-based reporting functions. All three run on Bull HN systems. Vendor: Magna Software Corp.

Magna Powermax Servers Midrange computer. Pentium II CPU. Operating system: Windows 95/98. Vendor: MAXIMUS COMPUTERS, Inc.

Magna SCSI Server Midrange computer. Server. Pentium, Pentium Pro CPU. Vendor: MAXIMUS COMPUTERS, Inc.

Magna Wide SCSI Server Midrange computer. Server. Pentium Pro CPU. Vendor: MAXIMUS COMPUTERS, Inc.

Magna-Artist,CAD,Media,NT2 Desktop computer. Pentium CPU. Operating systems: Windows 95/98. Vendor: MAXIMUS COMPUTERS, Inc.

Magna X Application development environment. Generates server code for Unix, MVS/CICS systems access mainframe data in client/server systems. Includes high-level language and visual design tools. Generates front-end Visual Basic code. Users do not have to know Unix or C. Vendor: Magna Software Corp.

Magnatronic 620 Desktop computer. Notebook. Pentium processor. Vendor: Ace Computers.

Magnets Communications, Internet software. Allows companies to broadcast messages to everyone in the company. Messages can appear as screen savers or in small on-screen windows. Broadcasts can contain information, graphics, and links to Intranet or Internet Web pages. Vendor: Wayfarer Communications, Inc.

Magnia 3000,5000 Midrange computer. Pentium II CPU. Vendor: Toshiba America Information Systems, Inc.

Magnifi Server Knowledge management software. Accessible through the Internet. Enables indexing, locating and retrieving multimedia data. Vendor: Magnifi, Inc.

MAI Computer vendor. Manufactures desktop computers.

MAI Origin ADS Relational database for large computer environments. Runs on Basic Four systems. Utilizes EQM. Vendor: MAI Basic Four, Inc.

Mail Link (STMP, Mac, MHS) Communications software. Gateway for e-mail systems. Runs on Macintosh systems. Vendor: Starnine Technologies, Inc.

mail merge The use of software packages to merge names and addresses on one file with text from another file in order to produce multiple letters. The function uses both database software and word processing software and most major word processing packages have mail merge capabilities that work with most databases.

Mail Server Communications software. E-mail package. Vendor: Microsoft Corp.

Mail/VE Communications software. E-mail system. Runs on CDC systems. Vendor: Control Data Corp.

Mailbag Communications protocol. Handles EDI. U.S. standard.

Mailbox Communications software. E-mail system. Runs on IBM systems. Vendor: I.P. Sharp Associates, Ltd.

MailMail Communications software. Gives access to e-mail, fax and data from any Web browser. Can use PCs, network computers, Web-TV, handheld computers, etc. Released: 1998. Vendor: Infonet Software Solutions.

Mailman Communications software. E-mail system. Runs on VINES, Windows systems. Vendor: Reach Software Corp.

Mailmate(/MM,QM,1B) Communications software. Gateway for e-mail systems. Runs on Appletalk, DECnet networks. Vendor: Alisa Systems, Inc.

MAINDS Application development system. Runs on large computer systems. Used to develop MAINSAIL programs. Vendor: XIDAK, Inc.

mainframe computer A large general-purpose computer, in particular one to which other computers can be connected so they can share the facilities of a centralized data center.

Mainframe Express Application development environment. Used to develop mainframe COBOL applications from PCs. Includes Animator debugging utility. Vendor: Originally Micro Focus, now MERANT.

Mainlan for Windows Communications software. Network operating system. Runs on Pentium type desktop computers. Vendor: Mainlan, Inc.

MainLink Communications software. Network connecting PICK systems. Link between desktop computer, midrange, and mainframes. Vendor: Pick Systems.

MAINPM Operating system add-on. Debugging/testing software that runs on DG, DEC VAX, IBM, HP, Unix systems. Vendor: XIDAK, Inc.

MAINSAIL Compiler language used in large computer environments. Vendor: XIDAK, Inc.

maintenance Any activity that keeps an existing piece of hardware or software functioning.

Maintenance Workbench See MWB.

MainView Operating system enhancement used by systems programmers and operations staff. Automates the data center. Vendor: Boole & Babbage, Inc.

MainView Desktop Operating system software. Off-loads mainframe activity to OS/2 workstations. Vendor: Boole & Babbage, Inc.

MainWin (Studio) Application development environment. Used for cross-platform development. Creates single source code module for multiple platforms. Creates Unix version of applications written for Windows. Used in client/server systems. Vendor: Mainsoft Corp.

Mambo Application development tool. Used to build Web applications that link to mainframe and relational databases. Runs on Unix, Windows NT systems. Released: 1996. Vendor: Inprise Corp.

MAN Computer network established for a specific locality. Wide area network limited in range to around a 30 mile radius. Stands for: Metropolitan Area Network.

Managed Client Backup Operating system software. Provides automatic backups for clients in client/server systems. Includes partial file backups and duplicate file detection. Released: 1998. Vendor: Seagate Technology, Inc.

managed PC PCs that can be managed from a central location. Includes electronic/upgradable ROM, remote power on and off, and built-in desktop management.

Managed Query Environment See MQE.

Managed Reporter Edition, Managed Reporter Administrator's Kit Application development tool. Report writer. Allows control of data access. Part of Focus Six for Windows. Vendor: Information Builders, Inc.

ManageIT Operating system enhancement. Optimizes, monitors and predicts performance of databases, e-mail and other applications. Released: 1999. Vendor: Computer Associates International, Inc.

ManagePro Desktop software. Project management package. Includes Individual Edition and Multi-User Network Edition. Vendor: Avantos Performance Systems, Inc.

Manager See CA-Manager.

Manager family CASE product. Automates all phases of the development cycle including project management functions. Used for client/server computing. Associated product names: DataManager, DesignManager, DictionaryManager, MethodManager, ControlManager, SourceManager, ManagerView. Vendor: MANAGER SOFTWARE PRODUCTS, Inc.

Manager's Portfolio EIS. Runs on Pentium type desktop computers. Vendor: Easel Corp.

ManagerView CASE product. Automates project management function. See Manager family.

ManageWise Communications software. Network management software. Monitors distributed LANs. Includes Intel's LANDesk tools. Vendor: Novell, Inc.

ManageWorks See Pathworks.

ManageX Operating system software. Monitors Windows NT systems across the enterprise. Has Unix-like feel. Vendor: NuView, Inc.

Manhattan family Midrange computer. Pentium, Pentium Pro CPU. Operating systems: Windows NT. Vendor: AST Research, Inc. (division of Samsung Electronics).

ManMan MRP, or CIM software for companies that deal with products that must be mixed or blended. Stands for: Manufacturing Management for Process. Runs on DEC VAX systems. Vendor: Ask Computer Systems, Inc.

ManMan HP, ManMan VAX MRP, or CIM, software. Runs on DEC, Hewlett-Packard systems. Vendor: Ask Computer Systems, Inc.

ManMan/X ManMan software that runs on Unix systems. See ManMan.

Mantext Word processor for IBM mainframes. Works with Mantis. Vendor: Cincom Systems, Inc.

Mantis Application development system. Includes Mantis 4GL, program generator, testing and debugging tools, and documentation writers. Includes desktop computer Mantis, Mantis SQL for DB2 Access. Runs on Bull HN, DEC, IBM, NCR, Wang systems. Vendor: Cincom Systems, Inc.

Manufacturing Manager Application software. Manufacturing software handling supply chain functions. Vendor: SynQuest, Inc.

Manufacturing Order Processing See MOP.

Manufacturing Resource Processing See MRP.

MAP Operating system enhancement used by systems programmers. Monitors and controls system performance in IBM, DEC, Unisys systems. Stands for: Modeling and Analysis Package. Vendor: Quantitative System Performance.

MAP/Administrator Application development tool. Automated administration tool that integrates methodologies and tools. Used with AD/Method methodologies in client/server environments. Vendor: Structured Solutions, Inc.

Map Discovery System Application development tool. Allows users to do geographic data mining to determine unsuspected geographic patterns. Runs on Unix, Windows systems. Vendor: Information Discovery, Inc.

MAPI API (Application Programming Interface). Messaging software. Used in client/server applications. Provides the interface between different servers. Used with Windows. Stands for: Message Application Programming Interface. Vendor: Microsoft Corp.

Mapics/(DB,XA) Application software. Object oriented ERP product. Runs on AS/400 systems. XA includes Java, euro support. Stands for: Manufacturing and Accounting Production Information Control System/(DataBase, eXtended Advantage.) Vendor: Mapics, Inc.

Mapmaker Application development tool. Includes geocoding software, an address dictionary, and street maps. Runs on Unix, Windows systems. Vendor: MapInfo.

MAPPER programs Application development tools. Used for analysis of client/server applications. Interfaces with DB2, Informix, Oracle, Sybase. Includes: MAPPER ADMIN, MAPPER C, MAPPER Kit Tools, MAPPER RDMS I/F, MAPPER 1100. Vendor: Unisys Corp.

MAPS/DB Database for large computer environments. Runs on DEC systems. Includes interactive query facility and report generator. Vendor: Ross Systems, Inc.

Marcam Software vendor. Products: : MAPICS family of MRP systems for midsize environments.

Marcon Plus Relational database for desktop computer environments. Runs on Pentium type desktop computer systems. Vendor: Interactive Support Systems, Inc.

Mariner Internet browser. Includes e-mail, FTP, Gopher, newsgroup access. Runs on Windows systems. Released: 1995. Vendor: FTP Software, Inc.

Marixx Usxx, SS Midrange computer. SPARC CPU. RISC machine. Operating Systems: Solaris. Vendor: Aries Research, Inc.

MARK IV,V Program generator that accepts input from fill-in-the-blanks forms. Runs on IBM mainframes. Obsolete. Vendor: Sterling Software (Answer Systems Division).

MarketFirst Application software. Provides marketing automation by responding to and tracking customer information requests, analyzing when a lead changes to a prospect, and analyzing which marketing programs are most effective. Released: 1998. Vendor: MarketFirst Software.

Market Focus Communications, Internet software. Tracks usage of Web sites/pages. Vendor: Interse.

MarketOne Application software. Provides marketing automation by responding to and tracking customer information requests, analyzing when a lead changes to a prospect, and analyzing which marketing programs are most effective. Released: 1998. Vendor: DataMind Corp.

Mars Computer vendor. Manufactures desktop computers.

Marshall Workflow software. Object-oriented suite of Windows applications. Works with manufacturing software. Vendor: Ramco Systems Corp.

Martin/Odell Object-oriented development methodology.

Marvel Communications service. Includes e-mail, bulletin boards, libraries, Internet news groups, Microsoft tools and product information. Runs directly from Windows95. Also called the Microsoft Network. Vendor: Microsoft Corp.

MASM Assembler language for Intel's 80x86 computer chips. Stands for: Macro AsseMbler. Vendor: Microsoft Corp.

Maspar Computer vendor. Manufactures large computers.

MasPar MP-1,2 Massively parallel processor. Supports up to 16,384 processors. Operating systems: Unix, Ultrix. Vendor: Maspar Computer Corp.

MASS-II Manager Relational database for midrange and desktop computer environments. Runs on DEC, PC-DOS systems. Utilizes SQL. Can be used in client/server computing. Includes report writer. Vendor: Microsystems Engineering Corp.

Massively Parallel Processing See MPP.

Master Certified Novell Engineer See MCNE.

master console In a system with multiple consoles, the console used for communication between the operator and the operating system.

Master Production Scheduling See MPS.

Masterfile Database for desktop computer environments. Runs on IBM desktop computer systems. Vendor: Evolution, Inc.

Master*IT* Internet software. Monitors and manages Internet and extranet sites, and Web servers. Automatically reports on problems and takes corrective action. Runs on Unix, Windows NT systems. Vendor: Computer Associates International, Inc.

Mastermind Operating system enhancement used by systems programmers and Operations staff. Job scheduler for IBM AS/400 and S/38 systems. Full name: MastermindCThe Operations Expert. Vendor: Pansophic Systems, Inc.

MaTrix Application development tool. Includes screen painter, data editor. Used to distribute applications on Unix systems. Vendor: Ampersand Corp.

Matryx Application development tool. DSS. Accesses multidimensional databases. Vendor: Stone, Timber, River.

Mattisse Object-relational database. Supports inheritance, polymorphism. Handles multimedia, time-series data. Runs on Macintosh, Solaris, Unix, VMS, Windows systems. Vendor: Mattisse Software, Inc.

Maui Code name for the beta version of a Web browser written entirely in Java. Vendor: Netscape Communications Corp.

Max Communications. Web agent that searches Web sites to analyze effectiveness of the sites. Simulates human browsing and used to evaluate site design and improve the usability of the site. Vendor: WebCriteria, Inc.

MAX 32 Operating system for midrange MODCOMP systems. Vendor: Modular Computer Systems, Inc. (MODCOMP).

Max/Enterprise Communications. Network management software. Manages networks that include legacy and client/server systems. Integrates with OpenView, NetView, Spectrum, Windows. Vendor: Maxm Systems Corp.

MAX IV (Real Time OS) Operating system for midrange MODCOMP systems. Vendor: Vendor: Modular Computer Systems, Inc. (MODCOMP).

MAX/SPF Application development tool. Screen editor for use with TSO/ISPF systems. Vendor: Integrity Solutions, Inc.

Maxess SNA Gateway Communications software. Network connecting IBM systems. Vendor: 3Com Corp.

MAXION Midrange computer. RISC machine. MIPS R4400 CPU. Operating systems: MAXION/OS, Unix. Vendor: Concurrent Computer Corp.

Maxm Communications software. Integrates network management systems and LAN systems. Vendor: Maxm Systems Corp.

MaxPro Midsize computer. Server. Pentium II, Pro processors Operating systems: OS/2, Unix, Windows. Vendor: CSS Laboratories, Inc.

Maya Desktop software. Graphics package. Vendor: Silicon Graphics, Inc.

Mazdamon See CA-Mazdamon.

Mbill Billing system based on M&D's Millennium technology. Vendor: Ramyk Consulting Group, Inc.

MBOS/5 See BOS/CBOS/5.

Mbps Millions of bits per second. Term used to define speeds during data transfer. Industry standard is 10Mpbs, but fast ethernet and 100VG-Anylan both provide 100Mpbs.

MCA Micro Channel Architecture. Design for microcomputers.

McAfee Enterprise Network management system. Provides security and management functions for Windows NT systems. Vendor: McAfee Associates, Inc.

McCabe Visual Testing Toolset Application development tool. Testing tool that assists software testers throughout the software life cycle. Runs on Unix, Windows systems. Released: 1996. Vendor: McCabe & Associates, Inc.

McCormack & Dodge See M&D.

McDesk Operating system enhancement. Provides Macintosh functions such as printing by dragging a file icon to the printer icon in Windows systems. Vendor: Granite Software, Inc.

MCIS Communications software. Messaging servers based on Exchange. Includes Mail, News, Chat, Merchant, and Conference servers. Stands for: Microsoft Commercial Internet System. Vendor: Microsoft Corp.

McMax Relational database for desktop computer environments. Runs on Macintosh systems. Vendor: Nantucket Corp.

MCNE Technical certification. Master Certified Novell Engineer. Applicants must be CNEs. Specialize in one of five skill areas: Management, Connectivity, Messaging, Internet/Intranet, Client/Network. May specialize in multiple categories.

MCP(/AS,VS) Operating system for large computer Unisys systems. Vendor: Unisys Corp.

MCP (+ Internet) Technical certification. Microsoft Certified Professional. Base certification. Requires passing one exam. Internet option requires passing two additional exams.

MCP601, MCP603, MCP604, MCP620 Computer chip, or microprosessor. The specific chips in the PowerPC family. Faster than the 486 processor. Produced in conjunction with IBM, Apple. Vendor: Motorola Computer Systems, Inc.

MCSD Technical certification. Microsoft Certified Solution Developer. Requires passing two required and two elective exams.

MCSE (+ Internet) Microsoft Certified Systems Engineer. Consists of four required and two optional exams. Internet option adds three more exams. Applicants can receive credit for the networking exam if they have Novell or Banyan certification.

MD-DRAW Graphics package for mainframes. Vendor: Maersk Data AS.

MDBS III,IV Relational database for midrange and desktop computer environments. Runs on DEC, IBM, Unix, Xenix systems. Includes: MDBS DDL/DMS, MDBS IDML, MDBS QRS, MDBS RDL. Vendor: mdbs, Inc.

MECO Operating system enhancement used by systems programmers and Operations. Provides remote control of a data center. Vendor: Software Engineering of America.

Medallion Desktop computer. Pentium processor. Vendor: Tangent Computer, Inc.

Media 1. Operating system add-on. Library management system that runs on DEC VAX systems. Vendor: International Structural Engineers, Inc.
2. Application development tool. DSS. Provides access to data stored in relational and multidimensional databases. Runs on Unix, Windows systems. Vendor: Speedware Corp.

Media Best Buy Desktop computer. Pentium CPU. Operating system: Windows 95/98. Vendor: Royal Electronics, Inc.

Media/Schedule/Vault Operating system enhancement used by systems programmers. Increases system efficiency in DEC VAX systems. Manages disk storage space. Vendor: International Structural Engineers, Inc.

MediaCUBE Midrange computer. Operating system: Transit. Vendor: nCUBE.

MediaGem Desktop computer. Pentium CPU. Operating system: Windows 95/98. Vendor: Tangent Computer, Inc.

MediaGo Notebook computer. Pentium CPU. Vendor: HyperData Technology Corp.

MediaNote CD Book xxx Notebook computer. Pentium CPU. Vendor: MAXIMUS COMPUTER, Inc.

MediaOrganizer Multimedia object management software. Allows users to store and retrieve text, still image, animation, audio and video data. Images can be displayed in multiple windows on a single screen. Runs on Windows systems. Vendor: Lenel Systems International, Inc.

MediaPro Desktop computer. Pentium processor. Vendor: Technology Advancement Group Inc.

Meeting Maker Groupware. Scheduling package. Runs on Apple systems. Vendor: ON Technology.

MeetingPlace WebPublisher Groupware. MeetingPlace is an electronic meeting system and WebPublisher provides an interface with the Internet. Allows a user to schedule conferences, access meeting materials and listen to recordings over the Internet. Released: 1996. Vendor: Latitude Communications, Inc.

Mega 43D,45D2,46D2xx Desktop computer. Pentium CPU. Operating system: Windows 95/98. Vendor: Megadata Corp.

Mega CD-Note Desktop computer. Notebook. Pentium processor. Vendor: Comtrade Computer, Inc.

Mega/Net Models 1000,2000,3000 Relational database machine. A dedicated computer that contains the DBMS necessary to process the database. These databases are compatible with most mainframe operating systems and databases. Vendor: Mega/Net Corp.

Megabook 880 Notebook computer. Pentium CPU. Vendor: Megaimage, Inc.

megabyte See M.

Megafiler Relational database for desktop computer environments. Runs on Macintosh systems. Vendor: Megahaus Corp.

MemMaker Operating system enhancement. Utility that manages computer storage by providing services such as data compression and virus protection. Runs on MS-DOS 6.0 systems. Vendor: Microsoft Corp.

Memo 1. Communications software. E-mail system. Runs on Prime systems. Vendor: Interpac Software, Inc. 2. Communications software. E-mail system. Runs on IBM systems. Full name: Memo Electronic Mail System. Vendor: Verimation, Inc.

Memo/EDI Communications software. EDI package. Runs on IBM mainframe systems. Vendor: Verimation, Inc.

Memphis Code name for next release of Windows 95/98. Merges the desktop interface with the Explorer Web browser. Provides common user interface with Windows NT. Vendor: Microsoft Corp.

Mentis Wearable computer. Similar in functionality to handhelds, but worn on body to free hands. Primarily used in manufacturing for such things as providing access to diagnostic manuals with which the user makes repairs. Vendor: Interactive Solutions, Inc.

menu A list of options for an applications system from which the user selects the function desired. Online applications systems are either menu-driven or command driven (although some have both options). Types of menus: Pull down, pop up, list, button.

MERANT Software vendor. Produced application development tools. Company formed by the merger of INTERSOLV and Micro Focus in 1999.

Mercator Communications software. Message oriented middleware, data transformation tool. EDI package. Runs on Windows systems. Released: 1995. Vendor: TSI International Software Ltd.

Mercator for R/3 Application development tool. Creates links between legacy systems and SAP's R/3 by automatically converting old files to R/3 format. Released: 1996. Vendor: TSI International Software, Ltd.

Merced See IA-64.

MerchantXpert See CommerceXpert.

Mercury 1. Application development tool. Extension to BusinessObjects. Lets end users manipulate and analyze enterprise data stored in relational databases using a mouse. Interfaces with Oracle, Sybase, Informix, Teredata, DB2. Used in data warehousing. Runs on most desktop system. Vendor: Business Objects, Inc. 2. Application development tool. Allows developers to create software to collect and analyze performance statistics on the application level for client/server systems. Vendor: IBM Corp. and Legent Corp.

3. Communications, Internet software. Allows users to download applications from a network, store them locally and use them off-line. Vendor: Netscape Communications Corp.

Mercury 2000 Communications software. E-mail system. Runs on DEC systems. Vendor: Cableshare, Inc.

Mercury Multimedia Notebook computer. Pentium CPU. Vendor: Xediom.

Merge Operating system software. Allows DOS and Windows applications to run on Unix systems. Vendor: Locus Computing Corp.

Merge 286,386 Desktop computer operating system for Pentium-type desktop computers, Unix systems. Vendor: Locus Computing Corp.

Merise Structured programming design methodology named for its developer. Accepted as a standard design methodology by some companies, and used by some CASE products.

Merlin Code name for version of OS/2 Warp. Includes voice interface with a vocabulary of over 20,000 words. Vendor: IBM Corp.

MERVA/370 Communications software. Network connecting IBM computers. Used in banking environment. Used with DSNL/370. Vendor: IBM Corp.

MES Manufacturing Execution System. Part of manufacturing systems. Software that provides a link between the company's corporate planning and its process control systems. The purpose of a Manufacturing Execution System is to enforce business rules pertaining to plant floor operations. Allows the operation of production plans based on actual capabilities which are based on such things as the status of current orders, work in-process and equipment availability.

Mesa/AD Application development tool. Automates design functions. Includes a formal methodology for the design step which creates preliminary design diagrams as structure charts. Runs on Unix systems. Released: 1994. Vendor: Mesa Systems Guild, Inc.

Mesa/Vista Application development tool. Manages the development process. Works with many legacy development tools. Provides a familiar and consistent user interface to all project data regardless of platform or operating environment. Runs on Unix, Windows systems. Released: 1997. Vendor: Mesa Systems Guild, Inc.

message 1. Data transmitted between units of an online system. Messages can be sent between users and programs, from user to user, and from program to program. 2. In object-oriented development, information that is passed to an object to direct its execution. The message contains the name of the object, what is to be done, and any necessary parameters.

MessagePad 130, 2000 Handheld, pen-based computer. Operating system: NewtOS. Vendor: Newton, Inc. (division of Apple Computer Inc.)

Message Application Programming Interface See MAPI.

Message Express Communications software. Middleware. Application Programming Interface (API) that provides four common communications verbs to allow programmers to develop client/server applications without knowing the different communications protocols and operating systems in the network.

Message Format Service See MFS.

Message-Oriented Middleware See MOM.

message passing Technology used in client/server computing to handle communications. Type of middleware. The client passes a message to the server and does not have to wait until the message is received. Opposite of remote procedure calls.

Message Router Communications software. Gateway for e-mail systems. Runs with DECnet, SNA networks. Includes: Message Router/Memo, VAX Message Router VMSmail Gateway, VAX Message Router/S Gateway, P Gateway, VAX Message Router X.400 Gateway, Message Router gateways for CC:Mail/MHS, Network Courier, 3Com 3+Mail. Vendor: Digital Equipment Corp.

MessagePad Personal digital assistant. Reads handwritten notes, automatically adds appointments to a calendar, dials a phone, or sends a fax. Part of the Newton family. Vendor: Apple Computer, Inc.

Messaging Server Communications, Internet software. Provides messaging services. Supports LDAP, SMTP. Part of Apollo. Vendor: Netscape Communications Corp.

Meta Systems Toolset CASE product. Automates analysis, programming, and design functions. Vendor: Meta Systems Ltd.

MetaCenter Application development tools. Used in data warehousing for both development and management of multi-tiered warehouses. Includes tools for data extraction, warehouse management, data quality analysis. Includes software from Intellidex Systems. Vendor: Carleton Corp.

MetaCOBOL See CA-MetaCOBOL.

metacomputer A computer system that is controlled through a network and can access memory, CPU, and storage resources from any computer in the network. A metacomputer was set up on the Internet with 600 desktop systems and used to decode encrypted information.

MetaCube 1. Application development tool. OLAP software used in data warehousing. Multidimensional database that allows users to view data in three or more dimensions such as time, place and product. Vendor: Informix Software, Inc. 2. Application development tool used in data warehousing. Provides a drag-and-drop interface for DSS software. Can work with existing relational databases. Released: 1994. Vendor: Stanford Technology Group.

metadata Data about data. For example, when used with e-mail, metadata describes who sent a message, to whom, when sent, and when received. Also used with data warehousing and typically contains basic information, summary information, and pointers to related information sources.

MetaData Manager See Warehouse Control Center.

MetaEdit CASE tool. Includes Personal, Workbench versions. Vendor: MetaCase Consulting.

Metafact EIS. Runs on IBM mainframes and desktop computers. Vendor: Integrated Data Architects.

MetaFrame Operating system software. Used to deploy, manage and access business applications for the enterprise. Supports heterogeneous platforms and networks in client/server environments. Provides applications management, performance and security functions. Originally called pICAsso. Runs on Windows NT systems. Released: 1998. Vendor: Citrix Systems Inc.

MetaH See ADL.

Metaphase (Enterprise) Application software. PDM package. Three-tier, Web-based architecture. Allows product design and manufacturing teams in large, decentralized enterprises to collaborate on the design, planning, implementation, and management of product development processes and information. Vendor: Metaphase Technology, division of SDRC (Structured Dynamics Research Corp).

Metaphor Application development tool used for business access and analysis. Used in heterogeneous desktop systems. Vendor: Metaphor, Inc.

MetaStar Enterprise System management software. Used to deploy information-based applications over the Web. Uses XML to allow users to access data records in SQLbase and Oracle databases from a Web browser. Released: 1999. Vendor: Blue Angel Technology.

Metastor See Huron.

MetaTest Operating system add-on. Debugging/testing software that runs on Unix systems. Vendor: Software Research Associates.

Metaview Developer's Workbench Application development system. Includes 4GL, relational DBMS, text/image processing functions, interactive debugging tools. Runs on IBM systems. Vendor: Metafile Information Systems, Inc.

MetaViewer Data management software used in data warehousing. Searches through reports stored on laser disks. Runs on Windows systems. Vendor: Metafile Information Systems Inc.

Metavision CASE product. Automates analysis, design, and programming functions. Includes code generator for C, COBOL. Design methodologies supported: Gane-Sarson, DeMarco, Yourdon. Databases supported: DB2, Oracle, SQL, VSAM, RMS, dBase. Runs on Pentium type desktop computers. Vendor: Applied Axiomatics, Inc.

MetaWeb 1.Communications, Internet software. Output management software that lets users of internal networks direct documents to Web pages. Released: 1996. Vendor: Structural Dynamics Research. 2. See Dazel Output Server.

Metaworks See PowerDesigner.

metering Measuring the use of software on a network. Metering programs tell which stations on the network use which software. Used to determine licensing agreements and usage.

method Term used in object-oriented development. A procedure or function specified for an object class. Also called routine.

Method/1 CASE product. See Foundation.

MethodManager CASE product. See Manager family.

methodology A guideline identifying how to develop a software system. It is an approach to problem solving and is defined by a set of rules and specific development tools.

Methods Automation CASE product. Automates design function. Works with dBase III database.

MethodWorks Application development tool. Automates design function and can be used by non-technical end-users. Runs on Solaris, Windows systems. Vendor: ParcPlace Systems, Inc.

Mewel Application development tool. Uses libraries to allow programmers to write one set of source code and then move the application to DOS, OS/2 Unix, VMS or Windows. Vendor: Magma Software Systems.

Mezzanine Document retrieval system. Works with Web servers. Runs on OS/2, Unix, Windows systems. Vendor: Saros Corp.

MFast Application development tool. Program that generates CICS maps. Vendor: H & M Systems Software, Inc.

MFC Application development tool. Library of GUI components used to develop Internet and desktop applications. Stands for: Microsoft Foundation Classes. Vendor: Microsoft Corp.

MFCbase Data management system. Runs on DEC, IBM desktop systems. Vendor: Peter A. Johnson & Associates, Inc.

MFE Operating system for midrange Motorola systems. Vendor: Motorola Computer Systems, Inc.

MFG/PRO Application software. Supply Chain management software. Includes manufacturing distribution, financial, and service (or support) management applications. Vendor: QAD, Inc.

MFS 1. Message Format Service. Part of IMS online environment. Used by programmers to define how screens will look in an online system. Vendor: IBM Corp. 2. CASE software. Automates design function. Used for financial applications. Vendor: Computrol, Inc.

MGRW Application development tool. Report generator for IBM systems. Stands for Matrix Generator Report Writer for MPSX. Vendor: IBM Corp.

MHG Gateway Communications software. Gateway for e-mail systems. Runs on Pentium type desktop computers. Vendor: Retix.

MHS 1. Stands for Message Handling Service. Communications service provided by Action Technology, Inc. and Novell, Inc. 2. Data communications protocol. Applications layer. Part of X.400. Stands for: Message Handling System.

MHS-6000 Communications software. Gateway for e-mail systems. Runs with X.25 networks. Vendor: Unisys Corp.

MHz Megahertz. Used to measure CPU speeds.

MI³MS 1000 Image processing system that runs on desktop systems. Vendor: Minolta Corp.

MIB See Madman MIB.

MIC II Operating system for IBM and compatible desktop computer systems. Stands for Management Information Center. Vendor: Develcon Electronics.

MIC III Operating system for midrange DEC VAX systems. Vendor: Develcon Electronics.

MICO Communications. Middleware, object request broker (ORB). Free, open source product. CORBA compliant. Uses GNU tools. Can work on any platform. Stands for: MICO is CORBA.

MICR Magnetic Ink Character Recognition. Printed characters that can be directly read into a computer system because of particles of magnetic material in the ink.

micro See PC.

Micro/Answer II Communications software. Network connecting IBM systems. Micro-to-mainframe link. Vendor: Sterling Software (Answer Division).

MICRO-CAPS Application development tool. Program generator for Unisys systems. Vendor: Software Research, Inc.

Micro Express Computer vendor. Manufactures portable computers.

Micro Focus AppMaster Builder See AppMaster Builder.

Micro Focus AppMaster Renovator See AppMaster Renovator.

Micro Focus BridgeWare See BridgeWare.

Micro Focus CICS OS/2 Application development tool. Allows user to develop applications for CICS OS/2 or mainframe CICS. Runs on OS/2 systems. Vendor: Micro Focus, Inc.

Micro Focus COBOL See COBOL/2.

Micro Focus DB2 Application development tool. Used to develop applications for DB2/2, DB2, SQL/DS. Runs on OS/2, Windows systems. Vendor: Micro Focus, Inc.

Micro Focus Dialog System See Dialog System.

Micro Focus NetExpress See NetExpress.

Micro Focus PL/1 Application development tool. Used to develop mainframe PL/1 applications on an OS/2 system. Vendor: Micro Focus, Inc.

Micro Focus Workbench See COBOL/2 workbench.

Micro-Kernel Operating System; control programs in OSF/1. Based on Mach kernel and designed for high performance systems such as parallel processors. Can run other Operating Systems, so users could run MS-DOS or OS.2 under OSF/1 (A Unix type Operating System). Vendor: OSF Research Institute.

Micro Palm Computer vendor. Manufactures handheld computers.

Micro Planner 500,2000 Desktop system software. Project management package. Runs on IBM, Macintosh systems. Vendor: Micro Planning International.

Micro PROLOG See PROLOG.

Micro RSTS Operating system for DEC systems. A subset of RSTS/E. Vendor: Digital Equipment Corp.

Micro/RSX Operating system for DEC MicroPDP-11/73 computers. Vendor: Digital Equipment Corp.

Micro Saint Application development tool. Provides process modeling functions. Runs on Macintosh, Windows systems. Vendor: Micro Analysis & Design, Inc.

Micro/VMS Operating system for DEC systems. Vendor: Digital Equipment Corp.

MicroARTE Operating system for Encore RSX series computers.

Microbase Data management system. Runs on Apple II systems. Includes information manager and report writer functions. Vendor: Compumax Associates, Inc.

MicroCOMPstation Midrange computer. TurboSPARC RISC CPU. Operating systems: Solaris. Vendor: Tatung Science & Technology Inc.

microcomputer See PC.

MicroFLEX Desktop computer. Pentium, AMD, Cx6x86MX processors. Vendor: Micro Express.

Microflex NPxxx Desktop computer. Notebook. Pentium, AMD processors. Vendor: Micro Express.

Microflex PC Handheld computer. Vendor: DAP Technologies.

Micrografx Draw Desktop system software. Graphics package. Vendor: Micrografx, Inc.

microkernel Operating system design feature that builds separate, independent layers for specialized operating system functions, especially those that are hardware dependent. This can make the operating system portable. Used in Mach, Windows NT, OS/2, OSF/1.

MicroMan Esti-Mate Desktop system software. Estimates time requirements for Information Systems projects. Vendor: POC-IT Management Services, Inc.

MicroMan II Desktop system software. Project management package. Assigns and tracks resources across multiple projects. Interfaces with MicroMan Esti-Mate. Runs on Pentium type desktop computers. Vendor: POC-IT Management Services, Inc.

MicroMPX Operating system for Encore RSX series computers.

Micron Computer vendor. Manufactures desktop computers.

MicroPages Application development tool. Query and report functions. Allows users to create yellow pages on the Web. Runs on Unix, Windows NT systems. Vendor: Microlytics.

micropayments Very small payments (under $1). Systems are being set up on the Internet where consumers can set up accounts and make numerous small payments for items such as purchasing a single song or article reprint. Also called microcash.

MicroPDP-11/73,93 Midrange computer. Operating systems: DSM-11, Micro/RSX, RSX, RSTS, RT-11, Ultrix. Vendor: Digital Equipment Corp.

Micropulse See Group Four Micropulse.

Microslate Computer vendor. Manufactures pen-based computers.

Microsoft Access Relational database for desktop environments. Runs on Windows systems. Uses SQL. Used in client/server computing. Provides visual programming environment. Interfaces with SQL Server, dBase, Excel, 1-2-3, Foxpro, Paradox. Has programming language, Access Basic. Supports Visual Basic. Vendor: Microsoft Corp.

Microsoft BackOffice Operating system software. Includes system management and communications software. Includes Exchange Server, SQL Server, Mail Server, SNA Server, Systems Management Server. Vendor: Microsoft Corp.

Microsoft certification See MCP (+ Internet) and MCSE (+ Internet), MCSD.

Microsoft Commercial Internet System See MCIS.

Microsoft Cluster Server See MSCS.

Microsoft Exchange See Exchange.

Microsoft Foundation Classes See MFC.

Microsoft Internet Information Server See Internet Information Server.

Microsoft Mail See Mail Server.

Microsoft Network Communications service that provides a wide variety of information and services. Includes user forums, news groups, Apple customer support, access to the Internet. Vendor: Microsoft Corp.

Microsoft Office Software suite. Desktop system software that includes Microsoft Word, Excel, Powerpoint, and Mail. Microsoft Access can also be included. Includes MOM (Microsoft Office Manager), a tool that serves as the program manager for the entire suite. Office 95, Office 97 versions. Vendor: Microsoft Corp.

Microsoft Plus Operating system utilities for Windows 94. Includes disk maintenance tools and Internet Explorer. Vendor: Microsoft Corp.

Microsoft Project Desktop system software. Project management package. Runs on Pentium type systems. Vendor: Microsoft Corp.

Microsoft Project for Windows Desktop system software. Project management package. Runs with Windows. Vendor: Microsoft Corp.

Microsoft Repository Application development tool. Used with data warehousing. Holds metadata and provides the interconnections between disparate data. Runs on Windows, Unix systems. Vendor: Microsoft Corp and PLATINUM Technology.

Microsoft Transaction Server See MTS.

Microsoft Visual Basic See Visual Basic.

Microsoft Word Desktop system software. Word processor. Runs on Pentium type systems. Vendor: Microsoft Corp.

Microsoft Works Desktop system software. Integrated package. Runs on Macintosh systems. Vendor: Microsoft Corp.

Microsoft Write Desktop system software. Word processor. Runs on Apple Macintosh systems. Vendor: Microsoft Corp.

MicroStation 1. Program development tool for Unisys mainframe systems. Part of Group Four Application development environment. Vendor: ESI.
2. CAD software. Runs on IBM, Macintosh, Unix systems. Vendor: Intergraph.

Microstep CASE product. Automates analysis, design, and programming functions. Includes code generator for C. Design methodologies supported: RAD. Databases supported: Btrieve, dBase. Runs on Pentium type desktop computers. Vendor: Syscorp International.

MicroTrak Desktop system software. Project management package. Runs on DEC, IBM, systemsUnix systems. Vendor: SofTrak Systems, Inc.

MicroVAX Midrange computer. Operating systems: Micro/VMS, Ultrix, Vaxeln, OpenVMS. Vendor: Compaq Computer Corp.

MICS/Net,SNA Communications software. Network connecting IBM computers. Vendor: Legent Corp.

MICS/ Operating system enhancements used by systems programmers. Monitors and controls system performance in IBM systems. Products include: MICS/VSE Power, MICS/CICS, MICS/DASD, MICS/DB2, MICS/IDMS, MICS/IMS, MICS/MVS, MICS/VM and CMS. Vendor: Legent Corp.

MIDAS Communications and application development tools. Middleware used to build multi-tiered distributed applications. Uses DCOM. Stands for: Multi-Tier Distributed Application Services Suite. Also called Borland MIDAS. Runs on Windows NT systems. Released: 1997. Vendor: Inprise Corp.

Midas/EDI Communications software. EDI package. Runs on DEC, Windows systems. Vendor: MKS, Inc.

middleware Communications software that connects heterogeneous computer environments. Supports multiple protocols. Used in client/server environments to connect the front-end client GUI systems with the back-end server database managers. Allows queries to multiple databases through a standard interface. Uses common APIs. There are several types of middleware: remote procedure call, message passing, conversation, legacy middleware, object-oriented middleware and transaction server.

MidPoint for IEF Communications software. Middleware. Allows access to IEF controlled applications from GUI tools, report writers, statistical and analytical tools. Vendor: MidCore Software, Inc.

midrange computer An intermediate-size computer. Midrange computers include small mainframes, servers and large desktop machines. Catchall term for multi-user systems that do not have the speed and capacity of a mainframe. Formerly called mini computers. Also called midsize computers.

Midwest Micro Computer vendor. Manufactures desktop computers.

MIF Management Information File. Database used in DMI. See DMI.

Migradata Data dictionary for DB2 databases. Vendor: Computer Horizons Corp.

Migration Architect Application development tool. Data conversion tool. Helps users profile legacy data before implementing ERP warehouse systems. Runs on Sun, Windows systems. Vendor: DB-Star, Inc.

MIIS/Magic Operating system for midrange DEC VAX and Data General systems. Stands for: Meditech Interpretive Information System. Vendor: Medical Information Technology, Inc.

MIKSolution Application development tool. DSS. Provides access to data stored in multidimensional databases. Vendor: MIK.

Milan P-1xx Notebook computer. Pentium CPU. Vendor: Fujitsu PC Corp.

Millennia XKU, XRU, MME, LXE Desktop computer. Pentium CPU. Operating system: Windows 95/98. Vendor: Micron Electronics, Inc.

Millennium 1. Mainframe computer. Operating systems: MVS. Vendor: Amdahl Corp.
2. Group name for financial and administrative programs. Runs on IBM mainframes. Programs include: AP:Millennium (accounts payable), AR:Millennium (accounts receivable), CM:Millennium (currency management), CP:Millennium (capital projects), GL:Millennium (general ledger), HR:Millennium Payroll, HR:Millennium Personnel, HR:Millennium Flexible Benefits. Vendor: Gaec Computer Corp. (formerly Dun & Bradstreet Software Services).

Millennium:SDT Application development system. Includes 4GL for designing and executing applications online using prototyping. Uses M:PDL as procedural access languages. Interfaces with DB2. Vendor: Gaec Computer Corp. (formerly Dun & Bradstreet Software Services.)

Millicent Communications. Security protocol for e-commerce. Used to handle micropayments (small payments, typically under $1). Developed by DEC.

MILnet The Department of Defense's network.

MIMD Method of programming parallel computers. Stands for: Multiple Instruction/Multiple Data. Each processor executes its own set of instructions so the programmer must break application code into multiple, parallel streams. Contrast with SIMD.

MIME Communications specifications that allow messages sent on SMTP e-mail systems (Internet, TCP/IP) to work with X.400 systems. Stands for: Multipurpose Internet Mail Extensions.

MIMEsweeper Communications software. Provides security for Internet applications. Runs on Windows. Vendor: Integralis, Inc.

MIND-data Design tool used to develop and manage communications networks. Stands for: Modular Interactive Network Designer. Associated products include: MIND-Inventory, MIND-Packet. Vendor: Contel Business Systems.

MINDOVER See CA-MINDOVER.

MindNet Application development tool. Document management system that works with natural language input to do such things as retrieve multiple documents, create executive summaries. Included with Office suite. Planned release date: 2000. Vendor: Microsoft Corp.

MindWrite Desktop system software. Word processor. Runs on Macintosh systems. Vendor: Access Technology, Inc. (CA).

MindSet Application development tool used for data mining. Used to determine which of hundreds of realtionships among variables are significant. Runs on Unix systems. Vendor: Silicon Graphics Inc.

MineShare Suite Application development tool. Used to build and manage data marts. Includes end-user query tools to analyze the information and distribute reports. Released: 1998. Vendor: MineShare, Inc.

mini, minicomputer See midrange computer.

MINISYS Database for large computer environments. Runs on HP systems. Designed for bibliographic information, but adaptable to other uses. Vendor: McLeod-Bishop Systems, Ltd.

Minitab Application development tool. Statistical forecasting tool. Provides summary information and determines significance of degree of relationship between two factors. Runs on MS-DOS, MVS, Unix systems. Vendor: Minitab, Inc.

Mink Communications software. Network connecting Unix systems. Micro-to-midrange link. Vendor: Corporate Microsystems, Inc.

Minxware MRP, or CIM software. Runs on IBM, Unix systems. Vendor: Minx Software, Inc.

mips Million instructions per second. Measurement used to state computer speeds.

Mips Computer chip, or microprocessor. 64-bit chip. Runs Irix operation system. Vendor: Silicon Graphics, Inc.

Mirage Application development tool. Lets developers migrate Windows applications to Unix systems. Vendor: Software Pundits, Inc.

Mirror DBMS Database for midrange and desktop computer environments. Runs on IBM, Stratus systems. Vendor: Data Systems for Industry.

mirror site Internet technology. A replica of an existing site. The purpose of a mirror site is to redact traffic on a single server. A mirror site can be set up to handle some of the traffic at busy times. Mirror sites can be set up in diverse locations to speed up Web access as users can download information more quickly from a server that is geographically closer to them, i.e. if a busy New York-based Web site sets up a mirror site in England, users in Europe can access the mirror site faster than the original site in New York.

Mirrors Application development tool. Allows users to convert 16-bit Windows applications to 32-bit applications to run under OS/2 Warp. Stands for: Source Migration Analysis Reporting Toolset. Vendor: Micrografx, Inc.

MIS Management Information System. Any system designed to aid in the performance of management functions. Term is also used as a synonym for IS.

MIS-Turbo Operating system enhancement. Increases system efficiency in IBM applications accessing VSAM files. Vendor: Leeds Associates, Inc.

MISC Computer chip, or microprocessor. Vendor: Cyrix Corp.

Mission Control Console Communications software. Network management software that lets administrators manage all aspects of the network including automatic software updates and client/server configuration information. Includes Java-based interface and works with the Internet and intranets and extranets. Released: 1998. Vendor: Netscape Communications Corp.

mission critical system A software application that is essential to the functioning of the corporation. If a mission critical system fails, the company could face bankruptcy.

MIT/400 Application development tool. DSS. Provides access to data stored in multidimensional databases. Vendor: SAMAC, Inc.

Mitchell Humphrey & Co. Software vendor. Products: FMS II financial systems for midsize environments.

MitemView Communications software. Transaction processing monitor. Used to build client/server networks which integrate data from multiple sources. Runs on Macintosh systems. Vendor: Mitem Corp.

Mitrol Application development tool. Includes data manager, terminal monitor, development language. Runs on IBM mainframe systems. Vendor: Mitrol.

MKS Code Integrity Application development tool. Used with programs written in C. Includes a source code analyzer which is used to identify potential problems. Runs on Unix systems. Released: 1996. Vendor: Mortice Kern Systems, Inc.

MKS Toolkit Application development tool. Creates scripts and automates Unix tasks to run under Windows NT. Vendor: Mortice Kern Systems, Inc.

MM/IOS Operating system for midrange Contel systems. Vendor: Contel Business Systems.

MMCX Groupware. Multimedia conferencing system. Includes e-mail, voice mail, faxes. Stands for: MultiMedia Communications eXchange. Vendor: Lucent Technologies.

MMOS Operating system for midrange Flexible systems. Vendor: Flexible Computer Corp.

MMS-1100 Communications software. E-mail system. Runs on Unisys systems. Vendor: Formula Consultants, Inc.

MMX Pentium chip. Includes multimedia processing enhancements.

MNDS Communications. LAN (Local Area Network) connecting desktop computers. Vendor: Connections Telecommunications.

mnemonic A name or symbol used for a code or function. Assembler languages are made up of mnemonics, symbols that represent the actual machine code.

MNI Comunications. Provides link for diverse communications channels including both wired and wireless channels. Stands for: Mobile Networks Integration. Vendor: Motorola, Inc.

MNP 10 Communications protocol for cellular systems. Written by Microcom, Inc.

MNP Communications protocol. Network protocol that works with V.42. CCITT protocol, used with X.25.

Moab Work in progress name for upgrade of IntranetWare. Includes virtual memory and memory protection to function as an application server. Supports Java, CORBA, ActiveX. Release date: 1998. Vendor: Novell, Inc.

Mobile Assistant II Wearable computer. Similar in functionality to handhelds, but worn on body to free hands. Primarily used in manufacturing for such things as providing access to diagnostic manuals which the user follows to make repairs. Vendor: Xybernaut Corp.

Mobile Best Buy Notebook computer. Pentium CPU. Vendor: Royal Electronics, Inc.

mobile computing Use of notebook computers in a business or department. In theory, workers take the entire office to the field. In some functions mobile computing is so dominant that workers reserve offices as they would hotel rooms for the few occasions they physically work in the office. Requires strong communications support.

Mobile Networks Integration See MNI.

MobilePro 200, 400 Handheld computer. Operating system: WinCE. Vendor: NEC Computer Systems.

Mobilon HE-4500 Handheld computer. Released: 1998. Vendor: Sharp Electronics, Corp.

Mobius 1. Computer vendor. Manufactures desktop computers.
2. Communications software. Network connecting desktop computers and DEC computers. Vendor: FEL Computing.

MobyBrick2 Desktop computer. Notebook. Pentium processor. Vendor: The Brick Computer Company, Inc.

MODCOMP Computer vendor. Manufactures large and desktop computers.

Model 1100E, 1200E, 1400E Mainframe computer. Operating systems: MVS/SP, MVS/XA. Vendor: Amdahl Corp.

Model 204 Relational database for large computer environments. Includes 4GL named User Language. Runs on IBM systems. Includes data dictionary. Full name: Model 204 Database Management System. Part of The Advantage Series. Vendor: Computer Corp. of America.

Model 300 Operating system enhancement used by systems programmers. Monitors and controls system performance in IBM systems. Creates a model for system performance. Vendor: Boole & Babbage, Inc.

Model 5990 Mainframe computer. Operating systems: MVS, Unix, VM, VSE, ACP, UTS, HPO. Vendor: Amdahl Corp.

Model 5995 Mainframe computer. Operating systems: MVS, VM, TPF, UTS. Vendor: Amdahl Corp.

Model 7300 Mainframe computer. Operating systems: Unix, UTS. Vendor: Amdahl Corp.

Model Transporter See TopTier.

ModelMart Application development tool. Provides model management for client/server, Web and data warehouse development. Enables all members of a workgroup to have access to most current models. Runs on Macintosh, Unix, Windows systems. Released: 1997. Vendor: Computer Associates International, Inc.

ModelPro Application development tool. Used for data modeling on Macintosh systems. Vendor: D. Appleton Co.

modem Modulator/demodulator. A device used in communications systems to allow data transmission over communications lines.

MODIMS Inventory manager for wholesale distribution businesses. Runs on IBM RS/6000 systems. Stands for: Modular Order Entry and Inventory Management System. Vendor: Briareus Corp.

Modula-2 Compiler language used in desktop computer environments. Used to develop real-time applications. Enhanced version of Pascal. Created in 1982.

Modula-3 Object-oriented programming language.

Modular Scalable ServerArray Midrange computer. Server. Pentium CPU. Vendor: Advanced Modular Solutions, Inc.

module See program.

MOLAP OLAP analysis provided by a system relying on dedicated, pre-calculated data sets. Stands for: Multi-dimensional On Line Analytical Processing.

MOM Communications. Middleware that lets applications on different platforms and networks exchange data. Works with mainframe, Unix and Windows by having each platform send data to message queues, where they are held until another application accesses them. Stands for: Message-Oriented Middleware.

MOMA Message Oriented Middleware Association. Consortium of middleware vendors.

Momenta 1/xx Handheld, pen computer. Operating System: Windows for Pen Computing, proprietary. Vendor: Momenta Corp.

Momentum Business Applications Software vendor. Spin-off from PeopleSoft in 1998. Develops intranet and data analysis software.

MON+ Operating system add-on. Debugging/testing software that runs on Apple II systems. Vendor: Byte Works, Inc.

Monarch(/ES) Application development tool. Query and reporting system. Vendor: Datawatch Corp.

Monarch for Windows Application development tool. Allows users to view, print, query, and extract reports from any system. Vendor: Personics Corp.

Money Desktop software. Checkbook and home financial management program. Vendor: Novell, Inc.

MONEY Operating system enhancement used by systems programmers. Provides system cost accounting for DEC systems. Vendor: Digital Equipment Corp.

Monitor Operating system enhancement used by systems programmers. Monitors and controls system performance in IBM MVS, DB2, VTAM, CICS systems. Full name: The Monitor for MVS, CICS, CICS(VSE), VTAM, DB2. Vendor: Landmark Systems Corp.

Monitor/Plus Operating system enhancement. Monitors and controls system performance in DEC VMS systems. Vendor: Data Center Software, Inc.

MONITROL Application software. Manufacturing software used to monitor plant productivity, provide quality assurance functions and evaluate and control industrial processes. Retrieves data from plant floor devices and stores in real-time database. Runs on Unix systems. Released: 1995. Vendor: Hilco Technologies Inc.

Monte Carlo Notebook computer. Pentium CPU. Vendor: Fujitsu PC Corp.

Montegro Notebook computer. Pentium CPU. Vendor: Fujitsu PC Corp.

Montery Project by IBM, Intel, SCO and Sequent to accelerate growth of Intel processor-based servers running Unix.

MOODS Object-oriented database design. Introduces a framework for the representation of multimedia objects. Groups objects according to media types. Stands for: Multimedia Object-Oriented Database schema.

MOP Applications software. Part of manufacturing systems. Stands for: Manufacturing Order Processing.

MOS Operating system for Apple Macintosh systems. Stands for: Martian Operation System. Vendor: dogStar Software.

MOSES Object-oriented development methodology. Stands for: Methodology for Object-Oriented Software Engineering. Associated with Henderson-Sellers and Edwards.

Mosaic Internet browser. Originally developed by NCSA at the University of Illinois and is public domain software. Also called NCSA Mosaic. Faster and easier to use commercial Mosaic browsers are available.

MOT Application development approach for component-based development. Includes IDE (integrated development environment to design manage and monitor distributed components. Uses CORBA, JavaBeans. Stands for: Morphous Object Technology. Runs on Unix, Windows systems. Released: 1998. Vendor: Ikonodyne.

Motif Graphics user interface. Used in many Unix systems. Vendor: Open Software Foundation.

Motorola Computer vendor. Manufactures desktop computers.

mouse A device that is used for pointing and drawing. A mouse is rolled across a desktop and thus controls the movement of the cursor. The device contains one or two buttons that can be used to select functions to be performed. Commonly used in desktop computers, and starting to be used in mainframe operations.

Move for Servers Application development tool. Moves data from one relational database to another while maintaining the data relationships. Works with diverse databases. Automatically creates tables in the receiving table when necessary. Works with Oracle, DB2 UDB, Sybase, SQL Server, Informix. Runs on MVS systems. Released: 1998. Vendor: Princeton Softech (Division of Computer Horizons Corp.).

Mozart (Composer) Application development environment. Runs on Pentium type desktop computers. Version 3.0 includes object-oriented applications builder that allows developers to build tailored graphic user interfaces to run under MS-DOS and windows. Screen scraper. Supports client/server computing. Vendor: Mozart Systems Corporation.

Mozart Performer Runtime environment for Mozart Composer development environment. Vendor: Mozart Systems Corporation.

MP 133A Portable computer. Pentium CPU. Vendor: Micro Express.

MP/OS Operating system for midrange DG systems. Vendor: Data General Corp.

MPD-8x00 Desktop computer. Pentium processor. Vendor: MicroX.

MPE Operating system for midrange HP systems. Versions include: MPE V, MPE XL. Vendor: Hewlett-Packard Co.

MPE/IX Operating system for midrange Hewlett-Packard systems. Unix-type system. Vendor: Hewlett-Packard Co.

MPEX/3000 Operating system add-on. Data management software that runs on HP systems. Vendor: VESOFT, Inc.

MPP Massively Parallel Processing computer system. Comparable in size and use to a supercomputer, but internally consists of hundreds (sometimes thousands) of RISC microprocessors rather than a single powerful processor.

MPS 1. Operating system add-on. Library management system that runs on IBM systems. Stands for: Management of Production Software. Vendor: Pansophic Systems, Inc.
2. Applications software. Part of manufacturing systems. Stands for: Master Production Scheduling.

MPSX Mathematical/statistical programs for IBM systems. Stands for: Mathematical Linear Programming Extended. Vendor: IBM Corp.

MPTN Communications software. Links local area networks to SNA. Stands for: Multi-Protocol Transport Networking. Vendor: IBM Corp.

MPW Application development system. Allows development in C, Pascal, Fortran. Builds applications from scripts. Vendor: Apple Computer Inc.

MPX100 Desktop computer. PowerPC RISC CPU. Operating system: AIX, MacOS. Released: 1997. Vendor: Motorola Computer Group.

MQE Managed Query Environment. Building a work environment that offers user-friendly access, analysis, and reporting of corporate data. Computer professionals maintain the data and access tools and provide security and efficiency controls.

MQI Message Queue Interface. Proposed standard from IBM for interfacing applications with message-oriented middleware.

MQSeries Communications software. Message-oriented middleware. Messaging passing tool. Stores data in message queues until the receiving application is ready. Allows access to CICS, IMS and OpenVMS. Runs on most platforms. Stands for: Message Queuing Series. Vendor: IBM Corp.

MQSeries Three Tier See Oak.

MQView Communications software. Manages message queues from middleware systems that handle communications from Unix and Windows to mainframe systems. Interfaces with MQSeries. Vendor: Apertus Technologies.

MRC Operating system enhancement used by systems programmers. Monitors and controls system performance in IBM systems. Vendor: Software Engineering of America.

MRC Productivity, Query CASE product. Automates analysis, design, and programming functions. Includes code generator for RPG. Design methodologies supported: RAD. Databases supported: OS/400 relational database (Productivity), CPF relational database (Query). Runs on IBM AS/400 (Productivity) and Sys/38 (Query) systems. Vendor: Michaels, Ross & Cole Ltd.

MRCS Application development tool. Report generator for IBM systems. Stands for: Multiple Report Creation System. Vendor: McDonnell Douglas Information Systems Co.

MRDS Relational database for large computer environments. Runs on Bull HN systems. Stands for: Multics Relational Data Store. Vendor: Bull HN Information Systems.

MRP Name given to application software designed to handle manufacturing resources, supplies and planning. Stands for: Manufacturer Resource Planning, Material Requirements Planning, or Manufacturing Requirements Planning.

MRPx MRP, or CIM software. Runs on IBM AS/400 systems. Vendor: J.D. Edwards.

MS-DOS Operating system for IBM and compatible desktop computer systems. Called PC-DOS for IBM systems. Vendor: Microsoft Corp.

MS-Mail See Microsoft Mail.

MS-Windows Operating system software. Window manager. Transfers data between programs. Runs on Pentium type desktop computers. Vendor: Wang Laboratories, Inc.

MSA Management Science America, Inc. Company that writes application software, concentrating on accounting and administrative systems. Term is often used to refer to the accounting and administrative software. Company now owned by Dun & Bradstreet Software.

MSCM Operating system enhancement used by systems programmers. Monitors and controls system performance in IBM MVS systems. Stands for: Multi-System Configuration Manager. Handles computer center configuration. Vendor: IBM Corp.

MSCS Communications. Network management software. Provides clustering capabilities so that if one server goes down, another will automatically take its place. Stands for: Microsoft Cluster server. Formerly known as Wolfpack. Vendor: Microsoft Corp.

MSHF2 Communications software. Network connecting IBM computers. Stands for: Matrix Switch Host Facility 2. Vendor: IBM Corp.

MSM/STAM Operating system enhancement used by systems programmers. Monitors and controls system performance in IBM MVS systems. Stands for: Multiple Systems Manager and Shared Tape Allocation Manager. Vendor: Duquesne Systems, Inc.

MSMQ Communications software. Message oriented middleware. Transmits messages between servers which can be on different (but connected) systems. Works with Windows NT systems. Vendor: Microsoft Corp.

MSNF/TCAM Communications software. Network connecting IBM computers. Vendor: IBM Corp.

MSS Mass Storage System. Online storage system that includes both disks and drums and extends online data storage capacity to almost 500 billion characters.

MSSE Operating system add-on. Data management software that runs on IBM systems. Stands for: Mass Storage System Extensions 3850. Vendor: IBM Corp.

MTOS Operating system for IBM and compatible desktop computer systems. Versions include: MTOS-UX, MTOS-68K, MTOS-86/PC. Vendor: Industrial Programming, Inc.

MTS 1. Communications software. Transaction processing monitor that runs on IBM RS/6000 systems. Stands for: Micro Focus Transaction System. Compatible with CICS on AIX, Unix, OS/2 and Windows NT systems. Vendor: Micro Focus, Inc.
2. Application server providing middleware and development functions. Allows developers to write applications that access data from diverse databases over a network. Includes DCOM to handle routing data over the network. Full name: Microsoft Transaction Server. Also called Viper. Part of Windows NT. Vendor: Microsoft Corp.

MUC-DOS Operating system for IBM and compatible desktop computer systems. Vendor: Haar Industries, Inc.

MUD Interactive game played on the Internet. Users enter the game through Telnet. Stands for: Multiuser Dialog or Dimension.

MULTI Application development system. Develops programs in Ada, C, C++, FORTRAN, Pascal. Includes optimizers, source level debuggers, version control. Runs on Unix, Windows NT systems. Vendor: Green Hills Software, Inc.

Multi/CAM CASE product. Automates analysis, design, programming, and project management functions. Includes license for Excelerator, another CASE product. Vendor: AGS Management Systems, Inc.

multi-dimensional aggregation tables Application development design tool. An aggregation that contains metrics calculated along multiple business dimensions, i.e. sales by region by quarter.

Multia VX40A-xx AXP Midrange computer. Alpha CPU. Used to access legacy, Unix, and PC applications. Vendor: Digital Equipment Corp.

Multicast Communications. Sending a single message to a subset of users on a network of communications system.

Multics Communications software. Transaction processing monitor. Runs on Bull HN systems. Includes: Multics Communications System, Multics File Transfer Facility, Multics HASP Facility. Vendor: Bull HN Information Systems.

multidimensional database Database built by consolidating multiple databases and letting users share information. This design provides fast response time for complex data queries. Holds data in layers based on a hierarchical structure and often allows viewing data in three or more dimensions such as time, place, and product. Usually includes tools to create front-end applications. Data is stored in cells that exist in multiple layers with a hierarchical structure, i.e. shampoo is part of health & beauty products, is part of consumer goods. Users can query all layers. Used in OLAP.

MultiFiler/Finder Image processing software. Interfaces with most SQL databases.Application development tool. Allows access to non-relational mainframe databases from desktop systems. Vendor: Optika Imaging Systems, Inc.

MultiFinder See Finder.

MultiGantt Desktop system software. Project management package. Runs on Pentium type desktop computer systems. Vendor: SofTrak Systems, Inc.

MultiLink/VB Application development tool. Provides a link for client applications developed through Visual Basic to SQL databases. Vendor: Q+E Software, Inc.

MultiMate Desktop system software. Word processor. Runs on Pentium type systems. Includes MultiMate Advantage II, MultiMate Advantage II LAN, MultiMate Professional Word Processor. Obsolete. Vendor: Ashton-Tate Corp.

multimedia Combining text, graphics, animation, audio and video on desktop computers/workstations.

Multimedia 3D Screamer Desktop computer. Pentium CPU. Operating system: Windows 95/98. Vendor: Comtrade Electronics, U.S.A. Inc.

Multimedia Notebook Notebook computer. Pentium CPU. Operating systems: Windows 95/98. Vendor: Comtrade Electronics U.S.A. Inc.

Multimedia System Desktop computer. Pentium processor. Vendor: ABS Computer Technologies, Inc.

Multimedia Toolbook Multimedia authoring tool. Mid-level, scripting tool. Runs on Windows systems. Vendor: Asymetrix Corp.

Multimedia(VL-Bus), Best Buy, Dream Machine Desktop computer. Pentium CPU. Operating systems: Windows 95/98. Vendor: Comtrade Electronics U.S.A., Inc.

MultimediaWide SCSI Powerstation Desktop computer. Pentium CPU. Operating system: Windows 95/98. Vendor: Comtrade Electronics, U.S.A. Inc.

Multimedia Workbench for Supra Server Multimedia authoring tool. High-level, scripting tool. Applications will run under Windows, Macintosh. Vendor: Cincom Systems, Inc.

MultiNet Communications software. Network connecting Novell users with DEC VAX and alpha systems. Vendor: TGV, Inc.

Multinet/400 Communications software. EDI package. Runs on AS/400 systems. Vendor: Fischer EDI, Inc.

Multinet/Unix Communications software. EDI package. Runs on Unix systems. Vendor: Fischer EDI, Inc.

Multiplan Desktop system software. Spreadsheet. Runs on Apple II, Apple Macintosh, IBM systems. Vendor: Microsoft Corp.

multiplexor A device which transmits data from multiple sources through a single channel. Used in communications.

Multiprint/VM Operating system software. Allows mainframe output to be printed on LAN printers. Runs on IBM VM systems. Vendor: BlueLine Software, Inc.

multiprocessing Combining more than one computer to execute programs simultaneously, or using computers with more than one processor.

multiprogramming A mode of operation that allows for concurrent execution of programs in the same computer. Most large computers provide multiprogramming capabilities. Also called multitasking.

Multiproject Desktop system software. Project management package. Runs on Pentium type systems. Vendor: Technisoft.

multisensory I/O See virtual reality.

multitasking See multiprogramming.

Multiterm/VM,VSE,MVS Communications software. Network connecting multiple terminals to IBM mainframes. Vendor: BlueLine Software, Inc.

multithreading See thread, threading.

Multitiered, server based processing Part of client/server computing. One type of application partitioning. Splits logic between server and client.

Multitrak See Enterprise-Wide Work Management System.

MultiTSO Operating system enhancement used in IBM TSO systems. Reduces TSO overhead time and allows access to CICS, DB2, SAS, QMF, OMEGAMON through TSO screens. Vendor: Technologic Software Concepts, Inc.

MultiUser Helix Relational database for desktop computer environments. Runs on Macintosh systems. Includes DSS. Vendor: Odesta Corp.

MultiView Communications. Terminal emulation software used to integrate PCs, thin clients and/or Unix character terminals to existing legacy applications. Family of applications that includes MultiView 2000, Desktop, and Mascot/ALPHABrowser. Released: 1998. Vendor: JSB Corp.

MultiWare Latest enhancement of MultiNet. See MultiNet.

Multiway Application development tool. Includes database, 4GL. Used to develop large scale analysis and reporting applications. Part of Acumate. Vendor: Kenan Technologies.

MUMPS 1. Operating system and compiler language. Stands for: Massachusetts General Hospital Utility Multi-Programming System. Developed explicitly to serve the needs of the medical community and manipulates large amounts of textual data. 2. Operating system for Smoke Signal systems.

MUMPS/VM Application development system. Runs on IBM systems. Vendor: IBM Corp.

MUSIC/SP Operating system for mainframe IBM systems. Stands for: Multi-User System for Interactive Computing/System Product. Vendor: IBM Corp.

MusicMax Desktop computer. Pentium CPU. Operating systems: Windows 95/98. Vendor: MAXIMUS COMPUTERS, Inc.

mv.ESx00i Midrange computer. Server. Pentium Pro CPU. Vendor: General Automation, Inc.

MV Manager Operating system enhancement used by systems programmers. Used to monitor performance of an application through systems such as CICS and JES2. Used in IBM MVS environments. Vendor: Boole & Babbage, Inc.

MV/UX Operating system for midrange Data General Eclipse systems. Unix-type system. Vendor: Data General Corp.

MVAP Operating system enhancement used by systems programmers. Monitors and controls system performance in Unisys systems. Functions as a predicting tool. Vendor: Unisys Corp.

MVD-5x00 Desktop computer. Pentium processor. Vendor: MicroX.

MVP Multimedia Video Processor. Vendor: Texas Instruments, Inc.

MVP D,V,VF,VP,VX Desktop computer. Pentium processor. Vendor: Directwave Inc.

MVP Development Kit Application development tool. Works with MVP. Supports image processing, two- and three-dimensional graphics, audio, and full-motion graphics. Vendor: Precision Digital Images Corp.

MVP-S,NT Midrange computer. Server. Pentium II, Pro CPU. Operating system: Windows NT. Vendor: Directwave, Inc.

MVS General-purpose operating system for IBM, Amdahl large computer systems. Stands for: Operating system/Multiple Virtual Storage. Versions include: MVS/XA, MVS/ESA. Synonym: OS/MVS. Vendor: IBM Corp.

MVS/AutoOPERATOR Operating system enhancement used by Operations staff and systems programmers. Provides operator console support in IBM MVS systems. Vendor: Boole & Babbage, Inc.

MVS/DCE Operating system enhancement. Layer of services that runs on MVS/ESA OpenEdition systems. Makes MVS look like Unix to programmers and operators. Vendor: IBM Corp.

MVS/ESA See MVS.

MVS/JCL See JCL.

MVS/Quick-Ref Operating system add-on. A quick reference resource that documents operating system software, development tools, and applications in IBM ISPF systems. Vendor: Chicago-Soft, Ltd.

MVS/SE Operating system enhancement used by systems programmers. Monitors and controls system performance in IBM MVS systems. Stands for: MVS Systems Extensions. Vendor: IBM Corp.

MVS/TSO See TSO.

MVS/XA See MVS.

MVS/XA DPF Operating system add-on. Data management software that runs on IBM systems. Stands for: MVS/XA Data Facility Product. Vendor: IBM Corp.

MWB Application development tool. Allows maintenance for mainframe systems to be done on desktop systems using visual tools. Stands for: Maintenance Workbench. Vendor: INTERSOLV, Inc.

MXP-6x00 Desktop computer. Pentium processor. Vendor: MicroX.

MX Edge Desktop computer. Notebook. Pentium processor. Vendor: MicroX

MX/IX Operating system for mainframe Prime systems. Vendor: Prime Computer, Inc.

Mystique Series P-xxxxx Desktop computer. Pentium processor. Vendor: Akia Corp.

n-parallel PROLOG Application development tool. Used to develop parallel processing applications. Includes testing/debugging functions. Generates C code. Runs on Unix systems. Vendor: Paralogic, Inc.

N3, N4/SXL Notebook computer. Vendor: Leading Edge.

NAM Public-domain software written by NASA which provides a GUI to access bulletin boards, libraries, lists of users, e-mail services across Internet . Runs on Macintosh, MS-DOS, Unix systems. Stands for: NASA Access Mechanism.

NameTag Communications, Internet software. Web search product used to find and classify key phrases in text documents. Vendor: IsoQuest, Inc.

NAMS Operating system enhancement used by systems programmers. Controls communications networks in DEC systems. Stands for: Network Analysis & Management System. Vendor: Digilog, Inc.

NAN Communications. Network management software. Provides hardware and software inventory over networks. Works with NetWare, LAN Server, LAN Manager, Vines. Stands for: Norton Adminis-

trator for Networks. Vendor: Symantec Corp.

NANOS Operating system enhancement used by systems programmers. Increases system efficiency in DG RDOS systems. Multiprogramming option. Vendor: Nanosecond Systems, Inc.

nanosecond One billionth of a second. Computer speeds can be measured in nanoseconds.

NAPA 1. Communications . Network management software. Monitors network activity, analyzes job terminations, reports on errors in NetView networks. Stands for: Network Automated Problem Application. Vendor: Peregrine Systems, Inc. 2. Operating system enhancement used by systems programmers. Monitors and controls system performance in NOMAD systems. Stands for: NOMAD Application Performance Analyzer Performance Monitor. Vendor: Aonix.

NAPA/AIM Communications software. Connects MVS systems to PNMS systems. Provides inventory configuration data for all devices on the network. Stands for: Network Automated Problem Applications/Automatic Inventory Manager. Vendor: Peregrine Systems, Inc.

NAS Set of software standards and products used to develop integrated applications running on different vendors' systems, mainly DEC and IBM. Stands for: Network Application Support. Vendor: Digital Equipment Corp.

National Advanced Systems Computer vendor. Manufactures mainframe computers.

NatStar Application development environment used for enterprise development. Runs on OS/2, Unix, Windows NT systems. Users build a model with visual tools and C code is automatically generated for all chosen platforms. Vendor: Nat Systems International, Inc.

Natural 4GL. Accesses Adabas, SQL/DS, IMS, DB2, and VSAM data files. Vendor: Software AG Americas Inc.

Natural Architect, Construct CASE product. Automates analysis, design, and programming functions. Includes code generator for Natural. Design methodologies supported: Chen, Gane-Sarson, Martin, DeMarco, Yourdon. Databases supported: Adabas, DB2, VSAM, SQL/DS. Construct runs on IBM mainframes, DEC VAX, and Wang; Architect on Macintosh desktop computers. Vendor: Software AG Americas Inc.

Natural Architect Workstation Application development system. Includes diagramming editors for data modeling, data flow analysis, and program specification. Used to develop applications in Natural, and interfaces with IBM mainframes, DEC VAX, and Wang VS systems. Runs on Macintosh systems. Vendor: Software AG Americas Inc.

Natural Console Operating system enhancement used by Operations staff and systems programmers. Provides operator console support in IBM systems. Vendor: Software AG Americas Inc.

Natural Construct Application development environment. Uses fill-in-the-blank panels to build applications. Runs on mainframe and midrange systems. Vendor: Software AG Americas Inc.

Natural for Windows Application development environment. Lets developers design, code, test, maintain applications that can run on multiple platforms. Used with NEW/Define. Vendor: Software AG Americas Inc.

Natural ND Application development environment. Adds event-driven extensions to Natural. Supports drag-and-drop development. Interfaces with Adabas. Includes version control, online testing tools, SQL generation. Stands for: Natural New Dimension. Runs on OS/2, Unix, Windows systems. Vendor: Software AG Americas Inc.

Natural Network Operating system enhancement used by systems programmers. Controls communications networks in IBM systems. Vendor: Software AG Americas Inc.

Natural Operations Operating system enhancement used by systems programmers. Job scheduler for IBM systems. Vendor: Software AG Americas Inc.

NaturallySpeaking Speech recognition software. Vendor: Dragon Systems, Inc.

Navigation Server Communications software. Transaction processing monitor. Divides and guides queries through SQL server databases running on MPP systems. Used in data warehousing. Also called SYBASE MPP. Vendor: Sybase, Inc.

Navigator 1. See TopTier.
2. See Netscape.

Navigator Gold Application development tool. Web browser and authoring tool. Inserts HTML commands in text and provides WYSISYG editing. Vendor: Netscape Communications Corp.

Navigator/MF Application development tool. Program utility that analyzes COBOL code for interactive programs. Vendor: Centura Software.

Navigator(/PM) Application development tool. Automates programming function. Analyzes source code for such things as adherence to standards and presence of logic flaws (unexecuted code). Languages analyzed: COBOL. Runs on IBM systems. Vendor: Compuware Corp.

Navigator Systems Series Application development environment. Contains methods, tools, and training for entire development life cycle. Includes HyperProject. Vendor: Ernst & Young LLP.

Navigraph Operating system enhancement. Downloads performance data from IBM CICS, MVS, VSE, and DB2 systems to a desktop computer system that graphically displays the performance of the entire system. Vendor: Landmark Systems Corp.

NaviPress Application development tool. HTML editor. Has WYSIWYG interface. Runs on Macintosh, Unix, Windows systems. Vendor: Navisoft, Inc.

NaviServer Communications software, Web server. Provides access to the Web from Windows NT, Unix systems. Lets users view documents. Vendor: Navisoft, Inc.

NBP Communications protocol. Transport Layer. Part of AppleTalk. Stands for: Name Binding Protocol.

NC See Network Computer.

NC2xx Network computer. Vendor: Tektronic Inc.

NCI(/XF) Communications software. Network connecting IBM computers. Stands for: Network Control Interface. Vendor: Westinghouse Electric Corp.

NCP(/SSP,VS,VTAM) Communications software for large IBM systems. Stands for: Network Control Program. Vendor: IBM Corp.

NCR 2990 Network computer. Vendor: NCR Corp.

NCR 386/IX Operating system for NCR desktop computer systems. Vendor: NCR Corp.

NCR-DMS Operating system add-on. Data management software that runs on NCR systems. Vendor: NCR Corp.

NCR Globalist Midrange computer. Pentium CPU. Operating system: OS/2, Unix, Windows NT. Vendor: NCR Corp.

NCR Teredata See Teredata.

NCR WorldMark Midrange computer. Pentium CPU. Operating system: OS/2, Unix, Windows NT. Vendor: NCR Corp.

NCS Application development tool. Stands for: Network Computing System. A system that allows development of software across networks of computers from various vendors. Vendor: Apollo Computer, Inc.

NCSA Mosaic See Mosaic.

NCSS Application development tool. Statistical forecasting tool. Provides summary information and determines significance of degree of relationship between two factors. Runs on Windows systems. Vendor: NCSS Statistical Software.

NCUBE Computer vendor. Manufactures large computers.

nCUBE 2,2s Massively parallel processor. Supports up to 8192 processors. Operating systems: Unix. Vendor: nCUBE.

NDM Communications software. Network connecting IBM, DEC, Tandem systems. Stands for: Network DataMover. Vendor: Systems Center, Inc.

NDM-MVS/SQL Operating System add-on. Data management routine. Transfers data between DB2 and flat files while insuring data integrity, availability, reliability and security. Vendor: Systems Center, Inc.

NDMS Communications software. Client/server management system which manages NetWare servers and clients. Stands for: NetWare Distributed Management Services. Vendor: Novell, Inc.

NDS (for NT) Communications software. Object-based. Directory function of NetWare. Provides interface with Windows NT so companies can manage both NetWare and NT networks as one. Can handle a billion objects. Stands for: NetWare Directory Services. Vendor: Novell, Inc.

NDX ST Desktop computer. Pentium CPU. Operating systems: Windows 95/98. Vendor: Tandem Computers, Inc. (a Compaq company).

Near & Far Designer Application development tool. Used to develop XML applications. Graphical DTD editing tool. Vendor: Microstar.

Nebula AMD Desktop computer. AMD processor. Vendor: ET Technology.

NEC Computer vendor. Manufactures large and desktop computers.

NekoTech MACH Midrange computer. RISC CPU. Operating systems: Windows NT. Vendor: *Neko*Tech Inc.

Nekotech SuperServer Midrange computer. Alpha CPU. Operating system: Unix, Windows NT. Vendor: *Neko*Tech Inc.

NEM/32 Communications software. E-mail system. Runs on Concurrent systems. Vendor: Concurrent Computer Corp.

Neo New name for DOE. See DOE.

NEONimpact Communications. Message-oriented middleware. Works with existing applications, files, databases, objects, hardcopy output, and Web clients. Platform independent. Vendor: NEON (New Era of Networks, Inc.)

NEONsecure Operating system software. Provides access to multiple applications from a single password. Combines Kerberos authentication and DES encryption. Vendor: NEON (New Era of Networks, Inc.)

Neptune Pentium Desktop computer. Pentium processor. Vendor: ET Technology.

Neostation Network computer. Vendor: Neoware Systems, Inc.

NerveCenter Communications. Network management software. Manages distributed networks by correlating events. Works with OpenView, Unix, Windows NT, NetWare. Vendor: Seagate Enterprise Management Software, Inc.

NEST Technology which used NetWare programs to add control functionality to devices such as fax machines and electrical meters. Allows users to transmit data to and from any NetWare network. The NetWare monitors could, i.e., eliminate the need for meter readers. Stands for: Novell Embedded Systems Technology. Vendor: Novell, Inc.

NEST SDK Application development tool. Allows developers to create applications that can be plugged into private and global public networks. Supports SNMP. Stands for: Novell Embedded Systems Technology SDK (Software Development Kit). Runs on OS/2, Windows systems. Released: 1996. Vendor: Novell, Inc.

Net 127 Communications software. Network operating system. Runs on Pentium type desktop computers. Vendor: Trans-M Corp.

Net.Analysis Communications software. Manages Web traffic. Allows users to track who visits site, what pages they see, etc. Runs on Windows systems. Released: 1996. Vendor: Net.Genesis Corp.

Net.Analysis Pro Communications. Internet software. Provides tracking and analysis of Web activity. Correlates information submitted in an online form and the users subsequent activity on the Web. Runs under Unix, Windows NT. Vendor: Net.Genesis Corp.

Net.Commerce (Pro) Application software for e-commerce. Includes tools for creating a virtual mall. Released: 1998. Vendor: IBM Corp.

Net.Data Application development tool. Makes DB2 databases accessible through the Web. Includes macro language. Versions available for MVS, Unix. Vendor: IBM Corp.

Net.db Application development tool. Used to design and deploy Web applications. RAD tool which lets developers design data driven pages from a browser. Reads database data and automatically creates HTML views for each table. Database content can then be browsed, updated and published from any Web browser. Released: 1998. Vendor: Centura Software Corp.

Net-It Central Application development software. Allows end-users to publish documents on the Internet. To publish, users drag-and-drop documents into online folders. The software can work with most documents and recognizes hypertext links. Vendor: Net-It Software Corp.

Net/Master Communications. Network manager linking multiple networks in IBM, DEC, Fujitsu, Tandem, and AT&T systems. Vendor: Systems Center, Inc.

Net/One Lan Manager Communications software. Network operating system which links systems from different vendors. Interfaces with LAN Manager, Novell, 3Com. Works with Token-Ring, Ethernet. Vendor: Ungermann-Bass, Inc.

Net Partitioner Communications software. Policy-based security management system. Java based. Allows users to define access lists. Runs on Unix, Windows systems. Released: 1996. Vendor: Solsoft Inc.

Net-Pass Communications software. Network connecting IBM computers. Vendor: Software AG Americas Inc.

Net-Works II Communications software. E-mail system. Runs on Apple II systems. Vendor: High Technology Software Products, Inc.

Net/WrkVMS, Net/WrkHP Communications software. Middleware. Provides integration among DEC VAX VMS, HP/Unix platforms, IBM AS/400, RS/6000, OS/2 platforms. Vendor: KnowledgeNet.

Net30 Communications software. Network operating system. Runs on Pentium type desktop computers. Vendor: Invisible Software, Inc.

NetAgent Communications software. Internet product used with e-commerce. Provides customer service over the Web. Vendor: Eshare Technologies Inc.

NetAnswer Application development tool. Query and retrieval system for the Web. Allows providers to distribute large volumes of data, text, and multimedia content over the Internet. Interfaces with all standard browsers and servers. Vendor: Dataware Technologies, Inc.

NetApp Midsize computer. Server. Alpha processor. Vendor: Network Appliance, Inc.

NetArchive Operating system enhancement. Provides systems management functions for enterprise-wide systems. Provides backup and hierarchical storage management. Vendor: PLATINUM Technology, Inc.

NetAttache Enterprise Server Communications, Internet software. Offline browser that searches Web and disseminates HTML from Web sites across a corporate network. Vendor: Tympani Development, Inc.

NetBackup Operating system enhancement. Provides storage backup and recovery for VMS systems. Released: 1998. Vendor: Veritas Software Corp.

NetBankPC Applications software. Provides a secure environment for banking over the Internet. Vendor: FiTech, Inc.

NetBasic Programming language. Used to write NetWare Loadable Applications. Based on Basic. Vendor: Novell, Inc.

Netbatch-Plus Operating system enhancement used by systems programmers and Operations staff. Job scheduler for Tandem systems. Vendor: Tandem Computers, Inc. (a Compaq company).

NetBEUI Communications. Protocol. Extension to NetBios. Stands for: NetBIOS Extended User Interface.

NetBIOS Communications software. Networking protocol and set of program modules used on IBM and compatible desktop computers. Performs network functions such as error control and data transmission.

Netcache Application development tool. Provides web caching functions. Vendor: Network Appliance, Inc.

Netcaster Communications, Internet software. Push component of Communicator browser. Vendor: Netscape Communications Corp.

NetCensus Operating system software used to migrate to Windows 2000. Surveys networks and reports on how many PCs are running and what software is running. Released: 1999. Vendor: Tally Systems Corp.

Netcenter 1. Communications. Network manager for IBM mainframe systems. Vendor: IBM Corp.
2. Web portal built by Netscape. Includes e-commerce functions.

NetChamp Network computer. Vendor: LG Electronics USA, Inc.

NetConMT Communications software. Network operating system (NOS) connecting Windows, Unix systems. Vendor: Netcon Business Systems.

NetControl Communications software. Network manager. Set of modules that follow SNMP. Manages NetWare, Vines, LAN Manager. Vendor: Xtree, Inc.

NetCool Communications. Network management system. Integrates management of entire environment under one platform. Interfaces with SunNet Manager, OpenView, NewView. Vendor: Micromuse USA, Inc.

NetCraft Application development tool. Visual development environment that generates Java applets. Runs on Windows systems. Released: 1996. Vendor: SourceCraft, Inc.

NetCruiser Communications software. Internet kit which facilitates the use of basic Internet services. Vendor: Netcom.

NetCube Application development tool. DSS. Provides access to data stored in multidimensional databases. Vendor: NetCube Corp.

NETDA Communications software. Network connecting IBM computers. Stands for: NETwork Design and Analysis. Vendor: IBM Corp.

NetDeploy System management software. Automates software distribution and installation over networks. Includes two parts: Packer (used by the network administrator to create a distribution package) and Launcher (used by end users to install the software). Vendor: Open Software Associates Ltd.

NetDirector (for Unix) Communications. Network manager for Unix and OS/2 systems. The Unix version interfaces with OS/2 systems. Supports multiple protocols for multi-vendor networks. Vendor: Ungermann-Bass, Inc.

NetDynamics Application server providing middleware and development tools. Used to build Web applications. Includes Application Server (handling Java applets), Platform Adapter Components (provides the framework for integrating enterprise systems), Command Center (real-time, local and remote, management of enterprise applications), and Java Object framework (includes over 400 classes and 4000 methods). Vendor: Sun Microsystems, Inc.

Netex Communications software. Network connecting computers of most of the major vendors. Stands for: Network Executive. Vendor: Network Systems Corp.

Netexpert Communications. Network manager that manages large-scale, multivendor network situations. Vendor: Objective Systems Integrators, Inc.

NetExpress Application development tool. Development environment for e-COBOL applications that takes legacy mainframe systems written in COBOL and extends them to the Web and other distributed platforms. Developers create components out of the existing code and create the new applications from the components. Converts existing COBOL programs to Unix and Windows NT and builds a Web-based front end. Vendor: MERANT.

Netfax Communications. LAN (Local Area Network) connecting IBM systems. Vendor: OAZ Communications, Inc.

Netfinity Midrange computer. Pentium Pro processors. Vendor: IBM Corp.

Netfinity Manager Communications. Systems management software. Suite of tools and utilities designed to manage both IBM and non-IBM networked desktop, and notebook systems Runs under OS/2, Windows systems. Vendor: IBM Corp.

NetFire Operating system software. JVM that runs on NetWare systems. Runs on Intel's 64-bit Merced processor. Vendor: Novell, Inc.

NetFire Pro xxx Midsize computer. Server. Pentium Pro processor. Operating system: Windows NT. Vendor: Reason Computer.

NetFRAME Desktop computer. Server with Pentium III Xeon processor. Vendor: Micron PC, Inc.

NetGain Systems management software. Caching tool. Runs on Windows systems. Released: 1996. Vendor: NetStream, Inc.

Netgain Application software. Sales software, providing customer information functions. Vendor: Firstwave Technologies, Inc.

NetGateway Software products that allow IBM mainframe systems to communicate with client/server workstation systems. Workstations must be running Open Client. Vendor: Sybase, Inc.

NetHelp2 The format used for Help screens by Netscape. This format must be used with Netscape browsers.

NetInstall Operating system software. Software distribution package for Windows applications. Provides a central point for application installation. Allows network administrators to manage and customize desktops from a central console. Runs on Windows systems. Released: 1999. Vendor: InstallShield.

Netis Best Buy, PowerStation Desktop computer. Pentium processor. Vendor: Netis Technology, Inc.

NetIT Central Application development software. Used to publish documents on the Internet or extranets. Can publish documents from spreadsheets, presentation software and word processors and retain the original look of the document. Documents can be retrieved by keywords chosen by the publisher. Released: 1999. Vendor: Allegis Corp.

NetLabs/Assist Reporting software that works with NetLabs/Manager. Vendor: NetLabs, Inc.

NetLabs/Discovery Communications. Network management software. Analyses components on TCP/IP networks. Runs on IBM RS/6000, Sun, Unix systems. Vendor: NetLabs, Inc.

NetLabs/Manager Communications. Network management software. Used with heterogeneous networks. Unix based. Latest versions called NetLabs/DiMons. Runs on IBM RS/6000, Sun, Unix systems. Vendor: NetLabs, Inc.

NetLabs/Vision Development Environment Application development environment. Creates Motif, OpenLook Unix applications. Runs on IBM RS/6000, Sun, Unix systems. Vendor: NetLabs, Inc.

NetMake Application development tool. Compiles large programs in parallel over a Unix network. Vendor: Aggregate Computing, Inc.

NetMaker XA Systems management software. Allows users to build models of networks and proposed changes to determine effect of changes. In addition to basic network monitoring functions, also measures the effect of Internet browsing on networks. Runs on AIX, Solaris systems. Vendor: Make Systems, Inc.

NETMAN See CA-NETMAN.

NetMap Application development tool. Visualization tool that can color-code groups or clusters of data and display trends and patterns as graphs. Used for analysis type queries and with data warehousing. Runs on Unix systems. Vendor: Software AG Americas Inc.

NetModeler Communications. Network management software. Allows users to build models of networks and proposed changes to determine effect of changes. Full name: Network Performance Modeler for OS/2. Vendor: IBM Corp.

NetMon Communications software. Network analyzer. Included in Windows NT Server. Stands for: Network Monitor. Vendor: Microsoft Corp.

NetObjects (Fusion) Application development tool used to build Web sites. Authoring tool. Includes visual site planner and library of site styles. Runs on Windows systems. Vendor: NetObjects, Inc.

NetOwl, NameTag Document retrieval system. Works with Web servers. Runs on Windows NT systems. Vendor: IsoQuest Inc.

NetPC Network computer type. Hybrid device that supports a hard drive but not floppy or CD-ROM drives. Operating system: Windows. Accesses applications from a server or a local hard drive. Vendor: Microsoft Corp and Intel Corp.

NetPlex 4xx Desktop computer. Pentium CPU. Operating systems: Windows 95/98. Vendor: Dell Computer Corp.

NetPro Communications software. Management and migration programs used to upgrade versions of NetWare. Vendor: NetPro, Inc.

Netpro 200 Desktop computer. Pentium CPU. Operating system: Windows NT Workstation. Vendor: Royal Electronics, Inc.

Netra i, j, nfs Midrange computer. TurboSPARC RISC CPU. Operating system: Solaris. Vendor: Sun Microsystems, Inc.

Netra Proxy Cache Communications. Internet product that allows customers to cache frequently accessed Web sites and information. Combination of hardware and software. Vendor: Sun Microsystems, Inc.

Netron Fusion CASE tool. Builds frames, which are software components that resemble objects and allow for reuse. Uses Fusion Workplace to provide access to design, analysis, program construction tools. Uses QuickStarts template programs. Formerly called Netron Cap. Vendor: Netron, Inc.

Netscape Internet browser. Available through retail sales. Allows users to choose among several national Internet access providers, sets up account. Name of browser actually Navigator. Vendor: Netscape Communications Corp.

Netscape Application Server Application server providing middleware and development tools. Assists in the deployment of business critical, transaction-oriented Internet and intranet applications and with integrating them with user's existing business based environments. Interfaces with Web servers and the enterprise databases. Runs on Unix, Windows systems. Released: 1998. Vendor: Netscape Communications Corp.

Netscape BillerXpert Communications. E-commerce application. Billing/payment software that allows customers to view bills online, establish flexible payment options and schedules future bill payments. E-mails customers with information about new bills, past-due payments and special offers. Runs on Sun systems. Released: 1998. Vendor: Netscape Communications Corp.

Netscape Collabra Server See Collabra Server.

Netscape Commerce Server See Commerce Server.

Netscape Enterprise Server See Enterprise Server.

Netscape EXCpert Communications. Internet software that enables business-to-business transactions over the Internet and EDI networks. Runs on Unix, Windows systems. Released: 1998. Vendor: Netscape Communications Corp.

Netscape One Application development environment. Used for component based development. Supports Java, JavaScript, IIOP, CORBA. Stands for: Open network Environment.

Netscape SuiteTools Application development tools used to develop and manage intranet applications. Includes Visual JavaScript Pro and Component Builder. Uses Java, JavaScript and CORBA to create and assemble JavaBeans components. Released: 1997. Vendor: Netscape Communications Corp.

Netscout Manager System management software. Monitors network traffic on both WAN and LAN setups. Runs on Unix, Windows systems. Vendor: NetScout Systems, Inc. (formerly Frontier Software Development, Inc.)

NetScript/6000 Communications software. Enables NewView/6000 to manage DEC, Tandem, Unisys, and AS/400 systems in addition to IBM mainframes and microcomputers. Vendor: Diederich & Associates, Inc.

NeTservices Application development tool. Data management software that manages data residing on both Unix and Windows NT systems. Vendor: Auspex Systems, Inc.

NetShare Communications software. Resource management system that will determine which computers are in an environment, what the availability of computers is, and remotely execute an application on the appropriate system. Includes: RIB (Remote Information BASE), DMI (Decision Making Interface), RES (Remote Execution Service). Runs on HP, IBM, Sun systems. Vendor: Aggregate Computing, Inc.

NetSolutions Application development tool used to develop Web applications. Runs on Unix, Windows systems. Released: 1996. Vendor: Oracle Corp.

Netspace HSM Operating system software. HSM. Vendor: Avail Systems Corp.

NetSpy Operating system enhancement used by systems programmers. Controls communications networks in IBM systems. Vendor: Legent Corp.

NetStation Desktop computer. Pentium processor. Vendor: Tri-Star Computer Corp.

NetStrada 5000, 7000 Midrange computer. Server. Pentium Pro CPU. Vendor: Olivetti North America.

NetThread Groupware package that runs over the Web. Provides Web-based discussion forms. Vendor: Web.Genesis Corp.

NetTools Operating system software. Configuration management package. Includes HelpPlus, a workstation monitor. Vendor: McAfee Associates.

NetView Systems management software. Network manager linking multiple networks in IBM systems. Interfaces with LAN Server. Includes various programs and links AIX, MVS, OS/2, Windows systems. Vendor: IBM Corp.

netViz Operating system software. Configuration manager. Provides graphical control of all equipment in multiple networks. Runs on Pentium type desktop computers. Vendor: Quyen Systems, Inc.

NetWare Communications software. Network operating system connecting IBM desktop computer, Unix. DEC, Macintosh systems. Includes many programs under NetWare title. Vendor: Novell, Inc.

NetWare Connect 2 Communications software. Provides access to all network services to remote PC and Macintosh users. Vendor: Novell, Inc.

NetWare Directory Services See NDS.

NetWare Distributed Management System See NDMS.

NetWare for SAA Communications software. Network manager that allows NewView (IBM) systems to function as a LAN. Vendor: Novell, Inc.

NetWare/IP NLM Communications software. Allows applications following IPX protocols to run under IP protocols. Stands for: NetWare/Internet Protocol NewWare Loadable Module. Vendor: Novell, Inc.

NetWare Lite Communications software. Peer-to-peer network operating system. Vendor: Microsoft Corp.

NetWare Loadable Module See NLM.

NetWare Management System See NMS.

NetWare SQL Relational database for desktop computer environments. Runs on NetWare systems. Vendor: Novell, Inc.

NetWare SunLink Communications software. Network connecting NetWare LANs with various TCP/IP networks used with Unix. Vendor: SunSelect.

NetWeave Communications. Message-oriented middleware. Allows developers to connect legacy systems. Vendor: NetWeave.

NetWizard Operating system software. Manages desktops by automating distribution and updating of software throughout a network. Runs on Windows systems. Released: 1997. Vendor: Attachmate Corp.

network A combination of hardware and software that connects two or more computers and/or assorted devices.

Network Communications software. Network connecting HP systems. Vendor: Operations Control Systems.

Network/3000 Communications software. Network connecting HP computers. Vendor: DIS International Ltd.

network administrator Support personnel. Monitors functioning of networks, usually LANs. Installs networks, adds new users, troubleshoots network. Title usually indicates experience in midrange, desktop computer systems. Can be any experience level.

network agent See agent.

network analyst, architect, designer, engineer Technical developer. Plans, installs, and supports the company's networks including both LANs and WANs. Knowledge of hardware, protocols, LANS, NOS. Certification available and often required or at least a plus. Senior or mid-level title.

network computer A computer costing under $1000 that provides GUI access to the Internet /intranets. The computer has no disk or CD drives so all data and programs must be loaded from a server machine. Except for the GUI functions the computer functions as a dumb terminal and does no other processing. Runs JAVA applications. Manufactured by many vendors.

Network Computer TC,XL Network computer. Vendor: Boundless Technologies, Inc.

Network Courier Gateway Communications software. Gateway for e-mail systems. Runs with 3+Open, Vines, MS-Net, Netware networks. Vendor: Consumers Software, Inc.

Network DataMover See NDM.

Network File System See NFS.

Network Health Communications. Web-based software that analyzes network performance by condensing thousands of network statistics into trends and problem situations. Can access reports from any Web browser. Vendor: Concord Communications, Inc.

Network Inspector System management tool designed for small businesses. Keeps track of inventories and manages PCs. Vendor: Fluke Corp.

network interface card See nic.

network layer In data communications, the OSI layer that finds the path in the network, or routes the data. The network layer is also responsible for packet switching and monitoring activity on the networks. X.25 and LU6.2 are network protocols.

network management systems Communications software that graphically displays network components, displays alarms, collects and graphs statistics, and allows for implementation of user-specific management tools. Controls network operations in a multi-network system. Usually works with diverse networks and operating systems.

Network Navigator Software distribution tool. Used to install and maintain multiple versions of a single application running on a network. Vendor: Novell, Inc.

Network News Transfer Protocol See NNTP.

Network Node Manager See NNM.

Network Notes Communications service. Internet style network service based on Lotus Notes. Business users maintain information on the system and decide what access other businesses have as Notes clients. Includes online catalogs for sales. The overall system provides the security. Vendor: AT&T Corp.

Network OLE Application development tool. Allows Windows applications to connect across a network. Vendor: Microsoft Corp.

network operating system Software package that provides both operating system and LAN (Local Area Network) functionality such as centralizing file and print services. Handles diverse systems. It is the system software that runs on the server in a local area network (LAN), and interfaces with other operating systems. Also called LAN operating system.

Network-OS (Plus) Communications software. Network operating system which links MS-DOS systems. Conforms to OSI standards. Vendor: CBIS, Inc.

Network PM Network management system. Tracks traffic problems and warns of potential problems. Vendor: 3DV Technology.

Network Scheduler Groupware. Allows scheduling of time, meeting rooms, equipment, etc. for individuals and/or groups. Runs on Pentium type desktop computers. Vendor: CE Software.

Network Server Midrange computer. Server. PowerPC CPU. Vendor: Apple Computer, Inc.

Network Station Network computer. Connects to OS/2, System/390, AS/400, RS/6000, Windows NT servers. Vendor: IBM Corp.

NetWorker Operating system enhancement. Provides backup functions for distributed systems. Runs on Unix systems. Vendor: Legato Systems, Inc.

NetWorker (Business Suite) Operating system enhancement. Provides network wide backup and recovery services for multivendor networks. Uses x-windows based graphics user interface. Vendor: Legato System, Inc.

Networker Remote Communications software. Provides backup for remote desktops and laptop computers. Includes disaster recovery functions. Runs on Windows NT. Released: 1999. Vendor: Legato Systems, Inc.

Network*IT* Pro Communications software. Network management system. Released: 1998. Vendor: Computer Associates International, Inc.

NetworkMCI Business Communications service. Internet style network service. Provides e-mail, teleconferencing. Business users maintain information on the system and decide what access other businesses have. Includes online catalogs for sales. The overall system provides the security. Vendor: MCI Communications Corp.

Networks Messaging Facility See NMF.

Neugents Network management system. Based on neural network technology, which means the software "learns" from activity and continues to add to its knowledge base. In networks, Neugents are included in CA-Unicenter TNG and are Windows NT server-based network agents which analyze the network environment, decide when its operating state has become abnormal and notify administrators. Neugents software is available separately and can be used to analyze conditions in business markets, predict changes in those conditions, and suggest action to avoid problems and take advantage of opportunities. Runs under Windows NT. Released: 1999. Vendor: Computer Associates International, Inc.

Neural Connection Application development tool used for data mining. Uses neural networks to discover and predict relationships in data. Runs on Windows 95/98 systems. Vendor: SPSS Inc.

Neural Network Utility Application development tool. Artificial intelligence software used for data mining. Uses neural networks to discover and predict relationships in data. Supports risk assessment, fraud detection, portfolio management and product testing. Runs on AIX, OS/2 systems. Released: 1991. Vendor: IBM Corp.

NeuroGenetic Optimizer Application development tool used for data mining. Uses neural networks to discover and predict relationships in data. Runs on Windows systems. Vendor: BioComp Systems Inc.

NeuralWorks Predict Application development tool used for data mining. Uses neural networks to discover and predict relationships in data. Runs on Macintosh, Unix, Windows systems. Vendor: NeuralWare Inc.

NevOS Real-time operating system for desktop computers. Vendor: Microprocessing Technologies.

NEW/Define Application development tool. Supports analysis of client/server systems. Used with Natural for Windows. Stands for: Natural Engineering Workbench. Vendor: Software AG Americas Inc.

New Technology See NT.

NewEra See Informix-NewEra.

NewFace Application development tool. Windows front-end for character-based applications running on Hewlett-Packard 3000 systems. Vendor: M.B. Foster Associates Ltd.

NewGeneration Software Software vendor. Products: Concert Series of financial, HR, payroll systems for AS/400 environments.

NewLook Application development tool. Used to develop applications using uni-Verse. Vendor: VMARK Software, Inc.

NeWS Platform for development and support of window-based applications. Runs on VMS, PC/MS-DOS, Unix, Xenix systems. Stands for: Network extensible Window System. Vendor: Sun Microsystems, Inc.

NewsEDGE Application development tool used for information filtering. Finds articles of interest according to user defined specifications. Runs on OS/2, Windows systems. Vendor: Desktop Data, Inc.

News-OS Operating system for SONY desktop computer systems. Vendor: SONY Corporation of America.

newsgroup Automated message in the Internet in which subscribers post messages to the entire group on a specific topic. Newsgroups are established for interest groups and are both personal and professional. Operated through Usenet.

Newt OS Operating system for handheld computers. Vendor: Apple Computer, Inc.

NEWT-SDK Application development tool. Used to build applications for the Internet. Lets developers include e-mail, file transfers in Windows-based applications. Vendor: NetManage, Inc.

Newton Personal digital assistant, or handheld computer. Family of computers, including MessagePad. Vendor: Apple Computer, Inc.

Newton Intelligence Operating system for the Newton family of handheld computers. Vendor: Apple Computer, Inc.

Newton MessagePad See MessagePad.

Newton Toolkit Application development tool. Includes NewtonScripts, object-oriented database. Used to develop applications for Newton family of handheld computers. Vendor: Apple Computer, Inc.

NewtonScripts Object-oriented programming language used in Newton Toolkit. Vendor: Apple Computer, Inc.

NewWave Application development environment for object-oriented program development. Includes GUI, object database. Includes office automation, word processing, e-mail, windowing, and other software. Vendor: Hewlett-Packard Co.

NewWave Office Groupware. Runs on various systems and provides office applications such as word processing, spreadsheets, e-mail, calendars. Vendor: Hewlett-Packard Co.

NEX/COM SNA Communications software. Network connecting IBM desktop computers. Vendor: DSC Nestar Systems, Inc.

NEX/OS Communications software. Network operating system which links MS-DOS systems. Vendor: DSC Communications Corp.

Nexpert Object Application development tool for AI systems that works in a client/server architecture. Accesses dBase III, DEC Rdb, Gemstone Ontos, DB2, SQL/DS, Informix, Ingres, Oracle Server, SQL Server. Includes application and report generators. Graphics oriented. Vendor: Neuron Data, Inc.

NeXT, Nextcube, Nextstation Midrange computer. Operating systems: NextOS, Mach, Unix. No longer manufactured.

Next Generation Internet See NGI.

NeXTOS Operating system for Next workstations. Unix-type system. Vendor: Apple Computer, Inc.

NextPoint S3 Harmony Operating system software. Measures performance in clients/server systems by simulating user activity to measure response time. Vendor: NextPoint Networks, Inc.

NeXTStep Application development environment. Used for building object-oriented applications. Includes Graphic user interface, windows functions, and comes installed on NeXT systems. Also runs on IBM's RS/6000 systems; version called OpenStep runs on Solaris systems. Includes: Interface Builder. Vendor: Apple Computer, Inc.

NFS Communications protocol and software system. Applications layer protocol that provides access to remote files without having to know where the files are. Developed by Sun Microsystems. Stands for: Network File System. Used with TCP/IP. Accepted file sharing standard by Unix vendors.

NFSaccess Communications software. Network connecting DEC, SUN, NCR, Bull HN, Harris, IBM desktop computer systems. Vendor: Syntax Systems, Inc.

NGI The Next Generation Internet. A multi-agency initiative for Federal research and development of advanced networking technologies. Systems are working on test sites that are 100 to 1000 times faster than today's Internet.

nic Communications device. Network interface card. Connects desktop computers to networks from a slot in the PC.

Nielsen Spotlight Application development tool. Visualization tool that can color-code groups or clusters of data and display trends and patterns as graphs. Used for analysis type queries and with data warehousing. Vendor: A.C. Nielsen Co.

Nielsen Visual Topline Application development tool. Visualization tool that can color-code groups or clusters of data and display trends and patterns as graphs. Used for analysis type queries and with data warehousing. Vendor: A.C. Neilsen Co.

Night Hawk Midrange computer. PowerPC CPU. Operating systems: PowerMAX OS, Unix. Vendor: Concurrent Computer Corp.

Night Operator See Supermon.

NII National Information Infrastructure. Network connecting homes to university, government, and business networks. Proposal by consortium of computer vendors. Extension of NREN.

NII Testbed National Information Infrastructure Testbed. Consortium of companies, universities, and national laboratories. Concerned with standards, functions, and building applications for the Internet.

Nile Code name for OLE DB. See OLE DB.

Nino Handheld computer. Operating system: Windows CE. Vendor: Philips Mobile Computing Group.

NIROS Operating system for Siemens Nixdorf systems. Vendor: Siemens Nixdorf Computer Corp.

Nirvana Application development tool. Query and report functions. Creates graphs. Runs on AS/400, Windows systems. Vendor: Synergy Technologies.

NIST National Institute of Standards and Technology.

NLM A program that can be loaded and executed by NetWare. NLMs facilitate basic functions, such as printing. Stands for: NetWare Loadable Module. Vendor: Novell, Inc.

NLM Server Object-oriented database. Stands for: NetWare Loadable Module Server. Vendor: Object Design, Inc.

nmake Application development tool. Manages software and documentation. Runs on Unix systems. Included with ADE. Vendor: NCR Corp.

NMAP Operating system software. Assists the network manager in controlling access to applications installed on a network. Stands for: Network Menuing Admin Pack. Runs with Norton Desktop Utilities. Vendor: Symantec Corp.

NMF Communications software. Middleware. Links OLE objects across multiple systems. Supports DDE protocols. Stands for: Networks Messaging Facility. Vendor: Symbiotics, Inc.

NMS Communications. Network management software. Stands for: NetWare Management System. Runs with Windows or OS/2. Distributed network manager. Follows SNMP protocols. Vendor: Novell, Inc.

NNM Communications software. Network management system. Map and monitor network devices. Includes a data warehouse, event-correlations services, Web server. Runs on Windows NT systems. Stands for: Network Node Manager. Released: 1998. Vendor: Hewlett-Packard Co.

NNTP Internet protocol. Governs discussion group messaging. Stands for: Network News Transfer Protocol.

node A connection point in a communications network. In some systems, notably DEC networks, a node is a computer. In other usage, notably IBM, a node can be a terminal as well as a computer.

Nokia 9000 Handheld computer. Operating system: GEOS. Vendor: Nokia.

NOMAD(2) Application development system. Includes relational database/4GL for large computer environments. Runs on DEC, IBM systems. Includes DSS, report generator. Interfaces to IMS, IDMS, DB2, SQL/DS, Teredata. Versions include DB2 NOMAD, SQL Nomad, Teradata Nomad. Vendor: Aonix.

Non-Uniform Memory Access See NUMA.

NonStop Himalaya Midrange computer. RISC machines. MIPS Rxxx CPUs. Operating systems: OS/2, Unix, Windows NT. Vendor: Tandem Computers, Inc. (a Compaq company).

NonStop SQL(/MX) Relational database/4GL for midrange and desktop computer environments. Can be used in data warehousing. Includes Object Relational Data Mining. Supports DataBlades. Runs on Tandem systems. Includes report generator, CASE tools. Runs on Windows NT systems. Released: 1997. Vendor: Tandem Computers, Inc. (a Compaq company).

NonStop (Kernel) Operating system for Tandem RISC computers. Vendor: Tandem Computers, Inc. (a Compaq company).

Norand Computer vendor. Manufactures pen-based computers.

normalization Term used in database design and processing. A type of database design that spreads data into tables that contain only unique and pertinent information about the subject of the table.

Northbridge NX801 Midrange computer. Pentium Pro multiprocessor. Operating systems: Windows NT. Vendor: Axil Computer, Inc.

Northstar Application development tool. Analyzes COBOL code and moves it to ADW for redevelopment as a client/server application. Vendor: Knowledgeware, Inc.

Norton AntiVirus Desktop software. Virus protection. Runs on NetWare networks. Released: 1998. Vendor: Symantec Corp.

Norton DiskLock Administrator See DiskLock.

Norton File Manager Operating system software. File management system. Includes file compression, uuencode and uudecode functions. Vendor: Symantec Corp.

Norton Utilities Desktop system software. Operating system add-on. Runs on Pentium type desktop computer systems. Vendor: Peter Norton Computing.

NOS See network operating system.

NOS Tools Operating system add-on. Data management software that runs on CDC systems. Vendor: Control Data Corp.

NOS/VE Operating system for large computer CDC systems. Vendor: Control Data Corp.

Note-able Tools Application development tool. Used to develop applications for Lotus Notes. Vendor: Workgroup Productivity Corp.

Notebook Application development system. Runs on Lotus Notes and interfaces with relational databases. Supports Visual Basic and LotusScript. Vendor: Lotus Development Corp.

notebook computer A portable desktop computer that weighs between three and six pounds. Most plug into full-size monitors and/or keyboards. Most are near the size of a standard notebook, 8 1/2″ x 11″.

Notebook Companion Operating system software. Utilities package for portable computers. Includes a security system. Vendor: WizardWorks Group.

NoteBrick3 Desktop computer. Notebook. AMD processor. Vendor: The Brick Computer Company, Inc.

Notes Groupware. In addition to standard groupware functions, provides E-mail capabilities, data management functions, and design tools. Interfaces with ProShare to provide EMS capabilities. Runs on Macintosh, OS/2, Unix, Windows systems. Vendor: Lotus Development Corp.

Notes Application Library Application development tools. Library of functions such as project tracking and budget planning that can be used as is or customized to a company's needs. Used to develop groupware systems. Vendor: Lotus Development Corp.

Notes Express Groupware. Scaled down version of Notes. Includes e-mail, some applications. Does not access Notes applications. Also called Notes lite. Vendor: Lotus Development Corp.

Notes lite See Notes Express.

Notes Mail Communications software. E-mail system. Client/server system. Interfaces with cc:Mail. Vendor: Lotus Development Corp.

Notes Reporter Application development tool. Report Writer used with Notes databases. Vendor: Lotus Development Corp.

NotesSuite Combination of Notes and SmartSuite. Vendor: Lotus Development Corp.

NotesView Network management software used with Lotus Notes. Works with OS/2, Windows NT, NetWare, Unix. Vendor: Lotus Development Corp.

Notework Communications software. E-mail system. Runs on local area networks. Vendor: ON Technology Corp.

Notework/MHS Gateway Communications software. Gateway for e-mail systems. Runs on Netware networks. Vendor: ON Technology Corp.

Notrix, Notrix Composer Application development tools. Allows users to manipulate Lotus Notes data using common programming languages. Handles relational database replication to Notes. Vendor: Percussion Software.

Nouveau Communication. Middleware, object request broker (ORB). Used with e-commerce, Internet applications and provides interoperability between COM, C++ and Java. Runs on Unix, Windows systems. Originally developed by NobelNet. Released: 1999. Vendor: Rogue Wave Software, Inc.

Nova 4GL used in midrange and desktop computer environments. Vendor: Uniq Digital Technologies, Inc.

NovaNet 7 Enterprise Communications. Network backup software. Works with Windows NT and NetWare. Released: 1998. Vendor: Novastor Corp.

NovaView Application development tool. Provides OLAP front-end to SQL Server. Includes "What If..." functions. Vendor: Cognos Inc.

Novell Software vendor. Produces data communications software (NetWare) and operating system software (UnixWare). Also markets communications hardware. See Netware.

Novell certification See CNE, CNA, MCNE, CIP.

Novell Embedded Systems Technology See Nest.

NPL Relational database for midrange and desktop computer environments. Runs on Apple, DEC, HP, IBM systems. Includes query facility and report generator. Full name: NPL Information Management System. Vendor: Database Applications, Inc.

NREN National Research and Education Network. Upgrade of American Internet and intended to combine NSInet, ESnet, NSFnet and MILnet. Focuses on NSFnet and is based on MCI's national network.

NRUNOFF CASE product. Part of CASE 2000 system. Vendor: Nastec Corp.

NS 7000 Midrange computer. Server. Pentium CPU. Vendor: Auspex Systems, Inc.

NS NetServer Midrange computer. RISC CPU. Operating systems: Unix, Solaris. Vendor: Auspex Systems, Inc.

NS200 Midrange computer. SPARC CPU. Operating system: Solaris. Vendor: Integrix, Inc.

NS7000,8000-xxxx Midsize computer. Server. Pentium II, Pro processors. Vendor: Gateway 2000.

NSCP National Scalable Cluster Project. Research investigation into metacomputers as an alternative to mainframe and/or MPP systems. Funded by the National Science Foundation.

NSFnet A network composed of leased lines (from MCI) to over a million commercial, academic and research users.

NSInet NASA's network.

NSSN National Standards Systems Network. A user-friendly view of information systems standards available over the Internet. Created by NIST and ANSI.

NT See Windows NT.

NT Advanced Server, NT AS See Windows NT Server.

NT Enterprise Operating system. Includes Wolfpack clustering technology. Supports up to eight processors. Full name: Windows NT Enterprise Edition. Released: 1997. Vendor: Microsoft Corp.

NT Server, NT Advanced Server See Windows NT Server.

NTNX Communications software. Network operating system. Interfaces with Novell LANs. Vendor: Alloy Computer Products, Inc.

NTrigue Communications software. Connects Windows NT servers with Unix workstations. Vendor: Citrix Systems, Inc.

NTX 2000 Desktop computer. Pentium Pro CPU. Operating system: Windows NT. Vendor: Sequent Computer Systems.

Nuance6 Speech recognition software. Vendor: Nuance Communications.

NUMA Architecture used to cluster computers. Allows clustering more than the standard 16 processors. Arranges the processors into groups (called nodes) to control communications. Each of these nodes has its own memory pool which is also shared. Stands for: Non-Uniform Memory Access.

NUMA-Q Computer design technology used in parallel processing. Stands for: Non-Uniform Memory Access for quads. Gives a single system image to parallel processors. Used to build scalable processors and can reach up to 252 processors. Vendor: Sequent Computer Systems, Data General Corp.

NUMA-Q 1000 Desktop computer. Operating systems: Unix, Windows NT. Released: 1999. Vendor: Sequent Computer Systems, Inc.

NUMA-Q 2000 Midrange computer. Parallel processor. Up to 252 Pentium Pro processors. NUMA architecture. Vendor: Sequent Computer Systems, Inc.

NUMARC Operating system enhancement used by systems programmers. Controls communications networks in IBM desktop computer systems. Vendor: ADC Telecommunications, Inc.

NuMedia xxxx Desktop computer. Pentium processor. Vendor: PC Importers, Inc.

NuMega DevPartner for Java Application development tool. Used to debug Java programs. Detects inefficient code and potential bottlenecks. Runs on Windows systems. Release date: 1998. Vendor: NuMega Technologies, division of Compuware.

NuMega DevPartner Studio 6 Application development tool. Used for testing. Automatically detects software errors and performance problems. Supports Microsoft's Visual Studio. Runs on Windows systems. Vendor: NuMega Technologies, division of Compuware.

Numetrix/3 Application software. Manufacturing software that provides supply chain management. Assists manufacturers in deciding quantities of each product that should be made at each production site. Enables users to consider what-if situations. Runs on Unix, Windows systems. Released: 1997. Vendor: Nemetrix, Ltd.

NuTCRACKER Application development tool. Converts Unix application to run under Windows NT. Vendor: Mortice Kern Systems, Inc.

Nuts & Bolts Operating system utilities. Provides file recovery, file security, disk protection, etc. Includes WebScan X. Vendor: Helix Software Co.

NUView Application development tool. Windows front-end for character-based applications running on Hewlett-Packard 3000 systems. Vendor: Advanced Systems, Inc.

NWS Communications software. Web server. Stands for: Novell Web Server. Vendor Novell, Inc.

nX Operating system for TC2000 MPP systems. Unix type system. Vendor: Bolt Beranek and Newman, Inc.

O/S200 Operating system for midrange PICK systems. Includes DBMS. Vendor: The Ultimate Corp.

O.S.5 Operating system for Hewlett-Packard desktop computer systems. Vendor: Hewlett-Packard Co.

O2 Object-oriented database. Runs on Unix, Windows systems. Supports OQL, SQL. Vendor: Ardent Software, Inc.

O2 Technology See Ardent Software.

O2Web Application development tool. Connects any O2 database to the Web. Translates objects to HTML format. Runs on Unix systems. Vendor: Ardent Software, Inc.

OAG Open Applications Group. Consortium of software vendors founded in 1994. Its mission is to provide multivendor client/server applications which can be integrated without additional software. Provides specifications for business functions by covering, i.e., how manufacturing functions (such as process control) interface with financial applications.

Oak Communications software. Middleware. Full name: MQSeries Three Tier. Provides links to GUI development tools. Used to combine existing mainframe applications with client/server programs. Supports Smalltalk, C++ versions of VisualAge. Vendor: IBM Corp.

Oberon Software vendor. Produces EAI software. Main product: Prospero.

Obex Communications software. Technology that encapsulates data into objects that can be transferred over e-mail and/or networks. Integrated with Paradox, Quattro Pro. Stands for: Object Exchange. Vendor: Inprise Corp.

OBI Standards for EDI over the Internet. Combines legacy and Internet technology. Stands for: Open Buying on the Internet.

ObjChart Application development environment. Object-based modeling tool. Generates object code (Smalltalk, C++) directly from design diagrams. Vendor: IBM Corp.

object Entity that contains data and procedures, or methods—descriptions of how to manipulate that data. Used in object-oriented development. Objects are defined by class definitions and have three characteristics: encapsulation, polymorphism and inheritance.

Object/1 Application development system. Object-oriented. Works in a client/server architecture. Accesses OS/2 EE Database Manager, MDBS IV, SQL Server, Oracle Server. Builds client applications with a GUI based front-end. Includes application and report generators. Runs on IBM OS/2 systems. Vendor: MDBS, Inc.

Object Administrator Application development tool. Estimates query's use of system resources so queries can be optimized, compiled or canceled. Vendor: PLATINUM Technology, Inc.

Object Bridge (for COM, CORBA) Communications and development tool. Allows combination of CORBA and OLE systems. Vendor: Visual Edge Software Ltd.

Object CM Application development tool. Object-oriented configuration management system. Integrates with PCTE. Vendor: The Alsys CASE Division.

Object COBOL Application development tool. Add-on to COBOL Workbench that provides class libraries and enables the development of object-oriented, GUI applications in COBOL. Vendor: Vendor: MERANT.

object database Object-oriented database, object-oriented database management system. Data collection that holds values and processing information. Must provide for inheritance, encapsulation, and polymorphism. Also called ODBMS, objectbase, object database. Often used with multimedia applications.

Object Development Facility See ODF.

Object Development Workbench Application development tool. Includes ODF, Object Repository, Gen/c++. Vendor: System Software Associates, Inc.

Object Factory Application development tool. Visual development tool that generates C++ code. Creates database independent applications. Uses the memento pattern to generate database code by setting up one class for data representation and a second class for interface information to keep the actual data separate from information about how to save, restore and update it. Runs on Unix, Windows systems. Released: 1997. Vendor: Rogue Wave Software, Inc.

Object GEODE CASE tool. Automates analysis, design and programming functions with object-oriented and real-time approaches to development. Includes graphical editors, C/C++ code generator, design-level debugger. Runs on Unix, Windows NT. Vendor: CS Verilog.

Object Linking and Embedding See OLE.

Object/LM Operating system software. Security package for object-oriented applications. Protect objects from unauthorized use and allows an object to be an entire application or a specific resource. Tracks object utilization and associated costs. Confirms to CORBA standards. Runs under Unix, Windows systems. Released: 1997. Vendor: Black & White Software, Inc.

Object management group See OMG.

Object Management Workbench See OMW.

Object Manager Application development tool. Repository for objects. Handles configuration management and version control. Interfaces with PowerBuilder, Visual Basic, TeamTest. Vendor: LBMS, Inc.

Object Master Application development tool. Screen editor. Provides multiple screen browse capabilities for object libraries. Vendor: ACI US, Inc.

Object Modeling Technique See OMT.

Object/Observer Operating system software. Monitors the communication of distributed, client/server objects. Follows CORBA standards. Monitors traffic from both the client and the server. Includes Inprise's Visibroker, Orbix, and OrbixWeb. Runs under Unix, Windows systems. Released: 1998. Vendor: Black & White Software, Inc.

object-oriented development An approach to programming that treats data and processing as independent objects which can be reused and interchanged between programs. Objects are stored in libraries and programs are build by combining object to accomplish a task.. Also called object-oriented, object-oriented programming.

object-oriented middleware Communications software that manages communications between objects. Includes ORBs (Object Request Broker).

Object Partner CASE tool. Object-oriented approach for mission critical applications. Based on OMT and C++. Supports UML. Runs on Unix, Windows NT. Vendor: Verilog SA.

Object Pascal Computer language used to develop object-oriented applications. Used with MacApp.

object program A program in machine code. Object programs are created by translator programs.

Object Relational Data Mining Application development tool. Used in data warehousing for data mining queries. Vendor: Tandem Computers, Inc. (a Compaq company).

object/relational database Database that stores both object and relational data. Multimedia data is stored in tables, but database can be indexed and searched based on the objects.

Object Repository Application development tool. Part of Object Development Workbench. Repository for business objects. Vendor: System Software Associates, Inc.

object request broker A form of middleware used in object-oriented systems that locates, loads and executes objects. Creates interoperability by letting objects connect regardless of platform. Also, objects created in different environments (C++, Smalltalk, Objective C) cannot work together without some form of intervention. Object request brokers supply the commonality. Called ORBs.

Object REXX Object-oriented programming language.

Object Series Family of products to develop applications to run under Windows. Manages and routes communications between objects on a network. Includes ObjectScript, ObjectView. Supports development of client/server systems. Includes debugging features, object data dictionary. Vendor: Matesys Corp.

Object Tracker Application development tool. Tracks and analyzes object usage to better maintain objects and resources. Vendor: PLATINUM Technology, Inc.

Object Windows Library See OWL.

ObjectAda Application development systems. Develops Ada applications in C/C++ and Java. Integrates with SoftBench. Runs on Unix systems. Vendor: Aonix.

objectbase See OODB, OODBMS.

ObjectBroker Communications, application development tool. Middleware, ORB (Object Request Broker). CORBA based. Used to integrate existing applications as well as used in new development. Works with diverse platforms. Originally written by DEC. Released: 1997. Vendor: BEA Systems, Inc.

ObjectCenter Application development environment. Used to develop C, object-oriented C++ programs. Runs on Unix systems. Vendor: Centerline Software, Inc.

ObjectCraft Application development environment. Visual tool that generates C++ code. Runs on Windows systems. Vendor: SourceCraft, Inc.

ObjectCycle Application development tool. Provides version control functions for multiple clients. Runs on Windows systems. Vendor: Powersoft (division of Sybase, Inc.).

ObjectFactory for C++ Application development tool. Visual tool that generates C++ code and stores it in text files that can then be customized. Vendor: Rogue Wave Software.

ObjectFinder Electronic Meeting System (EMS). Vendor: ObjectVision.

ObjectForms Application development tool. Connects ObjectStore databases to Web. Generates HTML Runs on Solaris, Windows NT systems. Vendor: Object Design, Inc.

ObjecTime CASE tool. Automates design, testing functions. Used graphical modeling to develop object-oriented real-time systems. Part of Object Series. Vendor: ObjecTime Ltd.

ObjectIQ Application development environment. Used in object-oriented environments. Includes GUI editor, C/C++ interface, workbench functions. Includes ObjectIQ-DF, ObjectReuser. Runs on most workstations. Vendor: Hitachi America Ltd.

Objective C Programming language. Object-oriented language that combines parts of Smalltalk into a C-type language. Used with NextStep.

Objectivity/DB Object-oriented database. Runs on DEC, HP, IBM, Sun, Unix systems. Vendor: Objectivity, Inc.

ObjectMaker Application development tool. Eases migration of legacy systems to object technology. Runs on OS/2, Unix, Windows systems. Vendor: Mark V Systems, Ltd.

ObjectMaker TDK Application development tool. Used by developers to customize methods, syntax rules, notations, etc. Manages all information in a repository. Stands for: Tool Development Kit. Vendor: Mark V Systems, Ltd.

ObjectModeler Application development tool. Part of PowerTools. Supports Rumbaugh, Coad/Yourdon, Booch, Jacobsen methodologies. Includes editors for C++, Smalltalk. Vendor: ICONIX Software Engineering, Inc.

ObjecTool Application development tool. Provides analysis and design model, C++ and Smalltalk code generators. Runs on Macintosh, OS/2, Unix, Windows systems. Vendor: Object International, Inc.

Objectory See Jacobsen.

ObjectPal Integrated programming language used with Paradox and dBase IV. Works with visual design and data modeling tools. Vendor: Inprise Corp.

ObjectPool Application development tool. Allows developers to encapsulate mainframe data stored in traditional databases into objects that can then be used in both client/server and object-oriented applications. Vendor: Sapiens International Corp.

ObjectPro Application development tool. Object-oriented language that allows programmers to build prototypes that can be translated into C code. Vendor: PLATINUM Technology, Inc.

ObjectQ Application development tool. Motif-based GUI builder. Includes object-based interfaces for Informix, Ingres, Oracle and Sybase databases. Runs on Windows systems. Vendor: Hitachi America, Ltd.

ObjectScript See Object Series.

ObjectStar Application development tool. Used for enterprise development. Migrates mainframe applications to client/server systems. Supports three-tiered application partitioning. Vendor: Antares Alliance Group.

ObjectStore Object-oriented database. Includes development tools used to develop Java, C++ and ActiveX applications. Has Internet interfaces. Runs on Solaris, Windows systems. Vendor: Object Design, Inc.

ObjectStudio Application development tool used to develop object-oriented applications. Allows developers to build large-scale client/server systems that will run on multiple hardware and software systems. Includes Enfin visual Smalltalk environment. Accesses all major relational DBMSs. Generates reusable Smalltalk. Includes TeamBuilder feature which enables groups of developers to simultaneously build object-oriented applications. Vendor: Runs on OS/2, Unix, windows systems. Released: 1996. Vendor: Ardent Software, Inc.

ObjectTeam See COOL:Jex.

ObjectView 1. Application development environment. Used to develop client/server applications. Builds client applications with a GUI-based front end. Includes 4GL, debugging tools, Personal SQL. Runs on OS/2, Windows systems. Interfaces with ADW. Vendor: KnowledgeWare, Inc.
2. See Object Series.

ObjectVision Application development environment. Allows development of applications without a programming language in non-procedural format so applications can be developed by non-programmers. Includes windowing functions. Works with client/server and builds client applications with a GUI based front-end. Vendor: Inprise Corp.

Objectworks(/C++,Smalltalk) Application development environment. Available for Smalltalk-80 and C++ environments and used to develop object-oriented applications. Includes set of class libraries containing over 5,600 reusable functions. Vendor: ParcPlace-Digitalk.

Obsydian See COOL:Plex.

OC/PS Communications software. Network connecting terminals, desktop computers, and mainframe systems. Stands for: OpenConnect/Presentation Services. Vendor: Mitek Systems Corp.

OC://WebConnect Communications software. Screen scraper. Emulates mainframe terminals from Web browsers. Converts data from Web browsers to terminal data. Vendor: OpenConnect Systems.

Occam Compiler language used in mainframe environments. Designed to facilitate parallel, or concurrent, processing taking advantage of the fact that new computers are not limited to sequential processing. Considered a difficult language.

OCCF Communications software. Transaction processing monitor. Runs on IBM systems. Stands for: Operator Communications Control Facility. Controls operator commands and messages from remote locations. Vendor: IBM Corp.

OCL Operation Control Language. Name given to operator commands in IBM environments.

OCR Optical Character Recognition. Printed characters that can be directly read into a computer system because of their graphic representation.

OCS/3000, Express Operating system enhancement used by systems programmers. Monitors and controls system performance in HP systems. Vendor: Operations Control Systems.

OCX Application development tool. Language used to create objects to appear on a tool palette in a visual programming environment. 32-bit controls that will replace 16-bit VBX (Visual Basic Controls). Lets developers write small modular programs that can stand alone or be merged in other programs. Used over the Internet. Stands for: OLE Control Extensions. Vendor: Microsoft Corp.

Odaptor Application development tool. Provides a bridge between Oracle's, Sybase's and Informix's relational database and C++, Smalltalk object-oriented programming languages. Runs on HP/UX servers, with AIX, DOS/Windows, HP/UX, Solaris, clients. Vendor: Hewlett-Packard Co.

ODB-II Object-relational database. Includes database engine, graphical development environment called ModelWorks, class libraries, interfaces to C, C++, Visual Basic, Oracle. Handles multimedia objects. Vendor: Fujitsu America, Inc.

ODB/Server Application development tool. Provides client access to DB2 systems in client/server environments. Follows ODBC. Supports OS/2 and Windows clients. Stands for: Open Database/Server. Runs on IBM/MVS systems. Released: 1996. Vendor: Aonix.

ODBC communications. Middleware. Simplifies connection between front-end (user interface) and SQL/relational databases and flat files. Used in Desktop environments. Stands for: Open DataBase Connectivity. Written by Microsoft.

ODBC Integrator Application development tool. Used to allow remote users access to data across platforms. Runs on Windows systems. Vendor: Dharma Systems, Inc.

ODBMS 2.0 Object-oriented database. Stores object created through Smalltalk. Vendor: VC Software Inc.

ODBS Relational database for desktop computer environments. Runs on IBM, Concurrent systems. Vendor: O'Hanlon Computer Systems.

Odds/MVS Operating system enhancement used by Operations staff and systems programmers. Provides operator console support in IBM systems. Vendor: Software Engineering of America.

ODE Toolkit Application development tool used for client/server development. Builds client applications with a GUI based front end. Vendor: Open Environment Corp.

ODF Application development tool. Part of Object Development Workbench. Allows definition of business objects. Stands for: Object Definition Facility. Vendor: System Software Associates, Inc.

ODMG Consortium of vendors setting standards for object-oriented data storage. Includes working with object, relational, and object-relational databases. Develops bindings (connections) from the databases to object-oriented languages including Java, C++, Smalltalk. Stands for: Object Data Management Group.

ODMS/ Communications software. Network connecting Apple Macintosh and DEC VAX. Includes ODMS/DocuShare, ODMS/Matrix. Vendor: Odesta Corp.

ODS Operational Data Store. Architectural approach to data storage that combines the size capabilities of a data warehouse with the accessibility of traditional transactional databases.

ODSI Set of four APIs that will allow users to access multi-vendor networks from a single logon. Stands for: Open Directory Service Interfaces. Vendor: Microsoft Corp.

ODWI Data warehouse framework. Strategy and product selection intended to offer smooth integration between the warehouse and the operational systems. Stands for: Open Data Warehouse Initiative. Vendor: Software AG Americas Inc.

Odyssey Communications software. Used to improve the speed of intranets. Converts Smartsuite and Office 97 data to HTML. Part of Lotus Smartsuite. Released: 1998. Vendor: Lotus Development Corp.

Odyssey Suite Communications, Internet software. Groupware. Includes calendar and reminder functions. Vendor: Ulysses Telemedia Networks, Inc.

OEA Application development architecture. Includes programs called Adaptlications that can be readily customized in client/server applications. Adaptlications available for manufacturing (PowerMan), financial (Intergy), sales (Xceed), customer-order-management (PowerCom), and help desk (Response) applications. Stands for: Object Extensible Architecture. Vendor: PowerCerv Corp.

off-line browser Internet browser that lets users save entire Web sites locally. These browsers monitor Web sites for new pages and retrieve pages at off-peak hours. Users could, for example, send entire Web catalogs to customers. Customers can then browse catalogs without worrying about connect time.

Off-Net Sniffer Communications software. Protocol analyzer used to debug network problems. Works with over 200 LAN and applications protocols. Vendor: Network General Corp.

Office, Office95, Office98 See Microsoft Office.

Office 2000 Software suite. Upgrade to Office 98. Five suite options available which includes part to all of the following: Excel 2000, FrontPage 2000, Outlook 2000, PhotoDraw 2000, PowerPoint 2000, Publisher 2000, Small Business Tools, Word 2000. Outlook 2000 is the only module with significant changes. Release date: 2000. Vendor: Microsoft Corp.

Office Automation Specialist Someone who works with desktop computers and off-the-shelf software packages, such as word processors, databases, and spreadsheets. Part of the administrative staff, not IS.

Office Workstation Desktop computer. Pentium processor. Vendor: ACE Computers.

OfficeLogic Groupware. Includes e-mail, scheduling, phone messaging, databases. Vendor: Lan-Aces, Inc.

OfficeServer Communications software. Messaging product. Includes file handling with attributes such as data and type attached to each document. Supports CC:Mail, MS-Mail. Runs on OpenVMS, Unix, Windows NT systems. Vendor: Digital Equipment Corp.

OfficeVision A software package that provides basic office functions such as phone, mail, and calendar support. Part of SAA and runs on MVS, OS/2, and OS/400, VM systems. Vendor: IBM Corp.

offline browser See off-line browser.

OHF Operating system enhancement used by systems programmers. Increases system efficiency in IBM CICS systems. Automatically creates CICS help windows. Vendor: Software Engineering of America.

OHMTools Application development tools. 4GL building blocks with subroutine library. Runs on Unix, Windows NT systems. Vendor: OHM Systems, Inc.

OL Object-oriented development methodology. Stands for: Object Lifecycles. Associated with Shlaer and Mellor.

OLAP On-Line Analytical Processing. Software used to help consolidate and analyze business information. Any query system that provides multidimensional data manipulation, display and visualization of data for reporting purposes. Includes EIS, GIS, DSS, data warehousing systems. Stands for: On-Line Analytical Processing.

OLE 1. Software that allows users to create an object (a spreadsheet, word processing memo, etc.) in one application and then move the object to another application which edits it. "Linking" simply means making object available to other applications, but all editing must be done in the application which created the object. "Embedding" allows the change to be made from any application. Has turned into ActiveX.
2. Application development tool. Object-oriented tool that allows two separate applications to be mixed. Used with Windows. Vendor: Microsoft Corp.

OLE Controls See OCX.

OLE DB Database software. Extension to OLE which provides a transparent interface allowing all types of data and platforms to be accessed from diverse systems. Previous code name Nile. Vendor: Microsoft Corp.

OLEnterprise Application development tool. Allows Windows applications to connect with MVS, Unix applications. Released: 1996. Vendor: Open Environment Corp.

OLGA Operating system enhancement used by systems programmers. Monitors and controls system performance in IBM MVS, CICS, and DB2 systems. Vendor: Systar, Inc.

OLIAS Internet browser. Displays both HTML and SGML DOCUMENTS. Runs on Unix systems. Vendor: HalSoft.

Oliver Application development tool. On-line testing and debugging tool for CICS applications. Runs on VSE systems. Vendor: Compuware Corp.

Olivetti Computer vendor. Manufactures desktop computers.

OLS Operating system add-on. Security/auditing system that runs on IBM systems. Full name: OLS C Online Security. Vendor: Softsystems, Inc.

OLTP An online system that gathers information and updates data in real-time. The file or database is changed when the new information is entered. The information and update process is called a transaction. Stands for: Online Transaction Processing.

OM-Axcess Communications software. Manages Web sites with multiple Web servers. Handles end-used authorization by individual and by group. Vendor: Open Market, Inc.

OM-Express Communications, Internet software. Off-line browser. Vendor: Open Market, Inc.

OM3 Message Broker Communications. Message-oriented-middleware, (MOM). Interfaces with MQSeries, integrates CORBA, DCOM. Works with any size system and allows requests to be handled from any requestor regardless of platform. Runs on Unix, Windows systems. Also called Cloverleaf OM3. Vendor: Healthdyne Information Enterprises Inc. (HIE).

OME See Open Messaging Environment.

OmegaCenter Operating system enhancement used by systems programmers. Provides enterprise-wide systems management for IBM mainframe systems. Handles MVS, CICS, IMS, DB2, VM systems. Vendor: Candle Corp.

Omegamon (II) Operating system enhancement used by systems programmers. Monitors and controls system performance in IBM systems. Includes: OMEGAMON for CICS, DB2, DBCTL, IMS, MVS, SMS, VTASM, VM. Vendor: Candle Corp.

OMG Object Management Group. A consortium of vendors whose goal is to develop standards for object-oriented applications. Developed CORBA standard.

Omni-8 Database for large computer environments. Runs on DEC systems. Vendor: Solutions Unlimited.

Omnibase(/SQL,4GL) Relational Database/4GL for large computer environments. Runs on DEC systems. Has program development tools, query and report generation capabilities. Utilizes SQL. Vendor: Signal Technology, Inc.

OmniBroker See ORBacus.

OmniConnect Communications. Connects multiple disparate data sources as though they were a single database. Vendor: Sybase, Inc.

OmniDesk Workflow software. Image processing system that runs on mainframe systems. Vendor: Sigma Systems, Inc.

Omnidex Application development tool. Indexing and warehousing tool used to search through millions of records to summarize data and download it to desktop systems. Used in EIS systems. Full name: Omnidex for Data Warehousing. Vendor: Dynamic Information Systems Corp.

Omnifile Relational database for desktop computer environments. Runs on IBM systems. Vendor: SSR Corp.

OmniGo See HP OmniGo.

Omniguard Operating system add-on. Security/auditing system that runs on IBM systems. Vendor: On-Line Software International.

OmniGuard/EAC Operating system add-on. Security package for open systems environments. Runs on AIX, HP-UX, Solaris systems. Stands for: OmniGuard/Enterprise Access Control. Vendor: Axent Technologies.

Omnilink/36 Communications software. Network connecting IBM computers. Micro-to-midrange link. Vendor: On-Line Software International.

OmniORB Communications. Middleware, object request broker (ORB). Freeware. CORBA compliant. Vendor: AT&T Laboratories Cambridge.

OmniPlex Desktop computer. Pentium CPU. Operating systems: Windows 95/98. Vendor: Dell Computer Corp.

OmniReplicator Application development tool. Data replicator. Allows two way replication. Interfaces with Oracle, DB/2. Sybase, Informix. Part of OmniWarehouse. Vendor: Praxis International, Inc.

Omnis(7) Application development environment that works in a client/server architecture. Used for enterprise development. Builds client applications with a GUI based front-end. Accesses DEC Rdb, DB2 Informix, OS/2 EE Database Manager, Ingres, Oracle Server. Includes application and report generators. Allows development of any SQL-based application. Vendor: Blythe Software, Inc.

OmniTrans Communications software. EDI system. Vendor: Release Management Systems.

OmniWarehouse Data warehouse. Runs on Unix systems. Includes OmniReplication. Vendor: Praxis International, Inc.

OMT Object-oriented development methodology. Stands for: Object Modeling Technique. Associated with Rumbaugh (Rumbaugh/OMT).

OMW Application development tool. Automates analysis and design functions. Lets users create object-oriented systems without knowing object-oriented languages. Includes C++ code generator. Based on Martin/Odell object-oriented methodology. Interfaces with Kappa. Runs on Unix systems. Stands for: Object Management Workbench. Vendor: James Martin & Co. and Intellicorp, Inc.

ON Guard Communications, Internet software. Firewall. Vendor: ON Technology Corp.

On/Q Application software. Supply chain management software designed for large companies and handle the relationships with multiple vendors. Vendor: QAD.

ON/X Application development tool. Integrated development environment and runtime platform for on-line systems. Vendor: Shared Financial Systems.

Once Express Communications software. Internet product used with e-commerce. Provides customer service over the Web. Vendor: Business Evolution Inc.

Ondemand ATM Campus Manager Communications software. Provides centralized control of all ATM (Asynchronous Transfer Mode) products. Based on Motif, X-Windows GUI. Vendor: Chipcom Corp.

One See Netscape ONE.

One-to-One Communications. Internet software. Allows Web site tuning and presentation rearrangement in real-time. Vendor: BroadVision Inc.

One-UP Desktop software. Multidimensional spreadsheet. Runs on MS-DOS systems. Vendor: ComShare, Inc.

OneView Application software. Manufacturing software handling supply chain functions. Runs on Unix, Windows systems. Vendor: Manugistics, Inc.

OneWorld Application software. Component based ERP system. Includes financial, HR, manufacturing, payroll, procurement sales, supply-chain functions. Vendor: J.D. Edwards & Co.

online Pertaining to a program or system that receives input from, and/or sends output to, terminals during execution. Also called communications, data communications, interactive, real-time, telecommunications.

OnLine See Informix-OnLine, Informix-OnLine 7.0.

OnLine English 4GL used in mainframe environments. Vendor: Cullinet Software, Inc.

Online Log Display See DBA Tool Kit.

OnLine Query 4GL used in mainframe environments. Accesses IDMS databases and VSAM files. Vendor: Cullinet Software, Inc.

Online Reference Data management software. Allows users to write textual data into DB2 databases. Vendor: Data Base Architects, Inc.

Online Test Operating system add-on. Debugging/testing software that runs on IBM systems. Vendor: DBMS, Inc.

OnLine XPS Parallel database. Can run client/server applications on a parallel server, or across a cluster of systems. Can be used for data warehousing or for operational transaction processing. Vendor: Informix Software, Inc.

OnNet Communications software. Provides all levels of interconnectivity between LANs and WANs. Follows TCP/IP protocols. Vendor: FTP Software.

OnSchedule Operating system enhancement. Job scheduler for Unix systems. Vendor: Paradigm Systems Corp.

Ontos Object-oriented database based on C++. Especially suited for CAD/CAM functions, but also used in CASE and office automation applications. Runs on Unix systems. Vendor: Ontos, Inc.

OO-COBOL Object-oriented programming language.

OOA/D Object-oriented development methodology. Stands for: Object-Oriented Analysis/Design. Associated with Coad & Yourdon.

OOAD Object-oriented development methodology. Associated with Martin and Odell.

OOADA Object-oriented development methodology. Associated with Booch.

OODB, OODBMS See object database.

OO, OOD, OOP See object-oriented development.

OOSD Object-oriented development methodology. Stands for: Object-Oriented System development. Associated with de Champeaux et al.

Opal Application development tools. Suite of tools which deploys communication and database applications distributed across IP networks. Integrates data from multiple mainframe and Unix databases. Enables users to access applications via Web browsers and includes screen scraper tools to put a GUI front-end on mainframe systems. Released: 1998. Vendor: Computer Associates International, Inc.

OPC Operating system enhancement used by Operations staff and systems programmers. Provides operator console support in IBM systems. Also helps plan operations activity. Stands for: Operations Planning & Control. Vendor: IBM Corp.

OpCon/Cross Platform Server Systems management software. Scheduling and batch processing software for client/server environments. Supports most mid-size, RISC, and desktop systems. Vendor: Software Management Associates.

Open 2200/500 Mainframe computer. Operating systems: Unix, OS 2200. Vendor: Unisys Corp.

Open Access II Relational database for desktop computer environments. Runs on Pentium type desktop computer systems. Vendor: Software Products International.

open architecture Designing and developing computer systems that will work in any hardware environment.

Open ClientConnect Application development tool. Used with client/server computing. Allows developers to turn a mainframe CICS application into a client for a Unix server. This means mainframe applications can access data from the remote workstations. Vendor: Sybase, Inc.

Open Data Warehouse Initiative See ODWI.

Open DataBase Connectivity See ODBC.

Open Desktop Graphical operating system built on Unix. Lets users create their own icon-driven workstations. One of two operating systems accepted as part of ACE standards. Vendor: The Santa Cruz Operation.

Open Directory Service Interfaces See ODSI.

Open Gateway for DB2 Software products that allow IBM mainframe systems to communicate with client/server workstation systems. Workstations must be running Open Client. Vendor: Sybase, Inc.

Open/Image for Notes Imaging software that interfaces with Notes. Vendor: Wang Laboratories, Inc.

Open/Image Windows Image processing system that runs on Wang, RS/6000 systems. Vendor: Wang Laboratories, Inc.

Open Interface Application development tool. Workbench used for cross-platform development. Object-oriented application generator for C programs. Generates code for Windows, OS/2, Motif, Open Look, and Macintosh. Runs on OS/2, Unix, Windows systems. Vendor: Neuron Data, Inc.

Open Look Graphics user interface for Unix systems. Includes windows functions. Jointly developed by Sun and AT&T.

Open Messaging Environment Enterprise messaging software. Built from Novell's communications software and WordPerfect's Office suite. Also called OME.

Open PGP Specification for security of e-mail functions over the Internet. Stands for: Open Pretty Good Privacy.

Open PL/1 Application development tool set. Includes PL/1, Code-Watch (source level debugger). Interfaces with VSAM, GUI tools, DEC systems. Migrates mainframe PL/1 applications to Unix. Runs on Unix systems. Vendor: Liant Software Corp.

Open Plan Desktop system software. Project management package. Runs on Pentium type desktop computers, Macintosh, Unix. Vendor: Welcom Software Technology.

Open Profiling Standard See OPS.

Open Server 400 Communications software. Gateway for e-mail systems. Runs on Netware, Vines, PC LAN networks. Vendor: Retix.

Open Server for CICS Software products that allow IBM mainframe systems to communicate with client/server workstation systems. Workstations must be running Open Client. Vendor: Sybase, Inc.

Open ServerConnect Data Management Software. Allows users to connect relational and nonrelational databases. Vendor: Sybase, Inc.

Open Software Foundation See OSF.

Open Solutions Architecture See OSA.

open source A tradition of open standards, shared source code, and collaborative development that has contributed to software such as the Linux and FreeBSD operating systems; the Apache Web server; Perl, Tcl, and Python languages; and much of the Internet's infrastructure. Open source is a trademark of the Open Source Initiative. Open source implies that software is available without restrictions or charge, and that modifications and derivative works are permitted. Officially, open source means that source code must be available for redistribution without restriction and without charge, and the license must permit the creation of modifications and derivative works, and must allow those derivatives to be redistributed under the same terms as the original work. Licenses that conform with the Open Source Definition include the **GNU** Public License (GPL), the BSD license used with Berkeley Unix derivatives, the X Consortium license for the X Window System, and the Mozilla Public License.

open systems A philosophy that both software and hardware vendors should make pertinent information about their products available to each other so that products from different vendors can work together. This extends to the agreement that vendors should follow the same standards. Open systems also states that all systems should be interoperable, scalable, and portable. Movement towards open systems is most obvious in the area of communications.

Open Systems Interconnect See OSI.

Open Text 5 Document retrieval system. Works with Web servers. Runs on Unix, Windows systems. Vendor: Open Text Corp.

Open Warehouse Data warehouse framework. Strategy and product selection intended to offer smooth integration between the warehouse and the operational systems. Vendor: Hewlett-Packard Co.

Open/workflow Workflow software. Includes GUI, report generator. Vendor: Wang Laboratories, Inc.

OpenAPI Application development tool. Automates programming function. Works with INFRONT/ development tools. Vendor: Multi Soft, Inc.

OpenBASIC Application development system. Includes report writer, data query tools, RDBMS interfaces. Runs on Unix, Windows systems. Vendor: MAI Systems Corp.

OpenBridge Data warehousing software. Builds a distributed warehouse by connecting datamarts. Vendor: Informatica Corp.

OpenCASE CASE product. Automates analysis, design, and programming functions. Works with Informix databases. Runs on Unix systems. Vendor: Informix Software, Inc.

OpenConnect Communications software. Network connecting IBM, DEC VAX, AT&T, HP, Apollo, and Sun systems. Vendor: Nitek Systems Corp.

OpenDNM Communications software. Manages DEC networks running on Hewlett-Packard systems with OpenView. Vendor: Ki Research.

OpenDoc Architecture for object-oriented system development. Standards to turn programs and routines into objects that can be accessed across diverse platforms. Apple, IBM, Novell, WordPerfect, Borland cooperate in the development. Conforms to CORBA standards. Competitive with Microsoft's OLE. Available for OS/2, MacOS, AIX, Windows, mainframes.

OpenEDI Communications software. EDI. Runs on Unix systems. Vendor: Ross Systems, Inc.

OpenEdition MVS Operating system for IBM mainframe computers. Version of MVS designed for client/server environments. Supports Unix application programming interfaces and has a Unix look and feel. Vendor: IBM Corp.

OpenGL Application programming tool. Software library of tools for texture mapping and special effects for 3D graphics. Included with most operating systems. Vendor: Silicon Graphics, Inc.

OpenImage Imaging software. Vendor: Wang Laboratories, Inc.

OpenIngres Application development tool. Provides object-oriented and connectivity enhancements to Ingres. Used to move applications from mainframes to client/server systems. Includes Ingres database, gateways that allow Unix users to retrieve data from IBM mainframe systems. Runs on OS/2, Unix, VMS, Windows NT systems. Used in Client/server systems. Vendor: Computer Associates International, Inc.

OpenIngres/ESQL Application development tool. Stands for: Embedded SQL. Allows users to query databases and also change table definitions. Vendor: Computer Associates International, Inc.

OpenIngres/ODBMS Integration of the relational database Ingres and the object-oriented database ODB-II. Allows users to search, query and manage objects through relational database commands. Vendor: Computer Associates International, Inc.

OpenIngres/Star Relational database for distributed computer environments. Can update data as a single database regardless of format or location. Runs on DEC, IBM, Unix systems. Vendor: Computer Associates International, Inc.

OpenInsight (for Notes) Application development tool that works in a client/server architecture. Object-oriented repository which supports derivatives of Basic, C, C++ languages. Used for team development. Builds client applications with a GUI-based front end. Interfaces with SQL Server, DB2. Vendor: Revelation Technologies, Inc.

OpenJ Application development tool. Designed for user with light programming experience. Used to create applications using JavaBeans components. Vendor: Corel Corp.

OPENjdbc Communications software. Connects Internet browsers to servers to access multiple databases. Stands for: Java Database Connectivity. Released: 1997. Vendor: I-Kinetics, Inc.

OpenLink ODBC Drivers Communications software. Middleware. Runs on OS/2, Unix. VMS, Windows systems. Vendor: OpenLink Software Inc.

OpenMail Groupware and e-mail system. Client/server system that runs on a Unix server and allows clients to keep existing e-mail software. Vendor: Hewlett-Packard Co.

OpenMapper Application development system. Combines Mapper tools with Designer Workbench. Runs on Windows, Sun systems. Vendor: Unisys Corp.

OpenMaster Communications. Systems management software. Provides all management functions including application and security management. Follows SNMP, CDE standards. Cross-platform support for Macintosh, OS/2, Unix, Windows. Vendor: Bull HN Information Systems, Inc.

OpenMind Groupware. Runs on Macintosh, Windows systems. Provides Internet access. Vendor: Attachmate Corp.

OpenNote 680xx Notebook computer. Pentium CPU. Vendor: Kiwi Computer, Inc.

OpenNT (SDK) Application development tool. Allows users to run software written for Unix on Windows NT platforms. Vendor: Softway Systems, Inc.

OpenODB Relational database with object-oriented extensions. Interfaces with Informix and AllBase. Runs on Hewlett-Packard systems. Vendor: Hewlett-Packard Co.

OpenROAD Application development environment. Graphical. Includes data templates and styles, ability to edit multiple applications, integrates with most databases. Runs on Unix, Windows systems. Vendor: Computer Associates International, Inc.

OpenScape Communications, Internet software. Provides middleware and development tools. Used to link Internet browsers to corporate databases and applications. Released: 1996. Vendor: OneWave, Inc.

OpenServer See SCO OpenServer.

OpenSNA Communications. Network manager that manages SNA networks from an SNMP based platform. Vendor: Peregrine Systems, Inc.

OpenStep Application development environment. Version of NextStep which runs on Solaris systems. Version also available for Windows NT. Vendor: SunSoft, Inc., Apple Computer, Inc.

OpenUI Application development environment. Visual tool with object-oriented functionality used in client/server systems. Interfaces all phases of the development cycle. Vendor: Open Software Associates, Inc.

OpenView System management software. Distributed network manager linking multiple networks. Follows SNMP, CMIP, DME protocols. Connects Unix systems. Vendor: Hewlett-Packard Co.

OpenView Desktop Administrator Communications package. Hardware and software management suite used to manage PCs. Released: 1997. Vendor: Hewlett-Packard Co.

OpenView Node Manager Communications. Network management software. Client/server management system which manages NetWare servers and clients. Vendor: Hewlett-Parkard Co.

OpenView Service Simulator Communications. Network management software. Allows users to simulate adding new applications, users and technologies to networks, thus predicting networks needs and avoiding failures. Developed jointly with MIL-3 who markets it under the name IT DecisionGuru. Vendor: Hewlett-Packard Co.

OpenVision System management software. Used with Unix systems. Vendor: OpenVision Technologies, Inc.

OpenVista Application development environment that lets developers convert mainframe green screens to GUI frontends. Automatically generates Java classes and code. Includes with OC://WebConnect and Web-Connect Pro. Vendor: OpenConnect Systems, Inc.

OpenVMS Operating system for DEC systems. Provides multi-user and real-time support. Vendor: Digital Equipment Corp.

OpenWarehouse Web Application development tool. Uses Web browsers to access data warehouses. Vendor: Hewlett-Packard Co.

OpenWeb Application development tool. Links Web front end systems to client/server systems. Creates applications that use a Web browser as a user interface. Full name: OpenWeb NetDeploy. Vendor: Open Software Associates, Inc.

OpenWindows (DevGuide) Application development environment. Used to develop applications using Open Look GUI. Runs on RISC systems. Vendor: Sunsoft, Inc.

OPENworkshop Application development tool. Object-oriented environment that has data objects designed first and then the methods that manipulate the objects are controlled by the data. A single set of code can produce both character and graphic presentations. Runs on DEC, Unix, Windows systems. Released: 1997. Vendor: Thoroughbred Software International, Inc.

OPERA See CA-OPERA.

operating system Collection of programs that manages the resources of the hardware system and controls the execution of programs.

operating system add-on A computer program that works with the operating system to provide additional support to the applications programming staff. Add-ons include debugging/testing packages, data management routines, library management systems, and security/auditing systems.

operating system enhancement An extension to the operating system. Systems software used to improve the operation of the data center. Usually transparent to applications programmers. Used by systems programmers and Operations.

operating system software Software which provides operating system type functions but is not associated with a specific operating system.

Operational Data Store See ODS.

Operations Coordinator Communications. Network management system. Used for full enterprise networking. Vendor: Nynex Allink Company.

Operations Environment Operating system enhancement used by systems programmers and Operations staff. Job scheduler for IBM AS/400 systems. Vendor: Informed Management Environment, Inc.

OperationsCenter Communications. Network management system. Used for full enterprise networking managing IBM, Sun, UP and Unix systems. Works with OpenView. Vendor: Hewlett-Packard Co.

operator Support personnel. Operates the equipment in a data center. Controls execution of computer programs by providing hardware and software support. Can be any experience level. Title usually means experience in mainframe computer installations.

OPMAN Operating system enhancement used by systems programmers. Monitors and controls system performance in IBM system. Concentrates on activity of Operations staff. Vendor: Compucept, Inc.

OPS Open Profiling Standard. Set of standards that defines storing personal information in software and lets users decide whether a Web site should be able to access that data.

OPS/MVS Operating system enhancement used by Operations staff and systems programmers. Provides operator console support in IBM systems. Vendor: Goal Systems International, Inc.

OPS/R2 Application development tool. Used to develop rules-based expert systems. Developed programs can run stand-alone or be imbedded in C/C++ programs. Runs on OS/2, Unix, Windows, Windows NT systems. Vendor: Production Systems Technologies, Inc.

OPS38 Artificial intelligence system. Runs on AT&T, DEC, HP, IBM, Sun systems. Rule-based, expert systems programming language. Vendor: Production Systems Technologies, Inc.

OPS5+ Programming language used in the development of artificial intelligence systems. Stands for: Official Production System 5.

OPTASM Compiler language used in desktop computer environments. Runs on IBM desktop systems.

Optegra Total Data Management Data management software. Gives on-line access and control over data to end-users regardless of data's location. Vendor: Computervision Corp.

optical card Card similar in appearance to regular credit cards but which includes optical data storage. Used to create a portable medical record.

optical disk A disk storage device that is read and written by light. Provides greater storage capacities than magnetic disks, thus allowing large amounts of data to be kept online. Used in imaging systems.

optical fiber Communications lines using light rather than electronic currents to transmit a signal. Very fast and very reliable and commonly used for long distance transmission. Expensive, so rarely used with LANs.

OptiGrowth System management software for SAP R/3 systems. Provides capacity planning functions for R/3 systems. Part of Energizer PME. Released: 1998. Vendor: OptiSystems Solutions, Ltd.

Optika Workflow software. Object-oriented package. Vendor: Optika Imaging Systems, Inc.

Optima Workflow software. Used for business process re-engineering. Vendor: AdvanEdge Technologies.

Optima++ See Power++

Optima for Java Application development tool. Converts C++ systems to Java for use on the Internet. Creates applications that use a Web browser as a user interface. Runs on Windows systems. Vendor: Powersoft (division of Sybase, Inc.)

Optima MT, SL, DT Desktop computer. Pentium processor. Operating systems: Windows 95/98. Vendor: Acer American Corp.

Optimal Manager EIS. Runs on Pentium type desktop computers. Vendor: Transpower Corp.

Optimal Performance Communications software. Automated network design tool used to set up LANs. Vendor: Optimal Network Corp.

Optimal Surveyor Communications software. Analysis tool that uses data from sniffer probes to monitor network traffic. Vendor: Optimal Network Corp.

Optimizer See CA-Optimizer.

Optimodel Operating system enhancement used by systems programmers. Increases system efficiency in IBM systems. Manages disk storage space. Vendor: Legent Corp.

OptiPlex Desktop computer. Pentium CPU. Operating systems: Windows 95/98. Vendor: Dell Computer Corp.

OptiTrak System management software for SAP R/3 systems. It identifies and analyzes likely causes of slow response time and correlates details for R/3, database and operating system components. Detects and diagnoses problems in real-time. Part of Energizer PME. Released: 1998. Vendor: OptiSystems Solutions, Ltd.

OptiWatch System management software for SAP R/3 systems. It identifies and analyzes likely causes of slow response time and correlates details for R/3, database and operating system components. Detects and diagnoses problems in real-time. Part of Energizer PME. Released: 1998. Vendor: OptiSystems Solutions, Ltd.

Optix Network System software. Document management, imaging, archival/retrieval and natural language search functions. Runs on AIX, A/UX, Macintosh, Solaris, Windows systems. Vendor: Blueridge Technologies, Inc.

Optix Workflow Workflow software. Add-on to Optix Network. Allows creation of multiple routes for electronic documents. Runs on AIX, A/UX, Solaris systems. Vendor: Blueridge Technologies, Inc.

OPUS Massively parallel processor. Stands for: Open Parallel Unisys Server. Operating system: Unix type. Vendor: Unisys Corp.

OQL Object-oriented programming language. Used with distributed computing, object technology and the Internet. Superset to SQL. Includes OM (Object Model), ODL (Object Definition Language) OML (Object Manipulation Language). Developed by ODMG. Stands for: Object Query Language.

OR-Compass Application development tool. Data modeling tool that allows designers to mix object and relational technologies. Supports object-enabled databases from Informix, Oracle, IBM. Released: 1998. Vendor: Logic Works, Inc.

Oracle Relational Database/4GL for mainframe and/or midrange environments. Client/server architecture. Runs on all major computer systems. Utilizes SQL. Can be used in client/server computing. Includes DSS, report generator, CASE tools. Vendor: Oracle Corp.

Oracle Application Server Application server providing both middleware and development functions. Integrates enterprise-wide Web platforms over most Web servers. CORBA compliant. Supports multiple operating systems including Unix and Windows NT. Vendor: Oracle Corp.

Oracle Applications Application software. Internet enabled. ERP system which includes financials, HR, manufacturing, supply chain, and sales force automation systems. Business process automation which includes supply-chain management, workflow. Version 11 (released 1998) includes support for European, Latin American currencies. Vendor: Oracle Corp.

Oracle APS Application software. Manufacturing system, supply chain software. Used for e-commerce applications. Includes: Oracle Demand Planning, Advanced Supply Chain Planning, Global ATP Server and Oracle Manufacturing Scheduling modules. Supports all manufacturing processes. Stands for: Advanced Planning and Scheduling. Released: 1999. Vendor: Oracle Corp.

Oracle Card Application development tool. Works with Windows and/or Macintosh computers to allow nontechnical users to query databases from a graphic interface. Generates SQL for accessing data and building reports. Includes: Query Builder, Table Builder, Stack Builder. Runs as client in any client/server architecture. Vendor: Oracle Corporation.

Oracle CASE CASE product. Automates analysis, design, and programming functions. Includes code generator for Oracle SQL. Design methodologies supported: Gane-Sarson, Information Engineering. Databases supported: DB2, Oracle SQL. Runs on DEC, IBM, and Unix systems. Vendor: Oracle Corp.

Oracle CDE Application development tools. Used for design and modeling of client/server applications. Includes graphical reporting tools. Includes: Oracle Forms, Oracle Reports, Oracle Designer. Stands for: Cooperative Development Environment. 1996 upgrade called Developer/2000. Vendor: Oracle Corp.

Oracle CPG Application software. ERP software for the consumer packaged goods industry. Includes financial and manufacturing software. Released 1997. Vendor: Oracle Corp.

Oracle Discoverer Application development tool. Query and analysis tool. Includes estimate of how long query will take. Released: 1997. Vendor: Oracle Corp.

Oracle EBU System software. Provides backup and restore to disk for Oracle7 databases. Stands for: Enterprise Backup Utility. Vendor: Oracle Corp.

Oracle Express Multidimensional/relational database. Includes query, graphics, reporting, forecasting, statistics, modeling, communications, client/server integration Web enabling and application development tools. Allows users to use Excel spreadsheet screens for input. Runs on DEC, OS/2, Unix systems. Released: 1997. Vendor: Oracle Corp.

Oracle Financials Name used to refer to application software from Oracle that includes many functions such as financial analysis, general ledger, government purchasing, etc.

Oracle Forms Application development tool. Program that develops programs accessing Oracle databases. Supports Windows, Presentation Manager, Motif, Open Look and Macintosh GUIs. Part of Oracle Tool Kit. Earlier versions called SQL*Forms. Vendor: Oracle Corp.

Oracle Glue API (Application Programming Interface). Provides a bridge between various databases and desktop operating systems. Supports ODBC. Interfaces with Oracle, DB2, NonStop SQL, Paradox, dBase. Vendor: Oracle Corp.

Oracle Groupware Groupware. Includes e-mail, access to corporate data, multimedia support. Formerely called Oracle Documents. Vendor: Oracle Corp.

Oracle InterOffice See InterOffice.

Oracle Lite Version of Oracle database for handheld systems. Released: 1998. Vendor: Oracle Corp.

Oracle*Mail Communications software. E-mail system. Runs on DEC, IBM desktop systems. Vendor: Oracle Corp.

Oracle Manufacturing MRP, or CIM software. Runs on Unix systems. Vendor: Oracle Corp.

Oracle Office Desktop software. Provides e-mail, scheduling, and planning functions. Vendor: Oracle Corp.

Oracle Quicksilver Relational database for desktop computer environments. Runs on Pentium type desktop computer systems. Vendor: Oracle Corp.

Oracle Reports Application development tool. Internet and intranet reporting tool. Accesses corporate data. Produces reports in HTML, PDF, ASCII and PCL PostScript formats. Translates into foreign languages and contains tools to imbed graphics into reports. Runs on Windows systems. Vendor: Oracle Corp.

Oracle RMAN System software. Provides backup and restore to disk for Oracle8 databases. Stands for: Recovery Manager. Vendor: Oracle Corp.

Oracle Server Database server software that works with Oracle databases. Client/server architecture. Interfaces with NetWare, Vines. Vendor: Oracle Corp.

Oracle Universal Server See Universal Server.

Oracle Workgroup/2000 Application development tools. Suite of products used to build and deploy scalable, client/server applications. Runs on OS/2, Unix, Windows systems. Vendor: Oracle Corp.

Oracle7 Version of Oracle database. Stores application code within the database. Allows operating of distributed databases. Also called Cooperative Server.

Oracle7 Parallel Server Version of Oracle 7 used in parallel systems. Released: 1992. Vendor: Oracle Corp.

Oracle8 Oracle 8 contains object-oriented support including support for non-relational data such as text and images. Also has parallel query and data warehousing features. Used for enterprise-wide processing. Uses Cartridges to provide access to non-relational data. Vendor: Oracle Corp.

Oracle8i Database designed specifically for Internet applications. Version of Oracle that includes IFS (Internet File System) which handles images and text documents, Java capabilities and XML support. Also handles relational data. Update to Oracle released in 1998. Vendor: Oracle Corp.

Oracle8i Lite Database designed for mobile computing. Provides interface with Java programs and any client hardware including laptops and PDAs. Released: 1998. Vendor: Oracle Corp.

Oracle8i ORB Plus Communications. Middleware, object request broker (ORB). Connects diverse objects. CORBA compliant. Stands for: Object Request Broker. Vendor: Hewlett-Packard Co.

OracleWare Software suite from Oracle and Novell. Consists of Oracle7, either NetWare or UnixWare, and Oracle Office. Support for all packages in the suite will be through Oracle.

Orange Systems Software vendor. Products: ALCIE IV financial, payroll systems for midsize environments.

ORB See object request broker.

Orb/Enable Application development tool. Used to develop CORBA compliant applications. Allows developers to visually browse, manipulate and manage CORBA information, thereby eliminating the need to know and remember coded interfaces. Includes: IDL Editor, Visual Interface Repository Browser and Server Manager. Runs under Unix, Windows systems. Released: 1998. Vendor: Black & White Software, Inc.

ORBacus Communications. Middleware, object request broker (ORB). Free for non-commercial use. CORBA compliant. Formerly known as OmniBroker. Vendor: Object Oriented Concepts, Inc.

ORBit Communications. Middleware, object request broker (ORB). Part of GNU project.

Orbix, OrbixWeb Communications, application development tool. Middleware. ORB (Object request broker) which integrates OLE and CORBA interface modules. Runs on Windows systems. Included in Forte application development environment. OrbixWeb provides Web interface. Vendor: Iona Technologies, Inc.

ORBIX Application development tool. JCL and/or SYSOUT maintenance program for HP systems. Vendor: Datamaster Computer Service.

Orbix+Isis Communications software. Uses Iona ORB and Isis technology for replication. Vendor: Isis Distributed Systems, Iona Technologies, Inc.

OrbixBuilder Application development tool. Used to develop CORBA compliant applications. Provides drag and drop editing of CORBA based code in Java and C++, automatic integration of user interface and distributed object code. Includes Orbix or OrbixWeb ORB, integrated graphical CORBA utilities, automatic code generation and visual CORBA tutorials. Runs under Unix, Windows systems. Released: 1999. Vendor: Black & White Software, Inc.

OrbixCOMet Application development tool. Provides a bridge between COM and CORBA technologies. This allows applications to access both Unix and Windows systems. Vendor: Iona Technologies Ltd.

ORBlink Communications. Middleware, object request broker (ORB). Full name: Allegro ORBlink. Integrates with Allegro CL and interfaces with Java and C++ applications. CORBA compliant. Vendor: Franz, Inc.

ORCA Operating system enhancement used by systems programmers. Monitors and controls system performance in IBM systems. Stands for: Online Resource Control Aid. Measures performance of online systems. Vendor: Software Diversified Services.

Orchestrate Application development tool. Used in data warehousing. Builds warehousing and data mining applications on parallel processors. Used to determine which of hundreds of realtionships among variables are significant. Runs on Unix systems. Vendor: Torrent Systems.

ORexx Scripting language based on Rexx. Adds object-oriented functions. Part of OpenDoc. Works with OS/2, MacOS, AIX, Windows.

Org Publisher for Intranets Application software. Turns data from HR/Payroll databases into charts that allow users to locate such things as people, job titles, phone numbers. Vendor: TimeVision, Inc.

Organizer 97 GS Communications software. Includes PIM features, group scheduling and calendaring features. Links to Web sites. Vendor: Lotus Development Corp.

Organizer II Handheld computer. Vendor: Psion.

Organon Word processor for DEC systems. Vendor: Orion Information Systems.

Orgware See Baan.

Origin200, Origin2000 Midrange computer. Server. Parallel processor with up to 128 processors. MIPS R1000 CPU. Vendor: Silicon Graphics, Inc.

Orion AMD Desktop computer. AMD processor. Vendor: ET Technology.

Orion/Itasca Object-oriented database. Runs on Unix systems. Supports Windows clients. Vendor: Ibex Systems, Inc.

Orlando Application development environment. Includes library of business objects and data query tools. Used to build EIS systems that execute with most databases. Runs on Windows systems. Vendor: SAS Institute, Inc.

Orlando II Data warehousing software suite. Provides analysis tools. Vendor: SAS Institute, Inc.

ORMS Communications software. E-mail system. Runs on IBM systems. Stands for: On-Line Report Management System. Vendor: CHI/COR Information Management, Inc.

OS 1100 Operating system for mainframe Unisys systems. Vendor: Unisys Corp.

OS/2 Operating system for IBM desktop systems. Provides 32-bit functionality. Supports DCE, one of two operating systems accepted by ACE standards. Vendor: IBM Corp.

OS/2 DBM Relational database included with OS/2. Runs on IBM PS/2 systems. Vendor: IBM Corp.

OS/2 EE Database Manager Database management system that is part of OS/2 operating system and runs on IBM desktop systems. Vendor: IBM Corp.

OS/2 LAN Manager See LAN Manager.

OS/2 LAN Server See LAN Server.

OS/2 (Warp) Operating system for desktop computers. Runs on Pentium type desktop systems. Smaller version of OS/2. Includes Launchpad, Java support. Vendor: IBM PC Corp.

OS/2 (Warp) Connect Communications software. Peer-to-peer network operating system (NOS) connecting Windows, OS/2 systems. Vendor: IBM Corp.

OS/2 (Warp) Server Operating system including full NOS. Combines OS/2 Warp with LAN Server. Supports Windows clients. Vendor: IBM Corp.

OS 2200 Operating system for mainframe Unisys systems. Vendor: Unisys Corp.

OS/286,386 Operating system for IBM and compatible desktop computer systems. Enhancement to OS/3. Vendor: A.I. Architects, Inc.

OS/3 Operating system for mainframe Unisys systems. Vendor: Unisys Corp.

OS/32 Operating system for Concurrent systems. Vendor: Concurrent Computer Corp.

OS/32 MTM Application development system. Runs on Concurrent systems. Program development timesharing monitor. Vendor: Concurrent Computer Corp.

OS/390 Operating system for IBM mainframes. Based on MVS and includes Unix APIs for running Unix applications. Adds Java, Internet support. Includes Web application server and encryption features for e-commerce. Vendor: IBM Corp.

OS/390 Unix Systems Services Operating system for IBM S/390 computers. Runs Unix applications. Formerly called OpenEdition. Vendor: IBM Corp.

OS/400 Operating system for IBM AS/400 systems. Includes DBMS, and SQL, RPG, COBOL compilers. Vendor: IBM Corp.

OS/400 V3R1 See V3R1.

OS/8 Operating system for DEC midrange computers. Vendor: Digital Equipment Corp.

OS/9 Real-time operating system. Runs on Motorola, Intel, PowerPC processors. Vendor: Microware Systems, Inc.

OS Connection Communications software. Network connecting IBM, Unix, DEC VAX. Vendor: TRW, Inc.

OS/JCL See JCL.

OS/MP Operating system for Solbourne microcomputers. Vendor: Solbourne Computer, Inc.

OS/MVS See MVS.

OS/Open Real-time operating system for PowerPC systems. Vendor: IBM Corp.

OS/TSO See TSO.

OS/VS1 Operating system for mainframe IBM systems. This operating system is no longer supported by IBM. Vendor: IBM Corp.

OSA Open Solutions Architecture. Architecture from Novell that allows developers to create applications for NetWare and NDS that are Java compliant and Web enabled. Vendor: Novell, Inc.

OSAM Operating system enhancement used by systems programmers. Monitors and controls system performance in Unisys systems. Stands for: Online Systems Activity Monitor. Vendor: Unisys Corp.

OSF Open Software Foundation. A nonprofit vendor-supported research and development consortium working to establish standards for Unix systems. Supports and develops OSF/1 (Operating System), Motif (GUI), and DCE (communications environment). Merging with COSE to avoid duplicate development efforts.

OSF/1 General-purpose operating system for midrange systems. Version of Unix. Vendor: Open Software Foundation.

OSF DCE Application development tools. Used to develop, run, and support distributed applications. Runs on Unix systems. Stands for: OSF Distributed Computing Environment. Vendor: The Open Group.

OSF/Motif See Motif.

OSI 1. Open Systems Interconnect. Specifications set by ISO (International Standards Organization) to allow communications between products from different vendors. Most computer vendors agree to follow these standards.
2. Communications software. Network connecting NCR computers. Stands for: Office System Interface.

OSL Operating system enhancement used by systems programmers. Stands for: Optimization Subroutine Library. Increases system efficiency by increasing speed of executing number-crunching operations in IBM systems. Vendor: IBM Corp.

Osmium See Exchange.

OSx Operating system for Pyramid systems. Vendor: Pyramid Technology Corp.

OTS-1100 Operating system add-on. Security/auditing system that runs on Unisys systems. Vendor: Formula Consultants, Inc.

OURS Open User Recommended Solutions group. Organization dealing with problems of handling different vendor software in client/server environments. Group deals with personnel issues such as education.

Outlook (98) Groupware. Includes calendaring, scheduling, e-mail, and PIM capabilities. Purchased separately (Outlook 98) or included with Office 97 and Exchange. Vendor: Microsoft Corp.

outsourcing Turning over some, or even most, information system functions to outside contractors.

Ovation Desktop software. Used to create presentation graphics from Unix systems. Vendor: Visual Engineering, Inc.

overhead The amount of time system software (operating systems, database management systems, communications systems, etc.) uses the computer system.

Overlord Communications. Network management system. Used for full enterprise networking. Vendor: NetLabs, Inc.

OverLord See Dimons 3G.

OverQuota Application software. Automates sales functions. Includes customer relationship management and instant messaging which lets users locate and ask questions of experts at a company. Released: 1999. Vendor: Relavis Corp.

OverVUE Relational database for desktop computer environments. Runs on Macintosh systems. Vendor: ProVue Development Corp.

OWL Object oriented class library. Stands for: Object Windows Library. Vendor: Inprise Corp.

P/A See programmer/analyst.

P/OS Operating system for DEC systems. Stands for: Professional's Operating System.

P-Rade Relational database for desktop computer environments. Runs on IBM, Unix systems. Vendor: P-Stat, Inc.

P-Stat Relational database for desktop computer environments. Runs on IBM, Unix systems. Vendor: P-Stat, Inc.

P+WorkBench/ILR Application development tool. Tools and methodology that cover the development cycle, including project management, system delivery, system ownership, and prototyping. Runs on OS/2 systems. Vendor: DMR Group, Inc.

P/X See Project/2 P/X.

P3 (for Windows) Desktop software. Project management package. Allows multiple users. Supports OLE, which allows users to attach data from spreadsheets, word processors, etc. to project plans. Stands for Primavera Project Planner. Vendor: Primavera Systems, Inc.

P³P Platform for Privacy Preferences. Set of standards being developed by the World Wide Web Consortium. Defines what information about users Web sites can access.

P5-xxx Desktop computer. Pentium II CPU. Operating system: Windows 95/98. Vendor: Gateway 2000, Inc.

P6 Computer chip, or microprocessor. Considered to be a new generation of chip and twice as fast as Pentium chips. Designed to be used in multiprocessor server systems. used in Pentium II, Celeron, Xeon systems. Released: 1995. Vendor: Intel Corp.

P6D-200 Desktop computer. Pentium Pro CPU. Operating system: Windows 95/98. Vendor: Royal Computer Systems, Inc.

P-75CX Desktop computer. Pentium CPU. Operating system: Windows 95/98. Vendor: Data General Corp.

PA-80000 64-bit chip. Vendor: Hewlett-Packard Co.

PA-RISC RISC chip from Hewlett-Packard.

PaBLO Application development tool. OLAP software. Provides access to data stored in multidimensional and relational databases. Runs on Macintosh, Windows systems. Vendor: Hummingbird Communications, Ltd.

PAC Reverse Application development tool. Provides reverse engineering. Extracts the data model and stores data descriptions in the PacBase repository. Vendor: CGI Systems, Inc.

Pacbase CASE product. Automates analysis, design, and programming functions. Includes code generator for COBOL. Design methodologies supported: Information Engineering, Yourdon, Merise. Databases supported: DB2, Oracle, RDMS, SQL/DS. SQL/400, Teradata, Rdb, CA-Datacom. Runs on Bull HN, IBM, Unisys systems. Includes: Paclan, Paclan/X. Vendor: CGI Systems, Inc.

Pace, Pace for Open Systems Application development system. Includes program generator, 4GL. Runs on Wang systems. Pace for Open Systems is for general development and not tied to Wang platforms. Vendor: Wang Laboratories, Inc.

Pace RDBMS Relational database for large computer environments. Runs on Wang systems. Vendor: Wang Laboratories, Inc.

Pacebase Relational database for desktop computer environments. Runs on IBM systems. Vendor: Watcom Products, Inc.

Pacer Post Communications software. Gateway for e-mail systems. Runs on DEC VAX, Macintosh systems. Vendor: Pacer Software, Inc.

PacerForum Groupware. Supports electronic conferencing. Vendor: Pacer Software, Inc.

PacerLink Communications software. Network connecting DEC VAX, Unix, DG, Stratus, IBM desktop computers, and Apple Macintosh computers. Micro-to-mainframe link. Vendor: Pacer Software, Inc.

PacerPrint Operating system software. Allows printers to be shared in Unix/Macintosh systems. Vendor: Pacer Software, Inc.

PacerShare Operating system software. Allows Unix data storage volumes and directories to be accessed as if they were Macintosh disks. Vendor: Pacer Software, Inc.

Pack-Mate Desktop computer. Pentium processor. Vendor: Packard Bell NEC Inc.

Package/It Application development tool. Provides conversion, analysis, and database maintenance functions by consolidating database request modules into packages for DB2 databases. Vendor: PLATINUM Technology, Inc.

packet The data and control characters that are transmitted along communications lines. Information is sent in messages which, if long, are split into packets for better transmission.

packet switching In data communications, the breaking apart of long messages into smaller, fixed-length packets. Packets are then routed to the destination as independent pieces and may not travel the same route. The packets must be reassembled in the correct sequence at the receiving system.

PacketShaper Communications software. Manages network bandwidths. Released: 1998. Vendor: Packeteer, Inc.

Packit Operating system add-on. Data management software that runs on Unisys systems. Vendor: Software Clearing House, Inc.

PackRat Desktop system software. Personal Information Manager (PIM). Runs on Windows systems. Vendor: Polaris Software.

PACLAN Application development tool. Automates design and programming functions. Concentrates on sharing of information so developers can work together. Runs on OS/2, MVS, VSE, Windows systems. Vendor: CGI Systems, Inc.

PacReverse CASE product. Reverse-engineering tool. Vendor: CGI Systems, Inc.

PACS Plus Operating system enhancement used by systems programmers. Provides system cost accounting for DEC VAX systems. Stands for: Process Accounting and Chargeback System. Vendor: Signal Technology, Inc.

PadBase Application development tool. Used to develop pen-based applications. Vendor: GRiD Systems Corp.

PadPlus Pen-based computer. Vendor: Fujitsu America, Inc.

PADS Operating system add-on. Debugging/testing software that runs on Unisys systems. Stands for: Programming Advanced Debugging System. Vendor: Unisys Corp.

page locking See data locking.

PageMaker See Adobe PageMaker.

PageMill Application development tool. Web authoring tool. Automatically inserts HTML tags into text to create Web pages. Available for Macintosh, Windows. Vendor: Adobe Systems, Inc.

PageVault Operating system enhancement. Provides security for electronic documents. Can tailor security so that only certain users can read even paragraphs and images on pages. Keeps documents unreadable until a defined date and time. Ensures that confidential information cannot be printed. Released: 1998. Vendor: Authentica Security Technologies, Inc.

Pakmanager Operating system enhancement used by systems programmers. Increases system efficiency in DEC VAX systems. Manages disk storage space. Vendor: Demax Software, Inc.

PAL The programming language used with Paradox. Stands for: Paradox Application Language. Vendor: Inprise Corp.

PALcode Privileged Architecture Library. Set of routines in DEC's Alpha computers that allows multiple operating systems to run. Alpha systems run VMS, OSF/1, NT, and Unix operating systems. Vendor: Digital Equipment Corp.

Palm VII Handheld computer. Provides e-mail and limited Web access. Released: 1999. Vendor: Palm Computing (subsidiary of 3COM Corp.).

PalmBook Handheld computer. Vendor: ProLinear Corp.

PalmOS Operating system for handheld computers. Vendor: Palm Computing (subsidiary of 3COM Corp.).

PalmPilot 1000, 5000, Pilot Professional, Palm III Handheld, pen-based computer. Operating system: PalmOS. Vendor: : Palm Computing (subsidiary of 3COM Corp.).

Palomar Applications development environment used over the Internet. Users can build applications from Java, HTML, JavaScript. Uses JavaBeans which lets developers access OpenDoc and ActiveX components. Vendor: Netscape Communications Corp.

Panagon Capture Document management software. Captures data and stores it in FileNET repositories so it is accessible through Panagon software. Data can be faxed, scanned, captured from EDI documents and Web pages. Runs on Windows systems. Vendor: FileNET Corp.

Panagon IDM,2 Document management software. Used in client/server, Internet systems. Integrates document management, document imaging, and workflow functions. Allows users to view, distribute, share, manage and revise any document regardless of where it is stored in the system. Stands for: Panagon Integrated Document Management. Vendor: FileNET Corp.

Panagon Visual WorkFlo See Visual Workflo.

Panapt Application development tool. Automates moving programs from testing to production libraries. Stands for: Pansophic Automated Production Turnover. Vendor: Pansophic Systems, Inc.

Panaudit Plus Operating system add-on. Security/auditing system that runs on IBM systems. Vendor: Pansophic Systems, Inc.

Panlink Communications software. Network connecting IBM computers. Micro-to-mainframe link. Vendor: Pansophic Systems, Inc.

Panorama Application development tool. Allows end-users to run queries against data warehouses. Released: 1999. Vendor: The Great Elk Co.

Panther Desktop computer. Pentium Pro CPU. Operating system: Lynx. Vendor: *Neko*Tech Inc.

Panvalet See CA-Panvalet.

PAP Communications protocol. Session Layer. Part of AppleTalk. Stands for: Printer Access Protocol.

Paperback Writer Desktop system software. Word processor. Runs on Pentium type systems. Vendor: Paperback Software International.

PaperClip Document management software. Runs on Windows NT systems. Vendor: PaperClip Imaging Software, Inc.

PAR Operating system enhancement used by systems programmers. Monitors and controls system performance in Unisys systems. Stands for: Performance Analysis Routines. Vendor: Unisys Corp.

Paradigm Operating system enhancement. Automates help desk. Integrates with CA-Unicenter TNG. Vendor: Computer Associates International, Inc.

Paradigm Plus CASE product. Automates analysis and design. Follows HP Fusion object-oriented methodology. Vendor: Computer Associates International, Inc.

Paradigm/XP Workflow software. Used to expand help desk into customer service center. Vendor: Sterling Software.

Paradise Application development tools. Includes visual debugger, shared memory technology, performance monitoring. Runs on Unix, Windows NT systems. Vendor: Scientific Computing Associates, Inc.

Paradox Relational database for desktop computer environments. Runs on IBM desktop systems. Utilizes SQL. Can be used in client/server computing. Has its own programming language, PAL. Vendor: Corel Corp (licensed from Inprise/Borland).

Paradox Engine Application programming interface for C, C++. and Pascal development of programs accessing data in Paradox databases. Vendor: Inprise Corp.

Paragon XP/S MPP computer. 4000 processors. Operating system: OSF/1. Vendor: Intel Supercomputer Systems (Division of Intel Corp).

parallel computer A computer that has multiple processors and can have each processor work on part of a problem. Considered to be the supercomputer of the future; parallel computers are faster than standard supercomputers. Used mostly in scientific and engineering applications, but quickly being implemented in the business world. Parallel computers usually have a few processors; massively parallel processers can have thousands.

Parallel Enterprise Server Mainframe computer. Air-cooled machine. Also called System/390, SP1, SP2. Parallel processor; from 4 to 128 processors. Vendor: IBM Corp.

Parallel Inference MPP computer. 16,384 processors. Operating system: Proprietary. Vendor: Flavors Technology.

parallel processing See cluster.

parallel sysplex IBM's mainframe clustering technology that lets up to 32 System/390 machines act as a single mainframe. Also used by Amdahl, Hitachi.

Parallel Visual Explorer Application development tool. Visualization tool that can color-code groups or clusters of data and display trends and patterns as graphs. Used for analysis type queries and with data warehousing. Vendor: IBM Corp.

Paramid Midrange computer. RISC CPU. Operating system: Unix. Vendor: Transtech Parallel Systems Corp.

Paramount Operating systems enhancement for Windows systems. Resource manager that graphically displays systems information. Interfaces with NetSpy, LanSpy, Monitor products. Vendor: Legent Corp.

Parasight Application development tool. Automates testing functions. Analyzes program structure for applications that run on Encore 90 RISC machines. Vendor: Encore Computer Corp.

PARC Palo Alto Research Center. Xerox Corporation's research and development center.

ParcPlace Smalltalk See Smalltalk-80.

Parity DBSS Application development system. Runs on IBM systems. Used to develop DB2 applications. Stands for: Parity Data Base Support System. Vendor: Paragon Software International.

Park City Work in progress name for upgrade of NetWare due to be released in 1998. Vendor: Novell, Inc.

Parlance Document Manager Document retrieval system. Works with Web servers. Runs on Unix systems. Vendor: Xyvision, Inc.

Parlay Application Server Application server providing middleware and development tools. Used for Java development and integrates applications, transactions, data and messaging with JavaBean components. The JavaBean Suite includes JavaBeans for CICS, IMS, MQSeries, and EDA middleware that accesses more than 80 databases and ERPs. Runs on DEC, OS/390, OS/400, Unix, Windows NT systems. Vendor: Information Builders, Inc.

PARSDOS Operating system for IBM and compatible desktop computer systems. Vendor: Gulf Data, Inc.

parser Language translator. Reads data from XML documents and checks it for validity. Also called XML parser, XML processor.

partition Fixed division of computer storage.

partitioning See application partitioning.

Parts Application development tool. Client/server tool that creates applications from prefabricated software components. Stands for: Parts Assembly and Reuse Tool Set. Vendor: Digitalk, Inc.

Parts CICS Wrapper Application development tool. Helps to connect existing CICS applications to new client/server tasks. Vendor: Digitalk, Inc.

Parts Communications Wrapper for EHLLAPI Application development tool. Allows developers to add GUI front-ends to legacy mainframe systems. Stands for: Emulator High-Level Language Application Programming Interface. Vendor: Digitalk, Inc.

Parts Workbench Application development tool. Workbench software. Object-oriented system that allows developers to choose and connect "parts" from a catalog to build an application for OS/2, Windows. Vendor: Digitalk, Inc.

Parts Wrapper Application development tool. Bridge between ADW (CASE tool) and Parts Workbench (application development environment). Full name: Parts Wrapper for ADW. Vendor: Digitalk, Inc. and Knowledgeware, Inc.

PAS Operating system enhancement used by systems programmers. Monitors and controls system performance in IBM systems. Monitors system software and warns of potential problems. Stands for: Problem Alert System. Vendor: Morino Associates, Inc.

Pascal Compiler language. Language based on ALGOL. Mainly used in desktop computer environments.

Passport 1. Application development environment. Visual tool used for enterprise development. Provides dynamic application partitioning for client/server systems. Includes RAD tool. Accesses Oracle, Sybase, Ingres, Rdb. Generates C, SQL code. Runs on Windows systems. Vendor: InSync Software Corp.
2. See Carelton Passport.

Passport Interconnect Data management tool that allows programmers to access multiple databases from a single application. Allows access to Oracle, Sybase, Ingres, and Rdb. Vendor: Passport Corp.

Passport IntRprise Application development tool. Used to develop client/server and Internet applications that use Java front ends. A single set of code can be used on standard networks and the Internet. Runs on DEC, Unix, Windows systems. Released: 1998. Vendor: Passport Corp.

Pastran Programming development tool. Programming utility that translates Pascal to Ada. Vendor: R.R. Software, Inc.

Pathfinder Application development tool. Captures and analyzes system data for applications running under IBM's MVS. Vendor: A+ Software, Inc.

Pathfinder Pentium Desktop computer. Pentium processor. Vendor: ET Technology.

Pathmaker Application development tool. Program generator for online applications. Runs on Tandem systems. Vendor: Tandem Computers, Inc. (a Compaq company).

Pathtracker Operating system software. Keeps track of equipment on client/server systems. Vendor: Distributed Technologies, Inc.

PATHVU Application development tool. Automates programming function. Analyzes source code for such things as adherence to standards and presence of logic flaws (unexecuted code). Language analyzed: COBOL. Runs on IBM systems. Vendor: Compuware Corp.

PathWay Communications software. Network operating system connecting DEC, Sun, Macintosh, and IBM desktop computers. Supports client/server computing. Provides e-mail services. Vendor: The Wollongong Group, Inc.

Pathworks Communications software. Network providing micro-to-mainframe links. Based on LAN Manager, links LANs to VAX servers. Works in a client/server environment. Interfaces with VMS, OSF/1, Apple networks, NetWare, LAN Manager. Includes ManageWorks which provides management of NetWare and LAN Manager Networks. Vendor: Digital Equipment Corp.

Patriot II Desktop computer. Pentium processor. Vendor: USA Flex, Inc.

Patrol Communications. Network management software. Monitors and controls machines, databases, applications. Used with client/server systems. Runs on most workstations. Used on Internet to direct execution of CGI scripts across multiple servers. Supports MQSeries, Exchange. Works with CA-Unicenter, NetView, SunNet Manager, Polycenter. Vendor: BMC Software, Inc.

Patrol for Red Brick Warehouse Operating system software. Monitors Red Brick relational databases and detects errors from both software and hardware. Runs on Unix, Windows NT systems. Vendor: BMC Software, Inc.

Patrol Knowledge Module Communications software. OLTP, or communications control systems. Interfaces with MQSeries, NetWare, Tuxedo. Released: 1997. Vendor: BMC Software, Inc.

PATROLWATCH Internet management software. Monitors both intranets and Internet applications. Released: 1996. Vendor: BMC Software, Inc.

PATS Application development system. Runs on Unisys systems. Stands for: Programming And Testing System. Vendor: ESI.

patterns A pattern is literature, not software. It is the written explanation of a solution to a common recurring problem. The pattern defines the context of the problem, the problem itself and the solution. Different kinds of patterns include design patterns, process patterns, analysis patterns, architectural and organizational patterns.

Pattern Recognition Workbench Application development tool used for data mining. Used to determine which of hundreds of realtionships among variables are significant. Runs on Windows systems. Vendor: Unica Technologies Inc.

Pax-2 Res Operating system enhancement. Balances batch job execution across networks. Sends all job requests to a central server which matches requests to resources. Runs on DEC, HP, IBM RS/6000, Sun systems. Vendor: VXM Technologies, Inc.

PayBase 32 Portable computer. Vendor: Bottomline Technologies, Inc.

PBwin/(Standard) Application development tool. Used for business process modeling. Runs on Windows systems. Vendor: Logic Works, Inc.

PBM Operating system enhancement used by systems programmers. Controls communications networks in IBM systems. Monitors and controls banking machines. Vendor: IBM Corp.

PBoss Operating system enhancement used by systems programmers and operations staff. Job scheduler for Datapoint systems. Vendor: Telmi Systems, Inc.

PBX Telephone networking. An internal telephone switching system that now often handles terminals in addition to both analog and digital telephones. Stands for: Private Branch eXchange.

PC Personal computer. A small computer that includes a terminal, keyboard, and disk and/or diskette storage. Designed primarily as stand-alone machines, but can be connected. Also called micro, microcomputer, laptop computer, and desktop computer. Single-user system, although PCs can be networked.

PC/204 Communications software. Network connecting IBM systems. Micro-to-mainframe link. Vendor: Computer Corp. of America.

PC 300,700 Desktop computer. Pentium CPUs. Operating system: Windows 95/98. Vendor: IBM Corp.

PC 300GL Desktop computer. Pentium II CPU. Operating system: Windows. Bundled with Microsoft Office and Lotus SmartSuite. Released:1999. Vendor: IBM Corp.

PC-30x0, 90x0, 9300 Notebook computer. Pentium CPU. Operating system: Windows 95/98. Vendor: Sharp Electronics, Inc.

PC-BLIS Operating system for IBM and compatible desktop computer systems. Vendor: Information Processing, Inc.

PC Card Solid state storage device. Used in most portable computers.

PC Companion Handheld computer. Vendor: Compaq Computer Corp.

PC Connect Communications software. Network connecting IBM systems. Micro-to-midrange link. Vendor: DecisionLink, Inc.

PC Contact Communications software. Network connecting IBM systems. Micro-to-mainframe link. Vendor: Cincom Systems, Inc.

PC Data Base Database for desktop computer environments. Runs on IBM desktop systems. Vendor: Wang Laboratories, Inc.

PC Docs Open Document management software. Provides keyword and phrase searching. Runs on Windows systems. Vendor: PC Docs, Inc.

PC-DOS Operating system for desktop systems. Includes some pen support. Similar to MS-DOS. Vendor: IBM Corp.

PC/Foccalc Desktop system software. Spreadsheet. Runs on Pentium type systems. Add on to PC/Focus. Vendor: Information Builders, Inc.

PC/Focus Relational database for desktop computer environments. Runs on IBM desktop systems. 4GL capabilities. Vendor: Information Builders, Inc.

PC/Host Autoware Communications software. Network connecting desktop computers and mainframes. Vendor: Attachmate Corporation.

PC INFO Relational database for desktop computer environments. Runs on IBM desktop systems. Vendor: Henco Software, Inc.

PC/IX Operating system for IBM desktop computer systems. Vendor: IBM Corp.

PC Lan Communications software. Network operating system. Network connecting IBM desktop computers. Vendor: IBM Corp.

PC Link Communications software. Network connecting IBM systems. Micro-to-mainframe link. Vendor: McCormack & Dodge Corp.

PC-LINT See Lint.

PC Mantis See Mantis.

PC-Metric Application development tool. Automates programming function. Analyzes source code for such things as adherence to standards and presence of logic flaws (unexecuted code). Languages analyzed: ADA, assembler, C, C++, COBOL, Fortran, Lisp, Pascal, PL/1. Other versions: UX-Metric, VX-Metric. Runs on Pentium type desktop computers, DEC desktop computers. Vendor: Set Laboratories, Inc.

PC-MOS/386 Operating system for IBM and compatible desktop computer systems. Vendor: The Software Link, Inc.

PC-NFS Communications software. Network connecting microcomputers to Unix workstations, IBM mainframes, and DEC VMS systems. Vendor: Sun Microsystems, Inc.

PC NOMAD Relational database for desktop computer environments. Runs on IBM desktop systems. 4GL capabilities. Vendor: MUST Software International.

PC/NOS Operating system for IBM and compatible desktop computer systems. Vendor: Corvus Systems, Inc.

PC Paintbrush Desktop system software. Graphics package. Runs on MS-DOS systems. Vendor: Zsoft Corp.

PC programmer, PC software specialist Application developer. Works with word processors, spreadsheets, PC databases, PC programming languages. Used for departmental, small company software development. Can be any experience level.

PC/RTX See RTX.

PC Server Midrange computer. Pentium Pro, II CPU. Operating systems: OS/2, Unix, Windows NT. Vendor: IBM Corp.

PC/TCP OnNet Communications. Network management software. Runs on Windows systems. Vendor: FTP Software, Inc.

PC technician Support personnel. Coordinates, controls, and maintains the personal computers within a company. Installs new hardware and upgrades. Often part of a help desk staff. Can be any experience level.

PC/VRTX Operating system for IBM and compatible desktop computer systems. Vendor: DYAD Technology Corp.

PC-Write Desktop system software. Word processor. Runs on Pentium type systems. Vendor: Quicksoft.

PC-X Software that connects desktop computer users with mainframe and Unix applications. Produced by different vendors.

PC97 Desktop computer. Microcomputer for individual use. Vendor: Microsoft Corp.

PCAnywhere Data communications software. Allows remote access control of desktop computers. Supports TCP/IP. Vendor: Symantec Corp.

PCDB Database for desktop computer environments. Runs on IBM desktop systems. Vendor: Wang Laboratories, Inc.

PCF Operating system add-on. Data management software that runs on IBM systems. Stands for: Pool Configuration Facility. Vendor: Chicago Soft, Ltd.

PCI Lan Workstation Desktop computer. Pentium CPU. Operating system: Windows 95/98. Vendor: Compu-tek International.

PCI ProServer Midrange computer. Server. Pentium CPU. Vendor: Netis Technology, Inc.

PCIOS Operating system add-on. Data management software that runs on Unisys systems. Vendor: Unisys Corp.

PCmainframe Communications software. Network connecting IBM computers. Micro-to-mainframe link for CICS environments. Vendor: CFsoftware, Inc.

PCMCIA Desktop computer card that contains hard disk drives, modems, or additional storage.

PCMS Application development tool. Software process management system. Tracks all components of a software project, including versions of code under development. Stands for: Process Configuration Management Software. Vendor: SQL Software, Ltd.

PCnet Communications. LAN (Local Area Network) connecting desktop computers. Vendor: Orchid Technology, Inc.

PCNX Operating system for Pentium-type desktop computers, Unix systems. Vendor: Wendin, Inc.

PCS System management software. Provides system performance data collection, performance analysis, workload scheduling, capacity planning, statistical analysis. Runs on HP, IBM, Sun, Unix systems. Stands for: Performance Collection Software. Vendor: Hewlett-Packard Co.

PCTE See Emeraude PCTE.

PCTE Applications development environment. Integrates use of diverse CASE tools. Incorporates ToolTalk. Stands for: Portable Common Tool Environment. Vendor: Sun Microsystems, Inc. and Emeraude.

PCVMS Operating system for IBM and compatible desktop computer systems. Vendor: Wendin, Inc.

PCworks Communications software. Network connecting IBM, Unix, DEC VAX. Micro-to-mainframe link. Vendor: Touch-Stone Software Corp.

PCX12 Communications software. EDI package. Runs on Pentium type desktop computers. Vendor: Advanced Communications Systems.

PDA 1. Personal Digital Assistant. Name for light-weight (1 pound) task specific computers. These machines are pen-based, offer voice recognition, fax and modem communication, and include a pager. Typical use: write in "lunch—John" and the system will send John a fax and enter the lunch date in the user's appointment book. Also called handheld computers.
2. Operating system enhancement used by systems programmers. Monitors and controls system performance in IBM DB2 systems. Stands for: PLATINUM Database Analyzer. Vendor: PLATINUM Technology, Inc.

PDB Operating system enhancement used by systems programmers. Monitors and controls system performance in IBM systems. Stands for: Planning Data Base. Vendor: Boole & Babbage, Inc.

PDF See ISPF/PDF

PDM Application software. Part of ERP systems. Automates tracking and updating of information about products, such as design specifications, parts and manufacturing records, etc. Stands for: Product Data Management.

PDO Application development tool. Server-based object model and messaging functions. Allows objects to send and receive messages regardless of whether they reside in the same application or in the same computer. Stands for: Portable Distributed Objects. Part of NextStep. Vendor: Apple Computer, Inc.

PDP Old Midrange computer systems. Operating systems: Unix, RSTS, RSX, DSM, ULTRIX. Vendor: Digital Equipment Corp.

PDQ 1. Data management software used in parallel processing. Splits complex queries into several parts to run simultaneously across multiple processors. Stands for: Parallel Data Query. Vendor: Informix Software, Inc. and Sequent Computer Systems.
2. 4GL used in mainframe environments. Stands for Personal Data Query. Used in Bull HN systems. Vendor: Bull HN Information Systems, Inc.

PdQ Handheld computer/smart phone. A device which combines a cellular digital telephone and a PDA. Provides voice and data transmission, wireless e-mail access and a personal organizer. Released: 1999. Vendor: Qualcomm Inc.

PDS Software vendor. Products: HR, payroll systems for midsize environments.

PDS/400 Application development system. Used to develop applications for AS/400 systems. Vendor: Applied Logic Corp.

PDS-C Source Generator Application development tool. Translates programs written in 4GL to C. Runs on Unix systems. Vendor: Parameter Driven Software, Inc.

PDSFAST Operating system enhancement used by systems programmers. Increases system efficiency in IBM systems. Improves speed of IBM data utility IEB-COPY. Vendor: Software Engineering of America, Inc.

PDSMan/MVS Operating system add-on. Data management software that runs on IBM systems. Vendor: Goal Systems International, Inc.

Pdstools Application development tool. Manages IBM mainframe datasets. Includes full screen edit and browse. Vendor: Serena International.

PDSUPDATE Application development tool. JCL and/or SYSOUT maintenance program for IBM systems. Vendor: Software Engineering of America, Inc.

PE See Process Engineer.

pedabyte One thousand terabytes (Terabyte = one trillion bytes).

Peer Planner Application development tool. Statistical forecasting tool. Allows user to make predictions based on past patterns. Runs on Windows systems. Vendor: Delphus, Inc.

peer-to-peer network operating system Communications software. Network operating system that allows systems on a LAN to share files, printers, and applications directly without going through a server. Provides centralized network management, user administration, and security functions. Used mostly with small (2-5 system) LANS.

PeerNet Communications software. EDI package. Runs on IBM mainframe systems. Vendor: SDM International, Inc.

PEF Operating system enhancement used by systems programmers. Increases system efficiency in Unisys systems. Stands for: Performance Enhancement Feature. Vendor: Unisys Corp.

Pegasus Operating system software. Measures performance in client/server systems by simulating user activity to measure response time. Measures available bandwidths. Vendor: Ganymede Software, Inc.

Pegasus Application Monitor System management software. Monitors performance of packaged and custom written software over corporate networks. Released: 1999. Vendor: Ganymede Software, Inc.

Pegasus Network Monitor System management software. Monitors performance of client/server networks. Includes a data analysis engine and performs trend analysis. Reports, scheduled at user-defined intervals, can help to pinpoint the source of performance problems. Vendor: Ganymede Software, Inc.

Pen Programming language. Scripting language designed to deal with free-form input. Used to create CGI scripts.

pen-based computer Handheld computer that accepts written input with a pen, or stylus. Can also be connected to a keyboard and to larger machines. Includes PDAs.

Pen Developer Assistant Application development tool. Used to develop applications for pen computers. Vendor: IBM Corp.

Pen for OS/2 Operating system enhancement. Provides pen support by including handwriting recognition and development tools for pen applications. Runs on OS/2 systems. Vendor: IBM Corp.

PEN/KEY Handheld, Pen-based computer. Vendor: Norand Corp.

Pendant TFT Desktop computer. Pentium CPU. Has flat-panel display screens which can rotate 90 degrees to provide a choice of landscape or portrait displays. Released: 1998. Vendor: Tangent Computer, Inc.

PENnet PC,Plus,X25PI Communications software. Network connecting Concurrent computers. Micro-to-midrange link. Vendor: Concurrent Computer Corp.

PennyLAN Communications software. Network operating system which allows desktop computers on a LAN to share peripherals. Vendor: American Research Corp.

PenPoint Operating System for pen computers. Vendor: Go Corp.

Pentium, Pentium II, Pentium Pro Family of computer chips, or microprocessors. 64-bit processor. At least twice as fast as 486 systems. Operating systems: DOS/Windows, Windows 95/98, Windows NT, OS/2, Unix. Vendor: Intel Corp.

Pentium III Computer chip, or microprocessor. Computer speeds range from 500 MHz to 900 MHz and are expected to reach 1GHZ. Released: 1999. Vendor: Intel Corp.

Pentium MMX See MMX.

Pentium Server Desktop computer. Pentium CPU. Operating system: Windows 95/98. Vendor: Compu-tek International.

PenView Pen-based computer. Vendor: Norand Corp.

PeopleSoft Software vendor. Produces ERP systems. Client/server human resource, payroll, financial, manufacturing, and supply chain systems for IBM, DEC, HP systems. PeopleSoft version7 uses three tiered architecture.

PeopleTools Application development tools. Work-flow application which includes e-mail and imaging support. Interfaces with SQLBase, SQL Server. Vendor: PeopleSoft, Inc.

PepperSeed Application development tool. Provides object-oriented analysis and design following Booch methodology. Generates C++ code. Runs on Macintosh, OS/2, Unix, Windows systems. Vendor: Cayenne Software, Inc.

Peregrine Network Communications. Network management software. Controls networks in IBM systems. Full name: Peregrine Network Management System 3. Vendor: Peregrine Systems, Inc.

Perfectdisk Operating system enhancement. Increases system efficiency in DEC VAX systems. Controls disk space to keep fragmentation down. Vendor: Raxco Software, Inc.

PerfectOffice Desktop software suite. Three versions are available, PerfectOffice Standard, PerfectOffice Professional, and a version on CD-ROM, PerfectOffice Select. Includes 6 basic applications plus Paradox and Visual Application Builder. Includes a scripting language (PerfectScript). Vendor: Novell, Inc.

Perfectsolution Document manager that works on LANs. Can find text files across the network. Vendor: Softsolutions, Inc.

Performa xxx Midrange computer. RISC PowerPC system, also 680X0 CPU models. Operating system: System 7. Vendor: Apple Computer, Inc.

Performance Architek Performance prediction software. Creates predictive models from design specifications. Allows companies to evaluate performance under DOS, Unix and OS/2 and to locate bottlenecks in applications systems. Vendor: Windtunnel Software, Inc.

Performance EDI Communications software. EDI. Runs on AS/400, Unix, Windows systems. Released: 1996. Vendor: Retail Vendor Software.

Performance Estimator Application development tool. Analyzes IBM DB2 applications during the design phase to identify performance problems. Runs on Windows systems. Vendor: PLATINUM Technology, Inc.

Performance xxx Desktop computer. Pentium CPU. Operating system: Windows 95/98. Vendor: Zenith Data Systems Direct.

PerformanceStudio Operating system software. Measures performance of client/server and Web systems. Generates performance tests. Runs on Unix, Windows systems. Vendor: Rational Software Corp.

PerformanceWorks Systems management software. Application management system used to predict performance bottlenecks. Released: 1996. Vendor: Landmark Systems Corp.

Performer Application development environment. Visual system used with client/server, Internet development. Used for enterprise development. Released: 1996. Vendor: Texas Instruments, Inc.

Performix Application development tool. Automates testing of Web-based applications. Vendor: Rational Software Corp.

PerfStat Operating system enhancement. Gathers and interprets performance data on Cray systems. Vendor: Instrumental, Inc.

peripheral device Any hardware device except the computer itself. Also called peripheral unit.

Perl Programming language. Scripting language used in Unix systems. Combines elements of C language and Unix shell commands. Stands for: Practical Extraction and Report Language. Popular in creating CGI scripts on the Internet.

Persistence Application development tool. Builds object-oriented applications that access data in relational databases. Generates C++, SQL code. Vendor: Persistence Software, Inc.

Persistence PowerTier See PowerTier.

persistent data Permanent data, or data that remains after a process is concluded.

Person to Person EMS (Electronic Meeting System). Vendor: IBM Corp.

personal communication system Cellular telephone and data link that combines voice and data communications, pen technology, computing, and functions such as calendaring and scheduling in a tablet size device.

Personal Consultant (Plus) Artificial intelligence system. Expert system building tool. Runs on desktop systems. Vendor: Texas Instruments, Inc.

personal digital assistant See PDA.

Personal Netware Communications software. Peer-to-peer network operating system. Interfaces with NetWare. Vendor: Novell, Inc.

Personal Oracle Lite Relational database. Version of Oracle7 for laptops. Runs on MacOS, Windows systems. Vendor: Oracle Corp.

Personal Partner See FH-2000.

Personal VM Operating system for IBM and compatible desktop computer systems. Desktop computer version of VM. Vendor: VM Labs, Inc.

PersonalJava Application development environment. Allows developers to build Java applets that run on a wide range of devices including game consoles, handheld computers and smart phones. Vendor: Sun Microsystems, Inc.

PersonalSQL Application development tool. Reporting tool. Part of ObjectView. Vendor: KnowledgeWare, Inc.

Persuasion Desktop system software. Graphics package. Used for presentation graphics. Vendor: Aldus Corp.

Pertmaster Desktop system software. Project management package. Runs on Pentium type systems. Versions include: Pertmaster Advance, Pertmaster 1500, Pertmaster 2500. Vendor: Pertmaster International, Inc.

Pervasive Solutions Technical framework for Extended Enterprise strategy. Technology includes using 64-bit computers, Unix type operating systems. Vendor: Hewlett-Packard Co.

Perwill*EDI Communications software. EDI package. Runs on DEC, Unisys, Unix systems. Vendor: Perwill EDI, Inc.

PES See Parallel Enterprise Servers.

Petra X-D Midsize computer. Server. Pentium II processor. Operating system: IRIX. Vendor: Tangent Computer, Inc.

pF/s Real-time operating system for desktop computers. Provides extremely fast response. Vendor: Forth, Inc.

Pf/x Operating system for Harris computers. Vendor: Forth, Inc.

PFS:First Choice Desktop system software. Integrated package. Runs on Pentium type systems. Vendor: Software Publishing Corp.

PFS:Professional File Database for desktop computer environments. Runs on IBM desktop systems. Vendor: Software Publishing Corp.

PFS:Professional Plan Desktop system software. Spreadsheet. Runs on Pentium type systems. Lotus lookalike. Vendor: Software Publishing Corp.

PFS:Professional Write Desktop system software. Word processor. Runs on Pentium type desktop computers. Vendor: Software Publishing Corp.

PFS:Window-works Desktop system software. Word processor. Runs on windows on IBM and compatible desktop computers. Vendor: Spinnaker Software.

PGP Encryption methodology. 128-bit key encryption methodology. Uses two keys, both public and secret. Stands for: Pretty Good Privacy.

PGP VPN Communications. Internet software. Firewall. Vendor: Network Associates, Inc.

PHASOR Application development tool. JCL and/or SYSOUT maintenance program for IBM systems. Vendor: Tone Software Corp.

Phenom Handheld computer. Operating system: Windows CE. Full name: Phenom Handheld PC Ultra. Vendor: LG Electronics.

PHD Relational database for desktop computer environments. Runs on IBM desktop systems. Full name: PHD C Relational Data Base. Vendor: Micro Business Applications, Inc.

PHIGS Programmer's Hierarchical Interactive Graphics System. Standard for graphics used in robotics, CAD, and other engineering applications.

PHIPS See CA-PHIPS.

PHLEX Application development tool. Program generator for IBM systems. Vendor: Cognos, Inc.

Photoshop See Adobe Photoshop.

PhotoStyler Image editing software. Runs on MS-DOS, Windows systems. Vendor: Aldus Corp.

physical layer In data communications, the OSI layer that defines the most basic communications functions; the handshake signals.

physical processing Term used for any processing that is hardware dependent. Deals with problem solutions that are specific to a certain computer and/or data storage device. Usually associated with systems programmers. Contrast with logical processing.

PI-7000,9000 Pen-based computer. Vendor: Sharp Electronics, Inc.

PI Systems Computer vendor. Manufactures pen-based computers.

PIC-1000, 2000 Handheld, pen-based computer. Operating system: Magic Cap. Vendor: Sony Electronics, Inc.

pICAsso See MetaFrame.

PICCL Parallel Implementation of Concurrent Common Lisp. Allows for parallel implementation of Lisp applications. Vendor: Top Level, Inc.

Piccolo Communications. Message-oriented middleware that connects application components regardless of the platform or the network protocol used. Runs on Tandem systems. Vendor: Cornerstone Software.

PICK General-purpose operating system/DBMS for midrange and desktop computer systems. Runs on Unix systems as a DBMS, and this is becoming its main function. Vendor: Pick Systems.

PIE/CICS,IMS,TSO,VM Operating system enhancement used by systems programmers. Controls communications networks in IBM systems. Vendor: Technology Software Concepts, Inc.

Pike's Peak Desktop computer. Pentium CPU. Operating systems: Windows 95/98, NT. Vendor: Aspen Systems, Inc.

Pilot Operating system enhancement used by systems programmers. Job scheduler for IBM AS/400, System/38. Vendor: Advanced Systems Concepts, Inc.

PILOT Operating system enhancement used by systems programmers. Monitors and controls system performance in IBM systems. Capacity planning and performance tuning system. Includes PILOT/CICS, PILOT/MVS, PILOT/SMF, PILOT/VM. Vendor: Axios Products, Inc.

Pilot Decision Support Suite Application development tool. OLAP software. Data mining tool. Creates multidimensional databases. Vendor: Pilot Software, Inc.

Pilot Internet Publisher OLAP software. Uses Internet architectures. Interfaces with Java, ActiveX. Part of Pilot Decision Support Suite. Released: 1996. Vendor: Pilot Software, Inc.

Pilot Software Software vendor. Products include EIS, OLAP, data warehousing software.

PIM 1. Personal Information Manager. Desktop computer software that assists individuals in managing personal schedules and projects. Keeps track of, and organizes, data from many sources.
2. See Parallel Inference Machine.

Pinnacle System software. Monitors performance of Lotus Notes applications. Released: 1998. Vendor: Candle Corp.

Pinpoint Application development tool. Analyzes existing programs to determine what applications need restructuring to fit into a new system. Vendor: Language Technology, Inc.

PIP Operating system add-on. Data management software that runs on DEC systems. Transfers data files from one peripheral device to another. Stands for: Peripheral Interchange Program. Vendor: Digital Equipment Corp.

Pipes Communications software. Message-oriented middleware. Allows information to be transmitted between different environments. Vendor: PeerLogic, Inc.

Pipes Platform Communications software. Middleware. Finds the best path between two applications running on diverse systems. Vendor: PeerLogic, Inc.

Pippin Operating system. Version of Macintosh OS to be used with a network computer. Vendor: Apple Computer, Inc.

Pixie Desktop system software. Graphics package. Interfaces with Lotus 1-2-3, Excel. Runs on Windows systems. Vendor: Zenographics, Inc.

PKEnable Operating system software. Provides security functions. Makes secure transactions less costly, faster to implement, and more manageable across multiple applications and public key infrastructure (PKI) standards. Links enterprise applications to digital certificate products and services from all the leading digital certificate authorities. Stands for: Public Key enable. Released: 1999. Vendor: SHYM Technology Inc.

PL/1 Compiler language used in large computer environments. Stands for Programming Language One. Language developed by IBM to handle both business and scientific applications.

PL/M Compiler language used in desktop computer environments.

PL/SQL 4GL. Extension to SQL. Runs with Oracle. Vendor: Oracle Corp.

Plan 9 Operating system for desktop computers. Unix-type system. Vendor: Lucent Technologies, Inc.

Plan Manager Application development tool. Program that allows DB2 users to define application plans across multiple DB2 systems. Part of DB Excel. Vendor: Reltech Products, Inc.

Planmacs See CA-Planmacs.

PlanMaster Desktop system software. Project management package. Runs on Pentium type systems. Vendor: Engineering Software Co.

Planner's Choice Desktop system software. Spreadsheet. Runs on Apple II systems. Vendor: Activision, Inc.

PlanPerfect Desktop system software. Spreadsheet. Runs on Pentium type systems. Vendor: WordPerfect Corp.

Plantrac Desktop system software. Project management package. Runs on Pentium type desktop computers, RS/6000, Sun, Apollo, Unix. Vendor: Computerline, Inc.

platform The combination of a computer and an operating system. All software is written for a platform; if either the computer or operating system is changed, a new version of the program must be written. For applications (business) programs often the term platform is used synonymously with operating system.

Platform for Privacy Preferences See P³P.

Platinum Code name for the latest release (4th) of Exchange Server. Integrates with Windows 2000 and Office 2000. Creates a single Web source for documents and applications such as word processors and spreadsheets. Release date: 2000. Vendor: Microsoft Corp.

PLATINUM Software family that provides system enhancements to IBM DB2 systems. Programs include: PLATINUM Database Analyzer, PLATINUM Report Facility, PLATINUM Plan Analyzer, PLATINUM Data Compressor, PLATINUM Detector, PLATINUM Rapid Reorg, and PLATINUM Pipeline for DB2. Vendor: PLATINUM Technology, Inc.

PLATINUM AutoAction Operating system enhancement. Monitors performance of internal and external application systems, including SAP R/3, AutoSys and NetArchive. Merges messages from multiple systems to a single display. Runs on DEC, DG, IBM mainframe, OS/400, Unix, Windows NT systems. Released: 1997. Vendor: PLATINUM Technology, Inc.

PLATINUM Compile/PRF Application development tool. Program, or code generator. Generates COBOL code from PLATINUM report facility or from SQL. Runs on IBM mainframes. Vendor: PLATINUM Technology, Inc.

Platinum ERA Application software. ERP suite to link sales and marketing, customer service, manufacturing, distribution and financials. Stands for: Platinum Enterprise Ready Applications. Released: 1998. Vendor: Platinum Software Corp.

PLATINUM Integrator Application development tool. Allows access to multliple legacy mainframe data systems (IMS, CICS, DB2) from client/server applications without rewriting code. Vendor: PLATINUM Technology, Inc.

PLATINUM Package/IT Operating System enhancement used in IBM DB2 systems. Used with conversions. Vendor: PLATINUM Technology, Inc.

Platinum ProVision See ProVision.

PLATINUM Report Facility Application development tool. Query and report functions. Interfaces with multiple databases, multiple platforms. Runs on MVS systems. Vendor: PLATINUM Technology, Inc.

PLATINUM Repository Application development tool used in data warehousing. Manages, maintains and provides access to enterprise data, applications and systems in a heterogeneous environment. Runs on MVS, OS/2. Vendor: PLATINUM Technology, Inc.

Platinum Series Financial software. Includes general ledger, accounts payable and receivable, currency manager, inventory, payroll, manufacturing, and banking software. Runs on Pentium type desktops. Vendor: Platinum Software Corporation.

Platinum SQL NT Application software. Financial systems. Vendor: Platinum Software Corporation.

Platinum Supreme, xxxx Desktop computer. Pentium processor. Vendor: Packard Bell NEC Inc.

Plato 1. Application development tool. OLAP software. Includes multidimensional database and provides access to data stored in multidimensional and relational databases. Released: 1998. Runs on Windows systems. Vendor: Microsoft Corp. 2. Development system used for creating CAI systems. Stands for: Programmed Logic for Automatic Teaching Operations. Runs on CDC systems. Vendor: Control Data Corp.

Playback See *QA*Center.

PLF Application development tool. JCL and/or SYSOUT maintenance program for IBM systems. Stands for: Procedure Library Facility. Vendor: Software Pursuits, Inc.

PlotTrak Desktop system software. Project management package. Runs on DEC VAX, IBM, Unix systems. Vendor: SofTrak Systems, Inc.

PLP Communications protocol. Network Layer. Part of X.25. Stands for: Packet Layer Protocol.

plug-and-play 1. Hardware specifications that define capability standards for desktop computers and peripherals and specifications for software that will automatically recognize what hardware is in a system and configure the equipment to work together. The software will be included with operating systems; OS/2 and Windows 95/98 both support plug-and-play.
2. In object-oriented technology, storing objects in an open, distributed environment that allows them to be used in any application.

plug-in Program added to a Web browser. Common plug-ins are to enable execution of programs providing sound and animation.

Plumtree Corporate Portal EIP software, corporate portal. Includes Portal Gadgets which are links to databases. Companies can develop personalized Portal Gadgets. Accesses Lotus Notes databases, DB2 databases, Word and Excel documents. In addition to standard internal access has established partnerships with external data sources to provide information on such things as weather and stock prices. Runs on Windows NT. Released: 1998. Vendor: Plumtree Software, Inc.

PM/Focus Application development tool. Used to develop GUI applications. Interfaces with EDA/SQL. Application development tool. Allows access to non-relational mainframe databases from desktop systems. Vendor: Information Builders, Inc.

PM/SS Application development tool. Program that analyzes the impact of implementing system changes, analyzes and documents programs and helps enforce programming and data-naming standards. Interfaces with IMS and IDMS. Vendor: Adpac Corp.

PMAT Application development tool. Automates programming and design functions. Vendor: XA Systems Corp.

PMB/MVX Operating system enhancement used by operations and systems programmers. Automatically creates a report of problems and transmits it to the service vendor. Keeps records of problems and reports service calls. Runs on IBM systems. Stands for: Problem Management Bridge. Vendor: IBM Corp.

PMD Operating system add-on. Debugging/testing software that runs on Harris systems. Stands for: Post Mortem Dump. Vendor: Harris Corp.

PMF Operating system enhancement used by systems programmers. Increases system efficiency in IBM systems. DASD storage manager. Vendor: Software Engineering of America, Inc.

PMF/VM Operating system enhancement used by systems programmers. Increases system efficiency in IBM systems. Stands for: Print Management Facility/Virtual Machine. Print spooler. Vendor: IBM Corp.

PMS-II Desktop system software. Project management package. Runs on Pentium type systems. Stands for: Project Management System—II. Vendor: North America MICA, Inc.

PMS80 Desktop system software. Project management package. Runs on Pentium type desktop computers, Unix. Vendor: Pinnell Engineering, Inc.

PNMS Communications. Network management system. Manages SNA, TCP/IP, Novell networks. Runs on MVS, Unix systems. Vendor: Peregrine Systems, Inc.

POEMS System management software. Set of tools that provide enterprise-wide system management functions. Used for data warehousing. Stands for: PLATINUM's Open Enterprise Management System. Includes: UniVision, NetArchive, SAFE, Xfer, Apriori, AutoSys/Altal, and Enterprise Console. Vendor: PLATINUM Technology, Inc.

POET Object-oriented database for C++ systems. Vendor: POET Software Corp.

POET Universal Object Server Object database. Provides system services, security and data integrity functions and GUI-based workbench for application development. Is both language and platform independent and includes SDKs (Software Development Kits) for Java and C++. Released: 1997. Vendor: POET Software Corp.

Point to Point Tunneling Protocol See PPTP.

PointCast Free service over the Internet that sends almost up-to-the-minute news information. Used push technology to push corporate data on intranets.

Poise DMS Relational database for large computer environments. Desktop computer version available. Runs on Pentium type desktop computer systems. Vendor: Campus America/POISE.

Polaris Desktop computer. Pentium processor. Vendor: ProGen Systems.

Polaris Pentium Desktop computer. Pentium processor. Vendor: ET Technology.

policy based networking Management system for networking that follows user defined policies such as limiting use of the network to the operational departments during prime work hours, or giving the sales applications top priority for bandwidth. Allows users to set priorities by defining policies.

Policy-based security management Security system based on user defines lists of accesses and permisssions.

polling Determining whether devices in a network are ready to transmit by checking each device in a predetermined order.

Poly xxx Midrange computer. Server. ALPHA CPU. Vendor: Polywell Computers, Inc.

PolyAlpha, Vision Desktop computer. Alpha processor. Vendor: Polywell Computers, Inc.

Polycenter Manager Communications. Network management software. Manages a multivendor data center. Vendor: Digital Equipment Corp.

Polycorder Handheld computer. Vendor: Omnidata International, Inc.

polymorphism Term used with object-oriented development. The principle that the same message can be sent to different objects which may respond in different ways.

Pool-DASD Operating system enhancement used by systems programmers. Increases system efficiency in IBM systems. Manages disk storage space. Vendor: Empact Software.

POP³ Communications protocol. Used for e-mail. Provides standards for offline e-mail, that is storing mail on servers and then having each user download the mail from the servers. The mail is deleted from the server when it is downloaded.

portable The principle that software must be movable at source code level among computers from different vendors and of different architectures. Part of open systems.

portable computer Desktop computer. Typically weighs between 20 and 30 pounds and is used for field operations such as surveying, geographic surveys.

Portable Distributed Objects See PDO.

Portable Netware See NetWare.

Portable Operating System Interface for Unix See POSIX.

Portable OS/2 New version of OS/2 that is based on Mach Micro-Kernel and will run on both microcomputers and workstations (Intel and RISC machines). Competitive with Windows NT. RISC, PowerPC versions are also called StarBase. Vendor: IBM Corp.

Portage Application development tool. Converts Unix application to run under Windows NT. Vendor: Consensys, Inc.

portal High level web site that allow browsers a one-stop location to start Web searches. Called gateway sites, as they provide a gateway to other sites. There are various kinds of portals. AOL, Excite, Netcenter (Netscape) and Yahoo are general interest portals. Other portals have been built around specific interests such as sports, women's issues, the stock market, etc. Companies are building their own portals called EIPs as gateways to their own applications, documents, reports, etc. as well as Internet sites.

Portal Gadgets See Plumtree Corporate Portal.

Ported NetWare Communications software. Network operating system (NOS) connecting Unix systems. Portable NetWare for Unix. Vendor: Sun Microsystems, Inc.

Portege 3010 Laptop computer. Sub-compact, under three pounds. Pentium processor. Vendor: Toshiba America Information Systems, Inc.

Portege T3400 Notebook computer. Vendor: Toshiba America Information Systems, Inc.

Portfolio Handheld computer. Vendor: Atari.

Portos Real-time operating system for desktop computers. Supports Java, OLE. Vendor: Oberon Microsystems, Inc.

Portus Communications, Internet software. Firewall. Vendor: Livermore Software Laboratories International.

POS Operating system for Apple II desktop computer systems. Full name: Piaser's Operating system. Vendor: The Mike Piaser Co.

Pose CASE product. Automates analysis, design, and programming functions. Includes code generator for COBOL. Design methodologies supported: Gane-Sarson, Yourdon. Databases supported: DB2, SQL/DS, Oracle, Focus, OS/400 relational database, Adabas. Runs on Pentium type desktop computers. Stands for: Picture Oriented Software Engineering. Vendor: Computer Systems Advisers, Inc.

POSIX Definition of standards by IEEE to control interfaces between application programs and Unix. Stands for: Portable Operating System Interface for Unix. Most vendors are adhering to this standard.

POST Application development tool. Utility that assesses the stability of application systems by analyzing the number and types of changes made to the system and the number of abnormal end that occur. Runs on IBM MVS systems. Stands for: Production and Operation Stability Tool. Vendor: Coopers & Lybrand.

Posta Communications, Internet software. Sends large, complex documents usually sent by fax or overnight delivery over the Internet's e-mail. Released: 1997. Vendor: Tumbleweed Software Corp.

PostgreSQL Object-relational DBMS. Runs on Unix systems. Vendor: developed by a team of developers over the Internet.

PostScript Used in desktop publishing. PostScript is a page description language used to describe text fonts and graphics. Requires a PostScript printer. Supports HTML for Internet interface. Vendor: Adobe Systems, Inc.

POWER Operating system enhancement used by systems programmers. Increases system efficiency in IBM VSE systems. Full name: VSE/SP POWER. Input and output spooler. Vendor: IBM Corp.

Power++ Visual programming system that uses C++. Runs on Windows 95/98, Windows NT systems. Used for database and Internet applications. Vendor: Powersoft (division of Sybase, Inc).

Power Base, Center, Tower Midrange computer. PowerPC RISC CPU. Operating systems: MacOS, System 7. Vendor: Power Computing Corp.

Power Center Operating system software. Provides system management functions. Used to deploy, configure and manage client/server applications. Detects and corrects network problems including problems with applications, peripherals and computers. Runs on DEC, Unix, Windows systems. Released: 1996. Vendor: Power Center Software, LLC.

Power Desktop Integrated package of Communications, Internet software. Includes Emissary, an HTML editor, Winsock. Vendor: The Wollongong Group, Inc.

Power EDI Communications software. EDI package. Runs on OS/2, Unix, Windows systems. Vendor: EDI Able, Inc.

Power Graph UV Desktop computer. Pentium CPU. Operating system: Windows 95/98. Vendor: Royal Electronics, Inc.

Power Macintosh Midrange computer. RISC PowerPC system. Operating system: System 7. Vendor: Apple Computer, Inc.

Power Media (MMX) Desktop computer. Pentium CPU. Operating systems: DOS/Windows, Windows 95/98. Vendor: Royal Electronics, Inc.

Power Media III Notebook computer. Pentium CPU. Vendor: Astro Research, Inc.

Power Objects Application development tool used to develop object-oriented programs. Visual programming tool using drag-and-drop technology. Based on Visual Basic. Provides cross-platform support. Formerly called Project X. Vendor: Oracle Corp.

Power Pen Pal Application development tool. Used to develop pen-based applications. Vendor: PenPal Associates.

Power System Desktop computer. AMD processor. Vendor: Austin Computer Systems, Inc.

PowerAdvantage Midrange computer. PowerPC RISC CPU. Operating systems: AIX, Windows NT. Vendor: General Automation.

Powerbase Relational database for desktop computer environments. Runs on Pentium type desktop computer systems. Vendor: Compuware Corp.

POWERBench Application development system. Versions available for COBOL, FORTRAN, C++. Runs on AIX systems. Vendor: IBM Corp.

PowerBook See Macintosh PowerBook.

PowerBrick See CD-PowerBrick.

PowerBroker Communications software. Middleware. Upgrade to Xshell Distributed ORB. Supports Smalltalk, C++, Visual Basic. See Xshell Distributed ORB. Runs on Unix, Windows systems. Follows CORBA standards. Vendor: Expersoft Corp.

PowerBrowser Internet browser. Can run Java applets. Includes database called Blaze. Vendor: Oracle Corp.

PowerBuilder Application development environment. Used to develop departmental client/server systems. Used for RAD. Uses object-oriented features but does not require developers to know object-oriented programming. Builds client applications with a GUI-based front end. Interfaces with Oracle, AllBase SQL, SQL Server, Ingres, DB2, Centura, Xdb, Informix. Runs on Unix, Windows systems. Vendor: Powersoft (division of Sybase, Inc).

PowerBuilder Desktop Application development tool. Creates client/server applications that link xbase (dBase, FoxPro) with SQL databases. Vendor: Powersoft (division of Sybase, Inc).

Powerburst Communications software. Speeds connections to remote hook-ups. Includes Powerburst Client which runs on the remote PC and Powerburst Agent. Vendor: AirSoft, Inc.

PowerCenter Application development tool. Used to build, manage and network incrementally-built data marts for distributed data warehousing. Integrates disparate data marts and allows for the dynamic sharing and re-use of data. Runs on Unix, Windows systems. Released: 1998. Vendor: Informatica Corp.

PowerCom See OEA.

PowerConnect Application development tools. Packaged software products used to extract data and metadata from hard-to-access ERP and other legacy applications. Three products available: PowerConnect for SAP R/3, PowerConnect for PeopleSoft and PowerConnect for IBM DB2. Interfaces with PowerCenter. Released: 1999. Vendor: Informatica Corp.

PowerDesigner Application development tool. Used in client/server environments for database design. Interfaces with most major database systems. Provides reverse engineering, data modeling, process modeling tools. Includes: Metaworks (manages data design information), AppModeler (generates application objects), ProcessAnalyst (provides data flow diagramming), PDM (Product Data Management), WarehouseArchitect (supports warehouse design), DataArchitect (database design functions) and Viewer (information browser). Runs on Windows systems. Vendor: Powersoft (Division of Sybase, Inc.)

Powerdigm XSU, XLI Desktop computer. Pentium II, Pro CPU. Operating system: Windows NT. Vendor: Micron Electronics, Inc.

PowerDynamo Communications. Web server. Distributes Web applications using existing database replication facilities. Vendor: Sybase, Inc.

PowerEdge Midrange computer. Pentium, Pentium Pro CPU. Operating system: Windows NT. Vendor: Dell Computer Corp.

PowerFlow Workflow software. Includes business modeling component to graphically design and automate work-flow steps. Runs on Windows systems. Vendor: Optika Imaging Systems, Inc.

PowerFrame 1. Application development tool. Class library used with PowerBuilder. Vendor: MetaSolv Software, Inc. 2. Midsize computer. Server. Pentium processor. Operating system: Windows NT. Vendor: Tricord Systems, Inc.

PowerHawk Desktop computer. PowerPC CPU. Operating system: PowerMAX OS. Vendor: Concurrent Computer Corp.

PowerHelp Help Desk software. Sends messages to remote pagers and handheld computers. Vendor: Astea International, Inc.

PowerHouse Dictionary Data management system. Runs on DEC, DG systems. Component of PowerHouse Application development environment. Creates databases and file structures from dictionary. Vendor: Cognos, Inc.

PowerHouse, PowerCASE CASE product. Automates analysis, design, and programming functions. Includes code generator for SQL, Powerhouse 4GL. Design methodologies supported: entity-relationship models, data flow diagrams. Databases supported: Rdb, Powerhouse Starbase, any standard SQL database. Runs on DEC VAX, OS/2 systems. Vendor: Cognos, Inc.

PowerHouse StarBase Database/4GL for large computer environments. Runs on DEC, DG, HP systems. Utilizes SQL. Includes report generator, CASE tools. Vendor: Cognos, Inc.

PowerHouse StarNet, StarGate Communications software. Networks connecting midrange systems. Vendor: Cognos, Inc.

PowerHouse Windows Application development tool that works in a client/server architecture. Builds client applications with a GUI-based front end. Interfaces with AllBase SQL, InterBase. Vendor: Cognos, Inc.

PowerJ Application development tools. RAD tool user for Java, Internet applications. Includes CORBA support. Vendor: Powersoft (division of Sybase Inc).

PowerKeys Application development tool. Part of Approach that provides interfaces to other relational databases. Vendor: Lotus Development Corp.

Powerlan Communications software. Network operating system. Runs on Pentium type desktop computers. Vendor: Performance Technology, Inc.

PowerLite Portable computer. Vendor: RDI Computer.

PowerMan See OEA.

PowerMart (Suite) Application development tool used in data warehousing. Extracts production data for data warehouses. Inventories and searches metadata. Includes PowerCenter. Vendor: Informatica Corp.

PowerMate Enterprise Series Desktop computer. Pentium II CPU. Operating system: Windows NT. Vendor: NEC Computer Systems.

PowerMate professional Desktop computer. Pentium II CPU. Operating system: Windows NT. Vendor: NEC Computer Systems.

PowerMAX Midrange computer. RISC CPU. Operating system: Windows NT. Vendor: CyberMax Computer, Inc.

PowerMax CD Book xxx Notebook computer. Pentium CPU. Vendor: MAXIMUS COMPUTERS, Inc.

Powermax DS, TFT Desktop computer. Pentium CPU. Operating system: Windows 95/98. Vendor: MAXIMUS COMPUTERS, Inc.

PowerMAX OS Real-time operating system. Unix based system. POSIX compliant. Vendor: Concurrent Computer Corp.

PowerMAXION Midrange computer. PowerPC SMP multiprocessor. Operating system: PowerMAX OS. Vendor: Concurrent Computer Corp.

PowerMedia MMX Desktop computer. Pentium CPU. Operating system: Windows 95/98. Vendor: Royal Electronics, Inc.

PowerModel Application development environment. Object-oriented. Visual tool used to develop applications for Unix systems. Runs on Unix, Windows systems. Vendor: Intellicorp, Inc.

PowerObjects Application development tool. Toolset used in object-oriented development based on Visual Basic. Runs on MacOS, OS/2, Windows systems. Supports Internet, mobile users. Vendor: Oracle Corp.

PowerOpen Operating system. Unix type system. Runs on PowerPC systems. Applications for AIX, A/UX will run under PowerOpen. Vendor: Apple Computer, Inc./IBM Corp.

PowerPak Professional Edition Application development tool. Used to develop Visual Basic applications. Vendor: Crescent (Division of Progress Software).

PowerPC Family of computer chips, or microprocessors. 64-bit chip. Operating systems: AIX 4.3, OS/400, OS/390. Vendor: Motorola Corp.

PowerPDL Application development tool. Part of Power Tools. Supports pseudocode design. Allows automatic generation of documentation from code comments. Vendor: ICONIX Software Engineering, Inc.

PowerPLUS Desktop computer. Pentium processor. Vendor: Austin Computer Systems, Inc.

PowerPoint (for Windows) Desktop system software. Graphics package. Interfaces with Lotus 1-2-3, Excel, dBase. Runs on Macintosh, Windows systems. Part of Office. Vendor: Microsoft Corp.

POWERportable Portable computer. Vendor: IBM Corp.

PowerQ Series Communications software. Manages message queues from middleware systems that handle communications from Unix and Windows to mainframe systems. Interfaces with MQSeries. Vendor: PowerQ Software.

PowerServer SMP-N Midrange computer. Server. Pentium CPU. Vendor: Micron Electronics, Inc.

Powersim Application development tool. Provides process modeling functions. Runs on Windows systems. Vendor: Powersim Corp.

PowerSite Application development tool. RAD tool used to build enterprise-wide applications for the Web. Manages and deploys large-scale systems. Vendor: Powersoft (division of Sybase, Inc.)

PowerStack Midrange computer. PowerPC RISC CPU. Operating systems: Unix, Windows NT. Vendor: Motorola Computer Group.

Powerstep Desktop system software. Spreadsheet. Runs on NeXT systems. Vendor: Ashton-Tate Corp.

PowerStudio Enterprise Application development tools. Integrated RAD tools used to develop C++, Java systems. Includes object-oriented functions and Web development. Vendor: Powersoft (division of Sybase, Inc.).

PowerTier Application server providing middleware and development tools. Used to develop and deploy Internet and intranet applications for heterogeneous environments including most databases, most Web servers, and multiple languages and platforms. Available for Java and C++. Vendor: Persistence Software.

Powertools Operating system add-on. Data management software that runs on IBM systems. Vendor: Software Diversified Services.

PowerTools 1. Application development tool. Suite of tools for analysis and design of object-oriented applications. Includes ObjectModeler, FastTask, DataModeler, FreeFlow, PowerPDL, AdaFlow, QuickChart, SmartChart, ASCIIBridge, CoCoPro. Vendor: ICONIX Software Engineering, Inc.
2. Set of library routines that provide parallel processing across multivendor Unix systems. Helps design small subroutines that can run simultaneously in a cluster of machines, thus providing fast execution for the task. Vendor: VXM Technologies, Inc.

PowerTrip Notebook computer. Tillamook CPU. Vendor: Power Computing Corp.

PPE Operating system add-on. Debugging/testing software that runs on IBM systems. Stands for: Problem Program Evaluator. Vendor: Boole & Babbage, Inc.

PPP Communications protocol. Used to transfer information through telephone circuits and modems. Used to connect Macintosh users to the Internet. Stands for: Point-to-Point protocol.

PPPI Applications software. Funnels data from shop floor machines to accounting systems. Provides an interface between the factory and laboratory systems. Provides controls for the entire manufacturing process. Interfaces with R3. Stands for: Production Planning Process Industries. Vendor: SAP AG.

PPTP Communications. Protocol used to set up open tunneling over the Internet. Used with VPNs. Stands for: Point to Point Tunneling Protocol. Supplanted by L2TP.

Praesidium/Security Communications framework covering both Intranets and the Internet. Includes: Praesidium/VirtualVault and Eagle Firewall. Vendor: Hewlett-Packard Co.

Praesidium/VirtualVault Operating system software. Security package that connects enterprise databases and applications to the Internet. Vendor: Hewlett-Packard Co.

PractiBase Relational database for desktop computer environments. Runs on IBM desktop systems. Includes programming language and report generator. Vendor: PractiCorp International, Inc.

Praxa family Application software. Financial, manufacturing, sales systems for DEC environments. Vendor: Unitronix.

PRC 3310 Handheld computer. Vendor: Symbol Technologies, Inc.

Prealert MVS/IDMS Operating system enhancement used by systems programmers. Monitors and controls system performance in IBM IDMS systems. Vendor: Allen Systems Group, Inc.

Precise/CPE Communications software. Middleware. Allows queries to multiple databases on different platforms. Runs on Unix systems. Vendor: Precise Software Solutions.

Precise/MPX, MQX Real-time operating system for desktop computers. Vendor: Precise Software Technologies, Inc.

Precise/Pulse System software. Monitors performance of multiple Oracle databases. Runs on Unix, Windows NT systems. Released: 1998. Vendor: Precise Software Solutions, Inc.

Precise/SQL Application development tool. Speeds up loading and querying databases. Used in data warehousing. Runs on Unix systems. Vendor: Precise Software Solutions.

Precision Workstation Desktop computer. Pentium processor. Vendor: Dell Computer Corp.

PrecisionBook Desktop computer. Notebook. Pentium processor. Vendor: RDI Computer Corp.

Preclass Application development tool used for data mining. Used to determine which of hundreds of relationships among variables are significant. Runs on Macintosh, Windows systems. Vendor: Prevision Corp.

Predict Data dictionary and operational repository. Works with Adabas. Part of Unix Productivity Pack. Vendor: Software AG Americas Inc.

Predict Gateway Application development tool. Transfers data design from CASE products to Natural language. Uses IEW, ADW, Excelerator. Vendor: Software AG Americas Inc.

Predictive Support XL Operating system enhancement used by systems programmers. Monitors and controls performance in HP systems. Analyzes peripherals and reports on potential failures. Can be set up to automatically call HP maintenance when finding a problem. Vendor: Hewlett-Packard Co.

PREDITOR Application development tool. Provides editing functions for program languages including COBOL, Java. Vendor: Compuware Corp.

Prelude 1. Application development system. Includes relational database, application generator, report writer, word-processor and spreadsheet. Used in scientific, office automation applications. Vendor: Phase II Software Corp. 2. Groupware. Electronic meeting system used over multiple locations. Lets users conduct sessions over most networks. Users share documents while using standard voice conference calling. Vendor: ConferTech International, Inc. 3. Code name for DOE. See DOE.

Premenos Software vendor. Writes EDI software for midsize, RISC systems.

Premio Apollo BX Desktop computer. Vendor: Premio Computers, Inc.

Premium xx Midsize computer. Server. Pentium II processor. Operating system: Windows NT. Vendor: AST Research, Inc.

PREP PowerPC Reference Platform. A document that can be used to produce PowerPC clones. Produced by IBM and Motorola to make PowerPC systems readily available quickly.

Presario 10x0 Notebook Notebook computer. Pentium CPU. Vendor: Compaq Computer Corp.

Presario xxxx Desktop computer. Pentium CPU. Operating system: Windows 95/98. Vendor: Compaq Computer Corp.

Present Operating system add-on. Data management software that runs on DG systems. Vendor: Data General Corp.

presentation layer In data communications, the OSI layer that is concerned with transforming data from the internal representation of one computer to the internal representation of another. The most common transformation would be from EBCDIC to ASCII character code systems. This layer was invented by OSI.

Presentation Maker Desktop system software. Graphics package. Interfaces with Lotus 1-2-3, Excel, dBase. Runs on Macintosh, MS-DOS, Windows systems. Vendor: System Generation Associates, Inc.

Presentation Manager Graphics user interface for OS/2 systems. Includes windowing functions. Vendor: Microsoft Corp.

Presentation Team Desktop system software. Graphics package. Interfaces with Lotus 1-2-3, Excel, Quattro, Supercalc. Runs on MS-DOS systems. Vendor: Digital Research, Inc.

Presentations Desktop software. Graphics package. Vendor: WordPerfect Corp, Novell, Inc.

Presenter Desktop software. Graphics package. Used to make presentation documents. Vendor: Corel Corp.

PresenterPC Desktop system software. Graphics package. Interfaces with Lotus 1-2-3, Excel. Runs on MS-DOS systems. Vendor: Dicomed, Inc.

Prestige Software Software vendor. Creates financial management software. Division of Computer Associates International, Inc.

Pretty Good Privacy See PGP.

Preview Plus, Gold Application development tool used for business access and analysis. Provides end-user access for queries, reports, charts. Vendor: Idea.

preVue Application development tool. Provides testing, debugging functions. Includes modules for client/server testing, regression testing. Runs on Unix systems. Vendor: Rational Software Corp.

PRF Application development tool. Mainframe-based reporting and data access tool for end-users. Also called Report Facility. Vendor: Computer Associates International, Inc.

Pride- CASE product. Automates all phases of the development cycle including project management functions. Includes: Pride-SDM, Pride-DBEM, Pride-PMS, Pride-ISEM. Vendor: M. Bryce & Associates, Inc.

Pride-IRM Data management system. Runs on DEC, IBM, Unisys systems. Includes data dictionary, documentation tools. Interfaces with Pride-ISEM, Pride-PMS. Full name: Pride-Information Resource Manager. Vendor: M. Bryce & Associates, Inc.

Prim-OS Real-time operating system for parallel systems. Vendor: Czech und Matzner.

Primavera Project Planner See P3 (for Windows).

Prime Information Relational database for midrange and desktop computer environments. Runs on Prime systems. Vendor: Prime Computer, Inc.

Prime Oracle Relational database for large computer environments. Runs on Prime systems. Vendor: Prime Computer, Inc.

PrimeLink Communications software. Network connecting Prime computers. Vendor: Prime Computer, Inc.

Primenet Communications. LAN (Local Area Network) connecting Prime computers. Vendor: Prime Computer, Inc.

PRIMERGY Midrange computer. Pentium Pro CPU. Operating systems: OS/2, Unix, Windows NT. Vendor: Siemens Nixdorf Information Systems Inc.

PRIMIX Operating system for Prime systems. Unix-type system. Vendor: Prime Computer, Inc.

PRIMOS Operating system for Prime systems. Vendor: Prime Computer, Inc.

Primrose Application development tool. Develops applications that run in a client/server or peer-to-peer network. Provides GUI tools. Vendor: Tesseract Corp.

Primrose Workplace Application development tool that works in a client/server architecture. Builds client applications with a GUI-based front end. Interfaces with SQL Server, OS/2 EE Database Manager. Vendor: Tesseract Corp.

Prioris Midrange computer. Pentium CPU. Operating systems: OS/2, Windows 95/98. Vendor: Digital Equipment Corp.

Prism CASE product. Part of Excelerator system. Vendor: Cullinet Software, Inc.

PRISM Application software. ERP system. Runs on AS/400 systems. Includes Production Model for process maufacturing. Vendor: Marcam Solutions.

Prism Scalable Data Mart, Enterprise, Warehouse, Conversion Application development tools. Used to build data warehouses. Runs on DEC's Unix, IBM mainframe. Vendor: Prism Solutions, Inc.

Private Operating system add-on. Security/auditing system that runs on HP systems. Vendor: Operations Control Systems.

Private Data Mover Operating system software. Transfers files between Unix and Windows NT servers. Released: 1998. Vendor: Morgan Network Software.

private key encryption Security technique. Uses a single key known to both sender and receiver that is used to encrypt and decrypt information. Also called symmetric keys.

PrivateNet Communications, Internet software. Firewall. Vendor: NEC Technologies, Inc.

Privileged Architecture Library See PALcode.

PRL See Insight 2+.

PRMS See CA-PRMS.

PRMS PRO/5, Visual PRO/5 Application development tool. Used to develop business applications. Includes data dictionary, file maintenance and housekeeping utilities, and two and three dimensional arrays. Works with PRO/5 Data Server to create client/server applications. Runs on Unix systems. Released: 1997. Vendor: BASIS International, Ltd.

Pro-2 Application development tool. Program generator for IBM CICS programs. Interfaces with IMS, IDMS, and DB2 systems. Vendor: Software Engineering of America, Inc.

Pro*Ada Programming interface from Oracle databases to programs written in Ada. Runs on DEC VAX systems. Interfaces also available for C, COBOL, Fortran, Pascal, PL/1 as Pro*C, etc. Vendor: Oracle Corp.

Pro-C (Lite, Professional) Application development tool. Program, or code generator. Generates C code, interfaces with most databases. Runs on desktop systems. Vendor: Pro-C, Ltd.

Pro_EDI Communications software. EDI. Runs on Unix systems. Vendor: Data Management Strategies, Ltd.

PRO/Enable Application development tool. Used to develop 3-tier object-oriented application accessing relational databases and using existing enterprise data. Performs the functions of a data broker between the clients and the server and keeps a pool of objects currently in use by client programs. Automatically loads objects into the pool whenever referenced by clients. Runs under Unix, Windows systems. Released: 1998. Vendor: Black & White Software, Inc.

Pro/ENGINEER Application development tool used for mechanical engineering. Provides total project development management. Includes support for open systems and customized data flow systems. Vendor: Parametric Technology Corp. (PTC).

PRO-IV Application development environment. Includes ProKit*Workbench. Provides fill-in-the-blanks method of program development. Runs on DEC, IBM, Unix systems. Vendor: McDonnell Douglas Systems Integration Co.

Pro-IV Workbench Application development tool. Creates data models for systems accessing MDBS databases. Vendor: mdbs, Inc.

Pro Manager Desktop system software. Project management software. Vendor: Holland Systems Corp.

PRO/MVS Operating system enhancement used by systems programmers. Monitors and controls system performance in IBM systems. Performance reporting tool add-on to EPILOG. Vendor: Candle Corp.

Pro/TSX-Plus Operating system for DEC desktop computer systems. Vendor: S & H Computer Systems, Inc.

Pro*xxx Programming interfaces from Oracle databases to programs written in different languages. Versions available: PRO*ADA, PRO*C, PRO*COBOL, PRO*FORTRAN, PRO*PASCAL, PRO*PL/1. Vendor: Oracle Corp.

Proalter/Plus Operating system enhancement used by systems programmers. Monitors and controls system performance in DB2 systems. Vendor: On-Line Software International.

ProAudit Operating system add-on. Security/auditing system that runs on IBM DB2 systems. Provides audit trail for DB2. Vendor: On-Line Software International.

Probase Relational database for desktop computer environments. Runs on IBM desktop systems. Vendor: Probase Group, Inc.

Probe/38 Application development tool. Program that maintains information about batch program development in IBM System/38 environments. Vendor: Advanced Systems Concepts, Inc.

ProBuild Application development tool. Programming utility that builds application prototypes. Vendor: On-Line Software International.

Probus EIS EIS. Vendor: Decision Technologies, Inc.

PROC Set of JCL statements that is stored and executed as a unit.

Procalc 3D Desktop system software. Spreadsheet. Runs on Pentium type desktop computers. Vendor: Parsons Technology.

Process and Modeling Application development tool. Used for business process modeling. Vendor: PTech, Inc.

PROcess Application Generator Application development system. Includes program definition database and dictionary, program generators. Runs on Unix systems. Vendor: Bytelyne Computer Systems, Inc.

Process Configuration Management Software See PCMS.

Process Continuum Application development tool. Project management software. Vendor: PLATINUM Technology, Inc.

Process Engineer Application development tool. Defines, deploys, and measures the development process. Includes project management, library functions. Provides iterative prototyping. Used for client/server development. Includes: Process Library, Process Manager. Vendor: LBMS, Inc.

Process Manager Communications. Used to create and manage workflow applications over extranets. Runs on Netscape's web server. Released: 1998. Vendor: Netscape Communications Corp.

process model Using computers to follow the work flow of manufacturing. For example, an item could be electronically created by a design engineer. The design would be automatically routed to engineering who would electronically refine and create specifications for the item. The specifications are then routed to the factory floor where the item is created.

processing modeling tool Application development tool. Users build a diagram of the steps in a process, and provide equations defining the movement between steps. Allows users to simulate the real world by providing different inputs to the model.

process patterns Written explanations of techniques, actions and tasks for developing software.

ProcessAnalyst See PowerDesigner.

ProcessIt Workflow management software. Used to help design, implement, execute, and manage business processes. Windows-based tool. Vendor: NCR Corp.

ProcessMaker Application development tool. Object-oriented. Develops graphic representations of processes, or work flows. Vendor: MarkV Systems.

ProcessModel Application development tool. Provides process modeling functions. Runs on Windows systems. Vendor: Promodel Corp.

Procode Application development tool. Automates programming function. Vendor: Clyde Digital Systems.

Prodas Relational database for midrange and desktop computer environments. Runs on IBM, Unix, Xenix systems. Vendor: Conceptual Software, Inc.

ProdeaBeacon Application development tool. End-user query tool. Interfaces with DB2, SQL Server, Sybase, Oracle, Red Brick. Used for data warehousing. Runs on Windows systems. Vendor: Prodea Software Corp.

Prodeveloper CASE product. Automates analysis and design functions. Vendor: Holland Systems Corp.

Prodigy An on-line utility that provides information and services to monthly subscribers. Services include stock market access, ticket purchases, and library access. Competitive with CompuServe. Vendor: IBM and Sears, Roebuck & Co.

Prodoc re/NySys CASE product. Automates analysis, design, and programming functions. Includes code generator for Ada, C, COBOL, Fortran, Pascal. Runs on Pentium type desktop computers, RS/6000, Sun Sparcstation. Full name: Prodoc RE/NU Sys Workbench. Vendor: Scandura Intelligent Systems.

ProDOS Operating system for Apple II desktop computer systems. Vendor: Apple Computer, Inc.

Product Data Manufacturing See PDM.

production library A dataset that contains programs which execute as part of the regular data center schedule.

ProEdit Operating system enhancement used by systems programmers. Increases system efficiency in IBM DB2 systems. Testing program and table editor. Vendor: On-Line Software International.

Professional Oracle Relational database for desktop computer environments. Runs on IBM desktop systems. Vendor: Oracle Corp.

Professional Support Tool See PST.

Professional Write Desktop system software. Word processor. Runs on Pentium type desktop computers. Vendor: Software Publishing Corp.

Professional Write Plus Desktop system software. Word processor. Runs on windows on IBM and compatible desktop computers. Vendor: Software Publishing Corp.

Profile Operating system enhancement used by systems programmers. Monitors and controls system performance in IBM systems. Vendor: Software Engineering of America, Inc.

Profiler 1. Database for desktop computer environments. Runs on Apple II systems. Vendor: Pinpoint Publishing.
2. See DistribuLink.

ProFinder Desktop computer. Pentium CPU. Operating system: Windows 95/98. Vendor: ProGen Technology Inc.

Proforma xxxx Desktop computer. Pentium processor. Vendor: Duracom Computer Systems.

PROFS Communications software. E-mail system. Runs on IBM systems. Stands for: PRofessional OFfice System. Vendor: IBM Corp.

ProGen Plus Application development tool. Generates RPG/400 programs and screens. Vendor: Business Computer Design.

Progeni GLE See GLE.

program Series of instructions that accomplish a task. Also called module, routine.

program generator A computer program that will generate all, or part, of the code necessary to build a program in a source language. Also called application generator, code generator.

program library A dataset that holds programs. Program libraries can hold source programs and/or object programs.

program utility A computer program that provides specific assistance to programmers. Utilities perform such functions as monitoring datasets and converting programs, data, and/or JCL to run on another system.

Programaker Application development tool. Produces business applications without writing code. Used in DEC VAX environments. Vendor: Iskra Software International.

Programmer Operating system add-on. Library management system that runs on IBM systems. Vendor: Software Controls, Inc.

programmer Developer. Could be either applications or technical. Analyzes specifications, designs logic, writes code, tests and debugs, and documents computer programs. Can be any experience level.

programmer/analyst Application developer. Title should mean experience working with users and indicate mid to senior level experience. Check for experience carefully, title often mis-used. Title is often abbreviated P/A.

Programmer Assistant CASE product. Part of Re-engineering Product Set. Vendor: Bachman Information Systems, Inc. (now Cayenne Software).

programmer trainee See Entry-level Programmer.

programming A phase in the system development cycle. Writing a sequence of instructions to be executed by a computer to solve a problem. This includes analyzing the problem to determine what must be done, designing the program logic, coding the program, testing the code, and documenting the finished program.

Prograph Application development environment. Object-oriented. Builds client/server applications with a GUI-based front end. Interfaces with Oracle, Informix, Ingres, SQL Server. Vendor: TGS Systems.

Progress (ADE) Application development environment. Used to develop departmental client/server systems. Includes relational database, 4GL, SQL, data dictionary, help system, compiler. Accesses DEC Rdb, Oracle Server, Progress. Utilizes SQL. Includes application and report generators. Runs on DEC, IBM, Unisys systems. Vendor: Progress Software Corp.

Pro*IV* Application development tool. Used for RAD development. Interfaces with diverse databases and across multiple platforms. Developer Studio includes a GUI interface and JavaSuite can Web enable any Pro*IV* application. Vendor: ProIV Inc.

Project 1. Application development tool. Program that provides cost accounting information for a project. Vendor: International Structural Engineers, Inc.
2. Desktop software. Project management package. Includes VDA. Vendor: Microsoft Corp.

Project/2 Series X Desktop system software. Project management package. Includes GUI and manages multiple projects. Runs on desktop systems. Vendor: Project Software & Development, Inc.

Project 95, 98 Desktop software. Project management system. With Project 98, managers can use Intranet technology. Runs on Windows 95/98 systems. Vendor: Microsoft Corp.

Project Cost Management Application development tool. Project management system. Interfaces with Walker financial systems. Runs on IBM mainframes, Unix systems. Vendor: Walker Interactive Systems, Inc.

Project DOE Application development tool. Tool kit for developing object-oriented systems. Also includes tools for wrapping code from old systems in a new layer of objects. Stands for: Project Distributed Objects Everywhere. Vendor: Sun Microsystems, Inc.

Project Eagle Name for IBM project to make its client/server software less expensive and easier to install. Includes multi-platform line of Software servers in integrated packages.

Project Engineer Desktop system software. Project management package. Runs on Pentium type desktops. Vendor: LBMS, Inc.

Project/I-80 Desktop system software. Project management package. Runs on Pentium type desktop computers. Vendor: Micro Research Systems Corp.

Project Kickstart Desktop software. Project management program designed to help users set up the initial stages of project management such as assigning resources and listing tasks. Runs on Windows systems. Vendor: Microsoft Corp.

project leader, team leader Application developer. Senior level supervisory position. Supervises the work done by mid-level and junior developers. Duties may include personnel management and project planning and scheduling. Sometimes acts as analyst or programmer.

project management software Desktop computer software that helps to plan, schedule, and control a project during its development.

project manager, programming manager Application developer. First level management. Manages personnel and does project planning and scheduling for a specific application or function area. Other duties could be to conduct performance appraisals, determine salaries and increases, hire and fire staff, and be accountable for the system budget.

Project Outlook Desktop system software. Project management package. Runs on Pentium type desktop computers. Vendor: Strategic Software Planning Corp.

Project Planner See P3 (for Windows).

Project Rescue Plan and software from Sun Microsystems that enables corporations to turn 486 microcomputers running MS-DOS into network computers to work as front-ends to client/server and/or Internet systems.

Project Scheduler Desktop system software. Project management package. Runs on Pentium type desktop computers. Vendor: Scitor Corp.

Project Workbench Desktop system software. Project management package. Runs on Pentium type desktop computers. Manages multiple projects simultaneously and gives a complete picture of all activity underway at any time. Vendor: Applied Business Technology Corp.

Project Workbench Professional Project management software. Includes re-engineering features. Vendor: Applied Business Technology Corp.

Project X See PowerObjects.

Projectbase Desktop system software. Project management package. Runs on Pentium type desktop computers. Vendor: Center for Project Management.

ProjectGuide Application development tool. Creates project models for scheduling systems. Runs on MS-DOS, Windows systems. Vendor: Marin Research.

ProKit*Analyst CASE product. Automates analysis function. Vendor: McDonnell Douglas Systems Integration Co.

Prokit*Workbench CASE product. Automates analysis, design and programming functions. Design methodologies supported: Chen, Gane-Sarson, Stradis. Databases supported: IMS, DB2. Runs on Altos, AT&T, DEC, IBM, Prime systems. Products include: PRO-IV. Vendor: McDonnell Douglas Systems Integration Co.

ProLiant Midrange computer. Pentium, Pentium Pro CPU. Operating system: Windows 95/98. Vendor: Compaq Computer Corp.

Prolifics Application development tool. Object-oriented. Creates three-tiered client/server and Web based applications. Embeds online OLTP middleware. Supports Windows 3.x, Windows 95/98, Macintosh clients. Runs on Unix, Windows NT systems. Vendor: Prolifics (a JYACC company).

Prolinc Communications software. Network manager that provides access to Banyan's VINES, Microsoft's LAN Manager, and AT&T's Star Group network operating systems. Allows users to access multiple hosts. Vendor: Hughes LAN Systems.

ProLinea Desktop computer. Pentium CPU. Operating system: Windows 95/98. Vendor: Compaq Computer Corp.

Prolinear Computer vendor. Manufactures handheld computers.

PROLOG Programming language used in the development of artificial intelligence applications. Stands for: PROgramming in LOGic. Versions include: PROLOG I, PROLOG II, Micro PROLOG.

PROLOGIC Management Systems Software vendor. Products: Manufacturing, distribution systems for midsize environments.

ProMacro Macro language. A superset of BASIC that will allow users to create their own enhancements to BASIC programs. Compatible with Visual Basic. Runs on Windows systems. Vendor: Netlogic, Inc.

ProMacs Application development tool. Program generator for IBM systems. Interfaces to CICS, IMS, VSAM, DB2. Includes: ProMacs/CICS, ProMacs/BMS, ProMacs/Batch. Vendor: Computer Associates International, Inc.

PROMEDIA-A,B Desktop computer. Pentium CPU. Operating system: Windows 95/98. Vendor: MAXIMUS COMPUTERS, Inc.

PROMAL Compiler language used in desktop computer environments.

ProMAX Desktop computer. Pentium processor. Vendor: Cybermax, Inc.

PROMIS Desktop system software. Project management package. Runs on Pentium type systems. Full name, PROMIS C Project Management Integrated System. Vendor: Strategic Software Planning Corp.

Promiselan Communications software. Network operating system. Runs on Pentium type desktop computers. Vendor: Moses Computers, Inc.

Promix Process Manufacturing Series MRP, or CIM software. Runs on DEC systems. Vendor: Ross Systems, Inc.

ProMod family CASE product. Automates analysis, programming, and design functions. Includes: Pro/Source, Pro/Cap, ProMod/CM, ProMod/DC, ProMod/MD, ProMod/RT, ProMod/SA, ProMod/TMS, ProMod/2167A. Vendor: ProMod, Inc.

ProMod SourcePilot Application development tool. Program, or code generator. Generates C, C++, Fortran code. Runs on OS/2, VMS, Unix systems. Vendor: G&E Systems, Inc.

Promulgate/(Vines,PC) Communications software. Gateway for e-mail systems. Runs on Vines, TCP/IP, 3+Open networks. Vendor: Management System Designers, Inc.

ProNote Notebook computer. Pentium CPU. Operating system: Windows 95/98. Vendor: ProLinear Corp.

ProntoWatch Communications. Software that analyzes network performance by condensing thousands of network statistics into trends and problem situations. Vendor: Proactive Networks, Inc.

ProOptimize Operating system enhancement used by systems programmers. Increases system efficiency in IBM DB2 systems. Vendor: On-Line Software International.

Propagation See replication.

ProReports Application development tool. Report generator. Users can create a report under Windows and then run the report from any supported client or server machine. Vendor: Software Interfaces, Inc.

ProSecure Operating system add-on. Security/auditing system that runs on IBM DB2 systems. Vendor: On-Line Software International.

ProSEED 4GL used in midrange environments. Works with SEED's database. Vendor: SEED Software Corp.

ProServa Midrange computer. Pentium CPU. Vendor: NEC Computer Systems.

ProShare Communications software. EMS software. Allows two people working on desktop computers to work on the same document simultaneously. At the end of the collaboration both documents are recaptured as one document. Works with add-in boards for videoconferencing. Interfaces with Lotus Notes. Includes ProShare Conferencing Video. Vendor: Intel Corp.

ProShare Conferencing Video System 200 Video conferencing system. Released: 1996. Vendor: Intel Corp.

ProSignia Server Midrange computer. Pentium CPU. Operating system: Windows 95/98. Vendor: Compaq Computer Corp.

Prosignia xxx Desktop computer. Notebook. Pentium processor. Vendor: Compaq Computer Corp.

ProSource, Pro Cap Application development tool. Program generator for DEC VAX systems. Vendor: ProMod, Inc.

Prospero Application development tools. EAI software that integrates disparate applications by allowing the end-user to create Building Blocks. Works with J.D. Edwards, SAP systems. Building Blocks will be created for the major ERP vendors. Released: 1996. Vendor: Oberon Software, Inc.

Prospero SDK Application development tool. Allows developers to create Prospero Building Blocks for any application. Building Blocks allow for the integration of any piece of software into an ERP system. Released: 1998. Vendor: Oberon Software, Inc.

ProStar 5200, 6200, 7200, 8200 Notebook computer. Pentium CPU. Vendor: Prostar Computer, Inc.

ProTeam Application development tool. Automates engineering, quality assurance, technical support and sales operations for software companies. Includes: SupportTeam, QualityTeam, ViewTeam. Utilizes Sybase's SQL Server. Links to PeopleSoft and SAP. Runs on mid-size systems. Vendor: Scopus Technology, Inc.

Protean Application software. Includes financials, logistics, manufacturing functions. Object-oriented software. Includes: Plant Planning, Quick Scheduler. Runs on Unix systems. Interfaces with Oracle, Informix. Released: 1997. Vendor: Marcam Corp.

ProtectIT Operating system software. Security package. Includes policy-based security management, audit tools. Interfaces with RACF. Vendor: Computer Associates International, Inc.

Protege Midrange computer. Pentium CPU. Operating systems: Unix, Solaris. Vendor: Mobius Computer Corp.

Protel Software vendor. Produces software for CAD/CAM applications.

protocol Set of rules agreed to for data communications. Used by communications specialists.

Protocop Communications software. Network management tool. Displays problems across networks as they occur. Vendor: Optical Data Systems, Inc.

Protogen+ Application development environment. Visual tools used to design and implement Windows applications. Used in client/server systems. Includes code generators for C, OWL, MFC, Pascal. Vendor: Protoview Development Corp.

Proton Application server providing middleware and development tools. Used to develop and deploy three-tier Java applications over the Internet. Available as Proton EJB with Enterprise JavaBean server, or as Proton Web without EJB support. Released: 1999. Vendor: Pramati Technologies Ltd.

Protool/Online Application development tool. Program generator for IBM systems. Vendor: Software Intelligent Systems.

Protos Application development system. Runs on HP systems. Vendor: Protos Software Co.

prototype A model of a program and/or system suitable for evaluation of the design, performance, and potential of the product. Ideally the prototype can be added to until it becomes the full working system.

Protracs Desktop system software. Project management package. Runs on Pentium type desktop computers. Vendor: Applied Microsystems, Inc.

ProUtility Operating system enhancement used by systems programmers. Increases system efficiency in IBM DB2 systems by automating the scheduling and administration of DB2 utilities. Vendor: On-Line Software International.

ProVision System management software. Suite of nine integrated tools that provides multiplatform database and systems management. Released: 1997. Vendor: PLATINUM Technology, Inc.

ProWorks Application development system. Used to develop C applications. Includes run-time libraries and development utilities. Generates code for Intel systems. Runs on Unix systems. Vendor: Sun Microsystems, Inc.

PROXMVS Application development tool. Allows programmers to develop COBOL programs on a desktop computer or workstation using mainframe JCL and utilities. Vendor: Micro Focus, Inc.

proxy server Communications. Internet software. An intermediate server that sits between the Web browser and an application server. Proxy servers save the Web pages they retrieve for a certain amount of time, and often can fulfill a later request for a specific page from saved information instead of having to return to the originating site. Proxy servers can also be used as filters and deny access to certain sites.

ProxyReporter Communications. Internet software. Creates reports of Internet usage. Runs on Solaris, Unix, Windows. Released: 1998. Vendor: Wavecrest Computing, Inc.

Proxy Server Operating system software. Provides security functions used to connect Windows NT systems to the Internet. Vendor: Microsoft Corp.

pruning The process of sorting through enormous numbers of facts and rules during the decision-making process to arrive at a conclusion. A computer program must be able to do this in order to be said to have artificial intelligence.

PS-1000,3000 Handheld computer. Vendor: Prolinear Corp.

PS/Financials Applications software. Accounting, human resource, payroll systems. Runs on OS/2, Unix, Windows systems. Interfaces with most relational/SQL databases. Client/server technology. Vendor: PeopleSoft, Inc.

PS/General Ledger Financial software. Client/server architecture. Interfaces with SQLbase and DB2. Vendor: PeopleSoft, Inc.

PS/nVision Application development tool. Used by end-users. Includes graphical reporting and development functions. Vendor: PeopleSoft, Inc.

PSAM Access method alternative to VSAM in IBM systems. Vendor: Universal Software, Inc.

pseudocode An artificial language used to design computer programs. Synonymous with Structured English.

Psion Computer vendor. Manufactures handheld computers.

PSL/PSA Application development system. Includes query system, report generator. Stands for: Problem Statement Language/Problem Statement Analyzer. Runs on IBM mainframe systems. Vendor: Meta Systems Ltd.

pSOS Operating system for TC2000 MPP systems. Unix type operating system. Uses pSOS+ kernel. Vendor: Bolt Beranek and Newman, Inc.

PSR High-level statements used as input to CA-ADS/Generator. Stands for: Program Specification Rules.

PSS Operating system enhancement used by systems programmers. Monitors terminal sessions in DEC VAX systems. Stands for: Performance Simulation System. Vendor: Advanced Systems Concepts, Inc.

PSS Gateway M400 Communications software. Gateway for e-mail systems. Runs with SNA networks. Vendor: IIS Technologies, Inc.

PST Professional Support Tool. Name given to software designed to support non-technical professionals and managers. Includes query and analysis software, financial modeling tools. Included in EIS systems.

PT4x00 Handheld computer. Vendor: Symbol Technologies, Inc.

Ptech Software and services vendor. Works with business process modeling. Major product is FrameWork.

PTF Program Temporary Fix. A temporary solution to a problem in one of IBM's programs. PTFs are sent to all companies using the software with the error, and must be applied by systems programmers. Permanent fixes are made with new releases of the software.

PTMS Accounting system that runs on IBM RS/6000 systems. Handles management tasks for accounting offices. Stands for: Practice Time Management System. Vendor: Briareus Corp.

public domain software Software that anyone can use. Non commercial software.

public-key encryption Security technique. Used both to establish confidentiality and authenticate that information does come from the sender. Two keys are used: a private key which is kept confidential and a public key which is sent to potential correspondents. A document is encoded with one key can only be decoded with the other. Encryption is handled totally by software. Called asymmetric keys.

Publicalc Desktop system software. Spreadsheet. Runs on Unix systems. Vendor: SSC, Inc.

publish and subscribe Communications technology used with object-oriented networks. Enables diverse systems and applications to communicate without even being aware of each other. Many sources can subscribe to published objects, i.e. an EDI request can obtain customer's name from a billing system that is unaware of the EDI functions. Developed by Sun Microsystems.

Publishers Paintbrush Desktop system software. Graphics package. Runs on Windows systems. Vendor: Zsoft Corp.

PublishingXpert See CommerceXpert.

Pure Java Application development tools. Suite of tools which includes an application modeler, database access tool, code repository, and an object-oriented framework. Works with Jbuilder, Visual Cafe, VisualAge for Java. Released: 1998. Vendor: Sun Microsystems, Inc.

PureCoverage Application development tool. Testing tool that verifies whether each line of code has been tested. Runs on Unix systems. Vendor: Rational Software Corp.

PureDimension Application development tool used in data warehousing. Converts data from warehouses into a format usable by leading communications, database and analysis tools. Interfaces with DecisionSuite, DSS Agent, Essbase, Express, Discover, Pablo, PowerPlay, BrioQuery. Runs on Unix, Windows NT. Released: 1999. Vendor: Carleton Corp.

PureVision Application development tool. Monitors the function and use of software being tested at Beta sites. Collects information over the Internet. Vendor: Pure Software, Inc.

Purify (NT) Application development tool. Provides runtime error-detection functions. Runs on Unix, Windows NT. Vendor: Rational Software Corp.

Purveyor Communications software, Web server. Provides access to the Web from Windows NT, Windows 95/98, Open VMS, NetWare systems. Lets users view documents. Vendor: Process Software.

push technology Push software collects information and automatically sends it to a user without specific request. The user defines some information categories and receives updates without having to search or ask for them. For example, investment services can automatically send stock updates to clients when stocks in their portfolios change, or a news service could automatically send news updates to subscribers every hour. Used over the Internet.

PUT Program Update Tape. Tape containing PTFs to correct program errors that is sent by IBM to companies using the applicable software. PUTs are processed by systems programmers.

PV-Wave Application development tool used for data mining. Provides graphs of different subsets or summaries of data. This allows user different perspectives of the data. Runs on Windows 95/98 systems. Vendor: Visual Numerics, Inc.

PV-Wave Point & Click Application development tool. Allows nonprogrammers to analyze large amounts of data through a GUI. Runs with Open Look and Motif on Sun workstations. Vendor: Precision Visuals.

PVCS Application development tool. Provides version control and configuration management for application development and maintenance. Runs on DOS, OS/2, Unix systems. Includes: Reporter, Version Manager, Configuration Builder, Production Gateway, Developer's Toolkit. Stands for: Polytron Version Control System. Vendor: MERANT.

PVCS Production Gateway Application development tool. Allows developers to synchronize LAN and mainframe libraries. Based on APPC protocols. Runs with MVS, OS.2, Windows. Vendor: MERANT.

PVCS Version Manager Application development tool. Debugging/testing programs. Vendor: MERANT.

PVM Parallel Virtual Machine. A technology developed to let networked systems perform as parallel machines. Developed by Oak Ridge National Laboratory and supported by supercomputer vendors.

PW2 Advantage Midrange computer. Pentium CPU. Operating system: OS/2Windows 95/98. Vendor: Unisys Corp.

PXRos Real-time operating system for desktop computers. Vendor: HighTec EDV Systems GmbH.

Pyramid Nile Midrange computer. Server. MIPS R4400 CPU. Vendor: Siemens Pyramid Information Systems, Inc.

Pythias Internet browser. Provides direct support for database access. Runs on Windows systems.

Python Compiler. Supports object-oriented programming. Can be used as a scripting language for Web applications.

Q&A Desktop system software. Flat-file database with word processing functions. Runs on Pentium type systems. Vendor: Symantec Corp.

Q/Artesian Application development tool. Automates conversion from COBOL to COBOL II. Vendor: Eden Systems Corp.

Q/Auditor (COBOL, PL/1) Application development tool. Automates programming function. Analyzes source code for such things as adherence to standards and presence of logic flaws (unexecuted code). Languages analyzed: COBOL, PL/1. Runs on IBM systems. Vendor: Eden Systems Corp.

Q-Calc Desktop system software. Spreadsheet. Runs on Unix systems. Vendor: Quality Software Products Co.

Q-Calc Standard Desktop system software. Spreadsheet. Runs on Unix systems. Vendor: Unipress Software, Inc.

Q Diagnostics Center Operating system software. Monitors performance of Oracle database applications. Provides graphical reports. Vendor: Savant Corp.

Q+E Application development tool. End-user query tool and report writer. Vendor: INTERSOLV, Inc.

Q:ManagerIMS Operating system add-on. Tracks data movement in online IMS systems. Runs on IBM mainframes. Vendor: BMC Software, Inc.

Q-Media for Windows Multimedia authoring tool. Mid-level, timeline-based tool. Vendor: Q-Media Software Corp.

Q'Reporter Operating system enhancement used by systems programmers. Increases system efficiency in DEC VAX systems. Manages disk storage space. Vendor: W. Quinn Associates.

QA Partner Application development tool. Debugging and testing tool that performs software quality assurance testing. Includes integrated test analysis and reporting. Used to test client/server applications. Runs on Macintosh, IBM and compatible desktops. Released: 1998. Vendor: Seque Software, Inc.

QACenter Application development tool. Testing tool that provides enterprise-wide testing functions for client/server, mainframe, and process management applications. Includes *QA*Batch (regression testing for mainframe applications), *QA*Director (organizes the testing), *QA*Hiperstation (testing mainframe applications), *QA*Load (database, Web testing), *QA*Playback (CICS-based testing), *QA*Run (client/server testing), *QA*TrackRecord (tracks discovered bugs). Released: 1997. Vendor: Compuware Corp.

QADirector Application development tool. Debugging/testing programs. Vendor: Compuware Corp.

QAHiperstation Application development tool. Creates test data and scripts for applications for CICS, IMS, IDMS and TSO environments. Debugging/testing programs. Vendor: Compuware Corp.

QASE Operating system support software. Provides capacity planning for client/server systems. Provides analytic modeling and network simulations. Runs on Macintosh, Unix systems. Vendor: Advanced System Technologies, Inc.

QC Operating system add-on. Data management software that runs on IBM systems. Stands for: Quick Compress. Vendor: Stockholder Systems, Inc.

QDB/Analyze, QDB/Connect System software. Identifies and measures data quality problems. Analyze works with legacy systems and data warehouses. Connect is a client/server environment that works with relational database environments. Vendor: QDB Solutions.

QDistrib Operating system software. Controls distribution of reports throughout the enterprise. Handles multiple destinations including devices, files and people. Runs on DEC, Unix, Windows systems. Released: 1996. Vendor: QMaster Software Solutions Inc.

QDMS-R Relational database for midrange and desktop computer environments. Runs on DEC systems. Vendor: Quodata Corp.

QDS 4GL used in mainframe environments. Stands for: Query Driven System. Used in Bull HN systems. Vendor: Bull HN Information Systems, Inc.

QED Application development tool. Screen editor for Harris systems. Vendor: Harris Corp.

QEMM Operating system enhancement. Utility that manages computer storage by providing services such as data compression and virus protection. Runs on MS-DOS systems. Stands for: Quarterdeck Expanded Memory Manager. Vendor: Quarterdeck Office Systems, Inc.

QIC-AIX Operating system enhancement used by systems programmers. Increases system efficiency in IBM systems. Reduces time needed to construct VSAM files. Vendor: Quantum International Corp.

QLP 1100 4GL used in mainframe environments. Stands for: Query Language Processor. Works with DMS databases. Vendor: Unisys Corp.

QMan Relational database for midrange and desktop computer environments. Runs on Unix systems. Vendor: Breakpoint Computer Systems, Inc.

QMaster Batch Operating system software. Manages the execution of batch programs across diverse Unix and Windows systems, even in different geographic regions. Runs on DEC, Unix, Windows systems. Released: 1997. Vendor: QMaster Software Solutions Inc.

QMaster Print Operating system software. Provides print management and spooling functions. Manages centralized printing. Runs on DEC, Unix, Windows systems. Released: 1997. Vendor: QMaster Software Solutions Inc.

QMF 4GL used in mainframe environments. Stands for: Query Management Facility. Designed for use by end-users. Includes report generator. Works with DB2 databases. Vendor: IBM Corp.

QNX Neutrino Operating system software. Real-time operating system kernel for desktop systems. Vendor: QNX Software Systems, Ltd.

QNX RTOS Real-time Unix operating system for desktop computers. Includes micro kernel and optional modules for file servers, GUIs etc. Vendor: QNX Software Systems, Ltd.

QP5/xxx, QP6/xxx Desktop computer. Pentium CPU. Operating systems: Windows 95/98, NT Workstation. Vendor: Quantex Microsystems, Inc.

QR See Quick restore.

QSAM Access method used in IBM mainframe systems.

QT Operating system add-on. Data management software that runs on IBM systems. Stands for: Quick Tran. Vendor: Stockholder Systems, Inc.

QTAM Access method used in IBM mainframe systems.

QTP Application development tool. Program that analyzes volume of online transactions. Runs on DEC, DG, HP systems. Part of PowerHouse development environment. Vendor: Cognos, Inc.

QuadStar Midrange computer. Pentium Pro CPU. Operating system: Windows NT. Vendor: Tangent Computer, Inc.

QualEDI Communications software. EDI package. Runs on Pentium type desktop computers. Vendor: The APL Group, Inc.

quality assurance The development of standards and the formal monitoring to make sure that the standards are followed. Often part of the testing process.

QualityTeam See ProTeam.

Quantify Application development tool. Testing tool. Identifies bottlenecks in software code. Vendor: Rational Software Corp.

quantum computer Computer of the future based on quantum mechanics. Can perform calculations a billion times faster than a Pentium III PC. Uses atoms instead of PC chips. Basic computers have been built, but the technology is twenty to thirty years away.

Quantum I/O,PM Operating system enhancement used by systems programmers. Increases system efficiency in DEC VAX systems. Improves I/O speed. Vendor: Computer Information Systems, Inc.

Quantum RS Operating system enhancement used by systems programmers. Provides system cost accounting for DEC VAX systems. Vendor: Computer Information Systems, Inc.

Quarterdeck Mosaic Internet browser. Displays both HTML and VRML documents. Runs on Windows systems.

Quattro Desktop system software. Spreadsheet. Runs on Pentium type systems. Full name: Quattro: The Professional Spreadsheet. Obsolete. Vendor: Inprise Corp.

Queman Operating system enhancement used by systems programmers. Increases system efficiency in DEC VAX systems. Controls print queues. Vendor: Data Center Software, Inc.

QUEO-IV,QUEO-V Database for large computer environments. Runs on Prime systems. Vendor: Computer Techniques Inc.

Quepro Application development tool. Report generator for Unix, NCR, IBM desktop computers. Vendor: Bytel Corp.

query A program that produces a report. A query could be simple "show me John Jones' starting salary" or complex "show me the average starting salaries of each department in the company summarized by department within division within branch." Query programs produce no permanent data. If the terminal is turned off, or the printed report discarded, the information is gone.

Query 4GL used in mainframe environments. Used in Bull HN systems. Vendor: Bull HN Information Systems, Inc.

Query/250 4GL used in mainframe environments. Works with HP databases. Vendor: Hewlett-Packard Co.

Query/36,AS/400 4GL environments. Vendor: IBM Corp.

Query Builder See Oracle Card.

Query.DL/1 4GL used in mainframe environments. Works with IMS databases. Vendor: IBM Corp.

Query-IV Application development tool based on SQL. Contains pre-defined queries, or will generate SQL statements from a "point-and-click" user interface. Runs on Unix, Windows systems. Vendor: Thoroughbred Software International, Inc.

query language An interactive language that can be used to access data (usually in databases). Query languages are used to retrieve information, update information, and even to write complete programs. Term is often used interchangeably with 4GL.

Query program See query.

QueryObject Application development tool used for data mining. Used to determine which of hundreds of realtionships among variables are significant. Runs on Windows systems. Vendor: Cross/Z International Inc.

QueryObject System Application development tool used in data warehousing. Builds data marts on IBM S/390 systems. Builds multidimensional analytical environments directly from typical mainframe data sources such as DB2 and VSAM. Vendor: QueryObjects Systems Corp.

Quest 1. Query language that allows non-technical people to retrieve information from databases without help from information system professionals. Accesses DB2 (IBM), Oracle, SQLserver (Microsoft), and OS/2 Extended Edition Database Manager (IBM). Vendor: Centura Software Corp.
2. Application development tool. Report generator for NCR systems. Vendor: Software Clearing House, Inc.
3. Multimedia authoring tool. High-level, icon-based, scripting tool. Runs on Windows systems. Vendor: American Training International, Inc.

Quest Server Application development tool used with the Web. Interfaces with all SQL database systems. Allows users to build and publish searchable databases by browsing existing SQL databases. Vendor: Level Five Research.

Questview Application development tool. Programming utility that allows programmers to view and update databases in IBM AS/400 and S/38 environments. Vendor: Questcomp, Inc.

queue Any list.

Quick Copy Operating System enhancement used by systems programmers. Increases system efficiency in IBM DB2 systems by copying tables faster than the standard copy. Vendor: PLATINUM Technology, Inc.

Quick-Mail Communications software. E-mail system. Runs on DEC systems. Vendor: Horizon Data Systems.

Quick-Plan Desktop system software. Project management package. Runs on Pentium type systems. Vendor: Mitchell Management Systems, Inc.

Quick Query A menu-driven language for nontechnical users that allows them to use Answer/Reporter. Vendor: Sterling Software, Inc. (Answer Software Division).

Quick Restore Operating system enhancement. Provides backup and restore routines for Unix networks. Vendor: Workstation Solutions, Inc.

Quick Schedule Plus Desktop system software. Project management package. Runs on Pentium type desktop computers. Vendor: Poser Up Software Corp.

Quick Tran Operating system enhancement used by systems programmers. Increases system efficiency in IBM systems. Improves speed of data transmission between computers. Vendor: Software Engineering of America.

QuickApp Application development tool. Used in RAD (rapid application development). Incorporated into many middleware products. Vendor: Digital Communications Associates, Inc.

QuickChart Application development tool. Part of Power Tools. Supports Yourdon/Constantine Structured Design. Vendor: ICONIX Software Engineering, Inc.

Quicken Desktop software. Checkbook and home financial management program. Vendor: Microsoft Corp.

QuickGen/DB2 Application development tool. Automates programming function. Generates DB2 SQL data objects, Presentation Manager window definitions, DB2 programs. Runs on IBM DB2 systems. Vendor: The Object Group, Inc.

Quickgraph Plus Desktop system software. Graphics package. Runs on MS-DOS, Windows systems. Vendor: Sumak Enterprises, Inc.

QuickObjects Application development environment. Includes visual language. Used to build object-oriented client/server systems. Runs with SQLWindows. Vendor: Centura Software Corp.

QuickPascal Object-oriented programming language.

QuickPlace Communications software. Teamware. Allows users to share information and files over the Web. Includes a common calendar and the ability to edit and trade documents. Released: 1999. Vendor: Lotus Development Corp.

Quicksilver Desktop system software. Operating system add-on. Runs on Pentium type desktop computer systems. Compiles dBase programs for both stand-alone and network use. Vendor: Wordtech Systems, Inc.

Quicksilver/SQL Application development tool that works in a client/server architecture. Accesses Sqlbase, Oracle Server. Includes report generator. Vendor: Wordtech Systems, Inc.

QuickSite Application development tool. Web authoring program. Automatically inserts HTML tags into text to create Web pages. Vendor: DeltaPoint, Inc.

Quickstart Operating system add-on. Data management software that runs on IBM MVS systems. Enables users to restart batch jobs from the last database checkpoint. Vendor: Sysdata International, Inc.

QuickStart Data Mart Data warehousing software. Used for rapid development of data marts. Integrates Sybase IQ with Passport and other query tools. Bundled with IBM's RS/6000 systems. Released: 1996. Vendor: Sybase, Inc.

QuickStarts COBOL program templates that can be used to develop standard applications. Used with Netron's Fusion. Vendor: Netron, Inc.

Quicktime Multimedia system. Integrates sound, video and animation. Part of Apple's System 7 operating system, but available for general personal computer usage and runs on MS-DOS and Windows systems. Vendor: Apple Computer, Inc.

Quikfacts Application development tool. Report generator for Unisys systems. Part of GLE development environment. Vendor: PROGENI Systems, Inc.

Quikjob Operating system add-on. Debugging/testing software that runs on IBM systems. Vendor: Goal Systems International, Inc.

Quiklib Application development tool. Programming utility that maintains source code libraries. Runs on Hewlett-Packard systems. Vendor: Computer and Software Enterprises, Inc.

Quikwrite Operating system add-on. Data management software that runs on IBM systems. Works with Quikjob. Vendor: Goal Systems International, Inc.

QUILL Application development tool. Report generator for DEC systems. Includes query facility. Vendor: Digital Equipment Corp.

Qwiknet Professional Desktop system software. Project management package. Runs on Pentium type desktop computers. Vendor: Project Software & Development, Inc.

QUIN-54, 55 Desktop computer. Pentium CPU. Operating system: Windows 95/98. Vendor: DTK Computer, Inc.

R/2 Application software. Runs on large computer systems. Support is due to end in 2004. Vendor: SAP America, Inc.

R/3 Application software. ERP system that includes manufacturing and financial packages. Client/server package that runs on Unix systems. More often referred to as SAP than as R/3. Uses ABAP programming language. Vendor: SAP America, Inc.

R:BASE (5000) Relational database for desktop computer environments. Runs on IBM desktop systems. Vendor: Microrim.

r-tree Application development tool. Multifile report generator. Allows developers to define any number of fields from both fixed and variable data. Runs on DEC, Unix, Windows systems. Released: 1998. Vendor: FairCom Corp.

R2LAN Communications. LAN (Local Area Network) connecting desktop computers. Vendor: Crosstalk Communications/DCA.

RA/2 Operating system add-on. Programming utility that retrieves data from RACF databases. Used by systems programmers. Vendor: Advanced Software Products Group, Inc.

Rabbit-x Operating system enhancement used by systems programmers. Increases system efficiency in DEC VAX systems. Series of programs that enhances assorted operating system functions. Vendor: Raxco Software, Inc.

Racal Management System Communications software. Manages diverse equipment. Integrates with OpenView, NetView. Vendor: Racal-Datacom, Inc.

RacerPC Desktop computer. Pentium processor. Vendor: Racer Computer Corp.

RACF(/VM) Operating system add-on. Security/auditing system that runs on IBM systems. Stands for: Resource Access Control Facility. Vendor: IBM Corp.

RAD Rapid Application Development. A methodology which uses iterative prototyping to develop systems rather than follow formal design and review requirements. Combines development tools such as visual/4GLs and/or CASE software with management techniques such as brainstorming to quickly develop prototypes that can be modified to become fully operational.

RADD Toolkit Application development tool. Used to develop Lotus Notes applications. Runs on Windows systems. Stands for: Rapid Application Deployment and Development Toolkit. Vendor: Workgroup Productivity Corp.

Radely EDI Communications software. EDI package. Runs on DEC, Unix systems. Vendor: Radley Corp.

Radial See RuleMaster.

Radiant Midrange computer. UltraSPARC CPU. Operating system: Solaris. Vendor: Aspen Systems, Inc.

RADIO Cluster Midrange computer. Pentium CPU. Operating system: Windows NT. Vendor: Stratus Computer, Inc.

Radius Computer vendor. Makes Macintosh clones.

RADSL Version of ADSL. Stands for: Rate Adaptive Digital Subcriber Line. See ADSL.

RAF Communications software. Network connecting IBM desktop computers and DEC VAX computers. Stands for: Remote Access Facility. Vendor: Datability Software Systems, Inc.

Rage Media Desktop computer. Pentium CPU. Operating system: Windows 95/98. Vendor: Royal Electronics, Inc.

Raid 1. Operating system add-on. Debugging/testing software that runs on CICS systems. Vendor: On-Line Documentation, Inc.
2. Redundant Array of Inexpensive Disks. An installation of duplicate disk storage to ensure data reliability and reduce downtime.

Raima Software vendor. See RDM++.

Raima Database Manager See RDM++.

Rajan Relational database for desktop computer environments. Runs on IBM desktop systems. Vendor: Anil, Inc.

RAM Random Access memory. Dynamic computer storage. Programs and data are read into RAM when executing and reside there only temporarily.

Rampal Database access software that accesses Oracle's database server and Borland's Paradox databases. Developed jointly by Oracle Corp and TSR Systems Ltd.

random access The ability to access data in a non-sequential manner. Also called direct access. Data can be accessed randomly, or any data record can be accessed directly.

RangeLAN Communications software. Wireless LAN. Operates within a range of 500 feet.

RAP Communications. Protocol. Data link layer. Stands for: Remote Data Protocol. Vendor: Microsoft Corp.

RAPID Operating system enhancement used by systems programmers. Increases system efficiency in IBM VM systems. Provides printed support. Vendor: VM/CMS Unlimited, Inc.

Rapid/3000 Application development system. Runs on HP systems. Includes: Transact/3000, Inform/3000, Report/3000. Vendor: Hewlett-Packard Co.

Rapid Application Deployment and Development Toolkit See RADD Toolkit.

Rapid Deployment Template See RDT.

Rapid Reorg Operating system enhancement used by systems programmers. Increases system efficiency in IBM DB2 systems by reorganizing tables. Vendor: PLATINUM Technology, Inc.

Rapide See ADL.

RapiDeploy Operating system software. Installs and configures operating systems and application systems on many computers. Can configure new PCs and migrate existing systems. Vendor: KeyLabs, Inc.

Rapport Script Object based word processing product. Runs on Unix systems. Used to create complex reports and documents. Supports slide presentations. Vendor: Clarity Software, Inc.

RAPS See CA-RAPS.

RAS Operating system enhancement used by operations staff and systems programmers. Disaster control package. IBM systems. Defines resources required for processing after loss of system. Stands for: Resource Analysis System. Vendor: CHI/COR Information Management, Inc.

RAS/3000 Operating system enhancement used by systems programmers. Provides system cost accounting for HP systems. Stands for: Resource Accounting System. Vendor: Computer Consultants & Service Center, Inc.

RAS Enterprise Systems management software. Security package that handles end-user security issues. Vendor: Technologic Software Concepts, Inc.

RASCAL Real-time version of Pascal.

RAServer 2000, 2500, 2900 Midrange computer. Server. Pentium CPU. Vendor: Rascom, Inc.

RATFOR Compiler language used in midrange environments. Fortran compiler for Harris systems. Includes RATMAC. Vendor: Harris Corp.

Rational Apex Application development environment. Used for design and programming. Includes GUI debugger. Versions available for C/C++, Ada. Runs on Unix systems. Vendor: Rational Software Corp.

Rational Rose CASE products. Automates analysis, design functions and programming functions. Includes appropriate code generators. Available: Rational Rose/C++, Forte, PowerBuilder, Smalltalk, SQLWindows, Visual Basic. Used for object-oriented development. Runs on Unix, Windows systems. Vendor: Rational Software Corp.

Rational Suite Application development tools. Allows software teams to create visual application models and to use those models throughout the entire lifecycle. Includes: AnalystStudio, DevelopmentStudio and TestStudio. Interfaces with Microsoft's Visual Studio. Runs on Unix, Windows systems. Released: 1999. Vendor: Rational Software Corp.

Rational TeamTest Application development tool. Testing tools for functional testing of Web, ERP, and client/server applications. Includes LoadTest, Manager, Robot, Process, SiteCheck. Originally developed by SQA Inc. and called SQA Suite. Used with Windows 3.x, Windows NT, Windows 95/98. Vendor: Rational Software Corp.

Rational Visual Test Application development tool. Provides automatic testing for cross-Windows applications. Vendor: Rational Software Corp.

Raven Midrange computer. RS/6000 system with 12 64-bit processors. Vendor: IBM Corp.

Raxmanager Operating system enhancement. Monitors and controls system performance in DEC VAX systems. Vendor: Raxco Software, Inc.

Raxmaster Operating system enhancement. Increases system efficiency in DEC VAX systems. Vendor: Raxco Software, Inc.

Razor/CM,PT Application management tools. Razor/CM is a software configuration manager providing version control and release management. Razor PT is a configurable problem tracking system. Runs on Unix systems. Vendor: Tower Concepts, Inc.

RC/Migrator Data management system that works with DB2 databases. Runs on IBM systems. Vendor: PLATINUM Technology, Inc.

RC/QUERY,UPDATE Operating system enhancement used by systems programmers. Increases system efficiency in IBM DB2 systems. Vendor: PLATINUM Technology, Inc.

RC/Secure Application development utility. Programming utility that provides security in IBM DB2 environments. Vendor: PLATINUM Technology, Inc.

RCOM/MF Communications software. Network connecting IBM computers. Micro-to-mainframe link. Vendor: Computer Vectors, Inc.

RCS Application development tool. Tracks files across multiple platforms. Integrates with Visual Basic. Stands for: Revision Control System. Vendor: Mortice Kern Systems, Inc.

RCS K5 xxx Desktop computer. AMD K5 CPU. Operating system: Windows 95/98. Vendor: Royal Computer Systems Inc.

RCS P6C-166 Midrange computer. AMD K5 CPU. Vendor: Royal Computer Systems, Inc.

RD/Share Operating system add-on. Library management system that runs on IBM systems. Vendor: RD Labs, Inc.

RDB Expert Application development tool. Automates database design for Rdb/VMS databases. Vendor: Digital Equipment Corp.

Rdb/VMS Relational Database. Runs on DEC systems. Also referred to as Rdb. Vendor: Digital Equipment Corp.

RDBMS Relational Data Base Management System. See DBMS.

RDC Application development tool. Automates programming and testing functions. Full name: Rand Development Center. Vendor: Rand Information Systems, Inc.

RDD-100 CASE tool. Automates analysis, design functions. Works from requirements statements. Stands for: Requirements Driven Development. Modules include report writer, behavioral modeling. Interfaces with Teamwork. Runs on Unix systems. Vendor: Ascent Logic Corp.

RDF Operating system add-on. Data management software that runs on Tandem systems. Vendor: Tandem Computers, Inc. (a Compaq company).

RDI Computer Computer vendor. Manufactures desktop computers.

RDM 1. Relational database for midrange and desktop computer environments. Runs on DEC, IBM systems. Full name: RDM: The Application Developer. Vendor: Interactive Technologies, Inc. 2. Workflow software. Object-oriented package. Concentrates on document management. Vendor: Interleaf, Inc.

RDM++ Database used in object-oriented systems. Combines relational and pointer-based methodologies. Includes class libraries and other object-oriented functionality including object identity, persistence, polymorphism and inheritance. Runs on OS/2, Unix, Windows systems. Stands for: Raima Database Manager. Vendor: Raima Corporation.

RDM/2 Relational database for large computer environments. Desktop computer version available. Runs on most major systems. Utilizes SQL. Vendor: Amperif Corp.

RDM/CE Version of Raima database for handheld systems. Stands for: Raima Database Manager. Runs under Windows CE. Released: 1998. Vendor: Raima Corporation.

RDOS Operating system for midrange Data General Eclipse systems. Vendor: Data General Corp.

RDS-Assort Application development tool. Query tool used for data mining. Used by retailers to analyze product assortment at stores. Released: 1998. Vendor: NeoVista Software, Inc.

RDT Application development tools. Tools available to build templates to use BusinessObjects for Oracle, PeopleSoft, and SAP applications. Stands for: Rapid Deployment Template. Released: 1996. Vendor: Business Objects, Inc.

RDT xxxx Handheld, pen-based computer. Vendor: Omnidata International, Inc.

RDW See Retek.

RE/Cycle Application development tool. Moves existing applications into multi-user repository environment. Then performs all maintenance and new development. Vendor: CGI Systems, Inc.

Re-engineering The process of analyzing and redesigning business systems to improve speed, service, and quality before automating the systems. It means designing the manual and non-computerized functions to take advantage of new technologies, not simply automating the existing procedures. Usually uses RAD techniques. Also called BPR (business process redesign).

Re-engineering Product Set CASE product. Automates analysis, programming, and design functions. Works with existing DB2 and IDMS databases. Associated product names: Data Analyst, Database Administrator, Information Model Library, Programmer Assistant, Shared Design Database, Systems Analyst, Workstation Manager. Runs on IBM 386 desktop computers. Interfaces with Excelerator, IEW. Vendor: Bachman Information Systems, Inc. (now Cayenne Software)

RE for IE CASE product. Reverse-engineering tool. Part of IEF. Vendor: Texas Instruments, Inc.

Re:Solution Image processing system that runs on large computer systems. Vendor: ABB Engineering Automation.

RE/Vision Application development tool. Measures COBOL code and re-engineers COBOL programs. Works with Q/Auditor, Q/Artesian. Vendor: Eden Systems Corp.

ReachOut Data communications software. Allows remote access control of desktop computers. Supports TCP/IP. Vendor: Ocean Isle Software.

React 1. Operating system enhancement used by systems programmers. Monitors and controls system performance in IBM DOS/VSE CICS systems. Vendor: Macro 4, Inc. 2. Utility program for DEC systems. Stands for: Remote Access Test system.

Readit Operating system add-on. Debugging/testing software that runs on Unisys systems. Vendor: ESI.

Ready Desktop computer. Pentium CPU. Operating system: Windows 95/98. Vendor: NEC Computer Systems.

REAL/IX Real-time Unix operating system for MODCOMP desktop machines. Vendor: Modular Computer Systems, Inc. (MODCOMP).

Real/Star family (1000, 9xxx) Desktop computer. RISC CPU. Operating system: REAL/IX. Vendor: Modular Computer Systems, Inc. (MODCOMP).

Real/Star family MODCOMP 97 Desktop computer. 680X0 CPU. Operating system: REAL/IX. Vendor: Modular Computer Systems, Inc. (MODCOMP).

real-time An online system that provides answers to queries, and/or processes data updates to a database, when the data or query is entered. Opposed to batch, which holds the data or query until an entire "batch" can be processed.

real-time operating system An operating system which works with real-time, not batch, programs.

Real370 Application development tool. Source level debugger for IBM S/370 assembler. Runs on Pentium type desktop computers. Vendor: Realia, Inc.

RealAudio Communications, Internet software. Lets users with multimedia PCs play sound found on the Internet, i.e. there is a Web site that plays the original "War of the Worlds" broadcast. Used to add sound to Web sites. Vendor: Progressive Networks.

RealCICS Mainframe-compatible version of CICS that runs on desktop systems. Vendor: Realia, Inc.

RealDL/I Mainframe-compatible version of DL/I that runs on desktop systems. Vendor: Realia, Inc.

Realia CICS, COBOL, DL/1, IMS See CA-Realia CICS, etc.

Reality DBMS OS Operating system for midrange McDonnell Douglas systems. Vendor: McDonnell Douglas Information Systems Co.

Realix Operating system for midrange MODCOMP systems.

Realizer See CA-Realizer.

RealObjects Application development tool. Generates three-tiered business components and wizards to graphically join these reusable components into specific application. Runs on Windows systems. Released: 1996. Vendor: Cognos Inc.

RealSecure Operating system software. Security program that detects intruders, or hackers. Vendor: Internet Security Systems, Inc.

REBOL Compiler language. Interpreter. Used to develop Internet applications. Stands for: Relative Expression-Based Object language.

Recital Open Information Environment See ROIE.

RECODER(/CICS) CASE product. Automates analysis function. Vendor: Language Technology, Inc.

record A collection of fields about a single instance. For example, each record in a human resource file or database would contain all the fields about a single employee.

record locking See data locking.

Recovery for CICS See CICS Integrity Series.

Recovery Plus Operating system add-on. Data management software that runs on IBM IMS systems. Vendor: BMC Software, Inc.

Recovery Plus for VSAM See CICS Integrity Series.

RECOVR Operating system add-on. Data management software that runs on DEC PDP systems. Vendor: Machine Intelligence and Industrial Magic.

Recycle-SF Application development tool. Used to reverse engineer COBOL/CICS/DB2 applications. Vendor: Virtual Software Factory, Ltd.

Red Brick Formation Application development tool. Visual programming tool used to build data-transformation applications. Used in data warehousing. Released: 1998. Vendor: Red Brick Systems, Inc.

Red Pepper Software vendor. Products include manufacturing planning and scheduling and supply chain software. Acquired by PeopleSoft, and Red Pepper software is included in PeopleSoft manufacturing system.

RedBack Application development tool used to build Web applications for UniData databases. Includes GUI form painter, repository for Web pages. Runs on Unix, Windows systems. Released: 1997. Vendor: Ardent Software, Inc.

Redi-Micro Communications software. EDI package. Runs on Pentium type desktop computers. Vendor: Control Data.

Redimaster EIS. Runs on Pentium type desktop computers. Vendor: American Information Systems, Inc.

Rediscovery Application development tool. Restructures COBOL code and manages the reuse of old applications. Helps developers reuse existing COBOL code as an object on desktop systems. Vendor: IBM Corp.

REDUCE Programming language based on ALGOL and designed for nonmathematical manipulation of mathematical expressions.

REELBackup, REELLibrarian Operating system enhancements. Used to manage tape backup functions. Vendor: SCH Technologies, Inc.

Reengineering Set CASE tool. Used to extract subsystems from existing code. Organizes and streamlines existing software. Runs on Unix, Windows systems. Vendor: Software Emancipation Technology, Inc.

referential integrity The function in relational databases that protects against deleting information as long as dependant information remains in the database. i.e., if a Sales database has a vendor table and an order table, vendors could be deleted only if there were no open orders.

REFINE Application development tool. Re-engineering tool for Ada, C, COBOL, and FORTRAN programs. Vendor: Reasoning Systems, Inc.

Reflection EnterView Communications. Terminal emulation software for Web servers. Allows an unlimited number of users to access IBM mainframe, Unix, and DEC systems from the Web. Runs on Windows systems. Released: 1998. Vendor: Walker Richer and Quinn Inc. (WRQ)

Reflection X for NT Communications software. Used in client/server computing. Links PC clients to Unix servers. Vendor: WRQ.

Reflex(Plus, Workshop) Database for desktop computer environments. Runs on IBM desktop systems. Vendor: Inprise Corp.

Regal Portable computer. Vendor: Micro Express.

region Non-fixed division of computer storage.

Registry Editor Operating system software. Stores configuration files for all applications and systems initialization files in one database. Part of Windows NT. Vendor: Microsoft Corp.

regression analysis Function used in database processing. Statistical operations that help to predict the value of the dependent variable from the values of one or more independent variables.

regression testing A form of testing to ensure that changes made to a program or system do not affect its performance. All paths of a program are tested before a change is made and the results are saved. The program is modified and the changes are tested. Then the original pre-change tests are rerun to ensure the program or system still functions correctly in all ways.

Regulus Operating system for IBM systems. Vendor: Alcyon Corp.

REGULUS Operating system for midrange Smoke Signal systems. Vendor: Smoke Signal.

rehosting Moving applications off a mainframe to a midrange machine. Does not encompass converting to client/server technologies.

REI-DCV, REI-TUA Operating system enhancement used by systems programmers. Monitors and controls system performance in DEC VAX systems. Vendor: Rioux Engineering, Inc.

Relate/3000,DB Relational database for midrange and desktop computer environments. Runs on DEC, DG, HP systems. Vendor: Computer Representatives, Inc.

relational database A database with a table and index structure that has proved to be the most popular type of database.

Relational Data Base Manager Relational database/4GL for midrange and desktop computer environments. Runs on IBM, Unix, Xenix systems. Has 4GL capabilities. Vendor: Ultrasoft, Inc.

Relational DBC 386 Relational database machine. A dedicated computer that contains the DBMS necessary to process the database. These databases are compatible with Bull HN Bull systems. Vendor: Bull HN Information Systems, Inc.

Relativity Application development tool. Moves data from mainframe VSAM files to client/server relational database environments. Vendor: Liant Software Corp.

Relay/ Communications software. Network connecting IBM systems. Micro-to-mainframe link. Includes: Relay/150, Relay/VM, Relay/3270. Vendor: Relay Communications, Inc.

Release/Shipment Communications Communications software. EDI package. Runs on Pentium type desktop computers. Vendor: Genzlinger Associates, Inc.

Reliance Relational database/4GL for large computer environments. Runs on Concurrent systems. Utilizes SQL. Includes report generator. Vendor: Concurrent Computer Corp.

Reliance Access Query language used with Reliance Plus databases. Includes RQL/32 and RUS/32.

Reliance Builder Application development system. Includes 4GL and interfaces with Reliance Plus. Modules included: Application Builder, Transaction Sequencer, Menu Builder. Runs on Concurrent systems. Vendor: Concurrent Computer Corp.

Reliance Plus Relational database for large computer environments. Runs on Concurrent systems. Vendor: Concurrent Computer Corp.

Reliant RM1000 MPP computer. Operating system: Unix. Vendor: Siemens Nixdorf Information Systems Inc.

remote Terminals and/or processors that are a distance away from the main computer facility.

Remote Computing Services Application development tool. Allows users to designate a job for processing on remote systems. Vendor: SAS Institute, Inc.

Remote IMS Application development tool. Provides testing functions for IMS systems from remote workstations. Vendor: Originally Micro Focus, now MERANT.

Remote Library Services Application development tool. Allows desktop computer users direct query access to SAS mainframe data. Vendor: SAS Institute, Inc.

Remote Network Monitoring See Rmon Plus.

remote procedure call See RPC.

RemoteWare Communications software. Middleware. Lets remote and mobile workers access enterprise networks. Includes session automation, software distribution functions. Automatically keeps software installations standard by removing unauthorized software from the remote machines. Supports MAPI. Released: 1998. Vendor: XcelleNet, Inc.

Renaissance 1. See VIA/Renaissance. 2. Financial applications (General Ledger, Fixed Asset, Project Accounting, etc.). Client/server software. Vendor: Ross Systems.

Renaissance Balanced Scorecard Application development tool used in data warehousing. Provides multidimensional data analysis with software that measures customer performance so users can rank their customers. Vendor: Gentia Software.

Renaissance Migration Operating system enhancement used by systems programmers. Increases system efficiency in Sun systems. Manages disk storage space. Vendor: Epoch Systems, Inc.

Renaissance Software Software vendor. Products: Financial, Sales systems for AS/400 systems.

Rendezvous Software Bus Communications software. Message-oriented middleware. Uses publish-and-subscribe. Interfaces with Visual Basic, PowerBuilder. Vendor: Teknekron Software Systems, Inc.

Renovator Plus Application development tool. Integrates legacy mainframe applications with client/server networks. Vendor: Easel Corp.

REORG PLUS See DB2 REORG PLUS.

repeater Communications device used to connect two networks that operate under the same physical layer protocols.

Replic-Action Application development tool. Handles relational database replication to Notes. Vendor: Casahl Technology, Inc.

Replica Desktop system software. Document exchange software that allows cross-platform use of documents created under any software. For example, a document created in WordPerfect under Windows 95/98 will be readable by a user on a Macintosh system. Vendor: Farallon Computing, Inc.

replication Automatically copying and synchronizing data from one environment to another making necessary changes so the data can be accessed in the new system. Usually means coping it from one database into another. Commonly used in distributed processing, (including client/server) applications, data warehousing. Also called propagation.

Replication Agents Application development tool. Replicates DB2 data updates to client/server distributed systems. Runs on MVS systems. Vendor: Sybase, Inc.

replication Server A dedicated computer system that executes a replication application.

Replication Server See Sybase Replication Server.

Replication Toolkit Application development tool. Allows users to develop replication modules working with DB2 databases for distributed systems. Runs on MVS systems. Vendor: Sybase, Inc.

report Data output from a computer program that goes to paper or a terminal screen. Reports are non-permanent data. After the terminal is turned off or the paper discarded, the program will have to be rerun to recreate the data if needed. Programs that only produce reports are called query programs.

Report/3000 See Rapid/3000.

Report/DB2 Application development tool. Report generator for IBM systems. Vendor: Informix Software, Inc.

report generator A computer program that can generate formatted reports from simple input statements thus eliminating the need to write complete programs to accomplish fundamental print tasks. Also called report writer.

Report-IV Application development tool. Report generator for DEC VAX, Unix systems. Part of Idol-IV. Vendor: Thoroughbred Software International, Inc.

report mining See data mining.

Report Server Application development tool. Schedules and automatically distributes reports. Interfaces with Lotus Notes. Part of Focus Six for Windows. Vendor: Information Builders, Inc.

report writer See report generator.

Report Writer Application development tool. Report generator for DEC DIBOL systems. Vendor: Federated Consultants.

Report Writer, Report Analyst Application development tool. Query and report functions. Runs on Unix, Windows systems. Vendor: Raima Corp.

ReportCast Application development tool. Lets users access reports from Web browsers. Vendor: Actuate Software Corp.

Reporter/32 Application development tool. Report generator for Concurrent systems. Works from fill-in-the-blanks terminal screens. Vendor: Concurrent Computer Corp.

Reporter Edition Application development tool. Report writer. Part of Focus Six for Windows. Vendor: Information Builders, Inc.

ReportMart Application development tool. Enables users to search, view, and execute reports accessing information objects in dozens of local or remote source systems, including relational databases, desktop applications, file servers, and packaged applications. Provides Java-based access over the Web. Users don't need to know the location of information objects or the application which created them. They simply click on the object, and it is downloaded either as HTML or a native MIME type. Vendor: SQRIBE Technologies, Inc. (acquired by Brio Technology in August, 1999).

ReportPacks See Trend.

ReportSmith Application development tool. Database reporting and query tool. Used with Delphi. Runs on Windows systems. Vendor: Inprise Corp

repository An organized, shared collection of information about data. Includes the information found in a data dictionary and can also keep track of specific data requests by subject or division. The actual data is pointed to by the repository. Some repositories keep track of development projects.

Repository-based code generation Part of client/server computing. One type of application processing. Stores all application requirements in a repository and lets the system generate to code for various platforms.

Repository Manager/MVS An IBM program that is part of SAA, Common Programming Interface. Provides a consistent means of storing and accessing data across different applications. Also acts as a database of rules about software development. Part of IBM's AD/Cycle. Vendor: IBM Corp.

REQL Query language used with Revolve. See Revolve.

reQuest Relational database for midrange environments. Runs on Convergent systems. Vendor: System Automation Software, Inc.

RequisitePro Application development tool. Handles project management functions. Creates requirements repository which includes change management information. Can handle multiple projects. Runs on Windows systems. Released: 1998. Vendor: Rational Software Corp.

RES See NetShare.

RescueWare Application development tool. Converts legacy COBOL applications to Java, C++, Visual Basic. Released: 1998. Vendor: Relativity Technologies, Inc.

Reserve Enterprise Communications, Internet software. Scheduler and office manager. Runs on Unix, Windows NT systems. Vendor: Amplitude software corp.

Resolve 2000 EIS. Runs on Pentium type desktop computers. Vendor: Metapraxis, Inc.

Resolve Plus Operating system enhancement used by systems programmers. Monitors and controls system performance in IBM systems. Online tool for detecting, diagnosing, and solving system performance problems. Vendor: Boole & Babbage, Inc.

Resolve Productivity Set Relational database for large computer environments. Runs on Unix systems. Vendor: Resolve Logic Systems.

resource People, equipment, and/or software needed to accomplish a task.

ResourceCenter Application development tool. Catalogs, locates, and retrieves software, Including source and object code, design documents, and debugging reports. Vendor: CenterLine Software, Inc.

Response See OEA.

response time Elapsed time between user input to an online system and the response by the system.

ResponseR Relational database for large computer environments. Runs on Wang systems. Vendor: Coyne Kalajian, Inc.

restart Resume execution of a computer job other than at the beginning.

Restart Operating system enhancement used by systems programmers. Increases system efficiency in IBM VSE Power systems. Vendor: Quantum International Corp.

Resumix Interact for Lawson Application development tools. Passes data between Resumix's systems and Lawson's HR package. Released: 1999. Vendor: Resumix, Inc.

Retail Performance Monitor Application software. OLAP software for the retail industry. Used by a chain of stores. Identifies unprofitable products by store and analyzes effect of price markdowns. Released: 1999. Vendor: Pilot Software Inc.

ReTarGet Application development tool. Uses point-and-click to extract, scrub and migrate legacy data to relational databases. Automatically converts data types. Runs on Unix systems. Released: 1997. Vendor: Rankin Technology Group, Inc.

Retek Data warehouse software. Used for large and midsize systems. Also called RDW. Runs on Unix systems. Released: 1997. Vendor: Retek Information Systems, Inc.

RETOOL Application development tool. Program that adds structure to COBOL programs. Vendor: Computer Data Systems, Inc.

Retrievalware Application development tool. Used to build systems to retrieve text, image, voice over LANs and the Internet. Search engine. Released: 1996. Vendor: Excalibur Technologies Corp.

Retrofit CASE product. Automates analysis function. Vendor: Peat Marwick Advanced Technology.

RETROFIT Application development tool. Converts unstructured COBOL code to structured COBOL according to company's standards. Vendor: Compuware Corp.

reusability Application development concept. Creating data and/or programs that can be used by many applications. Part of object-oriented design. Also part of Information Engineering methodology and enterprise-wide processing.

Reveal Data management software used in data warehousing. Searches through reports stored online. Runs on Unix, Windows NT systems. Vendor: O'PIN Systems.

reverse-engineering The process of going backwards in the system development cycle, i.e. from coding to design.

Review Operating system enhancement used by systems programmers. Monitors and controls system performance in Adabase systems. Vendor: Software AG Americas Inc.

Revolution xxxx Midrange computer. Server. Pentium CPU. Vendor: Advanced Logic Research, Inc.

Revolve Application development tool. Provides system wide data analysis. Vendor: MERANT.

REvolve Application development tool. Programming utility that analyzes COBOL, CICS, DB2 and JCL statements. Incorporates query language REQL to build custom reports. Runs on Windows systems. Vendor: Burl Software Laboratories, Inc.

Rex Handheld computer, or PDA. Uses TrueSync. Vendor: Franklin Electronic Publishers.

REX-80/86 Operating system for DEC VAX systems. Vendor: Systems & Software, Inc.

Rex/Guard Operating system add-on. Security routines that protect data and Exec procedures in Rexx systems. Vendor: Banner Software, Inc.

REXCOM Relational Database/4GL for large computer environments. Runs on DEC, DG, Unix systems. Vendor: Rexcom Systems Corp.

REXX Scripting language that can be used as a job control language and a programming language. Used in desktop computer and IBM VM/CMS environments.

RexxWare Communications software. Management and migration programs used to upgrade versions of NetWare. Vendor: Simware, Inc.

Rexxware Migration Toolkit System software. Automates upgrading NetWare systems. Vendor: Simware, Inc.

Rhapsody 1. Application development tool. Used for object-oriented analysis, design, and implementation. Generates C++ code. Includes debugging functions. Vendor: Ilogix, Inc.
2. Operating system. Runs on Macintosh systems. Combination of System 7, Copland, NeXTOS. Now called Mac OS X. Vendor: Apple Computer Inc.

Rhapsody Client Image processing system that runs on midrange and mainframe systems. Vendor: AT&T Computer Systems.

RHC-44,88 Handheld computer. Vendor: Paravant Computer Systems.

Rhythm Supply Chain Planner Application software. Manufacturing software handling supply chain functions. Runs on Unix systems. Released: 1996. Vendor: i2 Technologies, Inc.

RIB See NetShare.

ring topology A network configuration where each computer is connected only to two other computers. The connection of all the computers forms a ring.

RIO EIP software, corporate portal. Allows users to publish to Internet or intranets by any standard software including word processors, spreadsheets, etc. Sets up channels of information so different corporate users can define their own needs. Runs on Unix, Windows NT.. Vendor: Datachannel, Inc.

RISC machine Midrange computer. Stands for: Reduced Instruction Set Computing machine. A computer designed with a specific instruction set so it can be more efficient. RISC machines are large desktop (midrange) computer systems. They are often called servers as they function in that capacity most of the time. Dominate operating system: Unix.

RISC/OS 4.5I Operating system. Version of Unix that removes English-language commands so it can be used internationally. Vendor: Mips Computer Systems, Inc.

RiskAdvisor Applications software. Data warehouse software for the insurance industry. Includes a warehouse model of insurance industry specific tables. Includes DSS software based on Forest & Trees containing typical queries for the insurance industry. Runs on Windows systems. Vendor: PLATINUM Technology, Inc.

RiverView (Pro, for Unix, for Windows) Communications. Network management software which configures, monitors, and troubleshoots devices. Provides network maps and cut and paste creation of network topologies. Runs on Unix, Windows systems. Released: 1997. Vendor: Advanced Computer Communications, Inc.

RJE Communications software. Transaction processing monitor. Runs on IBM systems. Stands for: Remote Job Entry Workstation. Vendor: IBM Corp.

RL/1 Application development tool. Report generator for Unix, Wang systems. Vendor: Business Computer Solutions, Inc.

RLN Communications software. Allows remote users to link to a LAN. Stands for: Remote Local-area Network. Vendor: Intercomputer Communications Corp.

rlog Operating system software. Provides remote login and file transfer functions. Allows administrators and developers to perform administrative and maintenance tasks on remote systems from a centralized location. Runs on Unix systems. Released: 1995. Vendor: Firesign Computer Co.

RM/COBOL Application development tool. Used to transition legacy systems to client/server. Vendor: Liant Software Corp.

RM-COS Operating system for IBM, Stride, NCR, and compatible desktop computer systems. Vendor: Ryan McFarland Corp.

RM200 Desktop computer. MIPS CPU. Operating systems: SINUX, Windows NT Workstation. Vendor: Siemens Nixdorf Information Systems, Inc.

RM400, RM600 Midrange computer. Server. MIPS 4400 CPU. Vendor: Siemens Pyramid Information Systems, Inc.

RMDS Application development tool. Report generator for IBM systems. Stands for: Report Management & Distribution System. Vendor: IBM Corp.

RMF Operating system enhancement used by systems programmers. Monitors and controls system performance in IBM systems. Stands for: Resource Management Facility. Vendor: IBM Corp.

RMI Communications standards. ORB technology which is part of the Java virtual machine and allows Java objects to be executed remotely. Stands for: Remote Method Invocation.

RMON Communications management. Any software that gathers statistics from remote networks and reports across a wide area network to a central management site. Stands for: Remote MONitoring.

RMON Plus Communications software. Collects data on traffic and errors. Handles the data link layer of activity. Extension of SNMP protocol. Works with IP, IPX, DecNet. Stands for: Remote Network Monitoring. Vendor: Technically Elite Concepts, Inc.

RMS 1. Data management system. Runs on DEC systems. The collection of routines that process input/output functions. Stands for: Record Management Services. Vendor: Digital Equipment Corp.
2. Operating system enhancement used by systems programmers. Monitors and controls system performance in IMS IDMS systems. Stands for: Real-time Performance Monitor. Vendor: DBMS, Inc.

RMS/ONLINE Application development tool. Report generator for IBM systems. Stands for: Report Management System/Online. Vendor: Mantissa Corp.

RMS(/XA) Communications software. Network connecting Datapoint computers. Stands for: Resource Management System. Vendor: Datapoint Corp.

RMX-86, RMX-286 Operating system for Scientific Micro systems.

RNA Component framework used to develop and deploy scalable e-commerce and Internet applications that operate over different platforms. Includes: RW-Client (ActiveX, MFC, and Java components for building GUIs); RW-Tools (C++ components); RW-Connect (includes Nouveau ORB and RPC middleware); RW-DataServer (relational databases, C++ connectivity); RW-Architect (includes Visual Case UML modeling tool). Stands for: Rogue Wave InterNet Architecture. Vendor: Roque Wave Software, Inc.

RNO See RUNOFF.

Road Warrior See Versa.

Roadster Desktop computer. Notebook. Pentium processor. Vendor: AMS Technology, Inc.

Roaster (Professional) Application development tools. Used to create cross-platform applications based on Java. Runs on Macintosh systems. Vendor: Natural Intelligence, Inc.

RoboHelp Application development tool. Add-on to Microsoft Word 97. Used to create Help documents. Creates WinHelp, HTML Help (Microsoft) and NetHelp (Netscape) files from Word documents. Vendor: Blue Sky Software Corp.

robot See spider.

Robot See SQA Suite.

Robot38 Operating system enhancement used by operations staff and systems programmers. Provides operator console support in IBM S/38. Vendor: Help/38 Systems.

RoboMon Operating system enhancement. Provides enterprise-wide system administration functions. Monitors all systems in the enterprise and uses push technology to handle remote installations. Vendor: Heriox Corp.

Rochade Data repository. Interfaces with many CASE tools including Bachman, Excelerator, Teamwork, IEF, IEW/ADW. Builds catalog of metadata when used with data warehousing. Runs on DEC, IBM, Unix systems. Vendor: R&O, Inc.

Rock City Desktop computer. Pentium processor. Vendor: the Panda Project Inc.

Rocket/QMF Operating system enhancement used by systems programmers. Monitors QMF sessions in IBM DB2 systems. Vendor: Rocket Software, Inc.

Rodeo Desktop computer. Notebook. Pentium processor. Vendor: AMS Technology, Inc.

ROIE Development suite. Includes several modules. Accesses Oracle, Informix, OpenIngres, DB2, rdb databases. Runs on Unix systems. Stands for: Recital Open Information Environment. Vendor: Recital Corp.

ROLAP A computer application that provides OLAP functionality from data stored in a relational database. Stands for: Relational On-Line Analytical Processing.

Rollout Term used in data warehouse design and processing. Represents the activity of distributing the same data warehouse solution to a larger audience than the one initially served by the first implementation. Rollout involves concerns of scaling the decision support system to many additional users and standardization.

ROM Read Only Memory. Computer storage that hold firmware and/or embedded systems. The programs stored in ROM cannot be erased.

Roma Auto Bridge Applications development tool. Allows developers to use middleware from IBM and Microsoft at the same time. Released: 1998. Vendor: Candle Corp.

ROPES Operating system enhancement used by systems programmers. Increases system efficiency in IBM systems. Print Spooler. Stands for: Remote Online Print Executive System. Vendor: Axios Products, Inc.

Roscoe See CA-Roscoe.

ROSE Communications protocol. Applications layer. Developed by OSI. Stands for: Remote Operations Service Element.

Rosettanet Consortium of IT businesses working on creating standards for e-commerce for the IT industry. The consortium started with building catalog standards (defining standard names and part numbers for products and processes so buyers could use universal catalogs), defining standards for software specifications, for memory specifications and for laptop specifications.

Ross Systems Software vendor. Products include Renaissance CS family of manufacturing, accounting, and sales software. Runs on various midrange systems.

router Communications device used to connect two networks that operate under the same network layer protocols. Specifically, a computer configured to determine the best path between nodes in a packet switching network, reduces traffic congestion.

RouterPM Application development tool. Publishes network management reports on a server so they can be accessed by an Internet browser. Vendor: 3DV Technology, Inc.

routine See program.

Royal Mystique Notebook computer. Pentium CPU. Vendor: Royal Electronics, Inc.

RP/Server Communications. Middleware. Allows desktop users to access IBM mainframe data distributed throughout the enterprise. Interfaces with NOMAD. Stands for: Remote Processing/Server. Runs on IBM mainframe systems. Released: 1998. Vendor: Aonix.

RP/Web Application development tool. Used to build Web-enabled applications on a mainframe-based Web server. Supports distributed applications and allows developers to move portions of applications between servers. Stands for: Remote Processing/Web. Runs on IBM mainframes. Released: 1998. Vendor: Aonix.

RPC Remote procedure call. Technology used in client/server computing to handle communications. The client "calls" the server with a request and waits until the request is handled.

RPC Tool Communications software. Network connecting IBM mainframes and DEC, Sun, and IBM workstations and desktop computers. Version also available that allows Macintosh desktop computers to operate across multivendor systems. Vendor: Netwise, Inc.

RPCpainter Application development tool. Used to build three-tiered client/server systems. Includes graphical tools for building interfaces, remote procedure calls. Vendor: Greenbrier & Russell, Inc.

RPG Compiler language used in large computer environments. Stands for Report Program Generator and was designed for applications with heavy reporting responsibilities. Dominant use in midrange environments. Versions include RPG, RPG II, RPG III, RPG IV, RPG/400.

RPG V Application development tool. Program that allows programmers to code free-form RPG. Program generates RPG III code. Runs on IBM S/38 systems. Vendor: Help/38 Systems.

RPL Compiler language used in midrange environments. Runs on Prime systems. Vendor: Database Systems Corp.

RPM-RSTS Operating system enhancement used by systems programmers. Increases system efficiency in DEC PDP systems. Used to locate system bottlenecks. Vendor: Northwest Digital Software, Inc.

RPS Operating system for midrange IBM Series 1 systems. Vendor: IBM Corp.

RPS 1100 Communications software. Transaction processing monitor. Runs on Unisys systems. Stands for: Remote Processing System. Vendor: Unisys Corp.

RQE Operating system enhancement used by systems programmers. Increases system efficiency by managing resource sharing between IBM and DEC processing systems. Stands for: Remote Queuing Extended. Vendor: Advanced Systems Concepts, Inc.

RQL/32 Query language used with Reliance databases. Stands for: Relational Query Language. Part of Reliance Access.

RRDS Relative Record Data Set. A type of VSAM file.

RS/6000 Desktop computer. RISC machine. IBM Power CPU. Operating system: AIX. Vendor: IBM Corp.

RS/6000 N40 Notebook Notebook computer. Vendor: IBM Corp.

RS/6000 SP Computer technology. CPU design that can execute 10 trillion instructions per sec. A computer built following this technology could execute in 1 second what would take a hand-held calculator 10 million years. Vendor: IBM Corp.

RSA encryption Security function. Public key encryption technique named after three MIT professors (Rivest, Shamir, and Adleman). See public key encryption.

RSAF Communications software. Network connecting Wang systems. Stands for: Remote System Administration Facility. Vendor: Wang Laboratories, Inc.

RSAM Access method used in IBM midrange systems.

RSC-JC Operating system enhancement used by systems programmers. Provides cost accounting for DEC VAX systems. Stands for: Job Costing. Vendor: Resource Systems Corp.

RSC-MM Operating system add-on. Data management software that runs on DEC VAX systems. Stands for: Module Management and Services. Vendor: Resource Systems Corp.

RSC-RC Operating system enhancement used by systems programmers. Provides system cost accounting for DEC VAX systems. Stands for: Resource Chargeback. Vendor: Resource Systems Corp.

RSC-RW Application development tool. Report generator for DEC VAX systems. Stands for: Report Writer. Vendor: Resource Systems Corp.

RSCS Operating system enhancement used by systems programmers. Increases system efficiency in IBM systems. Stands for: Remote Spooling Communication Subsystem. Part of ESF package. Vendor: IBM Corp.

RSDS Access method used in IBM midrange systems.

RS1/170, RS2/xxx Midsize computer. Server. UltraSPARC processor. Vendor: Integrix, Inc.

RSTS General-purpose operating system for desktop computers and midrange systems. Versions include: Micro/RSTS, RSTS/E.

RSVP Communications protocol. Used with Internet applications. Stands for: Resource Reservation Protocol. Sets bandwidth reservations to ensure response time and quality of service for voice, video, and data communications.

RSX General purpose operating system for desktop computers and midrange systems. Versions include: Micro/RSX, RSX-11M, RSX-11M-Plus, RSX-11S.

RT-11 Operating system for DEC PDP systems. Most compact operating system for PDP-11 computers. Stands for: Real-time system. Vendor: Digital Equipment Corp.

RT-Ada/88k Operating system for RISC systems. Real-time system that was designed to support 88K Ada applications. Vendor: Ready Systems Corp.

RT PC SQL/RT Relational database for midrange and desktop computer environments. Runs on IBM RT systems. Full Name: Structured Query Language/RT. Vendor: IBM Corp.

RT-Mach Real-time operating system for desktop computers. Vendor: Carnegie Mellon University.

RT-VOS Operating system for Harris systems. Vendor: Harris Corp.

RTA/IMS,CICS Operating system enhancement used by systems programmers. Monitors and controls system performance in IBM systems. Vendor: Candle Corp.

RTE Operating system for Hewlett-Packard systems. Versions include: RTE-A, RTE-6/VM. Vendor: Hewlett-Packard Co.

RTEE See ISEE, TSEE, RTEE.

RTIRIM Relational database for large computer environments. Desktop computer version available. Runs on IBM, CDC, Prime systems. Vendor: RIM Technology, Inc.

RTL A library of run-time routines for common functions. Stands for: Run-Time Library. Used in DEC VMS systems. Vendor: Digital Equipment Corp.

RTMP Communications protocol. Transport Layer. Part of AppleTalk. Stands for: Routing Table Maintenance Protocol.

RTMX O/S Real-time operating system for desktop computers. Vendor: RTMX, Inc.

RTOS Operating system for IBM and compatible desktop computer systems. Vendor: MicroWay, Inc.

RTP Communications protocol. Network Layer. Part of Vines. Stands for: Routing Update Protocol.

RtPM Application software. MES (Manufacturing Execution System) software. Makes information regarding process parameters, materials, scheduling and production readily available and easy to use. Provides accurate and timely information about plant status. Stands for: real-time Production Management. Vendor: Hilco Technologies, Inc.

RTQ Communications software. Transaction processing monitor. Runs on IBM systems. Stands for: Remote Terminal Query. Allows access to remote terminal screens. Vendor: On-Line Documentation, Inc.

RTU Operating system for midrange, desktop systems. Vendor: Harris Corp.

RTworks Application development tool. Used to build real-time monitoring and control systems. Runs on most RISC systems. Vendor: Talarian Corp.

RTX Operating system for IBM and compatible desktop computer systems. Versions include: AT/RTX, PC/RTX, RTX286, RTX84. Vendor: Real-Time Computer Science Corp.

Rubix (Master) Relational database for midrange and desktop computer environments. Runs on Unix, Xenix systems. Vendor: Infosystems Technology, Inc.

Ruggedized PowerLite Portable computer. Pentium CPU. Vendor: RDI Computer Corp.

RuleChecker Application development tool. Allows developer to automatically check code against a set of predefined rules. Rules can be defined for each project. Part of Logiscope. Runs on Unix, Windows systems. Vendor: CS Verilog.

RuleMaster Artificial intelligence system. Expert system building tool. Includes RuleMaker, Radial. Vendor: Radian Corp.

RuleServer Application development tool. Used in client/server development. Handles application partitioning. Works with Visual Basic, PowerBuilder, 4GLs. Vendor: Trinzic Corp.

Rumba Data Access Application development tool. Provides access to IBM mainframe and midsize systems from PC applications. Vendor: Wall Data, Inc.

Rumba for NetWare Communications software. Connects AS/400 and mainframe systems to NetWare. Vendor: Wall Data, Inc.

Rumba for Profs, Office/Vision Communications software. E-mail system. Provides e-mail in client/server environments by providing the client software to Profs and Office/Vision host systems. Vendor: Wall Data, Inc.

Rumba Office 95/NT Application Development tool. Used to develop object-oriented Visual Basic systems. Vendor: Wall Data, Inc.

Rumbaugh Object-oriented development methodology. Provides analysis and design techniques. Used to create object models.

Run-Time Library See RTL.

RUNOFF Application development tool. Program that facilitates the preparation of manuscripts. Also called RNO. Vendor: Digital Equipment Corp.

RunTime Helix Relational database for desktop computer environments. Runs on Macintosh systems. Includes DSS. Vendor: Odesta Corp.

RUS/32 Update system for Reliance databases. Stands for: Reliance Update System. Part of Reliance Access.

RW-Metro Application development tool. Used to develop object-oriented applications accessing relational databases. The developer imports the source code for class definitions to build the relationships between objects. Works with any relational database. Released: 1998. Vendor: Rogue Wave Software, Inc.

RX/V Operating system for Ridge systems. Unix-type system. Vendor: Ridge Computers.

RXSQL Data management system. Runs on IBM VM systems. Transfers data between SQL/DS databases and files. Vendor: IBM Corp.

S/36,38 See System/36,38.

S/360,S/370 Mainframe computer. Operating systems: MVS, VSE. Vendor: IBM Corp.

S/390 ESO, G3 Mainframe computer. Parallel processor. CMOS technology. Operating systems: MVS, VM, VSE. Also called System/390, SP1, SP2, PES machines. Vendor: IBM Corp.

S/420 Midrange computer. SuperSPARC CPU. Operating system: Solaris. Vendor: Axil Computer, Inc.

S-Bridge Communications software. Gateway for e-mail systems. Runs with Netware, MS-NET networks. Vendor: Computer Mail Services, Inc.

S-Designor See PowerDesigner.

S/MIME (3) Specification for security of e-mail functions over the Internet. Uses encryption and digital signatures. Stands for: Secure Multipurpose Internet Extensions 3.

S-MON Operating system enhancement used by systems programmers. Increases system efficiency in DEC VAX systems. Performance monitor and debugging aid. Vendor: Bear Computer Systems, Inc.

S-MP Superserver Supercomputer. Operating system: Solaris. Vendor: Cray Research, Inc.

S-TCAT/C Operating system add-on. Debugging/testing software that runs on DEC VAX, IBM, Unix systems. Stands for: System Test Coverage Analysis Tool for C. Vendor: Software Research Associates.

S10, S16, S26, S40, S46 Midrange computer. Pentium Pro processor. Operating systems: Unix, OS/2, Windows NT. Vendor: NCR Corp.

SA Browser See System Architect.

SA-1100 Computer chip. Used for small notebook, handheld, wallet PCs. Vendor: Digital Equipment Corp.

SA/RT Application development tool. Models real-time systems in DEC VAX environments. Stands for: Structured Analysis/Real-Time. Vendor: Tektronix, Inc.

SAA Systems Application Architecture. A series of software standards, communications protocols, and interfaces from IBM. The purpose of SAA is to allow users to develop software that will run on all IBM computer systems. IBM has named certain software to conform to SAA standards, and is making any necessary changes to this software. New product development, both by IBM and other vendors who write software for IBM systems, will follow SAA.

Saber-C Application development tool for developing programs in C. Facilitates development under X-windows. Vendor: Saber Software, Inc.

Sablime Application development tool. Program tracking system that keeps track of program development and change. Included with ADE. Vendor: NCR Corp.

Safari InfoTOOLS Application development tools. Suite of tools providing data access, management and connectivity functions to provide two-tier systems with three-tier capabilities. Runs on DEC, Unix, Windows systems. Released: 1997. Vendor: Interactive Software Systems Inc.

Safari ReportWriter Application development tool. Query and report functions. Three-tiered client/server system. Runs on Unix, VMS systems. Vendor: Interactive Software Systems.

Safari UDMS Application development tools. Provides report writing and data management functions. Allows access to data across multiple databases, file structures and platforms including Rdb, RMS, VAX DBMS, Ingres, Oracle, C-ISAM, Sybase, Informix, Supra and RS/1. Stands for: The User Data Management System. Runs on Unix, Windows systems. Released: 1997. Vendor: Interactive Software Systems Inc

SAFE Operating system enhancement. Provides systems management functions for enterprise-wide systems. Security system for the entire enterprise. Vendor: PLATINUM Technology, Inc.

Safe/400 Operating system enhancement. Security package for AS/400 systems. Vendor: Millennium Systems Products, Inc.

Safe-T-Net Operating system add-on. Security/auditing system that runs on Tandem systems. Vendor: Tandem Computers, Inc. (a Compaq company).

Safeguard Operating system add-on. Security/auditing system that runs on Tandem systems. Vendor: Tandem Computers, Inc. (a Compaq company).

SafeKeyper Security package that uses digital certificates. Used over the Internet and browsers must be enabled to work with the product. Vendor: BBN Corp.

Saffron Application development tools. Saffron 4GL is a scripting language for database reporting. Interfaces with DB/2, Informix, CA-Ingres, Oracle, Sybase, SQL Server. Saffron Publisher is a graphical user interface for designing web pages. Saffron Designer allows the user to define the layout of Web pages. Runs on Windows NT systems. Vendor: WebSci Technologies.

SAG Consortium of vendors working on defining standards to enable multiple SQL-based relational databases and tools to work together. Stands for: SQL Access group. Participants include Hewlett-Packard, DEC, Oracle, Sun, and others.

SAG Dual Pentium, RAID, STF Midsize computer. Server. Pentium II, Pro, processors. Vendor: SAG Electronics.

SAGE Application development tool. Program that runs on Harris systems. Stands for: Software Analysis and Generation. Vendor: Harris Corp.

Sagent Application development tools used in data warehousing. Used to create and operate data marts. Includes: Data Mart Server, Admin, Information Studio, Design Studio, WebLink, Analysis, Reports. Vendor: Sagent Technology, Inc.

Sager NP8xxx Notebook computer. Pentium CPU. Vendor: Sager Midern Computer, Inc.

Sagister Midrange computer. RISC processor, based on PowerPC. Operating system: AIX. Vendor: Bull HN Information Systems, Inc.

SAIC-SDDL CASE product. Automates programming and design functions. Stands for: SAIC Software Design & Documentation Language. Vendor: Science Applications International Corp.

SAIC VUE Application development tool. GUI used with Unix. Stands for: SAIC Visual User Environment. Full name: SAIC VUE Developer's Toolkit. Vendor: Science Applications International Corp.

SAL 4GL used with Centura databases and development tools. Stands for: SQLWindows Application Language. Vendor: Centura Software Corp.

SalesLogix Application software. Sales management system that includes account and contact management, manages unlimited contacts, assigns accounts to sales people or sales teams, provides lead processing, calendar management. Client/server system with salespeople on mobile systems. Interfaces with any Web browser. Server runs on Windows NT systems. Vendor: SalesLogix, Inc.

SalesTrak Application software. Tracks customer needs, forecasts sales and handles team sales. Vendor: AurumSoftware, Inc. (division of Baan).

Salsa Application development tool. Object-oriented tool for end-users. Users choose prebuilt objects such as "Employee" (which could contain name, address, employee number, etc.) and build database models. Vendor: Wall Data, Inc.

Salvo Communications software. Provides a Web interface with existing mainframe systems. Translates terminal screens into HTML screens to be accessed through browsers, and then translates the Web user input into mainframe terminal input. Vendor: Simware, Inc.

SAM 1. Access method used in IBM mainframe systems. Stands for: Sequential Access Method.
2. Systems management software. Security package that handles end-user security issues. Stands for: Security Administration Manager. Vendor: Schumann Security Software, Inc.

SamePage Groupware package that runs over the Web. Provides project management for workgroups. Vendor: WebFlow.

Sammi Application development tool. RAD tool used to build applications that have to manage networked data. Provides a common user access to multiple applications, databases, and computing platforms. Runs on DEC, Unix, Windows systems. Released: 1997. Vendor: Kinesix, Division of Scientific Software-Intercomp.

SAMS:Allocate Operating system enhancement. Allocates and controls disk and tape storage in MVS systems. Formerly called: VAM. Vendor: Sterling Software, Inc.

SAMS:Compress Operating system enhancement. Optimizes use of disk space by compressing data files up to 20% of original size. Formerly called: Shrink. Vendor: Sterling Software, Inc.

SAMS:Main Operating system software. Provides management of all data in client/server environments. Includes: SAMS:Expert, SAMS:Vantage, SAMS:Control. Vendor: Sterling Software.

SAMS:Vantage for Distributed Platforms, Vantage for MVS Operating system enhancement. Manages data storage across NetWare, Unix, Windows or MVS systems. Formerly called: SAMS:Expert. Vendor: Sterling Software, Inc.

SAMS:Vtape Operating system enhancement. Stores information intended to go to tape in a disk buffer so it can be retrieved at disk speeds before it reaches its final tape destination. Released: 1999. Vendor: Sterling Software, Inc.

Samsung Computer vendor. Manufactures desktop computers.

San Francisco Application framework used for object-oriented, component based systems. Can be used for ERP systems. Designed to speed the development of server-side Java applications. The application components will be created by third party vendors. Runs on AS/400, mainframe, Unix, Windows NT systems. Objects and frameworks are assembled into "towers" which can be licensed. Available towers: general ledger, accounts payable, accounts receivable, sales order management and warehouse management. Vendor: IBM Corp.

SAP Software vendor. Full name SAP AG. Products include: R/2,R/3 financial and manufacturing systems. SAP HR can be purchansed separately and integrated with R/3. See R/3.

SAP Connect Communications software. Links IBM's MQSeries middleware with SAP systems. Vendor: Candle Corp.

Sapiens Application development environment for developing object-oriented mainframe systems. Uses artificial intelligence facts and rules. Runs on IBM mainframe systems. Vendor: Sapiens USA, Inc.

Sapiens Ideo Application development environment. Builds GUI based applications Object-oriented. Accesses most mainframe and SQL databases. Runs on Unix, Windows systems. Released: 1995. Vendor: Sapiens USA, Inc.

Sapiens Vision Application development tool. Object-oriented tool used in RAD development of client/server systems. Runs on AIX, OpenVMS, SunOS, Ultrix systems. Vendor: Sapiens USA, Inc.

Sapiens Workstation Application development tool. Converts textual mainframe applications into client/server applications running under Windows. Applications must be written with Sapiens development tools. Vendor: Sapiens USA, Inc.

Sapphire Net Desktop computer. Pentium processor. Vendor: Tangent Computer, Inc.

Sapphire/Web Application server providing development tools and middleware. Links Web front end systems to client/server systems. Creates applications that use a Web browser as a user interface. Visual development tool that generates C++ code. Interfaces with Oracle, Informix, Sybase, mainframe systems. Vendor: Bluestone, Inc.

SAR 1. Operating system enhancement used by systems programmers. Improves system efficiency in IBM MVS systems. Recovers system volumes without operating system. Vendor: Innovation Data Processing.
2. Application development tool. JCL and/or SYSOUT maintenance program for IBM systems. Stands for: Sysout Archival and Retrieval System. Vendor: Essential Software, Inc.

Saros Document Manager Document management software. Intended for large enterprises, has no limit to servers and will download work when a server gets overloaded. Runs on OS/2, Unix, Windows NT systems. Vendor: Saros.

SAS Software vendor. Original products included mathematical/statistical programs for all major computer systems. Products also include 4GL, EIS, data warehousing software. Accesses DB2, SQL/DS, IMS, IDMS/R, Datacom/DB, Adabase, Oracle databases. Vendor: SAS Institute, Inc.

SAS/ACCESS Software Application development tool. Allows users to access data from many sources including IMS, DB2, Oracle, Rdb, IDMS, Datacom/DB, Adabase. Runs on DEC, IBM mainframe, MacOS, Unix, Windows systems. Released: 1995. Vendor: SAS Institute, Inc.

SAS/ASSIST Software Application development tool. Point and click interface which provides access to SAS System functions. Includes ability to create EIS systems. Runs on DEC, IBM mainframe, MacOS, Unix, Windows systems. Released: 1998. Vendor: SAS Institute, Inc.

SAS/AF Application development tool. Can be used to create objects for object-oriented development. Stands for: SAS Application Factory. Vendor: SAS Institute, Inc.

SAS/AF Software Application development environment. Used to create windows applications, CBT courses. and online help systems. Runs on mainframe, midsize, PC systems. Released: 1995. Vendor: SAS Institute, Inc.

SAS/EIS Application development tool. Used to develop EIS systems. Runs on MVS, OS/2, VMS, Windows systems. Vendor: SAS Institute, Inc.

SAS/ETS Application software. Financial planner which provides forecasting, planning and financial modeling functions. Runs on DEC, IBM mainframe, MacOS, OS/2, Unix, Windows systems. Released: 1998. Vendor: SAS Institute, Inc.

SAS/INSIGHT Application development tool. Prepares graphic data displays including bar charts, scatter plots and 3D charts. Links data so that changes in one graph show immediately on the others. Runs on DEC, IBM mainframe, MacOS, Unix, Windows systems. Released: 1998. Vendor: SAS Institute, Inc.

SAS/IntrNet Application development tool used to develop Web applications. Allows users to create dynamic Web pages without knowing HTML. Can access data in Oracle, Sybase, DB2, SAS warehouses. Runs on DEC, IBM mainframe, MacOS, OS/2, Unix, Windows systems. Released: 1997. Vendor: SAS Institute, Inc.

SAS/MDDB Multi-dimensional database. Provides unlimited views of multiple relationships. Stands for: SAS/Multi-dimensional Database. Runs on DEC, Unix, Windows systems. Released: 1998. Vendor: SAS Institute, Inc.

SAS/Warehouse Administrator Application development tool used in data warehousing. Manages meta data, schedules warehouse activities, handles warehouse organization and data transformation. Runs on DEC, Unix, Windows systems. Released: 1998. Vendor: SAS Institute, Inc.

SAT Operating system enhancement. Monitors and controls system performance in NCR systems. Stands for: Syslog Analog Tool. Vendor: Software Clearing House, Inc.

Satellite Group name for subsets of Millennium products that run on desktop computers. Products include: GL:Satellite, AP:Satellite. Vendor: McCormack & Dodge Corp.

Satellite (Pro) xxxx Notebook computer. Vendor: Toshiba America Information Systems, Inc.

Sather Object-oriented programming language. Patterned after Eiffel. Created at Berkeley.

SATT Application development tool. Automates testing function for PL/1 and COBOL II applications. Stands for: Software Analysis Test Tool. Part of AD/Cycle. Vendor: IBM Corp.

SAVERS/TRMS Application development tool. JCL and/or SYSOUT maintenance program for IBM systems. Runs on MVS systems. Vendor: Software Engineering of America.

Savvy PC Artificial intelligence system. Runs on Pentium type desktop computers. Provides database, programming language, and query language. Vendor: Excalibur Technologies Corp.

SB+ See System Builder.

SC... See Selling Chain.

SCADE CASE tool. Used to develop safety critical real-time systems. Runs on Unix systems. Vendor: Verilog SA.

scalable, scalability The principle that software will work when moved from a desktop computer to a large computer system. Part of open systems.

SCALE See Enterprise Developer.

SCALEScript See Enterprise Developer.

Scan/Cobol Application development tool. Automates programming function. Analyzes source code for such things as adherence to standards and presence of logic flaws (unexecuted code). Languages analyzed: COBOL, CICS, IMS, IDMS. Runs on IBM mainframes. Vendor: Computer Data Systems, Inc.

ScanMail System software. Monitors desktops from a central location. Virus checker. Runs on OS/2, Windows NT, Unix. Released: 1998. Vendor: Trend Micro, Inc.

SCdraw CASE product. Automates design function. Vendor: McDonnell Douglas Information Systems Co.

Scenario Application development tool used for data mining. Used to determine which of hundreds of relationships among variables are significant. Runs on Windows systems. Vendor: Cognos Inc.

SCENIC Mobile Notebook computer. Pentium CPU. Operating systems: OS/2 Warp, Windows 95/98, NT Workstation. Vendor: Siemens Nixdorf Information Systems.

SCENIC Pro Midrange computer. Pentium CPU. Operating system: Windows NT. Vendor: Siemens Nixdorf Information Systems.

SCERT II Operating system enhancement used by systems programmers. Monitors and controls system performance in IBM DG, Unisys, HP, Bull HN, DEC systems. Performance predictor and capacity planning tool. Stands for: Systems & Computers Evaluation and Review Technique. Vendor: Performance Systems, Inc.

SCF Operating system add-on. Security/auditing system that runs on IBM systems. Stands for: Security Control Facility. Vendor: Software Technology, Inc.

Schedule Operating system enhancement used by systems programmers. Job scheduler for DEC VAX systems. Vendor: International Structural Engineers, Inc.

Scheduler 1. Operating system enhancement used by systems programmers and operations staff. Job scheduler for Wang VS systems. Vendor: Glosser Software.
2. Operating system enhancement used by systems programmers. Job scheduler for HP systems. Vendor: Operations Control Systems.
3. See DistribuLink.

SCHEDULER See CA-SCHEDULER.

schema Data design tool. Set of statements, expressed in a data definition language (DDL), that completely describes the logical structure of a database. The diagrammatic representation of the data storage aspects of a database system.

Schema Manager Application development tool. Used by DBAs to create, track and deploy schema changes to Oracle databases. Provides change documentation, schema comparison and schema version control. Vendor: Quest Software, Inc.

Schema Mapper Products. See DBA Tool Kit.

Scientist's Workbench GUI for scientific applications. Runs on supercomputers. Vendor: Cornell University's Theory Center.

SCM Application management tool. Manages the source code of multi-programmer projects. Keeps all source files available to all programmers at all times. Tracks revisions to make releases consistent across the programming team. Stands for: Source Code Manager. Runs on DEC, Unix systems. Released: 1995. Vendor: UniPress Software, Inc.

SCO Foxbase See Foxbase (+,Plus).

SCO Integra See Integra.

SCO Internet FastStart Internet server. Includes: OpenServer Enterprise, Netscape Navigator. Released: 1996. Vendor: The Santa Cruz Operation, Inc.

SCO OpenServer Operating system. Single user version of Unix for desktop systems. Has Internet interface, including an HTML tool. Interfaces with Netscape's Web browser. Vendor: The Santa Cruz Operation, Inc.

SCO OpenServer 5 Operating system. Unix based system. Runs on desktop systems. Code name for system: Everest. Vendor: The Santa Cruz Operation, Inc.

SCO Professional Desktop system software. Spreadsheet. Runs on Unix systems. Vendor: The Santa Cruz Operation, Inc.

SCO Profile Suite Desktop system software. Integrated package. Runs on Pentium type systems. Vendor: The Santa Cruz Operation, Inc.

SCO System V General purpose operating system for desktop computers. Version of Unix. Vendor: The Santa Cruz Operation, Inc.

SCO Unix, Xenix Operating systems for IBM and compatible systems. Vendor: The Santa Cruz Operation, Inc.

SCO UnixWare Operating system for desktop computers. Unix type system. Conforms to X/Open standards. Runs Unix applications on NetWare and shares Unix files with NetWare users. Vendor: SCO, Inc.

SCO XVision Operating system software. Provides transparent access to Unix applications from Windows. Vendor: The Santa Cruz Operation, Inc.

SCONS/1000,3000 Operating system add-on. Library management system that runs on HP systems. Vendor: Corporate Computer Systems, Inc.

SCOOP Application development tool. Component of Eiffel. Includes Internet, client/server functions. Stands for: Simple Concurrent Object Oriented Processing. Released: 1996. Vendor: Interactive Software Engineering, Inc.

Scope Application software. Provides supply chain functions for manufacturing operations. Stands for: Supply Chain Operations Planning Environment. Runs on Windows systems. Vendor: Distinction Software, Inc.

SCOPE Application software. Part of SAP's ERP system. Stands for: Supply Chain Optimization, Planning and Execution.

SCOR Model of supply-chain operations developed by the Supply Chain Council, a consortium of hundreds of manufacturing companies. Intended to be a blueprint to map supply-chains, benchmark business performance, and evaluate supply-chain software. Stands for: Supply-Chain Operations Reference.

Scoreboard Application development tool. Automates programming function. Analyzes source code for such things as adherence to standards and presence of logic flaws (unexecuted code). Languages analyzed: COBOL. Runs on IBM mainframes. Vendor: Travtech, Inc.

Scout ixxxxx Desktop computer. Notebook. Pentium processor. Vendor: Melard Technologies Inc.

screen editor A computer program designed to handle the input of programming statements from a terminal screen. Similar to word processing software. Also called text editor.

screen generator See report generator.

screen painter Application development tool. A program that will generate the code necessary to define a terminal screen for an application system from a sample of the screen. Often part of applications development systems used for online system development.

screen scraper Application development tool used to translate a character user interface (CUI) on a mainframe or midrange system to a GUI by adding pull down menus and mouse support. The screen scraper software runs in the user's PC.

ScreenStar Portable computer. Vendor: Bitwise Designs, Inc.

script 1. In Unix systems, a prewritten set of commands.
2. A program written in a scripting language.

Script-IV 4GL. Runs on DEC VAX, Unix systems. Part of Idol-IV. Vendor: Thoroughbred Software International, Inc.

scripting Multimedia term. Used to describe multimedia tools that require programming to build applications.

scripting language Specialized programming language that processes text strings, searches files, databases and indexes, and generates reports. Heavily used with Web development. Easier to use than compiler or assembler languages. HTML and Perl are examples of scripting languages.

scriptlet Reusable HTML routine. Similar to Java's applets and work across platforms, except scriptlets do no calculations and are intended for page presentation and layout only. Developed: 1997. Vendor: Microsoft Corp.

ScriptX Multimedia description language. Object-oriented. Allows users to build one title that will work on multiple diverse devices. Vendor: Apple Computer Inc.

SCSI, SCSI-2 Communications protocol. Handles communication in peer-to-peer LANS with up to eight systems. SCSI-2 also covers peripherals such as tapes, printers, disks. Stands for: Small Computer System Interface. Pronounced: "scuzzy."

SDDGen Application development tool. Allows users to store and reuse software design elements. Supports multiple users. Stands for: Software Design Documentation Generator. Runs on Unix systems. Vendor: Trident Systems, Inc.

SDE Application development system. Runs on NCR systems. Stands for: Software Development Environment. Vendor: NCR Corp.

SDE/Workbench/6000 Application development tool. Group of programs used to develop applications for IBM's AIX RS/6000 system. Includes tools from third party vendors. Vendor: IBM Corp.

SDF Communications software. Network connecting IBM systems. Micro-to-mainframe link. Vendor: Multi Soft, Inc.

SDIS Communications software. EDI package. Runs on Pentium type desktop computers. Stands for: Standards Driven Interface System. Vendor: Telecommunications Interface Corp.

SDK Software Development Kit. Acronym used in name of many development tools and environments.

SDL Application development system. Includes 4GL, report generator. Stands for: System Definition Language. Runs on Unisys systems. Vendor: Source Data Systems, Inc.

SDLC The primary data link protocol used in IBM's SNA networks. Stands for: Synchronous Data Link Control. Vendor: IBM Corp.

SDM/LINK Communications software. EDI package. Runs on IBM systems. Vendor: SDM International, Inc.

SDM/Standard Application development tool. Aid for developing information systems. Includes method, documentation, estimating, and management guides. Vendor: AGS Management Systems, Inc.

SDM/Structured Programming development methodology. Uses structured documentation and generic techniques. Vendor: AGS Management Systems, Inc.

SDOS Operating system for Smoke Signal systems. Vendor: Smoke Signal.

SDS 1. Communications. Network management system. Distributes software across networks. Stands for: Software Distribution Service. Part of DME. Vendor: OSF.
2. See XCOM\SDS.

SDSL See DSL.

SDT Application development tool used to design large real-time systems. Runs on Sun systems. Vendor: Telesoft.

SE/Open for PowerBuilder Application development tool. Links Select Software Tool's Systems Engineer design tools with back-end PowerBuilder. Provides two way link, so design models can be built from PowerBuilder applications. Vendor: LBMS, Inc.

Seagate Client Exec Operating system software. Provides backup functions to a Windows NT server from user systems. Released: 1998. Vendor: Seagate Software.

Seagate Holos Application development environment used to create OLAP systems. Analyzes multidimensional data from any database. Provides parallel processing of data and has no limits to the amount of data that can be processed so scalable systems can be built. Vendor: Seagate Software.

Seahorse Handheld, pen-based computer. Operating system: NewtOS. Vendor: Digital Ocean, Inc.

Search 97 Agent Server Application development tool used for information filtering. Monitors content changes at selected Web sites. Runs on Windows systems. Vendor: Verity, Inc.

search engine Program that will search files, databases for selected text and phrases. Databases, Internet browsers, data warehouses all contain search engines. Some search engines build indexes to shorten search time for large data collections.

SearchExpress Document retrieval system. Runs on Windows systems. Vendor: Executive Technologies, Inc.

SearchServer See Fulcrum SearchServer.

Seascape Storage Enterprise Family of storage management software and storage devices. Includes StorWatch (software that allows central management of storage throughout a global enterprise from a web browser), Versatile Storage Server (disk storage accessible from OS/400, Unix, Windows NT systems), and Virtual Tape Server (optimizes tape storage). Released: 1998. Vendor: IBM Corp.

SEC100 Midrange computer. SPARC CPU. Vendor: Integrix, Inc.

second-generation computer Computer built with transistors. Commercially available in the early 1960s.

Secura Network computer. Vendor: Advanced Modular Solutions, Inc.

Secure E-Check Electronic version of paper check. Created using public/private key encryption. Cannot be changed once created and authenticity can be traced and verified. Signed with digital signatures. Under testing to be used by federal government.

Secure Electronic Transaction See SET.

Secure Multipurpose Internet Extensions See S/MIME 3.

Secure Ruix Database for Unix environments. Utilizes SQL. Vendor: Infosystems Technology, Inc.

Secure SQL Server Version of Sybase database that provides security functions allowing users to be able to store and access data according to individual security clearances. Intended for use by federal government users. Vendor: Sybase, Inc.

Secure Topix Operating system enhancement. Provides data security for online transaction processing in Unix environments. Vendor: AT&T Bell Laboratories.

Secure Web Server Communications software, Web Server. Includes security functions and can be used with e-commerce. Vendor: Red Hat Software.

Secure WebServer See WebServer, Secure WebServer.

SecureZone Communications software. Firewall. Vendor: Secure Computing.

Secured Network Gateway Internet software. Firewall. Vendor: IBM Corp.

SecureMax Operating system add-on. Security/auditing package that runs on DEC VAX systems. Vendor: Demax Software.

SecureONE Security framework which represents a set of APIs to create compatibility among various security packages.

SecurIT FIREWALL Internet software. Provides administration and security functions. Includes 3 modules: Basic Firewall, Kerberos and Virtual Private Network. Runs on Unix, Windows NT systems. Released: 1996. Vendor: Milkyway Networks Corp.

Security/3000 Operating system add-on. Security/auditing system that runs on HP systems. Vendor: VESOFT, Inc.

security/auditing system Packages that make sure only authorized people can access programs and data.

Security Cloner See Toolset-DB2.

Security for Open Systems See SeOS.

Security Management System See SMS.

SecurityPac Operating system enhancement. Security package for AS/400 systems. Vendor: Computer Security Consultants, Inc.

SED Editor program in Unix systems. Stands for: Stream Editor. Used to edit Unix scripts.

Sedit Operating system add-on. Unix editor which emulates mainframe editors. Runs on most Unix systems. Vendor: Treehouse Software, Inc.

Sedit.DB Application development tool. Assists in developing, testing, and managing DB2 applications. Runs on IBM DB2 systems. Vendor: Allen Systems Group, Inc.

Sedit(/DB) Operating system add-on. Data management software that runs on IBM MVS systems. Vendor: CGCI, Inc.

SEEC/Care COBOL Analyst Application development tool. Re-engineering tool for maintaining mainframe COBOL applications. Vendor: SEEC, Inc.

Seed DBMS Database for Unix environments. Vendor: Seed Software Corp. Division of Mantech Software Solutions Corp.

Seer/7000 Application development environment. Used to develop enterprise-wide, client/server systems. Provides choice of methods—model-based, object-oriented, RAD. Runs on Windows NT systems. Released: 1996. Vendor: Seer Technologies, Inc.

Seer*HPS CASE product. Used for enterprise development. Automates analysis, design, and programming functions. Includes code generator for C, COBOL. Used for client/server computing and component-based development. Design methodology supported: Information Engineering. Databases supported: DB2, OS2/DBM. Runs on IBM systems. Vendor: Level 8 Systems, Inc.

SELCOPY Operating system add-on. Data management software that runs on IBM systems. Vendor: Compute (Bridgend), Ltd.

Select Application development tool. Programming utility that accesses data from midrange computer systems for desktop computer word processors, spreadsheets, and database systems. Vendor: California Software Products, Inc.

SellerXpert See CommerceXpert.

Selling Chain Applications software. Front office software suite of sales and marketing functions. Includes: SC Catalog, SC Config, SC Pricer, SC Quote, SC Commission, SC Proposal, SC for Web, SC Contract, SC Promotion. Vendor: Trilogy Software.

SemioMap Application development software. Extracts information from unstructured text and presents it in a logical format to the user. Automatically sifts out irrelevant information and pursue links or relationships that interest user. Released: 1998. Vendor: Semio Corp.

senior programmer, engineer Developer. Could be applications or technical. Analyzes systems, designs logic and testing scenarios. May or may not actually write code. Works on systems and sub-systems rather than on single programs. Writes specifications and supervises mid-level and junior developers. Responsible for system performance. Senior level.

SENS Pro 5xx Notebook computer. Pentium CPU. Vendor: Samsung Electronics America, Inc.

Sens810 Notebook computer. Pentium CPU. Vendor: Samsung Electronics America, Inc.

Sentinel Communications. Monitors on-line traffic across networks by simulating user activity at remote sites. Warns system about such things as unavailable applications and network congestion. Originally developed by Bell Atlantic Corp and called EnView. Vendor: Amdahl Corp.

Sentinel Information Catalog Data management software used in data warehousing. Information catalog tool that provides a table of contents to the information stored in a data warehouse. Manages the metadata. Runs on Windows systems. Vendor: MayFlower Software.

SentinelTrack System management software. Provides enterprise-wide tracking of Java, Unix, Macintosh, and PC applications through Web browsers. Vendor: Rainbow Technologies, Inc.

SeOS Operating system software. Security package that provides access control and protects resources such as files, programs, userIDs, servers. Provides single point management of multiple Unix systems. Runs on Unix systems. Released: 1998. Stands for: Security for Open Systems. Vendor: Memco Software, Inc.

SEQUEL Application development tool. Query and report functions. Runs on AS/400 systems. Vendor: Advanced Systems Concepts Inc.

SeQueL to Platinum Applications software. Accounting package. Runs on OS/2, Windows systems. Interfaces with NetWare, SQL Server. Client/server technology. Vendor: Platinum Software Corp.

SequeLink See DataDirect SequeLink.

Sequent Computer vendor. Manufactures desktop computers.

Serdb V4.1 Version of Rdb database that provides security functions allowing users to be able to store and access data according to individual security clearances. Intended for use by federal government users. Vendor: Digital Equipment Corp.

SEREP System Error Record Editing Program. Program that edits and prints hardware error conditions. Used by systems programmers.

Series/1 Model xxxx Midrange computer. Obsolete. Used in manufacturing. Operating systems: EDX, RPS. Vendor: IBM Corp.

Series 3 Palmtop Handheld computer. Vendor: Psion Inc.

Series 3000i,6000i,7000i Desktop computer. Pentium CPU. Operating system: Windows 95/98. Vendor: Wyse Technology, Inc.

Server Server is a broad term that is used for many things. First of all, it describes both hardware and software. Server software is any program that makes anything—data, other programs, services, devices—available to any other program. A server computer is any computer that runs server software. There are many type of servers including database servers, Web servers, file servers, application servers, etc. Also see client/server.

Server/300,700,8000 Databases for Unix environments. Utilize SQL. Vendor: Sharebase Corp.

Server Express Communications. Middleware. Used to deploy e-commerce and distributed COBOL applications. Runs on Unix systems. Vendor: Originally Micro Focus, now MERANT.

Server Performance Prediction Operating system enhancement. Monitors and predicts performance of servers. Used to determine optimal performance. Released: 1999. Vendor: Tivoli Systems, Inc.

ServerPack Application development tool. Provides run-time testing tools for server software. Part of Cyrano Suite. Vendor: Cyrano, Inc.

ServerWare Group of software products that run under Windows NT, Unix, NonStop. Includes an SQL database, communications software including middleware and messaging software. Used to connect Tandem systems with Windows NT and Unix. Vendor: Tandem Computers, Inc. (a Compaq company).

Service Desk Applications software. Provides customer support by allowing customers to post problems on a Web site and ask for online assistance. Runs on Windows NT servers. Vendor: Silknet Software, Inc.

ServiceIT Operating system software. Knowledge-based help desk for Windows NT and Web-based environment. Vendor: Computer Associates International, Inc.

Servlet A small program, or object, created in Java. Similar to applets, except servlets run on the Internet server.

SES/objectbench Application development tool. Object-oriented. Creates Shlaer-Mellor models, tests the models through simulation. Generates C++ code. Vendor: SES, Inc.

SES/strategizer Application development tool. Modeling tool used to project an application's performance before implementation. Runs on Windows NT systems. Vendor: Scientific and engineering Software, Inc.

SES Workbench Graphical modeling tool that allows users to build models of virtually any kind of system. Has been used to model systems as diverse as the internals of a computer system and the design of the lines in a fast food restaurant. Used in re-engineering. Runs on Unix systems. Vendor: Scientific and Engineering Software, Inc.

session layer In data communications, the OSI layer that provides tools for applications such as e-mail, checkpoint/recovery and inquiry/response. Passwords are verified and system usage is monitored. The user communicates directly with this layer.

Session Recorder Application development tool. Program utility used to develop interactive application test situations. Part of COBOL/2 Workbench. Vendor: Micro Focus, Inc.

SET Communications protocol. Standards to exchange credit card information over the Internet. Uses electronic signatures, public key encryption. Stands for: Secure Electronic Transaction.

Seth Application development tool. Report generator for IBM systems. Vendor: CRE8 UnLimited.

SFA Sales Force Automation. See front-office software.

SFT III NetWare Fault tolerant version of NetWare. Protects against server memory and hard disk failures. Stands for: System Fault Tolerant III. Vendor: Novell, Inc.

SFTRAN3 Compiler language used in mainframe environments. Used in Harris systems and translates into FORTRAN. Vendor: Harris Corp.

SFVME Midrange computer. SuperSPARC RISC CPU. Operating system: Solaris. Vendor: Solflower Computer, Inc.

SGI Software vendor. Full name: Silicon Graphics, Inc.

SGML Basic markup language. Most other markup languages including HTML based on it. Complex with few limitations. Developer has full control over the positioning of text and images. Stands for: Standard Generalized Markup Language.

SGxxxx Desktop computer. Pentium CPU. Operating system: Windows 95/98. Vendor: Unisys Corp.

Shadow for DB2 Operating System enhancement. Runs on IBM DB2 systems. Used to facilitate remote use of DB2. Vendor: Neon Systems, Inc.

Shadow II Communications software. Transaction processing monitor. Runs on IBM systems. Vendor: THORN EMI Computer Software.

Share 1. Groupware. Works with existing communications systems. Vendor: Collabra Software, Inc.
2. Users group for information technology professionals. Focuses on technology for large systems.

Sharebase I, II Relational database for large computer environments. Desktop computer version available. Runs on most major midrange systems. Vendor: Britton Lee, Inc.

Shared Design Database CASE product. Part of Re-engineering Product Set. Vendor: Bachman Information Systems, Inc. (now Cayenne Software).

Shared LAN Cache Systems management software. Caching tool. Runs on Windows systems. Vendor: SourceCraft, Inc. Released: 1997. Vendor: Measurement Techniques, Inc.

Shared Work Manager CASE product. Part of Re-engineering Product Set. See Bachman/Re-engineering Product Set.

Shareoption 5 Application development tool. Programming utility that allows users to access a VSAM file for batch processing. Vendor: On-Line Software International, Inc.

shareware Software available over the Internet on a trial basis. After the trial, users are required to register and pay, which usually gets them additional features.

Sharp Electronics Computer vendor. Manufactures desktop computers.

shell Part of the Unix operating system. The shell is the program that interfaces with the user by reading commands from terminals or prewritten scripts and carrying out requested actions such as executing programs and transferring data between programs. Shell language is the equivalent to JCL in other operating systems.

shell update release Version of Windows NT which contains a Windows 95/98 interface. Vendor: Microsoft Corp.

SherpaWorks Application software. PDM package. Scalable system that handles product and process lifecycle requirements. Web-enabled application that integrates with legacy, enterprise-resource-planning (ERP) and engineering systems. Runs on Unix, Windows systems. Released: 1998. Vendor: Sherpa Corp.

shift supervisor Support personnel. Person responsible for the completion of the production schedule in a data center during a shift. Senior level. Title used in mainframe installations.

ShipIT System software. Provides automatic software distribution. Released: 1998. Vendor: Computer Associates International, Inc.

SHL Transform Application development tool. Used for re-engineering. Walks managers through business process and application development changes. Runs on OS/2, Unix, Windows NT systems. Vendor: SHL Systemhouse.

Shlaer-Mellor Object-oriented development methodology. Requires following a strict discipline, but works well with code generators thus allowing developers to make changes at analysis/design levels.

Shockwave Internet software. Compresses and transfers sound, graphics, and animation over the Internet. Used to add sound to Web sites. Included in some browsers so users can access multi-media titles and pages. Vendor: Network Music, Inc.

SHOPMON Operating system enhancement used by systems programmers. Monitors and controls system performance in IBM systems. Works with IDMS and MVS systems. Vendor: Allen Systems Group, Inc.

ShowBIZ Notebook computer. Pentium CPU. Vendor: Wedge Technology, Inc.

ShowBOOK Notebook computer. Pentium CPU. Vendor: Wedge Technology, Inc.

ShowCASE Application development tool. Used to develop object-oriented applications following the Booch design method. Generates C++ code. Runs on HP, Macintosh, Sun, Windows systems. Vendor: MultiQuest Corp.

ShowCase Vista Application development tool. Client/server query tool for AS/400 systems. Runs on Windows systems. Vendor: ShowCase Corp.

Showman P2, S2 Desktop computer. Notebook. Pentium processor. Vendor: TJ Technology, Inc.

ShowMe Groupware. EMS (Electronic Meeting System). Vendor: Sun Microsystems, Inc.

Showtext Desktop system software. Graphics package. Runs on MS-DOS systems. Vendor: Timeware.

Shrink See SAMS:Compress.

SHRINK See EXSYS.

SHTP Internet protocol. Sets e-mail extensions to ensure privacy and authenticate data transferred over the Web. Used by Web servers from Netscape, Open Market. Stands for: Secure Hypertext Transport Protocol.

SHTTP Protocol used to provide security measure on the Internet. Encryption standard. Stands for: Secure Hypertext Transfer Protocol.

Shuttle Desktop computer. Notebook. Pentium processor. Vendor: Tangent Computer, Inc.

Side by Side Application development tool. Provides line by line comparison of programs, JCL streams, screen panels. Runs on IBM mainframe systems. Interfaces with Librarian, Panvalet. Vendor: Jensen Research Corp.

Sidewinder Communications software. Internet gateway designed to protect classified information. Blocks data from moving between Internet and private corporate networks. Used to protect against hackers and viruses. Vendor: Secure Computing Corp.

Siebel Enterprise Applications Application software. Front-office systems. Modules include: Sales Enterprise (handles sales functions), Service Enterprise (customer service and call center functions), InterActive (on-line access to sales information including customer, competition data), and Tools (allows users to customize queries and reports.) Runs on Windows systems. Vendor: Siebel Systems, Inc.

Siena Handheld computer. Operating system: EPOC/16. Vendor: Psion Inc.

Signature Desktop system software. Family of graphical word processors. Replaces IBM's Displaywrite and Xyquest's Xywrite. Jointly developed by IBM and Xyquest.

Silk... Application development tools for Internet and client/server applications. Includes: SilkTest (provides automated functional and regression testing); SilkPerformer (load and performance testing); SilkControl (enterprise Web site management); SilkMeter (CORBA-based access control); SilkPilot (provides unit testing of CORBA objects); SilkObserver (transaction management and monitoring); SilkRealizer (provides scenario testing and system modeling); SilkRadar (automated defect tracking): and SilkTest (provides regression testing functions). Vendor: Seque Software.

Silverado Relational database for desktop computer environments. Runs on IBM desktop systems. LOTUS 1-2-3 add-on. Vendor: Computer Associates International, Inc.

Silvernet Communications software. Network operating system. Runs on multiple desktop computers. Vendor: Net-Source, Inc.

SILVERRUN CASE product. Automates design, programming, and testing functions. Runs on Macintosh, OS/2, Windows systems. Vendor: Computer Systems Advisers, Inc.

Silverrun-Enterprise Workflow software. Lets users build and manage enterprise-wide client/server models of business processes and corporate data warehouses. Vendor: Computer Systems Advisers, Inc.

Silverrun for PowerBuilder Application development tool. Generates server-level database designs for various databases including Ingres, Oracle, Sybase, Watcom, Informix. Vendor: Computer Systems Advisers, Inc.

SilverStream Application Server providing middleware and development tools. Used to build, deploy and manage Web applications. Links Web front-end systems to client/server systems using three-tiered architecture. Connects to legacy databases. Java, HTML based. Vendor: SilverStream Software, Inc.

SIM Society for Information Management. Professional association for high-level IS executives.

SIMAN Operating system enhancement used by systems programmers. Monitors and controls system performance in Unisys systems. Stands for: SIte MANagement Control. Vendor: Unisys Corp.

SimbaEngine Application development tool. Allows users to verify correctness of SQL data, access up to 5000 items in a table in a single query, and access non-SQL data. Runs on unix, Windows systems. Vendor: Simba Technologies, Inc.

SimbaExpress Communications software. Middleware. Connects ODBC applications to databases without proprietary middleware. Runs on Unix, Windows NT systems. Vendor: Simba Technologies, Inc.

SIMD Method of programming parallel computers. Stands for: Single Instruction/Multiple Data. Each processor executes the same instruction at the same time, but on different data. Contrast with MIMD.

Simon 1. Application development tool. Screen editor for Unisys systems. Part of GLE development environment. Vendor: PROGENI Systems, Inc.
2. Personal digital assistant, or hand-held computer. Includes phone, fax, pager, electronic-mailbox. Vendor: BellSouth Cellular Corp.

SIMON Application development tool. Interactive testing and debugging tool. Runs on VSE systems. Vendor: Compuware Corp.

SIMPAC Application development tool. Interactively alters and monitors systems. Interfaces with FORTRAN, Ada, Pascal, C/C++ libraries. Runs on Windows NT systems. Vendor: SCS Engineering, Inc.

Simple Application development tool. Program that develops application prototypes. Runs on Prime systems. Vendor: Prime Computer, Inc.

Simple Concurrent Object Oriented Processing See SCOOP.

Simple Mail Transfer Protocol See SMTP.

Simple Workflow Access Protocol See SWAP.

SIMS Communications software. Micro to mainframe link. Passes data query to mainframe and returns data to micro through hierarchy of menus. Interfaces with Systematics mainframe financial and accounting systems. Stands for: Systematics Information Management System. Vendor: Systematics Information Services, Inc.

SIMSCRIPT Programming language used for discrete simulation.

SIMUL8 Application development tool. Provides process modeling functions. Runs on Windows systems. Vendor: Visual Thinking International Ltd.

Simula67 Computer language developed in 1967 that is the forerunner of object-oriented programming. Introduced the concept of classes.

simulation Software that represents a physical unit. An example is a program that will allow one computer to execute the commands of another machine.

Sina Object-oriented programming language.

Single Sign-On Operating system enhancement. Provides security functions for MVS systems. Uses public-key encryption and Kerberos. Vendor: CyberSafe Corp.

Sinix General-purpose operating system. Unix-type system. Uses Motif. Conforms to X/Open standards. Vendor: Siemens Nixdorf Information Systems.

SINUX Operating system for Siemens Nixdorf midrange systems. Unix based. Vendor: Siemens Nixdorf Information Systems, Inc.

SIP400 Desktop computer. Pentium II processor. Vendor: CTX International, Inc.

SIPC Desktop computer. Stands for: Simply Interactive PC. Microcomputer for individual use. Vendor: Microsoft Corp.

SIRF/IDMS Operating system enhancement used by systems programmers. Monitors and controls system performance in IBM systems. Works with IDMS systems. Vendor: Allen Systems Group, Inc.

Site Assure Suite Communications. Internet software. Agent-based middleware providing load balancing and resource management functions to manage traffic over the Web. Automates the management of spikes in Web traffic. Released: 1999. Vendor: Platform Computing Corp.

Site Server Internet software. Used to manage Intranets and Web sites. Automatically replicates content of sites across multiple Web servers. Used to build web sites to publish and deliver information to corporate employees over the Internet or intranets. Provides users with a structured content submission, posting and approval process. Runs on Windows NT systems. Vendor: Microsoft Corp.

Site Server Commerce Edition Application software for e-commerce. Used to set up a virtual store. Runs on Windows NT systems. Released: 1998. Vendor: Microsoft Corp.

SiteAnalyzer Application development tool. Used in data warehousing. Compiles statistics describing data usage. Vendor: Information Builders, Inc.

SiteMeter Operating system software. Metering product. Vendor: McAfee Associates.

SiteMinder Internet software. Provides security controls on a central server that interacts with many different firewalls. Released: 1996. Vendor: NeTegrity, Inc.

SiteStak TW-xxx Midsize computer. Server. Pentium Pro processor. Vendor: Data General Corp.

SiteSweeper Application development tool. Monitors Internet activity tracking bad links, modifications, and number of links to site pages. Vendor: Site Technologies.

SITGO-10/20 FORTRAN IV language processor for DEC systems. Vendor: Digital Equipment Corp.

SIX Operating system enhancement used by systems programmers. Increases system efficiency in IBM IMS systems. Vendor: Innovative Designs.

Six Sigma A term used by statisticians and engineers to describe a state of nearly zero defects. Has been used as a goal for information systems.

SJS Operating system enhancement used by systems programmers. Monitors and controls system performance in IBM systems. Stands for: Synthetic Jobstream. Vendor: Stockholder Systems, Inc.

Skipjack See clipper chip.

Skyline See HDS Skyline series.

SLAVE Artificial intelligence language used in speech recognition and simulation. Stands for: Symbolic Language for Automated Verification and Execution.

SLEUTH Operating system add-on. Debugging/testing software that runs on IBM COBOL systems. Vendor: Computer Data Systems, Inc.

Slice 'n dice Term used in data warehouse processing. The activity of data analysis along many dimensions and across many subsets. This includes analysis of the data warehouse from the perspective of fact tables and related dimensions.

slide-based Multimedia term. Used to describe multimedia tools that control all changes from slides.

Slidewrite Plus Desktop system software. Graphics package. Interfaces with Lotus 1-2-3, Excel, dBase. Runs on MS-DOS systems. Vendor: Advanced Graphics Software, Inc.

Slim-Control Desktop system software. Project management package. Runs on Pentium type desktop computers. Vendor: Quantitative Software Management, Inc.

Slimnote Notebook computer. Vendor: Twinhead Corp.

SLIP Communications protocol. Lets Internet users connect through telephone circuits and through modems. Essential for Windows systems. Stands for: Serial Line Internet Protocol.

SLS Operating system add-on. Library management system that runs on DEC VMS systems. Vendor: Digital Equipment Corp.

SM-EDA-xx Midsize computer. Server. SPARC Pentium processor. Operating systems: OS/2, Unix, Windows NT. Vendor: Filetek, Inc.

Small Computer System Interface See SCSI.

Smalltalk Object-oriented programming language. The original versions, Digitalk (Smalltalk/v) and ParcPlace (Smalltalk-80) have been merged under the name Smalltalk.

SMART Application development tool. Allows users to convert 16-bit Windows applications to 32-bit applications to run under OS/2 Warp. Stands for: Source Migration Analysis Reporting Toolset. Vendor: One Up Corp.

smart card Card similar in appearance to regular credit cards but which includes microprocessors for monetary transactions. Used for such transactions as pay phones, subways, and parking meters.

Smart DB Workbench Application development tool used with data warehousing. Provides visual database design. Includes legacy data migration functions. Runs on Unix, Windows NT. Vendor: Smart Corp.

Smart*Doc Operating system enhancement used by systems programmers. Reports on job status in NCR systems. Vendor: Software Clearing House, Inc.

Smart Elements Application development environment. Used to develop object-oriented systems with a GUI. Runs on DEC, HP, IBM RS/6000, Sun systems. Vendor: Neuron Data, Inc.

Smart/Restart Operating system enhancement. Restarts production jobs from point of failure or last checkpoint. Vendor: Relational Architects International.

Smart Sockets Communications. Message-oriented middleware (MOM). Works with Unix and Windows systems. Vendor: Talarian Corp.

Smart*Start Operating system enhancement used by systems programmers. Job scheduler for NCR systems. Vendor: Software Clearing House, Inc.

smart terminal A terminal with display options, such as blinking cursors and reverse imaging (dark on white). Sometimes used as a synonym for intelligent terminal.

Smart Warehouse Data warehouse framework. Strategy and product selection intended to offer smooth integration between the warehouse and the operational systems. Vendor: Pyramid Technology.

SmartBase Application development tool used with data warehousing. Manages metadata. Interfaces with SmartMart and Microsoft Repository. Released: 1998. Vendor: Information Builders, Inc.

SmartBatch Operating system enhancement. Partitions a large job into pieces and distributes them to different processors depending on resource availability. Includes: Batch Accelerator, BatchPipes, Data Accelerator. Vendor: IBM Corp.

Smartbook Notebook computer. Vendor: Commax Technologies, Inc.

SmartBuilder Application development tool. Rapid application development software that interfaces with CASE tools and builds models that can be turned into working prototypes. Vendor: SmartStar Corp.

SmartCheck Application development tool used for testing. Provides automatic run time error analysis of Visual Basic programs. Part of DevCenter. See DevCenter.

Smartcheck/Smartview Application development tool. Automates programming function. Analyzes source code for such things as adherence to standards and presence of logic flaws (unexecuted code). Language analyzed: C. Runs on DEC, IBM, Sun systems. Vendor: Procase Corp.

SmartClient GUI used in financial and manufacturing application software from Oracle. Vendor: Oracle Corp.

Smartcom I,II,III CASE product. Automates project management function. Vendor: Hayes Advanced Systems.

SmartContent Application development tool. Graphically displays information on the Internet and across Intranets. Vendor: Perspecta, Inc.

SmartMove Program which includes both software and services from IBM, Lotus and other business partners. Used to assist users in moving from host-based and LAN e-mail systems to Notes and Domino.

SmartSockets Communications. Message-oriented middleware toolkit. Allows programs to communicate across platforms. Runs on DEC, Unix, Windows systems. Released: 1996. Vendor: Talarian Corp.

SmarText Electronic book software. Provides users with on-line access to manuals. Users input from standard word processors to create the electronic books. Vendor: Lotus Development Corp.

SmartForecasts Application development tool. Statistical forecasting tool. Allows user to make predictions based on past patterns. Runs on Windows systems. Vendor: Smart Software Inc.

SmartLook Application development tool. Allows users to view data on legacy systems from desktop computers using a GUI. Runs on Windows systems. Vendor: STS Systems, Inc.

SmartMart Data warehouse software. Used to build data marts. Includes tools for data extraction, transformation, storage, reporting and analysis. Includes Web access. Released: 1996. Vendor: Information Builders, Inc.

SmartMatch Application development tool used in data warehousing. Used for data cleansing. Detects duplicate names and addresses. Vendor: Group 1 Software, Inc.

SmartMode Application development tool used in Focus environments. Allows users to predict the resource utilization of Focus requests. Used for planning and interfaces with many databases including Oracle, Sybase, Informix. Vendor: Information Builders, Inc.

SmartMode for DB2 Works with DB2 databases. See SmartMode.

Smartreportpainter Application development tool. Report generator. Runs on DEC VAX systems. Vendor: SmartStar Corp.

SmartStar Vision, VMS, for Windows Application development environment. Used to develop object-oriented systems. Integrates GUI objects with database items. Versions available for Unix, VMS, Windows. Vendor: SmartStar Corp.

Smart*Stream* Series of client/server applications packages. Includes: SmartStream Decision Support (middleware that moves mainframe data to Sybase and provides queries and reporting functions based on Access); Financial Stream (accounting software); HR Stream (benefits, payroll, personnel). Runs on OS/2, Unix systems. Vendor: Gaec Computer Corp, Ltd. (formerly Dun & Bradstreet Software.)

SmartSuite (,97) Software suite. Includes 1-2-3 for Windows, AmiPro, Organizer, Freelance Graphics, Approach, Organizer, ScreenCam and CC:Mail. All applications in SmartSuite 97 are 32-bit applications. Vendor: Lotus Development Corp.

SMARTsystem Application development tool. Includes routines for re-engineering C programs, object-oriented database. Vendor: Perspective Software, Inc.

SmartTest Operating system add-on. Debugging/testing software that runs on IBM systems. Vendor: VIASOFT.

Smartview EIS. Runs on IBM desktop systems. Vendor: Dun & Bradstreet Software Services.

SmartWatch Operating system software. Measures performance in client/server systems by monitoring actual response time experienced by the user. Vendor: Landmark Systems Corp.

SMbasic Compiler language used in midrange environments. Emulates Business BASIC for DG systems. Vendor: Data General Corp.

SMDS A standard communications network for which a variety of products is available. Stands for: Switched Multimegabit Data Services. Alternate technologies are frame relay and ATM.

SME Application development system. Runs on Unisys systems. Stands for: System Maintenance Support Environment. Vendor: PROGENI Systems, Inc.

SMERFS Application development tool. Evaluates reliability by estimating number of errors in software. Works on historical data. Stands for: Statistical Modeling and Estimation of Reliability Functions for Software. Provided at no charge by U.S. Navy (Naval Surface Warfare Center, Dahlgren, VA).

SMF Operating system enhancement used by systems programmers. Monitors and controls system performance in IBM systems. Stands for: System Management Facility. Vendor: IBM Corp.

SMGL Development tool used to connect multimedia content (i.e. graphics, text images). The coding that allows files to appear as formatted pages on the World Wide Web. Stands for: Standard Generalized Markup Language. Similar to HTML but provides more formatting and tables.

SMIL Communications. Internet standards that reduce bandwidth for animation. Describes the layout and behavior of a screen presentation. Based on EML. Stands for: Synchronized Multimedia Integration Language.

SMIS See Lightship SMIS.

SMP 1. Symmetric multiprocessing. Design feature that lets all processors in a multiprocessing environment operate as equals rather than in a client-server mode. The processors share a common memory pool. 2. Communications protocol. Upgrade to SNMP. See SNMP.

SMR/JMR Application development tool. JCL and/or SYSOUT maintenance program for IBM systems. Stands for: SYSLOG Management and Retrieval/JOBLOG Management and Retrieval. Vendor: Duquesne Systems, Inc.

SMS 1. Communications standard. Interfaces storage management systems with a variety of file and database management systems. Stands for: Storage Management Service. Used with NetWare 4.01. Vendor: Novell, Inc.
2. See System Management Server.
3. Operating system add-on. Security/auditing system that runs on Prime systems. Stands for: Security Management System. Vendor: Q.S., Inc.

SMT Application development system. Used to develop applications for real-time monitoring of physical processes. Includes reusable classes. Can be specialized by industry. Stands for: System Management Template. Runs on Unix systems. Vendor: Template Software, Inc.

SMTF Operating system software. Moves data between mainframe and Unix systems. Used with data warehousing. Stands for: Symmetrix Multi-Host Transfer Facility. Vendor: EMC Corp.

SMTP Communications protocol. Applications layer protocol that deals with e-mail. Stands for: Simple Mail Transfer Protocol. Used with TCP/IP, the Internet , Unix.

SMU Operating system add-on. Library management system that runs on IBM systems. Stands for: Source Management Utility. Vendor: COSMIC.

SNA Communications protocols. Seven layer protocol used for WANs. Overall name of IBM communications systems. Stands for: Systems Network Architecture. Vendor: IBM Corp.

SNA Gateway Communications software. Transaction processing monitor. Runs on IBM CICS systems. Vendor: Interact.

SNA Server Communications software. Network connecting Desktops to host systems running SNA. Full name: SNA Server for Windows NT. Vendor: Microsoft Corp.

SNAP Application development environment used to develop distributed applications. Includes GUI, object technology. Stands for: Strategic Networked Application Platform. Vendor: Template Software, Inc.

SNAPpack Data Migrator Application development tool. Converts data to and from SAP R/3 and over 70 DBMS and application sources. Reformats data, edits fields, generates Abap/4 code for customization. Runs on IBM mainframe, Unix systems. Released: 1997. Vendor: Information Builders, Inc.

Snapshot Data management tool. Moves data from DB2 databases to OS/2 formats. Allows applications to download mainframe data and work on desktop computers. Vendor: Bridge Technology.

SnapShot Systems management software. Manages Web traffic. Vendor: Tinwald Networking Technologies, Inc.

SNiFF+ Application development environment. C/C++ environment that includes graphical browsing, configuration management, documentation tools, and support for development teams. Runs on Unix, Windows systems. Vendor: Integrated Systems, Inc.

SNiFF+ for Java Application development tool. Used to generate Java code. Runs on Windows systems. Released: 1996. Vendor: Integrated Systems, Inc.

sniffer Hardware and/or software used to debug network problems.

snippet JavaScript programs. Includes functions such as scrolling messages (also called ticker tapes), warning boxes, cookie counters, and bozo filters.

SNMP Communications protocol. Standards for monitoring and configuring both devices and applications. Part of TCP/IP protocol suite. Stands for: Simple Network Management Protocol. Provides a common format for devices such as modems, routers, and bridges. Runs on MS-DOS, Unix systems. Part of TCP/IP.

SNOBOL Compiler language used in large computer environments. One of the first list processing languages. Stands for: StriNg Oriented symBOlic Language.

Snoop Operating system add-on. Debugging/testing software that runs on IBM systems. Vendor: Interactive Solutions, Inc.

Snoopit Operating system software. Provides terminal control and security functions. Saves terminal sessions to play back later. Provides control of remote terminals and keyboards. Allows a single session to appear on multiple terminals simultaneously to conduct training sessions. Runs on Unix systems. Released: 1995. Vendor: Cactus International, Inc.

snowflake schema Data design tool. An extension of the star schema design where each of the points of the star further radiates out into more points. In this form of schema, the star schema dimension tables are more normalized. The advantages provided by the snowflake schema are improvements in query performance due to minimizing disk storage for the data and improving performance by joining smaller normalized tables rather than large denormalized ones. The snowflake schema also increases flexibility of applications because of the normalization and therefore the lower granularity of the dimensions.

SNS Communications software. Network connecting IBM and DEC computers. Allows communication between mainframes, midsize and micro systems. Includes: SNS Connect, SNS/LU6.2, SNS/Netconnect, SNS/SNA Gateway, SNS Printq, SNS SNApath, SNX/937X. Vendor: Interlink Computer Sciences, Inc.

SNS/NFS Communications software. Allows a mainframe to act as a file server to a LAN. Vendor: Interlink Computer Sciences, Inc.

SNS/Program Application development tool. Allows developers to write programs that exchange data between IBM mainframes and DECnet systems. Vendor: Interlink Computer Sciences, Inc.

SNS/TCPaccess Communications software. Connects IBM mainframes and TCP/IP networks. Vendor: Interlink Computer Sciences, Inc.

socket Communications term. As hardware, a socket is a plug, or connection. In the Internet, a section of a network computer that is dedicated to a specific task.

SOCKS Framework Communications. Internet protocol. Provides secure data feeds across firewalls. Vendor: NEC Systems Inc.

Socrates Application development tool. Provides a data analysis front-end to Microsoft's Plato, which is a multidimensional database and OLAP tool. Runs on Windows systems. Released: 1998 Vendor: Cognos Inc.

SoDA Application development tool. Generates documentation from development sources. Vendor: Rational Software Corp.

Soft-Switch See Lotus Soft-Switch.

SoftAudit/ONE Operating system enhancement. Monitors installation's use of software products, locates programs, tracks program use and modification. Runs on IBM MVS systems. Released: 1997. Vendor: Isogon Corp.

SoftBench Application development tool. Provides interface for multi-vendor development tools. Analyzes C++ code. Part of CASEdge. Runs on Solaris, UP-UX systems. Vendor: Hewlett-Packard Co.

SoftBench OpenStudio Application development tools. Enables developers to write C++ applications for client/server systems with Windows clients and HP-UX servers. Vendor: Hewlett-Packard Co.

Softbol Compiler language used in midrange and desktop computer environments.

SoftMail Communications software. E-mail system. Runs on IBM systems. Vendor: H & M Systems Software, Inc.

SoftNet Communications software. Allows Unix workstations to become integrated with NetWare servers and/or clients. Vendor: UniPress Software, Inc.

SoftProbe/86 Operating system add-on. Debugging/testing software that runs on DEC VAX, Unix systems. Vendor: Systems & Software, Inc.

SoftSolutions Groupware. Document management system that interfaces with Lotus Notes and work-flow software. Vendor: Novell, Inc.

SoftTest Application development tool. Provides testing, debugging functions. Formats test data. Generates decision tables summarizing testing showing what system functions have been tested. Runs on Windows systems. Vendor: Bender & Associates.

software Programs, procedures, rules, and documentation used in a computer system.

Software 2000 Software vendor. Produces object-oriented applications for IBM's AS/400 environments. Includes human resource software.

Software Backplane CASE product. Provides a framework for CASE technology. Includes development tool integration functions. Runs on DEC, Unix systems. Vendor: Altherton Technologies.

Software Design Documentation Generator See SDDGen.

Software Development Kit See SDK.

software engineer Software developer. Title originally was used in the desktop world and was associated with client/server, GUI and Unix systems, but now is used by developers throughout the industry and has no definitive definition. Can be any experience level.

Software Engineering Workbench CASE product. Automates analysis and design functions.

Software Foundry Application development tool. Program generator for DEC VAX, Sun Apollo systems. Vendor: Quantitative Technology Corp.

Software Meter Operating system enhancement used by systems programmers. A group of products that monitors IBM MVS/ESA systems. Vendor: Proginet Corp.

software process management system Software that controls all aspects of software development. Tracks code development, change requests, provides version control. Automates entire project development plan.

Software Quality Management System See SQMS.

Software Refinery Application development tool. Used to convert mission-critical systems to Unix systems. Includes source code analyzer, workbench functions. Vendor: Reasoning Systems.

software specialist Developer. Could be applications or technical. Used for a variety of job skills and levels, so has no real meaning.

software suite Desktop system software bundled together. Includes any combination of word processor, spreadsheet, graphics, e-mail, and database software.

Software TestWorks Application development tool. Testing and debugging tool. Interfaces with STW/Advisor. Vendor: Software Research.

Software through Pictures CASE product. Automates analysis, design, and programming functions. Component based, visual modeling tool. Includes object-oriented analysis and design tools, code generator for Java, IDL and C++. Databases supported: DB2, Informix, Ingres, Interbase, Oracle, Sybase, any standard SQL database. Supports Booch, UML, OMT methodologies. Runs on Unix, Solaris, Windows NT systems. Vendor: Interactive Development Environments, Inc.

SoftWindows Emulation software. Runs on PowerPC Macintosh, Unix systems. Allows users to run Macintosh/Unix and Windows applications on the same machine. Users still need to have Windows, as this software emulates the Intel 486 processor rather than the software. Vendor: Insignia Solutions, Inc.

SOGA Communications software. Used to connect PCs with IBM host systems. Stands for: SNA Open Gateway Architecture. Part of SNA Server, bundled with BackOffice. Runs on Windows NT systems. Vendor: Microsoft Corp.

Solaris Operating system for workstations. Unix-type system. Mostly used on workstations, but is scalable from 486 laptops to supercomputers. Can run software developed for Windows, Macintosh. Originally named SunOS. Vendor: Sun Microsystems, Inc.).

Solaris Internet Server Application development tool. Sends Web sites and groupware business applications over corporate networks and the Internet . Used to build intranets. Released: 1996. Vendor: Sun Microsystems, Inc.

SolarNet Communications software. Used in client/server computing. Links Windows clients to Unix servers. Vendor: Sun Microsystems, Inc.

Solo xxxx Notebook computer. Pentium chip. Operating system: Windows 95/98. Vendor: Gateway 2000, Inc.

SOLOMAN/MRS Operating system add-on. Library management system that runs on Bull HN systems. Stands for: Source/Object Library Online MANager.

Solstice System management software. Provides both network and systems management functions. Predicts potential failures. Object oriented. Runs on Sun systems. Includes: AdminTools, AutoClient, Backup, Cooperative Consoles, Enterprise Manager, FireWall-1, SunNet Manager. Vendor: Sun Microsystems, Inc.

Solstice Internet Mail Server See Internet Mail Server.

Solstice PC-Cache FS Systems management software. Caching tool. Reduces network traffic by copying frequently used files into user's hard drive. Runs on Windows systems. Vendor: Sun Microsystems, Inc.

Solution Frameworks Application development tool. Data mining software designed for banking applications. Vendor: HyperParallel, Inc.

Solution Series/ST See Cyborg Systems.

Solution-IV Application development tool. Utility that provides accounting information for IDOL-IV environments. Runs on MS-DOS, Unix, VMS systems. Vendor: Thoroughbred Software International, Inc.

Solve:Automation Operating system enhancement used by systems programmers. Automates IBM's MVS mainframe operations by providing functions to, i.e., automatically establish communications lines between branches during work hours. Vendor: Sterling Software, Inc.

Solve:Central Operating system enhancement used by systems programmers. Object oriented modules that provide problem, change, configuration, and asset management. Used with IBM mainframe, Unix systems. Vendor: Sterling Software, Inc. (Operations Management Division).

Solve:LAN LAN manager for NetWare that runs on a mainframe. Can manage multiple NetWare LANs. Vendor: Systems Center.

Solve:Netmaster Communications. Network management software. Links multiple SNA networks and MVS systems operations. Runs on MVS systems. Vendor: Sterling Software, Inc.

Solve:Operations Operating system enhancement. Automates management of the data center. Versions available for MVS, SNA and CICS or OpenView and NetView. Formerly called: SOLVE:Automation, SOLVE:Commander. Vendor: Sterling Software, Inc.

SOM System Object Management. Technology from IBM which uses an object-oriented framework for defining and managing binary class libraries. CORBA compliant. Middleware that links all IBM platforms. Developers can write modules as if they were all on the same system as long as they communicate through SOM. Used to develop client/server systems. Includes DSOM, which supports distributed computing.

SOMobjects Communications, Middleware, object request broker (ORB). CORBA compliant. Includes development kit. Creates objects independent of programming language. Used with OS/2, AIX, Windows systems. SOM stands for: System Object Module. Vendor: IBM Corp.

SONET Communications network associated with ISDN. Stands for: Synchronous Optical NETwork. Provides a standard way of transmitting digital information using fiber-optic circuits.

SONY-OS Operating system for SONY systems. Vendor: SONY Corporation of America.

sort To arrange records in order according to one or more key fields. Sort programs are part of the operating system, and are also available from outside vendors.

SORT See CA-SORT.

Sound Bytes Software that makes synthesized speech more lifelike. Adds a natural rhythm that includes pitch variations, word stress patterns, and punctuation. Vendor: Emerson & Stern Associates.

Source Code Manager See SCM.

Source Integrity Application development tool. Used for software management and monitors the complete development lifecycle. Select Edition and Professional Editions available. Vendor: Mortice Kerns Systems, Inc.

source language The language computer programs are written in, i.e., COBOL. Source language must be translated into machine language in order to be executed.

Source Navigator Application development tool. CASE tool that directs user through analysis, change, compile and debug cycles. Provides code analysis and reverse engineering for C/C++, Fortran, Java and Tcl programs. Produces the design definition from the code. Runs on Unix, Windows systems. Released: 1997. Vendor: Cygnus Solutions.

SourcePoint Application development tool used in data warehousing. Automates extraction, transport and loading of data warehouses. Interfaces with Natural, COBOL, C. Vendor: Software AG Americas, Inc.

source program A program written in a computer language, such as COBOL.

SourceManager CASE product. See Manager family.

Sourcemaster Operating system add-on. Library management system that runs in IBM MVS environments. Vendor: International Business Information Systems.

Sourcerer I Application development system. Runs on Bull HN systems. Vendor: Scientific and Business Systems, Inc.

SourceTools Operating system add-on. Library management system that runs on DEC VAX systems. Vendor: Oregon Software, Inc.

Sourceview Application development tool. Automates programming and design functions.

Soundx 3000, 4000 Notebook computer. Pentium CPU. Vendor: Sceptre Technologies, Inc.

SP-Xmotif Programming utility that works with X-Windows to provide a graphical user interface. Runs on AT&T Unix systems. Vendor: Concurrent Computer Corp.

SP1,SP2 See Parallel Enterprise System.

SPA Software Publishers Association. Purpose is to prevent software piracy.

Space Recovery System See SRS.

SpaceOLAP Application development tool. DSS. Provides access to data stored in relational databases. Runs on Unix, Windows systems. Vendor: InfoSpace Inc.

SpaceSQL Application development tool. Query and report functions. Java based. Creates queries to run from a Web browser. Generates SQL. Runs on Windows NT systems. Vendor: InfoSpace Inc.

Spacial Analysis System See Span.

spam, spamming Sending mass mail over the Internet . Firewalls often include functions to block spamming as a mass mailing can cause lengthy delays or even crash a server.

Span Geographical information system. Stands for: SPatial ANalysis System. Vendor: Tydac Technologies Corp.

SPARC family Midrange computers. RISC machines. SPARC, microSPARC, SuperSPARC TurboSPARC, UltraSPARC CPUs. Operating systems: Solaris. Includes: SPARCcenter, SPARCclassic, SPARCcluster, SPARCserver, SPARCstation (IPX, Voyager), UltraSPARC. Stands for: Scalable Performance Architecture. Vendor: Sun Microsystems, Inc.

Sparc OS Operating system for Tatung desktop computer systems. Vendor: Tatung Science & Technology, Inc.

SPARCplug Desktop computer. HyperSPARC processor. Vendor: Ross Technology, Inc.

Sparcus Operating system enhancement. Provides online file management in Unix systems. Allows online access to data stored on optical cartridges. Vendor: Hiarc, Inc.

SPARCworks Application development tools. Used to develop C,. C++, FORTRAN, Pascal applications. Includes browse, debug functions. Runs on Unix systems. Vendor: Sun Microsystems, Inc.

SPARCworks/iMPact Application development tool. Used to develop multi-processing systems. Vendor: Sun Microsystems, Inc.

SPARCworks/TeamWork Application development tool. Monitors source code and reports on modifications. Distributes and coordinates group projects. Runs on Sun systems. Vendor: Sun Microsystems, Inc.

SPARCworks/Visual Application development tool. GUI builder for OSF/Motif compliant applications. Runs on Sun systems. Vendor: Sun Microsystems, Inc.

SPARTA Midrange computer. Pentium Pro CPU. Operating system: Windows NT. Vendor: NeTpower, Inc.

SPD See TRW SDP.

SPDS Application development tool. Accesses corporate data when gathering information for DSS. Stands for: Scalable Performance Data Server. Runs on Solaris systems. Released: 1998. Vendor: SAS Institute, Inc.

SPEC 1170 Standards for Unix. Set of over 1100 system and library calls. Adhered to by leading Unix vendors. Defined by X/Open.

Specif-X CASE product. Automates analysis and design functions.

Spectra 4GL used in large computer environment. Runs on DEC, IBM systems. Vendor: Cincom Systems, Inc.

Spectrum Communications. Network management system. Distributed network manager that handles large inter-networks running on multivendor equipment. Manages OpenView, NetView/6000, NetWare, SunNet networks. Vendor: Cabletron Systems, Inc.

Spectrum Writer Application development tool. Query and report functions. Generates reports from COBOL, ASM statements. Runs on MVS, VM, VSE systems. Vendor: Pacific Systems Group.

spEDI Communications software. EDI package. Runs on Unix systems. Vendor: St. Paul Software, Inc.

SpeechWorks Speech recognition software. Vendor: Applied Language Technologies, Inc.

Speed 1 Database for Unix environments. Vendor: Tom Software, Inc.

Speed II 4GL used in midrange environments. Runs on Wang systems. Vendor: Tom Software, Inc.

SpeedServer, SpeedSeeker Communications. Internet software. Improves speeds on both the server and the browsers. Vendor: Sitara Networks, Inc.

Speedware Environment Communications software. Network connecting HP systems. Micro-to-midrange link. Vendor: Infocentre Corp.

Speedware Media EIS. Includes accounting and reporting tools. Vendor: Speedware Corp.

Sperry Old computer vendor. Merged with Burroughs to form Unisys.

SPF See ISPF.

SPF/2 Version of ISPF/PDF that runs on IBM OS/2 systems. Interfaces with REXX. Vendor: Command Technology Corp.

SPF/PC(,2) Application development tool. Screen editor for IBM desktop computers that is functionally equivalent to ISPF/PDF. Vendor: Command Technology Corp.

SPF/Professional Application development tool. Used to develop CICS, IMS, DB2 applications from a PC. Runs on Windows 95/98, Windows NT systems. Vendor: Command Technology Corp.

SPF/Win Application development tool. Allows Windows users to interface with ISPF screens. Vendor: Command Technology Corp.

SPF/Workbench Application development tool. Allows developers to use ISPF commands in client/server development. Includes: SPF/PC, SPF Shell for Micro Focus, SPF Shell for Xdb, SPF Shell for DB2/2, SPF/Shell for MVS. Vendor: Command Technology Corp.

Sphinx 1. Database. Development name for release of SQLServer. Also called SQLServer7.0. Includes parallel querying and data warehousing tools. Vendor: Microsoft Corp.
2. Operating system for Data General Eclipse systems. Unix-type system. Vendor: Data General Corp.

SPI-TAB Application development tool. Program that works with online applications. Vendor: Axios Products, Inc.

spider A program that prowls the Internet, attempting to locate new, publicly accessible resources such as WWW documents and files available in public archives. Also called wanderers or robots (bots), spiders store the information they find in a database, which Internet users can search by using an Internet-accessible search engine. Used to, i.e., search the Web for e-mail addresses of doctors.

Spider Application development tool. Links Web front end systems to client/server systems. Creates applications that use a Web browser as a user interface. Interfaces with Oracle, Sybase, Informix databases. Vendor: Spider Technologies, Inc.

Spider-Man Application development tool. Integrates information from accounting packages into HTML pages so they can be accessed through Web browsers. Vendor: Hyperion Software.

Spiffy Enhancement to ISPF. Vendor: Isogon Corp.

spiral development Developing software through an iterative method where functions are delivered in stages and improvements are identified by providing software with increasing functionality. Each stage is rapidly produced from changing design specifications. Contrast with waterfall development.

SPL An assembler language for Hewlett-Packard systems. Stands for: Systems Programmers Language. Vendor: Hewlett-Packard Co.

SPL/3000 Systems programming language for HP systems. Vendor: Hewlett-Packard Co.

SPM Operating system enhancement used by systems programmers. Monitors and controls system performance in DEC systems. Collects and reports performance statistics for the operating system. Stands for: Software Performance Monitor. Vendor: Digital Equipment Corp.

SPMS Operating system add-on. Data management software that runs on IBM systems. Stands for: Storage Pool Management System. Vendor: Chicago Soft, Ltd.

Spocgen Operating system add-on. Security/auditing system that runs on IBM systems. Vendor: SPOC.

spooling Using disk storage to temporarily hold data being sent to external devices, most often printers. Stands for: Simultaneous Peripheral Operations OnLine.

SPOT Operating system enhancement used by operations staff and systems programmers. Provides operator console support in IBM systems. Stands for: Systems Programmers & Operators Transaction. Vendor: THORN EMI Computer Software.

Spotfire 4 Application development tool used for data mining. Specific functions for biological, medical and pharmaceutical research. Released: 1999. Vendor: Spotfire, Inc.

SPOX Real-time operating system for desktop computers. Vendor: Spectron Microsystems, Inc.

SPP 1. Massively parallel processor. Stands for: Scalable Parallel Processing. Operating system: Unix. Vendor: Unisys.
2. Communications protocol. Transport layer. Part of Vines. Stands for: Sequenced Packet Protocol.

SPP-UX Operating system for PA-RISC systems. Scalable Unix-type system that runs on parallel computers. Vendor: Convex Technology Center.

SPQR/20 CASE product. Automates project management function. Supplies cost accounting information for project development and maintenance. Vendor: Software Productivity Research.

spreadsheet A desktop computer program to help with management planning and budgeting using a columnar pad. Spreadsheets have also been written for large computer systems.

Sprint Desktop system software. Word processor. Runs on Pentium type systems. Emulates other popular word processors. Vendor: Inprise Corp.

SPS Operating system add-on. Data management software that runs on IBM systems. Stands for: Systems Programmer Series. Vendor: A+ Software, Inc.

SPSS, SPSS for Windows Mathematical and statistical package. Versions available for large and desktop computer systems. Vendor: SPSS, Inc.

SPUFI Name used for routine using SQL by itself to perform a function rather than imbedding SQL statements in a program. Stands for: SQL Processing Using File Input.

SQA Suite See Rational TeamTest.

SQL 4GL. Stands for: Structured Query Language. Has become the standard 4GL for data access to relational databases in most environments from mainframe to desktop computer. SQL is a standard language but many vendors have added proprietary features.

SQL+ Relational database for midrange and desktop computer environments. Runs on HP, IBM systems. Utilizes SQL. Vendor: Advanced MicroSolutions.

SQL* Applications software. Financial, sales software for midsize environments. Vendor: Design Data Systems.

SQL/200 Relational database for large computer environments. Runs on Stratus systems. Vendor: Stratus Computer, Inc.

SQL/400 Relational database for midrange systems. Runs on IBM AS/400 systems. Provides access to database that is part of OS/400. Utilizes SQL. Can be used in client/server computing. Vendor: IBM Corp.

SQL Ad Hoc Application development tool. Program and report generator. Runs with Tandem's Nonstop SQL. Vendor: Software Professionals, Inc.

SQL Anywhere (Professional) Relational database. Runs on laptop systems. Runs on Windows, OS/2. New version of Watcom SQL. Vendor: Sybase, Inc.

SQL:Attach Application development tool. Program that works with batch jobs in IBM DB2 systems. Vendor: Intex Solutions, Inc.

SQL:Bridge Application development tool. Programming utility that downloads DB2 or SQL/DS data to a desktop computer. Vendor: Intex Solutions, Inc.

SQL*Calc Desktop system software. Spreadsheet. Runs on Pentium type desktops, AT&T, DEC VAX, Unix systems. Vendor: Oracle Corp.

SQL Capture CASE product. Reverse-engineering tool. Creates data models from existing database structures. Part of Visible Analyst Workbench. Vendor: Visible Systems Corp.

SQL Central Database software. Administrative tool used with SQL Server and Sybase IQ. Vendor: Sybase, Inc.

SQL Debugger Application development tool. Debugging tool included with Visual C++ Enterprise Edition. Runs on Windows NT systems. Vendor: Microsoft Corp.

SQL*Design Dictionary Application development tool. Program that assists in design and development of applications using Oracle databases. Vendor: Oracle Corp.

SQL/DS Relational database for large computer environments. Runs on IBM VM and VSE systems. Utilizes SQL. Can be used in client/server computing. Full name: Structured Query Language/Data System. Vendor: IBM Corp.

SQL/EXEC Application development tool. Interpretive language for developing SQL/DS programs. Runs on IBM systems. Vendor: VM Systems Group, Inc.

SQL Financials International Software vendor. Products: financial systems for midsize environments.

SQL*Forms Application development tool. Creates form layouts and automatically generates the database access. Interfaces with CASE Dictionary, Oracle7. Runs on DEC VAX, IBM mainframe, OS/2, Sun, Unix systems. Included with CASE Generator. Vendor: Oracle Corp.

SQL/Link Application development tool. Allows users to create applications that can access both SQL and Lotus Notes databases. Full name: SQL\Link for Lotus Notes. Will be incorporated in Centura's SQLWindows. Vendor: Brainstorm Technologies, Inc. for Centura Software Corp.

SQL/Menu Application development tool. Program generator for end-users with little or no SQL experience. Runs on IBM systems. Vendor: VM Systems Group, Inc.

SQL*Menu Application development tool. Creates menu layouts. Interfaces with CASE Dictionary, Oracle7. Runs on DEC XAX, IBM mainframe, OS/2, Sun, Unix systems. Included with CASE Generator. Vendor: Oracle Corp.

SQL Monitor Operating system software. Provides database performance monitoring and tuning for client/server systems. Includes SQL Monitor Server, which runs on Unix, and SQL Monitor Client which runs on Unix, Windows systems. Vendor: Sybase, Inc.

SQL Navigator Application development tool. GUI tool for developing PL/SQL. Used to develop and test stored procedures, schemas, and SQL scripts, Runs on Windows systems. Released: 1998. Vendor: Quest Software, Inc.

SQL*Net Communications software. Network connecting diverse systems. Vendor: Oracle Corp.

SQL NOMAD See NOMAD(2).

SQL*Plus 4GL. Used for ad hoc data queries and report generation. Interfaces with CASE Dictionary, Oracle. Included with CASE Generator. Runs on DEC VAX, IBM mainframe, OS/2, Sun, Unix systems. Vendor: Oracle Corp.

SQL/PM Operating system enhancement. Performance monitor for SQL/DS systems. Runs on IBM systems. Vendor: VM Systems Group, Inc.

SQL Power Plan Analyzer Application development tool. Speeds up loading and querying databases. Used in data warehousing. Runs on Solaris, Windows systems. Vendor: The Development Group for Advanced Technology Inc.

SQL Pump Application development tool. Used for data replication. Provides automatic system shutdown from data integrity errors, scheduled replication points and real-time data conversion and validation. Released: 1998. Runs on Windows NT systems. Vendor: DataMirror Corp.

SQL*QMX 4GL. Compatible with IBM's QMF. Vendor: Oracle Corp.

SQL*ReportWriter Application development tool. Report generator. Creates report layouts and automatically generates the database access. Interfaces with CASE Dictionary, Oracle. Included with CASE Generator. Runs on DEC VAX, IBM mainframe, OS/2, Sun Unix systems. Vendor: Oracle Corp.

SQL<>SECURE Operating system enhancement. Provides security functions at the database level. Includes the following components: Password Manager, Database Security Manager, Audit Manager, Policy Manager. Works with Oracle, Sybase, SQL Server databases. Vendor: Braintree Security Software, Inc.

SQL Server Relational database for desktop systems. Uses SQL and works with client/server computing. Vendor: Microsoft Corp.

SQL Server 10 Database management engine for Sybase DBMS. See Sybase, SQL Server.

SQL Source Generator See Toolset-DB2.

SQL-Station, SQL-Station/Team Edition Application development tool. Helps create, edit, test, tune and manage database code for server systems. Includes: Code Manager, Coder, Debugger, Plan Analyzer. Vendor: PLATINUM Technology, Inc.

SQL Toolset Application development system. Runs on DEC, Unix systems. Includes APT Workbench (containing program generator, 4GL, and edit routines) and Data Workbench (containing query language and data manipulation routines). Vendor: Sybase, Inc.

SQLab Tuner Application development tool used to tune applications using Oracle databases. Resolves I/O bottlenecks and identifies poorly performing SQL statements. Vendor: Quest Software, Inc.

SQLassist Application development tool. Programming utility that builds SQL queries for use with Oracle databases. Vendor: Software Interfaces, Inc.

SQLBase (for Windows) Relational database for desktop computers on a LAN. Works with client/server computing. Interfaces with Microsoft's Windows and includes screen editor and debugging facilities. Also called SQLBase Server. Runs on OS/2, Unix. Vendor: Centura Software Corp.

SQLDBM Application development tool. Programming utility that provides catalog management services. Runs with Tandem's Nonstop SQL. Vendor: Software Professionals, Inc.

Sqlfile System Application development tool that works in a client/server architecture. Accesses SQL Server, Novell Netware SQL, Oracle Server. Includes application and report generators. Vendor: Vinzant, Inc.

SQLHost Application development tool. Includes middleware to integrate Visual Basic applications with DB2 and other legacy data. Full name: SQLHost for Visual Basic. Vendor: Centura Software Corp.

SQLizer Application development tool for DB2 environments. Vendor: Centura Software Corp.

SQLNetwork Communications software. Links desktop computer and/or workstation applications with databases residing on large computer systems. Vendor: Centura Software Corp.

SQLWindows Application development environment. Used for developing client/server applications. Builds client applications with a GUI based front-end. Accesses Teredata, SQLbase, OS/2 EE Database Manager, DB2, Oracle Server. Includes program generator. Runs on Unix, Windows systems. Released: 1996. Vendor: Centura Software Corp.

SQLWindows SQLXpert Application development tool used to tune applications using Oracle databases. Traces system-wide SQL activity and evaluates each SQL statement. SQL statements are then compared to a knowledge base and the statements are rewritten based on actual database content. Released: 1998. Vendor: Quest Software, Inc.

SQMS CASE product. Automates project management function. Supplies cost accounting information for project development and maintenance. Stands for: Software Quality Management System. Vendor: Software Quality Tools Corp.

SQR, SQR3 Workbench Application development tool. Report writer. Utilizes SQR 4GL. Includes SQR-Execute for creating run-time applications. Runs on Unix, Windows systems. Vendor: SQRIBE Technologies (acquired by Brio Technology in August, 1999).

Square 1,5,6 Desktop computer. Pentium processor. Vendor: Reason Computer.

Square 5H, 5MX Desktop computer. Notebook. Pentium processor. Vendor: Reason Computer.

SRAM See CA-SRAM.

SRS Operating system enhancement. Manages disk space for IBM MVS systems. Eliminates space abends. Stands for: Space Recovery System. Vendor: DTS Software, Inc.

SRTM Operating system add-on. Debugging/testing software that runs on Concurrent systems. Vendor: Concurrent Computer Corp.

SSF Operating system for mainframe Unisys systems. Stands for: System Software Facility. Vendor: Unisys Corp.

SSH Communications, Internet software. Security system that allows users to log into remote systems. Runs under Unix. Both free and commercial versions available. Stands for: Secure Shell.

SSL Protocol used to provide security measure on the Internet . Encryption standard. Uses public and secret key encryption. Stands for: Secure Sockets Layer.

SSP Operating system for IBM System/34,36. Vendor: IBM Corp.

SSSLIB Operating system add-on. Library management system that runs on Unisys systems. Vendor: Schamburg & Schamburg Systems, Ltd.

SSSTEX Application development tool. Screen editor for Unisys systems. Vendor: Schamburg & Schamburg Systems, Ltd.

SSX/VSE Operating system for IBM systems. Vendor: IBM Corp.

Stabilize Operating system enhancement used by systems programmers. Increases system efficiency in IBM CICS systems. Detects and repairs system problems. Vendor: On-Line Software International.

Stack Builder See Oracle Card.

Staffware Workflow software. Object-oriented package. Runs on Unix, Windows systems. Vendor: Staffware Corp.

stage-based Multimedia term. Used to describe multimedia tools that present objects which are given paths to follow and a number of frames in which action should occur.

STAIRS(/VS) Text editor and retrieval system for IBM systems. Stands for: STorage And Information Retrieval System. Vendor: IBM Corp.

Standard Generalized Markup Language See SMGL.

Star Application development system. Runs on NCR systems. Vendor: Century Analysis, Inc.

STAR Artificial intelligence system. Runs on DEC systems. Programming language for the development of expert systems. Stands for: Simple Tool for Automated Reasoning. Vendor: COSMIC.

STAR-1100 Operating system add-on. Tape management system that runs on Unisys systems. Vendor: Formula Consultants, Inc.

STAR:Flashpoint See VISION:Flashpoint.

star schema Data design tool. A modeling technique that has a single table (fact table) in the middle of the schema connected to a number of tables (dimension tables) encircling it. The star schema design is optimized for end user business query and reporting access. This design will contain a fact table such as sales, compensation, payment, or invoices qualified by one or more dimension tables (month, product, time, and/or geographical region). The table in the center of the star is called the fact table and the tables that are connecting to it in the periphery are called the dimension tables.

star topology A network configuration where each computer is connected to a central hub.

StarWatch Desktop system software. Project management package. Runs on Pentium type desktop computers. Vendor: Pathfinder, Inc.

StarBase 1. See Portable OS/2.
2. See PowerHouse StarBase.

Starfire Midrange computer. SPARC CPU. Vendor: Sun Microsystems, Inc.

StarGate See PowerHouse StarGate.

Stargroup LAN Manager Communications software. Network operating system. Unix based version of LAN Manager. Runs on AT&T, IBM, NCR systems. Vendor: AT&T/NCR Corp.

StarGroup Server Communications. LAN (Local Area Network) connecting DOS, OS/2, Macintosh and Unix workstations. Vendor: AT&T Computer Systems.

Starion Desktop computer. Pentium CPU. Operating system: Windows 95/98. Vendor: Digital Equipment Corp.

StarLAN Communications. LAN (Local Area Network) connecting desktop computers. Vendor: AT&T Information Systems, Data Systems Group.

StarMax Desktop computer. PowerPC RISC CPU. Operating systems: AIX, MacOS. Vendor: Motorola Computer Group.

StarNet See PowerHouse StarNet.

StarOffice Applications suite. Includes word processing, spreadsheet, graphics, e-mail. Vendor: Network Computer, Inc.

StarServer RAID, SMP Midsize computer. Server. Pentium II processor. Operating system: Windows. Vendor: TriStar Computer Corp.

StarSQL Pro Communications software. Middleware. Allows Web front-ends to mainframe systems. Provides ODBC access to DB2 databases. Runs on OS/2, Windows systems. Vendor: StarQuest Software Inc.

StarTeam Application development tool. Provides tools to manage project development by teams. Includes configuration management functions such as keeping track of which files are being worked on and by whom, managing bug histories and providing version control. Vendor: StarBase Corp.

Starter Suite Multimedia kit. Used to create multimedia networks with Macintosh and IBM and compatible desktop computers. Vendor: Starlight Networks, Inc.

StarView Application development tool. Workbench. C++ multi-platform toolset. Runs on most desktop systems. Vendor: Star Division Corporation.

StarWorks Video networking software. Supports up to 40 DOS/Windows, Windows 95/98, Macintosh users accessing video and/or audio. Used with multimedia applications. Vendor: Starlight Networks, Inc.

Stata Application development tool. Statistical forecasting tool. Provides summary information and determines significance of degree of relationship between two factors. Runs on Macintosh, Unix, Windows systems. Vendor: Stata Corp.

Statemate CASE product. Automates programming, testing, and project management functions. Used for design of realtime systems. Includes: Kernal, Analyzer, Prototyper, Document. Vendor: I-Logix, Inc.

Station Classic+ Midrange computer. RISC machine. microSPARC CPU. Operating system: Solaris. Vendor: DTK Computer, Inc.

Statistica Application development tool. Statistical forecasting tool. Provides summary information and determines significance of degree of relationship between two factors. Runs on Windows systems. Vendor: StarSoft Inc.

Statit Application development tool. Statistical forecasting tool. Provides summary information and determines significance of degree of relationship between two factors. Runs on Unix, Windows systems. Vendor: Statware, Inc.

Stats from Knowware Desktop system software. Graphics package. Runs on MS-DOS systems. Vendor: Knowware.

Status/38 Operating system enhancement used by systems programmers. Provides system cost accounting for IBM System/38. Vendor: Advanced Systems Concepts, Inc.

Statview Application development tool. Statistical forecasting tool. Provides summary information and determines significance of degree of relationship between two factors. Runs on Macintosh, Windows systems. Vendor: Abacus Concepts, Inc.

Stealth GS, SE Desktop computer. Pentium processor. Vendor: USA Flex, Inc.

Step/DP/Pro Desktop computer. Pentium CPU. Operating system: OS/2, Windows 95/98, NT Workstation. Vendor: Everex Systems, Inc.

Step DP,QP/Pro RAID Midrange computer. Pentium Pro CPU. Operating system: Unix. Vendor: Everex Systems, Inc.

Step DPe/3000 Desktop computer. Pentium CPU. Operating system: OS/2, Unix, Windows NT. Vendor: Everex Systems, Inc.

Step Premier Desktop computer. Pentium CPU. Operating system: Windows 95/98. Vendor: Everex Systems, Inc.

StepClient Desktop computer. Pentium CPU. Operating system: Windows NT. Vendor: Everex Systems, Inc.

StepNote Notebook computer. Pentium CPU. Vendor: Everex Systems, Inc.

StepPremier Desktop computer. Pentium CPU. Operating system: Windows NT. Vendor: Everex Systems, Inc.

StepServer 2, MDP/Pro Midrange computer. Pentium Pro CPU. Operating systems: OS/2, Unix. Vendor: Everex Systems, Inc.

StepStation Desktop computer. Pentium CPU. Operating system: Windows NT. Vendor: Everex Systems, Inc.

Stereo 800x600 Notebook computer. Pentium CPU. Vendor: Comtrade Computer, Inc.

STL ToolKit Object library. Portable version of Systems ToolKit. Vendor: ObjectSpace, Inc.

Stop-X37 Operating system enhancement used by systems programmers. Increases system efficiency in IBM systems. Manages disk storage space. Vendor: Empact Software.

Storage Management Service See SMS.

Storage Manager Operating system add-on. Manages file backup and storage for Windows NT. Vendor: Palindrome Corp.

Storage-Server 100 Operating System enhancement used by systems programmers. File server that utilizes optical and magnetic storage. Vendor: Digital Equipment Corp.

StorageCenter Systems management software. Provides storage management functions. Vendor: Software Partners/32, Inc.

stored procedure Routine that manipulates data but resides on a database server rather than in an application. Used to build standard routines that implement business rules. Popular with Oracle, Sybase, Informix databases.

Storm Application development tool. Query environment that allows programmers to prepare views of an Informix database that users customize with a mouse and icons. Works with Motif and Microsoft's Windows. Vendor: Informix Software, Inc.

StorWatch See Seascape Storage Enterprise.

Storyboard Desktop system software. Graphics package. Runs on Pentium type desktop computer systems. Vendor: IBM Corp.

Storyboard Live Multimedia authoring tool. Entry-level, stage-based tool. Vendor: IBM Corp.

StoryServer Application development tool. Used to build e-commerce applications. Supports team-based development. Creates dynamic Web sites that change depending on the users actions so a focused presentation can be made to each Web visitor. Vendor: Vignette Corp.

STP See Software through Pictures.

STP for BRP Application development tool. Generates SQL, C++, Smalltalk code. Stands for: Software Through Pictures for Business Process Reengineering. Vendor: Interactive Development Environments, Inc.

STP/IM Application development tool. Information modeling tool set. Supports client/server development. Runs on HP, RS/6000, Sun systems. Stands for: Software through Pictures/Information Modeling. Vendor: Interactive Development Environments, Inc.

Stradis Structured programming design methodology. Accepted as a standard design methodology by some companies, and used by some CASE products.

Stradis/Draw CASE product. Automates design function. Vendor: McDonnell Douglas Information Systems Co.

Strategic Networked Application Platform See SNAP.

Strategy Application development software. Data warehousing software that includes centralized management functions and interfaces with DB2 databases. Vendor: ShowCase Corp.

Stratium Java Framework Application framework. Used to build thin client applications using Oracle8i. Supports Java, SQL, XML, CORBA/IIOP. Vendor: Stratium Consulting.

Stratus Computer vendor. Manufactures midsize computers.

Stratus FXT Operating system for Stratus mainframe systems. Vendor: Stratus Computer, Inc.

StreamBuilder Application development tool. Allows users to customize Smart-Stream applications and migrate the changes through new releases of Smart-Stream. Based on PowerBuilder. Vendor: Geac Computer Corp, Ltd. (formerly Dun & Bradstreet Software.).

Stream*Line* Applications software. ERP software that includes Smart*Stream* HR and financial software. Includes supply chain functions. Runs on Windows NT. Vendor: Geac Computer Corp, Ltd. (formerly Dun & Bradstreet Software.).

StreetTalk Communications software. Distributed directory. Defines attributes of objects attached to the network. Included with VINES, can be used with NetWare. Called Universal StreetTalk as stand-alone package. Vendor: Banyan Systems, Inc.

Strobe Operating system enhancement used by systems programmers. Monitors and controls system performance in IBM CICS, DB2, IDMS/DC systems. Vendor: Programart.

StrongARM Specialized processor to handle handheld computers and smart phones. Vendor: Intel Corp.

Structure/4 Relational database for large computer environments. Desktop computer version available. Runs on DEC. IBM, Prime systems. Has SQL-like query language. Vendor: Command Business Systems, Inc.

structure charts Analysis and design tool used to show the hierarchical structure of a system. Used with structured development methodologies.

Structure(s) CASE product. Automates programming and design functions. Vendor: Optima, Inc.

Structured English See pseudocode.

StructureBuilder Application development tool. Can be used alone or with other development environments. Allows developers to bring in existing Java files. Builds class diagrams from Java code. Vendor: Tendril Software Inc.

structured programming A development methodology that states that all programs are made up of a few definitive logic structures and should be analyzed, designed, and programmed using only these structures.

StudioJ Application development tool. Libraries of Java classes and components used for GUI development, data analysis, charting and database access. Vendor: Rogue Wave Software, Inc.

StudioStation Desktop computer. Pentium processor. Vendor: Tri-Star Computer Corp.

STW/Advisor Application development tool. Analyses source code, generates test data. Interfaces with Software TestWorks. Vendor: Software Research.

STW/Coverage Application development tool. Provides testing, debugging functions. Measures untested code. Runs on Unix systems. Vendor: Software Research, Inc.

STW/Regression Application development tool. Provides testing, debugging functions. Executes and verifies tests for both GUI and character systems. Runs on Unix systems. Vendor: Software Research, Inc.

STW/Web Application development tool. Testing and debugging software that lets Web developers create test scripts and simulate hundreds of interactive users. Released: 1996. Vendor: Software Research, Inc.

STX(12) Communications software. EDI system. Vendor: Supply Tech, Inc.

STX (SNA to X.25) Communications software. Network connecting IBM computers and non-IBM data sources including public networks. Vendor: Legent Corp.

Stylistic 2300 Laptop computer. Operating systems: Windows. Pen tablet computer. Vendor: Fujitsu Personal systems, Inc.

Stylistic 500, 1000, 1200 Handheld, pen-based computer. Vendor: Fujitsu Personal Systems, Inc.

SUDS Communications. Network management system. Uses all major local area networks to provide automated software distribution and automated file updating and replacement. Stands for: Software Update and Distribution System. Vendor: Frye Computer Systems, Inc.

SuiteDome Communications software. Middleware. Vendor: Suite Software, Inc.

SuiteSpot Communications, E-mail, Internet software. Used to build Intranets. Includes Java, search functions, messaging server, mail and messaging functions. Interfaces with Oracle, Sybase, Informix. Latest version called Apollo. Supports LDAP protocols. Vendor: Netscape Communications Corp.

SuiteTalk Communications software. Network connecting DEC VAX systems. Vendor: Suite Software.

SuiteValet Communications. Middleware, object request broker (ORB). CORBA compliant. Runs on Unix, DEC, Windows NT. Vendor: Suite Software.

Summit 1. Desktop computer utility package that lets Lotus 1-2-3 users access Emerald Bay. Vendor: Migent, Inc. 2. Midrange computer. Alpha CPU. Vendor: Aspen Systems, Inc.

Summit D System development methodology used by Coopers & Lybrand. Includes a defined project management definition, interfaces to project management tools, and software support tools.

Sun Computer vendor. Manufactures desktop computers.

Sun Connect Application software. Architecture based on Java and JavaBeans that ties together banks and brokers through the Web. Vendor: Sun Microsystems, Inc.

Sun ConsoleServer See Sun Enterprise Toolset.

Sun DataCenter Scripts See Sun Enterprise Toolset.

Sun Enterprise Toolset Application development tools. Suite of tools that provide interfaces between IBM mainframe and Unix systems. Includes: SunDANS, SunRAI, Sun Paperless Reporter, Sun ConsoleServer, and Sun DataCenter Scripts. Vendor: Sun Microsystems, Inc.

Sun INGRES Relational database for midrange and desktop computer environments. Can be used in client/server computing. Runs on Sun systems. Vendor: Sun Microsystems, Inc.

Sun Paperless Reporter See Sun Enterprise Toolset.

Sun UNIFY Relational database for midrange and desktop computer environments. Can be used in client/server computing. Runs on Sun systems. Vendor: Sun Microsystems, Inc.

Sun Web Server Communications. Web server. Runs on Solaris systems. Vendor: Sun Microsystems, Inc.

SunDANS See Sun Enterprise Toolset.

SunNet Manager Systems management software. Distributed network manager linking multiple networks. Follows SNMP, CMIP protocols. Vendor: Sun Microsystems, Inc.

SunNet MHS Communications software. Gateway for e-mail systems. Runs with TCP/IP networks. Vendor: Sun Microsystems, Inc.

SunOS Old operating system for Sun systems. Unix-type system. Most recent version is called Solaris. Vendor: Sun Microsystems, Inc.

SunPro WorkShop Application development tools. Used in Unix systems for C, C++, Fortran development. Includes SPARCworks/iMPact, SPARCworks/TeamWare. Runs on Sun systems. Vendor: Sun Microsystems, Inc.

SunRAI See Sun Enterprise Toolset.

SunScreen Internet software. Firewall. Released: 1997. Vendor: Sun Microsystems, Inc.

SunSoft WorkShop Application development tools. Suite of tools used to develop object-oriented systems for Unix. Vendor: Sun Microsystems, Inc.

Super Communications software. Terminal emulator. Works with Unisys systems. Vendor: Formula Consultants, Inc.

Super 3D Desktop computer. Pentium processor. Vendor: ABS Computer Technologies, Inc.

Super COMPserver Midrange computer. RISC machine. SuperSPARC CPU. Operating system: Solaris. Vendor: Tatung Science & Technology, Inc.

Super Crunch Desktop system software. Spreadsheet. Runs on Macintosh systems. Vendor: VisiCorp.

Super-Link Application development system. Runs on DEC VAX, IBM systems. Vendor: Multi Soft, Inc.

Super Nova Desktop computer. Pentium processor. Vendor: ET Technology.

Super Show and Tell Multimedia authoring tool. Entry-level, slide-based tool. Runs on Windows systems. Vendor: Ask Me Multimedia Center, Inc.

Super-UX Operating system for SX3 supercomputers. Vendor: HNSX Supercomputers, Inc.

Superbase (for Windows) Application development environment. Contains DBMS, 4GL. Interfaces with dBase, Lotus, Excel. Vendor: Software Publishing Corp.

SuperBOOK Notebook computer. Pentium CPU. Vendor: Wedge Technology, Inc.

SuperCalc See CA-Supercalc.

SuperCede Java/ActiveX Edition, SuperCede Database Edition Application development tools. RAD tool user for Internet applications. Released: 1997. Vendor: Asymetrix Corp.

supercomputer Computers designed to process complex scientific applications; speed is the most important feature in their design. Not commonly used for business applications.

Supercycle CASE product. Automates analysis, design, and programming functions. Includes code generator for CA-Ideal. Design methodologies supported: Information Engineering, DeMarco, Yourdon. Databases supported: DB2, DA-Datacom. Runs on IBM systems. Vendor: MFJ International.

SuperDOS Operating system for IBM and compatible desktop computer systems. Vendor: Bluebird Systems.

SuperEnglish Database for large computer environments. Runs on DG systems. Includes 4GL and report generator. Vendor: ACS.

SuperExpert Artificial intelligence system. Expert system building tool. Runs on desktop systems. Vendor: Softsync, Inc.

SuperMac Midrange computer. PowerPC CPU. Operating system: System 7. Vendor: UMAX Computer Corp.

Supermon Operating system enhancement used by operations staff and systems programmers. Provides operator console support in System/38 and System/400. Also called Night Operator. Vendor: Adollar Software.

SuperNOVA Application development environment. Used for enterprise development. Includes program generator, 4GL. Works with client/server computing. Builds client applications with a GUI based front-end. Allows access to DB2/6000 databases from non-IBM clients. Runs on Unix systems. Vendor: SuperNova, Inc.

SuperNova/Visual Concepts Application development tools. Suite of visual tools used to develop component based systems. Accomodates multiple languages and tools including COBOL, PowerBuilder, Visual Basic and Java. Released: 1998. Vendor: SuperNova, Inc.

SUPEROPTIMIZER Operating system enhancement used by systems programmers. Increases system efficiency in IBM CICS, IMS, TSO, and VM installations. Improves response time on 3270 networks. Vendor: BMS Software, Inc.

SuperPDL CASE product. Automates design function. Vendor: Advanced Technology International, Inc.

SuperProject See CA-Superproject.

superserver computer Computer that acts as the server in large client/server environments. Can be considered as a high-powered workstation machine. Provides mainframe capabilities, but does not require a controlled environment.

SuperStor Operating system utility. Compresses data for storage. Used in PC-DOS. Vendor: AdStor, Inc.

SuperStructure Application development tool. Automates programming function. Vendor: Computer Data Systems, Inc.

Supertask! Real-time operating system for desktop computers. Vendor: U.S. Software.

Supertracs Communications software. Transaction processing monitor. Runs on IBM systems. Vendor: Sterling Software (Systems Software Marketing Division).

SuperWeb Communications software, Web server. Provides access to the Web from Windows NT systems. Lets users view documents. Vendor: Frontier Technologies.

SUPERZAP Operating system add-on. Data management software that runs on IBM systems. Vendor: Software Engineering of America.

Supply Chain Advantage Application software. ERP, e-commerce and supply chain applications bundled together. Includes J.D. Edwards' manufacturing, financial and distribution software, SynQuest's production planning and scheduling software and IBM's Net.Commerce software.

supply chain collaboration Application function. Part of manufacturing systems. Process of retailers and suppliers sharing information such as sales forecasts and product designs, and work together to develop the information.

Supply Chain Collaborator Applications software. Manufacturing software handling supply chain functions. Runs on Unix systems. Released 1997. Vendor: PeopleSoft, Inc.

Supply Chain Council See SCOR.

Supply Chain Designer Application software. Manufacturing software handling supply chain functions. Runs on Windows systems. Released: 1997. Vendor: Caps Logistics, Inc.

Supply Chain Navigator Application software. Manufacturing software handling supply chain functions. Runs on Unix, Windows systems. Vendor: Manugistics, Inc.

supply-chain operations Term used in manufacturing. Refers to the process of managing materials from buying supplies through distributing products. Includes cost analysis and management of inventory, warehousing, transportation functions. Supply-chain software automates these functions.

Supply Chain Strategist Application software. Manufacturing software handling supply chain functions. Runs on Windows systems. Vendor: Released: 1997. Vendor: InterTrans Logistics Solutions, Inc.

SupportMagic SQL Systems software. Help desk and asset management package. Interfaces with Web. Released: 1997. Vendor: Magic Solutions, Inc.

SupportTeam See ProTeam.

Supra (Server) Relational database for large computer environments. Runs on most computer systems. Includes development tools. Has Mantis query language. Utilizes SQL. Vendor: Cincom Systems, Inc.

Supre/Daisys CASE software. Set of tools to be used in client/server architecture. The software splits an application and decides which tasks should be performed by the server and which by the clients. Decisions are based on business processing functions and technical information provided by the company. Vendor: S/Cubed, Inc.

Supreme DVD Desktop computer. Pentium processor. Vendor: ABS Computer Technologies, Inc.

Sure-Server E-commerce, Internet technology. Digital certificate offered by Wells Fargo and GTE CyberTrust.

Sureloc Operating system add-on. Security/auditing system that runs on IBM systems. Interfaces with Dun & Bradstreet's Millennium software. Vendor: Phase 2 Consulting, Inc.

SureSync Operating system software. Provided backup, disaster recovery, and software distribution functions for Windows systems. Vendor: Software Pursuits, Inc.

Suretrak Project Scheduler Desktop system software. Project management package. Runs on Pentium type desktop computers. Vendor: Primavera Systems, Inc.

SUREVSAM Operating system enhancement used by systems programmers. Increases system efficiency in IBM systems. Optimizes VSAM performance. Vendor: Chicago Soft, Ltd.

Surge Communications software. Code name for 1998 upgrade of GroupWise messaging and groupware system. Vendor: Novell, Inc.

SurviveIT Operating system software. Provides Windows NT replication. Has a second server automatically take over in case of system failure. Released: 1999. Vendor: Computer Associates International, Inc.

SVCommands Application development tool. Reverse engineering tool for legacy systems. Provides such functions as documenting relationships, providing functional analysis for programs and applications systems, and finding unused code. Vendor: Adpac.

SVGA CD-Note Desktop computer. Notebook. Pentium processor. Vendor: Comtrade Computer, Inc.

SWAP Communications. Protocol intended to let workflow products from different vendors work together. Stands for: Simple Workflow Access Protocol.

Swat Operating system add-on. Debugging/testing software that runs on DEC VAX, DG, Unix systems. Vendor: Data General Corp.

Swiftcalc Desktop system software. Spreadsheet. Runs on Pentium type desktop computers. Vendor: Timeworks, Inc.

Switched Token Ring In communications, backbone technology.

switches Communications device. Connect parts of networks and direct data packets to their appropriate destination. Handle multiple connections. Provide each connection with the necessary bandwidth.

SWS Midrange computer. RISC machine. SPARC CPU. Operating system: Solaris. Vendor: Integrix, Inc.

SX-3R Series Mainframe computer. Operating system: Super-UX. Vendor: NEC Computer Systems.

SX-4 Model xxR Supercomputer. Operating systems: Unix, Super-Ux. Vendor: HNSX Supercomputers, Inc.

SX-5 Mainframe computer. Parallel processor. Supports up to 512 Pentium processor. Operating system: Unix. Vendor: HNSX Supercomputers Inc.

SX OS Operating system for mainframe HNS systems.

Sybase Relational database/4GL for midrange and desktop computer environments. SQL Server is the database engine. Runs on DEC, Unix systems. Utilizes SQL. Includes SQL Toolset. Can be used in client/server computing. Versions include System 10, System 11. Latest architecture is Adaptive Server. Includes application and report generators. Vendor: Sybase, Inc.

Sybase IQ See Adaptive Server IQ.

SYBASE MPP See Navigation Server.

Sybase Replication Server Application development tool. Synchronized replications of data on heterogeneous systems throughout a client/server environment. Vendor: Sybase, Inc.

Sybase SQL Anywhere See SQL Anywhere.

SybaseWare Combination of Sybase and NetWare. Bundled software from Sybase and Novell.

SYLVA Foundry CASE product. Automates programming and design functions. Includes: SYLVA System Developer, SYLVA Picture Programmer. Vendor: Cadware, Inc.

Symbiator Application development tool. Provides bidirectional replication of data from SQL databases in AS/400 systems. Vendor: ExecuSoft Systems, Inc.

Symbolic Debug/1000 Operating system add-on. Debugging/testing software that runs on HP systems. Vendor: Hewlett-Packard Co.

Symbolics Genera Operating system for Symbolic systems. Vendor: Symbolics, Inc.

Symbolics LISP Operating system for Symbolic systems. Vendor: Symbolics, Inc.

SymDump Operating system add-on. Debugging/testing software that runs on IBM CICS systems. Part of InterTest. Vendor: On-Line Software International.

SYMETRA(2) Midrange computer. Pentium Pro CPU. Operating system: Windows NT. Vendor: NeTpower, Inc.

Symix Computer Systems Software vendor. Products: Financial, HR, payroll systems for midsize environments.

symmetric multiprocessing See SMP.

Symmetrix Multi-Host Transfer Facility See SMTF.

Symmetry 5000, S200 Desktop computer. Pentium CPU. Operating systems: DYNIX/ptx, Unix. Vendor: Sequent Computer Systems, Inc.

SyMON Operating system software. System monitoring tool for Sun computers. Solstice SyMON monitors enterprise servers and Sun Enterprise SyMON monitors all Sun products on a network including servers, desktops, storage and clustering devices. Stands for: System MONitor. Vendor: Sun Microsystems, Inc.

Symphony Desktop system software. Integrated package. Runs on Pentium type systems. No longer marketed. Vendor: Lotus Development Corp.

Symphony family Communications. Family of peer-to-peer wireless products including a modem and PC card. Vendor: Proxim, Inc.

Symposium Application development tool. Lets users create a live classroom situation to present instructor-led training over the Web. Vendor: Centra Software, Inc.

Synbol Technologies Computer vendor. Manufactures handheld computers.

Synchronicity Application development tool used in object-oriented development. Design and analysis tool based on Smalltalk. Integrates with Enfin. Runs on Unix, Windows systems. Vendor: Easel Corp.

Synchronous Optical Network See SONET.

Synchroworks Workbench software. Application development software used to develop client/server systems. Works with Solaris, Motif. Interfaces with Oracle, Sybase. Vendor: Oberon Software.

Syncsort Operating system software. Fast, efficient sort program. Runs on DEC, IBM, Unix systems. Vendor: Syncsort, Inc.

Synctrac Operating system enhancement used by systems programmers. Coordinates and reports on multiple software environments in IBM MVS systems. Vendor: Serena International.

Synergist Application development tool. Creates online applications that can access desktop computer or host database. Runs on DEC VAX systems. Vendor: Gateway Systems Corp.

Synergy 1. Application development tool. Software process management system. Runs on OS/2, Windows systems. Includes GUI, CASE functions. Interfaces with IEW, ADW, IEF. Vendor: CASE Methods Development Corp.
2. Desktop system software. Project management package. Image processing system that runs on Pentium type desktop computers, Sun, Unix. Vendor: Bechtel Software, Inc.

Synergy ADE Application development tool. Provides portability among operating systems. Runs on Unix systems. Vendor: Synergex International Corp.

Synergy xxx Desktop computer. Pentium CPU. Operating system: Windows 95/98. Vendor: Royal Electronics, Inc.

SynerJ Application development tools. Software suite used to develop and deploy Enterprise JavaBeans (EJBs). Includes three modules: SynerJ Developer (used to build EJBs, applets and servlets), SynerJ Server (stand-alone EJB server that can run any Java component) and SynerJ Deployer (graphical tool used to import Java archive files and connect file components). Released: 1999. Vendor: Forte Software, Inc.

syntax The rules governing the structure of a computer language.

SyntheSystem Application development tools. Includes tools for design, coding, and testing of applications. Incorporates menus, documentation, help screens, library control, and data security. Runs on IBM mainframe systems. Vendor: Logicon/Fourth Generation Technology, Inc.

Synthetic Jobstream Operating system enhancement used by systems programmers. Increases system efficiency in IBM systems. Functionally verifies operating system and hardware changes before implementation. Vendor: Software Engineering of America.

SYS:A7 Model x11 Desktop computer. Pentium CPU. Operating system: Windows 95/98. Vendor: Unisys Corp.

SYSB-II Operating system add-on. Data management that runs on IBM CICS systems. Allows CICS VSAM files to be simultaneously updated by both online and batch application programs. Vendor: H&W Computer Systems, Inc.

SYSD Application development system. Runs on IBM CICS mainframe systems. Similar to ISPF/PDF. Vendor: H&W Computer Systems, Inc.

sysgen The process of selecting optional parts of an operating system and of creating a specific operating system tailored to the requirements of a company. Includes testing the system. Part of the job of systems programmers.

SYSI Operating system enhancement. Increases system efficiency in IBM CICS systems. Provides an interface between TSO and CICS. Vendor: H&W Computer Systems, Inc.

SYSLOG A data set that contains information about the activity of the system. Used by operations staff and systems programmers.

SYSM Communications software. E-mail system. Runs on IBM systems. Vendor: H&W Computer Systems, Inc.

SYSMAN Operating system add-on. Data management software that runs on DEC VMS systems. Controls disk space. Vendor: Digital Equipment Corp.

SYSmarkx System software. Benchmark tools for client/server environments. Allows users to evaluate hardware by simulating work loads. Vendor: Business Applications Performance Corp. (BAPCo).

SYSOUT A data set that contains messages from the operating system pertaining to the execution of each job.

SYSOUT maintenance program A computer program that keeps track of SYSOUT datasets so they can be easily retrieved.

Sysplan Operating system enhancement used by systems programmers. Provides trend analysis and forecasting of system resource usage in Hewlett-Packard systems. Vendor: Carolian Systems International, Inc.

Systat Application development tool. Statistical forecasting tool. Provides summary information and determines significance of degree of relationship between two factors. Runs on Windows systems. Vendor: SPSS, Inc.

Systel Application development system. Includes 4GL, interactive dictionary, forms generator. Interfaces with RMS Rdb, Adabas. Vendor: Performance Software, Inc.

System 10 Relational database. Supports client/server computing. Includes Replication Server. Database engine is SQL Server 10. Vendor: Sybase, Inc.

System 1022 Relational database for large computer environments. Runs on DEC systems. Vendor: CompuServe Data Technologies.

System 1032 Relational database for midrange and desktop computer environments. Runs on DEC systems. Also provides application development tools. Vendor: CompuServe Data Technologies.

System 11 Application development system.Includes relational database (Sybase, with SQL Server database engine). Supports multiprocessing systems, parallel data access. Vendor: Sybase, Inc.

System 21 Application suite. ERP software which includes manufacturing, financial, logistical, and customer service functions. Provides unique solutions for specific industries including food, automotive, drinks, and service management. Vendor: JBA International Inc.

System 310AP PC, or micro computer. Operating systems: Xenix, iRMX. Vendor: Intel Corp.

System/36,38 Midsize computers. Operating system: CPF. Replaced by the AS/400. Vendor: IBM Corp.

System/390 See S/390.

System 7.0 Operating system for Apple Macintosh systems. Vendor: Apple Computer, Inc.

System 7000 R3 Midrange computer. RISC machine. MIPS R3000 CPU. Operating system: Unix. Vendor: NCR Corp.

system administrator Support personnel. System administrator is an official title in a Unix environment, but is also used in other midrange and desktop systems. The system administrator is responsible for such things as installing new software, adapting software to the system, running system backups, recovering lost data and maintaining security. Monitors functioning of computer systems, hardware, and/or networks. Can be any experience level.

System Analyzer Application development tool. Analyzes existing programs to determine what applications need restructuring to fit into a new system. Vendor: XA Systems Corp.

system architect Technical developer (occasionally used by application developers). Designs computer systems and handles software integration. Builds infrastructures. Usually a senior title, sometimes mid-level. Title is used in mainframe and midrange systems.

System Architect CASE product. Automates analysis, design, and programming functions. Includes data modeling, object-oriented, and workflow reengineering techniques. Generates Access, SQL Server databases from models. Runs on OS/2, Windows systems. Links to Power-Builder. Includes SA Browser. Used for client/server system development. Vendor: Popkin Software and Systems, Inc.

System Builder Application development environment. Consists of two parts: SB+ Server which is a 4GL development environment for both host-based and client/server development; and SBClient, which adds GUI features to host-based applications. Includes automated documentation, version control, and debugging functions. Applications can run on Unix and Windows systems. Vendor: Ardent Software Inc.

System Definition Language See SDL.

System Developer I, II CASE product. Automates analysis, design, and programming functions. Includes code generator for COBOL. Design methodologies supported: Information engineering, object-oriented analysis and design, Constantine DeMarco, Gane-Sarson, entity-relationship models. Databases supported: dBase III and higher. Runs on IBM mainframe and desktop computer systems. Vendor: Cadware, Inc.

System Management Program Operating system enhancement used by systems programmers. Increases system efficiency in Unix systems. Vendor: Taskforce Software Corp.

System Management Server Communications. Network management system. Uses SQL Server to keep track of software running on every Desktop in a network. Manages Windows, Windows NT, NetWare, and DOS servers and clients. Will also run on NetView/6000. Also called Systems Management Server (SMS). Vendor: Microsoft Corp.

system management software Software that provides functions of both operating systems and network management systems to provide enterprise-wide management of communications systems. Provides higher level support to network management systems. Also called enterprise system management software.

System Management Template See SMT.

System Object Model Application development tool. Tool kit for developing object-oriented systems. Vendor: IBM Corp.

System Policy Editor Operating system software. Runs with Chicago (Windows 4.0). Provides graphical monitoring of system, manages what resources each user can access. Vendor: Microsoft Corp.

system software Any software used directly by technical people (programmers, technical support, analysts) in order to perform IT functions. This includes operating systems, data management systems and communications systems. Opposed to applications software.

System Software Associates Software vendor. Products: BPCS family of financial, manufacturing systems for midsize environments.

system test A test of a complete software system.

System Vision Year 2000 Application development tool. Used to analyze size of century-compliant projects. Vendor: Adpac Corp.

System W/Datman Data management system. Runs on IBM systems. Also: System W/Easytrieve, System W/SQL/DS, and DB2. Vendor: Comshare, Inc.

System Z Application development system. Includes DBMS, 4GL, text editor, screen painter. Runs on Unix systems. Vendor: Zortec, Inc.

Systematics Information Management System See SIMS.

SystemPro/XL Desktop computer. Pentium CPU. Operating system: Windows 95/98. Vendor: Compaq Computer Corp.

systems analyst Application developer. Senior person skilled and experienced in the analysis phase of the system development cycle. Strong interpersonal skills are required, as analysts spend much of their time with the users determining needs and processing functions. Business knowledge is also important and often systems analysts are required to know a specific industry such as banking, or a specific application such as human resources.

Systems Analyst CASE product. Part of Re-engineering Product Set. Vendor: Bachman Information Systems, Inc. (now Cayenne Software).

Systems Application Architecture See SAA.

systems engineer Developer, probably technical. Used for a variety of job skills and levels, so has no real meaning.

Systems Engineer CASE product. Automates all phases of the development cycle. Used for client/server development. Vendor: Select Software Tools.

Systems Management Server Communications. Systems management software. Used with OS/2, DOS, Windows, MacOS, Unix. Provides software distribution, configuration management, performance monitoring. Vendor: Microsoft Corp.

systems management software Software that monitors, troubleshoots, tunes and generally manages heterogeneous computer environments. Typically includes problem management, software distribution, license management, configuration management, security management, performance and operations management. Used in client/server environments.

systems manager Technical developer. Manages other technical developers and is responsible for the performance of the computer systems. Title is usually used in mainframe and midrange systems.

systems network architecture See SNA.

systems programmer Technical developer. A systems programmer maintains the operating system programs and environment. In addition to the operating system itself, systems programmers work with communications systems, DBMS, and operating system enhancements and add-ons. Provides technical support to application developers. Plans and evaluates hardware and software purchases. Ensures system efficiency and security. Can be any experience level.

systems tester Usually technical developer. See tester.

Systems ToolKit C++ object library. Vendor: ObjectSpace, Inc.

SystemView (for AIX, for OS/2) A multi-vendor network management system from IBM. Provides standard user interface and data structure. Provides centralized control of decentralized data centers. Vendor: IBM Corp.

SystemWatch Communications. Network management software that allows user to monitor multiple host systems simultaneously. Manages printers, file systems, backups, performance, security. Runs on DEC, Unix systems. Released: 1996. Vendor: OPENService, Inc.

Sysview(/XL) Operating system enhancement used by systems programmers. Monitors and controls system performance in HP systems. Vendor: Carolian Systems International, Inc.

SyteLine Application software. Manufacturing software handling inventory control, supply chain functions. Interfaces with Progress databases. Runs on Unix, Windows systems. Released: 1996. Vendor: Symix Systems, Inc.

T, TB Tera, or terabyte, approximately one trillion bytes.

T-ask 4GL used in mainframe environments. Works in conjunction with Total databases. Vendor: Harris Corp.

T-base Relational database for desktop computer environments. Runs on IBM systems. Vendor: Traveling Software, Inc.

T-Fast/VSE See TD-Fast/VSE.

T/M Online Operating system enhancement used by systems programmers. Monitors and controls system performance in IBM systems. Monitors TSO performance. Vendor: Morino Associates, Inc.

T1/T3 Type of circuits used in communications. Uses telephone company circuits for data communications. Being replaced in many systems by ATM.

T3D, T3E Supercomputer. MPP. Supports up to 1024 processors. T3E is air-cooled. Operating systems: Unix, Unicos. Vendor: Cray Research, Inc.

T5000 Handheld computer. Vendor: Itronix.

T66xxx Portable computer. Vendor: Toshiba America Information Systems, Inc.

T90 Mainframe computer. Operating systems: UNICOS. Vendor: Cray Research, Inc.

Tabasco Code name for release of NetWare which will synchronize user data between NetWare and Windows NT. Vendor: Novell, Inc.

table A way of structuring data in arrays so it can be processed by one or more arguments. Relational databases are made up of tables.

Table Builder See Oracle Card.

table locking See data locking.

tablespace Database technology. Term used with DB2 to represent the definition of a data table.

Tadpole Technology Computer vendor. Manufactures desktop computers.

Tags CASE Product. Automates all phases of the development cycle including project management functions. Vendor: Teledyne Engineering, Inc.

Tahoe Development architecture for knowledge management systems. Includes document management and workflow features. Supports XML. Vendor: Microsoft Corp.

TAL Compiler language used in midrange environments. Runs on Tandem systems. Vendor: Tandem Computers, Inc. (a Compaq company).

TAL AE See CommonPoint.

Taligent 1. Application development framework with an object layer on top of a Unix operating system, a distribution method for objects, and an application development environment. Handles both development and run-time environments. Runs on AIX, HP-UX, OS/2, System 7. Vendor: Hewlett-Packard Co. 2. Company funded by IBM, Apple, Hewlett-Packard to develop object-oriented software.

TalkShow Groupware. Conferencing software that automatically moves changes made on a terminal screen with pen or keyboard to the other user's screens. Vendor: Future Labs, Inc.

Talladega Desktop computer. Pentium processor. Vendor: Racer Computer Corp.

TAM Operating system enhancement used by systems programmers. Controls communications networks in IBM systems. Stands for: TSO Access Manager. Vendor: Tone Software Corp.

Tamaris CS Application software. Financial systems for mainframe environments. Vendor: Walker Interactive Systems.

Tandem Computer vendor. Manufactures large and desktop computers. Purchased by Compaq in 1997.

Tangent Computer vendor. Manufactures desktop computers.

Tangent EISA/VL, NetRun, PCI, VL Desktop computer. Pentium CPU. Operating systems: Windows 95/98. Vendor: Tangent Computer, Inc.

Tangent MediaGem See MediaGem.

Tanglewood System management software. Suite of graphical products designed to track and troubleshoot network problems. Used in client/server, Unix environments. Includes Sybase's SQL Server database and Tivoli's Management Environment (TME). Vendor: Sybase, Inc. and Tivoli Systems, Inc.

Tango (Enterprise) Application development tool. Internet tool. Allows users to create Web pages that access SQL databases without writing SQL or HTML. Vendor: EveryWare Development Inc.

Tango for Windows Application development tool. Connects Windows, Unix and Macintosh Web servers to standard databases. Generates HTML and SQL. Runs on Windows NT and Windows 95/98 systems. Vendor: EveryWare Development Inc.

Tape Operating system enhancement used by systems programmers. Increases system efficiency in IBM systems. Manages tape storage. Runs on HP systems. Vendor: Operations Control Systems.

Tape+ Operating system add-on. Tape management system that runs on NCR systems. Vendor: Software Clearing House, Inc.

Tape Handler See Entry-level Operator.

tape management system A system that keeps track of all magnetic tapes being used in the data center. The system will record and report on such things as what data files are on what tapes, expiration dates of data files, and location and identification of backup files.

Tape Mount Management See TMM.

TapeBox Operating system enhancement. Provides backup services for systems on Unix networks. Vendor: BoxHill Systems Corp.

Tapelib CMS Operating system add-on. Tape management system that runs on IBM systems. Vendor: Boole & Babbage, Inc.

Tapeman Operating system enhancement used by systems programmers. Increases system efficiency in Wang systems. Manages tape storage. Vendor: Software Extraordinaire, Inc.

Tapes Operating system add-on. Tape management system that runs on HP systems. Vendor: Unison Software.

Tapestry Application development tool. Rule-driven software used to load data warehouses. Transforms data from relational databases and PeopleSoft, SAP systems. Runs on Unix, Windows systems. Vendor: D2K, Inc.

Tapestry II Communications software. Network operating system. Runs on multiple desktop computers. Vendor: Torus Systems, Inc.

Tapesys VAX/VMS Operating system add-on. Data management software that runs on DEC VAX systems. Vendor: Software Partners/32, Inc.

TAPI Communications interface that supports services from a wide range of telephone systems. Allows users to merge phone and desktop computer functions. Stands for: Telephony Application Interface. Vendor: Microsoft Corp.

Tarantella Communications technology. Middleware . Server-based technology that lets users access Unix and other proprietary server applications through the Internet using any Java-enabled client. Web-enables existing applications without rewrites. Options available to access IBM mainframe data. Released: 1997. Vendor: SCO (Santa Cruz Operation).

Vendor: SCO, Inc.

Target-Batch Operating system enhancement used by systems programmers. Job scheduler for DEC VAX systems. Vendor: Target Systems Corp.

TargetLink Systems management software. Java based software that automates software distribution to Macintosh, Unix, Windows or network computer systems. Released: 1998. Vendor: Internet Image, Inc.

Tarpon Handheld computer. Vendor: 3Com Corp.

TAS Communications. NOS (Network Operating System). Allows Unix servers to function as file, print and application server. Also functions as Internet server. Runs on DEC, MacOS, OS/2, Unix, Windows. Released: 1997. Stands for: TotalNET Advanced Server. Vendor: Syntax, Inc.

TAS-Plus Relational database for desktop computer environments. Runs on IBM desktop computer systems. Includes data dictionary, program generator, and utilities. Vendor: Business Tools, Inc.

TASC Compiler language used in desktop computer environments. Runs on Apple systems.

Task Master Groupware. Workflow system. Runs on NetWare systems. Vendor: Datacap, Inc.

Task Monitor Desktop system software. Project management package. Runs on Pentium type desktop computers, Macintosh. Vendor: Monitor Systems, Inc.

TASM Assembler language for Intel's 80x86 computer chips. Stands for: TurboASseMbler. Vendor: Borland International, Inc.

Tatung Computer vendor. Manufactures desktop computers.

Taylor II Application development tool. Provides process modeling functions. Runs on Windows systems. Vendor: F&H Simulations Inc.

TC2000 Parallel processor. Supports up to 128 processors. Operating systems: nX, pSOS. Vendor: Bolt, Teranek and Newman, Inc.

TC486(E,V) Desktop computer. Pentium CPU. Operating system: Windows 95/98. Vendor: Unisys Corp.

TCAM Access method used in IBM mainframe systems. Used by system programming communications specialists. Stands for: Telecommunications Access Method.

TCAT Application development tool. Automates testing function. Stands for: Test Coverage Analysis Tool. Vendor: Software Research Associates.

Tcl Programming language. Scripting language. Looks like a combination of Bourne shell and C language statements. Can be used for any scripting function, but was created specifically for component based architecture and is used to assemble components to create applications. Stands for: Tools Command Language. Vendor: Scriptics Corp.

TCL/TK Application development tool. TCL is a scripting language included as a C library, while TK is a toolkit used to develop x-windows applications. Vendor: Infomagic.

TCO Total Cost of Ownership. Used to define what PCS actually cost a corporation when support is factored in.

TCP Operating system enhancement used by systems programmers. Job scheduler for Nixdorf, IBM systems. Controls online jobs. Vendor: Siemens Nixdorf Computer Corp.

TCP/Connect II Communications software. Internet kit which facilitates the use of basic Internet services. Vendor: InterCon.

TCP/IP Transaction Control Protocol/Internet Protocol. One of the transmission standards used in communications software. Popular in Unix, DEC environments. Developed by DARPA (Defense Advanced Research Projects Agency). Provides packet switching technology. Stands for: Transmission Control Protocol/Internet Protocol.

TCP/NET Communications software. Network connecting MODCOMP systems. Vendor: MODCOMP.

TCP-V, TCP-R Operating system for mainframe Telefile systems. Vendor: Telefile, Inc.

TCRF Operating system add-on. Data management software that runs on Bull HN systems. Stands for: Transactional Contex Restart Facility. Vendor: Bull HN Information Systems.

TCS 1. Application development tool. Moves data files from mainframe systems to client/server environments. Vendor: Micro Tempus, Inc.
2. Communications software. Transaction processing monitor. Runs on IBM systems. Stands for: Telecommunications Control System. Includes: TCS-ACF, TCS-AF. Vendor: IBM Corp.

TCS-5xxx,6xxx Desktop computer. Pentium CPU. Operating system: Windows 95/98. Vendor: Tatung Company of America.

TD-Fast/VSE Operating system add-on. Data management software that runs on IBM VSE systems. Includes: T-Fast/VSE (tape manager) and D-Fast/VSE (disk manager). Vendor: Tower Systems International.

TDGEN Application development tool. Automates testing function. Stands for: Test Data GENerator System. Vendor: Software Research Associates.

TDL1 Application development tool. Allows programs written for TOTAL databases to access IMS databases. Released: 1996. Vendor: Business Information Systems, Inc.

TDMF Operating system enhancement. Manages data transfers and backups between mainframe storage systems. Stands for: Transparent Data Migration Facility. Vendor: Amdahl Corp.

TDMS Data management system. Runs on DEC systems. Handles input/output of information on terminal screens. Stands for: Terminal Data Management System. DECforms is newer version. Vendor: Digital Equipment Corp.

TDZ 2000 Desktop computer. Pentium II CPU. Released: 1998. Vendor: Intergraph Corp.

Team Developer See Centura Team Developer.

Team Enterprise Developer See Enterprise Developer.

Team-Errico CASE product. Automates design, programming, and project management functions. Vendor: Boole & Babbage, Inc.

Team-Flow4: Work Processor Application development tool. Used with re-engineering for process design. Runs on Windows systems. Vendor: CFM, Inc.

team leader See project leader, team leader.

Team Manager 97 Project management software. Used to coordinate and consolidate team activities. Integrates with Outlook.

Team Mate Database for midrange and desktop computer environments. Runs on IBM systems. Includes report generator. Vendor: CMDS.

Team/V Application development tool used to develop object-oriented systems. Actually is an extension of Smalltalk. Allows developers to define, store, browse and share Smalltalk objects. Vendor: Digitalk, Inc.

TeamBuilder Application development tool. Works with Enfin. Allows groups of developers to simultaneously build applications. Vendor: Easel Corp.

TeamConnection Application development tool. Builds object-oriented repository. Runs on OS/2, Windows systems. Vendor: IBM Corp.

teamCreator Application development tool. Builds GUI front-ends for Unix and mainframe systems. Runs on Unix, Windows systems. Released: 1997. Vendor: Pericom Software, Inc.

TeamFocus Groupware. Electronic Meeting System (EMS). Also called GroupSystems. Vendor: IBM Corp.

TeamLinks Groupware. Includes conferencing, document routing, reference libraries. Provides the client interface to ALL-IN-1. Vendor: Digital Equipment Corp.

TeamNet Application development system. Runs on Unix-based Sun systems. Is an upgrade from Teamone. Vendor: Team One Systems, Inc.

teamOffice Groupware. Includes Team-Flow (work-flow software), TeamMail. Runs on Windows systems. Vendor: ICL.

TeamPad Handheld computer. Vendor: Fujitsu.

TeamTalk Groupware. Supports electronic conferencing. Vendor: Trax Softworks.

TeamTest See Rational TeamTest.

TeamTrack Application development tool. Web-based debugging tool that tracks software problems across multiple platforms. Used for both departmental and enterprise-wide development. Vendor: TeamShare, Inc.

Teamware Software developed to aid collaboration among workers. Allows workers to share and edit information, but does not offer full functionality of groupware. Unlike groupware which is set up and managed by IT professionals, teamware is controlled by the end-users.

TeamWare Flow Workflow software. Works with TeamWare Office groupware. Vendor: TeamWare.

Teamwork See COOL:Teamwork.

tech writers Support personnel. Writes the user documentation for software systems. This documentation explains how to use the programs and defines the user interface. Usually need to know a desktop publishing system. Can be any experience level.

Techmate TM6 Portable computer. Vendor: Consultronics, Ltd.

technical certification Programs offered by many vendors. Vendor supplies (and often authorizes other vendors to supply) training courses in various skill areas. Tests are then offered to verify knowledge. When a defined number of tests are passed, certification is granted. Some certificate programs are accepted by companies as proving knowledge and are included as a job requirement. Most popular certificate programs for Novell and Microsoft networking knowledge.

Technote Sxxxx Notebook computer. Pentium CPU. Vendor: Techmedia Computer Systems Corp.

TECO Application development tool. Screen editor for DEC systems. Vendor: Digital Equipment Corp.

Tecra xxxx Notebook computer. Pentium CPU. Vendor: Toshiba America Information Systems, Inc.

Tecsys Distribution, manufacturing, and financial software. Multilingual (English, Spanish, French). Uses Informix. Tecsys is the vendor name.

TED Version of CDE that can be purchased separately from Unix. Runs with AIX, HP-UX, SunOS, Solaris, Irix, Ultrix. CDE will be bundled in some versions of Unix. Vendor: TriTeal Corp.

TEK/LANDS Application development system. Runs on DEC, Sun, Unix systems. Vendor: Tektronix, Inc.

TekBook Desktop computer. Notebook. Pentium processor. Vendor: TJ Technology, Inc.

Teknekron Information Bus Communications software. Messaging middleware. Uses publish-and-subscribe architecture and works with TCP/IP networks. Vendor: Teknekron Software Systems, Inc.

TEL-EDI Communications software. EDI package. Runs on Pentium type desktop computers. Vendor: Piedmont Systems, Inc.

TELCON Communications software. Transaction processing monitor. Runs on Unisys systems. Vendor: Unisys Corp.

telecommunications See communications.

Telemark Operating system enhancement used by systems programmers. Controls communications networks in DEC systems. Vendor: Tellabs, Inc.

Telenet Communications network set up in the early 1980s. Packet switching network that uses X.75 protocols. Considered a public network, companies can subscribe to Telenet.

TelePad SL Pen-based computer. Vendor: Telepad.

telepresence See virtual reality.

Telescape Communications software. Terminal emulator for Apple Macintosh desktop computers. Vendor: Mainstay.

Telescript Communications software. Communications protocol used with hand-held systems. Vendor: General Magic, Inc.

teletex Part of ISDN. E-mail for home and business.

TeleUSE Application development tool. Used to build object-oriented Motif GUIs in C or C++. Runs on DEC, Unix systems. Released: 1995. Vendor: Aonix.

TeleUse/Win Application development tool. Develops and migrates Unix applications with a Motif GUI to Windows. Vendor: Thompson Software Products.

Teleview Application development tool. Program generator for IDMS programs. Vendor: Telemap Corp.

Telink Communications software. EDI package. Runs on Pentium type desktop computers. Vendor: EDI, Inc.

Telluride Midrange computer. Alpha CPU. Vendor: Aspen Systems, Inc.

Telnet Communications protocol. Applications layer protocol that handles terminal logons to a communications system. Lets the user connect to another machine on a network as if logged on directly. Used with TCP/IP.

Telon See CA-Telon.

Telxon PTC Handheld, pen-based computer. Vendor: Telxon Corp.

Templar Communications software. EDI system that works through a direct Internet connection. Vendor: Premenos Corp.

TempleMVV Application development tool used for data mining. Provides graphs of different subsets or summaries of data. This allows user different perspectives of the data. Runs on Unix, Windows systems. Vendor: Mihalisin Associates, Inc.

Tempo Operating system. Runs on Macintosh systems. Code name for development version of MacOS. Used on smaller, single-user machines. Vendor: Apple Computer, Inc.

Tempo Client Network computer. Vendor: Everex Systems, Inc.

Tempo K Series Desktop computer. Pentium CPU. Operating system: Windows NT. Vendor: Everex Systems, Inc.

Tempus Communications software. Network connecting IBM computers. Includes: Tempus-Link, Tempus-Access, Tempus-Share. Micro-to-mainframe link. Vendor: Micro Tempus, Inc.

TENEX Operating system for DEC systems. Vendor: Bolt Beranak and Newman, Inc.

terabyte Approximately one trillion bytes.

Teradata Relational database and data warehousing software. Includes data mining functions. Vendor: NCR Corp.

Teradata Nomad See Nomad (2).

terminal emulator See emulator.

Terminal Server See WTS.

Terminal Server Edition Operating system. Version of Windows NT used in this client processing where all processing occurs on the server. Clients can be many platforms including MacOS, UNIX, and Windows. Terminal Server provides access to 32-bit Windows-based applications from virtually any desktop. Vendor: Microsoft Corp.

TERMS Operating system enhancement used by systems programmers. Provides system cost accounting for IBM systems. Stands for: Total EDP Resource Management System. Vendor: 4GL Software Systems.

Terrain See COOL:DBA.

Terranova Express Software distribution system. Warehouses programs, distributes them, and monitors usage. Released: 1997. Vendor: Lucent Technologies, Inc.

TESS/3000 Application development tool. Screen editor for HP systems. Vendor: Computer Consultants & Service Center, Inc.

Tesseract Software vendor. Products include human resource, payroll systems. Software runs on IBM mainframe systems. Full name: Tesseract Corp.

Test/IMS Operating system add-on. Debugging/testing software that runs on IBM systems. Vendor: Consumer Systems Corp.

test library A dataset that contains programs which are currently in the test phase of the development cycle.

Test Suite Application development tool. Provides fragment testing, application testing, test reports and test management functions. Includes TouchPoint, WinRunner. Vendor: MERANT.

TestBed Application development tool. Provides testing/debugging functions. Verifies programming standards adherence, analyzes what code is tested. Runs on Unix, VMS, Windows systems. Vendor: Eastern Systems, Inc.

TestBytes Application development tool used for testing. Allows user to generate test scenarios and test data for any database application. Runs on Windows systems. Vendor: Mercury Interactive Corp.

TestChecker Application development tool used during testing. Measures test coverage and detects untested source code. Part of Logiscope. Runs on Unix, Windows systems. Vendor: CS Verilog.

TestDirector Application development tool. Testing tool used to generate test plans. Vendor: Mercury Interactive Corp.

tester Developer. Could be either applications or technical. Designs test data, test scenarios. Conducts system, or integration testing. Specific skills can refer to such things as regression testing, black box testing, QA (quality assurance). Can be any experience level.

Testgen CASE product. Automates design, programming, and testing functions. Used for Ada program development. Vendor: Software Systems Design, Inc.

testing A phase in the system development cycle. Running a computer program, or system, with planned data and matching the results against expected output.

Testmanager Operating system add-on. Debugging/testing software that runs on IBM systems. Vendor: MANAGER SOFTWARE PRODUCTS, Inc.

TestPlan Application development tool. Provides testing/debugging functions. Plans, manages and documents testing throughout project. Runs on Windows systems. Vendor: Eastern Systems, Inc

TestStudio Application development tool. Provides testing tools for functional testing to ensure that software meets the business requirements. Creates test scripts and manages test plans. Part of Rational Suite. Runs on Unix, Windows systems. Released: 1999. Vendor: Rational Software.

TestWare Operating system add-on. Debugging/testing software that runs on DEC PDP, VAX systems. Vendor: PRIOR Data Sciences Product Sales, Inc.

TestWeb Application development tool. Provides testing/debugging functions. Performs regression testing, benchmarking. Runs on Windows systems. Vendor: Eastern Systems, Inc

Texas Instruments Software/hardware vendor. Known as TI. Purchased by Sterling Software in 1997.

Texis Text retrieval system. Fully integrates documents into strategic business applications Allows developers to load documents or text fields into a TEXIS database, or leave it in its native format. Runs on Macintosh, OS/2 and Windows systems. Vendor: Thunderstone/EPI, Inc.

text editor See screen editor.

TextBridge OCR software. Provides document and data table handling. Lets users scan documents and convert to HTML for Web use. Vendor: Xerox.

TextDBMS Document retrieval system. Runs on MVS, Windows NT systems. Vendor: Data Retrieval Corp.

Textor See CA-Textor.

TextWare Document retrieval system. Runs on Windows systems. Vendor: TextWare Corp.

TFP Communications protocol. Part of AppleTalk. Stands for: TOPS Filing Protocol.

TFS Communications, Internet software. Firewall. Fullname: TurnStyle Firewall System. Vendor: Atlantic Systems Group.

TFTP Communications protocol. Mini version of FTP used to transfer files between two host computers. Stands for: Trivial File Transfer Protocol.

TGAL Application development tool. Screen editor for Tandem systems. Vendor: Tandem Computers, Inc. (a Compaq company).

The Application development environment See ADS.

the Connection See VMS/Ultrix Connection.

The Daily Planit Operating system enhancement used by systems programmers and operations staff. Job scheduler for Unisys systems. Vendor: Planit Systems, Inc.

The Meeting Room Groupware. Electronic meeting software. Runs on standard networks. Vendor: Eden Systems Corp.

The Networker Communications software. Network connecting Ethernet, Novell, and Token Ring networks. Vendor: Wyse Technology, Inc.

The Organizer Desktop system software. Personal information manager (PIM). Runs on Pentium type systems under Windows. Vendor: Lotus Development Corp.

the Poet Application development tool. Automates programming function. Generates ESQL-C code. Full name: Embedded SQL Poet. Vendor: Hellenic Systems, Inc.

The Relational Tools Application development tool. Debugging/testing programs. Vendor: Princeton Softech Inc.

The Reporter 1. Communications software. Network performance monitor. Vendor: Network General Corp. 2. Application development tool. Query and report functions. Runs on Windows systems. Vendor: Sea Change Systems Inc.

The Scheduler Operating system enhancement used by systems programmers and operations staff. Job scheduler for Unisys systems. Vendor: Software and Management Associates.

The Solution Application software. Includes HR, Payroll. Interfaces with DB/2, Oracle, Rdb. Runs on Unix systems. Client/server technology. Vendor: Cyborg Systems, Inc.

The Virtual Notebook EMS (Electronic Meeting System). Vendor: The Forefront Group.

Theo+DOS Operating system for desktop computers. Multiuser version of DOS. Vendor: THEOS Software Corp.

Theos Operating system for IBM and compatible desktop computer systems. Versions include: Theos 286, Theos 286-V, Theos 386, Theos 86. Vendor: THEOS Software Corp.

Theta Object-oriented programming language. Created at MIT.

thin client In client/server computing, a system where most of the processing is done on the server system. "Thin client" systems can use very basic desktop client machines, as they really just provide a graphic interface to a server system. Network personal computers always function as thin clients.

ThinBrick Desktop computer. Notebook. Pentium processor. Vendor: The Brick Computer Company, Inc.

Thinking Machines Computer vendor. Manufactures large computers.

ThinkPad xxx Notebook computer. Vendor: IBM Corp.

Thinx Desktop system software. Personal Information Manager (PIM). Runs on Windows systems. Vendor: Bell Atlantic.

third-generation computer Computer built with integrated circuits. Commercially available in the mid-to-late 1960s.

Thoroughbred Idol-IV See Idol-IV.

thread, threading A thread is a process within a program that does a specific task. A single-threaded program does one thing at a time, while a multi-threaded application performs many tasks at the same time. Multi-threaded applications can, i.e., update a database, display a graph and print a report at the same time. A multi-threaded program can process more than one transaction at a time. Used with multiprocessing.

Threads.h++ Application development tool. Allows developers to develop multi-threaded programs in C++. Runs on Unix, Windows systems. Vendor: Rogue Wave Software, Inc.

ThreadX Real-time operating system for desktop computers. Vendor: Express Logic, Inc.

Three Dimensional Architecture See 3DA.

three-tiered Architecture used in client/server development which partitions an application into three parts: data access, user interface, application logic. The data access will run in the server, the user interface will run in the client, and the application logic can run in either system, or even in a third system. Location can be decided dynamically.

throughput A measure of the amount of work processed by a computer system over a given period of time. Also spelled thruput.

Thruway Operating system enhancement used by systems programmers. Increases system efficiency in DEC VAX systems. Handles remote devices. Vendor: Software Partners/32, Inc.

ThunderBox II Portable computer. Vendor: Lightning Computers.

ThunderBrick Desktop computer. Pentium processor. Vendor: Ergo Computing Inc.

TI See Texas Instruments.

TI System V Operating system for TI systems. Vendor: Texas Instruments, Inc.

TiaraLink LanWare Communications. LAN (Local Area Network) connecting desktop computers. Vendor: Tiara Computer Systems, Inc.

TIB/ActiveEnterprise Communications. Middleware. EAI system which integrates diverse applications. Moves data transparently across local networks, throughout the corporate enterprise, or over the Internet. Includes TIB/Rendezvous, a publish and subscribe technology. Connects all applications, databases, and networks. Provides event-driven technology so all business functions can share the same data in real-time. Released: 1998. Vendor: TIBCO Software, Inc.

TIB/Hawk System management software. Monitors and manages distributed applications across the enterprise. Runs on Solaris, Unix, VMS, Windows NT. Released: 1998. Vendor: TIBCO Software, Inc.

TIB/ObjectBus Communications software. Middleware, object request broker (ORB). CORBA compliant. Used to develop object-oriented applications. Released: 1997. Vendor: TIBCO Software, Inc.

TicToc Application development tool. Analysis tool used to find date dependencies. Used for year 2000 work. Vendor: Isogon Corp.

TIE See VEST.

TIF Application development tool. Program that includes report generation, analysis, and data selection functions. Runs on IBM systems. Stands for: The Information Facility. Vendor: IBM Corp.

Tillamook Computer chip. MMX chip for notebook and portable computers. Vendor: Intel Corp.

Timberline Midrange computer. Alpha CPU. Vendor: Aspen Systems, Inc.

Time/1000 Operating system enhancement used by operations staff and systems programmers. Disaster control package. HP systems. Provides recovery procedures after power failure. Vendor: Corporate Computer Systems, Inc.

Time and Place/2 Groupware. Scheduling software. Includes meeting scheduler and individual memo capabilities. Runs on LAN Server, NetWare systems. Vendor: IBM Corp.

Time Line Desktop system software. Project management package. Runs on Pentium type systems. Includes Time Line Graphics 1.1, Time Line 2.0, Time Line 3.0. Vendor: Symantec Corp.

Time Machine 1.Desktop system software. Project management package. Runs on Pentium type desktop computers. Vendor: Diversified Information Services, Inc. 2. Data warehouse software. Includes storage, query and load functions. Runs on OS/2, MVS, Unix, VMS, Windows systems. Vendor: Data Management Technologies, Inc.

Timebase Time-series database. The database understands the concepts of days, weeks, months etc. so programming for these functions is not necessary. Vendor: Pilot Executive Software.

timeline-based Multimedia term. Used to describe multimedia tools that present objects which are controlled by a timeline with start/stop and synchronization capabilities.

TimesTen Data Manager Relational database. In-memory database. Uses main memory as primary data store thus achieves high performance without having to access disks. Runs on Unix, Windows systems. Released: 1998.Vendor: TimesTen Performance Software.

Timetable Desktop system software. Project management package. Runs on Pentium type desktop systems. Vendor: Accuratech, Inc.

Timetrap Application development tool. Debugging/testing programs. Vendor: MiraSoft, Inc.

TIMM Artificial intelligence system. Expert system building tool. Runs on Amdahl, DEC, IBM systems. Stands for: The Intelligent Machine Model. Desktop version can handle up to 500 rules; mainframe version can handle up to 5000 rules. Vendor: General Research Corp.

TIMS Text Information Management System. Generic term for any software that manages documents.

TinyTERM Communications. Connects PCs to Unix hosts and/or the Internet. Series of products including: TinyTERM Plus, TinyTERM+NFS and TinyTERM Pro. Vendor: Century Software, Inc.

TIP family CASE product. Automates analysis, programming, and design functions. Includes: TIP Plan, TIP Define, TIP Create. Vendor: Technology Information Products Corp.

TIRS Artificial intelligence system. Is part of IBM's SAA. Stands for: The Integrated Reasoning Shell. Vendor: IBM Corp.

TIS/XA Communications software. Transaction processing monitor. Runs on IBM systems. Stands for: The Information System/Extended Architecture. Vendor: Cincom Systems, Inc.

Titan II Desktop computer. Pentium processor. Vendor: ET Technology.

Titana Desktop computer. Pentium processor. Vendor: Unicent Technologies.

Titanium Data warehouse software. Provides data security functions. Includes data dictionary. Runs on OS/2, Unix, Windows systems. Vendor: Micro Data Base Systems, Inc.

Titanium Advanced Notebook computer. Pentium CPU. Vendor: Xediom.

Tivoli/TME System management software. Includes operating system and network management functions for Unix systems. Latest release synchronizes management information among multiple servers, handles network devices, uses a browser interface. Provides software distribution, operations management, backup control. Includes: Tivoli/TME NetView, Inventory, NetFinity, GEM. Runs on HP, IBM Sun systems. Stands for: Tivoli Management Environment. Vendor: Tivoli Systems, Inc.

TKAM VSAM-type access method for IBM systems.

Tlist Data management system. Runs on Apple II systems. Vendor: Eclipse Systems.

TM1 Application development tool. DSS used in data warehousing. Provides access to data stored in multidimensional databases. Vendor: Applix, Inc.

TMDS Operating system add-on. Debugging/testing software that runs on Tandem systems. Vendor: Tandem Computers, Inc. (a Compaq company).

TME(10) See Tivoli Management Environment(10).

TMM Operating system software. Performance monitor. Stores as many data files on a single tape as possible. Runs on IBM systems. Stands for: Tape Mount Management. Vendor: IBM Corp.

TMON 1. Operating system add-on. Debugging/testing software that runs on Apple II systems. Vendor: ICOM Simulations, Inc. 2. Operating system enhancements used by systems programmers. Performance monitors for IBM systems. Available: TMON for MVS, TMON for DB2, TMON for VTAM, TMON for CICS. Stands for: The Monitor. Vendor: Landmark Systems Corp.

TN3270 Communications software. Graphical terminal emulation software that lets Unix systems access mainframe and AS/400 systems. Vendor: Apertus Technologies, Inc.

TNB-5xxx Notebook computer. Pentium CPU. Vendor: Tatung Company of America, Inc.

TNC-1xxx Network computer. Vendor: Tatung Company of America, Inc.

TNCE Communications software. Suite of security software. Interfaces with Net-Ware. Stands for: Trusted Network Computing Environment. Software being developed by multiple vendors.

Tnet Communications software. Network connecting DEC and Apollo computers. Vendor: Calma Co.

TOAD Application development tool. Contains a browser to give developers access to database objects to build and test PL/SQL. Stands for: Tool for Oracle Application Developers. Released: 1998. Vendor: Quest Software, Inc.

TOADS Application development system. Includes 4GL, data dictionary. Stands for: Total Online Application development environment. Runs on DEC, IBM RS/6000, Sun systems. Vendor: USC Software Systems.

TODAY 4GL Application development system. Used for enterprise development. Develops applications for most computers, operating systems, and databases. Includes screen builder, report generator. Runs on Unix systems. Vendor: TODAY Systems, Inc.

Together/C++(Pro) Application development tool. Manages object modeling, generates documentation, and provides configuration management. Vendor: Object International, Inc.

Together/J Application development tool. Object-oriented modeling tool that supports UML. Includes reverse engineering tools. Works with JavaBeans. Runs on Solaris, Windows systems. Released: 1998. Vendor: Object International, Inc.

Token-Ring Communications. LAN (Local Area Network) connecting IBM desktop computers. Based on ring topology. A piece of data called a token is passed throughout the network. The computer with the token may send data to another computer. Vendor: IBM Corp.

Tolas family Applications software. Includes financial, sales, inventory products for midsize environments. Vendor: ADP-GSI.

Tone Application development environment. Runs on IBM VS1 systems. Vendor: Tone Software Corp.

Tone 4 Operating system enhancement used by systems programmers. Controls communications networks in IBM systems. Vendor: Tone Software Corp.

ToolBuilder Application development tool. Used to develop other development tools. Vendor: Lincoln Software.

Toolbus Application development tool. Integrates CASE tools from different vendors. Vendor: The Barton Group.

Tools.h++ Application development tool. C++ class library containing objects for, i.e., time and date conversions. Runs on Macintosh, Windows systems. Vendor: Rogue Wave Software.

Toolset-DB2 Operating system add-on. Data management software that runs on IBM DB2 systems. Includes: SQL Source Generator, Data Dumper, Security Cloner. Vendor: On-Line Software International.

Toolset(/XL) Application development tool. Runs on HP systems. Vendor: Hewlett-Packard Co.

ToolTalk Communications software. Allows applications to communicate across heterogeneous networks. Object-oriented program that provides interface for multivendor development tools. Runs on DEC, HP, Sun systems. Vendor: SunSoft, Inc.

TOP Communications protocol. OSI protocol stack for office automation. Stands for: Technical Office Protocol.

top down development See functional decomposition, analysis, design.

Top Secret See CA-Top Secret.

TopEnd Communications software. Middleware. Connects heterogeneous hardware, databases and operating systems. Provides distributed transaction management, workload balancing, message routing. Used to move mainframe systems to Unix systems. Interfaces with Internet and includes Java support. Originally developed by NCR Corp, and acquired by BEA Systems in 1998. Vendor: BEA Systems, Inc.

Topic Text retrieval system that runs on various computer systems and accesses data in various databases. Includes Topiql, an SQL access to networked databases. Vendor: Verity, Inc.

Topiql See Topic.

TOPIX Operating system for Sequoia systems. Vendor: Sequoia Systems, Inc.

topology The structure, or arrangement, of the devices in a communications network. Types of topologies: Star, mesh, hierarchical, ring. Networks can be built combining topologies.

TOPS-10,20 Operating system for DEC, SC Group systems. Vendor: Digital Equipment Corp.

TOPS/DOS Operating system for IBM and compatible desktop computer systems. Vendor: TOPS.

TOPS/Macintosh Operating system for Apple Macintosh systems. Vendor: TOPS.

TOPS/Sun,DOS,Mac Communications software. Network connecting MS-DOS, Macintosh and Unix workstations. Vendor: Sitka Corporation.

TOPS/VMS Operating system for DEC VAX systems. Vendor: TOPS.

TopTier Application development tool. Allows users to access disparate databases through the Internet or a LAN without having to know where or in what format the data resides. Includes: Enterprise Builder, Navigator, Model Transporter. Vendor: TopTier, Inc.

Topview Desktop system software. Windowing package. Runs on IBM desktop systems. Obsolete. Vendor: IBM Corp.

Torch/PMS Operating system enhancement used by systems programmers. Monitors and controls system performance in Unisys systems. Vendor: Datametrics Systems Corp.

Tornado 1. System management software. Object-oriented. Provides both network and systems management functions. Next version of OpenView. Will contain a common data repository that can be shared by multiple network and systems management applications. Vendor: Hewlett-Packard Co.
2. Desktop computer. Pentium CPU. Operating systems: Windows 95/98, Windows NT Workstation. Vendor: Tangent Computer, Inc.
3. Desktop computer. Notebook. Pentium processor. Vendor: Akia Corp.

Tornado NetStar Midrange computer. Pentium Pro CPU. Operating system: Windows NT. Vendor: Tangent Computer, Inc.

TOS 1. Operating system for Nixdorf systems. Vendor: Nixdorf Computer Corp.
2. Operating system for Atari systems.

TOSC1,4,6 Operating system add-on. Library management system that runs on Bull HN systems. Vendor: TOSC International, Inc.

Toshiba Computer vendor. Manufactures desktop computers.

TOSS/Mainframe Communications software. E-mail system. Runs on IBM CICS systems. Vendor: NBS Systems, Inc.

Total 1. Database for large computer environments. Runs on IBM systems. Vendor: Cincom Systems, Inc.
2. Database for large computer environments. Runs on Harris systems. Vendor: Harris Corp.

Total Central Database for large computer environments. Runs on Bull HN systems. Vendor: Bull HN Information Systems.

TOTAL Framework Application software. Manufacturing environment which includes workflow, assembly, storage, and system integration linking the components. Vendor: Cincom Systems, Inc.

Total ORDB Object-oriented database. Runs on Unix systems. Vendor: Cincom Systems, Inc.

Total Virus Defense Virus protection. Runs on Windows systems. Vendor: McAfee Associates, Inc.

TotalNET Advanced Server See TAS.

Toto Application development tool. DSS. Provides access to data stored in multidimensional databases. Vendor: Decision-Works Ltd.

Touchdown Communications software. Provides the client software for e-mail in client/server systems. Interfaces with Microsoft's EMS. Handles Windows, Macintosh, MS-DOS, Unix clients. Windows version formerly called Capone. Vendor: Microsoft Corp.

TouchPoint Application development tool. Used during testing. Provides testing for small sections of code so the actual new code can be tested without testing the entire application. Vendor: Originally Micro Focus, now MERANT.

Toughbook Desktop computer. Notebook. Pentium processor. Vendor: Panasonic Personal Computer Co.

TowerEiffel System for OS/2 Application development environment. Develops object-oriented systems. Vendor: Tower Technology Corp.

TowerJ Application development tool. Java compiler that compiles (creates machine code) class files. This creates executable class files that can execute on any platform. Used to develop server applications. Runs on Solaris, Unix, Windows systems. Released: 1998. Vendor: Tower Technology Corp.

TP-IX/88K Version of Unix operating system designed to run on RISC machines. Vendor: Tadpole Technologies, Inc.

TP monitor See transaction processing monitor.

TP5000 Network Operating system enhancement used by systems programmers. Controls communications networks in Prime systems. Full name: TP5000 Network Control Center. Vendor: Telenet Communications Corp.

TPF Operating system that concentrates on online transaction systems. Stands for: Transaction Processing Facility. Vendor: Stratus Computer, Inc.

TPNS Communications software. Network connecting IBM computers. Stands for: Teleprocessing Network Simulator. Used by system programming communications specialists. Vendor: IBM Corp.

TPS Application development environment for IBM Series/1 systems. Stands for: Transaction Processing System. Vendor: Software Consulting Service, Inc.

TPS-6 Operating system enhancement used by systems programmers. Monitors and controls system performance in Bull HN systems. Vendor: Bull HN Information Systems.

TPS QR6 Application development tool. Report generator for Bull HN systems. Vendor: Bull HN Information Systems.

TPX Communications software. Network connecting IBM computers. Stands for: Terminal Productivity Executive. Includes: TPX Access. Vendor: Legent Corp.

TRACE Operating system add-on. Debugging/testing software that runs on IBM systems. Vendor: A.K., Inc.

Tracer Operating system add-on. Debugging/testing software that runs on Unix systems. Vendor: Ready Systems.

Track 1. EIS software. Based on IBM's Executive Decisions. Includes ObjectTrack programming language. Accesses multivendor multiproduct environments. Provides OLAP functions. Works with relational databases. Vendor: Intelligent Office Co.
2. Operating system add-on. Debugging/testing software that runs on IBM systems. Vendor: Performance Software, Inc.

Track I CASE product. Automates design function. Full name: Track 1 C The Systems Manager. Vendor: National Systems, Inc.

TrackRecord Application development tool used for testing. Works with ActiveLink to provide management functions to development teams. Part of DevCenter. See DevCenter.

TRACS Database for large computer environments. Runs on DEC systems. Vendor: Casher Associates, Inc.

Trading Partner Communications software. EDI package. Runs on IBM mainframes. Also available, Trading Partner PC Kits which runs on desktop computers and can be customized to interface with any EDI program. Vendor: TSI International.

TRAN Application development system. Used to build distributed OLTP systems. Supports Windows and Motif GUIs. Full name: TRAN Distributed Systems Environment. Runs on Unix systems. Vendor: BEA Systems, Inc.

Trans-sync Communications software. EDI package. Runs on Pentium type desktop computers. Vendor: Dynamic Business Systems.

Trans4M MRP, or CIM software. Runs on most RISC systems. Vendor: CMI Competitive Solutions, Inc.

TransAccess Communications software. Middleware. Integrates applications running on MVS mainframes with Windows NT client/server systems. Vendor: Proginet Corp.

Transact Application software for ecommerce. Includes tools for taking and tracking orders, handling shipping and processing payments and taxes. Supports both business-to-business and business-to-consumer sales. Supports multiple languages, currencies, payment methods and business models simultaneously. Runs on Solaris systems. Released: 1998. Vendor: Open Market.

Transact/3000 See Rapid/3000.

transaction In an online system, the cycle of activity that includes sending input from a terminal to an online program, the processing of the input by the program, and sending output to the terminal user after the processing is complete.

Transaction Control Protocol/Internet Protocol See TCP/IP.

Transaction Express Communications software. Monitors transaction processing and ensures that information reaches its destination. Runs with Teknekron Information Bus. Vendor: Teknekron Software Systems, Inc.

transaction processing monitor Communications software that controls online business applications. Directs commands and data requests over a network and provides features such as load balancing, security routines, and error detection. Can handle large volumes of transactions. Also called TP monitor.

transaction queuing Function included with databases that allows users to run multiple queries against a database.

transaction server Technology used in client/server systems to handle data transfer. The transaction server protects the data, maintains performance, and provides load balancing functions. Used in high speed, high volume applications needing security features. Typical applications include e-commerce, stock market trading, airline reservations and credit card verification. Transaction server software can be included in either the operating system or in middleware.

Transaction Server for Windows NT Communications software. Included in IBM's Software Server suite. Used to create mainframe type middleware from Windows NT systems. Vendor: IBM Corp.

TransactSQL 4GL. Extension to standard SQL. Used with SQL Server. Vendor: Sybase, Inc.

Transcend Communications software. Network management system. Provides Web support. Vendor: 3Com Corp.

TranscendWare Communications software. Manages policy based networking. Vendor: 3Com Corp.

TransCentury Data Logic Generator, TransCentury Calendar Routines Application development tool. Used to convert data processing for year 2000. Vendor: Vendor: Platinum Technology, Inc.

TransferPro Operating system add-on. Allows Unix systems to read and write to MS-DOS and Macintosh systems. Vendor: Digital Equipment Corp.

Transform See SHL Transform.

Transform(/DB2) CASE product. Automates all phases of the development cycle including project management functions. Works with IMS and DB2 databases. Vendor: Transform Logic Corp.

Transformation Server Data Management software. Provides data replication and conversion functions. Used to distribute enterprise data over mixed databases. Interfaces with IBM AS/400 databases, SQL Server, Oracle and Sybase. Runs on DEC, Unix and Windows systems. Released: 1998. Vendor: DataMirror Corp.

Transit Operating system for MediaCUBE server systems. Unix based. Vendor: nCUBE.

TranSlate Communications software. EDI package. Runs onHP, IBM systems. Vendor: TranSettlements, Inc.

Translator*MVS Communications software. EDI package. Runs on IBM systems. Vendor: Sterling Software (Ordernet Services Division).

translator program A compiler, interpreter or assembler program that takes a source program and converts it to an object program.

Transparent Data Migration Facility See TDMF.

transport layer In data communications, the OSI layer that governs multiple network connections. The transport layer is responsible for insuring reliable data transfer even though several different networks may be involved.

TransPort Trek Portable computer. Vendor: Micron Electronics, Inc.

TransPort xxx Notebook computer. Pentium CPU. Vendor: Micron Electronics, Inc.

Transrelate Workbench Operating system enhancement used by DBAs. Monitors and controls DB2 environments in IBM systems. Vendor: Compuware Corp.

TransRING Operating system enhancement used by systems programmers. Controls communications networks in IBM systems. Connects multiple Token Ring networks. Vendor: Vitalink Communications Corp.

Transtar Repository Application development tool. Creates open, repository-based environment that interfaces with Emeraude PCTE. Runs on Unix systems. Vendor: Transtar Software, Inc.

TransTracker Internet management software. Monitors both Intranets and Internet applications. Vendor: PLATINUM Technology, Inc.

Trapeze Desktop system software. Spreadsheet. Runs on Macintosh systems. Vendor: Access Technology, Inc. (CA)

TRAPS Operating system add-on. Debugging/testing software that runs on IBM systems. Full Name: TRAPS Testing System. Vendor: Travtech, Inc.

TravelMate Notebook computer. Pentium CPU. Vendor: Acer America Corp. (formerly Texas Instruments).

TravelPro See AMS TravelPro.

Traverse Business accounting software for Windows. Interfaces with Office. Vendor: Microsoft Corp.

Tree4C, Tree4Fortran, Tree4Pascal Application development tool. Reverse-engineering tools for the appropriate language. Run on Sun workstations. Vendor: 1 Software Engineering.

Trend System management software. Web based network monitor. Web-based software that analyzes network performance by condensing thousands of network statistics into trends and problem situations. Collects data from SNMP and RMON to help analyze network performance. Can access reports from any Web browser. Runs on Solaris, Unix, Windows systems. Reports can be accessed from any Web browser. Includes ReportPacks (report generator). Vendor: DeskTalk Systems.

Trend Micro Desktop software. Virus protection. Vendor: Trend Micro Inc.

Tri-Dimensional 92x0 Midrange computer. Operating systems: REAL/IX, MAX 32, MAX IV. Vendor: Modular Computer Systems, Inc. (MODCOMP).

Tri-Star Computer vendor. Manufactures desktop computers.

Tri-Star DesignBook Notebook computer. Vendor: Tri-Star Computer Corp.

TriBase Database product used with Baan's application software. Vendor: Baan Co.

triggers A program that initiates an action when an event such as updating a record occurs. The trigger usually causes a program (often a stored procedure) to be executed. Used for replication. Common with some databases and data warehouses.

TriniCom Video conferencing software. Runs on Windows NT Workstation. Released: 1998. Vendor: Sony Electronics, Inc.

Trinzic Software vendor. Formed by a merger of AICorp, Inc. and Aion Corporation. Produces development tools for artificial intelligence and client/server computing. Full name: Trinzic Corporation.

Trinzic RuleServer See RuleServer.

Triple DES Security technique. Encrypting a DES message three times to make it that much harder to break.

Tritus SPF Application development tool. Desktop version of IBM's mainframe ISPF. Runs with AIX systems. Vendor: Tritus, Inc.

Trivial File Transfer Protocol See TFTP.

TRMS Application development tool. End-user report manager. Part of Savrs/TRMS. Runs on IBM MVS systems. Stands for: Total Report Management Solution. Vendor: Software Engineering of America.

Trojan Horse A destructive program that is disguised as something benign, such as a directory lister, archiver, or game. Also a program that contains a virus. Other destructive programs are called viruses, worms, backdoors and logic bombs.

TRON The Real-time Operation system Nucleus. A design strategy from Japan that intends to provide standard operating systems for all computers and all applications.

troubleshoot To detect and eliminate errors in programs and/or hardware.

Trove for Manuals Document retrieval system. Works with Web servers. Runs on OS/2, Unix, VMS, Windows systems. Vendor: Ringwood Software Inc.

TRPS Operating system enhancement used by operations staff and systems programmers. Disaster control package. Stands for: Total Recovery Planning System. Vendor: CHI/COR Information Management, Inc.

Tru64 Unix Operating system for 64-bit systems. Formerly DEC Unix. Runs on DEC Alpha systems. Vendor: Compaq Computer Corp.

TrueCoverage Application development tool used for testing. Provides code coverage analysis for Visual C++, Visual Basic and Java programs. Part of DevCenter. See DevCenter.

TrueSecure Service provided to protect networks from attacks over the Internet. Defends against bad passwords, undocumented dial-up lines, and insecure Web servers. Available: 1998. Vendor: International Computer Security Association, Inc.

TrueSync PDA software. Downloads appointments or telephone numbers from PIM software to PDAs. Vendor: Starfish Software, Inc.

TrueTime Application development tool used for testing. Analyzes and optimizes programs written in Visual C++, Visual Basic and Java. Part of DevCenter. See DevCenter.

TrustBroker See CyberSafe TrustBroker.

TrustBroker Security Suite Operating system software. Provides security for global networked organizations. Supports both public key and Kerberos authentication. Consists of: TrustBroker Security Server, TrustBroker Client for MS Windows, TrustBroker Client for Unix, TrustBroker for MVS, and TrustBroker Developer's Pack, all of which may be used independently. Runs on many platforms. Vendor: CyberSafe Corp.

TrustBroker Web Agent Operating system software. Internet software. Provides a single password for users to have access to all authorized Web pages. Runs on Sun systems. Vendor: CyberSafe Corp.

Trusted Oracle7 Version of Oracle database that provides security functions allowing users to be able to store and access data according to individual security clearances. Intended for use by federal government users. Vendor: Oracle Corp.

TrustedLink Communications software. EDI. Versions available for mainframe, midrange and desktop systems. Vendor: Harbinger Corp.

TRW MTP Operating system add-on. Debugging/testing software that runs on DEC systems. Vendor: TRW Technical Training Center.

TRW SDP Operating system enhancement used by systems programmers. Monitors and controls system performance in DEC VAX systems. Stands for: TRW Series Diagnostic Package. Vendor: TRW Technical Training Center.

TS-DOS Operating system for Tandy and NEC systems. Vendor: Traveling Software, Inc.

TS Network DataServer Communications. Interface to networked data servers. Maintains data integrity for all data files in the network. Runs on Unix, Windows NT systems. Released: 1997. Vendor: Thoroughbred Software International, Inc.

TS-PRINT Operating system enhancement used by systems programmers. Increases system efficiency in IBM systems. Handles print distribution. Vendor: Tone Software Corp.

TSANet Network application for technical support. Members are computer system vendors including 3Com, Apple, IBM, Hewlett-Packard, Microsoft, Novell, etc. Stands for: Technical Support Alliance Network. Part of WorldCom.

TSAPI Communications software. Interface between telephone and data communications. Stands for: Telephony Services Application Programming Interface. Vendor: AT&T & Novell, Inc.

TSEE See ISEE, TSEE, RTEE.

TSO Communications software. Provides environment for application development. Includes programming and testing functions. Vendor: IBM Corp.

TSO/SPF, TSO/ISPF See ISPF.

TSreorg Operating system software. Allows data managers to restructure tables and indexes without having to rebuild the database structure. Used with Informix, Oracle, sybase databases. Vendor: PLATINUM Technology, Inc.

Tss-Archive Operating system add-on. Data management software that runs on Bull HN systems. Vendor: Scientific and Business Systems, Inc.

TSS Operating system add-on. Security/auditing system that runs on Unisys systems. Stands for: Terminal Security System. Vendor: Unisys Corp.

Tsunami Development name for rewrite of TME to be released in 1998. See Tivoli Management Environment.

TSX(-32) Real-time operating system for midrange computer systems. Vendor: S & H Computer Systems, Inc.

TTE Operating system add-on. Debugging/testing software that runs on Unix systems. Vendor: Lore, Inc.

TUBES Communications software. Network connecting IBM systems. Vendor: Macro 4 Systems Software.

Tun NET Communications. Provides network and resource sharing applications for TCP/IP networks. Including e-mail and fax capabilities. Runs on Windows systems. Vendor: Esker, Inc.

Tun Plus Communications. Used to access UNIX hosts, run multi-user applications, share peripherals and exchange messages from Windows desktops. Includes Tun TCP, Tun EMUL, Tun SQL, Tun MAIL and Tun NET. Runs on Unix systems. Released: 1997. Vendor: Esker, Inc.

TUXEDO Communications. Transaction processing monitor. Used to build three tier client/server applications in heterogeneous environments. Applications can be developed independently of the networks, hardware and database environments. Includes Jolt (Java interface to Tuxedo applications), Manager (manages the applications) and Builder (development tool) and Internet interface. Runs on DEC, OS/400, MVS, Unix, Windows NT. Released: 1997. Vendor: BEA Systems, Inc.

Tun SQL Communications software. Middleware. Connects ODBC applications to databases without proprietary middleware. Runs on Unix systems. Vendor: Esker, Inc.

Turbo BASIC Compiler language used in desktop computer environments. Version of BASIC. Turbo BASIC utilities also available. Vendor: Inprise Corp.

Turbo C, C++ Compiler language used in desktop computer environments. Vendor: Inprise Corp.

Turbo Debugger See C++Builder.

Turbo Image Database for large computer environments. Runs on HP systems. Vendor: Hewlett-Packard Co.

Turbo Pascal Compiler language used in desktop computer environments. Version of Pascal. Turbo Pascal utilities also available. Vendor: Inprise Corp.

Turbo Pascal 5.5 Object-oriented version of Pascal.

Turbo PROLOG Compiler language used in desktop computer environments. Version of PROLOG. Used to build artificial intelligence applications. Vendor: Inprise Corp.

Turbo/RX Operating system for Ridge systems. Vendor: Ridge Computers.

Turbo Vision Application development tool. Combination of C++ compiler and development aids such as mouse support and debugging tools. Vendor: Inprise Corp.

TurboBPR Application software. Set of tools for strategic planning, operational cost and performance tracking and investment analysis for businesses. Developed by the Pentagon to decide which project to fund and used throughout the government.

TurboLaser Desktop computer. Alpha CPU. 64-bit server. Vendor: Digital Equipment Corp.

turnaround time The amount of time it takes between submitting a job for execution and receiving the output from that job.

TurnKey Data Mart Application development tool used in data warehousing. Includes a starter kit of data mining rules to be able to predict customer buying habits. Vendor: Broadbase Information Systems, Inc.

turnkey system A complete system, hardware, software, and operating system, that provides the computing system for a specific application.

TurnOver Application development tool. Controls change management for the entire development life cycle. Runs on AS/400 systems. Vendor: SoftLanding Systems, Inc.

TurnStyle Firewall System See TFS.

TUXEDO Communications software. Transaction processing monitor for Unix systems. Unifies multiple databases. Interfaces with Internet, Java. Full name: Tuxedo Enterprise Transaction Processing System. Originally from AT&T. Vendor: BEA Systems, Inc.

Twaice Artificial intelligence system. Runs on CDC, DEC, HP. IBM, Unix systems. Provides expert system shell. Vendor: Logicware, Inc.

TWEED See VEST.

Twin Application development tool. Creates Unix version of programs written for Windows. Runs on Unix systems. Vendor: Willows Software.

Twin Peaks Midrange computer. Alpha CPU. Vendor: Aspen Systems, Inc.

Twin/UX Desktop system software. Spreadsheet. Runs on Unix systems. Vendor: Mossaic Software, Inc.

TwinAccess Communications software. Network connecting Macintosh desktop computers to IBM midrange systems. Vendor: KMW Systems Corp.

Twinhead Computer vendor. Manufactures desktop computers.

twisted pair cable Communications lines commonly used in data transmission because of its low cost.

TWister Desktop computer. PowerPC Pentium processor. Vendor: Mactell Corp.

Two-tiered client based processing Part of client/server computing. One type of application partitioning. Puts all logic on the client systems with server providing just database access.

Tx00 Handheld computer. Pen-based. Vendor: Toshiba America Information Systems, Inc.

Tymnet Communications network set up in the early 1980s. Public network that uses X.75 protocols.

Typhoon Application development tool. Based on WebDNA and used to build dynamic Web pages. Allows developers to, i.e., automatically insert current date or time, and hide or show portions of text based depending on who the user is. Allows the developer to include banner ads, e-mail forms, mass e-mailing, page counters, password protection in web sites. Vendor: Pacific Coast Software.

TYphoon Desktop computer. PowerPC Pentium processor. Vendor: Mactell Corp.

TyView/SNMP Communications software. Manages large WANS. Vendor: TyLink Corp.

U/ACR Mathematical/statistical software that runs on IBM systems. Balances report figures from application software. Vendor: Unitech Systems, Inc.

U10-300 Midrange computer. SPARC CPU. Operating systems: Solaris. Released: 1998. Vendor: Tatung Science & Technology, Inc.

U6000/x00 Desktop computer. Pentium CPU. Operating system: Windows 95/98. Vendor: Unisys Corp.

UBackup Operating system add-on. Library management system that runs on Unix systems. Vendor: Unitech Software, Inc.

UBASIC/BBASIC Compiler language used in mainframe environments. BASIC compiler for Unisys systems. Vendor: Unisys Corp.

UCA Communications framework for integrating diverse environments in client\server computing. Stands for: Universal Communications Architecture. Includes QuickApp. Vendor: Digital Communications Associates, Inc.

UCANDU See CA-UCANDU.

UCF Communications software. Transaction processing monitor. Runs on IBM systems. Stands for: Universal Communications Facility. Interface to IDMS/R. Vendor: Computer Associates International, Inc.

UControl Operating system enhancement used by systems programmers. Monitors and controls system performance in Unix systems. Performs backups, security administration, and batch job scheduling. Vendor: Unitech Software, Inc.

UCOS III Operating system for Bull HN systems. Vendor: Bull HN Information Systems.

UCSD Operating system for Stride desktop computer systems. Vendor: Stride Micro.

UDB See Universal Server.

UDMS Relational database for large computer environments. Runs on DEC systems. Full name: The User's Data Management System. Design concentrates on end-users. Integrates with Ingres. Vendor: Interactive Software Systems, Inc.

UDP Communications protocol. Transport protocol used with TCP/IP. Stands for: User Datagram Protocol.

UDS Communications software. Links applications built with AppWare to network resources. Part of AppWare. Stands for: Universal Directory Services. Vendor: Novell, Inc.

UDS RDMS 1100 Relational database for large computer environments. Runs on Unisys systems. Vendor: Unisys Corp.

UETP Operating system add-on. Debugging/testing software that runs on DEC systems. Used to verify the integrity of the operating system and used during installation. Stands for: User Environment Test Package. Vendor: Digital Equipment Corp.

UFAS Operating system add-on. Data management software that runs on Bull HN systems. Stands for: Unified File Access System. Vendor: Bull HN Information Systems.

UFO Productivity System Application development tool. Program that develops models of online applications. Runs on IBM systems. Vendor: On-Line Software International.

UIB Application development tool. Creates graphic user interfaces that are dynamically switched between Motif and Open Look. Stands for: User Interface Builder. Works with C++, X Window applications. Vendor: Solbourne Computer, Inc.

UIM/X Application development tool used in object-oriented development. GUI builder, generates C++ code. Used to develop custom icons. Stands for: User Interface Management System. Vendor: Bluestone Software.

UIM/X Builder Application development tool. Creates customized GUI screens. Interfaces with UIM/X. Vendor: Bluestone Software.

UIM/Xmove Application development tool. Enables developers to work in a single integrated environment and define the layout and behavior of user interfaces. Can generate code that is portable across numerous platforms. Runs under Unix systems. Released: 1997. Vendor: Black & White Software, Inc.

UIMS Application development tools. Family of programs that provide a multi-language solution for developing GUI applications. Supports C, C++, Ada. Stands for: User Interface Management System. Vendor: The Alsys CASE Division.

UIS-Manager Operating system enhancement used by systems programmers and Operations. Monitors and controls Data Center operations. Runs on DEC VMS systems. Vendor: UIS, Inc.

ULTIM Operating system add-on. Data management software that runs on IBM systems. Vendor: XL Software, Inc.

Ultimail Communications software. Client/server messaging system. Vendor: IBM Corp.

ULTIMAX Handheld computer. Vendor: S.T. Research, Corp.

Ultimedia Multimedia software. Includes Ultimedia Builder (development tool), Ultimedia Perfect Image (image-editing package), Ultimedia WorkPlace (edits images to allow sorting and searching of images). Vendor: IBM Corp.

Ultinet Communications software. Network connecting PICK systems. Vendor: The Ultimate Corp.

Ultra Relational database for large computer environments. Runs on DEC systems. Full name: Ultra Interactive Data Base System. Vendor: Cincom Systems, Inc.

Ultra 30 Midrange computer. UltraSPARC RISC CPU. Operating system: Solaris. Vendor: Sun Microsystems, Inc.

Ultra-Base Relational database for midrange and desktop computer environments. Runs on Datapoint systems. Vendor: Downey Data.

Ultra Enterprise x000 Midrange computer. UltraSPARC RISC CPU. Operating system: Solaris. Vendor: Sun Microsystems, Inc.

Ultra HPC Midrange computer. UltraSPARC RISC machine. 64-bit processor. Operating systems: Solaris. Vendor: Sun Microsystems, Inc.

Ultra/wide NetPro, Omega Midrange computer. Pentium CPU. Operating system: Windows 95/98. Vendor: Royal Electronics, Inc.

UltraBook Desktop computer. Notebook. Pentium processor. Vendor: RDI Computer Corp.

UltraCalc Desktop system software. Spreadsheet. Runs on DEC, IBM, Unix systems. Vendor: Olympus Software, Inc.

Ultraopt/IMS Operating system enhancement used by systems programmers. Improves response time in IMS systems. Vendor: BMC Software, Inc.

UltraQuest Application development tools. Provides access to mainframe legacy data from the Web. Allows developers to produce ad hoc and catalogued reports. Includes UltraQuest Reporter, UltraQuest Menus, UltraQuest Applications and RP/web. Vendor: Aonix.

UltraSPARC Computer chip. 64-bit chip. Vendor: Sun Microsystems, Inc.

Ultrix/SQL Relational database for Ultrix environments. Runs on DEC systems. Vendor: Digital Equipment Corp.

ULTRIX(-11,32,32M) Operating system for DEC systems. Unix-type system. Conforms to POSIX, OSF and X/Open specifications. Derived from BSD. Vendor: Digital Equipment Corp.

Ultrix Worksystem Software See UWS.

UM See unified messaging.

UMAX Operating system for Encore systems. Unix-type system. Vendor: Encore Computer Corp.

UmaxPC Desktop computer. Pentium processor. Vendor: UMAX Technologies, Inc.

UML Unified Modeling Language. Is both a methodology and a notation format used in object analysis and design. Works with all object-oriented methodologies. Using UML, developers define a three-tiered model of the application: user interface, business logic and database.

UN/System V Operating system enhancement used by systems programmers. Monitors and controls system performance in Unix systems. Vendor: Charles River Data Systems, Inc.

UN6000/xx Desktop computer. Pentium CPU. Operating systems: Windows 95/98, Unix. Vendor: Unisys Corp.

UNAS Communications. Middleware, object request broker (ORB). CORBA compliant. Includes both development tools and runtime environment. Stands for: Universal Network Architecture Services. Vendor: TRW.

UNI-DOS Operating system for IBM and compatible desktop computer systems. Vendor: Link Data, Inc.

Uni-File Relational database for desktop computer environments. Runs on IBM desktop computer systems. Vendor: Univair, Inc.

Uni-Power Relational database for midrange and desktop computer environments. Runs on Unix systems. Vendor: Uni-Concepts, Inc.

Uni-Rexx Version of Rexx that runs on Unix systems. Vendor: The Workstation Group.

Uni-Xedit Version of Xedit that runs on Unix systems. Vendor: The Workstation Group.

Unicenter/ICE Systems management software. Manages Windows NT and Unix servers on the Internet and other TCP/IP networks. Stands for: Internet Commerce Enabled. Vendor: Computer Associates International, Inc.

UNICENTER/II See CA-UNICENTER II.

Unicenter TND Communications. Latest release of Unicenter. Stands for: The Next Dimension. Allows users to view historical events and predict future ones (including potential problems) from the central console. Release date: 1999. Vendor: Computer Associates International, Inc.

Unicenter TNG System management software. Includes both operating system and network management functions. Provides security, storage management, accounting, report management, and scheduling functions. Supports all major platforms from mainframe to desktop. Internet interface available. Versions available for client/server environments managing OS/2, Windows NT, NetWare, Windows for Workgroups, LANtastic, and Windows CE clients. Stands for: The Next Generation. TNG version released: 1997. Vendor: Computer Associates International, Inc.

Unicode A consortium working on establishing standards for multiple national languages. Members of the consortium include IBM, DEC, Apple, Zerox, Microsoft.

UniCon See ADL.

Unicos Operating system for Cray supercomputer systems. Unix-type system. Vendor: Cray Research, Inc.

Unidas Operating system add-on. Data management software that runs on Unisys systems. Vendor: Unisys Corp.

UniData Relational database. Includes client/server development environment. Can be used with data warehousing. Interfaces with Informix, Ingres, Oracle. Vendor: Ardent Software, Inc.

UNIFACE Application development environment. Includes 4GL. Used to develop departmental client/server applications with a GUI based front-end. Used in midrange and desktop computer environments. Accesses Sybase, Oracle, Ingres, Informix, Rdb. Uniface for OS/2 and Uniface for WorkPlace Shell access DB/2, DB2/2, DB2/6000. Runs on MS-DOS, OS/2, Unix, VMX, VOS systems. Vendor: Compuware Corp.

Uniface Personal Series Application development tools. Products that allow end-user to access data. Access data in 60 different databases, and from 35 platforms. Includes: Uniface Personal Query, Uniface Personal Access, Uniface Business Graphics. Vendor: Compuware Corp.

Unified Messager Communications, Internet software. Provides unified messaging services. Vendor: Octel Communications Corp.

unified messaging Technology that delivers e-mail, voice mail and faxes to one mailbox. Messages can be retrieved through phone or computer.

Unified Modeling Language See UML.

UniFlex/DA,MP,MS,RT General-purpose operating system for Unix-compatible systems. Vendor: Technical Systems Consultants, Inc.

UNIFY Relational database for large computer environments. Runs on Concurrent, DG, IBM, Unix systems. Utilizes SQL. Can be used in client/server computing. Vendor: Concurrent Computer Corp.

Unify Vision Application development environment. Object-oriented graphical tool for creating departmental client/server applications. Based on Galaxy tool set. Runs on Unix systems. Accesses most relational databases. Vendor: Unify Corp.

UniKix 1. Integration software. Implements IBM CICS mainframe applications in a Unix environment. Integrates with Internet, Java. Vendor: UniKix technologies. 2. Communications software. Transaction processing monitor. Runs on desktop, RISC systems. Vendor: Groupe Bull.

Unipac Computer Computer vendor. Manufactures desktop computers.

Uniplex II Plus Database for Unix environments. Utilizes SQL. Vendor: Uniplex, Inc.

UniPlus+ General-purpose operating system for Unix-type systems. Vendor: UniSoft Corp.

UniPrise Access/DAL Communications software. Middleware. Allows queries to multiple databases on different platforms. Runs on AS/400, MVS, Unix systems. Vendor: UniPrise Systems Inc.

UniQBatch Operating system enhancement. Sets up a mainframe batch environment for Unix systems. Includes automated job scheduling and operator control of the job stream. Runs on DEC, Unix systems. Released: 1999. Vendor: Macro 4, Inc.

Uniquest Application software. Financial, payroll systems for midsize environments. Vendor: DataStar.

UniSQL/X Object-relational database with object-oriented extensions. Runs on Sun, IBM systems. Vendor: UniSQL, Inc.

Unisys Computer vendor. Manufactures large and desktop computers.

Unite Application development tool. Allows Unix applications to run under Windows NT. Can run Unix commands under NT, and NT commands under Unix. Vendor: Consensys Corp.

Unitree Operating system enhancement used by systems programmers. Increases system efficiency in Unix systems. Manages disk storage space. Vendor: Discos Division of General Atomics.

UniTree Operating system enhancement. File and storage manager for large-scale networked client/server systems. Vendor: Amdahl Corp.

Unitronix Software vendor. Products: Praxa family of financial, manufacturing, sales systems for DEC environments.

UniV/SQL Name of SQL used with Uni-Vision database.

Universal Communications Architecture See UCA.

Universal Directory Application development tool. Provides an information directory that provides an index to a data warehouse. Speeds warehouse development and management. Centrally stores data definitions including origin and physical location. Software that inventories and searches metadata. Provides users with metadata descriptions. Includes ActiveX interface. Runs on Windows NT systems. Released: 1997. Vendor: Logic Works, Inc.

Universal Directory Services See UDS.

Universal Financials Applications software. Financial systems including purchasing, decision support, contract processing. Integrates with Microsoft's Office and SmartSuite. Vendor: Global Software, Inc.

Universal-Link Communications software. Network connecting IBM systems. Micro-to-mainframe link. Vendor: Universal Software, Inc.

Universal-Link/EDI Communications software. EDI package. Runs on IBM mainframe systems. Vendor: Universal Software, Inc.

Universal Management Agent Communications. System management software. Combination of software and firmware used to manage heterogeneous clients in a client/server environment. Includes components from LANDesk and Tivoli. Vendor: IBM Corp.

Universal Server Relational object-oriented DBMS. Different versions: 1. Informix Universal Server. Combines Informix's scalable relational database with Illustra's object management handling of video, audio, images, and two- and three-dimensional modeling. Includes DataBade technology to access non-relational data. Vendor: Informix Software, Inc. 2. Oracle Universal Server. Relational database that handles complex data types including multimedia text and messaging. Family of servers and tools that manage multiple data types including video, voice, character, and numeric data. Uses Cartridge technology to access non-relational data. Part of Oracle 7, Oracle 8. Vendor: Oracle Corp. 3. DB2 Universal Database. Multimedia, parallel database. Allows user to mix images, sound objects with traditional data. Includes DB2 Parallel Edition. Provides access to non-relational data through Relational Extenders. Used for enterprise wide processing and data warehousing. Runs on OS/2, OS/390, Unix, VSE, VM, Windows systems. Vendor: IBM Corp.

Universal StreetTalk See StreetTalk.

Universal Virtual Machine See UVM.

Universal Warehouse Application development tool. Combines Universal Server, MetaCube and a parallel database. Used to create data warehouses. Vendor: Informix Software, Inc.

UniVerse Application development system. Can be used in data warehousing. Includes 4GL, relational DBMS, file management and backup functions. Runs on Unix systems. Vendor: Ardent Software, Inc.

Universe See CA-Universe.

UniVision 1. Operating system enhancement. Provides systems management functions for enterprise-wide systems. Provides performance monitoring and resource utilization for Unix systems. Part of POEMS. Vendor: Platinum Technology. 2. Database management system. SQL database. Runs an extensive library of applications available for the Pick data model. Allows user to move legacy applications to Unix systems. Includes data dictionary functions. Vendor: Objectware, Inc.

UniWorks Operating system add-on. Debugging/testing software that runs on Unix systems. Vendor: Integrated Solutions, Inc.

Unix General-purpose operating system. Versions of Unix, such as AIX, Ultrix, HP/UX, Xenix have been written for specific computer systems. Most common operating system used with RISC machines. Many vendors provide Unix, but it was originally developed by AT&T. Unix is written in C, so UNIX/C is commonly used to represent both skills.

Unix 1100 MRP, or CIM, software. Runs on Unisys systems. Vendor: Cimcase International, Inc.

Unix/DSM Systems management software. Includes configuration and administration functions. GUI interface. Manages multiple versions of Unix. Vendor: Enlighten Software Solutions, Inc.

Unix/FT Unix shell for networks running SCO Unix. Vendor: Tricord Systems, Inc.

Unix International A consortium of over 250 companies working on standards for Unix.

Unix Productivity Pack Application development tools. Includes Adabas, Natural, Entire Net-Work, Natural Construct, Predict. Used to develop Unix applications. Vendor: Software AG Americas Inc.

Unix-Safeword Operating system add-on. Security/auditing system that runs on Unix systems. Vendor: Enigma Logic, Inc.

Unix to Unix Copy Programs See UUCP.

UnixWare See SCO UnixWare.

UNLOAD PLUS See DB2 UNLOAD PLUS.

UNMA Communications software. Network manager linking multiple networks. Vendor: AT&T Information Systems.

UNOS Real-time Unix operating system. Vendor: Charles River Data Systems, Inc.

UPC*Premier Communications software. EDI package. Runs on AS/400, Windows systems. Vendor: EDI Support, Inc.

UPSU Operating system add-on. Security/auditing system that runs on IBM systems. Stands for: User Profile Security Utility. Vendor: Scott-Kennard & Associates, Inc.

UpToDate See CA-UpToDate.

UPX1000 Midrange computer. Ultra-SPARC CPU. Operating systems: Solaris. Vendor: Axil Computer, Inc.

UQueue Operating system enhancement used by systems programmers. Job scheduler for Unix systems. Vendor: Unitech Software, Inc.

URL Communications. Standard addressing system for Internet files and functions, especially on the World Wide Web. URLs contain the address of the server, location of the file in the server and what protocol must be used to access the file. Stands for: Universal Resource Locator.

US-DOS I,II Operating system for IBM desktop computer systems. Vendor: Ultrasoft, Inc.

USA.Net Free e-mail service. Accessible through wireless modems.

usability tests Term used for a testing scenario in which users are observed performing typical tasks with new hardware or software. Differs from system or beta testing in that users are not trying to crash the system, but work assuming that the system is bug-free. Usability tests are conducted to see whether the system makes sense to the user.

Usage/Reporter Application development tool. Works with Lotus Notes. Produces reports of Notes database activity. Vendor: DSSI.

USAM Operating system add-on. Data management software that runs on IBM systems. Vendor: Universal Software, Inc.

USecure Operating system add-on. Security/auditing system that runs on Unix systems. Vendor: Unitech Software, Inc.

Usenet Communications. A networking system on the Internet that holds most of the popular newsgroups.

user The person who provides the data for a computer system, updates that data, and uses reports from the system in his or her daily work. Reports are both regular, scheduled reports and ad hoc queries. The users of application systems are business people and are also called end-users, clients, and customers. The users of system software such as operating systems, data management systems and communications systems are computer professionals and have a technical background.

User-11 Database for large computer environments. Runs on DEC PDP systems. Is a DBMS for program development and includes program generator, high-level language, and run-time processor. Vendor: UserWare International, Inc.

User Datagram Protocol See UDP.

User Language 4GL used to build large-scale applications. Component of Model 204. Runs on IBM mainframe systems. Vendor: Computer Corp of America.

UserBase Application development system. Runs on DEC systems. Includes screen generator and data manipulation language. Vendor: Ross Systems, Inc.

Usernet Communications software. Network connecting Unisys, IBM desktop computers. Includes: Usernet Software/86, Usernet 3270 SNA. Vendor: Unisys Corp.

USF Operating system for Stratus computer systems. Unix-type system. Vendor: Stratus Computer, Inc.

USI/200E Midrange computer. Server. Pentium Pro CPU. Vendor: Integrix, Inc.

USoft Developer Application development tool. GUI builder used in client/server development. Generates applications and database structures from business logic contained in a repository. Formerly known as TopWindows. Runs on Unix systems. Vendor: Usoft.

Utah Application development tool. Object-oriented GUI builder that allows users to create GUIs with color, etching, and three-dimensional appearances. Runs on Windows systems. Vendor: ViewSoft, Inc.

UTIL Operating system add-on. Data management software that runs on IBM systems. Vendor: A.K., Inc.

utility A computer program that supports functions of the operating system. For example, sort programs are utility programs as they support the data management function.

UTLD1ENU Operating System enhancement used by systems programmers. Increases system efficiency in IBM DB2 systems by increasing space savings during compression. Vendor: PLATINUM Technology, Inc.

UTLX Operating system add-on. Debugging/testing software that runs on IBM CICS systems. Vendor: Software Technology, Inc.

Utopia Desktop software. Help desk package. Tracks problems and training classes and keeps inventory of hardware and software. Vendor: Hammersly Technology Partners, Inc.

Utopia (LSF) Communications software. Network operating system. Stands for: Utopia Load-Sharing Facility. Vendor: Platform Computer Corp.

UTS/580 Operating system for mainframe Amdahl systems. Unix-type system. Vendor: Amdahl Corp.

UTS Gateway Communications software. Network connecting Novell Lan to Unisys mainframe. Vendor: Computer Logics, Ltd.

UTX/32 Operating system for PowerNode systems. Stands for: Universal Time-sharing Executive. Versions include: 1467 UTX/32, 1470 UTX/32, 1490 UTX/32S Secure Unix. Vendor: Gould, Inc.

UUCP Unix-to-Unix Copy Programs. Permits communications between Ultrix or any other UUCP systems. Used in CompuServe.

UTS(/V) Operating system for mainframe Amdahl, IBM systems. Versions include: UTS, UTS/V, UTS/580. Vendor: Amdahl Corp.

Uuencode, uudecode Code system used on the Internet. Information can be transmitted to unlike systems by this code system.

UVM Application development tool. Enables developers to build platform-independent applications. Used to create programs in the appropriate language that can then run under any operating system. Works with VisualAge for Java, Basic, and Smalltalk. Supports MacOS, OS/2, Windows, Unix. Stands for: Universal Virtual Machine. Vendor: IBM Corp.

UWS Operating system for DEC workstations. Unix type system. Stands for: Ultrix Worksystem Software. Implements DECwindows. Vendor: Digital Equipment Corp.

UWS Desktop Midrange computer. UltraSPARC CPU. Operating system: Solaris. Vendor: Integrix, Inc.

UX See DG/UX.

UX-Metric See PC-Metric.

UXP/M Operating System for Fujitsu supercomputers. Unix type system. Vendor: Fujitsu Ltd.

V.42(bis) Communications protocol. Data Link protocol for the transfer of data between microcomputers and mainframes. CCITT protocol, used with X.25.

V-Bridge(/MHS) Communications software. Gateway for e-mail systems. Runs with Vines networks. Vendor: Computer Mail Services, Inc.

V/Copy II Operating system add-on. Data management software that runs on IBM systems. Vendor: VM Systems Group, Inc.

V-Drive Communications software. Network connecting DEC VAX and IBM desktop computers. Vendor: Virtual Microsystems, Inc.

V/Forms Application development tool. Form and screen painter. Interfaces with standard languages so developers do not have to learn a new language. Vendor: Visigenic Software, Inc.

V/Spool Operating system enhancement used by systems programmers. Increases system efficiency in IBM systems. Enhances spooling facility. Vendor: VM Systems Group, Inc.

V/Temp Operating system add-on. Data management software that runs on IBM systems. Vendor: VM Systems Group, Inc.

V3R1 Shorthand for Version 3 Release 1 of OS/400. Upgrade to OS/400 operating system for the AS/400 machine that includes DB2/400, PC connectivity software, TCP/IP implementation, Visual RPG, development tools, and support for Unix applications. Used to establish AS/400 as server in client/server systems. Vendor: IBM Corp.

V41 MK II, III Notebook computer. Pentium CPU. Vendor: Panasonic Communications & Systems Co.

VADS Application development system. Used to develop Ada systems. Includes debugger, library utilities, programming language. Stands for: Verdix Ada Development System. Vendor: Rational Software Corp.

VAIO 505G Notebook computer. Under three pounds with a 10.4 inch screen. Released: 1998. Vendor: Sony Electronics.

Validator/Req Application development tool. Used to model requirement statements. Generates tests and scripts from the requirements. Vendor: Aonix.

ValueMAX Desktop computer. Pentium processor. Vendor: Cybermax, Inc.

ValuePoint Desktop computer. Pentium CPU. Operating systems: Windows 95/98. Vendor: IBM Corp.

VAM/DS,GEN,VSAM See SAMS:-Allocate.

VanGogh Application development environment. Merger of VisualWorks and Visual Smalltalk Enterprise. Runs on Windows 95/98 systems. Vendor: ParcPlace-Digitalk, Inc.

VanGui for RM/COBOL Application development tool. Allows COBOL developers to create Windows applications using Windows and Visual Basic controls. Vendor: Ryan McFarland Corp.

VantageTeam for (Informix, Ingres) Application development tool. Allows users to build graphical models. Automates project management and documentation. Runs on Unix, Windows systems. Vendor: Cayenne Software, Inc.

Vantive Software vendor. Produces applications software. Front office software suite of sales and marketing functions. Includes: Vantive Sales, Vantive Enterprise, Vantive Support, Vantive FieldService, Vantive Quality, Vantive HelpDesk, VanWeb, VanDesign, Vantive On-The-Go.

Vantive Quality Application development tool. Interfaces with Emeraude PCTE. Provides testing/debugging functions. Manages and documents error handling. Attaches test cases to product change requests. Runs on Unix systems. Vendor: Vantive Corp.

VantiveVista Application software. Includes call-center, sales and marketing automation. Allows users to access data collected from front- and back-office applications, corporate intranets and the Internet. Vendor: Vantive Corp.

Vaps Application development tool. Builds real-time graphical interfaces. Uses object-oriented approach. Generates C code from graphical prototypes. Vendor: Virtual Prototypes, Inc.

VAR Value Added Retailer. A company that is authorized by a vendor to sell its computers and/or peripherals. The VAR typically provides some level of training and/or maintenance. Term is sometimes used to refer to authorized sellers of software products.

VarServer Midsize computer. Server. Pentium Pro processor. Operating system: Linux. Vendor: VA Research, Inc.

VarStation Desktop computer. Pentium processor. Vendor: VA Research Inc.

Vast-2 Application development tool. Program that precompiles Fortran programs. Vendor: National Advanced Systems.

Vault Operating system add-on. Data management software that runs on DEC systems. Vendor: International Structural Engineers, Inc.

VAX 10000-xxx Mainframe computer. Operating systems: OpenVMS. Vendor: Digital Equipment Corp.

VAX 4000,6000,7000 Midrange computer. Operating systems: OpenVMS, VMS, OSF/1, Ultrix. Vendor: Digital Equipment Corp.

VAX ADE Application development environment. User friendly. Runs on DEC systems. Stands for: VAX Application Development Environment. Vendor: Digital Equipment Corp.

VAX DBMS Database. Runs on DEC VAX systems. Vendor: Digital Equipment Corp.

VAX Debug Operating system add-on. Debugging/testing software that runs on DEC VAX systems. Vendor: Digital Equipment Corp.

VAX LISP See LISP.

VAX LSE/SCA Application development tool. Automates programming function. Analyzes source code for such things as adherence to standards and presence of logic flaws (unexecuted code). Languages analyzed: Ada Basic, Bliss, C, COBOL, Fortran, Pascal. Runs on DEC VAX systems. Vendor: Digital Equipment Corp.

VAX Message Router See Message Router.

VAX Performance Advisor Operating system enhancement used by systems programmers. Monitors and controls system performance in DEC VAX systems. Vendor: Digital Equipment Corp.

VAX Rdb/ELN See Rdb/ELN.

VAX/RMS See RMS.

VAX Software Performance Monitor Operating system enhancement used by systems programmers. Monitors 2and controls system performance in DEC VAX systems. Vendor: Digital Equipment Corp.

VAX VMS See VMS.

VAXcluster A connected group of VAX computers and/or VAXstation workstations that share processing, job queues, print queues and disk storage under a single VMS management system. Although all computers in the VAXcluster are managed as one system, each runs VMS independently.

VAXcluster Console System Operating system enhancement used by systems programmers and Operations. Allows all the connected VAX systems in a VAXcluster to be managed from a single console. Vendor: Digital Equipment Corp.

VAXELN Real-time development system for DEC systems. Assists in developing real-time applications. Vendor: Digital Equipment Corp.

VAXft x10 Midrange computer. Operating systems: OpenVMS, VAX/VMS. Vendor: Digital Equipment Corp.

VAXlink Communications software. Network connecting IBM and DEC systems. Vendor: Sterling Software (Answer Division).

VAXnotes Groupware. Vendor: Digital Equipment Corp.

VAXTPU Text-processing utility used in DEC VAX systems Vendor: Digital Equipment Corp.

VB See Visual Basic.

VB/Link, VB/Link for Notes Application development tool. Provides interface from development tools such as Visual Basic to Lotus Notes database so an application can access both Notes and SQL databases. Interfaces with Navigator, Mosaic, Explorer for development of Web applications. Vendor: Brainstorm Technologies, Inc.

VBA Application development environment. Proprietary edition of Visual Basic used in Microsoft's own products. Not licensed to other vendors. Stands for: Visual Basic Applications.

VBAssist Application development tool. Add-on to Visual Basic. Includes forms generator. Vendor: Sheridan Software Systems, Inc.

VBNet Application development tool. Builds a graphical front end for generated Visual Basic applications that run on the Web. Used to convert Visual Basic applications to run on the Internet. Vendor: TVObjects Corp.

VBOMP IBM CICS interface (DBOMP under VSAM).

Vbox See ZipLock.

VBScript Scripting language used to develop interactive applications for the Web. Builds Web pages with components from Java and ActiveX. Vendor: Microsoft Corp.

VBUG Operating system add-on. Debugging/testing software that runs on Harris systems. Vendor: Harris Corp.

VBX Application development tool. Allows users to write standard code for common functions, i.e. inserting a spreadsheet into an application. 16-bit controls that will be replaced by OCX. Stands for: Visual Basic Controls. Vendor: Microsoft Corp.

VC2 Desktop system software. Spreadsheet. Runs on Unix systems. Vendor: Software Innovations, Inc.

VCF/D,L,M Operating system add-on. Data management software that runs on IBM systems. Vendor: Software Engineering of America.

VCON Operating system enhancement used by systems programmers. Increases system efficiency in IBM systems. Tunes CICS VSAM files. Vendor: D & E Software, Inc.

VCQ See Visual CyberQuery.

VCS Application development tool. JCL and/or SYSOUT maintenance program for DEC systems. Vendor: Viking Software Services, Inc.

VdB Application development system. Generates applications that can run on intranets, includes Intranet Tools for Internet access. Full name: Visual dBase. Vendor: KSoft, Inc.

VDM Operating system enhancement used by systems programmers. Increases system efficiency in DEC VAX systems. Manages disk storage space. Vendor: Saiga Systems, Inc.

VDSL See DSL.

Vector:Connexion (for DB2) Communications software. EDI package. Runs on IBM systems. Used principally by banks. Vendor: Sterling Software, Inc. (Banking Systems Division).

Vectra DOS Operating system for Hewlett-Packard desktop computer systems. Vendor: Hewlett-Packard Co.

Velo 1 Handheld computer. Operating system: WinCE. Vendor: Philips Mobile Computing Group.

Velocis Database management software. Used in handheld computers. Runs under Windows CE. Developed by Raima Corp which was purchased by Centura Software in 1999.

Velociti Communications software. Middleware. Users subscribe to receive certain types of data. Runs on Unix, Windows NT systems. Vendor: Vitria Technology, Inc.

velOSity Real-time operating system for desktop computers. Vendor: Green Hills Software, Inc.

VENIX Operating system for IBM and compatible desktop computer systems. Version of Unix. Vendor: Software Kinetics, Ltd.

VENIX System V Operating system for IBM and compatible desktop computer systems, DEC, NCR desktop computer systems. Version of Unix. Vendor: Unisource Software Corp.

Venix Version 3.2 Operating system for IBM and compatible systems. Version of Unix. Vendor: Venturcom, Inc.

Ventana Desktop system software. Spreadsheet. Runs on Pentium type desktop computers. Vendor: Datamax Computer Systems, Inc.

Ventura Publisher Desktop system software. Desktop publisher. Runs on Pentium type systems. Vendor: Xerox Corp.

Venture II, Pro Desktop computer. Pentium CPU. Operating system: Windows 95/98. Vendor: ProGen Technology Inc.

Venturis Desktop computer. Pentium CPU. Operating systems: Windows 95/98. Vendor: Digital Equipment Corp.

Veranda Enterprise Messaging Reporter Desktop system software. Monitors e-mail, network fax and Internet activity. Provides audit of usage and consolidates billing. Vendor: Tally Systems Corp.

Verification Host X.25 Communications software. Security system for public access networks. Vendor: Plantronics Futurecomms, Inc.

VERIFY A utility program that verifies file structure in DEC systems. Also called VFY. Vendor: Digital Equipment Corp.

Verify xxxxDT Desktop computer. Notebook. Pentium processor. Vendor: Mag Portable Technologies, Inc.

VeriServ Operating system software. Measures performance in client/server systems by simulating user activity to measure response time. Vendor: Response Networks, Inc.

VeriSign Certificate authority. (NASDAQ: VRSN)

VERITAS Operating system software. Storage management system that includes integrated back-up functions and enterprise-wide storage management. Vendor: VERITAS Software Corp.

Veritas Desktop Management Suite Operating system software. Provides distribution functions for operating system and application software. Provides Windows NT backup. Released: 1999. Vendor: Veritas Software Corp.

Verity Knowledge management software. Accessible through the Internet. Searches data in documents and converts to HTML. Vendor: Verity, Inc.

Veronica Communications software. Search tool that handles keyword searches. Used with Gopher programs on the Internet.

Versa Notebook computer. Pentium CPU. Operating systems: Windows 95/98. Also called Road Warrior. Vendor: NEC Computer Systems.

VersaForm Relational database for desktop computer environments. Runs on Apple II systems. Full name: VersaForm Business Form Data Base. Vendor: Applied Software Technology.

Versant Object-oriented database. Supports multiprocessing and multithreading. Vendor: Versant Object Technology, Inc.

VersantACE Object-oriented database. Integration of Versant and NetDynamics. Applications built for NetDynamics can access objects in Versant ODBMS as well as data in legacy systems, client/server and/or Net-based applications. Released: 1998. Vendor: Versant and NetDynamics.

VersantWeb Application development tool. Used to build Internet applications. Runs on Unix, Windows systems. Released: 1997. Vendor: Versant Corporation.

Versatile Storage Server See Seascape Storage Enterprise.

VerSecure Hardware based encryption scheme implemented through plug-in boards. Released: 1998. Vendor: Hewlett-Packard Co.

version control Keeping track of revisions and versions of programs. Especially important in client/server systems because there can be several versions (one for each platform) of client programs, all of which must be updated when a change occurs.

Version Merger Application development tool. Upgrades new releases of vendor-supplied software that has been customized. Runs on MVS, VSE systems. Released: 1998. Vendor: Princeton Softech, Inc. (Division of Computer Horizons Corp.).

Versit Communications standards. Set of specifications that standardize the integration of telephone and data communications. Uses TSAPI. Issued by AT&T, IBM, Apple, and Siemens AG.

Vertigo Application development tool. Authoring tool used to produce video and audio applications for the Web. Vendor: Adobe Systems, Inc.

VESA Local Bus Best Buy, Professional, Win-Station, WinXpress Desktop computer. Pentium CPU. Operating systems: Windows 95/98. Vendor: Comtrade Electronics U.S.A., Inc.

VESA Multimedia, Value Line Desktop computer. Pentium CPU. Operating systems: Windows 95/98. Vendor: EPS Technologies, Inc.

VEST Application development tool. Translates VAX machine code into Alpha code. Stands for: VAX Executable Software Translator. Used with TIE (Translated Image Environment) and TWEED (Tool Which Evaluates Executable Dependencies). Vendor: Digital Equipment Corp.

Vetix MXI, LXI Midrange computer. Pentium Pro CPU. Operating systems: Windows NT. Vendor: Micron Electronics Inc.

VFY See VERIFY.

Via Application development tool used in data warehousing. Eliminates redundant directory information. Vendor: Zoomit Corp.

VIA/Alliance Application development tool. Analyzes COBOL code and JCL to give a picture of how applications are put together. Used in re-engineering, converting legacy applications to client/server, etc. Vendor: Viasoft, Inc.

VIA/Center Application development tool. Analyzes existing programs to determine what applications need restructuring to fit into a new system. Vendor: Viasoft, Inc.

VIA/DRE Relational database for a desktop computer environment. Runs on Pentium type desktop computers. Utilizes SQL. Can be used in client/server computing. Vendor: Via Information Systems Corp.

Via II Wearable computer. Similar in functionality to handhelds, but worn on body to free hands. Primarily used in manufacturing for such things as providing access to diagnostic manuals which the user follows to make repairs. Vendor: Via, Inc.

VIA/Insight Application development tool. Analyzes existing programs to determine what applications need restructuring to fit into a new system. Vendor: Viasoft, Inc.

VIA/Renaissance CASE product. Analyzes existing COBOL programs and integrates them into an applications development environment. A reverse-engineering tool. Vendor: Viasoft, Inc.

VIA/Smartdoc CASE product. Automates programming function. Analyzes source code for such things as adherence to standards and presence of logic flaws (unexecuted code). Languages analyzed: COBOL. Runs on IBM mainframes. Vendor: Viasoft, Inc.

VIA/SmartTest See SmartTest.

VIA/Visual Information Assistant Application development tool. Object-oriented GUI that does not require programming. Used to integrate data from diverse sources. Vendor: Applied Logic Programming.

ViaDuct Communications. Provides connectivity between Windows and Pick systems. Vendor: Via Systems, Inc.

ViaVoice Speech recognition package. Recognizes 32,000 words, accepts 140 words per minute. Translates spoken words into Word documents. Vendor: IBM Corp.

Vibe Application development environment. Visual, Java based. Provides extensive class libraries and visual control tools to build business applications in Java. Released: 1997. Vendor: Visix Software, Inc.

Vibrant xxx Notebook computer. Pentium CPU. Vendor: Transmonde Technologies, Inc.

VICE Application development tools. Allows developers to edit, browse, analyze, test and debug C programs. Stands for: Visual C Environment. Runs on Unix systems. Vendor: Lucent Technologies, Inc.

Video Remote Video conferencing software. Vendor: Corel Corp.

videotex Communications. Part of ISDN. Interactive access to a remote database by a person at a terminal. Used for such things as accessing an online telephone directory.

VIEW Database interface. Allows multiple users to access multiple sources of data. Interfaces with VSAM files, IMS, and DB2 databases. Stands for: Virtual Interface Engineering Window. Vendor: Developed by Ford's Car Product Development Division.

ViewCenter for Motif Application development tool. Used to create, test, modify and generate code for the user interface part of programs. Generates C++ code. Vendor: CenterLine Software, Inc. and Visual Edge Software, Ltd.

VIEWCOM Operating system enhancement used by systems programmers. Increases system efficiency in IBM systems. Handles print distribution. Vendor: Startech Software Systems, Inc.

ViewDirect Operating system enhancement. Provides an on-line interface to reports on optical disk and tape. Runs on IBM mainframe, Unix, Windows systems. Released: 1995. Vendor: Mobius Management Systems, Inc.

Viewer See PowerDesigner.

ViewKit ObjectPak Application development tool. Used to develop C++ programs. Includes C++ libraries and classes. Runs on RISC systems. Vendor: Integrated Computer Solutions, Inc.

Viewmax Graphics user interface for DR DOS 5.0 systems. Vendor: Digital Research.

Viewplex Communications. Network manager linking multiple networks. Vendor: Synernetics, Inc.

Viewpoint 1. Operating system enhancement used by systems programmers. Monitors and controls system performance in Unisys systems. Monitors online systems. Vendor: Datametrics Systems Corp.
2. Desktop system software. Project management package. Runs on Pentium type desktop computers. Vendor: Computer Aided Management, Inc.

Viewpoint TC Network computer. Vendor: Boundless Technologies, Inc.

ViewStar Workflow system. Object-oriented product which stores all workflow items in a reusable object library. Vendor: Viewstar Corp.

ViewTeam See ProTeam.

Vigor-100, 120, 133 Notebook computer. Pentium CPU. Vendor: Transmonde Technologies, Inc.

VIM API for messaging in client/server systems. Provides the interface between different servers. Stands for: Vendor Independent Messaging. Other common messaging APIs are MAPI, CMC. Vendor: Lotus Development Corp.

VINES Communications software. Network operating system connecting IBM and DEC desktop computers. Supports TCP/IP, X.25 protocols. Stands for: VIrtual NEtworking Software. Vendor: Banyan Systems, Inc.

VIO Operating system add-on. Data management software that runs on DEC VAX systems. Stands for: Virtual I/O. Vendor: Dave Froble Enterprises, Inc.

VIP Professional Desktop system software. Spreadsheet. Runs on Apple II systems. Lotus 1-2-3 lookalike. Vendor: Applied Engineering, Inc.; ISD Marketing, Inc.

VIP/VM Application development tool. Program that optimizes assembler code. Runs on IBM VM systems. Stands for: Vastly Improved Programs for VM. Vendor: BlueLine Software, Inc.

Viper 1. Operating system enhancement used by systems programmers. Increases system efficiency in Gould systems. Vendor: Datamax Computer Systems, Inc. 2. See MTS.

Virgil Operating system add-on. Debugging/testing software that runs on Gould systems. Vendor: Datamax Computer Systems, Inc.

Virgo Operating system enhancement used by systems programmers. Monitors and controls system performance in Gould systems. Measures performance of applications running in the system. Vendor: Datamax Computer Systems, Inc.

Virtual Data Warehouse Application development software. Used in data warehousing. Includes DataDirect Explorer (presentation component that allows end-users to build queries), DataDirect SmartData (Database interface that conforms to ODBC specifications). Vendor: INTERSOLV, Inc.

Virtual File Cabinet Groupware. Document management system. Provides Web access to document systems. Released: 1997. Vendor: Infodata Systems, Inc.

Virtual Library Manager Operating system enhancement used by systems programmers. Increases system efficiency in DEC VAX systems. Manages disk storage space. Vendor: Micro Technology, Inc.

Virtual Loadable Module See VLM.

Virtual Machine for Java Application development tool. Enables developers to build platform-independent applications. Includes J/Direct. Vendor: Microsoft Corp.

Virtual Notebook System See VNS.

Virtual Private Network See VPN.

virtual reality A technology that enables users to enter computer-generated worlds and interface with them three-dimensional through sight, sound, and touch. Users wear gloves, goggles, and earphones that are equipped with fiber-optic sensors that can interact with the computer program. Experiments with troubleshooting, data modeling, and product designs are current business uses. Synonyms: telepresence, artificial worlds, multisensory I/O, cyberspace.

Virtual Replica System software. Dynamic directory. No release date schedules. Vendor: Novell, Inc.

virtual storage A technique that simulates more storage than actually exists in a computer. The letters "VS" in an acronym usually refer to virtual storage.

Virtual Storage Manager Operating system enhancement used by systems programmers. Increases system efficiency in DEC VAX systems. Manages disk storage space. Vendor: Software Partners/32, Inc.

Virtual Tape Server See Seascape Storage Enterprise.

Virtual Vault Operating system enhancement. Security software that runs on top of existing operating systems and allows users to install a Web server outside of a corporate firewall so that external users can access business services. Vendor: Hewlett-Packard Co.

VirtualBranches Operating system enhancement. Data storage and retrieval software for DEC's OpenMVS systems. Vendor: Acorn Software, Inc.

Virtuoso 1. Real-time operating system for desktop systems. Versions include: Virtuoso Classico, Virtuoso Micro, Virtuoso Nano. Vendor: Eonic Systems, Inc.
2. Desktop software. Graphics package. Incorporates PostScript. Vendor: Altsys Corp.

Virus Computer program that attaches code to other programs. When these infected programs run, the unsuspected attached code can do very damaging things throughout the entire computer system. Entire systems can be deleted through viruses. A virus infects other programs within the computer system but cannot affect another system unless a person copies or downloads the affected program. Programs that have been affected by viruses are called "Trojan Horses." Other destructive programs are called worms, backdoors, and logic bombs.

VirusScan Virus protection. Vendor: McAfee Associates Inc.

VirusWall Virus detection software used for e-mail systems. Runs on Windows NT systems. Vendor: Trend Micro, Inc.

VIS Communications software. E-mail system. Runs on DEC systems. Stands for: Voice-mail Information System. Vendor: Voice-mail International, Inc.

Visaj Application development tool. Used to build cross-platform Java applications from a point-and-click environment. Runs on Unix systems. Vendor: DataViews Corp.

Visara Network computer. Vendor: Affinity Systems.

Visi Operating system add-on. Debugging/testing software that runs on Gould systems. Vendor: Datamax Computer Systems, Inc.

Visibility MRP, or CIM software. Runs on DEC, HP systems. Vendor: Visibility, Inc.

Visible Advantage Application development tool. Automates business planning, business process modeling, object-oriented design, and reverse engineering. Runs on Windows systems. Released: 1997. Vendor: Visible Systems Corp.

Visible Analyst Workbench CASE product. Automates analysis and design functions. Runs on Pentium type desktop computers. Vendor: Visible Systems Corp.

VisiBroker for C++, Java Communications, application development tool. ORB (Object Request Broker). Used to create applications that can operate over multiple platforms. Internet interface. CORBA compliant. Vendor: Inprise Corp.

VisiBroker Integrated Transaction Service Communications. Middleware, TP monitor. Supplies all the functions of a TP monitor, but works with distributed object applications. Includes VisiBroker ORB. CORBA compliant. Vendor: Inprise Corp.

VisiBroker ORB See VisiBroker for C++, Java.

VisiBroker SSL Pack Communications. Middleware. Provides security measures that allow developers to add authentication and encryption functions to applications. Available for Java and C++, and works with VisiBroker. Optionally available VisiBroker ORB that provides an introductory level of security. By using SSL Pack, developers can add authentication, encryption and digital certificate support capabilities to their distributed applications. Based on industry standards such as RSA's BSafe libraries, SSL comes in two flavors: Java and C++ to work with the corresponding versions of VisiBroker. Vendor: Inprise Corp.

Visimage Application development tool. Query and report functions. Produces ad hoc printed reports. Runs on Unix, Windows systems. Vendor: Vital Soft, Inc.

Visinet Communications. Network management system. Works with LAN Manager, LAN Server, NetWare, Pathworks. Vendor: VisiSoft, Inc.

Vision 1. Database for large computer environments. Runs on DEC, HP 9000 systems. Can be used in data warehousing. Vendor: Innovative Systems Techniques, Inc.
2. See Unify Vision.
3. See Sapiens Vision.

Vision 2000 EIS. Vendor: Software 2000.

Vision 4000 Application software. Manufacturing software handling workflow and supply chain functions. Runs on Windows systems. Released: 1998. Vendor: Manugistics, Inc.

VISION:Assess Application development tool. Graphically displays complexity of an application based on things like code violations, standards violations, and size. Vendor: Sterling Software, Inc.

VISION:Audit Operating system add-on. Uses query techniques to provide audit functions. Formerly called: DYL-AUDIT, ANSWER:Audit. Vendor: Sterling Software, Inc.

VISION:Bridge Application development tool. Allows batch and online queries to combinations of data from multiple and diverse databases. Formerly called: ANSWER/Reporter, ANSWER/Online, ANSWER:Bridge. Vendor: Sterling Software, Inc.

Vision Builder Application development environment. Allows developers to create client/server applications by writing business rules rather than code. Runs on Unix, Windows NT systems. Vendor: Vision Software Tools, Inc.

VISION:Clearaccess Application development tool. Visual query and reporting tool. Includes charting and multidimensional analysis functions. Runs on Macintosh, Windows systems. Works with multiple databases. Formerly called: CLEAR:Access. VISION:Data. Clear Access. Vendor: Sterling Software, Inc.

VISION:Clearmanage Application development tool. Software that manages and controls ad hoc queries from various platforms and databases. Works with Clear Access. Includes: Catalog Server, Profiler, Monitor. Runs on Macintosh, Windows systems. Formerly called: CLEAR:Manage, Clear Manager. Vendor: Sterling Software, Inc.

VISION:Distribute Application development tool. Automates distribution of client software for VISION:Flashpoint applications. Formerly called: STAR:Distribute. Vendor: Sterling Software, Inc.

VISION:Encrypt Operating system add-on. Provides security functions through encryption. Formerly called: DYL-Security, ANSWER:Encrypt. Vendor: Sterling Software, Inc.

VISION:Flashpoint Application development tool. Screen scraper. Used to create GUIs and integrate applications from multiple environments. Formerly called: Flashpoint, STAR:Flashpoint. Vendor: Sterling Software, Inc.

Vision:Inspect Applications development tool. Debugging aid which shows impact of program changes, detects dead code and overlapping and recursive procedures. Runs on OS/2, Windows systems. Vendor: Sterling Software (Applications Engineering Division).

Vision Jade Developer Studio Application development environment used to build HTML and Java Web applications. Allows developers to develop applications from rules and continuously evolve business processes. Runs on Unix, Windows. Released: 1998. Vendor: Vision Software Tools, Inc.

Vision:NorthStar Applications development tool. Reverse engineering tool used to migrate mainframe systems to client/server environments. Runs on OS/2, Windows systems. Vendor: Sterling Software (Applications Engineering Division).

Vision/Recital Application development environment. Object-oriented, visual. Includes application partitioning functions to support multi-tier architectures. Includes ROIE. Runs on Unix systems. Vendor: Recital Corp.

VISION:Results Application development tool. Data manager and report generator for IBM mainframe systems. Formerly called: DYL-280, ANSWER:Results. Vendor: Sterling Software, Inc.

Visionary Application development tool. Used to develop customized data analysis applications for data stored in Informix Dynamic Server databases. Includes a rapid development environment (called Studio). Released: 1999. Vendor: Informix Corp.

VisionBase Midsize computer. Server. Pentium II processor. Operating system: Windows NT. Vendor: Hitachi Data Systems Corp.

VisionBook Plus, Pro, Traveler Desktop computer. Notebook. Pentium processor. Vendor: Hitachi PC Corp.

VisionDesk Desktop computer. Pentium processor. Vendor: Hitachi PC Corp.

VisionQuest Groupware. Includes electronic meeting system (EMS). Includes exercises to help users define and prioritize topics prior to meetings. Vendor: Collaborative Technologies Corp.

VisiSchedule Desktop system software. Project management package. Runs on Pentium type systems. Vendor: VisiCorp.

VisiWord Plus Desktop system software. Word processor. Runs on Pentium type systems. Vendor: VisiCorp.

Visor/V 4GL used in mainframe environments. Works with HP databases. Vendor: Hewlett-Packard Co.

VISP Operating system add-on. Data management software that runs on Harris systems. Stands for: VULCAN Indexed Sequential File Package. Vendor: Harris Corp.

VisPro/C, VisPro/C++ Application development tool. Provides visual programming in C, C++. Work with C/Set, C/Set++. Vendor: Hockware, Inc.

VISTA Operating system add-on. Data management software that runs on Harris systems. Vendor: Harris Corp.

VISTA Plus Data management software used in data warehousing. Searches through reports stored online. Runs on Unix, Windows NT systems. Versions available for SAP R/3, Oracle Applications. PeopleSoft. Released: 1998. Vendor: Quest Software, Inc.

Visual Ada Application development tool. GUI based design tool and Ada code generator. Runs on Windows systems. Vendor: Aetech.

Visual AppBuilder Application development tool. Used to build ALMs (Application Loadable Modules) for client/server systems. The developer can call up existing ALMs as icons and merge functions by simply drawing lines between them. Vendor: Novell, Inc.

Visual Baler Application development tool. Allows users to create applications for diverse spreadsheet systems including Lotus 1-2-3, Excel, Quattro Pro. Once an application is created, it can be run on any system using Windows without running the spreadsheet. Vendor: Techtools, Inc.

Visual Basic Application development environment. Object-oriented system used with RAD development. Builds departmental client applications with a GUI-based front end. Interfaces with SQL Server, Ingres, Oracle, Access, DB2. Three versions available: Standard Edition (student and hobbyists), Professional Edition (departmental corporate use), Enterprise Edition (enterprise level corporate use). Vendor: Microsoft Corp.

Visual Basic Applications See VBA.

Visual Basic Script See VBScript.

Visual C Environment See VICE.

Visual C++ Application development environment. Enhancement of C++ which uses a visual interface and includes the compiler itself and AppWizard, AppStudio, ClassWizard. Produces code for Windows NT. Vendor: Microsoft Corp.

Visual C++ Cross Development Edition Application development tool. Workbench. Creates Macintosh version of Windows applications. Vendor: Microsoft Corp.

Visual C++ Enterprise Edition Application development environment. Includes DataView and SQL Debugger. Runs on Windows NT systems. Vendor: Microsoft Corp.

Visual Cafe See Cafe, Visual Cafe.

Visual Café Enterprise Suite Application development environment. Allows developers to write software across different Unix and Windows platforms. Provides a single user interface to view and debug work done on the different platforms as if it were done on a single system. Integrates with JServer. Runs on Windows systems. Vendor: Symantec Corp.

Visual CASE Application development tool. UML modeling tool. Integrates with C++. Allows developers to change code by clicking on UML diagrams. Part of the RW-Architect family of products, is the only UML modeling tool that is fully integrated with Visual C++. Vendor: Rogue Wave Software, Inc.

Visual COBOL Compiler language that includes productivity tools such as Visual Debug and Visual SMS (a screen editor). Vendor: mbp Software and Systems Technology, Inc.

Visual Companion Object Developers Kit Application development tool. Builds GIS features into software. Used with Smalltalk. Vendor: Object/FX Corp.

Visual CyberQuery Application development tool. Query and report functions. Runs on Unix systems. Vendor: Cyberscience Corp.

Visual DataFlex Application development tool. Suite of tools built around an object-oriented 4GL. Runs on Windows systems. Released: 1997. Vendor: Data Access Corp

Visual dBase See VdB.

Visual EDI Communications software. EDI package. Runs on Windows systems. Vendor: EDI Able, Inc.

Visual Enterprise DSS. Used for financial analysis and order processing. Vendor: FourGen Software, Inc.

Visual Express See CA-Visual Express.

Visual FoxPro Application development environment. Provides tab-oriented access to databases, forms, queries, reports and code for xBase applications. Vendor: Microsoft Corp.

Visual Insights Application development tool used for data mining. Provides graphs of different subsets or summaries of data. This allows user different perspectives of the data. Runs on Unix systems. Vendor: Lucent Technologies Inc.

Visual InterDev Application development environment. Used to develop Internet applications. Released: 1997. Vendor: Micrososft Corp.

Visual Interface Application development tool. Web authoring tool. Unix text editor.

Visual J++ Application development tool. Converts C++ systems to Java for use on the Internet. Runs on Windows systems. Formerly called Jakarta. Vendor: Microsoft Corp.

Visual JavaScript, Visual JavaScript Pro Application development tools for component based development. Used to build Web pages by assembling Java and JavaScript components. Users can drag and drop applications onto Web pages and customize each page's behavior. Runs on Windows systems. Released: 1997. Vendor: Netscape Communication Corp.

visual language See visual programming environment.

Visual Object COBOL Application development environment. Builds GUI front end applications in COBOL. Vendor: Micro Focus, Inc.

Visual Objects See CA-Visual Objects.

Visual PRO/5 See PRO/5.

Visual Programmer Application development tool. Visual programmer interface. Included with Watcom C/C++. Vendor: Blue Sky Software Corp.

visual programming environment Development software that allows user to use icons and point-and-click rather than writing code statements.

Visual Programming Environment See VPE.

Visual Pure Coverage Application development tool. Testing tool that verifies whether each line of code has been tested. Runs on Windows NT systems. Vendor: Rational Software Corp.

Visual RPG Application development system. Runs on Windows systems. Used to develop systems for AS/400. Vendor: Cozzi Research.

Visual Smalltalk Application development environment. Used to create scalable applications. Includes catalog of both visual and non-visual components. Includes Win32, OS/2 components. Vendor: ParcPlace-Digitalk, Inc.

Visual Smalltalk Enterprise Application development environment. Used to create scalable applications. Includes version control, configuration management, multi-user functions. Supports development teams. Runs on OS/2, Windows systems. Vendor: ParcPlace-Digitalk, Inc.

Visual SQL Applications development tool. Generates SQL data access statements as programs are developed. Generates Visual C++ code for the application design. Interfaces with Oracle, SQL Server, Integra, dBase. Runs on Windows NT, Windows 95/98 systems. Vendor: Blue Sky Software, Corp.

Visual Studio Application development suite. Packages Visual Basic, Visual C++, Visual J++, Visual FoxPro and Visual InterDev. Provides common libraries and access to third party tools. Allows users to access data from mainframe, Unix systems. Supports HTML and Web browser access. Vendor: Microsoft Corp.

Visual Test Application development tool. Used for testing Windows and Web applications. Integrates with Microsoft's Developer Studio and Visual C++.

Visual Thought Application development tool. Produces diagrams and flowcharts for software, network, business process, organizational and block diagrams. Runs on Unix, Windows systems. Released: 1997. Vendor: Confluent, Inc.

Visual Warehouse Warehouse software that includes DB2, tools for extracting data for OS/2, Unix systems. Supports the modular warehouse concept that builds department size warehouses that can be incorporated into a total enterprise warehouse at a later date. Vendor: IBM Corp.

Visual WorkFlo Workflow software. Used to model, re-engineer and automate business processes. Uses object-oriented technologies and optionally includes imaging services. Vendor: FileNet Corp.

Visual XML See XML Suite.

VisualAge (C++) Application development environment. Used for enterprise development. Object-oriented system development system based on Smalltalk. GUI builder used in client/server environments. Used to create front-end applications that can interface with COBOL programs. Both SmallTalk and C++ versions available. Known under code name "Camelot." Runs on AIX, OS/2, Windows systems. Vendor: IBM Corp.

VisualAge for Java Application development tools. RAD tool user for Internet applications. Professional and enterprise editions available. Enterprise edition includes Enterprise Access Builder. Incorporates JavaBeans. Released: 1997. Vendor: IBM Corp.

VisualAge for Smalltalk Web Connections Application development environment. Enhancement to VisualAge that builds applications that users can access over intranets or the Web. Vendor: IBM Corp.

VisualAge WebRunner Application development tool. Used to build JavaBeans components. Vendor: IBM Corp.

VisualBuilder for C++ Application development tool. Generates C++ applications. Vendor: IBM Corp.

VisualGen Application development environment. Visual programming environment that generates C/C++, COBOL code. Runs on AIX, OS/2, OS/400, MVS, VSE, Windows systems. Vendor: IBM Corp.

VisualGen (Team Suite) Application development environment. Includes 4GL, testing tools, GUI builder. Builds client/server applications for OS/2, Windows. Visual programming environment that runs on RISC systems but can be used to develop DB2, CICS, IMS mainframe applications. Generates C,C++ code. Vendor: IBM Corp.

VisualInfo Imaging software. Includes repository which links documents across OS/2, MVS, AIX systems. Vendor: IBM Corp.

Visualization Application development technology. Technique used to complement both data mining and OLAP systems. Used to detect patterns.

Visualization Data Explorer Application development tool used for data mining. Provides graphs of different subsets or summaries of data. This allows user different perspectives of the data. Runs on Unix systems. Vendor: IBM Corp.

Visualizer Application development tool. Web interface. Visual reporting tool that displays large amounts of data with multiple data dimensions on a single screen. For example, a screen could show sales by product, by salesperson, by time period, etc. Interfaces with PowerPlay and Impromptu. Runs on Windows systems. Released: 1999. Vendor: Cognos Inc.

Visualizer Query, Charts Application development tool. Query and report functions. Runs on AIX, OS/2 systems. Vendor: IBM Corp.

VisualWare Application development environment. Includes GUI, Visual Basic tools. Supports OLE. Vendor: Intergroup Technologies, Inc.

VisualWave Application development tool. Used to develop client/server applications for Internet Web. Uses Smalltalk. Vendor: ParcPlace-Digitalk, Inc.

VisualWeb Application development tool. Links Web front end systems to client/server systems. Creates applications that use a Web browser as a user interface. Vendor: ParcPlace-Digitalk, Inc.

VisualWorks Application development environment. Workbench software. Used to develop client/server applications. Supports multiple GUIs including Open Look, Motif, Presentation Manager, Macintosh, NeXT, and Windows. Can be used as an upgrade to Objectworks/Smalltalk. Interfaces with Oracle, Sybase, EDA/SQL. Vendor: ParcPlace-Digitalk, Inc.

Vital A design document which covers the integration of Apple Macintosh computers into a mixed-vendor client/server computing environment. Stands for: Virtually Integrated Technical Architecture Lifecycle. Written by Apple Computer.

VITAL Communications software. Connects Macintosh computers to mainframe networks. Stands for: Virtually Integrated Technical Architecture Lifecycle. Vendor: Apple Computer, Inc.

Vital Signs Operating system enhancement used by systems programmers. Increases system efficiency in IBM VM systems. Eliminates bottlenecks and improves I/O response times. Vendor: BlueLine Software, Inc.

Vital Signs 2000 EIS. Runs on AS/400, IBM and compatible desktop systems. Vendor: Software 2000.

VitalAnalysis Operating system software. Measures performance in client/server systems by monitoring actual response time experienced by the user. Vendor: VitalSigns Software, Inc.

Vitria Software vendor. Produces EAI software. Full name: Vitria Technology Inc. Main product: BusinessWare.

Vivante Desktop computer. Notebook. Pentium processor. Vendor: Transmonde Technologies, Inc.

Vivid NetDirector Operating system software. Manages networks and monitors performance from a browser interface. Vendor: Newbridge Networks Corp.

VizControls Application development tool. Provides visual solutions to understanding large amounts of data. Includes hyperbolic tree, table lens, cone tree, and perspective wall presentations. Vendor: InXight Software (subsidiary of Xerox Corp).

VJS See Visual JavaScript.

VLAN Virtual LAN. Logical, not physical, local area network. Enable network managers to create workgroups of users who need to communicate regardless of the physical network they use.

VLIW Type of computer architecture proposed by Hewlett-Packard. Technology for a 64-bit, billion instructions per second chip. Stands for: Very Long Instruction Word. Vendor: Hewlett-Packard Co.

VLM Communications software. Stands for: Virtual Loadable Module. Provides a client shell that allows a desktop computer to migrate to NetWare 4.01. Vendor: Novell, Inc.

VLOCK/VM Operating system add-on. Data management software that runs on IBM systems. Vendor: BMS Computer, Inc.

VLT Customizer Communications software. EDI package. Runs on MS-DOS, Unix systems. Vendor: RMS.

VM/386 Operating system for Pentium-type desktop computers. Vendor: IGC.

VM/AS Application development system. Includes data management, information retrieval, reporting, text processing, and project control functions. Runs on IBM VM systems. Stands for: VM/Application System. Vendor: IBM Corp.

VM/Batch/VMSchedule Operating system enhancement used by systems programmers and Operations staff. Job scheduler for IBM, Amdahl systems. Vendor: Systems Center, Inc.

VM/CMS General-purpose operating system for IBM, Amdahl. System allows other operating systems, including MVS, VSE, Unix, PICK, and VM itself, to be run subordinate to VM, so is often used by companies that are going through an operating system conversion and by companies that produce software that must run under different operating systems. Versions include: VMS/370, VM/CMS, VM/SP(HPO), VM/XA(SP), VM/ESA. Stands for: Virtual Machine/Conversational Monitoring System. Vendor: IBM Corp.

VM/Entry Operating system enhancement used by systems programmers. Increases system efficiency in IBM VM systems. Works with CMS applications in a VM environment. Vendor: IBM Corp.

VM:Migrate Operating system enhancement for MVS/ESA systems. Reduces storage costs and backup time by automatically migrating unused and fragmented files to secondary storage devices such as tape. Vendor: Sterling Software, Inc.

VM/MS Operating system enhancement used by systems programmers. Increases system efficiency in IBM systems. Allows multiple VM sessions. Vendor: Software Technology, Inc.

VM/PC Operating system for microcomputers. Version of VM/CMS that runs on Pentium type desktops. Vendor: IBM Corp.

VM/SP HPO Operating system enhancement used by systems programmers. Increases system efficiency in IBM VM systems. Stands for: VM/System Product High Performance Option. Vendor: IBM Corp.

VM/SP SSI Operating system enhancement used by systems programmers. Increases system efficiency in IBM VM systems. Handles multiprocessing. Vendor: VM/CMS Unlimited, Inc.

VMAccount Operating system enhancement used by systems programmers. Provides system cost accounting for IBM VM systems. Vendor: VM Software, Inc.

VMAIL Communications software. E-mail system. Runs on IBM systems. Vendor: Maxcom, Inc.

VMArchive Operating system add-on. Data management software that runs on IBM systems. Vendor: VM Software, Inc.

VMark Software See Ardent Software.

VMBackup Operating system add-on. Data management software that runs on IBM systems. Vendor: VM Software, Inc.

VMBACKUP-MS Operating system add-on. Data management software that runs on IBM systems. Vendor: IBM Corp.

VMBatch Operating system enhancement used by systems programmers. Monitors and controls system performance in IBM VM systems. Monitors batch program execution. Vendor: VM Software, Inc.

VMCenter (II) Operating system enhancement used by systems programmers. Monitors and controls system performance in IBM VM systems. Vendor: Systems Center, Inc.

VMCICS/DS Application development tool. Program that supports CICS program development. Stands for: Development System. Vendor: Unicorn Systems Co.

VMCICS/ES Operating system enhancement used by systems programmers. Controls communications networks in IBM systems. Stands for: Execution System. Vendor: Unicorn Systems Co.

VME Operating system for ICL mainframe computers. Vendor: ICL North America Business Systems.

VMEstation Midrange computer. SPARC CPU. Operating system: Solaris. Vendor: Advanced Laboratories, Inc.

VML Application development tool. Modeling language used with VODAK databases. Stands for: VODAK Modeling Language. Vendor: GMD-IPSI (Darmstadt, Germany).

VMLib Operating system add-on. Library management system that runs on IBM systems. Vendor: Pansophic Systems, Inc.

VMMAP, VMPPF Operating system enhancement used by systems programmers. Monitors and controls system performance in IBM VM systems. Vendor: IBM Corp.

VMOperator Operating system enhancement used by Operations staff and systems programmers. Provides operator console support in IBM VM systems. Vendor: VM Software, Inc.

VMOS Communications software. Network operating system. Runs on multiple desktop computers. Vendor: Starpath Systems, Inc.

VMPPF Operating system enhancement used by systems programmers. Monitors and controls system performance in IBM systems. Stands for: VM Performance Planning Facility. Vendor: IBM Corp.

VMS Operating system for DEC VAX systems. Stands for: Virtual Memory System. Vendor: Digital Equipment Corp.

VMS/Ultrix Connection Network software that allows a VMS system to communicate to any other system using TCP-IP protocols. Also called: the Connection. Vendor: Digital Equipment Corp.

VMSchedule Operating system enhancement used by systems programmers. Job scheduler for IBM VM systems. Vendor: VM Software, Inc.

VMSecure Operating system add-on. Security/auditing system that runs on IBM VM systems. Vendor: VM Software, Inc.

VMSQL/Edit Application development tool. Program that allows users to maintain data stored in SQL/DS databases. Vendor: VM Software, Inc.

VMSQL/Report Application development tool. Report generator for IBM VM systems. Works with SQL/DS databases. Vendor: VM Software, Inc.

VMTape Operating system add-on. Data management software that runs on IBM systems. Vendor: VM Software, Inc.

VMTAPE Operating system add-on. Tape management system that runs on IBM systems. Full name: VMTAPE Management System. Vendor: IBM Corp.

VNAT(B) Application development tool. Program that allows CICS users to access VSAM files. Runs on IBM systems. Works with Natural programs. Vendor: MB & Associates.

VNS Groupware. Object-oriented client/server system that runs on Macintosh, OS/2, Unix, Windows systems. Follows TCP/IP protocol. Stands for: Virtual Notebook System. Vendor: ForeFront Group, Inc.

VODAK Object-oriented database. Distributed, multimedia system. Includes VML, VQL. Interfaces with C++. Includes libraries, development tools. Runs on Unix systems. Released: 1992. Vendor: GMD-IPSI (Darmstadt, Germany).

Voice Xpress Speech recognition software. Released: 1998. Vendor: Lernout & Hauspie Speech Products.

VoiceType 2 Voice recognition software. Runs on MS/PC-DOS systems. Vendor: IBM Corp.

VoiceType Dictation Voice recognition system. Runs on OS/2, Windows systems. Vendor: IBM Corp.

Vollie See CA-Vollie.

VOMAD Operating system add-on. Data management software that runs on IBM systems. Vendor: Scientific and Business Systems, Inc.

Vortex 1. Database management utilities for client/server and host based environments. Includes Vortex Interface, Vortex Accelerator, and Vortex compilers. Interfaces with Oracle, Informix, Sybase. Runs on Unix, VMS systems. Vendor: Trifox, Inc. 2. Scripting language used by designers to easily create, deploy, and maintain Web based applications. An extension of HTML which can be compiled for fast execution. Vendor: Thunderstone/EPI, Inc.

Vortex01 Object-oriented database. Pure Java database. Stands for: Virtual Object Run Time Expository. Released: 1998. Vendor: Vortex01.

VOS Operating system for Stratus systems. Vendor: Stratus Computer, Inc.

Voyager 1. Database. Latest generation of dBase family. Part of xBase environment. Uses visual tools. Vendor: Inprise Corp. 2. Applications software. Manufacturing software handling supply chain functions. Vendor: Logility, Inc. 3. Communications. Middleware, object request broker (ORB). Connects diverse objects. CORBA and RMI compliant. Included with IBM's VisualAge, Symantec's Café and Novell's NDK. Vendor: ObjectSpace, Inc 4. Desktop computer. Notebook. Pentium processor. Vendor: Unicent Technologies.

Voyager Application Server Application server providing middleware and development tools. Develops, deploys and administers Web-based enterprise-wide systems. Includes Voyager Studio, a development environment for EJB development. Includes Voyager ORB Professional. Vendor: ObjectSpace, Inc.

voysKernel Operating system for Apple Macintosh systems. Vendor: Voysys Corp.

VP 2000 Supercomputer. Operating Systems: UXP/M. Vendor: Fujitsu Ltd.

VP Expert Artificial intelligence system. Runs on IBM systems. Rule-based expert system development tool. Vendor: Paperback Software International, Inc.

VP Graphics Desktop system software. Graphics package. Interfaces with Lotus 1-2-3, VP Planner. Runs on MS-DOS systems. Vendor: Paperback Software International, Inc.

VP-Info Relational database for desktop computer environments. Runs on IBM desktop computer systems. Vendor: Paperback Software International, Inc.

VP/ix Operating system enhancement. Executes MS-DOS applications while maintaining a Unix environment. Vendor: Sunsoft, Inc.

VP-MTI Programming utility that translates entire documents to machine-readable form while retaining charts and graphics in their original form. Stands for: ViewPoint Machine Translation Interface. Vendor: Xerox Corp.

VP-Planner Plus Desktop system software. Spreadsheet. Runs on Pentium type systems. Vendor: Paperback Software Internationa, Inc.

VP/XA Operating system for mainframe Amdahl systems. Allows companies to run MVS/XA on Amdahl systems. Vendor: Amdahl Corp.

VPE Application development environment. Used to develop client/server applications. Includes visual DBMS applications generator. Runs on Unix, Windows systems. Stands for: Visual Programming Environment. Released: 1996. Vendor: VPE, Inc.

VPN 1. Communications software. Allows a company to use the Internet to give remote users access to internal corporate networks. Includes encryption and encapsulation functions. Creates virtual tunnels across the Internet. Can replace WANs. Stands for: Virtual Private Network. 2. Communications, Internet software. Security system that provides encryption to set up Virtual Private Networks. Provides secure tunnels. Released: 1997. Vendor: Intel Corp.

VPP500 MPP computer. Operating systems: Unix, VMS. Vendor: Fujitsu Ltd.

VPS Operating system enhancement used by systems programmers. Increases system efficiency in IBM systems. VTAM print spooler. Vendor: Levi, Ray & Shoup, Inc.

VQL Query language used with multiversion databases including both object databases and multiversion relational. Used with Sybase, Versant, VODAK databases. Stands for: Version Query Language.

VR 4400 Computer chip, or microprocessor. Vendor: NEC Electronics.

VRC 3910 Handheld computer. Vendor: Symbol Technologies.

VRF Operating system enhancement used by systems programmers. Increases system efficiency in IBM systems. Stands for: VSAM Recovery Facility. Vendor: IBS Corp.

VRML Virtual Reality Modeling Language. Developers describe virtual worlds on the Web and users move through the worlds in the same manner as that used with computer games. A Web site could be turned into a Web world. Also used to build Web pages that could, i.e. demonstrate in 3-D how to assemble a product you just purchased. Based on SGM.

VRTX Real-time operating system for Sun, DEC systems. Vendor: Microtec (subsidiary of Mentor Graphics).

VRX(/E) Operating system for NCR systems. Stands for: Virtual Resource Executive. Vendor: NCR Corp.

VS 1. See Virtual Storage.
2. Operating system for Wang systems. Vendor: Wang Laboratories, Inc.

VS-1 Obsolete operating system for IBM mainframe computers. Also called OS/VS1.

VS/Presto-400 Application development tool. Migrates Wang/VS applications to AS/400 systems. Vendor: Financial Technologies, Inc.

VS-Toolbox Operating system add-on. Data management software that runs on DG systems. Vendor: Eagle Software, Inc.

VS6000, VS12000 Midrange computer. Server. Proprietary CPU. Vendor: Wang Laboratories, Inc.

VSAM Access method used in IBM mainframe systems. File manager that accesses data stored in three types of files: ESDS, KSDS, RRDS. Allows keyed, or indexed, access to data in addition to sequential. Used in online as well as batch systems. Stands for: Virtual Storage Access Method.

VSAM ANALYZER Operating system enhancement used by systems programmers. Data management software that runs on IBM mainframe systems. Improves space utilization and reorganizes VSAM files. Vendor: Design Strategy Corporation.

VSAM Assist Operating system add-on. Data management software that runs on IBM systems. Vendor: Softworks, Inc.

VSAM Data Compressor Operating system add-on. Data management software that runs on IBM systems. Vendor: Softworks, Inc.

VSAM-Lite Operating system add-on. Data management software that runs on IBM VSE/SP systems. Vendor: Universal Software, Inc.

VSAM Space Manager Operating system add-on. Data management software that runs on IBM systems. Vendor: Softworks, Inc.

Vsamtune Application development tool. Program that allows "what if" modeling. Vendor: Macro 4, Inc.

VSAMVIEW Operating system add-on. Data management software that runs on IBM mainframe systems. Manages and reports on VSAM files. Vendor: Design Strategy Corporation.

VSAT A shared satellite network. Users pay monthly charges based on use. Established for retail industry. Stands for: Very Small-Aperture Terminal. Vendor: Hughes Network Systems, Inc.

vsDesigner, vsSQL, vsObject maker Application development tools. Automates analysis and design functions. Vendor: Visual Software, Inc.

VSE General-purpose operating system for IBM large computer systems. Versions include VSE/SP, SSX/VSE. Used to be called DOS. Vendor: IBM Corp.

VSE/AF Operating system enhancement used by systems programmers. Increases system efficiency in IBM VSE systems. Stands for: VSE Advanced Functions. Vendor: IBM Corp.

VSE/ICCF See ICCF.

VSE/NMPF Communications software. Transaction processing monitor. Runs on IBM VSE systems. Stands for: VSE Network Management Productivity Facility. Vendor: IBM Corp.

VSE/POWER See POWER.

VSE/TAPE Operating system add-on. Tape management system that runs on IBM systems. Stands for: VSE/Tape Automation for the Production Environment. Vendor: IBM Corp.

VSE/VM Operating system enhancement. Monitors and controls system performance in IBM VM, VSE, and MVS systems. Used with VTAM, CICS, TSO, and ISPF. Vendor: Banner Software, Inc.

Vselect Operating system add-on. Data management software that runs on DEC MAX systems. Vendor: Evans, Griffiths & Hart, Inc.

VSLAN (II, III) Operating system software. Security package for midrange and desktop networks. Works with Ethernet and Token Ring. Vendor: Verdix Corp.

VSM/VM Operating system enhancement. Increases system efficiency in IBM systems. Improves response time for online systems. Vendor: Banner Software, Inc.

VSMR Operating system add-on. Data management software that runs on IBM systems. Stands for: Virtual Storage Measurement Reporter. Vendor: IBM Corp.

VSOS Operating system for mainframe ETA, Wang systems.

VSTimer Operating system enhancement used by systems programmers and Operations staff. Job scheduler for Wang VS systems. Vendor: Software Business Applications, Inc.

VSUM Operating system add-on. Data management software that runs on IBM systems. Stands for: VSAM Space Utilization Monitor. Vendor: On-Line Software International.

VT-52/100/102 Emulator Communications software. Terminal emulator. Vendor: Computer Logics, Ltd.

VTAM Access method used in IBM mainframe systems. Stands for: Virtual Telecommunications Access Method and is used in communications systems. Used by systems programmers communications specialists. Vendor: IBM Corp.

VTERM/220,4010,4105,4208 Communications software. Terminal emulator for DEC systems. Vendor: Coefficient Systems Corp.

VTOC Volume Table of Contents. The data on DASD that describes the contents of the volume.

VTPack Application development tool. Provides run-time testing tools for large heterogeneous server-based applications. Part of Cyrano Suite. Vendor: Cyrano, Inc.

VTS Virtual Tape Storage. See Seascape Storage Enterprise.

VUE Operating system for Harris systems. Vendor: Harris Corp.

VX-Metric See PC-Metric.

VX/VE Operating system for mainframe CDC systems. Vendor: Control Data Corp.

VX-REXX Development environment for OS/2 REXX applications. Interfaces with the Workplace Shell. Vendor: IBM Corp.

VxSim Application development tool. Builds simulations and prototypes. Simulates Sun workstations so programs can be developed for new hardware before it's available. Vendor: Wind River Systems, Inc.

VxWorks Real-time operating system for DEC and Sun workstations. Includes an applications development environment for real-time applications. Vendor: Wind River Systems, Inc.

Vycor Enterprise Operating system software. Automated help desk. Vendor: McAfee Associates, Inc.

VZ Programmer for OS/2 Application development environment. Used for object-oriented programming. Includes GUI toolkit, object database, C, C++ compilers. Runs on OS/2 systems. Vendor: VZ Corp.

W3 See Web.

W3 Connection See WWW Connection.

W3C World Wide Web Consortium. Founded in October 1994 to develop common protocols that promote the evolution of the Web. The consortium has defined XML as a Web standard and is working on XML's development.

WABI Application development tool. Software that allows users to run Windows applications on Unix systems. Stands for: Windows Application Binary Interface. Vendor: Sun Microsystems, Inc.

Wadsworth Communications technology. Will allow users to put applications on the World Wide Web without caring about the differences between Web browsers. Vendor: ParcPlace-Digitalk, Inc.

WAIS Communications. Search program designed to make searching user-friendly, efficient, and cumulative. Used with the Internet. Stands for: Wide Area Information Servers.

WALDO Operating System enhancement used by Operations staff and systems programmers. Provides console support and allows for automatic and remote IPL of MVS systems from desktop computers. Vendor: Software Engineering of America.

Walkabout Notebook computer. Vendor: Data General Corp.

Walker Interactive Systems Software vendor. Products: Tamaris family of financial, payroll, project management systems for mainframe environments.

WAN Wide Area Network. Communications network covering long distances.

WAN administrator Technical developer. Evaluates, selects, installs, and maintains both hardware and software for wide-area-networks. Writes middleware, works with protocols. Provides networking support for on-line applications systems. Senior or mid-level title.

Wanderer See spider.

Ward-Mellor Structured programming design methodology named for its developers. Based on event response. Concentrates on real-time application design. Accepted as a standard design methodology by some companies, and used by some CASE products.

warehouse See data warehouse.

warehouse analyst, architect, designer, engineer Developer. Could be application or technical. Designs the metadata and builds the indexing algorithms. Senior level title.

WarehouseArchitect See PowerDesigner.

Warehouse Control Center Application development tool used in data warehousing. Synchronizes metadata and provides central administration for the warehouse. Vendor: Intellidex Systems, Inc.

Warehouse for Workgroups Data warehouse database and programming. Intended for smaller systems. Supports up to 30 users. Vendor: Red Brick Systems.

Warehouse Manager/Change Manager Application development tool. Converts legacy applications and data into integrated information that can be used in Unix and/or OS/2 client/server environments. Also used in data warehousing. Vendor: Prism Solutions, Inc.

Warehouse Studio Application development tool. Complete package for developing data warehouses. Includes a database management system, software tools for warehouse design, metadata management and warehouse management, and a repository of metadata. Released: 1998. Vendor: Sybase, Inc.

Warehouse Technology Initiative See WTI.

Warehouse WORKS Data warehouse framework and line of products. Strategy and product selection intended to offer smooth integration between the warehouse and the operational systems. Includes database, query, management, and system functions. Vendor: Sybase Inc.

Warehouse XPP Data warehouse software. Runs on MPP systems. Automatically routes queries to least-used node. Vendor: Red Brick Systems.

WarehouseArchitect See S-Designor.

Warnier-Orr Structured programming design methodology named for its developers. Accepted as a standard design methodology by some companies, and used by some CASE products.

Warp See OS/2 Warp.

Warp Connect See OS/2 Warp Connect.

Warp Server See OS/2 Warp Server.

Wasserman/Pircher Object-oriented development methodology. Provides analysis and design techniques.

Watcom C/C++ Application development tool. Cross platform compiler used to develop applications for Windows, OS/2. Includes Visual Programmer and development tools. Vendor: Powersoft (division of Sybase, Inc).

WATCOM SQL Relational database. Bundled with PowerBuilder in PowerBuilder Desktop. Vendor: Powersoft (division of Sybase, Inc).

waterfall development Classical software development where software is developed from requirements and analysis to system development and delivery. Thorough analysis and design define the finished product from the beginning. The methodology mandates that every step of the process must be fully completed before moving on to the subsequent step. Waterfall methodologies are not appropriate for data warehouse or data mart due to their very slow speed. Contrast with spiral development.

WaveStar, WaveStation Desktop computer. Pentium processor. Vendor: Directwave Inc.

Wayfarer Application development tool. Allows users to develop applications with push technology to automatically distribute information from Web sites and business applications. Runs on Windows systems. Vendor: Wayfarer Communications, Inc.

WBEM Initiative from DMTF (Desktop Management Task Force) to create one set of data definitions for network management software. Built around CIM (Common Information Model) which provides common descriptions of network systems management data from static desktop hardware and software to dynamic data such as traffic levels on routers. Uses XML to structure CIM data for presentation, and HTTP to send data from one system to another. Stands for: Web-Based Enterprise Management.

Web Hypertext-based interface to the Internet. System for linking information on Internet computer systems that allows users access to sound, graphics and text not available through traditional Internet connections. The Web consists of HTML documents. Also called World Wide Web, WWW, W3.

Web 3.0 Communications software. Network operating system. Runs on Pentium type desktop computers. Vendor: Webcorp.

Web Agent Communications software. Synchronizes Web screens between users and call centers so both parties are seeing the same thing. Allows either party to draw circles around text and/or pictures so both can see. Released: 1998. Vendor: Aspect Telecommunications, Inc.

Web applications Applications that use a Web browser as a user interface.

Web browser Software that accesses the Web. The browser interprets the code and tags on a Web page and displays the information so Web pages will look different when viewed through different browsers.

Web-caching Storing frequently accessed Web sites or information on local servers. See caching.

Web Commander Communications. Internet software. Web server used for small to mid-size sites. Includes an HTML editor, other Web development tools, a mail server an automated credit card-verification and -clearing program for commercial sites. Vendor: Luckman.

Web DataBlade Application development tool. Links Web front end systems to client/server systems. Creates applications that use a Web browser as a user interface. Enables Illustra DBMS to act as a Web server. Vendor: Illustra Information Technology, Inc.

Web DB Application development tool. RAD tool which lets developers design data driven pages from a browser. Reads database data and automatically creates HTML views for each table. Database content can then be browsed, updated and published from any Web browser. Released: 1998. Vendor: Oracle Corp.

web designer See webmaster.

Web.Designer Application development tool. Web authoring tool. Creates Web pages for Windows platforms. Vendor: Corel Corp.

Web Defender Communications. Internet software. Allows administrators to manage access to Windows NT networks. Can log in with Windows NT id and password and access any corporate Web server. Vendor: Axent Technologies, Inc.

Web Developer Suite Application development environment used to build Web applications. Includes tools to model, create, deploy and manage a scalable Web-based transaction system. Can be used to build traditional LAN applications. Released: 1997. Vendor: Oracle Corp.

Web Development Kit Application development tool. Converts PowerBuilder applications to Java applications that can run on a Java virtual machine including Web browsers. Released: 1998. Vendor: PowerSoft (division of Sybase, Inc.)

Web DSS Application development tool. Lets users analyze marketing information stored in relational databases via Web browsers. Vendor: Microstrategy, Inc.

Web Element Application development tool. Links Web front end systems to client/server systems. Creates applications that use a Web browser as a user interface. Lets developers customize and embed Web browsers into applications. Runs on Windows. Released: 1995. Vendor: Neuron Data, Inc.

Web/Enable Application development tool. Creates applications in Java or C++. Vendor: Black & White Software.

Web for Windows Communications software. Peer-to-peer network operating system. Network operating system. Interfaces with MS-DOS. Works with Novell, Token-Ring, Ethernet, ARCnet. Vendor: Webcorp.

Web Force Communications software. Suite of programs that includes a Web server, Web browser, management utilities, HTML, and hardware options. Used to build an intranet. Vendor: Silicon Graphics, Inc.

Web Information Server Knowledge management software. Provides access to document management systems over the Internet. Vendor: Altris Software, Inc.

Web Integrity Communications, Internet software. Project management tool designed to support team building of Web sites and management of Web projects. Works with Java. Runs on Unix, Windows NT. Vendor: Mortice Kern Systems, Inc.

Web Link Communications software. EDI. Uses Web browsers. Runs on Unix systems. Released: 1997. Vendor: Sterling Commerce, Inc.

Web-Olap Communications software. Provides Internet accessibility to relational databases and data warehouses. Works with DecisionSuite Server. Vendor: Information Advantage, Inc.

web programmer, internet developer Developer, usually applications. Develops interactive programs using Internet skills such as HTML, Java, CGI scripts using languages such as Perl, VBScript, JavaScript. Can be any experience level.

Web Reports Application development tool. Lets users access reports from Web browsers. Vendor: Zanza Software, Inc.

web server Operating system software. Returns HTML pages to requesting browser. Provides security and configuration functions. Provide CGI tools to connect to corporate databases.

Web Server Web server. Users can publish and access information on intranets and the Internet. bundled with Netware. Vendor: Novell, Inc.

Web Server Benchmark System software. Analyzes Web server performance. Runs on Unix systems. Vendor: Neal Nelson & Associates.

web specialist Developer. Used for a variety of job skills and levels, so has no real meaning other than requiring Internet skills such as Java, CGI scripts, HTML, etc.

Web.sql Application development tool. Uses Web browsers to access data warehouses. Vendor: Sybase, Inc.

Web Toolkit Application development tool. Web authoring tool. C++ class library of HTML elements including test, graphics, frames, etc. Runs on DEC, Unix, Windows systems. Released: 1998. Vendor: ObjectSpace, Inc.

Web Transporter Communications, Internet software. Allows system managers to update files and programs on remote systems. Push technology delivers the updates. Vendor: Megasoft Online, Inc.

Web View Communications, Internet software. Interface with the Internet which has same interface as Windows 95/98 and lets users toggle between the operating system and the Internet. Vendor: Microsoft Corp.

WEB390 Communications. Web server designed for IBM's MVS. Provides secure Web browser access to CICS, IMS, and DB2 applications. Runs on OS/390 systems. Released: 1998. Vendor: Information Builders, Inc.

WebAccess See GroupWise WebAccess.

WebBase Application development tool. Links Web front end systems to client/server systems. Creates applications that use a Web browser as a user interface. Can access several relational databases at one time. Vendor: Expertelligence, Inc.

WebBuilder Internet software. Authoring tool. Used to design applications that access data in SQL databases. Includes a visual HTML editor. Runs on AS/400, DEC, Unix, Windows systems. Released: 1996. Vendor: VPE, Inc.

WebCatalog Application software used for e-commerce. Provides an online store with multiple storefronts, shopping cart processing, a database for product prices, custom shipping and invoicing, and e-mail integration. Vendor: Pacific Coast Software.

WebCD Knowledge Network Knowledge management software. Accessible through the Internet. Vendor: Fulcrum Technologies, Inc.

WebCenter Express Communications software. Internet product used with e-commerce. Provides customer service over the Web. Vendor: Acuity Corp.

WebClient Communications. Accesses mainframe data through a Java-enabled Web browser. Released: 1999. Vendor: Cisco Systems.

WebConnect (Pro) Communications/Internet software. Connects any Java-enabled Web browser to mainframes. Includes Open Vista development software to build GUI front-ends to mainframe systems. Released:1997. Vendor: OpenConnect Systems, Inc.

WebCraft Application development tool. Used to generate Java code. Creates applications that use a Web browser as a user interface. Runs on Windows. Vendor: SourceCraft, Inc.

Webcube P,Wxxxx Midsize computer. Server. Pentium Pro processor. Operating system: Linux. Vendor: Pacific Internet.

WebDAV Communications. Internet protocol. Provides Web servers with the equivalent of a network file system for the exchange of data. Reduced need for proprietary methods of interacting with Web servers. Accepted by W3C. (World Wide Web Consortium). Stands for: Web Distributed Authoring and Versioning.

WebDB Application development tool. Used for development and deployment of Internet applications. Included with Oracle8i and available independently. Runs on Unix, Windows NT. Released: 1999. Vendor: Oracle Corp.

WebDNA Application development tool. Internet technology. String of text inserted in web pages that contains information about the content and location of the web page. This allows search engines to find sites easily. Used in WebCatalog.

WebEIS Application development tool. Allows users to create interactive documents that are published as applets. There's no code to write as the software creates the HTML. Uses drag-and-drop to build the document. Documents can then be accessed through any browser. Released: 1997. Vendor: SAS Institute, Inc.

WebEx Communications, Internet software. Off-line browser. Vendor: Traveling Software, Inc.

WebExplorer Internet browser. Runs on OS/2. Vendor: IBM Corp.

WebFlow Groupware that runs on the Internet. Vendor: WebFlow Corp.

WebFOCUS Application server providing middleware and development tools. Allows users to display charts and graphs and show the results of predefined queries with any standard Web browser. Works with most relational databases including DB2, Informix, Oracle, Sybase. Full name: WebFOCUS Application Server. Runs on IBM mainframe, Unix, Windows NT systems. Released: 1997. Vendor: Information Builders, Inc.

WebFocus Suite Application development tool. A set of Java applets that adds report scheduling and distribution functions to WebFocus Application Server. Vendor: Information Builders, Inc.

WebForce Origin 200 Midrange computer. Server. MIPS R1000 CPU. Vendor: Silicon Graphics, Inc.

WebGalaxy Application development tool. Generates Web pages and Java applets. Runs on Windows 95/98 systems. Vendor: Allen Systems Group, Inc.

WebHelp Application software. Used for e-commerce. Provides a Web-based customer service system. Designed for small to midsize businesses. Released: 1998. Vendor: Lotus Development Corp., IBM Corp. and GWI Software, Inc.

WebIntelligence Application development tool. Suite of querying, reporting, and online analytical processing tools. Used to build intranets and business-to-business extranets. Includes InfoView, which is a single Web entry point for both WebIntelligence and BusinessObjects. Runs on Unix, Windows NT systems. Released: 1998. Vendor: Business Objects, Inc.

WebKit Application development tool. Allows developers to make current and new applications accessing UniVerse databases available via the Internet. Allows user to create HTML documents dynamically. Runs on Unix systems. Vendor: Pulsar Systems, Inc.

WebLogic Application development tool used to create, deploy, and manage Web applications. Interfaces with Enterprise JavaBeans and integrates with many Java development environments including Visual Café, Jbuilder, Visual J++ , and Visual Age. Vendor: BEA Systems, Inc.

WebLogic Enterprise Application server providing middleware and development tools. Used to build, deploy and manage component-based applications. Provides Java development server. Includes M3. Released: 1999. Vendor: BEA Systems, Inc.

webmaster Application developer. Does the graphic design, updates, and maintains Web sites. Requires Internet skills such as HTML, Java, Perl, VBScript, etc. and also database and business knowledge. Senior or mid-level title.

WebMate/Foundation Communications, Internet software. Stores pages from a Web site in a database for faster access. Includes scripting language based C and Perl and built-in debuggers. Released: 1996. Vendor: WebMate Technologies, Inc.

WebMerchant Application software used for e-commerce. Provides credit card verification, order notification through e-mail, immediate delivery of electronic goods (software, graphic images, etc.). Vendor: Pacific Coast Software.

WebObjects Application server providing middleware and development tools. Used to build Java applications for the Web. Adds an HTML layer on top of corporate data or applications. Accesses objects, relational databases. Supports Java. Runs on MacOSX, Unix, Windows NT systems. Vendor: Apple Computer, Inc.

WebPack Application development tool. Allows users to develop dynamic, searchable Web sites through a point-and-click interface. Lets users convert Windows documents into HTML. Vendor: InText Systems, Inc.

WebPatch Application development tool. Lets developers update software over the Web by downloading only changed programs. Released: 1996. Vendor: NetSync Corp.

WebPresenter Application development tool. Enables PowerPoint and Presenter documents to be viewed from any Java-enabled Web browser. Vendor: Contigo Software.

WebPublisher See MeetingPlace WebPublisher.

WebRecruiter Application software. Creates user profiles for online job applications and matches them with job requirements defined by Human Resources. Runs with Oracle WebServer. Vendor: Consortium LLC.

WebScan X Application development tool. Part of Nuts & Bolts. Security package. Detects viruses, Malicious Java applets and ActiveX components. Vendor: Network Associates.

Webserver, Secure Webserver Communications software. Provides access to the Web from Unix systems. Vendor: Open Market.

WebServer, WebServer 400 Communications software, Web server. Provides access to the Web from Windows NT, Solaris, OS/400 systems. Lets users view documents. Vendor: Oracle Corp.

WebSession Applications development tool. Internet software used to Web-enable legacy applications running on mainframe systems. Includes WebSession Java API (Java class library) and WebSession Servlet (used to develop automated logon scripts, customized screen settings, and user preferences, without modifying the existing legacy application). Released: 1999. Vendor: I/O Concepts, Inc.

WebShare Application server providing middleware and development tools. Used to build business-to-business applications. Runs on Unix, Windows systems. Released: 1997. Vendor: Radnet, Inc.

WebShieldX Communications software. Firewall. Blocks Java and ActiveX code from entering corporate systems from the Internet. Vendor: McAfee Associates, Inc.

WebSite (Professional) Communications software, Web server. Provides access to the Web from Windows NT, Windows 95/98 systems. Lets users view documents. Vendor: O'Reilly and Associates.

WebSpace Navigator Internet VRML browser. Runs on DEC, HP, Macintosh, OS/2, Unix, Windows systems. Vendor: Template Graphics Software.

WebSpective Communications. Internet software. Web management tool which handles traffic, performance monitoring, data collection, and centralized administration. Includes a content manager which distributes information across multiple Web sites. Vendor: Atreve Software, Inc.

WebSpeed Application server providing middleware and development tools. Includes WebSpeed Workshop for development and provides scaling to support increasing volumes of Web activity. Supports multiple levels of password protections. Runs on Unix, Windows systems. Released: 1996. Vendor: Progress Software Corp.

WebSpeed (Transaction Server) Application development tool. Includes a transaction processing system and a scripting language that links to HTML pages to create Web-based applications. Released: 1996. Vendor: Progress Software Corp.

WebSphere Application server providing middleware and development tools. Used to build Java-based e-commerce applications. Includes Java servlets, Java Server Pages, and Apache HTTP server. Includes support for EJBs. Full name: WebSphere Application Server. Released over time starting: 1998. Vendor: IBM Corp.

WebSphere Studio 3 Integrated development environment. Java based. Includes a workbench which is used to organize Web projects around a single user interface. Also includes ScriptBuilder (to build HTML and JSP), NetObject's Fusion (to develop interfaces), BeanBuilder, and VisualAge for Java (visual language). Released: 1999. Vendor: IBM Corp.

WebStar Communications software, Web server. Provides access to the Web from Macintosh systems. Lets users view documents. Vendor: StarNine.

WebStation Desktop computer. Pentium processor. Vendor: Tri-Star Computer Corp.

WebSurfer Internet browser. Runs on Windows systems. Vendor: NetManage.

WebSystem Communications, Internet software. Web server and gateway between Oracle databases and Web browsers. Vendor: Oracle Corp.

Webtropolis OrderNet Application software. Order management system that lets a business create a virtual store at any Web site. Interfaces with Oracle, SQLServer. Vendor: WebVision, Inc.

WebWhacker Communications, Internet software. Offline browser used to speed access. Collects several Web pages and presents them to the user at the same time. Runs on Macintosh, Windows systems. Vendor: ForeFront Group, Inc.

WebWorks Application development tool. Used with Framemaker publishing package. Lets users translate documents into HTML. Runs on Macintosh, Unix, Windows systems. Vendor: Quadralay Corp.

Wedge Technology Computer vendor. Manufactures desktop computers.

Wen 486 Notebook computer. Vendor: Wen Technology.

WFC Application development tool. Object-oriented framework for class libraries. Part of Visual J++. Stands for: Windows Foundation Classes. Used to develop software to run on Windows systems. Released: 1998. Vendor: Microsoft Corp.

Whew! Communications software. Search engine used on the Web. Released: 1998. Vendor: WordCruncher Publishing technologies, Inc.

white box testing Type of testing that builds test data and scenarios from the structure of the developed software. Consists Examples: Execute every statement at least once, execute each branch at least once. Must have the actual program, object or component. Also called structural testing, glass-box testing. Contrast with black box testing.

WIIS Imaging system. Runs on Wang systems. Stands for: Wang Integrated Image System. Vendor: Wang Laboratories, Inc.

WIMP Stands for Windows, Icons, Menus, Pointers. Describes necessary parts of a GUI.

Win32 Application development standards and tool. Allows developers to write applications for Windows that will run under both Windows NT, Windows 3.2, and Chicago. Two subsets available: Win32S (for Windows, Windows NT), and Win32c (for Windows 95/98). IBM follows Win32 in OS/2, Sun follows them in WABI. Vendor: Microsoft Corp.

WinBook Notebook computer. Vendor: Winbook Computer Corp.

WinCE Operating system for handheld computers. Based on Windows 95/98. Includes e-mail, Internet access. Code name Pegasus. Vendor: Microsoft Corp.

WinCenter (Pro) Communications software. Manages networks of network computers. Provides network management from a single point. Vendor: Network Computing Devices.

Wincite Knowledge management system. Interfaces with Oracle client/server systems. Vendor: Wincite Systems.

WinComm Pro Communications software. Lets users access a variety of on-line services including MCI Mail, CompuServe, Dow Jones. Can also cut and paste among multiple live sessions. Vendor: Delrina Corp.

WINconnect Business-to-business extranet set up for the insurance industry.

Wind/U Application development tool. Creates Unix version of applications written for Windows. Used in client/server systems. Vendor: Bristol Technology, Inc.

Windchill Application software. PDM package. Integrated suite of programs for product and process lifecycle management. Includes new Web-based workflow capabilities. Allows companies to manage across functions all phases in a product's lifecycle, from concept and definition to production, service, maintenance and retirement. Released: 1998. Vendor: Parametric Technology Corp.

WinDD Operating system software. Allows Windows NT screens to be displayed on X-terminals thus allowing desktop computer applications to run on Unix and mainframe systems. Vendor: Tektronix, Inc.

window A division of a screen in which one of several programs running concurrently can display information. Most common in desktop computer environments, but some availability on large computer systems.

Windows, Windows 95/98 Operating system. Includes multitasking, GUI. Runs on Pentium type desktop computers. Windows 95/98 is a 32-bit version which runs alone, prior versions require MS-DOS to provide a complete operating system. Windows 95/98 was also called Chicago, Windows 4.0. Vendor: Microsoft Corp.

Windows 2000 Operating system, major upgrade to Windows NT 4.0. Will include public-key security (Kerberos) and new directory measures. Existing applications will have to be revised to run under Windows 2000. Migration will require a great deal of training and utility programs to assist in the migration are available. Release date: 2000. See Windows NT.

Windows 98 Upgrade to Windows 95/98. Allows users to toggle between the Internet and the desktop. Release date: 1998. Vendor: Microsoft Corp.

Windows CE See WinCE.

Windows DNA Catch-all term for Microsoft's many Web application technologies including Visual InterDev, Visual Studio. Used to tie together different applications in a Windows network. Based on COM standards.

Windows for Pen Computing Operating System for pen computers. Vendor: Microsoft Corp.

Windows for Workgroups Peer-to-peer network operating system. Used for groupware. Includes e-mail, DDE and OLE. Vendor: Microsoft Corp.

Windows Notepad Application development tool. Text editor. Used to insert HTML tags into text. Vendor: Microsoft Corp.

Windows NT Operating System used primarily on workstations and servers. Utilizes full 32 bit operations. Runs on various systems. Supports Windows, MS-DOS, POSIX, LAN Manager, Net-Ware. Can act as either client or server in client/server environment. Includes all the features of Windows for Workgroups. Runs on desktop computers, DEC VAX systems. Stands for: New Technology. Also called NT. Vendor: Microsoft Corp.

Windows NT Embedded Operating system software. Embedded in large devices such as fax machines, copiers, and medical monitors. Allows these devices to be linked with corporate systems. Complements Windows CE which runs in smaller devices such as smart phones, handhelds, and pagers. In development. Vendor: Microsoft Corp.

Windows NT Enterprise Edition See NT Enterprise.

Windows NT Server Communications software. Network operating system connecting desktops running Windows NT. Includes wizards which automate routine network tasks such as adding users, and handle system configuration and management tasks. Full Name: Windows NT Advanced Server. Also called NT AS. Vendor: Microsoft Corp.

Windows NT Workstation Operating system. 32-bit version of Windows NT that runs on personal computers. Includes multi-processing, fault tolerant capabilities. Vendor: Microsoft Corp.

Windows Power Desktop computer. Pentium CPU. Operating system: Windows 95/98. Vendor: Compu-tek International.

Windows Terminal Server See WTS.

WindowsMaker Professional Application development tool. Object-oriented. Generates class libraries from OWL, MFC or C code. Allows developers to go back and forth between Borland and Microsoft libraries. Vendor: Blue Sky Software Corp.

Windtunnel See Bachman/Windtunnel.

WinEDI Communications software. EDI package. Runs on Windows systems. Vendor: Digit Software, Inc.

WinFrame Communications software. Allows remote users to access any application regardless of platforms. Runs on Windows NT systems. Vendor: Citrix Systems, Inc.

Wings/PC Desktop system software. Project management package. Runs on Pentium type desktop computers. Vendor: AGS Management Systems, Inc.

Wingz 1. Spreadsheet. Runs on Next, Unix systems. Vendor: Informix Software, Inc. 2. EIS. Graphical spreadsheet. Runs on Macintosh, Unix, Windows systems. Vendor: Investment Intelligence Systems Group.

WinHelp Office Application development tool. Used to create Help screens. Includes several help authoring tools and Robo-Help. Vendor: Blue Sky Software Corp.

WinInstall Operating system software. Installs desktop computer applications over LANs. Runs on Windows systems. Vendor: OnDemand Software, Inc.

WinLink Communications. Middleware. Connects Windows systems to legacy systems. Vendor: Via Systems, Inc.

WinMax Desktop computer. Pentium CPU. Operating systems: Windows 95/98. Vendor: MAXIMUS COMPUTERS, Inc.

WinNET Communications software. Network operating system (NOS) connecting Concurrent systems. Vendor: Concurrent Controls, Inc.

WinPad Operating system for PDA (handheld) computers. Includes application development environment. Also called At Work for Hand-helds. Vendor: Microsoft Corp.

WinPlus Application development environment. Object based. Used to develop GUI front ends for legacy systems, computer-based training systems, and process control. Vendor: ObjectPlus Corp.

WinPort Graphical user interface development tool. Vendor: Interactive Engineering Corp.

Winpro Desktop computer. Pentium CPU. Operating system: Windows 95/98. Vendor: Royal Electronics, Inc.

WinPro 486, Multimedia Desktop computer. Pentium CPU. Operating systems: Windows 95/98. Vendor: Leading Edge Products, Inc.

WinRunner Application development tool. Testing tool that automatically generates an initial set of scripts for GUI testing. Scripts are generated from the keystrokes during the original testing. Runs on OS/2, Windows systems. Vendor: Originally Micro Focus, now MERANT.

WinServer 5000, Pro Midrange computer. Pentium CPU. Operating systems: Windows 95/98. Vendor: Sequent Computer Systems, Inc.

WinSock Communications software needed to access the Internet through Windows. Mosaic, Cello, Gopher, Eudora are all Winsock programs.

Winsurf See WMA.

wIntegrate Application development tool. Allows developers to add Windows GUI features to applicati ons in a phased approach. Runs on Windows systems. Released: 1997. Vendor: Ardent Software, Inc.

Wintel Name used for the platform of Windows operating systems and Intel processing chips.

Winterm Network computer. Vendor: Wyse Technology, Inc.

WinTerm Operating system enhancement. Allows users to run Windows applications from network computers. Vendor: Citrix Systems, Inc.

Wintext Desktop system software. Word processor. Runs on Windows on IBM and compatible desktop computers. Vendor: Palsoft.

WinTower 486, Multimedia Desktop computer. Pentium CPU. Operating systems: Windows 95/98. Vendor: Leading Edge Products, Inc.

WinView Communications software. Middleware. Allows multiple users to access DOS, Windows, and OS/2 applications from diverse systems. Full name: WinView for Networks. Vendor: Citrix Systems, Inc.

Wireless OS-9 Operating system for handheld computers. Vendor: Microware Systems Corp.

WireTap Communications. Internet software that continuously monitors and manages network, intranet and Internet performance. Does traffic analysis and monitors end-to-end Web and SQL transaction response times. Vendor: PLATINUM Technology, Inc.

Wirfs-Brock Object-oriented development methodology. Provides analysis and design techniques.

WiseWan Communications. Network management software which handles enhanced band-width management. Released: 1999. Vendor: NetReality, Inc.

WITT Application development tool. Restructures COBOL code and manages the reuse of old applications. Automates testing function. Stands for: Workstation Interactive Test Tool. Part of AD/Cycle. Vendor: IBM Corp.

wizard Term used for an automated set of menus that leads the user through a series of steps to accomplish a specific task. Used in software such as desktop publishing where a wizard could move the user through the original setup of a document.

Wizard Mail Communications software. E-mail system. Runs on IBM systems. Vendor: H&W Computer Systems, Inc.

WizWhy Application development tool used for data mining. Used to determine which of hundreds of realtionships among variables are significant. Runs on Windows systems. Vendor: WizSoft, Inc.

WLFS Operating system software. Provides mainframe backup services to desktop systems. Runs on IBM VM and MVS systems, and supports DOS, AIX, OS/2, Macintosh, and SunOS desktops. Stands for: Workstation LAN File Services. Vendor: IBM Corp.

WMA Communications software. Allows Web front-end to mainframe systems. Stands for: Winsurf Mainframe Access. Also called Winsurf. Runs on Windows NT and works with Windows 95/98, 98 or NT clients accessing IBM, DEC or UNIX hosts. Uses Active Server Pages (ASP) and Active X. Released: 1999. Vendor: Data Interface.

WNDX Application development tool. Object-oriented, cross-platform development toolkit. Used to create cross platform GUIs. Allows users to develop code that will work in diverse environments including Windows, Presentation Manager, Motif, Macintosh. Vendor: The WNDX Group, Inc.

Wolf Mountain Communications technology from Novell. Defines clustering functions. Vendor: Novell, Inc.

Wolfpack See MSCS.

Word See Microsoft Word.

Word/3270 Mainframe word processor that works like WordStar. Vendor: Chicago Soft, Ltd.

Word Internet Assistant Application development tool. Text editor. Used to insert HTM tags into text. Vendor: Microsoft Corp.

Word Manager Desktop system software. Word processor. Runs on Pentium type systems. Vendor: Bluebird Systems.

Word Pro Desktop software. Word processor. Included with SmartSuite office suite. Vendor: Lotus Development Corp.

word processor 1. A desktop computer program that handles text input, editing and printing. Word processors have been written for large computer systems. 2. A clerk, or typist, who works with desktop computer word processing software. Part of the administrative staff, not IS.

WordPerfect Desktop system software. Word processor. Runs on Pentium type systems. Versions include: WordPerfect jr., WordPerfect Executive, WordPerfect, WordPerfect for Windows. Vendor: WordPerfect Corp.

WordPerfect Office See GroupWise.

WordPerfect Presentations Desktop software. Graphics package. Vendor: WordPerfect Corp.

Wordpower Word processor for IBM System/34. Vendor: Pansophic Systems, Inc.

Words, Graphs & Art Desktop system software. Graphics package. Interfaces with Lotus 1-2-3, Excel, dBase. Vendor: International Microcomputer Software, Inc.

WordScan Plus OCR software. Provides optical word recognition. Vendor: Caere Corp.

WordStar Desktop system software. Word processor. Runs on Pentium type systems. Versions include: WordStar, WordStar Professional, WordStar 2000, WordStar 2000 Plus, WordStar for Windows. Vendor: MicroPro International Corp.

work flow See workflow.

Work Management System Desktop system software. Project management package. Runs on Pentium type desktop computers. Vendor: Multitrak Software Development Corp.

Workbench 1. Application development tool. Program that allows data manipulation via commands. Works with Panvalet, Librarian. Vendor: Systech Software Products, Inc.
2. Application development tool. Allows users to customize SQLWindows applications and migrate the changes through new releases of SQLWindows. Based on PowerBuilder. Vendor: SQL Financials International, Inc.

Workbench Manager Application development tool. Allows users to tailor SQL Financials client/server financial software by changing software front-end. Vendor: SQL Financials International, Inc.

workbench software Application development software used in client/server computing. Allows developers to write code once and then compile for multiple platforms.

Workflo/ Workflow software. Includes imaging, document handling functions. Groupware. Includes imaging, document handling functions. Runs on Unix, Windows systems. Vendor: FileNet Corp.

WorkfloSystem Development Kit Application development tool. Integrates with FileNet's imaging and workflow systems. Allows users to create scripts to automate business processes. Vendor: FileNet Corp.

workflow Generic term describing how work is accomplished in a company or specific department or function. Analyzing work flow is part of re-engineering. Often used with imaging. The routines which manage the routing of images between people, programs, and processes. Also used as two words: work flow.

Workflow Analyzer Workflow software. Translates existing business processes into simulation models to analyze and test effects of change. Runs on Macintosh, Sun systems. Vendor: Meta Software Corp.

Workflow*BPR Workflow software. Vendor: Holosfx.

Workflow Connector Groupware. Workflow software. Versions available to interface with the Web or Microsoft's Exchange. Released: 1998. Vendor: Eastman Software, Inc.

Workflow Factory Workflow software. Used by non-technical people to develop workflow processes through predefined templates. Vendor: Delphi Consulting Group.

Workflow Manager See Actionworkflow.

Workflow Metro Workflow software. Allows Internet Web users to start a work process or track ongoing work from a Web browser. For example, companies can track the processing of support of sales information requests. Vendor: Action Technologies, Inc.

workflow software Software that incorporates data into forms or documents and then defines the path these documents should follow and monitors progress and changes by noting what revisions were made, when, and by whom. The software can automatically move data from databases into reports, spreadsheets, etc.

Workflow Template Application development environment. Object-oriented. Used to develop workflow systems. Integrates with legacy systems, imaging, and document management systems. Runs on top of SNAP. Vendor: Template Software, Inc.

WorkGroup Groupware. Client/server software which includes calendaring, scheduling, e-mail, fax, directory, and agent technology. Runs with an OS/2 server and OS/2, Windows clients. Vendor: IBM Corp.

workgroup software See groupware.

Workgroup Server Database system designed for workgroups. Version of Oracle7 that works with up to 50 users. Runs on OS/2, Windows NT systems. Vendor: Oracle Corp.

Workgroup Server 60,80,95 Midrange computer. PowerPC RISC CPU. Operating Systems: System 7, A/UX. Vendor: Apple Computer, Inc.

WorkMan Groupware. Workflow management software. Routes and tracks forms and data through an organization. Runs on NetWare, Vines systems. Vendor: Reach Software Corp.

WorkManager Workflow software. Object-oriented package. Vendor: Hewlett-Packard Co.

WorkPlace OS Operating system microkernel. Derived from Mach. Allows the PowerPC to run OS/2, AIX, and Windows applications. Will run programs written for Taligent. Vendor: IBM Corp.

WorkPlace Shell Operating system software. Object-oriented front end user interface to DOS, OS/2, Taligent, and AIX. Vendor: IBM Corp.

Works See Microsoft Works.

Workshop/204 Application development tool. Program that works with applications in Model 204 environment. Part of The Advantage Series. Runs on IBM mainframe systems. Vendor: Computer Corp. of America.

WorkShop NEO Application development tool. Creates object-oriented applications that can be accessed over the Internet. NEO is new name for DOE; see DOE.2. Vendor: Sun Microsystems, Inc.

@workStation Network computer. Vendor: Neoware Systems, Inc.

workstation A desktop computer, usually a RISC machine that offers high performance and better graphics functions than regular desktop computers. First used in scientific and engineering (CAD/CAM) applications, but growing more and more popular in business. Term is also used to refer to any terminal connected to a computer.

Workstation 10/xx Midrange computer. RISC CPU. Operating systems: Solaris. Vendor: Sanar Systems, Inc.

Workstation 400 Desktop computer. Pentium CPU. Operating system: Windows. Vendor: Dell Computer Corp.

Workstation LAN File Services See WLFS.

Workstation Manager CASE product. Part of Re-engineering Product Set. Vendor: Bachman Information Systems, Inc. (now Cayenne Software).

World UP Application development environment. GUI interface, used to develop virtual reality applications. Runs on Windows NT systems. Vendor: Sense8 Corp.

World Wide Web See Web.

WorldCom Network service that provides dial-up or Internet access to Lotus Notes or CC:Mail mailboxes. Allows subscribing companies to provide wide-area services without setting up a wide-area network. Vendor: Wolf Communications Co.

WorldDesk Commuter Midrange computer. Server. Pentium CPU. Vendor: Cubix Corp.

WorldMark Desktop computer. Pentium CPU. Vendor: NCR Corp.

Worldmark 5100M MPP computer. Vendor: NCR Corp.

WorldMart Application software. Allows users to analyze customer and sales information. Released: 1998. Vendor: Information Builders, Inc.

Worldtalk 400 family Communications software. Gateway for e-mail systems. Runs with Netware, 3+Open, Vines, LAN Manager networks. Vendor: Touch Communications, Inc.

Worm A program that propagates itself over a network, reproducing itself as it goes. A destructive program that replicates itself throughout disk and memory, using up the computer's resources and eventually putting the system down. A worm can affect many systems without any human action. Other destructive programs are called viruses, Trojan Horses, backdoors, and logic bombs.

WOSA Development architecture. Stands for: Windows Open Services Architecture. An umbrella concept from Microsoft that lets Windows provide the connections between applications and other services such as E-mail. Inlcudes specifications for ODBC, MAPI and TAPI.

WOSbase Database for desktop computer environments. Runs on IBM desktop systems. Multi-user. Vendor: WOS Data Systems, Inc.

WPS-PC Desktop system software. Word processor. Runs on Pentium type systems. Full name: WPS-PC Word Processing with List/Mail and Spell Checking. Also, WPS-DOS, WPS-Plus for DEC systems. Vendor: Exceptional Business Solutions, Inc.

wrapping Transitional step towards object-oriented systems. Uses object-oriented code such as C++, SmallTalk, to access data from traditional systems. "Wraps" traditional programs with object-oriented code.

Write See Microsoft Write.

WSAPI Application development tool. API in WebSite web server used to develop applications to run on the server. Stands for: WebSite API. Vendor: O'Reilly and Associates.

WSF2/ Operating system enhancement used by systems programmers. Increases system efficiency in IBM systems. Full name: WSF2/Extended Report Distribution. Vendor: RSD America, Inc.

WTI Data warehouse framework. Strategy and product selection intended to offer smooth integration between the warehouse and the operational systems. Stands for: Warehouse Technology Initiative. Vendor: Oracle Corp.

WTS Communications. Software that allows you to run Windows client applications on the server rather than the client systems. This allows changes to the applications to be made at the server rather than at each client. Interfaces with Picasso which works with non-Windows client software. Stands for: Windows Terminal Server. Used to be called TSE (Terminal Server Edition). Released: 1998. Vendor: Microsoft Corp.

WWW See Web.

WWW Connection Application development tool. Makes DB2 databases accessible through the Web. Includes macro language. Versions available for MVS, Unix. Also called W3 Connection. Vendor: IBM Corp.

Wylbur Application development environment. Runs on IBM online systems. Includes editing, file management, and remote job management. Runs on IBM systems. Vendor: ACS Commercial Services, Inc.

Wyse Computer vendor. Manufactures desktop computers.

WYSISYG Stands for: What You See Is What You Get. Software that allows user to work with the image that will appear on the screen when the program runs. Most commonly used with work processors. Also used with Web authoring tools.

Wysiword Desktop system software. Word processor. Runs on Windows on IBM and compatible desktop computers. Vendor: Microsystems Engineering Corp.

X.21(bis) Communications protocol. Physical protocol used with LAPB. CCITT protocol, used with X.25.

X.224, X.225, X.226 Communications protocols. Transport Layer. Part of X.25.

X.25 Communications protocol. Network layer. Developed by CCITT. Defines standards for packet-switching networks.

X.400 Communications protocol. Handles e-mail. CCITT protocol, used with X.25.

X.400 Gateway Communications software. Gateway for e-mail systems. Runs on Prime systems. Vendor: Prime Computer, Inc.

X.435 Communications protocol. Handles EDI. Works with X.400 e-mail protocol. Accepted international standard.

X.500 Communications protocol. Sets standards for e-mail directories. Used with X.25, X.400.

X.75 Communications protocol. Network protocol that is an extension of X.25 and is used in public networks. All X.75 networks conform to X.25 standards.

X-ample Applications development tool. Program generator used in DEC VAX systems. Contains DBMS, artificial intelligence query tool, report generator, screen and forms generator, and data dictionary. Vendor: Landmark Software Systems, Inc.

X-C 6000 Cross Country Portable computer. Pentium CPU. Vendor: Intronix.

X-C 6xxx Notebook computer. Pentium CPU. Operating system: Windows 95/98. Vendor: ProLinear Corp.

X-Change Communications software. EDI package. Runs on Pentium type desktop computers.

X.desktop Graphics user interface for Unix systems. Specifically designed for workstation use and adheres to Posix, OSF, and X/Open standards. Used with OSF/1 and DEC systems. Vendor: IXI Ltd.

X-Designer Application development tool. Lets users create interfaces for Unix and Windows applications from a single design. Runs on Unix, VMS systems. Vendor: Imperial Software Technology, Inc.

X-IPC Communications software. Message oriented middleware. Vendor: Level 8 Systems.

of data should be viewed, i.e. all brand names should be blue and large type. Stands for: Extensible Style Language.

X/Motif GUI used with X terminals.

X/Open Consortium of international computer vendors that certifies industry-standard Unix operating systems. X/Open's Spec 1170 is the accepted standard. Vendors must prove compliance with the standard and pay licensing fees to use the name Unix.

X/PTR Operating system enhancement used by systems programmers. Increases system efficiency in IBM systems. Handles print distribution. Vendor: Systemware, Inc.

X terminal Terminal that can execute GUI code to provide graphical front end to Unix and legacy applications.

X-Terminal A terminal designed to work with windowing applications. A terminal that can access many applications from host, or server, computers through windows. Designed to be a cheaper alternative to using desktop computers or workstations.

X User Interface See XUI Toolkit.

X-Windows Public domain software that allows development of programs and systems to run under a windows environment. Also allows desktop computers and workstation terminals to access applications on multiple hosts and display in windows on the screen. Used with multi-tasking systems such as Unix, MVS, VMS, etc. Includes GUI (graphic user interface). Developed by Massachusetts Institute of Technology.

X12 Set of EDI (Electronic Data Interchange) standards approved by ANSI.

X12/DISA Association created to set standards for EDI. Stands for: X12-Data Interchange Standards Association.

XA Stands for Extended Architecture. See MVS.

XA/R, R-S Model xx Midrange computer. RISC machine. i860 CPU. Operating systems: FTX, VOS. Vendor: Stratus Computer, Inc.

XA-RELO Operating system enhancement used by systems programmers. Increases system efficiency in IBM CICS systems. Vendor: Quantum International Corp.

XA-VSAM Operating system enhancement used by systems programmers. Increases system efficiency in IBM systems. Reduces VSAM file I/O time. Vendor: Quantum International Corp.

XA2000 Model 2xxx Mainframe computer. Supports up to 12 processors. Operating systems: VOS, FTX. Vendor: Stratus Computer, Inc.

XA2000 Model xxx Desktop computer. 680X0 CPU. Operating systems: DOS/Windows, Windows 95/98. Vendor: Stratus Computer, Inc.

XAMAP Operating system enhancement. Performance monitor for IBM VM systems. Vendor: Velocity Software, Inc.

Xamon Operating system enhancement used by systems programmers. Monitors and controls system performance in IBM VM systems. Vendor: Velocity Software, Inc.

Xample Relational database for large computer environments. Runs on DEC systems. Vendor: Landmark Software Systems, Inc.

XASSD Operating system enhancement used by Operations staff and systems programmers. Controls disk usage. Runs on IBM VM/XA systems. Vendor: Velocity Software, Inc.

xBase A database language standard created by a consortium of vendors. Products covered include dBase, FoxPro, Clipper, Paradox.

XBOL Application development environment. Includes program generator, report generator, data dictionary, screen painter. Interfaces with any database. Runs on IBM mainframe systems. Vendor: Excel Consulting & Programming, Inc.

Xceed See OEA.

XCOM/SDS Operating System software. Automatically distributes applications to different platforms including mainframe, midrange and desktop systems. Vendor: Legent Corp.

XD/Java, JavaDesigner Application development tool. Converts C++ systems to Java for use on the Internet. Runs on Windows. Vendor: Imperial Software Technology, Inc.

XDB Operating system add-on. Debugging/testing software that runs on Apollo, DEC VAX, IBM , Sun, Unix systems. Full name: XDB, Source Level Debugger. Vendor: Intermetrics, Inc.

XDB-DB2 Workbench Application development tool. Program that allows desktop computers to be used for DB2 program development. Interfaces with XDB and DB/2. Vendor: XDB Systems, Inc.

XDB-QMT 4GL used in desktop computer environments. Allows development of DB2 QMF reports from a desktop computer. Vendor: XDB Systems, Inc.

XDB-SQL Relational database for a desktop computer environment. Runs on Pentium type desktop computers. Utilizes SQL. Can be used in client/server computing. Vendor: XDB Systems, Inc.

XDE Application development system which includes a program editor, pseudo-code compiler, code manager, XDL (a 4GL), and a predefined set of object classes. Vendor: Micro Design International, Inc.

Xdebug Operating system add-on. Debugging/testing software that runs on IBM systems. Vendor: Kolinar Corp.

Xdesigner Application development tool. Used to build Motif GUIs through point-and-click screens. Runs on Unix systems. Vendor: DataViews Corp.

XDL 4GL. Includes object-oriented extensions. Part of XDE. Stands for Express Development Language. Vendor: Micro Design International, Inc.

xDSL Reference to all Digital Subscriber Lines, which includes ADSL, HDSL, SDSL, VDSL. See DSL.

Xecutive Office System Operating system enhancement used by Operations staff and systems programmers. Provides operator console support in DEC VMS systems. Vendor: Russell Information Sciences, Inc.

XEDIT Application development tool. Screen editor for IBM VM/CMS systems. Vendor: IBM Corp.

XELOS Operating system for Concurrent systems. Vendor: Concurrent Computer Corp.

Xenix General-purpose operating system for Pentium-type desktop computers, DEC PDP, Unix systems. Adaptation of Unix. Vendor: Microsoft Corp.

Xentis Application development tool. Report generator for DEC systems. Includes security features, sort capabilities, and totaling/subtotaling. Vendor: Gray Matter Software Corp.

Xentis/ Operating system add-on. Data management software that runs on DEC VAX systems. Includes Xentis/Dictionary, Xentis/Edit, Xentis/Report. Vendor: Park Software, Inc.

Xeon Computer chip, or microprocessor. 400 and 450 MHz Pentium machines for use in mid-range enterprise servers and workstations. Xeon replaces the Pentium Pro as Intel's main enterprise microchip. Xeon is the high end of the Pentium line (Celeron is the low end). Vendor: Intel Corp.

Xerox Parc Research center. Part of Xerox Corp. Developed original GUI, WYSIWYG editing, laser printing, object-oriented languages, notebook computers and enterprisewide e-mail. Latest development is the hyperbolic tree, a new interface for Internet access. InXight Software markets Parc's products.

XFaceMaker Application development tool. GUI builder that lets developers design X-window systems and C++ class libraries. Includes debugger. Vendor: Non-Standard Logics.

Xfer Operating system enhancement. Provides systems management functions for enterprise-wide systems. Manager distribution, installation and upgrading of software. Vendor: PLATINUM Technologies, Inc.

XGA CG-Note Desktop computer. Notebook. Pentium processor. Vendor: Comtrade Computer, Inc.

XGEN Application development tool. Program generator for Unisys systems. Vendor: Software Clearing House, Inc.

XJCF Operating system enhancement used by systems programmers. Increases system efficiency in IBM systems. Integrates printing with Xerox print systems. Stands for: Xerox Job Control Facility. Vendor: Xenos Computer Systems, Inc.

XL JCL generator. Runs on DEC VAX systems. Reduces DCL overhead by as much as 50%. Vendor: Boston Systems Office.

XL/Recover Application development tool. Analyzes existing COBOL programs and integrates them into an applications development environment. A reverse-engineering tool. Vendor: Index Technology Corp.

XL/SuperCASE CASE product. Automates analysis and design functions. Runs on DEC VAX systems. Interfaces with Excelerator. Vendor: Advanced Technology International, Inc.

XL/Windows CASE product. Windows-based version of Excelerator. See APS, PVCS, Excelerator, Design Recovery.

Xlib Library of routines used with DECwindows. Vendor: Digital Equipment Corp.

XLIB-86 Operating system add-on. Library management system that runs on DEC VAX, Unix systems. Full name: XLIB-86 Cross Librarian. Vendor: Systems & Software, Inc.

XLT12 Communications software. EDI package. Runs on IBM systems. Vendor: American Business Computer.

Xmath Object-oriented mathematical analysis and scripting language. Runs on Windows. Vendor: Integrated Systems, Inc.

XMENU Application development tool. Screen editor for IBM systems. Vendor: VM Systems Group, Inc.

XMetaL Application development tool. XML editor intended to be used by writers, not programmers. Includes word processing functions. Vendor: SoftQuad.

XML Programming language. An extension to HTML that gives more control of content. Used to define the content of a document (Web page) rather than the presentation of it. Used to exchange information and documents between diverse systems. XML is text-based and can be used on any platform. Formats data by using document tags to catalog information. Key elements in a document can be categorized according to meaning. Instead of a search engine selecting a document by the metatags listed in its header, a search engine can scan through the entire document for the XML tags that identify individual pieces of text and images. Accepted as a standard by W3C. Used in e-commerce to enable business partners to standardize specific XML syntax to describe i.e. purchase orders. Also used to define metadata, and in EDI systems, WBEM. Stands for: Extensible Markup Language.

XML parser, XML processor See parser.

XML Pro Application development tool. XML editor that easily integrates with any XML toolkit. Vendor: Vervet Logic.

XML Suite Application development tool. Used to develop and deploy e-commerce, EDI, application integration, and supply chain management applications. Includes XML-server (dynamic XML server), Visual-XML (developer's toolkit which generates Java and XML from drag-and-drop programming techniques), XwingML (development tool that merges XML and Java to allow users to generate Java Swing classes to create GUIs), and XML-Contact (Uses XML to connect handheld systems to enterprise systems). Released: 1999. Vendor: Bluestone Software, Inc.

XModem Communications protcols. Applications layer. Used to transfer files from Unix to other systems. Provides same functions as YModem, ZModem and is the most common of the three.

XNS Communications protocols. Set of protocols initiated as an open architecture. Stands for: Xerox Network Systems. Used in NetWare, Vines, 3+Open.

XoftWare Communications software. Used in client/server computing. Links PC clients to Unix servers. Vendor: AGE Logic.

XPAC Operating system enhancement used by systems programmers. Monitors and controls system performance in IBM VM systems. Provides early detection of system bottlenecks. Vendor: Macro 4 Systems Software.

XPACK Application development environment. Runs on DEC systems. Includes editor, file processor, and library routines. Vendor: MBA-Systems Automation, Inc.

XPE Cross-Platform Environment. Strategy from Legent to manage diverse computing environments. Will consist of a suite of software products from Legent and other vendors. Software will include middleware products, development tool kits, application program interfaces, and systems management products. Will work with MVS, Unix, OS/2.

XPEDITER(+) Operating system add-on. Debugging/testing software that runs on IBM systems. Supports TSO, IMS/DC, BTS, CICS, ROSCOE, COBOL. Vendor: Compuware Corp.

Xpense Management Solution Application software. Automation of travel and expense reporting. Employees report expenses in electronic forms which are routed to the payroll system. Reimbursement appears in the next paycheck. Vendor: Portable Software Corp.

XperCASE CASE product. Automates design, programming functions. Used in re-engineering. Provides graphical representation of code. Runs on Windows. Vendor: Siemens AG Austria.

Xpert Rule Analyser Profiler Application development tool used for data mining. Used to determine which of hundreds of relationships among variables are significant. Runs on Windows systems. Vendor: Attar Software Ltd.

XPERT Series Operating system add-on. Data management software that runs on IBM systems. Includes: DATA-XPERT, IMS-XPERT, DB2-XPERT. Vendor: Compuware Corp.

XpertGen, XpertRule DSS. Rules based, knowledge management systems. XpertGen is code generator. Developers can create decision trees in graphical form. Includes truth tables. Runs on Windows 95/98 systems. Vendor: Attar Software, Ltd.

XPF/Assembler, XPF/COBOL Operating system add-on. Debugging/testing software that runs on IBM systems. Vendor: Pansophic Systems, Inc.

XPG X/Open Portability Guide. Standards for developing Unix operating systems. Developed by the X/Open Consortium.

XPG3 Operating System for ICL Series 39 mainframe computers.

XPL System programming language for desktop computers.

XPL0 Compiler language used in desktop computer environments.

Xport Application development tool. Allows programs on desktop computers to access data from DEC and DG systems. Vendor: R.B. Zack & Associates, Inc.

XPS Relational database. Runs on parallel processors. Stands for: Extended Parallel Server. Vendor: Informix Software, Inc.

XPU4 Integration software. Allows a Unix system to act as an IBM host on an SNA network so IBM terminals can access applications on a Unix system. Vendor: Integris.

XQL 4GL used in desktop computer environments. Runs on Pentium type desktop computers. Works with Xtrieve. Vendor: Novell, Inc.

XR-VSAM Operating system enhancement used by systems programmers. Increases system efficiency in IBM systems. Reduces VSAM file I/O time. Vendor: Quantum International Corp.

XRAY Application development tool. Programming utility that allows source level debugging for programs written in C. Runs on CICS. Vendor: Prisym, Inc.

Xrefplus Operating system enhancement used by systems programmers. Includes a job scheduler exit point, data set reporting and ISPF interface. Runs on IBM MVS systems. Vendor: Jensen Research Corp.

XRunner Application development tool. Automates program testing for X-Windows based applications. Used to test client/server applications. Runs on HP Unix workstations. Vendor: Mercury Interactive Corp.

XShell Distributed ORB Application development tool. Used to develop systems by treating programs as objects. Objects can exist across computer networks. Runs on AT&T systems. Vendor: Expersoft Corp.

Xsight Software that allows the development of window applications. Runs on Unix, MS-DOS. Based on X-windows.

XSL Application development tool. Displays XML data. Allows users to design a template defining how each type

Xsoft Document retrieval system. Runs on Windows. Vendor: Zsoft.

XSYS/EXSYS Artificial intelligence system. Runs on IBM, DEC, Unix systems. Builds expert systems. Vendor: California Intelligence.

XTAL Real-time operating system for desktop computers. Vendor: Axe, Inc.

Xtrieve(/N) 4GL used in desktop computer environments. Runs on Pentium type desktop computers. Works with XQL. Vendor: Novell, Inc.

XTS/SNA Backbone Communications software. Network connecting DG and IBM computers. Vendor: Data General Corp.

XUI Toolkit Routines that work with DECwindows to provide standard programming. Stands for: X User Interface Toolkit. Vendor: Digital Equipment Corp.

XVision See SCO Xvision.

XvM Application development tool. Creates Motif version of OpenLook applications. Vendor: Qualix Group, Inc.

XVT Application development tool. API that builds GUIs. Used for cross-platform development. Includes windowing functions. Can develop applications that will run unchanged under Open Look, Motif, Presentation Manager, and the Macintosh. Stands for: Extensible Virtual Toolkit. Versions available for Ada, C, C++. Vendor: XVT Software.

XwingML See XML Suite.

XWORD Word processor for DEC systems. Vendor: MBA-Systems Automation, Inc.

XyWrite Desktop system software. Word processor. Runs on Pentium type systems. Vendor: XyQuest, Inc.

Y-MPxx Supercomputer. Operating systems: Unicos. Vendor: Cray Research, Inc.

Y2K, Y2K compliant Reference to the change of century. Y2K compliant systems will handle the year 2000 and in date comparisons the year 1999 will compare as less than (or sooner) than the year 2000. This is a problem because older systems maintained only two character dates, so 99 would compare as greater than (or later) than 00.

Yahoo Internet access. Free Web guide and search software. Carries advertising from some vendors offering online shopping; accesses Reuters NewsMedia, Inc. Vendor: Yahoo Corp.

Yellow Box Application development technology. Part of Rhapsody that supports Java applications and OpenStep APIs. Applications written to Yellow Box APIs can be recompiled to run on other platforms. Vendor: Apple Computer, Inc.

YESBOOK xxxx Desktop computer. Notebook. Pentium processor. Vendor: Maxtech Corp.

YModem Communications protocol. Applications layer. Transfers files from one system to another. Provides same functions as XModem, ZModem.

Yourdon Structured programming design methodology named for its developer. Based on functional decomposition. Accepted as a standard design methodology by some companies, and used by some CASE products.

Z-3D-PowerPro Desktop computer. Pentium II CPU. Operating systems: Windows 95/98, NT. Vendor: Zenon Technology, Inc.

Z-80, Z-8000, Z-80000 Assembler language for old desktop systems.

Z-Business NT Server Midsize computer. Server. Pentium II processor. Operating system: Windows NT. Vendor: Zenon Computer, Inc.

Z-Four Relational database/4GL for midrange and desktop computer environments. Runs on Unix, Wang systems. Includes development tools. Vendor: Business Computer Solutions, Inc.

Z-Millenium Desktop computer. Notebook. Pentium processor. Vendor: Zenon Computer, Inc.

Z-OptimaPro Midrange computer. Pentium II CPU. Operating systems: Windows 95/98, NT. Vendor: Zenon Technology, Inc.

Z-Platinum Desktop computer. Pentium processor. Vendor: Zenon Computer, Inc.

Z-Power Note (3000) Notebook computer. Pentium CPU. Vendor: Zenon Technology, Inc.

Z-Server Midrange computer. Pentium CPU. Operating systems: Windows 95/98. Vendor: Zenith Data Systems Direct.

Z-Station Desktop computer. Pentium CPU. Operating systems: Windows 95/98. Vendor: Zenith Data Systems Direct.

Z-Zenon Desktop computer. Pentium processor. Vendor: Zenon Computer, Inc.

Z-Star Notebook computer. Pentium CPU. Operating systems: Windows 95/98. Vendor: Zenith Data Systems Direct.

Zack (PC) Operating system enhancement used by Operations staff and systems programmers. Provides operator console support in IBM systems. Automated operator commands. Full name: ZackCThe Operator's Operator. Vendor: Altai Software.

zApp Application development tools. C++ generator. Developers Suite used for cross-platform development and can generate versions for multiple platforms including Windows NT, OS/2, Unix. Runs on DOS/Windows, OS/2, Unix. Includes: Developers Suite, Application Framework, Factory. Vendor: Rogue Wave Software.

ZDOS Operating system for IBM and compatible desktop computer systems. Vendor: Zenith Data Systems Direct.

Zeke Operating system enhancement used by systems programmers. Job scheduler for IBM systems. Automates online job scheduling. Full name: ZekeCThe Scheduler That Works. Vendor: Altai Software.

Zeon Computer chip, or processor. Based on Pentium architecture, is an interim chip release intended to provide increased speeds before the release of the 64-bit Merced chip due in 1999. Released: 1998. Vendor: Intel Corp.

Zephyr Massively parallel processor. Operating systems: Unix. Vendor: Wavetracer, Inc.

Zero Administration Kit Operating system software. Allows network administrators to configure desktops from a central station. It keeps users from changing system configuration. Also called ZAK. Runs on Windows NT Workstation systems. Vendor: Microsoft Corp.

Zetalisp See LISP.

ZIM Application development system. Includes relational database/4GL for midrange and desktop computer environments. Runs on DEC, IBM mainframes and desktop computers, Unix, and Xenix systems. Utilizes SQL. Can be used in client/server computing. Interfaces with SQL Server, Oracle. Includes report generator. Vendor: Sterling Software (Zanthe Systems Division).

Zinc Application Framework Application development tool. Used for cross-platform development. Allows developers to create one C++ program that can be compiled to run on Windows, Windows NT, OS/2, Macintosh, Unix systems. Contains C++ object-oriented library. Vendor: Zinc Software, Inc.

ZIP Communications protocol. Session Layer. Part of AppleTalk. Stands for: Zone Information Protocol.

ZipLock Internet software used in e-commerce. Used to deliver digital goods over the Internet. Three parts: Vbox (packages, provides security and marketing options for digital products); ZipLock Server (provides real-time management of inventory, licenses, and delivery and also tracks sales and customer data); and ZipLock Gateway (handles deliverys for multiple resellers on different platforms). Released: 1999. Vendor: Preview Systems, Inc.

ZMagic Application development tool. Includes application, report generators. Runs on midrange, desktop computer systems. Vendor: Zortec, Inc.

ZModem Communications protocol. Applications layer. Used to transfer files from one system to another. More efficient version of XModem and provides same functions as YModem.

Zoomer Personal digital assistant, or handheld computer. Discontinued. Vendor: Tandy.

ZooWorks Application development tool. Creates personal indexes from Web usage so users can look up sites of past searches by keyword and subject. Released: 1996. Vendor: Hitachi Software.

Zope. Application server providing development tools and middleware. Open source toolkit used to build dynamic Web sites. Based on Python and supports CORBA, COM, XML and most major databases. Also supports WebDAV protocol. Vendor: Digital Creations.

ZR 5700, 5800 Handheld, pen-based computer. Operating system: Synergy. Vendor: Sharp Electronics Corp.

ZR-3xxx Handheld computer. Operating system: Synergy. Vendor: Sharp Electronics Corp.

zShare Application development tool. Allows users to post and view corporate information over corporate intranets. Handles text, audio and video files. Uses XML. Released: 1998. Vendor: zBridge Software, Inc.

ZyImage Image processing software. Runs on Windows. Vendor: ZyLab Corp.

1,2...

+1Environment Application development environment. Used to develop object-oriented applications in many languages. Supports team development. Includes +1Base (Supports multiple projects), +1CM (configuration management functions), +1CR (report management), +1DataTree (Data storage used by +1Reports), +1Reports (report generator), +1Reuse (supports reuse of design, documentation, code, and test files), +1Test (unit, integration and regression testing functions). Runs on Unix systems. Vendor: +1Software Engineering.

1-2-3 See Lotus 1-2-3.

100BASE-T See Fast Ethernet.

100VG-Anylan In communications, network technology that transfers data at 100Mbps per second. Data is transferred over voice grade lines, and can be used with both ethernet and token ring LANS.

10Net Communications software. Network operating system. Runs on Pentium type PCs. Works with Ethernet, ARCnet, NetBIOS. Vendor: Tiara Computer Systems, Inc.

1394 See FireWire.

1st-Class Artificial intelligence system. Expert system building tool. Runs on desktop systems. Interfaces with Lotus 1-2-3. Vendor: 1st-Class Expert Systems, Inc.

1stFile Data management system. Runs on Macintosh systems. Vendor: 1st Desk Systems.

1stKey Relational database for Desktop system environments. Runs on Macintosh systems. Vendor: 1st Deck Systems.

1stTeam Data management system. Runs on Macintosh systems. Provides multi-user database. Vendor: 1st Desk Systems.

1View:Workflow Application development tool. Suite of client/server object-oriented modules providing icons and menu scripts for drag-and-drop program construction for routing/process applications. Runs on Unix systems. Released: 1994. Vendor: Network Imaging Corp.

20/20 Desktop system software. Integrated package. Runs on Unix systems. Includes: 20/20 Database Connection, 20/20 WideWriter. Vendor: Access Technology, Inc. (MA).

200LX palmtop PC Handheld computer. Operating system: MS-DOS. Vendor: Hewlett-Packard Co.

2200/xxx Mainframe/midrange computer. Operating system: OS 2200. Vendor: Unisys Corp.

220CSD Notebook computer. Vendor: Toshiba America Information Systems, Inc.

286 Descriptive term used for Desktop systems, as in "a 286 machine." Refers to machines built with a 16-bit chip and speeds from 5 to 20 times faster than the original PCs.

2Share EIP, corporate portal. Manages both Intranet and extranet systems. Browser-based. Integrates information from desktop files, databases and legacy systems onto a single web page. Allows users to create and publish documents based on corporate information. Runs on Solaris, Windows systems. Released: 1998. Vendor: 2Bridge Software.

3-D Charts to Go Desktop system software. Graphics package. Interfaces with Lotus 1-2-3, Excel. Runs on Windows systems. Vendor: Bloc Publishing Co.

3+Open Communications. LAN (Local Area Network) connecting Apple, Unix systems. Vendor: 3Com Corp.

3+Open LAN Manager Communications software. Network operating system connecting IBM and compatible PCs. 3Com's version of LAN Manager. Vendor: 3Com Corp.

3+Share Communications software. Network operating system which links MS-DOS, Macintosh systems. No longer actively marketed. Vendor: 3Com Corp.

300LX, 320LX palmtop PC Handheld computer. Operating system: WinCE. Vendor: Hewlett-Packard Co.

3090 Mainframe computer. Operating systems: MVS, VM, VSE. Models numbered over 180 can be considered supercomputers. Vendor: IBM Corp.

3270 Optimizer/CICS,IMS,VM Operating system enhancement used by systems programmers. Increases system efficiency in IBM systems. Vendor: BMC Software, Inc.

3270-PC Communications software. Network connecting IBM computers. Micro-to-mainframe link. Vendor: IBM Corp.

3270 Superoptimizer/CICS Operating system enhancement used by systems programmers. Increases system efficiency in IBM CICS systems. Vendor: BMC Software, Inc.

34-Front System development methodology used by Deloitte & Touche. Covers all phases of the development life cycle.

360,370 See S/360,S/370.

386 Descriptive term used for PCs, as in "a 386 machine." Refers to machines built with a 32-bit chip and speeds from 20 to 35 times faster than the original PCs.

386/ix Operating system for IBM and compatible PCs. Unix type system. Vendor: Interactive Systems Corp.

386/Multiware Communications software. Network operating system. Runs on multiple desktop systems. Vendor: Alloy Computer Products, Inc.

3DA Operating system. Merging of Unix systems from Hewlett-Packard and Santa Cruz Operations that will run on IA-64 bit systems. Stands for: Three Dimensional Architecture.

3dMMX Desktop computer. Pentium CPU. Operating system: Windows 95/98. Vendor: Royal Electronics, Inc.

43xx Mainframe computer. Operating systems: VM, MVCS, VSE. Vendor: IBM Corp.

486 Descriptive term used for PCs, as in "a 486 machine." Refers to machines built with a 32-bit chip and speeds two to three times as fast as a 386 PC.

486DX2 Computer chip, or microprocessor. Vendor: Texas Instruments, Inc.

4C Classic,Brief Application development tool. Automates programming function. Analyzes source code for such things as adherence to standards and presence of logic flaws (unexecuted code). Languages analyzed: C, Pascal, Modula-2. Runs on Pentium type PCs. Vendor: Tri-Technology Systems, Inc.

4D Desktop Application development tool. Used to develop database applications for the 4D database. Runs on Macintosh, Windows NT systems. Vendor: ACI US, Inc.

4D Passport Application development tool. Includes GUI development tools, report generators, and connects 4th Dimension Macintosh environments to SQL databases. Vendor: Acius, Inc.

4D SDK Application development environment. used to develop client/server applications. Runs on Macintosh, Windows systems. Released: 1995. Vendor: ACI US, Inc.

4GL A free-form, non-procedural LANguage with few rules and little predefined vocabulary. Designed to be user-friendly and heavily used to query databases and/or generate reports.

4th Dimension Relational database for Desktop system environments. Runs on Macintosh Windows systems. Vendor: Acius.

4Thought Application development tool. DSS (Decision Support System) used to develop models for business analysis, forecasting and measurement. Runs on Windows systems. Released: 1998. Vendor: Cognos Inc.

5d-xxxx Notebook computer. Pentium CPU. Vendor: 5D Technology.

6200 Mint Pentium Notebook computer. Pentium CPU. Vendor: Micro-International Corp.

6400 Superserver Supercomputer. Supports up to 64 processors. Operating system: Solaris. Vendor: Cray Research, Inc.

6x86 (MX), Media GX Computer chips, or microprocessors. Vendor: Cyrix Corp.

7051 POWER Desktop computer. RISC machine. IBM Power CPU. Operating system: AIX. Vendor: IBM Corp.

8086, 8088 language Assembler language for PCs.

9076 SP1 Scalable Mainframe computer. Operating systems: AIX/6000. Vendor: IBM Corp.

91,93 Series Midsize computer. Operating systems: Umax, Unix. Some models can be configured as the server in a client/server system. Clients can run DOS, Unix, Macintosh. Vendor: Encore Computer Corp.

9230A-xxx, 9250A Midrange computer. CMOS processor. Operating systems: MAXIV, MAX32, CLASSIC REAL/IX. Vendor: Modular Computer Systems, Inc. (MODCOMP).

9260A-xx Midrange computer. CMOS processor. Operating systems: MAX32, CLASSIC REAL/IX. Vendor: Modular Computer Systems, Inc. (MODCOMP).

937x-xx Midrange computer. Operating systems: VM, VSE, AIX, DPPX, MUMPS, MVS/370. Vendor: IBM Corp.

Part Three

Summary Tables

The following tables summarize computer system products by type. While most commonly used software is listed, in no way do these tables provide a complete listing of all the products available. They are a summarization of the products listed in this book. Specific information about each product can be found in the Glossary.

Computers By Vendor

Vendor	Large Mainframes, Supercomputers, Parallel Processorss	Midrange Midsize, Workstations, RISC, Servers	Desktop Microcomputers, Pcs, Notebooks
ABS Computer		Business Server SCSI Powerhouse	ABS Notebook Business Station,Server, Workstation BX,LX Supreme Multimedia System Super 3D Supreme DVD
Accton Technology			LanStation Pro
ACD Computers		APEX Mini-Server, RAID	
ACE Computers			ATX Powerstation Dreamedia Station Dual NT Enterprise Server Home Office Pro Magnatronic Office Workstation
Acer America		AcerAltos Altos System xxx	AcerEntra AcerNote AcerPower Aspire Extensa Optima MT, SL, DT TravelMate
ACMA			ACMA 7200 sPower, zPower
Advanced Logic Research			Evolution Optima DT, MT, SL Revolution
Advanced Modular		Modular Scalable Server Array	
Adv. Processing Labs		VMEstation	
Akia			Alcam XL Fusion Onebook Mystique Series Tornado
Altura			BOLT xx
Amdahl	Millennium Model 1100E,1200E,1400E Model 5990 Model 5995 Model 7300	EnVista	
American Multisystems		InfoGOLD	
Amrel Technology			Amrel Maverick Amrel Rocky 2000 Amrel Symphony
AMS Tech			AMS TravelPro Roadster Rodeo
Apple		Macintosh Network Server Performa (xxx) Power Macintosh Workgroup Server 60,80,95	iMac Macintosh Macintosh PowerBook Workgroup Server

Vendor	Large Mainframes, Supercomputers, Parallel Processorss	Midrange Midsize, Workstations, RISC, Servers	Desktop Microcomputers, Pcs, Notebooks
APS Technologies			M-Power
Aries Research		Marixx US xx, SS	
ARM Computer			ARMNote
Aspen		Alpine Avalanche Durango(II) Radiant Summit Telluride Timberline Twin Peaks	Columbine Glacier (II) Pike's Peak
AST Research		Centralan Manhattan family Premium xx	Advantage!xxx Ascentia Bravo LC,MS
Astro Research			Power Media III
Auspex		NS 7000 NS NetServer	
Austin Computer			C.A.P. System Edge xx Entrada 1000 Power System PowerPLUS
Axil		Axil Ultima family AxilNet! AxilServer Axilxxx NorthbridgeNX801 S/420 UPX1000	
BOS (Better On-line Solutions)			BOSaNOVA
Bottomline			PayBase 32
Brick Computer			BigScreen3 CD PowerBrick Ergo PowerBricks MobyBrick2 NoteBrick3 ThinBrick
Brother International			GeoBook NB-xx
BTG		BTG AXP275	
Bull HN		DPS/7000 DPS/9000/xxx DPX/20 Escala PowerCluster Estrella 300-xxx Sagister	
Carrera		Cobra(200,275), EV5 Hercules 200	
Census Computer			Elan Traveler Essentia Pxxx
Chaplet Systems			FLUFA 770 iLUFA 770

Vendor	Large Mainframes, Supercomputers, Parallel Processorss	Midrange Midsize, Workstations, RISC, Servers	Desktop Microcomputers, Pcs, Notebooks
Chatcom		ChatterBox	
CHEM USA			ChemBook xxxx
Colorbus		Cyclone II Cyclone Office	
Commax			Smartbook
Commercial Data Services	CDS-1 CDS-2000		
Commodore			Amiga 3000,4000
COMPAQ		ProLiant ProSignia Server	Armada 1550DMT Deskpro DPWa, DPWau LTE Elite, 5000 Presario xxxx Presario 10x0 Notebook ProLinea ProSignia System Pro/XL
Compu-Tek		Compu-Tek family	Deluxe Multimedia PCI Lan Workstation Pentium Server Windows Power
Computer Tech. Link		CTL	

Comtrade Electronics			Artist Dream Machine
			BusinessMachine
			Mega CD-Note
			Multimedia 3D Screamer
			Multimedia Notebook
			Multimedia (VL-Bus),
			    BestBuy, DreamMachine
			MultimediaWide SCSI
			    Powerstation
			Stereo 800x600
			SVGA CD-Note
			VESA Local Bus Best Buy,
			    Professional, WinStation,
			    WinXpress
			XGA CG-Note

Concurrent		MAXION NightHawk PowerMAXION	PowerHawk
Control Data Corp	Cyber xxx, CyberPlus	4680 InfoServer Cyber 920C	
Convex	C4/XA Model C46xx C3 Series Exemplar SPP1000, SPP1200		
Corollary		CBII/6000P	
Cray	6400 Superserver C9xx,EL9xx,J9xx Origon2000(with Silicon Graphics) J90 S-MP Superserver T3D, T3E T90 Y-MPxx		

Vendor	Large Mainframes, Supercomputers, Parallel Processorss	Midrange Midsize, Workstations, RISC, Servers	Desktop Microcomputers, Pcs, Notebooks
CSS Laboratories		MaxPro	
CTX Int'l			EzBook SIP400
Cubix		WorldDesk Commuter	BC Server CubixConnect DP 6200
CyberMax			BusinessMAX FamilyMAX PowerMAX ProMAX ValueMAX
Data General	Eclipse MV/xxx	AV20000 AviiON AV xxx Dasher II Eclipse MV/xxx SiteStak	AV2650(R) DG/Vision P-75CX Walkabout
Daystar Digital		Genesis LT, MP	
Dell			Dimension Inspiration Inspiron Latitude OmniPlex OptiPlex PowerEdge Precision Workstation NetPlex 4xxx Workstation 400
DFI			DFI 66xx Landmarq IPX
Directware		MVP-NT	Business 1,2 Wavestation MVP D,V,VF,VP,VX WaveStar, WaveStation
DTK Computer		Station Classic+	APRI-0031, 0032, 74 DBN54xxA DTK Quin DTK Quin-38 DTN 5xxxx Grafika 4xxx
Duracom Computer			Proforma
Encore	Infinity 90 Series Infinity R/T Series	91, 93 Series Encore RSX	
EPS Technologies			EPS Apex Evolution Green Power Workstation VESA Multimedia, Value Line
Ergo Computing			ThunderBrick
ET Technology			Atlas AMD Galilelo Pentium Nebula AMD Neptune Pentium Orion AMD Pathfinder Pentium Polaris Pentium SuperNova Titan II

Vendor	Large Mainframes, Supercomputers, Parallel Processorss	Midrange Midsize, Workstations, RISC, Servers	Desktop Microcomputers, Pcs, Notebooks
Everex Systems		Step DP/Pro Raid StepServer 2, MDP/Pro	eXplora Step/DP/Pro Step DPe/3000 Step Premier StepClient StepNote StepPremier StepStation Tempo K Series
Filetek		SM-EDA	
First Computer			FCS AM, DS FCS Pentium
Flavors	Parallel Inference		
Fujitsu	VP 2000 VPP500		LifeBook Milan P-1xx Monte Carlo Montegro
Futuretech			FutureMate
Gateway 2000		ALR xxxx NS7000,8000	Destination E1000 200 G-Series Gateway family Gateway Professional P5-xxx Solo S90
General Automation		mv.ESx00i PowerAdvantage	
Hertz Computer		Hertz NT WebServer	Hertz PCI Dual P-xxx Hertz Z-Pentium
Hewlett-Packard	HP 9000 EPS21, EPS30 HP Exemplar X-Class	HP 3000,9000 HP Domain Enterprise HP Exemplar S-Class HP NetServer	HP Brio HP NetVectra HP OmniBook HP Pavilion HP Vectra 500 HP Windows Client
Hiquality Computer		HiQ xxxxx	Chembookxxx
Hitachi	HDS EX HDS GX HDS Skyline Series	VisionBase	C090, C100, C120 E100D, E133T M100x, M120T, M133T VisionBook Pro VisionDesk
HNSX	SX-4 Model xxR SX-5		
HyperData			MediaGo
IBM	3090 43xx CMOS-390 ES/9000 Model xxx G4 Parallel Enterprise Server S/360, 370 S/390 ESO, G3 SP1 SP2	937x-xx AS/400 ES/9000 Model xxx Netfinity PC Server Raven RS/6000 Series/1 Model xxxx System 36, 38	7051 Power Aptiva PC 300, 700 RS/6000 N40 Notebook ThinkPad ValuePoint

Vendor	Large Mainframes, Supercomputers, Parallel Processorss	Midrange Midsize, Workstations, RISC, Servers	Desktop Microcomputers, Pcs, Notebooks
Impulse Computer			CompuBook 650
Integrated Business		ADP-P5	ADP-5xxxx, Pxxxxx
Integrix		HA1000 ISG NS200 RS1/170,RS2/xxx SEC100 SWS USI/200E UWS Desktop	
Intel			Celeron System 310AP
Intel Supercomputer	iPSC/860 Paragon XP/S		
Intergraph		InterServe	TDZ 2000
JC Information Systems			JC/Cheeta JC/Lion
Jetta Computer			JetBook
Keydata			Keynote xxxx
Kiwi Computer			OpenNote 680xx
Leading Edge			Fortiva N3, N4/SXL WinPro 486, Multimedia WinTower 486, Multimedia
Legato Systems			Legato Networker
Mactell			TWister Typhoon
Mag Portable			Verify xxxxDT
Marner International			CheetaRack
Maspar	MasPar MP-1,2		
MAXIMUS		Magna Powermax Server Magna SCSI Server Magna Wide SCSI Server	CD Notebook TFT Magna-Artist, CAD, Media, NT2 MediaNote DC Book xxx MusicMax PowerMAX Powermax DS, TFT PROMEDIA-A,B WinMax
Maxtech			YESBOOK
MegaComputer		Impact PCI Double Impact	Horizon MPC
Megadata			Mega 43D, 45D2, 46D2
Megaimage			Megabook 880
Melard Technologies			Scout ixxx
Micon			M-Note Pentium

Vendor	Large Mainframes, Supercomputers, Parallel Processorss	Midrange Midsize, Workstations, RISC, Servers	Desktop Microcomputers, Pcs, Notebooks
Micro Express			MicroFLEX Microflex NPxxx
MicroX			MPD-8x00 MVD-5x00 MX Edge MXP-6x00
Micron Electronics		NetFrame Vetix MXI, LXI	Best Buy ClientPro GoBook Millinnia Powerdigm PowerServer SMP-N TransPort xxx
MIPS		ARCSystem Magnum, Millennium	ARCSystem
Mitsubishi			Amity CN
Mobius		Atlantra AS/xxxx Protege	
MODCOMP		CLIII/95-xx Tri-Dimensional 92x0	Real/Star family (1000,9xxx) Real/Star family MODCOMP 97
Motorola		PowerStack	MPX100 StarMax
NCR	Worldmark 5100M	NCR Globalist NCR WorldMark System 7000 R3 S16, S26, S40, S46	WorldMark
nCUBE		MediaCUBE	
NEC	Sx-3R	Express5800 ProServa	Direction PC Express Server PowerMate Enterprise Series PowerMate Professional Ready Versa
Nekotech		AlphaServer AlphaStation NekoTech MACH Nekotech SuperServer	Bobcat Cougar Panther
Netframe Systems		Cluster Server Nxxxx	
Netis Technology		PCI ProServer	Netis Best Buy, PowerStation
NeTpower		Calisto SPARTA SYMETRA(2)	
Network Appliance		NetApp	
Olivetti		NetStrada 5000, 7000	Echos Pxxx, Pro
Pacific Internet		Webcube	
Packard Bell			Legend Supreme Pack-Mate Platinum Supreme

Vendor	Large Mainframes, Supercomputers, Parallel Processorss	Midrange Midsize, Workstations, RISC, Servers	Desktop Microcomputers, Pcs, Notebooks
Panasonic			CF-M31 V41 MK II, III Toughbook
Panda Project		Archistrat 4s	Rock City
PC Importers			Extreme NuMedia
Perifitech		Aerial xxx	
Periphonics			CallSPONSOR
Polywell Computers		Poly xxx	Poly Alpha, Vision
Power Computimg		Power Base, Center, Tower	PowerTrip
Premio			Premio Apollo BX
ProGen			AtlasX Discover 3D Discover MPEG II GameBreaker 3D Polaris ProFinder Venture II, Pro
ProLinear			ProNote X-C 6xxx
Prostar Computer			ProStar 5200, 6200, 7200, 8200
Quantex Microsystems			QP5/xxx
Racer Computer			Daytona Le Mans RacerPC Talledaga
RAScom		RAServer 2000, 2500, 2900	
RDI Computer			PrecisionBook UltraBook
Reason Computer			Netfire Pro Square 1,5,6 Square 5H
Ross Technology			SPARCplug
Royal Electronics		Ultra/wide NetPro, Omega	Cyber Graf Family Pak Graph Pro MMX Media Best Buy Mobile Best Buy Netpro 200 Power Graph UV Power Media MMX Rage Media Royal Mystique Synergy xxx Winpro
SAG Electronics		SAG Dual Pentium	
Sager Midern			Sager NP8xxx
Samsung Electronics			SENS Pro 5xx Sens810
Sanar		Workstation 10/xx	

Vendor	Large Mainframes, Supercomputers, Parallel Processorss	Midrange Midsize, Workstations, RISC, Servers	Desktop Microcomputers, Pcs, Notebooks
Sceptre Technologies			Soundx 3000, 4000
Sequent		NUMA-Q 2000	NTX 2000 NUMA-Q 2000 Symmetry 5000, S200 WinServer 5000, Pro
Sharp Electronics			PC-30x0, 90x0, 9300
Siemens Nixdorf	Reliant RM1000	PRIMERGY SCENIC Pro	RM200 SCENIC Mobile
Siemens Pyramid		Pyramid Nile RM400, RM600	
Silicon Graphics	Origin2000(with Cray)	Indigo IRIS Indigo Origin200 WebForce Origin 200	
Solflower		SFVME	
Sony Electronics			VAIO 505G
Stratus	Continuum xxxx XA2000 Model 2xxx	Continuum xxxx RADIO Cluster XA/R, R-S Model xx	XA2000 Model xxx
Sun		Darwin ftSPARC Netra i, j, nfs SPARC family Starfire Ultra 30 Ultra Enterprise x000 Ultra HPC	
Talkto Computers			Chatterbook
Tandem (owned by COMPAQ)		CS-150 Integrity family NonStop Himalaya	NDX ST
Tangent		Enterprise X-DQuadStar Tornado NetStar	Cyclone Hurricane Hurricane NetStar Medallion MediaGem Pendant TFT Petra X-D Sapphire Net Shuttle Tangent EISA/VL, NetRun, PCI, VL Tornado
Tatung Science & Techology		COMPstation MicroCOMPstation SuperCOMPserver U10-300	TCS-5xxx,6xxx TNB-5xxx
Techmedia Computer			Technote Sxxxx
Technology Advance		Galaxy Model xxx Constellation	MediaPro

Vendor	Large Mainframes, Supercomputers, Parallel Processorss	Midrange Midsize, Workstations, RISC, Servers	Desktop Microcomputers, Pcs, Notebooks
Thinking Machines	CM-2 CM-2a Model 4,7 CM-5		
TJ Technologies			Showman P2,S2 TekBook
Toshiba America		Magnia 3000, 5000	220CSD Equium Infinia Libretto Protege T3400 Satellite (Pro) Tecra xxxx
Transmonde			Vibrant xxx Vigor-100, 120, 133 vivante
Transtech		Paramid	
Tricord Systems		PowerFrame	
Tri-Star Computer		StarServer RAID,SMP	Tri-Star DesignBook NetStation StudioStation WebStation
Twinhead			Slimnote
UMAX Computer		SuperMac	ActionBook UmaxPC
UniCent Technologies			Avanta Titana Voyager
Unisys	2200 A11,14,18 ClearPath (2200) Open 2200/500 SPP	A2x00 Aquanta ClearPath HMP ix,mx ClearPath SMP PW2 Advantage	Aquanta Aquanta EL, LN (Notebook) SGxxxx SYS:A7 Model x11 TC486(E,V) U6000/x00 UN6000/xx
USA Flex			Flex xxx Patriot II Stealth GS,SE
VA Research		VarServer	VarStation
Wang		VS6000, VS12000	
Wavetracer	Data Transport Computer Zephyr		
Wedge Technology			ShowBOOK ShowBIZ SuperBook
Wen Technology			Wen 486
Winbook			WinBook
Wynn Data			FT-5xx Active, Passive Color

Vendor	Large Mainframes, Supercomputers, Parallel Processorss	Midrange Midsize, Workstations, RISC, Servers	Desktop Microcomputers, Pcs, Notebooks
Wyse			Series 3000i,6000i,7000i
Xediom			Mercury Multimedia Titanium Advanced
Zenith			Performance xxx Z-Server Z-Star Z-Station
Zenon		Z-Business Z-OptimaPro	Z-3D-PowerPro Z-Millenium Z-Platinum Z-PowerNote (3000) Z-Zenon

Small Computers

Handhelds

Computer	Vendor	Computer	Vendor
20xx TRAKKER	Intermec	PalmBook	ProLinear Corp.
200LX palmtop PC	Hewlett-Packard Co.	PC Companion	Compaq Computer
300LX, 320LX palmtop PC	Hewlett-Packard Co.	PEN/KEY	Norand
Cassiopeia	Casio	PenView	Norand
CE Pro	Microsoft	Phenom	LG Electronics
Concerto 4/25,4/33	COMPAQ Computer Corp.	PI-7000,9000	Sharp Electronics
CruisePad	Zenith Data Systems	PIC-1000, 2000	Sony Electronics Inc.
GriDPAD	AST Research, Inc.	PIC-2000	Sony Electronics Inc.
Gulliver	Hyundai Electronics America	Pilot	Palm Computing
HC 100,110,120	Psion, Inc.	Polycorder	Omnidata International, Inc.
HP 620-LX	Hewlett-Packard Co.	Portfolio	Atari
HP 660LX Palmtop PC	Hewlett-Packard Co.	PS-1000,3000	Prolinear
HP OmniGo	Hewlett-Packard Co.	PT4x00	Symbol Technologies, Inc.
HPW10E2	Hitachi Home Electronics	RDT xxxx	Omnidata International, Inc.
Husky FC-486, FS/2	Husky Computers, Inc.	Rex	Franklin Electronic Publishers
Hyperbook	HyperData Technology Corp.	RHC-44,88	Paravant Computer Systems
IBM 2488 Model 300, 800	IBM Corp.	Seahorse	Digital Ocean, Inc.
Infolio 160	PI Systems	Series 3 Palmtop	Psion
K2100, K2500	Kalidor	Siena	Psion
LGHPC	LG Electronics US	Simon	BellSouth Cellular Corp.
Magic Link	Sony Electronics, Inc.	Stylistic 500, 1000	Fujitsu
MessagePad 130, 2000	Newton, Inc.	T5000	Itronix Corp.
Microflex PC	DAP Technologies	Tarpon	3Com
MobilePro 200, 400	NEC Computer Systems	TeamPad	Fujitsu
Mobilon HE-4500	Sharp Electronics	TelePad SL	Telepad
Momenta 1/xx	Momenta Corp.	Telxon PTC	Telxon Corp.
Newton	Apple Computer	Tx00	Toshiba America
Nino	Philips Mobile	ULTIMAX	S.T. Research Corp.
Nokia 9000	Nokia	Velo 1	Philips Mobile
Organizer II	Psion	VRC 3910	Symbol Technologies
PadPlus	Fujitsu America, Inc.	Zoomer	Tandy
Palm VII	Palm Computing	ZR 5700, 5800	Sharp Electronics Corp.
		ZR-3xxx	Sharp Electronics Corp.

Portables

Computer	Vendor
ActionNote	Epson America
ALPHAbook	Tadpole Technology, Inc.
Armada xxxx	Compaq Computer Corp.
AS/400 Portable	IBM
BitWise 433, 466	Bitwise Designs
BriteLite	RDI Computer
Dauphin 1050	Dauphin Technology
EuroCom 3500	Europak International
FW5xxx, FW7000	Fieldworks, Inc.
Galaxy 1100	Science Applications
Guardian, Guardian Plus	Modgraph, Inc.
Hyperbook 700 Pro	HyperData Technology
IntelliView MPS-110,210	Intellimedia Corp.
Legato Networker	Legato Systems
LTE 5400	Compaq Computer Corp.
MP 133A	Micro Express
PayBase 32	Bottomline Technologies
PowerLite	RDI Computer
POWERportable	IBM
Regal	Micro Express
Ruggedized PowerLite	RDI Computer Corp.
ScreenStar	Bitwise Designs, Inc.
T66xxx	Toshiba America
Techmate TM6	Consultronics, Ltd.
ThunderBox II	Lightning Computers
TransPort Trek	Micron Electronics
X-C 6000 Cross Country	Intronix

Network Computers

Computer	Vendor
@workStation	Neoware Systems
Explora	Network Computing
HMX(Pro)	NCD
HP Net Vectra	Hewlett-Packard Co.
I-O Netstation	I-O Corp.
Internet Client Station	Idea Associates
JavaStation	Sun Microsystems
LanStation Pro	Accton Technology
NC2xx	Tektronix
NCR 2990	NCR Corp.
Neostation	Neoware Systems, Inc.
Netchamp	LG Electronics
NetPC	Microsoft and Intel
Network Computer TC,XL	Boundless Technologies
Network Station	IBM Corp.
Secura	Advanced Modular Solutions
Tempo Client	Everex Systems, Inc.
TNC-1xxx	Tatung
Viewpoint TC	Boundless Technologies
Visara	Affinity Systems
Winterm	Wyse Technology, Inc.

UNIX Systems

OS	Vendor	OS	Vendor
386/ix	Interactive Systems	PRIMIX	Prime Computer
3DA	HP & SCO	pSOS	n/a*
A/UX	Apple	QNX	QNX Software Systems
AD	Dascom	REAL/IX	MODCOMP
AIX	IBM	RISC/OS 4.5I	Mips Computer Systems
BSD (V4.3)	n/a*	RX/V	Ridge Computers
BSD/OS	Berkeley Software	SCO OpenServer	SCO
Caldera	Caldera	SCO OpenServer 5	SCO
Coherent	Mark Williams	SCO System V	SCO
CubixNet	Cubix	SCO Unix, Xenix	SCO
D-NIX	Diab Data	Sinix	Siemens Nixdorf
Destiny	Unix System Laboratories	SINUX	Siemens Nixdorf
DG/UX	Data General	Solaris	Sunsoft
Dynix(/ptx)	Sequent Computer Systems	Sphinx	Data General
Enix	Everex Systems	SPP-UX	Convex Technology Center
ETA System V	Compuware	SunOS	Sun Microsystems
Gemini	HP, SCO & Novell	TP-IX/88K	Tadpole Technologies
HP-UX	HP	Transit	nCUBE
IDRIS	Whitesmiths	Tru64 Unix	Compaq Computer
IN/ix	Interactive Systems	ULTRIX	Digital Equipment
Irix	Silicon Graphics	UMAX	Encore
IRIX OS	Silicon Graphics	Unicos	Cray
IS/3	Interactive Systems	UniFlex	Technical Systems Consultants
IX/370	IBM	UniPlus+	UniSoft
Linux	n/a*	UnixWare	SCO
Lynx	Lynx Real-Time Systems	UNOS	Charles River Data Systems
LynxOS	Lynx Real-Time Systems	USF	Stratus
MACH	n/a*	UTS/580	Amdahl
MPE/IX	HP	UTX/32	Gould
MV/UX	Data General	UWS	Digital Equipment
NeXTOS	NeXT	UXP/M	Fujitsu
nX	Bolt Beranek and Newman	VENIX	Software Kinetics
Open Desktop	SCO	VENIX System V	Unisource Software
OSF/1	Open Software Foundation	Venix 3.2	Venturcom
Plan 9	Lucent Technologies	VP/ix	Sunsoft
PowerMAX OS	Concurrent Computer	Xenix	Microsoft
PowerOpen	Apple & IBM		

* General type Unix system, not a product from a
 specific vendor

Languages

Compilers

Ada
ALGOL
APL
Aztec
BASIC
BLISS
C
C/XL
C7
Clipper
COBOL 2
COBOL 2 (MicroFocus)
COBOL II

CommonLISP
Coral 66
C3ADA
DG/L
DIBOL
EBASIC
FOCAL
FORGO-77
FORTH
FORTRAN
JOVIAL
LISP
LOGO

Magic/L
MAINSAIL
MetaCOBOL
Modula-2
MUMPS
Occam
OPTASM
Pascal
PL/1
PL/M
PROLOG
PROMAL

RATFOR
REBOL
RPG
RPL
SFTRAN3
SITGO
SNOBOL
Softbol
TAL
TASC
Visual COBOL
XPL0

4GL, QUERY Languages

Accent R[1]
ADS[1]
ASK/Windows
CA-Easytrieve
Dataflex(2.3B)[1]
DBQ
DXT
DYL-Inquiry
DYL-IQ Express(/OEM)
Easytalk
EQL
EQM
EZ/IQ
Focus[1]

Group Four Datapulse[1]
IDOL-II[1]
IDS I,II[1]
IIS/DESTINY[1]
IM/QUICK
Informix-SQL
Inquire
IQL
Mantis
Natural
NOMAD[1]
Nova
OnLine Query
PDQ

PowerHouse StarBase[1]
PROGRESS[1]
ProSEED
QDS
QLP
QMF
Query
Query/36,AS/400
Query.DL/2
RAMIS[1]
Reliance Access
RQL/32
SAL
Script-IV

SDL
Spectra
Speed II
SQL
SQL*Plus
SQR
System Z[1]
T-ask
User Language
VISION:Clear Access
Visor/V
VQL
XDL

Object Oriented

Advance
Agora
Beta
Blue
Borland C++
C++
COBOL for OS/2
Dylan
Flavors

Java
LENS
Liana
Loops
LotusScript
Modula-3
NewtonScripts
Object Pascal

Object REXX
Objective C
OO-COBOL
OQL
ORexx
Python
QuickPascal
Sather

ScriptX
Simula67
Sina
Smalltalk
Theta
Turbo Pascal 5.5
Xmath
XDL

Miscellaneous Languages (Scripting, Special Purpose, Etc.)

Abap
Access Basic
ADROIT
AWK
DAL
DHTML
ELF
G-Logis

GPSS
GQL
HTML
JavaScript
LotusScript
MASM
ObjectPAL
OPS5+

ORexx
PAL
Perl
Quest
REDUCE
REQL
REXX
ScriptX

SIMSCRIPT
Smbasic
TASM
Tcl
VBScript
Vortex
VQL
XPL0

---

[1]Database management system included

Application Development Environments

+1Environment
4D SDK
Actor Professional
AdaWorld
ADE
Advanced Revelation
Aion
Allegris Workshop
AM
ART*Enterprise
ArtBASE
BaseWorX
Borland C++
Build Momentum
BuildProfessional
C++ Builder
C++ Professional
C Set++
C4
CA-Ingres Windows
CA-OpenRoad
CA-Realizer
CA-Visual Objects
Centura
Centura Web Developer
DataEase
DB-UIM/X
Delphi,Delphi32
Developer/2000
Distributed Smalltalk
DOE
Easel
Eiffel

Elements
Encompass
Enfin
Entark
Entera
Enterprise Developer
Facets
Formida Fire
Forte
Gain (Momentum)
Galaxy(/C++)
GemStone DeODE
Gen/X
HarborView
ICPL
Informix(-NewEra)
IntelligentPad
Jade
JAM
JAM/Web
Jasmine SDK
Java 2
JdesignerPro
JDK
JFactory
Kappa
Lansa
LispWorks
LiveModel
MacApp
Magic(7)
Magna X
Mozart Composer

NatStar
Netlabs/Vision
NewWave
NeXTStep
ObjChart
ObjectCenter
ObjectCraft
ObjectIQ
ObjectStudio
ObjectView
ObjectVision
Objectworks
Omnis (7)
OpenROAD
OpenStep
OpenUI
OpenWindows
Optima++
Orlando
Passport
Performer
PersonalJava
PowerBuilder
PowerModel
PRO-IV
Prograph
Progress
Protogen+
QuickObjects
Rational Rose
Sapiens
Seer/7000
Smart Elements

SmartStar
SNAP
SQLWindows
SuperNova
System Builder
TRAN
Uniface
Unify Vision
VanGogh
VBA
Vibe
Vision (Unify)
Vision Builder
Vision/Recital
Visual Basic
Visual C/C++
Visual Cafe
Visual FoxPro
Visual Jade
Visual Object COBOL
Visual RPG
Visual SQL
VisualAge
VisualGen
VisualWare
VisualWorks
VPE
VZ Programmer
WinPlus
Workflow Template

Data Mining Tools

Axum
BrainMaker
CART
Clementine
CrossGraphs
Cubicalc
Data Desk
DataBase Mining Marksman
DataMind
Datasage
dbProphet

Decision Series
Discovery
FuziCalc
Geneva V/T
Intelligent Miner
Level5 Quest
Mineset
NetMap
Neural Connection
Neural Network Utility
NeuralWorks Predict

NeuroGenetic Optimizer
Object Relational Data
  Mining
Orchestrate
Pattern Recognition
  Workbench
PATTERN
Preclass
PV-Wave
QueryObject
RDS-Assort

SAS/Enterprise Reporter
Scenario
Solution Frameworks
TempleMVV
Visual Insights
Visualization Data Explorer
WizWhy
Xpert Rule Analyser Profiler

Relational Databases for
Large and Midrange Systems

ADABAS
Adaptive Server IQ
Allbase/(SQL)
Alpha/Four
Analect/RIM
ARES
Boeing Rim
C-A-T
CA-Datacom
CA-IDMS
CA-RAMIS
CA-Universe
Cdb Toolkit
CFMS
Cirrus
Condor 3
Data Base-Plus
Data+Trac
DATAMAT
Datascan
DB Magic
DB2
DBExpert
Decision DB
DG/SQL
DM

DNA-4
Empress(/32)
Enterprise:DB
filePRO 16 (Plus)
FOCUS
Harris INFO
Helix VMX
IMPRS
INFO
Information(Plus)
Informix (Dynamic Server)
Ingres II
Integra
InterBase
Interel
MAI Origin ADS
MASS-II Manager
MDBS III,IV
Model 204
MRDS
NOMAD(2)
NonStop SQL(/MX)
NPL
Omnibase(/SQL,4GL)
OpenIngres/Star
OpenODB

Oracle7
Oracle8
Oracle8i
Pace RDBMS
Poise DMS
Prime Information
Prodas
Progress (ADE)
RDM++
QDMS-R
QMan
Rdb/VMS
RDM
RDM/2
Relate/3000,DB
Reliance
reQuest
Resolve Productivity Set
ResponseR
REXCOM
RT PC SQL/RT
RTIRIM
Rubix (Master)
Sharebase I, II
SQL+
SQL/200

SQL/400
SQL/DS
Structure/4
Sun INGRES
Sun UNIFY
Supra (Server)
Sybase
System 10
System 1022
System 1032
Teradata
Trusted Oracle
UDMS
UDS RDMS 1100
Ultra
Ultra-Base
Ultrix/SQL
Uni-Power
UniData
UNIFY
Universal Server
Xample
XPS
Z-Four

Desktop Systems

1stKey
4th Dimension
Aladin
Approach(96)
Brock Activity Manager
Data-Ace
DATASTORE:lan,pro
dBXL
Double Helix II
Easy Base
Emerald Bay
FileMaker Pro (for
    Windows)
FoxBASE (+,Plus), FoxPRO

Goldatabase
IM/Personal
Information/pc
InterBase
Knowledgeman/2
LOGIX
M/SQL
Magic PC,LAN
Marcon Plus
McMax
Megafiler
Microsoft Access
MultiUser Helix
NetWare SQL

Omnifile
Oracle Quicksilver
OS/2 DBM
OverVUE
P-Rade
P-Stat
Pacebase
Paradox
PC/Focus
PC INFO
PC NOMAD
PHD
PractiBase
Probase

Silverado
SQL Anywhere
    (Professional)
SQL Server
SQLBase for Windows
T-base
TAS-Plus
Uni-File
VersaForm
VIA/DRE
VP-Info
XDB-SQL

Object, Object-relational Databases

Cache
CLOSQL
ConceptBase
DEC Object/DB
EXODUS
GemStone
Hybase
IDB Object Database

Illustra
Informix Internet
   Foundation 2000
Jasmine
JeeVan
KE Texpress
Mattisse
NLM Server

O2
Objectivity/DB
ObjectStore
ODB-II
ODBMS 2.0
Ontos
Orion/Itasca
Poet

PostgreSQL
RDM++
Total ORDB
UniSQL/X
Universal Server
Versant
VODAK
Vortex01

Multidimensional Databases

Acumate(ES)
Commander OLAP
CrossTarget
Essbase

FOCUS Fusion
Fusion
GentiaDB

Holos
MetaCube
Oracle Express

Plato
SAS/MDDB
TM_1

Applications Software (ERP Software)

Product	Vendor
IFS applications	IFS Systems
System 21	JBA International
Mapics	Mapics, Inc.
Oneworld	J.D. Edwards
Oracle CPG	Oracle Corp.
Prism	Marcam Solutions
R/3	SAP AG
Baan	Baan Company
Oracle Applications	Oracle Corp.
PeopleSoft	PeopleSoft, Inc.
Stream*Line*	Geac Computer Corp.

Communications

LANS (Local Area Networks)

3+Open	Ethernet	Netfax	StarLAN
Appletalk	LAN/3000,9000	PCnet	TiaraLink LanWare
Arcnet	LANLink	Primenet	Token-Ring
DaynaNetDNDS	MNDS	R2LAN	

NOS (Network Operating System)

10Net	Enhanced Spectrum	Net/One Lan Mgr	Powerlan
3+Open LAN Mgr	Great OS	NetconMT	Promiselan
3+Share	HP Lan	NetWare	QNX
386/Multiware	IntranetWare	NetWare Lite	Silvernet
Advanced Server	Invisible LAN	Network-OS	Stargroup LAN Manager
AppleShare	IOS	NEX/OS	Tapestry II
Aviator Wireless Network	LAN Manager	NT Server	TAS
BOS/LAN	LAN Server	NTNX	Utopia (LSF)
CCI Net	LANsmart	OS/2 (Warp) Connect	VINES
Chosenlan	Lansoft	OS/2 (Warp) Server	VMOS
CitrixMultiuser	LANstep	Pathway	Web 3.0
Datalan	LANtastic	PC Lan	Windows NT AS
DNS 300XA	Mainlan	PennyLAN	WINnet
DOSTOPS 3.0	Net 127	Personal NetWare	
Easynet NOS.2 Plus	Net30	Ported NetWare	

Network Management Systems

Accumaster Integrator	Kaspia Automated	NetCool	PNMS
ARCserve Replication	Monitoring System	NetDirector (for Unix)	Polycenter Manager
Assistance Center	LANDesk Manager	Netexpert	SDS
Balans	LANfocus	NetFinity	Solve:Netmaster
BBN/StatsWise	LANVIEW/Window	NetLabs/Discovery	Spectrum
BindView	Lassisnet	NetLabs/Manager	SUDS
BlueVision	LattisWare	NetModeler	System Management Server
BW-MultiConnect	LMU	NNS	System Watch
CA-Mazdamon	ManageWise	NMS	Transcend
CA-VMAN	Max/Enterprise	OpenSNA	UNMA
CEX	Mission Control Console	OpenView Node Manager	Viewplex
CiscoWorks	MSCS	Operations Coordinator	Visinet
Comnet III	NAN	OperationsCenter	Wisewan
Cross Platform Svcs	NAPA	Overlord	Wolfpack
EcoNet	NerveCenter	Patrol	
EcoSYSTEMS	Net/Master	PC/TCP OnNet	
Enterprise/Solver	Netcenter	Peregrine Network	

Systems Management Software

CA-Agentworks
CCC
GEM
ISM
Karat
Luminate for R/3
MetaStar Enterprise
Millicent

NetDeploy
NetFinity Manager
NetView
Newscout Manager
OpenMaster
OpenView
Openvision PCS
PCS

Pegasus
POEMS
ProVision
Soltice
SunNet Manager
Tanglewood
TIB/Hawk
Tivoli/TME

Tornado
Trend
Unicenter TNG
Universal Management
   Agent

Transaction Monitors (OLTP)

Access
Bachman/Ellipse
Calout Plus
CANDE
CICS(/ESA)
CLIP
Com-plete
Com-pose
COMS
CONNECT:Direct

Contrl
DASCOMP/B,M DB2
   Parallel Edition
DBOMP
Decintact
DFDS
DMS/CMS
DPM
DS/1000-IV DSN/MRJE
GEMCOS

IDMS-DC
IMS-DC
IMX 700 MitemView
Multics
OCCF
Patrol Knowledge
RJE
RPS 1100
RTQ
Shadow II

SNA Gateway Supertracs
TCS
TELCON
Tuxedo
TIP/30
TIS/XA
UCF
UniKix
VSE/NMPF

EDI (Electronic Data Interchange)

ACS X12
Adams-EDI
AT&TEDI
AMTrix Intellligent
   Messaging Agent
BPCs EDIPath
Commerce:Doculink
Data Interchange/2
Data Mail
Datainterchange
Datatran
DEC/EDI
Descrypt/EDI
E-Z Order
Easy-Connect
ECON
EDE-PC
EDI/36,38,400
EDI 400
EDI/ANK,INK,SCK,WDK
EDI-Answer
EDI Application Integrator
EDI*Asset
EDI BIZIBIX
EDI Business Partner
   Translator

EDI/comm
EDI/Entry, EDI/Developer
EDI Excel
EDI*Expert
EDI/Edge
EDI Link
EDI-Link-M,S,SB
EDI Manager
EDI*Net
EDI/Open
EDI*PC
EDI/Synapse
EDI*T System
EDI-Ware
EDI Windows
EDI/WINS-3000
EDIeasy
EDIpl Translator
EDISIM
EDItran,EDIfast
ESP II
Etoolkit/400
ExpEDIte
Extol EDI Integrator
EZ-EDI
FasTran

FEDI
FrEDI,FastFrEDI
Gateway*Express
Gentran
Hi-Life EDI
Hypertracs
I*B*T*S
Interconn
Intouch*EDI
JBA System 21 EDI
LAN II
LDJ Messenger
Mac-EDI
Memo/EDI
Mercator
Midas/EDI
Multinet
OmniTrans
OpenEDI
PCX12
PeerNet
Performance EDI
Perwill*EDI
Power EDI
Pro_EDI
QualEDI

Mac-EDI
Radely EDI
Redi-Micro
Release/Shipment
   Communications
SDIS
SDM/LINK
spEDI
STX12
TEL-EDI
Telink
Trading Partner
Trans-Sync
TranSlate
Translator*MVS
TrustedLink
Universal-Link/EDI
UPC*Premier
Vector:Connexion
Visual EDI
VLT Customizer
Web Link
WinEDI
X-Change
XLT12

Groupware

Action Plus
AOCE
Aspects
CA-UpToDate
Caucus
CCE
Change Agents
Channels
CM/1
Collabra Share
Common Knowledge
Cooperation
Coordinator
Corporate Online
    Repository
Desktop Conferencing
ERoom

Facilitator
FirstClass
ForComment
GroupSystems
GroupWise
Higgins
IDM
InForms
Innovation Team Project
InPerson
Instant Update
InterConnect for Notes
InterOffice (Suite)
Involv Intranet
LinkWorks
LiveMeeting Suite
Meeting Maker

MeetingPlace WebPublisher
MMCX
NetThread
Network Scheduler II
NewWave Office
Notes (Lotus)
Notes Express
Odyssey Suite
Office Logic
OpenMail
OpenMind
Oracle Groupware
Outlook
PacerForum
SamePage
Share
ShowMe

SoftSolutions
TalkShow
Teamfocus
TeamLinks
TeamOffice
TeamTalk
The Meeting Room
Time and Place/2
WebShare
Workflow Connector
WorkGroup
WorkMan
VAX Notes
VisionQuest
VNS

Workflow

ActionWorkflow
Approval
Conductor
Discovery Suite
Documentum
DSS Agent
Edify
Electronic Workforce
FloWare and Map/Builder
FlowMark
FlowModel

FormFlow
Framework
IBS Flowmaster
ImageMover
InConcert
Keyfile
Keyflow
Marshall
NovaManage
Omnidesk
Open/Workflow

Optika
Optima
Optix Workflow
Paradigm/XP
PowerFlow
ProcessIt
RDM
Silverrun:Enterprise
Staffware
Task Master
TeamWare Flow

Viewstar
Visual Workflo
Workflo/
WorkMAN
Workflow*BPR
Workflow Analyzer
Workflow Factory
Workflow Template
WorkManager

Protocols

ADCCP
ADSP
AEP
AFP
APdesktop computer
APPC
APPN
ARP
ASP
ATP
BACP
Bisync
CDCDLC
CDPD
CLNP
CMIP
CONS
CSMA/(CD)
DAP
DDCMP
DDP

DNA
DNS
FTAM
FTP
HDLC
ICA
ICMP
ICP
iKP
IMAP
IP
IPC
IPSec
IPX
IPX/SPX
Kermit
L2F
L2TP
LAPB, LAPD, LAPM
LDAP
Mailbag

MHS
MNP
MNP 10
NBP
NetBEUI
NFS
PAP
PLP
POP$_3$
PPP
PPTP
RAP
ROSE
RTMP
RTP
SCSI, SCSI-2
SMP
SMTP
SNA
SNMP
SOCKS

SPP
SWAP
Telescript
Telnet
TFP
TFTP
TOP
UDP
V.42(bis)
X.21(bis)
X.224, X.225, X.226
X.25
X.400
XModem
XNS
YModem
ZIP
ZModem

Index

Boldface entries indicated glossary page numbers.

OSI (Open Systems Interconnect) 63–65, **243**

P
partitioning. *See* application partitioning
PCs. *See* microcomputer
patterns 32–33, 34, 54, **246**
PDA (Personal Digital Assistant) 10–11, **247**
peer-to-peer network 61, 72, **248**
peripheral devices 3, 5, 40, 42, 55, 61, 95
Perl 38, 78, **249**, 346
Personal computer. *See* microcomputer
Personal Digital Assistants. *See* PDA
physical design 50, 51, 92, 93
physical layer 64, **249**
polymorphism 29, 31, **252**
presentation layer 65, **255**
program generators 16, 23, 24, 26, **257**
program specs 16
programming phase 13, 16, **257**
project management 11, 19, 23
protocol 55, 63–65, 76, 94, 96, **259**
pseudo-code 21, **260**
public key encryption 82, **260**

R
R/3. *See* SAP R/3.
RAD (Rapid Application Development) 22–23, 25, 32, 96, **264**
Reduced Instruction Set Computer. *See* RISC
regression testing 17–18, **266**
relational databases 38, 49, 52, 71, 73, **267**, 348
remote procedure call. *See* RPC
Report generators 16, 23, 26, 74, **267**
repositories 21, 51–52, 72–73, 87, **268**

RISC (Reduced Instruction Set Computer) 6, 7, 8–9, 10, 11, 44, **269**, 332–342
RPC 66, **270**
Rumbaugh 30, **272**

S
Sales Force Automation. *See* SFA
SAP R/3 86–87, **273**
screen editors 26, **275**
screen painters 26, **275**
scripting language 32, 38, 45, 78, **275**
secret key encription 81
servers 7, 8, 11, 55, 66, 68–70, 71, 82, **277**, 332–342
session layer 65, **278**
SFA (Sales Force Automation) 85, **278**
shell, shell language 39, 45–46, 78, **278**
Smalltalk 9, 29, 30, 67, 157, **280**, 346
SMP (symmetrical multiprocessing) 11, **281**
SNA (Systems Network Architecture) 63, 65, **282**, 353
SQL (Structural Query Language) 38, 52, 73, **286**, 346
structured programming 20–21, 27, **289**
supercomputers 6–7, **290**, 332-342
Supply Chain Management 84, 86, **291**
support personnel 6, 9, 41, 88, 94–96
symmetrical multiprocessing. *See* SMP
Systems Network Architecture. *See* SNA

T
TCP/IP 63, 65, 76, **295**
technical developers 36, 41, 42, 43, 50, 51, 55, 56, 60–61, 82, 86–87, 88, 89, 91, 92–94, 96
Telnet 76, **297**, 353
testing phase 17–18, **297**

three-tiered 72, 86, **298**
top-down development 21, **300**
Transmission Control Protocol/Internet Protocol. *See* TCP/IP
transport layer 65, **302**
two-tiered 72–73, **304**

U
Unix 8, 11, 38, 44–46, 63, 73, 78, 95, 96, **307**, 345
utility 25, 41, 45, **308**

V
vertical software 83, 85
visual languages 25, 38, **314**

W
WAN 60–61, **318**
Web 23, 31, 38, 69, 70, 76, 77–82, 94, **319**
web server 69, 77, **320**
wide area network. *See* WAN
white box testing 17, **322**
Wintel 46, **323**
workflow 58, 59, **324**, 352
World Wide Web. *See* Web

X
X.25 63–65, **325**, 353